Surface Treatment and Mechanical Properties of Sustainable Pavement Materials

Surface Treatment and Mechanical Properties of Sustainable Pavement Materials

Guest Editors

Hui Wang
Ge Zhang
Ling Qin

MDPI

Basel • Beijing • Wuhan • Barcelona • Belgrade • Novi Sad • Cluj • Manchester

Guest Editors
Hui Wang
School of Civil &
Environmental Engineering
and Geography Science
Ningbo University
Ningbo
China

Ge Zhang
College of Materials Science
and Engineering
Xi'an University of
Architecture and Technology
Xi'an
China

Ling Qin
School of Civil Engineering
Qingdao University of
Technology
Qingdao
China

Editorial Office
MDPI AG
Grosspeteranlage 5
4052 Basel, Switzerland

This is a reprint of the Special Issue, published open access by the journal *Coatings* (ISSN 2079-6412), freely accessible at: https://www.mdpi.com/journal/coatings/special_issues/Sustainable_Pavement.

For citation purposes, cite each article independently as indicated on the article page online and as indicated below:

Lastname, A.A.; Lastname, B.B. Article Title. *Journal Name* **Year**, *Volume Number*, Page Range.

ISBN 978-3-7258-6648-9 (Hbk)
ISBN 978-3-7258-6649-6 (PDF)
https://doi.org/10.3390/books978-3-7258-6649-6

Contents

About the Editors

Hui Wang

Hui Wang is from the School of Civil and Environmental Engineering, Ningbo University, Ningbo 315000, China. His research interests focus on pavement concrete, asphalt, cement-based materials, concrete durability, and steel corrosion. As a Guest Editor, he has organized the Special Issue "Surface Treatment and Mechanical Properties of Sustainable Pavement Materials" in the journal *Coatings*, committed to promoting the academic exchange and technological development of sustainable pavement materials. He has in-depth research experience in the performance optimization and durability improvement of pavement materials, and his research results provide important technical support for the application of eco-friendly materials in road engineering.

Ge Zhang

Ge Zhang is affiliated with the College of Materials Science and Engineering, Xi'an University of Architecture and Technology, Xi'an 710055, China. As a co-Guest Editor of the Special Issue "Surface Treatment and Mechanical Properties of Sustainable Pavement Materials", he focuses on the mechanical properties of concrete structures, advanced cement-based materials, and the hydration mechanism of cement. His research is dedicated to exploring the performance enhancement and sustainable development of pavement materials, and he has rich experience in the intersection of material science and civil engineering. Through organizing this Special Issue and compiling it into a Reprint, he aims to share cutting-edge research findings and promote the innovation and application of advanced pavement materials.

Ling Qin

Ling Qin works at the School of Civil Engineering, Qingdao University of Technology, Qingdao 266033, China. Her research interests include concrete durability, magnesium-based low-carbon cementitious materials, carbon sequestration of cement-based materials, solid waste resource utilization, and asphalt. As a co-Guest Editor, she has actively participated in the organization and compilation of the Special Issue "Surface Treatment and Mechanical Properties of Sustainable Pavement Materials". Her research focuses on low-carbon, environmental protection and resource recycling in the field of pavement materials, and her work contributes to the sustainable development of road engineering. She is committed to promoting academic communication and technological cooperation among researchers in related fields through this Reprint.

Preface

This Reprint is developed from the Special Issue "Surface Treatment and Mechanical Properties of Sustainable Pavement Materials" of the international journal *Coatings* (ISSN 2079-6412), which belongs to the section "Surface Characterization, Deposition and Modification". The mechanical properties and durability of pavement materials are crucial to their application in road engineering, and asphalt and cement-based materials are the main types used in current road projects. To promote the innovation, development, and practical application of sustainable pavement materials, we launched this Special Issue and compiled it into this Reprint. The included papers cover key research areas such as the preparation of sustainable pavement materials, surface treatment technology, mechanical property testing, durability evaluation, and mechanism analysis, reflecting the latest research progress and application practices in the field. This Reprint is intended for researchers, academic scholars, engineering technicians, and relevant professionals engaged in road engineering, material science, sustainable development, and other related fields. It aims to provide a platform for academic exchange and technical communication, share cutting-edge research results, and further promote the development of the sustainable pavement materials industry. We would like to express our sincere gratitude to all contributing authors for their valuable research outputs, as well as to the editorial team and reviewers for their rigorous work in supporting the publication of this Reprint.

Hui Wang, Ge Zhang, and Ling Qin
Guest Editors

coatings

MDPI

Article

The Influence of Modifiers on the Performance of Recycled Asphalt Mixtures

Xuejie Wang [1], Hui Zhang [1,*], Chenxi Gao [2], Qi Xue [3], Jia Yu [1], Feiting Shi [2,*], Shuang Lu [4] and Hui Wang [2]

[1] School of Intelligent Manufacturing, Nanjing University of Science and Technology ZiJin College, Nanjing 210023, China; wangxuejie5860@163.com (X.W.); yujiazijin@163.com (J.Y.)
[2] School of Civil Engineering and Geographic Environment, Ningbo University, Ningbo 315000, China; 15957895939@163.com (C.G.); huiwang123@aliyun.com (H.W.)
[3] Nanjing Municipal Design and Research Institute Co., Ltd., Nanjing 210012, China; xueqi0405@163.com
[4] School of Civil Engineering, Harbin Institute of Technology, Harbin 150001, China; lus@hit.edu.cn
* Correspondence: zhanghui219810@163.com (H.Z.); shifeiting@ycit.cn (F.S.)

Abstract

To enhance the performance of reclaimed asphalt pavement and evaluate its suitability under severe environmental conditions, this study systematically assessed the reinforcing effects of two modifiers: styrene–butadiene rubber (SBR) and polyurethane (PU). The stability and mechanical properties of mixtures with varying recycled asphalt contents were tested, with a focus on the effects of water immersion and high-temperature (150 °C) aging. The underlying mechanisms were elucidated using scanning electron microscopy with energy-dispersive X-ray spectroscopy (SEM-EDS). Results indicate that recycled asphalt incorporation significantly degraded the mixture's performance; stability and mechanical strengths decreased with recycled asphalt content following a cubic function, with maximum deterioration rates of 37.3%, 67.2%, 73.6%, and 72.1%. Water and thermal aging further accelerated performance decay, with deterioration rates reaching up to 61.2% and 63.9%, respectively. Modifiers effectively counteracted this trend: SBR and PU enhanced the aforementioned properties by up to 102.6%, 50%, 60.3%, and 56.9% and 139.5%, 81.4%, 113.2%, and 120.3%, respectively. Microstructural analysis revealed that RAP led to decreased C and O contents, increased Si and Fe presence, and a more porous structure. In contrast, the modifiers increased binder content and interfacial density. This research provides a theoretical foundation and technical pathway for the engineering application of high-performance recycled asphalt mixtures in complex environments.

Keywords: stability; recycled asphalt mixture; water action; high-temperature action; butadiene–styrene rubber; polyurethane

1. Introduction

The growing demand for transportation infrastructure has made asphalt pavement one of the most widely used road surfaces worldwide [1–3]. This pavement has the advantages of convenient construction and comfortable use. However, the exploitation of natural aggregates is limited, and the material cost of asphalt is relatively high [4,5]. The use of recycled asphalt pavement materials to prepare recycled asphalt mixtures has become a current research hotspot [6–8]. In order to provide affordable and sustainable materials for road construction, research and application of recycled asphalt have been carried out [9].

Asphalt recycling is the process of mixing a waste asphalt mixture with new asphalt and reusing it for road paving. The recycled asphalt is coated with asphalt binder, mineral filler, and fine aggregate on the surface and is composed of multiple particles of different thicknesses bonded together [10]. When preparing recycled mixtures, the asphalt film on the recycled asphalt surface softens with increasing temperature, yet it remains difficult to completely separate the old aggregates from the aged binder during hot-mix construction. This difficulty results in significant differences in the performance characteristics between recycled and fresh asphalt mixtures. For these reasons, the performance of a recycled asphalt mixture is worse than that of a conventional asphalt mixture [11–13]. Prior research has pointed out that a 30% content of recycled asphalt can reduce the water resistance coefficient by up to 37.1% [14,15]. Meanwhile, the incorporation of recycled asphalt adversely affects the temperature stability of the mixture. In addition, the compressive, splitting tensile, and flexural strengths are decreased by up to 43.1%, 45.2%, and 44.7%, respectively, with the addition of recycled asphalt [16,17]. Although some research on the performance of recycled asphalt mixtures has been reported, no systematic investigation of methods for improving recycled asphalt mixtures under various aging factors has been conducted.

Amino formate ester modifier and solid powder styrene–butadiene rubber modifier are frequently used as asphalt modifiers [18–20]. These asphalt modifiers can improve the bonding ability of asphalt binders. Additionally, the asphalt modifiers can increase the asphalt mixture's water stability, temperature stability, and mechanical strengths. The amino formate ester modifier can improve high-temperature dynamic stability of an asphalt mixture by 80%–150% [21,22]. Moreover, the splitting tensile strength, compressive strength, and flexural strength of the asphalt mixture modified by amino formate ester are improved by up to 80%, 45%, and 61% [23,24]. Meanwhile, the recycled asphalt mixture using the amino formate ester modifier shows excellent high-temperature stability, maintaining structural rigidity below 180 °C. The solid powder styrene–butadiene rubber modifier can increase the high-temperature dynamic stability, splitting tensile strength, compressive strength, and flexural strength of the asphalt mixture it modifies by up to 100%, 65%, 33.1%, and 52.3% [25]. However, little attention has been paid to comparing different modifiers in recycled asphalt mixtures under aging conditions. The novelty of this study is the investigation of recycled asphalt mixtures with modifiers under environmental conditions of water and high-temperature activity.

This paper aims to investigate the influence of recycled asphalt binder on the stability, splitting tensile strength, flexural strength, and compressive strength of asphalt mixtures. Asphalt modifiers, namely, amino formate ester and solid powder styrene–butadiene rubber, are employed to enhance the performance of asphalt mixtures. Some specimens are immersed in water for 48 h, while other samples are dried in an oven at 150 °C for 4 h. Scanning electron microscopy–energy-dispersive spectroscopy (SEM-EDS) is conducted to reveal the mechanism underlying the performance changes. This study will offer a novel approach for the utilization of recycled asphalt mixtures in road paving.

2. Experimental Section

2.1. Raw Materials

Maoming Petrochemical Donghai Brand No. 70A road petroleum asphalt, supplied by Maoming Zhengcheng Petrochemical Co., Ltd., Maoming city, China, was used in this study as the new asphalt. Waste asphalt recovered from the milled recycled asphalt mixture on Guanghui Expressway in Guangzhou, China, was used to manufacture a recycled asphalt mixture. The technical performance of the new asphalt was obtained from the manufacturer in accordance with the Chinese standard JTG E20-2011 [26]. The key properties include a penetration of 70.3 mm, a softening point of 49.1 °C, a rotational viscosity of 481 mPa·s, and a ductility of 11.7 cm. Butadiene–styrene rubber, provided by Shandong Gaoshi Technology Industry and Trade Co., Ltd., Zibo, China, was used as a modifier. The butadiene–styrene rubber was in the form of white powder, and the particle size distribution ranged from 0.3175 mm to 0.423 mm. The acrylonitrile mass ratio of butadiene–styrene rubber was 33.1%. Thermoplastic polyurethane, provided by the Hongtai Machinery Factory (Wenzhou, China) in the form of white powder with a particle size of 0.56 mm and a density of 1.11 g/cm^3, was employed as another asphalt modifier. In the experiment, the recycled asphalt was prepared in the lab by blending the waste asphalt. The detailed preparation procedure began by heating the waste asphalt to 165 °C and maintaining it at this temperature for 45 min until fully fluid. Subsequently, the rejuvenator was incorporated at a dosage of 6% by mass of the aged asphalt, followed by the addition of the No. 70A road petroleum asphalt, with continuous stirring at the same temperature for 20 min. The rejuvenator was a brown transparent liquid with a dynamic viscosity of 65 mm^2/s at 60 °C and a density of 0.95 g/cm^3. The recycled asphalt exhibited a penetration of 1.33 mm, a softening point of 78.6 °C, and a ductility of 13 cm. The coarse and fine aggregates used in this study were sourced from Zijin Linjiang Jianhua Furong Stone Field in Heyuan, China. Mineral powder from Boluo County Xiongmao Building Materials Co., Ltd., Huizhou, China, was applied in this research. Tables 1 and 2 present the chemical compositions of the new asphalt and aggregates, respectively. In Table 1, x represents the values 30 and 35, and y represents 44 and 46. Table 3 shows the cumulative passing rate of the recycled asphalt mixture.

Table 1. Chemical composition of new asphalt (%).

Type	Chemical Composition/%									
	Saturates	Aromatics					Asphaltenes			Cycloalkyl Aromatics
	C_xH_{62}	$C_{36}H_{57}N$	$C_{29}H_{50}O$	$C_{40}H_{59}N$	$C_{40}H_{60}S$	$C_{18}H_{10}S$	$C_{66}H_{81}N$	$C_{51}H_{62}S$	$C_{42}H_{56}O$	$C_{35}H_y$
New asphalt	11.13	5.63	7.03	5.62	5.65	21.1	4.48	2.98	4.48	31.90

Table 2. Chemical compositions of aggregates (%).

Type	Chemical Compositions/%						
	SiO_2	Al_2O_3	Fe_2O_3	CaO	MgO	Na_2O	K_2O
Coarse aggregate	62	17	5	6	5	3	2
Fine aggregate	62.8	15.5	6.2	4.4	4.9	3.3	2.9
Mineral Filler	34	4	3	56	2	0.6	0.4

Table 3. Cumulative passing rate of the recycled asphalt mixture (%).

Types	Particle Size/mm											
	26.5	19.0	16.0	13.2	9.5	4.75	2.36	1.18	0.6	0.3	0.15	0.075
Recycled asphalt	100	100	100	98.2	98.8	95.4	76.7	65.0	45.4	31.6	23.9	13.1
Coarse aggregate	100	100	100	94.0	19.7	0.0	0.0	0.0	0.0	0.0	0.0	0.0
Fine aggregate	100	100	100	100	100	99.9	87.8	71.1	42.2	25.4	14.9	0.0
Mineral Filler	100	100	100	100	100	100	100	100	100	100	98.8	84.5

2.2. Preparation of Asphalt Mixture

The asphalt modifiers were weighed on an electronic scale. After that, the asphalt modifiers were added to the base asphalt, including different types of recycled asphalt. The samples were dried at a temperature of 150 °C and at a speed of 1000 r/min for 10 min. After this stirring, the samples were mixed at a temperature of 160 °C and at a speed of

4000 r/min for another 60 min. When the stirring was finished, the samples were poured into molds to form specimens with sizes of Φ101.6 mm × 63.5 mm, 30 mm × 35 mm × 250 mm, and Φ100 mm × 100 mm. Six replicate specimens were prepared for each group. A Marshall compaction instrument purchased from Wuxi Huanan Experimental Instrument Co., Ltd., Wuxi, China, with a standard compaction hammer mass of 4.5 kg and a drop distance of 457 mm, was used to compact the asphalt mixture. The compaction speed is 60 times per minute, and the motor specifications are 370 W, 380 V, and 50 Hz. The mixing proportions of the recycled asphalt mixture are shown in Table 4. Figure 1 shows the recycled asphalt mixture. The modifier dosage (butadiene–styrene rubber or polyurethane) remained at 5% by mass of the asphalt mixture in this study, referring to prior research [27–29]. The mass ratios of recycled asphalt by the total mass of asphalt were 5%~95%.

Table 4. The mixing proportions of the recycled asphalt mixtures (kg/m^3).

Recycled Asphalt	Coarse Aggregate	Fine Aggregate	New Asphalt	Butadiene–Styrene Rubber	Polyurethane	Mineral Filler	Recycled Asphalt Mass Ratio (%)
103.5	264.5	126.5	1564	120.06	120.06	69	5
308.2	236.9	112.7	1400.7	120.06	120.06	69	15
515.2	209.3	98.9	1235.1	120.06	120.06	69	25
719.9	181.7	85.1	1071.8	120.06	120.06	69	35
926.9	154.1	71.3	906.2	120.06	120.06	69	45
1131.6	126.5	59.8	740.6	120.06	120.06	69	55
1338.6	98.9	46	575	120.06	120.06	69	65
1543.3	71.3	32.2	411.7	120.06	120.06	69	75
1750.3	41.4	20.7	246.1	120.06	120.06	69	85
1955	13.8	6.9	82.8	120.06	120.06	69	95

Figure 1. Asphalt mixture used for extracting recycled asphalt.

2.3. Measuring Process

2.3.1. The Measurement of Stability

The stability was measured by the following process. Specimens with a size of Φ101.6 mm × 63.5 mm were incubated in a 60 °C constant-temperature water bath for 30 min. Then, specimens of size Φ101.6 mm × 63.5 mm were loaded into the Marshall testing machine at a rate of 50 mm/min until failure. The maximum force value during the loading process is a measure of the stability of the asphalt mixture.

2.3.2. The Measurement of Mechanical Strengths

A loading testing machine with a temperature control box (provided by Jinan Touching Group, Jinan, China), whose temperature ranges from −40 °C to 900 °C, was used to measure the mechanical strengths. The temperature accuracy, loading range, and loading rate are 0.5 °C, 0 kN~100 kN, and 0.1 mm/min~500 mm/min, respectively. Before the measurement of mechanical strengths, specimens were cooled to room temperature and then cured in the standard environment (20.5 °C and 96.4% relative humidity) for 28 days. Specimens with a size of Φ101.6 mm × 63.5 mm were used for the measurement of splitting tensile strength. The loading rate and measuring temperature were 50 mm/min and 15 °C. Specimens with sizes of 30 mm × 35 mm × 250 mm and Φ100 mm × 100 mm were used for the measurement of flexural and compressive strengths. The loading rate was 50 mm/min, and the measuring temperature was 15 °C.

In order to study the effects of water and high temperature on the performance of asphalt mixtures, some specimens were immersed in water for 48 h. Some other specimens were subjected to a short-term oven-aging protocol at 150 °C for 4 h in a high-temperature chamber, provided by Shanghai Hanyu Technology Co., Ltd., Shanghai, China. This aging temperature was selected to effectively simulate the short-term thermal aging of the recycled asphalt mixture during construction stages (mixing, transportation, and paving), thereby enabling a more rigorous evaluation of its anti-aging capabilities and thermal stability [30–32]. After these treatments, the specimens were used for the measurement of mechanical strengths. The preparation process and macroscopic performance experiment of the recycled asphalt mixture are shown in Figure 2.

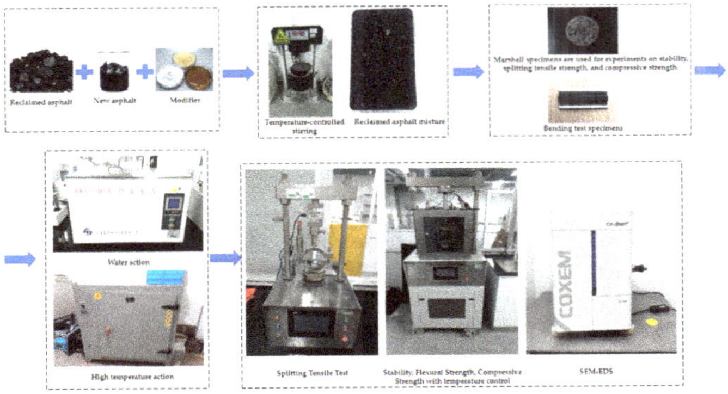

Figure 2. The preparation process and performance measurement of the recycled asphalt mixture.

2.3.3. The Measurement of Microscopic Properties

Ellipsoid samples with a diameter not exceeding 3 mm were used for the scanning electron microscopy–energy spectrum (SEM-EDS) analysis. The Hitachi S-4800 model scanning electron microscope, produced by Hitachi Corporation, Tokyo, Japan, was used for the measurement of SEM-EDS. Samples with gold-plated surfaces were placed at the top of the Hitachi S-4800 sample holder for cold field emission scanning electron microscopy. The acceleration voltage and the emission current of the SEM-EDS instrument are 15.0 kV and 10 μA.

3. Results and Discussion

3.1. The Stability of Asphalt Mixtures

The stability of asphalt mixtures containing different dosages of recycled asphalt is shown in Figure 3a. As observed in Figure 3a, the stability decreases in a cubic-function manner as a function of the mass ratio of the recycled asphalt mixture. This is attributed to the fact that the discarded asphalt has aged, and the performance of various parts has also deteriorated [33]. Therefore, the bonding strength between fresh asphalt and recycled asphalt mixtures decreases, leading to a decrease in stability. However, the asphalt modifier shows an improving effect on the stability of the asphalt mixture. The asphalt mixture with the modifier made of polyurethane exhibits higher stability than the asphalt mixture with the modifier made of butadiene–styrene rubber. When the recycled asphalt mixture is not treated with modifiers, the stability decreases at a rate of up to 37.3%. Meanwhile, when the polyurethane and butadiene–styrene rubber modifiers are added, the stability decreases at rates of up to 32.9% and 29.9%. The butadiene–styrene rubber and polyurethane modifiers can increase the stability at rates of 5.2%~11.8% and 9.1%~21.9%. This can be explained by the fact that the modifiers improve the bonding performance and compactness of asphalt [34]. Therefore, the modifiers exhibit an increasing effect on the stability of the asphalt mixture.

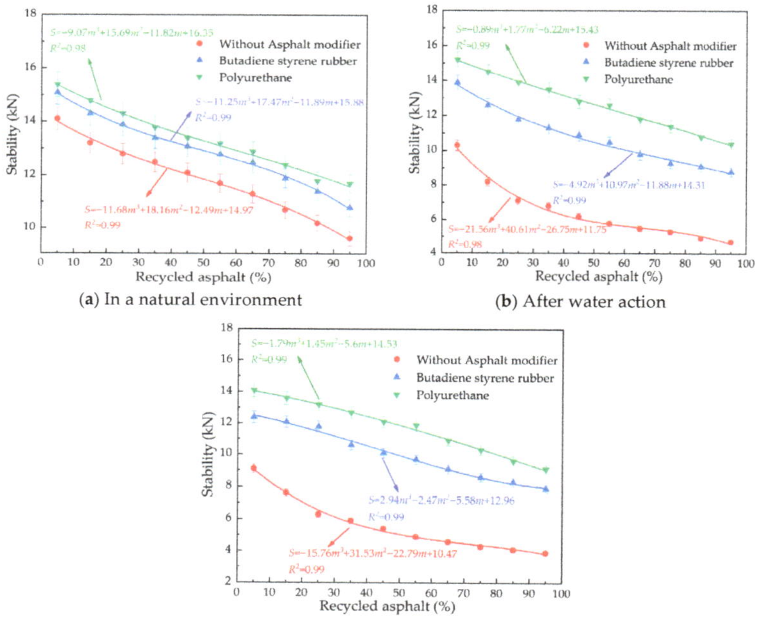

(**a**) In a natural environment

(**b**) After water action

(**c**) After high-temperature action

Figure 3. The stability of recycled asphalt mixtures.

Figure 3b shows the stability of the asphalt mixtures after being immersed in water for 48 h. As depicted in Figure 3b, the stability declines in a cubic-function manner with the mass ratios of recycled asphalt. As shown in this figure, the added modifiers can reduce the attenuation of stability. After being immersed in water for 4 h, the stability of the untreated asphalt mixture, butadiene–styrene rubber-modified asphalt mixture, and polyurethane-modified asphalt mixture decreases at rates of 20.9%~51.9%, 7.9%~21.8%, and 1.2%~11.1% after the untreated asphalt mixture, the butadiene–styrene rubber-modified asphalt mixture, and polyurethane-modified asphalt mixture are immersed in water for 48 h. After 4 h′ immersion, the stability decreases at rates of up to 61.2%, 38.5%, and 32.5% when adding 95% untreated asphalt, butadiene–styrene rubber-modified asphalt mixture, and polyurethane-modified asphalt mixture. After the asphalt mixture is immersed in water for 4 h, water invades the interior of the asphalt mixture through pores, gradually squeezes open the asphalt film, and occupies the surface of the aggregate, causing the asphalt to peel off from the aggregate. The asphalt film on the surface of the aggregate falls off, and the aggregate becomes loose [35]. The addition of butadiene–styrene rubber-modified asphalt mixture and polyurethane-modified asphalt mixture can increase the stability at rates of up to 87.2% and 121.3%, respectively. Asphalt modifiers can enhance the compactness of asphalt mixtures, and at the same time, asphalt modifiers can increase the adhesion between asphalt and aggregates, thus improving the water stability of asphalt mixtures [36].

The stability of asphalt mixtures after drying in an oven at a temperature of 150 °C for 4 h is shown in Figure 3c. As shown in Figure 3c, the relationship between the stability and the mass ratio of the recycled asphalt was fitted with a cubic function. In this figure, the stability decreases at rates of up to 63.9%, 39.7%, and 38.1% for the untreated asphalt mixture, butadiene–styrene rubber-modified asphalt mixture, and polyurethane-modified asphalt mixture. The high-temperature action reduces the stability at rates of up to 59.8%, 27.7%, and 22.2% for the untreated asphalt mixture, butadiene–styrene rubber-modified asphalt mixture, and polyurethane-modified asphalt mixture. High temperature can cause asphalt to soften and lose rigidity, thus decreasing the adhesion between asphalt and aggregate, as well as the compactness of the mixture [37]. In addition, high temperature can accelerate the oxidation of asphalt, causing changes in asphalt components, with an increase in asphaltene and a decrease in resin, which results in harder and more brittle asphalt [38]. Hence, the stability is decreased by high temperature. The stability of the modified asphalt mixtures with butadiene–styrene rubber and polyurethane can increase at rates of up to 102.6% and 142.9%, respectively. This can be attributed to the fact that the butadiene–styrene rubber and polyurethane can improve the compaction of and adhesion to asphalt [39], thus increasing the high-temperature stability of the asphalt mixture. In summary, short-term high-temperature exposure induces a thermal softening effect on asphalt, reducing the adhesion and cohesion of the asphalt binder [40]. The addition of modifiers can counteract these detrimental effects through their specific polymer structures, forming elastic networks or cross-linking points within the asphalt. Then, the thermal–oxidative aging pathways are effectively hindered. Therefore, the asphalt mixture's resistance to deformation at elevated temperatures is improved by the modifiers. Compared with prior research, the stability of the recycled asphalt mixture with polyurethane is 95.3% that of the ordinary asphalt mixture [41].

3.2. *The Mechanical Strengths of Asphalt Mixtures*

Figure 4a presents the splitting tensile strength of the recycled asphalt mixtures. As depicted in Figure 4a, the relationship between the splitting tensile strength and the mass ratio of the recycled asphalt mixture adheres to a cubic function. The increasing dosages of the recycled asphalt mixture decrease the splitting tensile strength at rates of up to 67.2%, 58.5%, and 51.3% for the recycled asphalt mixture without an asphalt modifier, the recycled asphalt mixture modified by butadiene–styrene rubber, and the recycled asphalt mixture treated with polyurethane. This phenomenon occurs because the aged asphalt experiences diminished adhesion and structural density over time, which consequently reduces its splitting tensile strength [40]. The asphalt modifiers butadiene–styrene rubber and polyurethane can increase the splitting tensile strength of the recycled asphalt mixture at rates of 9.9%~39.3% and 15.2%~71.4%. This is ascribed to the improved compactness and bonding performance, leading to increased splitting tensile strength [42].

Figure 4b,c illustrate the flexural and compressive strengths of the recycled asphalt mixtures. As depicted in Figure 4b,c, the flexural and compressive strengths vary in the form of cubic functions with the mass ratio of recycled asphalt. With the increase in the recycled asphalt's mass ratio, the flexural strength decreases at rates of up to 73.6%, 72.8%, and 72.1%. The corresponding compressive strength decreases at rates of up to 63.3%, 56.8%, and 50.1%. This can be attributed to the fact that the aging recycled asphalt reduces the bonding performance between asphalt and aggregate, leading to decreased flexural and compressive strengths [43]. The addition of the asphalt modifiers butadiene–styrene rubber and polyurethane increases the flexural strength at rates of up to 41.1% and 87.6%. With the addition of asphalt modifiers, the corresponding compressive strength is increased at rates of up to 56.9% and 73.8%. This can be explained by the improved bonding properties and compactness with the addition of the asphalt modifier, resulting in increased flexural and compressive strengths. When the mechanical strengths of the recycled asphalt mixture are compared with those of the ordinary asphalt mixture, the recycled asphalt mixture shows splitting tensile strength, flexural strength, and compressive strength that are 93.4%, 95.1%, and 90.3% those of the ordinary asphalt mixture [44,45].

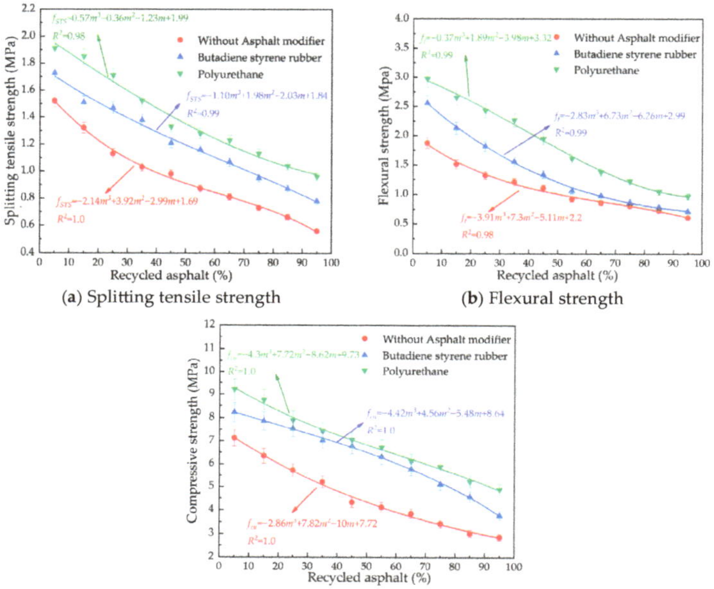

(**a**) Splitting tensile strength (**b**) Flexural strength

(**c**) Compressive strength

Figure 4. The mechanical strengths of recycled asphalt mixtures.

Figure 5a shows the splitting tensile strength of asphalt mixtures after immersion in water for 48 h. As shown in Figure 5a, the splitting tensile strength of the recycled asphalt mixture decreases with the mass ratio of recycled asphalt. The relationship between the splitting tensile strength and the mass ratio follows a cubic-function pattern. With the increasing mass ratio of recycled asphalt, the splitting tensile strength decreases at rates ranging from 66.3% to 78.6%. This can be attributed to the softening effect of water on asphalt, which results in a decrease in the splitting tensile strength [46]. The added asphalt modifiers butadiene–styrene rubber and polyurethane increase the splitting tensile strength at rates of up to 39.2% and 86.9%. Water action can decrease the splitting tensile strength at rates ranging from 2.3% to 42.3%. This is ascribed to the modifier's role in improving the structural integrity and internal cohesion of the mixture [47].

Figure 5b,c exhibit the flexural and compressive strengths of recycled asphalt mixtures after water action. The relationships between the flexural and compressive strengths follow cubic functions. The flexural and compressive strengths decrease at rates of up to 72.1% and 77% as the mass ratio of the added recycled asphalt increases. The flexural strength decreases at rates of 3.3%~35.1% after 48 h immersion in water. Meanwhile, the corresponding compressive strength decreases at rates of 3.8%~46.3% after 48 h immersion in water. The water action reduces the adhesion between the asphalt and aggregate interface, leading to a decrease in flexural and compressive strengths of the recycled asphalt mixture. The asphalt modifiers butadiene–styrene rubber and polyurethane increase the flexural strength at rates of 1.5%~40.7% and 16.2%~81.6%. The compressive strength increases at rates of 21.2%~69.5% and 40.2%~81.8%. This can be ascribed to the improved adhesive performance and higher compactness caused by the asphalt modifiers, thus increasing the flexural and compressive strengths [48].

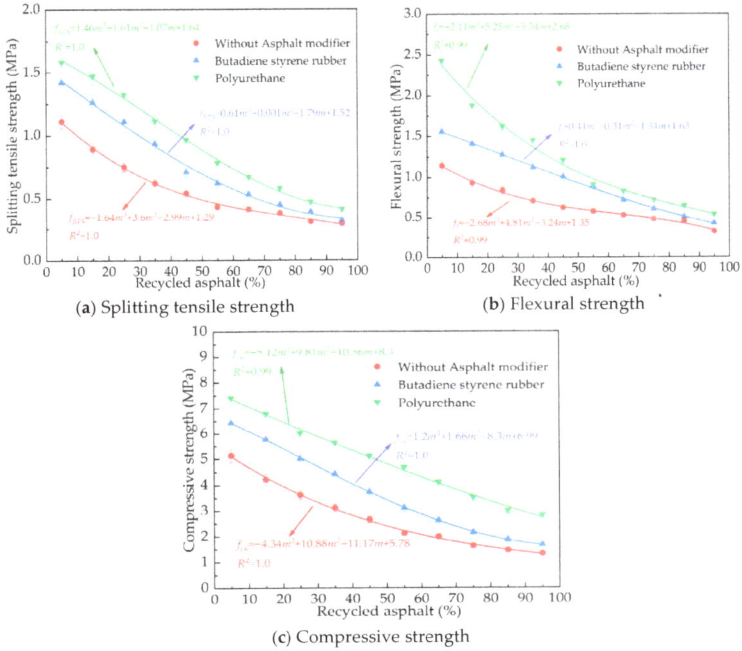

(c) Compressive strength

Figure 5. The splitting tensile, flexural, and compressive strengths of asphalt mixtures with different dosages of recycled asphalt after water action.

The splitting tensile strength of recycled asphalt mixtures is demonstrated in Figure 6a. The specimens were exposed to a temperature of 150 °C for 4 h. The splitting tensile strength follows a cubic function with the mass ratio of recycled asphalt. There are three types of recycled asphalt mixtures: the recycled asphalt mixture without surface modification, the recycled asphalt mixture modified by butadiene–styrene rubber, and the recycled asphalt mixture treated with polyurethane. The splitting tensile strength decreases at rates of up to 78.1%, 76.1%, and 74.8% for these three mixtures, respectively. This is because high temperature softens the asphalt, thereby reducing the splitting tensile strength [49]. High temperature reduces the splitting tensile strength at rates of 16.6%~59.1%. The asphalt modifiers butadiene–styrene rubber and polyurethane can increase the splitting tensile strength of recycled asphalt mixture at rates of up to 50% and 81.4%. This can be explained by the fact that modifiers improve the high-temperature resistance of asphalt mixtures [50]. Therefore, the attenuation effect on splitting tensile strength after high-temperature exposure is delayed.

The flexural and compressive strengths of recycled asphalt mixtures after being exposed to 150 °C temperature for 4 h are illustrated in Figure 6b,c. The flexural and compressive strengths decrease in the form of cubic functions with the mass ratio of the recycled asphalt mixture. The flexural strength decreases at rates of up to 77.8%, 76.7%, and 66.9% with the addition of the untreated asphalt mixture, butadiene–styrene rubber-modified asphalt mixture, and polyurethane-modified asphalt mixture. The corresponding compressive strength is reduced at rates of up to 77.3%, 73.6%, and 70.1%. When the recycled asphalt mixture is exposed to a high-temperature environment, the softening of the asphalt and component aging are accelerated. Then, the structural hardening and interfacial bonding deteriorate. Consequently, the mechanical strengths are decreased by high temperature [51]. The asphalt modifiers butadiene–styrene rubber and polyurethane can increase flexural strength by up to 60.3% and 113.2%, respectively. The corresponding compressive strength increases at rates of up to 47.2% and 120.3%. High temperature decreases the flexural and compressive strengths at rates of 12.8%~60.3% and 16.8%~47.6%. This can be ascribed to the fact that the modifiers can improve the recycled asphalt mixture's thermal stability and aging resistance [52]. Consequently, the flexural and compressive strengths are increased by the asphalt modifiers after exposure to 150 °C for 4 h. It can be concluded from this study that the mechanical properties of the recycled asphalt mixture and recycled asphalt satisfy cubic functions. The mechanical properties include the cohesive force of asphalt emulsion, the adhesion force at the asphalt–aggregate interface, and the interlocking friction between aggregate particles, which follow the shear strength theory of viscoelastic materials and the Mohr–Coulomb criterion [53]. Therefore, the mechanical properties roughly follow a cubic-function relationship. Moreover, the cubic-function fitting yields a higher value than the quadratic, linear, and other functions tested. Hence, cubic-function fitting is selected.

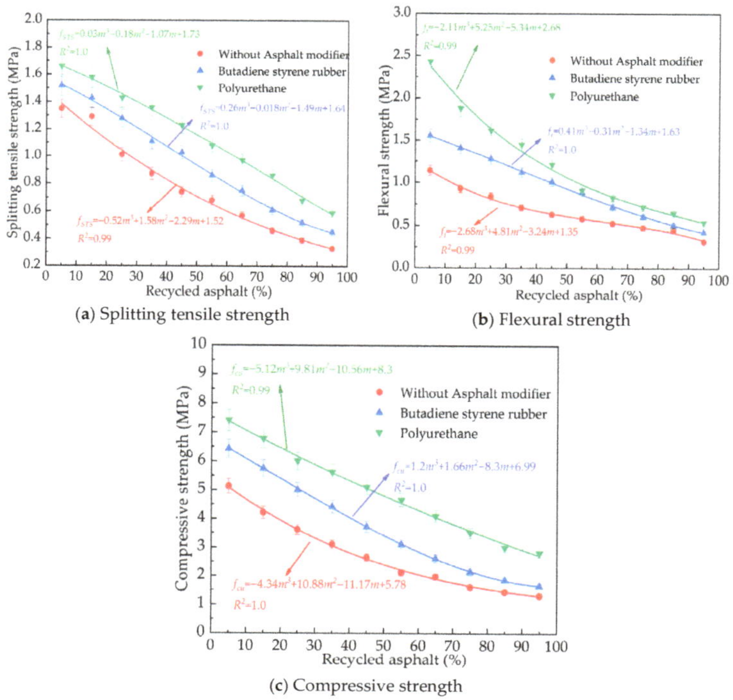

Figure 6. The splitting tensile, flexural, and compressive strengths of asphalt mixtures with different dosages of recycled asphalt after high-temperature action.

3.3. Microscopic Results

The scanning electron microscope–energy-dispersive spectroscopy (SEM-EDS) photos are shown in Figure 7. As illustrated in Figure 7, the elements C, O, Mg, Al, Si, Cl, Ca, and Fe are found. As can be observed in Figure 7, the added recycled asphalt leads to a decrease in the compactness of the microstructures and the contents of C and O. The elements Si and Fe are increased by the addition of the recycled asphalt mixture. When the asphalt modifiers butadiene–styrene rubber and polyurethane are added, the contents of C and O increase. Meanwhile, the asphalt modifiers reduce the looseness of the structures. The microstructure becomes denser due to the addition of modifiers. The microstructures of recycled asphalt mixtures with the polyurethane asphalt modifiers show higher compactness than the recycled asphalt mixture with the butadiene–styrene rubber asphalt modifier.

The SEM-EDS results elucidate the origins of the macroscopic behavior. The decline in C and O signifies the aging and degradation of the binder in the recycled asphalt mixture, compromising its adhesive and cohesive properties [54]. Concurrently, the increase in Si and Fe indicates a more pronounced contribution from the mineral aggregates. This signal enhancement is attributed to the thinning and discontinuity of the aged asphalt binder film, which reduces its masking effect and thereby enhances the detection of aggregate-characteristic elements [55]. These compositional changes collectively degrade the interface between the aged asphalt binder and the aggregates, thereby reducing microstructural compactness. This microstructural degradation directly accounts for the inferior stability and mechanical strengths of the unmodified recycled asphalt mixture. In contrast, the modifiers improve the properties of the aged binder. The butadiene–styrene rubber and polyurethane modifiers enhance the cohesion and adhesion of the binder, as evidenced by the restored C and O contents and the significantly denser microstructure [56,57]. This microstructural enhancement, facilitated by the improved binder, is the fundamental reason for the superior performance recovery of the modified recycled asphalt mixtures.

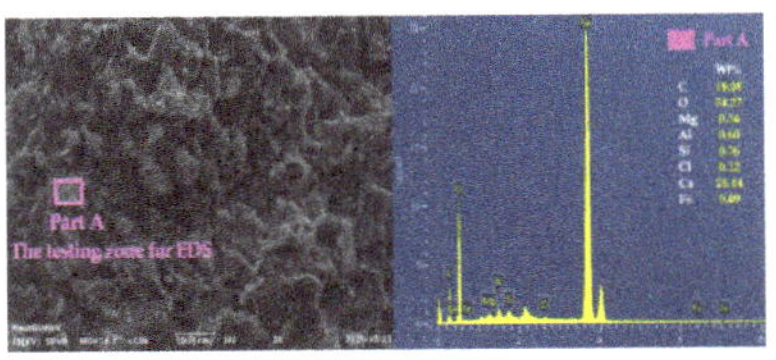

(a) With 15% recycled asphalt

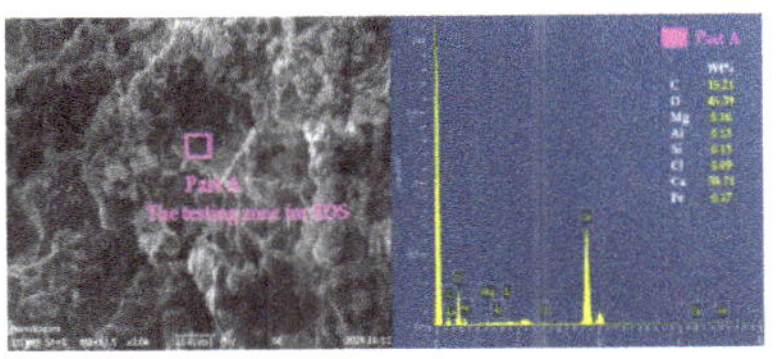

(b) With 35% recycled asphalt

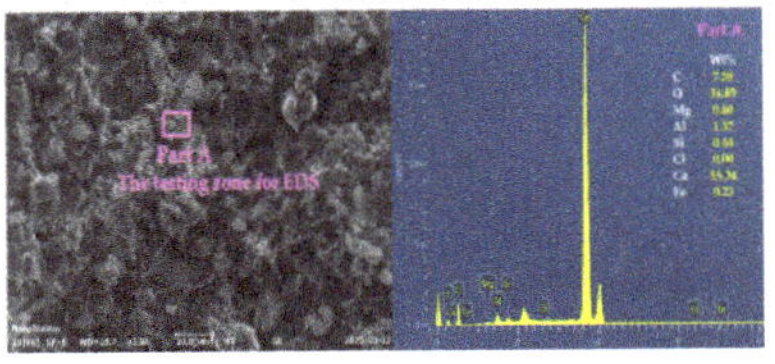

(c) With 65% recycled asphalt

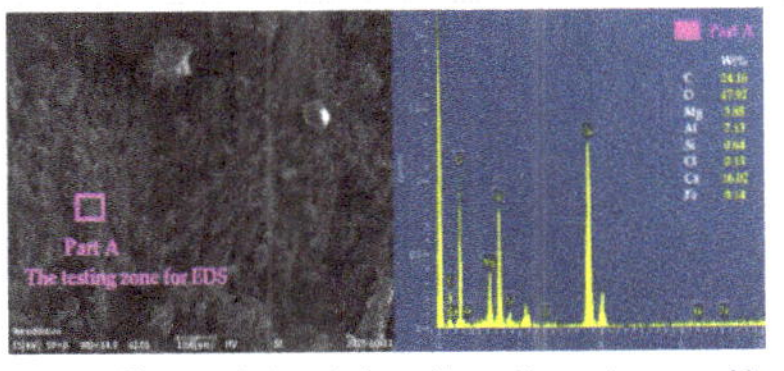

(d) With 65% recycled asphalt and butadiene–styrene rubber

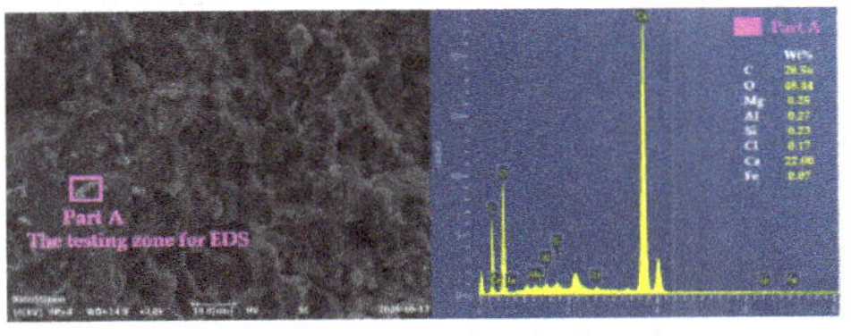

(e) With 65% recycled asphalt and polyurethane

Figure 7. SEM-EDS results of specimens with recycled asphalt and modifier.

4. Conclusions

This study systematically evaluated the performance evolution of recycled asphalt mixtures under varying incorporation ratios and aging conditions. It revealed the mechanisms and microstructural basis of enhancement provided by styrene–butadiene rubber and polyurethane modifiers. The principal conclusions are as follows:

(1) The incorporation of recycled asphalt significantly impairs the stability and mechanical properties of asphalt mixtures, with the degree of deterioration increasing as a cubic function of the recycled asphalt content. Both water immersion and high-temperature (150 °C) aging substantially exacerbated this performance degradation, highlighting the poor moisture and thermal stability of unmodified recycled asphalt mixtures.

(2) Modifiers effectively enhanced the overall performance of recycled asphalt mixtures. Both styrene–butadiene rubber and polyurethane significantly improved the stability and mechanical strengths. Polyurethane demonstrated superior performance in enhancing high-temperature stability and resistance to water damage, confirming its potential for application in demanding environments.

(3) Microstructural analysis indicated that the addition of recycled asphalt reduced the contents of C and O, increased the relative concentrations of mineral elements such as Si and Fe, and led to a more porous structure, uncovering the intrinsic reason for the macroscopic performance decline. The introduction of modifiers effectively restored the continuity of the binder, strengthened the interfacial bonding, and improved the compactness and integrity of the material.

In summary, employing suitable modifiers, particularly polyurethane, can significantly enhance the road performance and environmental durability of high-RAP-content mixtures. This work provides crucial technical support for the resource-efficient and high-value utilization of waste asphalt pavement materials.

Author Contributions: Conceptualization, X.W. and H.Z.; methodology, C.G. and H.W.; validation, F.S. and H.W.; formal analysis, S.L. and Q.X.; investigation, C.G., J.Y. and S.L.; resources, J.Y.; data curation, Q.X. and H.Z.; writing—original draft preparation, X.W. and C.G.; writing—review and editing, X.W., H.Z., C.G., Q.X., J.Y., F.S., S.L. and H.W.; visualization, F.S.; supervision, H.W.; project administration, X.W. and J.Y.; funding acquisition, X.W. and H.Z. All authors have read and agreed to the published version of the manuscript.

Funding: This work is sponsored by the Basic Science (Natural Science) General Research Project of Jiangsu Higher Education Institutions [No. 24KJD580003]. Research Project: Study on Mix Proportion Design of Hot In-Place Recycling for Asphalt Mixtures in Different Traffic Lanes.

Institutional Review Board Statement: Not applicable.

Informed Consent Statement: Not applicable.

Data Availability Statement: The data used to support the findings of this study are available from the corresponding authors upon request.

Conflicts of Interest: Author Qi Xue was employed by the company Nanjing Municipal Design and Research Institute Co., Ltd. The remaining authors declare that the research was conducted in the absence of any commercial or financial relationships that could be construed as a potential conflict of interest.

References

1. Russo, N.; Filippi, A.; Carsana, M.; Lollini, F.; Redaelli, E. Recycling of RAP (Reclaimed Asphalt Pavement) as Aggregate for Structural Concrete: Experimental Study on Physical and Mechanical Properties. *Discov. Civ. Eng.* **2024**, *1*, 99. [CrossRef]

2. Mariyappan, R.; Palammal, J.; Balu, S. Sustainable use of reclaimed asphalt pavement (RAP) in pavement applications—A review. *Environ. Sci. Pollut. Res.* **2023**, *30*, 45587–45606. [CrossRef]
3. Su, Y.; Luo, B.; Luo, Z.; Xu, F.; Huang, H.; Long, Z.; Shen, C. Mechanical characteristics and solidification mechanism of slag/fly ash-based geopolymer and cement solidified organic clay: A comparative study. *J. Build. Eng.* **2023**, *71*, 106459. [CrossRef]
4. Yao, L.; Leng, Z.; Lan, J.; Chen, R.; Jiang, J. Environmental and economic assessment of collective recycling waste plastic and reclaimed asphalt pavement into pavement construction: A case study in Hong Kong. *J. Clean. Prod.* **2022**, *336*, 130405. [CrossRef]
5. Da Costa, L.F.; Mendes, L.P.T.; de Medeiros Melo Neto, O.; de Figueiredo Lopes Lucena, L.C.; de Figueiredo Lopes Lucena, L. Economic and environmental benefits of recycled asphalt mixtures: Impact of reclaimed asphalt pavement and additives throughout the life cycle. *Environ. Sci. Pollut. Res.* **2025**, *32*, 11269–11291. [CrossRef]
6. Tarsi, G.; Tataranni, P.; Sangiorgi, C. The challenges of using reclaimed asphalt pavement for new asphalt mixtures: A review. *Materials* **2020**, *13*, 4052. [CrossRef] [PubMed]
7. Azadgoleh, M.A.; Mohammadi, M.M.; Ghodrati, A.; Sharifi, S.S.; Palizban, S.M.M.; Ahmadi, A.; Vahidi, E.; Ayar, P. Characterization of contaminant leaching from asphalt pavements: A critical review of measurement methods, reclaimed asphalt pavement, porous asphalt, and waste-modified asphalt mixtures. *Water Res.* **2022**, *219*, 118584. [CrossRef] [PubMed]
8. Min, Y.; Zhang, Y.; Qian, Z.; Zhang, H.; Hu, B.; Xie, Y.; Zhang, X. Performance evaluation and feasibility study on reuse of reclaimed epoxy asphalt pavement (REAP). *Constr. Build. Mater.* **2024**, *450*, 138653. [CrossRef]
9. Shams, H.; Ullah, H.; Qiu, Y.; Abdrhman, H.; Fu, J.; Yang, E. Investigating the Performance of Asphalt Pavements Modified with Reclaimed Asphalt and Crumb Rubber. *Sci. Rep.* **2025**, *15*, 32194. [CrossRef]
10. Cai, Y.; Cao, L.; Guo, J.; Tan, Z. Restoring Road Performance of Recycled Asphalt Mastic through Filler Asphalt Ratio Adjustment. *Sci. Rep.* **2025**, *15*, 17775. [CrossRef]
11. Gholami, M.; Khodaii, A.; Hajikarimi, P. A Multifaceted Purpose-Oriented Approach to Evaluate Material Circularity Index for Rejuvenated Recycled Asphalt Mixtures. *Sci. Rep.* **2025**, *15*, 12213. [CrossRef]
12. Nascimento, F.A.C.; Guimarães, A.C.R.; Castro, C.D. Comparative study on permanent deformation in asphalt mixtures from indirect tensile strength testing and laboratory wheel tracking. *Constr. Build. Mater.* **2021**, *305*, 124736. [CrossRef]
13. Costa, J.; Borges, P.; Santos, F.; Bezerra, A.; Van, W.; Blom, J. Cementitious binders and reclaimed asphalt aggregates for sustainable pavement base layers: Potential, challenges and research needs. *Constr. Build. Mater.* **2020**, *265*, 120325. [CrossRef]
14. Yi, X.; Chen, H.; Wang, H.; Shi, C.; Yang, J. The feasibility of using epoxy asphalt to recycle a mixture containing 100% reclaimed asphalt pavement (RAP). *Constr. Build. Mater.* **2022**, *319*, 126122. [CrossRef]
15. Pasandín, A.R.; Pérez, I.; Gómez-Meijide, B. Performance of High RAP Half-Warm Mix Asphalt. *Sustainability* **2020**, *12*, 10240. [CrossRef]
16. Pradhan, S.K.; Das, U.; Patra, A.R. Utilization of Reclaimed Asphalt Pavement (RAP) Materials in HMA Mixtures for Flexible Pavement Construction. *Mater. Today Proc.* **2023**. [CrossRef]
17. Prosperi, E.; Bocci, E.; Marchegiani, G. The Influence of Bitumen Nature and Production Conditions on the Mechanical and Chemical Properties of Asphalt Mixtures Containing Reclaimed Asphalt Pavement. *Materials* **2025**, *18*, 3713. [CrossRef] [PubMed]
18. Sun, J.; Cheng, Y.; Zou, G.; Wang, W. Study on the Water Stability and Safety Mechanism of Warm Mixed Recycled SBS Modified Asphalt Mixture. *Sci. Rep.* **2025**, *15*, 31188. [CrossRef] [PubMed]
19. Yu, D.; Luo, J.; Lang, H.; Hua, F.; Yang, Y.; Xie, M. Study on the Micro-Surfacing Properties of SBR Modified Asphalt Emulsion with Reclaimed Asphalt Pavement. *Mater* **2025**, *18*, 802. [CrossRef] [PubMed]
20. Fang, Y.; Zhang, Z.; Zhang, K.; Li, Z. Rheological property evaluation and microreaction mechanism of rubber asphalt, desulfurized rubber asphalt, and their composites. *J. Mater. Civ. Eng.* **2021**, *33*, 04021100. [CrossRef]
21. Zhang, Z.; Wang, Z.; Zhang, S.; Liu, H.; Guo, Y.; Liu, X.; Tian, P.; Yang, Y.; Xia, J. Experimental Studies and Molecular Dynamics Simulation of the Compatibility between Thermoplastic Polyurethane Elastomer (TPU) and Asphalt. *Constr. Build. Mater.* **2024**, *411*, 134316. [CrossRef]
22. Gong, M.; Jiao, B.Z. Thermodynamic properties analysis of warm-mix recycled a sphalt binders using molecular dynamics simulation. *Road Mater. Pavement Des.* **2023**, *25*, 239–258. [CrossRef]
23. Zhang, Z.; Sun, J.; Jia, M.; Ban, X.; Wang, L.; Chen, L.; Huang, T.; Liu, H. Effects of Polyurethane Thermoplastic Elastomer on Properties of Asphalt Binder and Asphalt Mixture. *J. Mater. Civ. Eng.* **2021**, *33*, 04020477. [CrossRef]
24. Zeng, Z.; Sudarsanan, N.; Underwood, B.S.; Kim, Y.R.; Guddati, M. Reflective Cracking Performance Evaluations of Highly Polymer-Modified Asphalt Mixture. *J. Transp. Eng. Part B Pavements* **2024**, *150*, 04024039. [CrossRef]
25. Jin, X.; Sun, S.; Guo, N.; Huang, S.; You, Z.; Tan, Y. Influence on Polyurethane Synthesis Parameters Upon the Performance of Base Asphalt. *Front. Mater.* **2021**, *8*, 2296. [CrossRef]
26. Liu, Y.; Dong, J.; Yuan, S.; Yu, J.; Guan, Z.; Chai, H.K.; Wang, Q. Impact resistance and enhancing mechanism of a novel mortar integrated with shear thickening fluid. *Constr. Build. Mater.* **2025**, *498*, 144036. [CrossRef]
27. Zhang, Y.; Jiang, C.; Li, M.; Qi, Z.; Yang, X.; Lin, Y.; Cao, S. A review on curve edge based architectures under lateral loads. *Thin. Wall. Struct.* **2025**, *217*, 113849. [CrossRef]

28. Sun, J.S.; Liao, T.; Ji, Y.X.; Li, H.; Qu, Y.Z.; Huang, X.B.; Li, J. Biomimetic inspired superhydrophobic nanofluids: Enhancing wellbore stability and reservoir protection in shale drilling. *Pet. Sci.* 2025, *in press*. [CrossRef]
29. Li, J.; Ji, Y.X.; Ni, X.X.; Lv, K.H.; Huang, X.B.; Sun, J.S. A micro-crosslinked amphoteric hydrophobic association copolymer as high temperature-and salt-resistance fluid loss reducer for water-based drilling fluids. *Pet. Sci.* **2024**, *21*, 1980–1991. [CrossRef]
30. Li, J.; Fan, J.; Sun, J.; Wang, Z.; Lv, K.; Qu, Y.; Ma, W. A Hyperbranched Copolymer as High-Temperature and Salt-Resistance Fluid Loss Reducer for Water-Based Drilling Fluids: Preparation, Evaluation, and Mechanism Study. *SPE J.* **2025**, 1–17. [CrossRef]
31. Wang, G.; Hao, J.; Wang, B.; Han, W. Comparative study on thermal and mechanical performances of liquid lead thermocline heat storage tank with different solid filling material layouts. *Case. Stud. Therm. Eng.* **2025**, *73*, 106754. [CrossRef]
32. Zong, Y.; Xiong, R.; Tian, Y.; Chang, M.; Wang, X.; Yu, J.; Zhang, Y.; Feng, B.; Wang, H.; Li, C. Preparation and Temperature Susceptibility Evaluation of Crumb Rubber Modified Asphalt Applied in Alpine Regions. *Coatings* **2022**, *12*, 496. [CrossRef]
33. *JTG E20-2011*; Standard Test Methods of Bitumen and Bituminous Mixtures for Highway Engineering. MOT: Beijing, China, 2011.
34. Agarwal, B.; Behl, A.; Kumar, R.; Dhamaniva, A. Fatigue Failure Assessment of Asphalt Mixtures Containing 40% Reclaimed Asphalt Pavement (RAP) Under Short-Term Aging Conditions. *J. Fail. Anal. Preven.* **2025**, *25*, 1831–1849. [CrossRef]
35. Hong, Z.; Yan, K.; Wang, M.; Yuan, J.; Ge, D.; Liu, J. The Laboratory Performance of Asphalt Mixture with Thermoplastic Polyurethane (TPU) and Amorphous Poly Alpha Olefin (APAO) Compound Modified Asphalt Binder. *Constr. Build. Mater.* **2022**, *349*, 128742. [CrossRef]
36. Wang, Y.; Zhang, W.; Wei, J.; Xiong, Y.; Qiao, Y.; Wang, X. Quantifying Moisture Susceptibility in Asphalt Mixtures Using Dynamic Mechanical Analysis. *Coatings* **2025**, *15*, 1109. [CrossRef]
37. Guo, T.; Li, Z.; Chen, Y.; Xu, Q.; Wang, J.; Jin, L. Study on the Performance and Mechanism of Graphene Oxide/Polyurethane Composite Modified Asphalt. *Sci. Rep.* **2025**, *15*, 2482. [CrossRef] [PubMed]
38. Chen, C.; Wu, H.; Zhang, W.; Mao, Y.; Zhang, C. Investigation of High Temperature Performance and Viscosity Characteristics of Modified and Unmodified Color Asphalt. *Exp. Tech.* **2022**, *46*, 317–333. [CrossRef]
39. Liu, B.; Zhang, K.; Fan, X.; Tu, C. Performance Evaluation of Desulfurized Rubber Powder and Styrene-Butadiene-Styrene Composite-Modified Asphalt. *Coatings* **2025**, *15*, 607. [CrossRef]
40. Gulzar, S.; Fried, A.; Preciado, J.; Castorena, C.; Underwood, S.; Habbouche, J.; Boz, I. Towards sustainable roads: A State-of-the-art review on the use of recycling agents in recycled asphalt mixtures. *J. Cleaner. Prod.* **2023**, *406*, 136994. [CrossRef]
41. Hu, J.; Qian, Z.; Huang, Q.; Liu, P. Investigation on high-temperature stability of recycled aggregate asphalt mixture based on microstructural characteristics. *Constr. Build. Mater.* **2022**, *341*, 127909. [CrossRef]
42. Hu, Y.; Cheng, X.; Sreeram, A.; Si, W.; Li, B.; Pipintakos, G.; Airey, D. Enhancing Fatigue Resistance and Low-Temperature Performance of Asphalt Pavements Using Antioxidant Additives. *Mater. Struct.* **2025**, *58*, 46. [CrossRef]
43. Qian, C.; Fan, W. Evaluation and characterization of properties of crumb rubber/SBS modified asphalt. *Mater. Chem. Phys.* **2020**, *253*, 123319. [CrossRef]
44. Al-Saffar, Z.H.; Yaacob, H.; Katman, H.Y.; Mohd Satar, M.K.I.; Bilema, M.; Putra Jaya, R.; Radeef, H. A review on the durability of recycled asphalt mixtures embraced with rejuvenators. *Sustainability.* **2021**, *13*, 8970. [CrossRef]
45. Yu, J.; Lin, Z.; Zou, G.; Yu, H.; Leng, Z.; Zhang, Y. Long-term performance of recycled asphalt mixtures containing high RAP and RAS. *J. Road. Eng.* **2024**, *4*, 36–53. [CrossRef]
46. Wang, M.; Li, J.; Chen, L.; Shi, C.; Ge, J. Research on Aging Characteristics and Interfacial Adhesion Performance of Polyurethane-Modified Asphalt. *Coatings* **2025**, *15*, 1194. [CrossRef]
47. Hamedi, G.H.; Dehnad, M.H. Effects of Dodecyl Amide, Nano Calcium Carbonate, and Dry Resin on Asphalt Concrete Cohesion and Adhesion Failures in Moisture Conditions. *Sci. Rep.* **2025**, *15*, 14451. [CrossRef]
48. Luan, Y.; Ma, T.; Wang, S.; Ma, Y.; Xu, G.; Wu, M. Investigating Mechanical Performance and Interface Characteristics of Cold Recycled Mixture: Promoting Sustainable Utilization of Reclaimed Asphalt Pavement. *J. Clean. Prod.* **2022**, *369*, 133366. [CrossRef]
49. Khan, A.M.; Elahi, A.; Waqas, R.M.; Kirgiz, M.S.; Nagaprasad, N.; Ramaswamy, K. Influence of Bentonite, Silica Fume, and Polypropylene Fibers on Green Concrete for Pavement and Structural Durability. *Sci. Rep.* **2025**, *15*, 29567. [CrossRef] [PubMed]
50. Ghaleh, M.; Asadi, P.; Eftekhar, M.R. Life Cycle Assessment Based Method for the Environmental and Mechanical Evaluation of Waste Tire Rubber Concretes. *Sci. Rep.* **2025**, *15*, 10687. [CrossRef] [PubMed]
51. Xie, S.; Cheng, Z.; Zhou, Y.; Cao, Y.; Wang, T.; Zhang, Z.; Dai, Y.; Zhang, W. Performance of Asphalt Mixtures Modified with Desulfurized Rubber and Rock Asphalt Composites. *Buildings* **2024**, *14*, 3026. [CrossRef]
52. Yang, M.; Chang, H.; Li, W.; Wang, H.; Lin, J.; Tong, Z.; Zhang, W. High-Temperature Deformation and Skid Resistance of Steel Slag Asphalt Mixture Under Heavy Traffic Conditions. *Buildings* **2024**, *14*, 3990. [CrossRef]
53. Orosa, P.; Pérez, I.; Pasandín, A.R. Evaluation of the shear and permanent deformation properties of cold in-place recycled mixtures with bitumen emulsion using triaxial tests. *Constr. Build. Mater.* **2022**, *328*, 127054. [CrossRef]
54. Li, Y.; Zhao, C.; Li, R.; Zhang, H.; He, Y.; Pei, J.; Lyu, L. Dry-Process Reusing the Waste Tire Rubber and Plastic in Asphalt: Modification Mechanism and Mechanical Properties. *Constr. Build. Mater.* **2025**, *458*, 139759. [CrossRef]

55. Ma, X.; Wang, D.; Liu, S.; Jiang, J.; Kan, J.; Tu, M. Analysis of asphalt microscopic and force curves under water-temperature coupling with AFM. *Case. Stud. Constr. Mat.* **2024**, *20*, e03071. [CrossRef]
56. Zhang, E.; Shan, L.; Tan, Y. Utilization of atomic force microscopy (AFM) in characterizing microscopic properties of asphalt binders: A review. *J. Test. Eval.* **2024**, *52*, 754–775. [CrossRef]
57. Zhao, S.; Zhang, H.; Gao, M.; Zhang, Q.; Sun, Q.; Dong, Q. Nano-microscopic analysis on the interaction of new and old asphalt mortar in recycled asphalt mixture. *Chem. Phys. Lett.* **2023**, *825*, 140593. [CrossRef]

coatings

MDPI

Article

Research on Fatigue Performance of Fast Melting Styrene-Butadiene-Styrene-Modified Asphalt with High Viscosity and Elasticity

Hao Zhang [1], Fei Guo [2], Xiaoyu Li [2], Shige Wang [1,*] and Jinchao Yue [2]

1 School of Architecture and Civil Engineering, Xinyang Normal University, Xinyang 464000, China
2 School of Water Conservancy and Transportation, Zhengzhou University, Zhengzhou 450001, China
* Correspondence: wangshige1979@163.com

Abstract

To overcome the limitations of conventional high-viscosity high-elasticity modified asphalt, including high production costs, phase separation, and thermal degradation, this study introduces a novel fast melting Styrene-Butadiene-Styrene modifier (SBS-T) for asphalt modification. The primary novelty of SBS-T lies in its ability to mitigate phase separation and thermal degradation while simplifying the production process, thereby offering a more robust and cost-effective alternative. The viscoelastic properties of SBS-T-modified asphalt were characterized through frequency sweep tests under varying loading conditions, while its fatigue behavior was quantitatively assessed using the Simplified Viscoelastic Continuum Damage (S-VECD) model. The results indicate that the SBS-T-modified asphalt exhibits outstanding viscoelastic performance across a broad range of temperatures and loading frequencies, and can better adapt to the temperature and load changes in complex pavement environments. Among them, the influence of long-term aging on the linear viscoelastic characteristics of SBS-T-modified asphalt is greater than that of ultraviolet aging. The SBS-T-modified asphalt also shows better stiffness and resistance to shear deformation. The fatigue life of asphalt gradually decreases with the deepening of the aging degree, among which the impact of long-term aging on fatigue life is greater than that of ultraviolet aging. Under different aging conditions, SBS-T-modified asphalt has shown good fatigue performance and is suitable for practical engineering applications.

Keywords: fast melting SBS; high viscosity and elasticity; linear amplitude sweep test; viscoelastic behavior; fatigue performance

1. Introduction

Vehicle overloading and channelized traffic exacerbate asphalt pavement damage. Water infiltration under traffic loads causes surface deterioration (pitting, loosening, shedding, grouting, cracking, potholes), compromising driving comfort, safety, and pavement lifespan [1,2]. Traditional asphalt exhibits limitations: it softens and ruts at high temperatures (>60 °C) while oxidizing and embrittling; conversely, it hardens and cracks at low temperatures (<0 °C) [3,4]. These deficiencies hinder its ability to meet modern highway demands for enhanced high-temperature stability, low-temperature crack resistance, and durability [4,5]. Consequently, developing high-viscosity, high-elasticity modified asphalt binders presents a promising solution.

As an increasingly important material in pavement engineering, high-viscosity and high-elasticity modified asphalt binder has demonstrated outstanding performance in highway and heavy-load road applications. Its superior characteristics, including remarkable thermal stability against rutting, exceptional flexibility at low temperatures, and sustained resistance to aging degradation, have made it a subject of extensive research in recent years [6,7]. Due to the excellent road performance of high-viscosity and high-elasticity asphalt and its strong adhesion to aggregates, it can significantly enhance the water damage resistance of asphalt pavement and is widely used in drainage pavement [8,9]. Currently, high-viscosity elastomeric modified asphalt exhibits several technical limitations: (1) thermodynamic instability manifests as SBS phase separation during storage and high-temperature conditions (>160 °C), particularly in low-aromatic-content binders; (2) processing demands stringent parameters (180–190 °C, 5000 rpm shear for 4–6 h) yet still yields limited storage stability. The root cause of these problems lies in their preparation process. To avoid these problems from the source, dry modification can be attempted. The key to the dry modification process lies in reducing the melting time of the modifier and accelerating the swelling rate of the modifier in asphalt [10–12]. Based on this, Guolu Gaoke optimized the structure of the SBS modifier and added components that enhance the melting capacity, achieving a balance between the melt index and the modification effect. A fast melting Styrene-Butadiene-Styrene (SBS-T) modifier was prepared—chemically, it differs from traditional SBS in that its styrene-butadiene block ratio is adjusted, and compatible components are grafted onto the molecular chain, which strengthens the interfacial adhesion between the modifier and asphalt. SBS-T is pre-ground into fine powder and then formed into granules. This drastically increases its surface area, enabling it to melt and disperse within the asphalt mix in seconds under the shear force of aggregates, unlike conventional SBS which requires prolonged, high-shear milling in asphalt. Physically, it has a reduced molecular weight distribution width and a more uniform particle size, lowering the energy barrier for melting and swelling. With a melt index equivalent to 100 times that of ordinary SBS [13–15], it can be directly put into the mixing plant for use and achieve rapid melting within 1 min. This structural and physical optimization enables it to perfectly solve problems such as performance attenuation of traditional high-viscosity and high-elasticity modified asphalt [16–18].

Performance assessment of high-viscosity, high-elasticity modified asphalt binders typically relies on fundamental parameters (penetration, ductility, softening point) to evaluate viscoelastic behavior and thermal susceptibility [19,20], supplemented by metrics like dynamic viscosity and the elastic recovery ratio to characterize adhesion and deformation resistance [21]. These quantitative measures effectively demonstrate performance enhancements from modifiers at varying concentrations. However, current studies often limit evaluation to qualitative interpretations of these parameters, lacking comprehensive analytical frameworks [22]. Fatigue cracking, a primary failure mode and cause of highway damage [23,24], involves the gradual expansion of pavement cracks, ultimately compromising surface integrity and significantly reducing pavement durability and service life [25–27]. Vehicle loads initiate micro-cracks, while temperature fluctuations (diurnal/seasonal) accelerate crack propagation, leading to macroscopic fatigue damage [28,29]. Consequently, asphalt's fatigue resistance critically determines its overall durability. Understanding the fatigue crack formation mechanism and developing effective improvement strategies are therefore vital for extending road service life and reducing maintenance costs.

This study employs a novel SBS-T to simultaneously compare the differential effects of thermo-oxidative and UV aging, and applies the S-VECD model to quantify fatigue performance. This approach breaks from conventional methods that isolate single aging types and rely solely on rheological characterization. Existing studies have proved that the addition of SBS-T modifiers can significantly improve the water stability performance, anti-aging performance, high-temperature rutting resistance, and low-temperature cracking resistance of asphalt. However, regarding the comparative study of the fatigue performance of SBS-T after short-term aging, ultraviolet aging, and long-term aging, there is still a lack of systematic exploration and analysis at present. The in-depth exploration in this direction is not only conducive to a comprehensive understanding of the long-term performance of SBS-T-modified asphalt, but can also provide a theoretical basis for its application in practical engineering. In this study, the stress–strain response laws of high-viscosity and high-elasticity SBS-T-modified asphalt with different dosages were obtained through LAS tests. Based on the S-VECD (Simplified Viscoelastic Continuous Damage) model, the definition of virtual strain energy fatigue failure, and the fatigue failure criterion, the fatigue life of high-viscosity and high-elasticity SBS-T-modified asphalt was quantitatively analyzed. We established the damage characteristic curve (DCC curve) of high-viscosity and high-elasticity SBS-T-modified asphalt, and compared and analyzed the fatigue performance of high-viscosity and high-elasticity SBS-T-modified asphalt under different aging environments. This study provides a theoretical basis for the application of high-viscosity and high-elasticity SBS-T-modified asphalt in drainage pavement. Its novelty lies in integrating multi-aging factor comparison with S-VECD-based fatigue quantification, addressing the gap of unsystematic fatigue analysis under diverse aging conditions. For application, it offers targeted guidance for SBS-T dosage optimization in drainage pavement, directly promoting the material's reliable use in durable pavement engineering.

2. Materials and Methods

2.1. Materials

2.1.1. Asphalt Binder

The selection of base asphalt significantly influences the performance of both asphalt binder and asphalt mixtures. In this study, 70# penetration grade base asphalt was selected, and its fundamental properties were characterized through standardized testing procedures. The technical specifications of the base asphalt are presented in Table 1.

The SBS-T modifier and conventional SBS modifier used in this study were supplied by Guolu Gaoke Engineering Technology Research Institute Co., Ltd. (No. 30 College Rd., Haidian District, Beijing, China), China. Figure 1 presents their morphological characteristics, while Table 2 summarizes their technical specifications.

Figure 1. SBS-T (**left**) and SBS (**right**) modifier [30].

Table 1. Basic technical specifications for 70# asphalt.

Technical Indicators	Unit	Measured Value	Test Methods
Penetration (25 °C)	0.1 mm	66.9	T0604-2011
Ductility (15 °C, 5 cm/min)	cm	>100	T0605-2011
Softening point	°C	52.8	T0606-2011
Brinell rotational viscosity (135 °C)	Pa·s	0.60	T0625-2011

Table 2. Technical specifications of modifiers.

Modifier	Technical Indicator	Unit	Measured Value
SBS	Structure type	-	Linearity
	Block ratio S/B	-	40/60
	Fracture elongation	%	≥700
	Melt flow rate	g/10 min	0.1–5.0
SBS-T	Appearance	-	Granular, uniform
	SBS content	%	≥50
	Ash content	%	≤5.0
	Melt index	g/10 min	≥2.0
	Dry mix dispersibility	-	No particle residue

2.1.2. Sample Preparation

The conventional SBS-modified asphalt and SBS-T-modified asphalt were prepared using the melt blending method. This process relies on thermal fusion at elevated temperatures to promote intermolecular interactions and achieve a homogeneous composite. SBS-T undergoes rapid physical melting and dispersion upon mixing with hot aggregate, immediately followed by an ultra-fast, catalyzed chemical crosslinking reaction within the asphalt binder. This entire process—achieving full polymer network formation through controlled crosslinking—is completed during a short mixing cycle, eliminating the prolonged maturation required by conventional SBS. Three SBS-T-modified asphalt formulations were prepared with modifier contents of 5 wt%, 6 wt%, and 7 wt%, alongside a control group comprising 6 wt% conventional SBS-modified asphalt. For each formulation, a minimum of three replicate samples were fabricated and tested to ensure statistical reliability. The preparation process, outlined in Figure 2, consisted of the following steps:

(1) The 70# penetration grade base asphalt was heated at 135 °C for approximately 2 h until fully liquefied, then transferred to a temperature-controlled hot plate maintained at 175 ± 2 °C.

(2) The predetermined quantity of SBS or SBS-T modifier was gradually introduced into the molten asphalt. Continuous manual stirring with a glass rod ensured uniform dispersion and prevented particle agglomeration.

(3) The mixture was subjected to high-shear mixing using an FM300 shear mixer (5000 rpm, corresponding to a shear rate of ~18.3 m/s, 30 min) under strict temperature control (175 ± 2 °C) to guarantee consistent polymer dissolution and blend homogeneity.

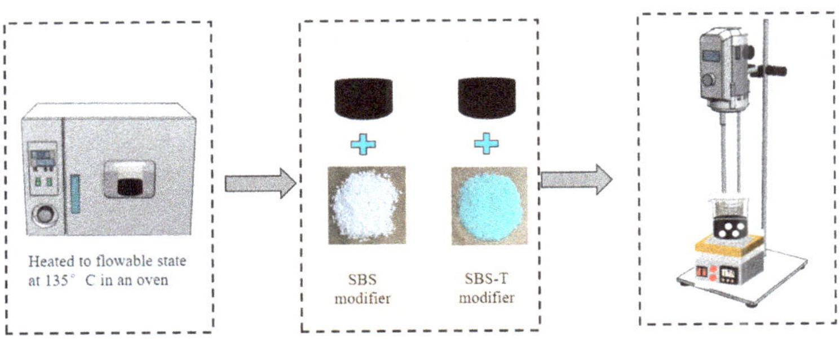

Figure 2. Preparation of modified asphalt.

According to the requirements of "Bitumen with high viscosity and elastic recovery for pavement" (GB/T 30516-2014), basic performance tests for SBS- and SBS-T-modified asphalt were carried out. The test samples were divided into the control group (6% SBS-modified asphalt) and the experimental groups (5%, 6%, and 7% SBS-T-modified asphalt). The test results are shown in Table 3. The asphalts with different modifier dosages all meet the technical standards of high-viscosity and high-elasticity modified asphalts.

Table 3. Technical specifications for modified asphalt.

Technical Indicators	Unit	6% SBS	5% SBS-T	6% SBS-T	7%SBS-T	Standard Values
Penetration (25 °C)	0.1 mm	50.1	53.7	49.3	48.1	40–80
Softening point	°C	76	77.5	86.7	89	≥70
Ductility (5 °C)	cm	28.1	31.3	33.7	36.3	≥20
Dynamic viscosity (60 °C)	Pa·s	25,370	23,100	26,880	29,630	≥20,000
Elastic recovery	%	92	94	96.5	98.1	≥85

2.2. *Aging Method*

2.2.1. Short-Term Aging

In compliance with JTG E20-2011 specifications, short-term aging was simulated using the Rolling Thin Film Oven Test (RTFOT) to evaluate the oxidative degradation of 70# base asphalt and SBS-T-modified asphalts during production and construction phases. The standardized protocol commenced with heating virgin asphalt to 160 ± 5 °C until complete fluidity was achieved, followed by precise measurement of 35.0 ± 0.5 g aliquots into pre-cleaned borosilicate glass bottles. Samples were then aged in a preheated RTFOT chamber at 163 ± 0.5 °C for 85 min under strictly controlled conditions, including 4000 ± 200 mL/min filtered dry air flow and 15 ± 0.2 rpm rotational speed. Immediately after aging, samples were collected for comparative testing, with the process maintaining <2% mass loss tolerance to ensure valid simulation of field aging mechanisms. This methodology (illustrated in Figure 3) effectively replicates thermo-oxidative binder hardening during mix production and paving operations while guaranteeing reproducible results through precise parameter control.

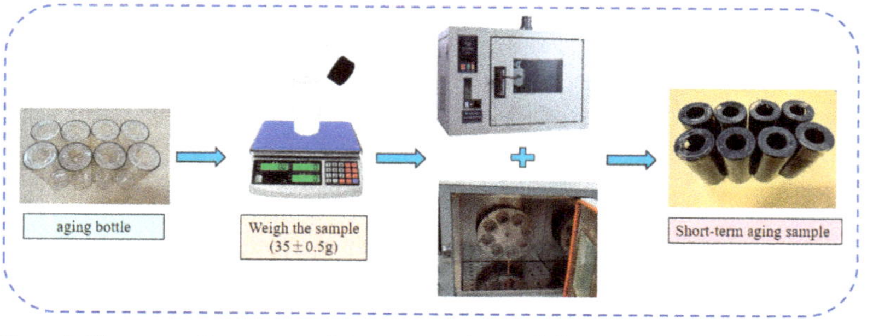

Figure 3. The preparation process of short-term aging asphalt.

2.2.2. UV Aging

As a viscoelastic composite, asphalt undergoes progressive degradation under combined thermal and solar radiation. While thermo-oxidative aging mechanisms are well characterized, photodegradation by ultraviolet (UV) radiation—constituting only 2% of the solar spectrum—remains poorly understood. The high-energy short-wavelength band (295–400 nm) drives complex photochemical reactions that systematically degrade asphalt's molecular structure over months to years of service.

Prior studies show that UV aging most significantly impacts asphalt performance within the first 194 h, with diminishing effects until stabilization. Accordingly, this study implemented a controlled 8-day (192 h) UV aging protocol using a 1000 W mercury arc lamp (primary emission: 365 nm). Temperature was maintained at 45 ± 0.5 °C via precision air cooling to isolate UV effects from thermal degradation.

For exposure, 20 g asphalt samples formed uniform 1 mm thick films in standardized trays (Ø140 mm × 10 mm depth). Trays were mounted on a rotating platform (5 rpm) within the aging chamber to ensure homogeneous irradiation. Figure 4 details the UV aging preparation.

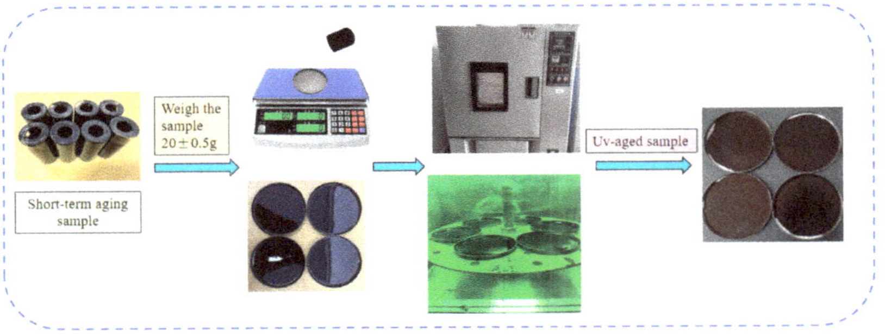

Figure 4. The preparation process of UV-aged asphalt.

2.2.3. Long-Term Aging

Asphalt undergoes sequential aging processes, beginning with short-term aging during construction (mixing and paving) followed by progressive long-term aging under environmental exposure during service life. To evaluate oxidation resistance under controlled conditions, this study conducted Pressure Aging Vessel (PAV) tests on 70# base asphalt and modified variants according to JTG E20-2011 specifications, using a precision PAV system (Tianjin Port Original Inspection Instrument Co., Ltd., Tianjin, China). The standardized protocol involved the following: (1) preparing 50 ± 0.5 g aliquots of RTFOT-aged asphalt in stainless steel pans (150 × 25 × 10 mm); (2) aging samples at 100 ± 0.1 °C and 2.1 ± 0.05 MPa for 20 ± 0.5 h to simulate approximately 5–10 years of field aging; and (3) immediate extraction of aged samples for subsequent analysis. The complete preparation process is illustrated in Figure 5, demonstrating our rigorous approach to replicating long-term oxidative aging in laboratory conditions while maintaining precise environmental controls (±0.1 °C temperature stability). This methodology provides reliable simulation of in-service aging mechanisms critical for performance evaluation of modified asphalts.

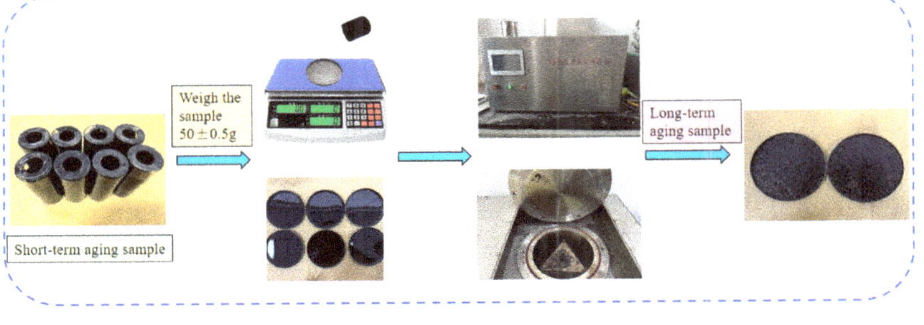

Figure 5. The preparation process of long-term aged asphalt.

2.3. Laboratory Testing

2.3.1. Frequency Sweep (FS) Test

Frequency sweep tests were performed using a dynamic shear rheometer in strain-controlled mode (0.1% strain) within the linear viscoelastic region (0.1–100 rad/s). This measured complex modulus, storage modulus, and loss modulus to characterize inherent viscoelastic properties. Tests were conducted on SBS-T-modified asphalt under four conditions: unaged, short-term aged, UV-aged, and long-term aged. Triplicate testing minimized sample variability.

Following the time–temperature superposition principle (TTSP) [31,32], master curves were constructed to unify data across temperatures. These curves delineate distinct behavioral regions, facilitating rheological analysis per Equation (1):

$$\log \omega_r = \log \omega + \log \varphi_T \tag{1}$$

where ω is the angular frequency in radians per second (rad/s); ω_r is the reduced angular frequency in radians per second (rad/s); and φ_T is the shift factor.

2.3.2. Linear Amplitude Sweep Test (LAS)

The LAS test measures asphalt's stress response under progressively increasing strain at a fixed frequency (10 Hz), identifying linear/nonlinear viscoelastic regions. At low strains, materials exhibit linear behavior; high strains induce nonlinearity with energy dissipation and strain hardening.

Tests at 25 °C applied three constant strain rates (0.1%/s, 0.06%/s, 0.0375%) across a 0.1%–30% strain, corresponding to 300 s, 500 s, and 800 s loading durations (Figure 6). Triplicate testing ensured repeatability.

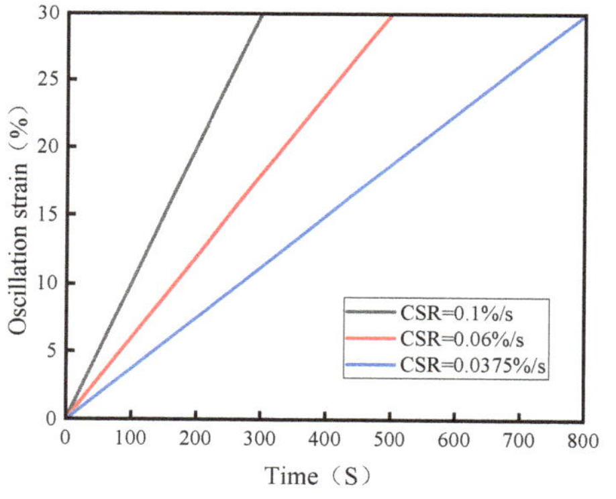

Figure 6. LAS test loading program.

This study evaluates the fatigue performance of modified asphalt based on the Simplified Viscoelastic Continuum Damage (S-VECD) theory. This theory overcomes the limitations of traditional testing conditions by introducing a damage variable to characterize the evolution of the material's internal state under loading. The research establishes a dynamic relationship model between the damage variable S and time t, enabling quantitative analysis of the material's fatigue damage behavior. The peak pseudo-strain of modified asphalt under cyclic loading can be expressed by Equation (2):

$$\gamma_p^R = \frac{1}{G_R} \cdot |G^*|_{LVE} \cdot \gamma_p \tag{2}$$

where γ_P is the virtual strain; G_R is the reference modulus of the material, taken as 1 MPa in this paper; and $|G^*|_{LVE}$ is the dynamic shear modulus of the asphalt material in linear viscoelasticity.

The relationship between the complex modulus C and the internal damage variable S of the asphalt material is expressed by Equation (3):

$$S(t) = \sum_{i}^{N}\left[\frac{DMR}{2}\left(\gamma_p^R\right)^2(C_{i-1} - C_i)\right]^{\frac{\alpha}{1+\alpha}}[t_{Ri} - t_{Ri-1}]^{\frac{1}{1+\alpha}} \quad (3)$$

$$t_{Ri} = \frac{t_i}{\varphi_T} \quad (4)$$

where DMR is the ratio of the initial dynamic shear modulus of the asphalt material to the dynamic shear modulus of the asphalt material in linear viscoelasticity; t_i is the loading time; and φ_T is the shift factor.

Based on the experimental data, the complex modulus C and damage variable S are fitted to derive the C-S curve of the material. The fitting formula is given in Equation (5):

$$C(t) = 1 - C_1 \cdot S(t)^{C_2} \quad (5)$$

where C_1 and C_2 are the fitting parameters of the material.

In the S-VECD theoretical model, the maximum stored peak pseudo-strain energy (PSE) is defined as the fatigue failure point of asphalt materials. The calculation formulas for stored energy, total pseudo-strain energy, and released energy are given by Equations (6)–(8), respectively:

$$W_s^R = \frac{1}{2} \times \tau_p \times \gamma_p^R / DMR = \frac{1}{2} \times C \times \left(\gamma_p^R\right)^2 \quad (6)$$

$$W_t^R = \frac{1}{2} \times \tau_{undamage} \times \gamma_p^R = \frac{1}{2} \times \left(\gamma_p^R\right)^2 \quad (7)$$

$$W_r^R = W_t^R - W_s^R = \frac{1}{2} \times (1 - C)\left(\gamma_p^R\right)^2 \quad (8)$$

The relationship between the stored energy and the released energy during the LAS test loading process is illustrated in Figure 7.

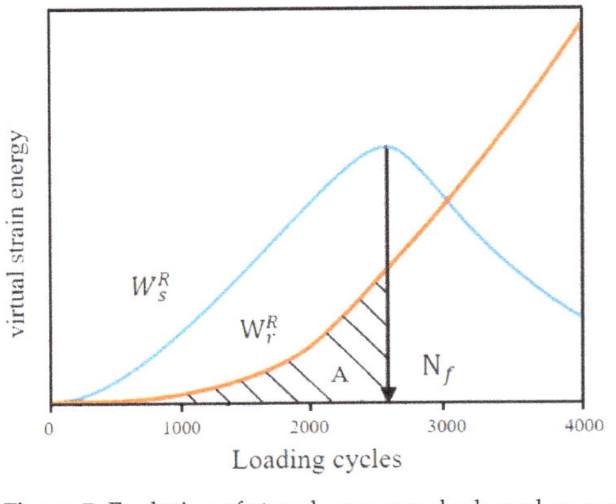

Figure 7. Evolution of stored energy and released energy in LAS test.

The peak value of ${W_s}^R$ is selected as the fatigue failure point, representing the state when the material reaches its maximum stored pseudo-strain energy. Once the failure point is determined, the average PSE from test initiation to fatigue failure can be calculated, and the pseudo-strain energy release rate (G^R) can be derived using Equation (9). This calculation method provides a more accurate description of the energy dissipation rate during dynamic loading, offering a more reliable theoretical basis for evaluating the fatigue performance of asphalt mixtures.

$$G^R = \frac{\overline{W_r^R}}{N_f} = \frac{A/N_f}{N_f} = \frac{A}{\left(N_f\right)^2} \tag{9}$$

where A is the area under the curve of the released pseudo-strain energy before the material reaches fatigue failure, as indicated by the shaded area in Figure 7. N_f denotes the number of loading cycles.

By varying the test methods and loading conditions, Wang et al. [33] plotted the calculated G^R (average PSE release rate) and corresponding N_f (fatigue life) on a double-logarithmic coordinate system. Their results demonstrated that the relationship between G^R and N_f can be well fitted by a power function. For the same material, the relationship between the average PSE release rate (G^R) and fatigue life (N_f) constitutes the fatigue failure criterion, which is capable of reflecting the inherent fatigue failure characteristics of the material itself, which can be fitted using Equation (10):

$$G^R = aN_f^b \tag{10}$$

where a and b are fitting parameters.

The fatigue life prediction equation for asphalt, based on the fatigue failure criterion, is presented in Equation (11):

$$N_f = \left[\frac{K}{a} \times (\gamma)^{2+2\alpha\left(\frac{C_2}{Q}\right)}\right]^{\frac{1}{b+1-\frac{C_2}{Q}}} \tag{11}$$

where K and Q are parameter combination values, which can be calculated using Equations (12)–(14):

$$K = \frac{1}{2} \times C_1 \times \left(|G^*|_{LVE}\right)^2 \times P^{(-C_2/Q)} \times \frac{1}{(C_2/Q)+1} \tag{12}$$

$$Q = 1 - \alpha \times C_2 + \alpha \tag{13}$$

$$P = \frac{f \times 2^{\alpha}}{(1-\alpha \times C_2 + \alpha)(C_1 \times C_2)\left(|G^*|_{LVE}\right)^{2\alpha}} \tag{14}$$

Equation (11) can be utilized to predict the fatigue life of asphalt materials under varying strains, thus offering a comprehensive analysis of their fatigue performance.

3. Results and Discussion

3.1. Dynamic Modulus Master Curve

The dynamic modulus represents a key parameter that characterizes the stiffness and deformation resistance of asphalt materials, where increased values correspond to enhanced load-bearing capacity. Through frequency sweep testing, the complex moduli of SBS-T-modified asphalt were measured across varying temperatures and loading frequencies, enabling construction of the dynamic modulus master curve through application of the time–temperature superposition principle. Figure 8 presents the following significant findings:

(1) In the unaged state, with increasing content of the SBS-T modifier, the complex modulus of the high-viscosity high-elasticity SBS-T-modified asphalt shows an upward trend, particularly in the low-frequency range (under low-shear-stress conditions). This indicates that the high-viscosity high-elasticity SBS-T-modified asphalt exhibits stronger elasticity and deformation resistance in its unaged condition. The reason lies in the optimized molecular structure and improved melting efficiency of the SBS-T modifier, which ensures its thorough dissolution and uniform dispersion within the asphalt matrix. Consequently, this enhances the overall rigidity and elastic recovery capability of the modified asphalt material. Compared with the traditional 6% SBS-modified asphalt with the same dosage, the 6% high-viscosity and high-elasticity SBS-T-modified asphalt shows a higher complex modulus in the unaged state, proving that it can provide better crack resistance and fatigue performance under low-temperature or low-stress conditions.

(2) Under short-term aging conditions, the complex modulus of high-viscosity and high-elasticity SBS-T-modified asphalt is generally increased compared with that of unmodified asphalt. The high-viscosity and high-elasticity modified asphalts of 5% SBS-T, 6% SBS-T, and 7% SBS-T all show good high-temperature stability. Short-term aging usually causes crosslinking and hardening of asphalt molecular chains, thereby enhancing its rigidity. During this process, high-viscosity and high-elasticity SBS-T-modified asphalt exhibits superior anti-aging ability compared to base asphalt, indicating that it can still maintain good elasticity and viscosity under high-temperature conditions. High-viscosity and high-elasticity SBS-T-modified asphalt can maintain better fluidity and elasticity during the short-term aging process, reducing cracks and fatigue damage caused by aging. In contrast, the modulus increase of SBS-modified asphalt with the same dosage was relatively small, and the extent of rigidity enhancement during the short-term aging process was limited, further verifying the superiority of high-viscosity and high-elasticity SBS-T-modified asphalt.

(3) During the ultraviolet aging process, the complex modulus of all high-viscosity and high-elasticity SBS-T-modified asphalts showed a significant increase, especially in the high-frequency band (under greater shear stress). Ultraviolet radiation can cause oxidation and crosslinking of the molecular chains in asphalt, resulting in an increase in the rigidity of the material. However, compared with SBS-modified asphalt and base asphalt of the same dosage, high-viscosity and high-elasticity SBS-T-modified asphalt shows a better recovery ability and lower hardening trend after ultraviolet aging. This indicates that high-viscosity and high-elasticity SBS-T-modified asphalt has obvious advantages in resisting ultraviolet aging.

(4) Under long-term aging conditions, the complex modulus of modified asphalt increased, particularly for 6% SBS-T and 7% SBS-T high-viscosity high-elasticity modified asphalt. Long-term aging enhances oxidation and crosslinking reactions in asphalt, thereby improving its stiffness and fatigue resistance. The high-viscosity high-elasticity SBS-T-modified asphalt demonstrates superior durability compared to both base asphalt and SBS-modified asphalt with the same dosage during this process. Notably, the 7% SBS-T-modified asphalt exhibits the highest complex modulus values, indicating that it maintains excellent mechanical performance and durability even under prolonged loading and environmental variations.

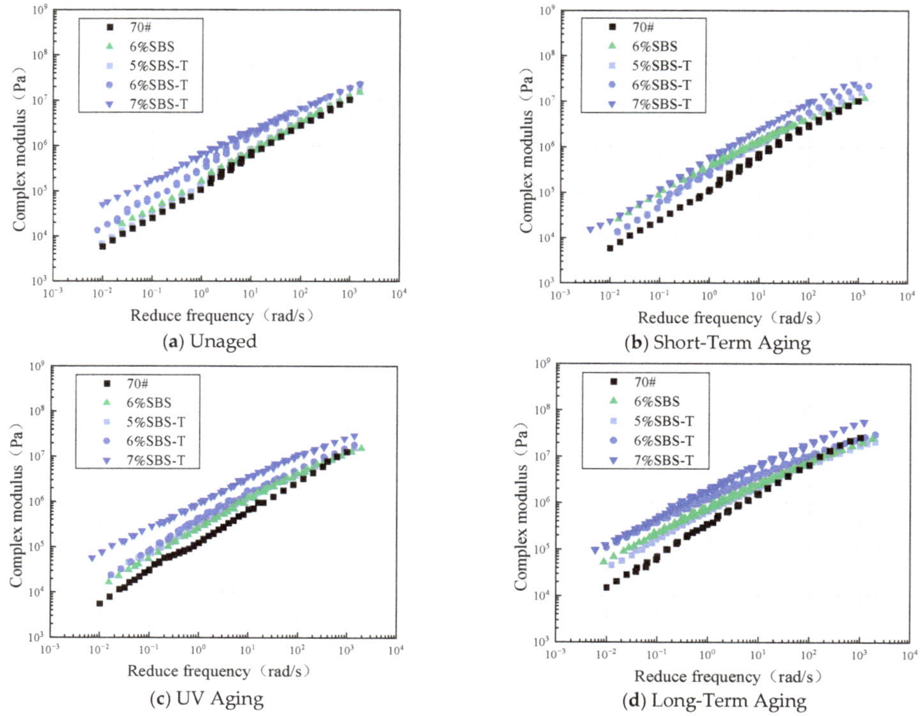

Figure 8. The main curves of the dynamic modulus under different aging effects.

3.2. Analysis of Stress–Strain Relationship

Figure 9 lists the stress–strain relationships of asphalt in the LAS test under four conditions (unaged, short-term aged, UV aged, and long-term aged). The following can be known from Figure 9:

(1) The increasing stress amplitude with the higher SBS-T dosage indicates a significant enhancement in stiffness. This is attributed to the formation of a more robust and continuous crosslinked polymer network within the asphalt matrix, which effectively restricts the mobility of asphalt molecules and bears a greater portion of the applied load. This microstructural reinforcement is consistent with the observed increase in complex modulus.

(2) Compared with 70# base asphalt and SBS-modified asphalt at the same dosage, high-viscosity and high-elasticity SBS-T-modified asphalt shows a larger stress peak width under the same aging conditions. Studies show that the larger the peak stress width, the greater the degree of shear deformation that asphalt can withstand. Therefore, the SBS-T modifier can enhance the stiffness and anti-deformation ability of asphalt.

(3) In the non-aging and short-term aging states, the stress–strain curves of high-viscosity and high-elasticity SBS-T-modified asphalt show a relatively clear elastic stage. All asphalt samples exhibit a higher stress response in the area with less strain, indicating that it can effectively store elastic energy during the initial strain. With the increase in strain, the curve tends to be stable, indicating that the material has entered the plastic deformation stage.

(4) Under ultraviolet aging conditions, the stress–strain curve of high-viscosity and high-elasticity SBS-T-modified asphalt shows more obvious changes. The molecular chain

breakage or degradation caused by ultraviolet aging leads to a reduction in the elastic properties of the material. In contrast, the stress response change of 7% SBS-T high-viscosity and high-elasticity modified asphalt after ultraviolet aging is relatively small, and it still maintains a relatively high stress value, showing its stronger ability to resist ultraviolet aging.

(5) The peak shear stress of 70# base asphalt, SBS-modified asphalt, and high-viscosity and high-elasticity SBS-T-modified asphalt after long-term aging is greater than that after ultraviolet aging, indicating that the long-term aging effect has a greater influence on the hardening and stiffness of asphalt materials than ultraviolet aging.

(6) Compared with 5% SBS-T and 6% SBS-T, 7% SBS-T asphalt shows stronger crack resistance and rutting resistance at each aging stage. Especially under high strain, its stress response is significantly stronger, reflecting that a higher content of SBS-T can effectively improve the durability of asphalt.

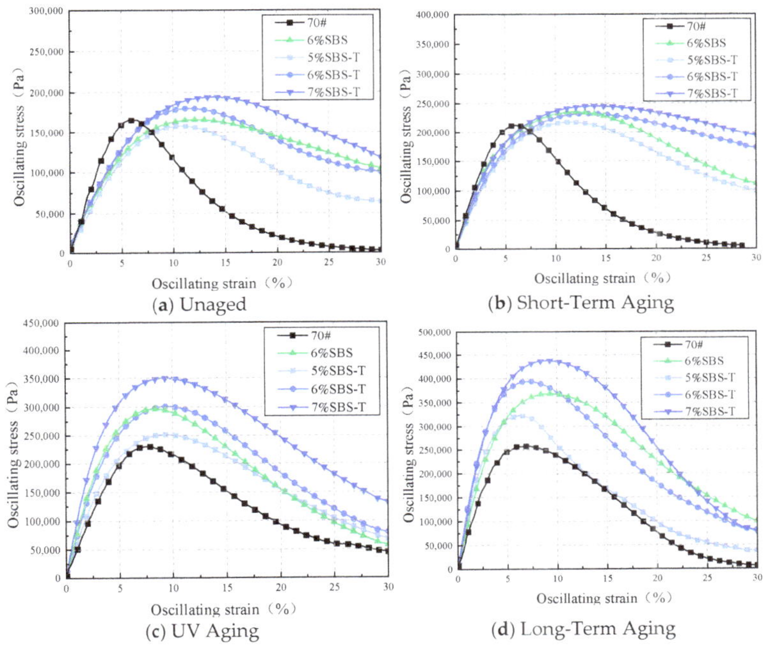

(a) Unaged (b) Short-Term Aging (c) UV Aging (d) Long-Term Aging

Figure 9. Stress–strain relationship graph.

3.3. Fatigue Damage Curve Analysis

By using the PSE theory to analyze the relationship between the value of C and S, and constructing the damage characteristic curve (DCC), the dynamic evolution law of the internal damage of the material can be visually presented. The DCC is helpful in explaining the damage development trend of asphalt binder and provides an important quantitative reference index for material performance evaluation. Figure 8 shows the DCCs under four aging modes: non-aging, short-term aging, UV aging, and long-term aging. By comparing the curve shapes and changing trends, the durability differences of different types of asphalt are evaluated, and the degree of performance decline is quantified. The end point of each DCC corresponds to the fatigue failure point of the asphalt binder.

The following can be seen from Figure 10:

(1) After short-term aging, ultraviolet aging, and long-term aging, the damage curve of 7% SBS-T high-viscosity and high-elasticity modified asphalt is always at the top, indicating that 7% SBS-T high-viscosity and high-elasticity modified asphalt can still maintain high stiffness after experiencing complex environmental effects, and therefore has better anti-aging performance.
(2) With the deepening of the aging degree, the downward trend of the DCC gradually slows down. After long-term aging, the DCCs of asphalt with different SBS-T dosages vary greatly and are significantly different from the samples after ultraviolet aging. This indicates that the influence of the photo-oxidation aging mechanism on the damage characteristics of asphalt binders is different from that of the thermal oxidation aging mechanism. Different aging mechanisms have different influences on the internal state characteristics of asphalt materials.
(3) For 6% SBS-T high-viscosity and high-elasticity modified asphalt and 6% SBS-modified asphalt at the same dosage, under long-term aging conditions, the high-viscosity and high-elasticity SBS-T-modified asphalt shows a lower damage accumulation than the traditional SBS asphalt. The fatigue failure point of 6% SB-T high-viscosity and high-elasticity modified asphalt is located after the curve of 6% SBS-modified asphalt, which indicates that SBS-T-modified asphalt is superior to ordinary SBS-modified asphalt in terms of fatigue durability.

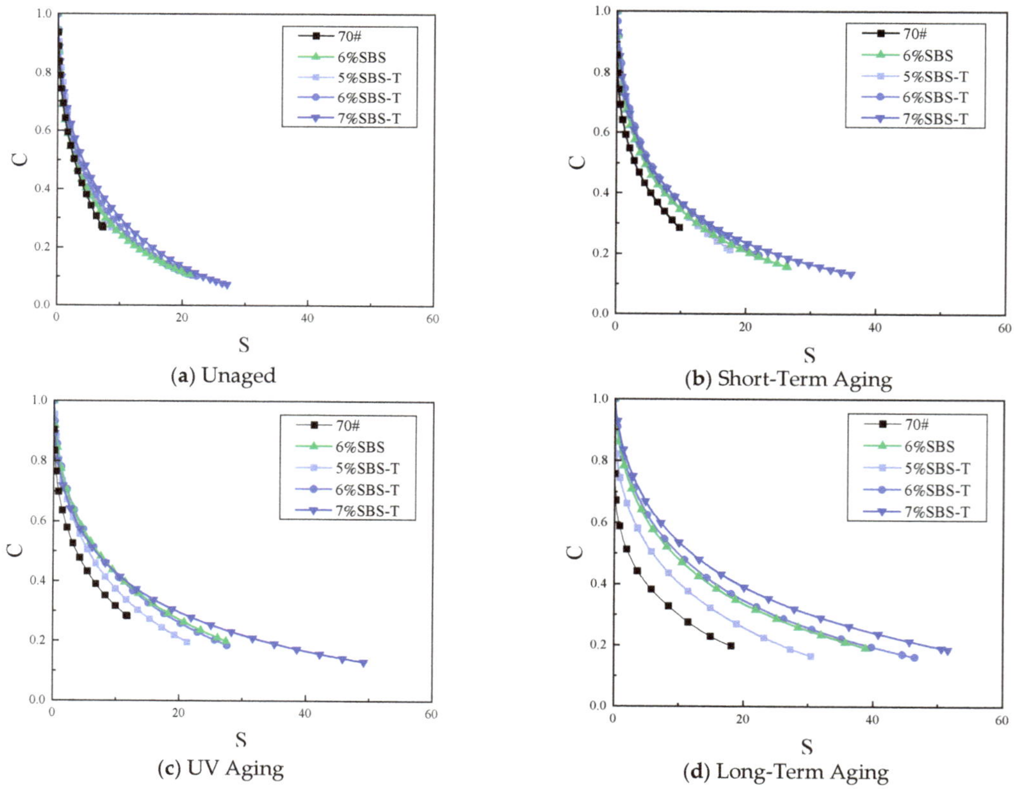

Figure 10. C-S curve graph.

3.4. Fatigue Life Analysis

Once the fatigue parameters a and b are determined, the fatigue life of asphalt can be predicted at different strain levels. Figure 9 shows the fatigue life of various types of asphalt within the strain level range of 1% to 10%. These prediction results are helpful for evaluating the performance of asphalt materials under different usage conditions, assisting in optimizing the proportion and application strategy of asphalt to enhance the durability and fatigue resistance of the pavement.

It can be known from Figure 11 that in the unaged state, SBS-T-modified asphalt (particularly the 6% and 7% dosage specimens) demonstrates significantly longer fatigue life compared to 70# base asphalt. The decline rate of its fatigue life with increasing strain levels is more gradual, indicating superior fatigue resistance even under high-strain conditions. After short-term aging, while the fatigue life of SBS-T-modified asphalt decreases, the performance degradation of 6% and 7% dosage specimens remains relatively minor, demonstrating excellent anti-aging properties. In contrast, the base asphalt shows significant fatigue performance deterioration after aging, particularly being more susceptible to fatigue damage under high-strain conditions.

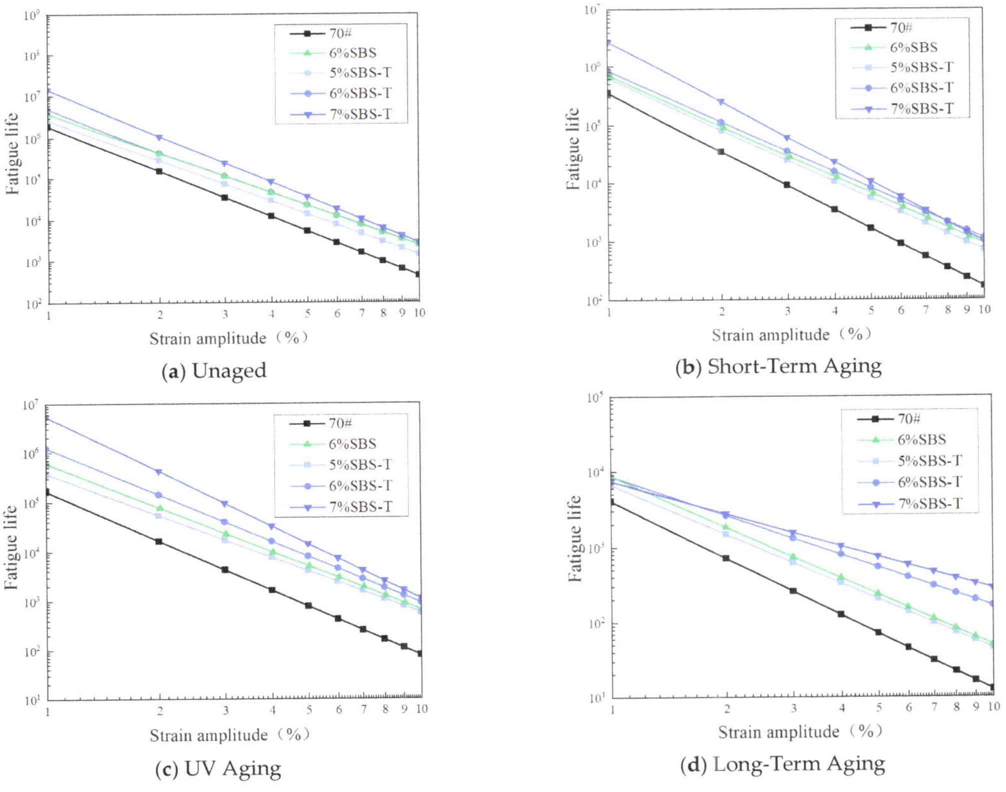

Figure 11. Fatigue life prediction curve.

Under UV aging conditions, SBS-T-modified asphalt exhibits superior aging resistance compared to conventional SBS-modified asphalt, owing to its stable molecular structure and optimized network system. Notably, the 6% and 7% SBS-T-modified asphalt specimens show comparable fatigue lives after UV aging, both significantly outperforming conventional modified asphalt. Long-term aging tests further validate the durability advantages of SBS-T-modified asphalt—even after extended aging, the 7% SBS-T modified asphalt maintains a relatively high fatigue life, demonstrating exceptional long-term service performance.

Comprehensive analysis reveals that the high-viscosity high-elasticity SBS-T-modified asphalt, through optimized material composition and microstructure, shows remarkable advantages in both fatigue and aging resistance. This material not only effectively delays the initiation and propagation of fatigue cracks but also maintains stable performance under various environmental conditions, which is crucial for extending pavement service life and reducing maintenance requirements. In particular, the 7% SBS-T-modified asphalt exhibits optimal performance across all aging conditions, providing valuable reference for developing high-performance pavement materials.

Figure 12 presents the fatigue life of SBS-T-modified asphalt under different strain levels. The results demonstrate the following:

(1) Across all aging conditions, SBS-T-modified asphalt exhibits superior fatigue life compared to both 70# base asphalt and conventional SBS-modified asphalt at equivalent dosages, with a pronounced increasing trend observed with higher SBS-T content.
(2) Short-term aging causes a systematic reduction in fatigue life for all specimens compared to the unaged state, while UV aging significantly deteriorates the fatigue performance of asphalt materials; nevertheless, the 7% SBS-T-modified asphalt maintains exceptional fatigue resistance.
(3) As the most severe service condition, long-term aging leads to dramatic fatigue life reduction for all asphalt types, yet the high-viscosity high-elasticity SBS-T-modified asphalt still demonstrates the most outstanding durability advantages.

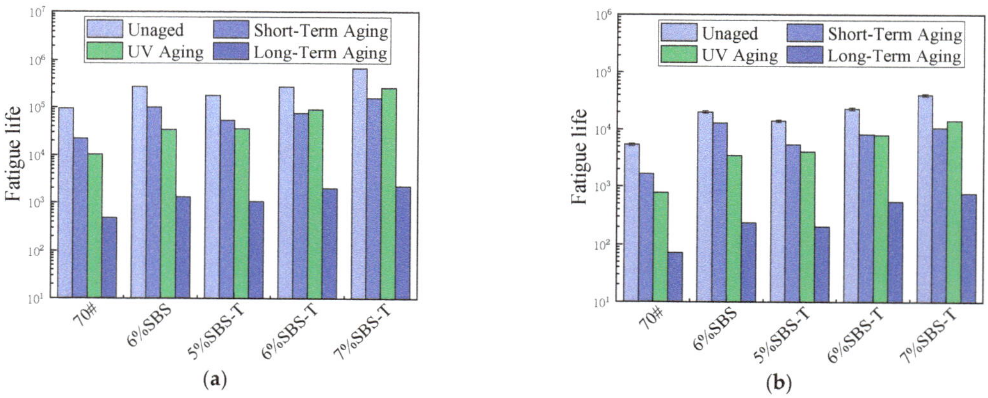

Figure 12. Fatigue life of asphalt under different strain conditions: (**a**) 2.5% strain; (**b**) 5% strain.

4. Conclusions

This study conducted an in-depth investigation into the fatigue characteristics of high-viscosity high-elasticity SBS-T-modified asphalt. Utilizing test methods based on the S-VECD theoretical model, the research systematically analyzed the cumulative fatigue damage and fatigue life of this modified asphalt. The influence of environmental conditions (such as aging and loading frequency) on fatigue life was also examined, yielding the following key conclusions:

(1) The high-viscosity high-elasticity SBS-T-modified asphalt demonstrates excellent viscoelastic properties across various temperatures and loading frequencies. Its dynamic modulus and phase angle exhibit greater stability with frequency variations,

particularly showing enhanced viscoelastic performance under high-frequency and high-temperature conditions. Compared to 70# base asphalt and conventional SBS-modified asphalt, this material maintains superior viscoelasticity during long-term service, demonstrating better aging resistance and more stable performance.

(2) Under identical strain conditions, the peak shear stress of SBS-T-modified asphalt is significantly higher than that of 70# base asphalt and SBS-modified asphalt at equivalent dosages, with a broader peak shear stress range. This indicates that SBS-T-modified asphalt possesses greater strength and enhanced resistance to shear deformation. After long-term aging, the peak stress of SBS-T-modified asphalt exceeds that of UV-aged samples, suggesting that long-term aging has a more pronounced effect on its strength and hardness than UV aging.

(3) The superior pseudo-strain energy characteristics and maintained rigidity under various aging conditions confirm SBS-T's exceptional durability. This suggests that infrastructure incorporating SBS-T-modified asphalt may require less frequent maintenance and have an extended service life, potentially offering better life cycle cost efficiency despite potentially higher initial costs.

(4) The fatigue resistance of SBS-T-modified asphalt improves significantly with increasing SBS-T content, indicating that the SBS-T modifier effectively enhances the asphalt's fatigue performance for extended service life under cyclic loading. However, fatigue life decreases noticeably with progressive aging, primarily due to increased hardness and brittleness during the aging process, which leads to material degradation and reduced fatigue strength.

(5) Laboratory simulation was used to mimic the UV aging of asphalt during pavement service, with only one irradiation duration selected. In subsequent research, the UV aging protocol can be optimized to investigate the performance changes of warm-mix polymer-modified asphalt under different UV aging degrees.

Author Contributions: Conceptualization, H.Z. and F.G.; methodology, H.Z.; software, X.L.; formal analysis, F.G.; data curation, S.W.; writing—original draft preparation, H.Z.; writing—review and editing, J.Y.; visualization, S.W.; supervision, X.L. and J.Y. All authors have read and agreed to the published version of the manuscript.

Funding: This research was funded by the Science and Technology Program of Henan Housing and Urban-Rural Development: HNJS-2024-K28.

Institutional Review Board Statement: Not applicable.

Informed Consent Statement: Not applicable.

Data Availability Statement: The original contributions presented in this study are included in the article. Further inquiries can be directed to the corresponding author.

Conflicts of Interest: The authors declare no conflicts of interest.

References

1. Yan, L.; Li, M.; Li, Q.; Li, H. Modified asphalt based on polyethylene with broad molecular weight distribution. *Constr. Build. Mater.* **2020**, *260*, 119707. [CrossRef]
2. Li, H.; Han, Y.; Guangxun, E.; Sun, Y.; Wang, L.; Liu, X.; Ren, J.; Lin, Z. Recycling of waste polyethylene in asphalt and its performance enhancement methods: A critical literature review. *J. Clean. Prod.* **2024**, *451*, 142072. [CrossRef]
3. Zheng, C.; Li, R.; Hu, M.; Zou, L. Determination of low-temperature crack control parameter of binding asphalt materials based on gray correlation analysis. *Constr. Build. Mater.* **2019**, *217*, 226–233. [CrossRef]
4. Li, Y.; Hao, P.; Zhao, C.; Ling, J.; Wu, T.; Li, D.; Liu, J.; Sun, B. Anti-rutting performance evaluation of modified asphalt binders: A review. *J. Traffic Transp. Eng.* **2021**, *8*, 339–355. [CrossRef]

5. Bai, Z.; Pei, Z.; Qiao, Y.; Cheng, P.; Guo, J.; Liu, Z.; Abdukadir, A.; Xiong, Y.; Yi, J. Study on the development and performance optimization of gussasphalt containing high percent of fine reclaimed asphalt pavement. *Constr. Build. Mater.* **2024**, *420*, 135557. [CrossRef]
6. Luo, Y.; Zhang, Z.; Zhang, H.; Zhang, M.; Zhang, K.; Zhao, Y. Performance optimization of high viscosity modified asphalt with SBS composite modifier and comparison of different high viscosity modified asphalts. *Int. J. Pavement Res. Technol.* **2021**, *15*, 1386–1398. [CrossRef]
7. Chen, L.; Zhang, Z.; Wang, Z.; Liu, H.; Zhang, D.; Sun, J. Using SBS Terpene-styrene Resin Blends as a Novel High-viscosity Asphalt Modifier. *J. Appl. Polym. Sci.* **2023**, *140*, e53355. [CrossRef]
8. Hu, M.; Sun, G.; Sun, D.; Zhang, Y.; Ma, J.; Lu, T. Effect of thermal aging on high viscosity modified asphalt binder: Rheological property, chemical composition and phase morphology. *Constr. Build. Mater.* **2020**, *241*, 118023. [CrossRef]
9. Zhu, J.; Zhang, K.; Liu, K.; Shi, X. Adhesion characteristics of graphene oxide modified asphalt unveiled by surface free energy and AFM-scanned micro-morphology. *Constr. Build. Mater.* **2020**, *244*, 118404. [CrossRef]
10. Luo, H.; Zheng, C.; Yang, X.; Bao, C.; Liu, W.; Lin, Z. Development of technology to accelerate SBS-modified asphalts swelling in dry modification mode. *Constr. Build. Mater.* **2022**, *314*, 125703. [CrossRef]
11. Luo, H.; Zheng, C.; Liu, B.; Bao, C.; Tian, B.; Liu, W. Study on SBS modifier of bio-oil/sulfur compound under dry modification mode. *Constr. Build. Mater.* **2022**, *326*, 127059. [CrossRef]
12. Zhuang, Y.; Yer, J.; Men, B.; Tang, G.; Wang, R. Experimental Study on Mechanism, Aging and Fatigue Performance of Warm Mixing Speed Melting SBS Modified Asphalt Binders. *Coatings* **2023**, *13*, 311. [CrossRef]
13. Yue, J.; Fu, J.; Wang, Y.; Ming, Y.; Tang, G.; Li, R. Effects of Complex Environmental Conditions on Fatigue Self-Healing Properties of Fast-Melting SBS-Modified Asphalt. *Materials* **2025**, *18*, 2157. [CrossRef] [PubMed]
14. Feng, L.; Zhang, X.; Sha, T.; Wang, D.; Niu, B.; Wang, R.; Hou, X. A Rheological Study of the High-Temperature Properties of Fast-Melting SBS/Epoxy-Modified Asphalt Binders. *Polymers* **2025**, *17*, 581. [CrossRef] [PubMed]
15. Li, R.; Wang, Y.; Zhu, W.; Chen, Y.; Yue, J. Research on the Rheological Performance of Fast-Melting SBS-Modified Asphalt under Complex Environmental Factors. *Coatings* **2024**, *14*, 1241. [CrossRef]
16. Men, B.; Li, X.; Sha, T.; Guo, F.; Tang, G.; Yue, J. Research on Fatigue Performance of Fast-Melting SBS/Epoxy Resin Composite-Modified Asphalt. *Coatings* **2024**, *14*, 789. [CrossRef]
17. Men, B.; Guo, F.; Kang, X.; Yue, J. Research on the Adhesion Properties of Fast-Melting SBS-Modified Asphalt–Aggregate Based on Surface Free Energy Theory. *Materials* **2023**, *16*, 7601. [CrossRef]
18. Huo, W.; Zhuang, Y.; Wang, Z.; Kang, X.; Wang, R. The Microscopic Mechanism and Rheological Properties of SBS-Modified Asphalt with Warm Mixing Fast-Melting. *Materials* **2023**, *16*, 5690. [CrossRef] [PubMed]
19. Zhou, Z.; Chen, G. Preparation, Performance, and modification mechanism of high viscosity modified asphalt. *Constr. Build. Mater.* **2021**, *310*, 125007. [CrossRef]
20. Tan, Y.; Zhang, H.; Cao, D.; Xia, L.; Du, R.; Shi, Z.; Dong, R.; Wang, X. Study on cohesion and adhesion of high-viscosity modified asphalt. *Int. J. Transp. Sci. Technol.* **2019**, *8*, 394–402. [CrossRef]
21. Zhang, F.; Hu, C. Preparation and properties of high viscosity modified asphalt. *Polym. Compos.* **2017**, *38*, 936–946. [CrossRef]
22. Wang, H.; Huang, Y.; Jin, K.; Zhou, Z. Properties and mechanism of SBS/crumb rubber composite high viscosity modified asphalt. *J. Clean. Prod.* **2022**, *378*, 134534. [CrossRef]
23. Zhang, L.; Hoff, I.; Zhang, X.; Liu, J.; Yang, C.; Wang, F. A Methodological Review on Development of Crack Healing Technologies of Asphalt Pavement. *Sustainability* **2023**, *15*, 9659. [CrossRef]
24. Yang, S.; Park, H.; Baek, C. Fatigue Cracking Characteristics of Asphalt Pavement Structure under Aging and Moisture Damage. *Sustainability* **2023**, *15*, 4815. [CrossRef]
25. Shi, C.; Cai, X.; Yi, X.; Wang, T.; Yang, J. Fatigue crack density of asphalt binders under controlled-stress rotational shear load testing. *Constr. Build. Mater.* **2021**, *272*, 121899. [CrossRef]
26. Wang, C.; Chen, Y.; Song, L. Investigating the Crack Initiation and Propagation of Asphalt Binder in Linear Amplitude Sweep Test. *J. Mater. Civ. Eng.* **2020**, *32*, 04020362. [CrossRef]
27. Zhang, Z.; Oeser, M. Energy dissipation and rheological property degradation of asphalt binder under repeated shearing with different oscillation amplitudes. *Int. J. Fatigue* **2021**, *152*, 106417. [CrossRef]
28. Yan, Y.; Hernando, D.; Park, B.; Allen, C.; Roque, R. Understanding asphalt binder cracking characterization at intermediate temperatures: Review and evaluation of two approaches. *Constr. Build. Mater.* **2021**, *312*, 125163. [CrossRef]
29. Jiao, L.; Elkashef, M.; Harvey, J.T.; Rahman, M.A.; Jones, D. Investigation of fatigue performance of asphalt mixtures and FAM mixes with high recycled asphalt material contents. *Constr. Build. Mater.* **2022**, *314*, 125607. [CrossRef]
30. Zhang, H.; Li, H.; Guo, F.; Wang, S.; Yue, J. Research on the Adhesion Performance of Fast-Melting SBS-Modified Emulsified Asphalt–Aggregate Based on the Surface Free Energy Theory. *Materials* **2025**, *18*, 3523. [CrossRef]
31. Schwarzl, F.R.; Staverman, A.J. Time-temperature dependence of linear viscoelastic behavior. *J. Appl. Phys.* **1952**, *23*, 838–843. [CrossRef]

32. Findley, W.N.; Lai, J. A modified superposition principle applied to creep of nonlinear viscoelastic material under abrupt changes in state of combined stress. *Trans. Soc. Rheol.* **1967**, *11*, 361–380. [CrossRef]
33. Wang, C.; Castorena, C.; Zhang, J.; Kim, Y.R. Unified failure criterion for asphalt binder under cyclic fatigue loading. *Road Mater. Pavement Des.* **2015**, *16*, 125–148. [CrossRef]

coatings

MDPI

Article

Effect of Basalt Fibers on the Performance of CO_2-Cured Recycled Aggregate Concrete Composite Slab–Column Assemblies with Bolted Connections Under NaCl Erosion

Di Wang [1], Yuanfeng Wu [2], Zhiqiang Xu [1,*], Na Xu [1], Chuanqi Li [1], Xu Tian [1], Feiting Shi [3] and Hui Wang [2,*]

[1] School of Chemical Engineering and Machinery, Liaodong University, Dandong 118000, China
[2] School of Civil Engineering and Geographical Environment, Ningbo University, Ningbo 315000, China
[3] School of Civil Engineering, Yancheng Institute of Technology, Yancheng 224051, China
* Correspondence: xuzhiqiang@liaodongu.edu.cn (Z.X.); huiwang123@aliyun.com (H.W.)

Abstract

Basalt fibers possess high tensile strength and excellent corrosion resistance, properties that may enhance the chloride resistance of recycled aggregate concrete (RAC) structures. Nevertheless, the effects of basalt fibers on RAC structures under chloride attack remain poorly understood. This study investigates mass loss and the deterioration of key mechanical properties in basalt fiber-reinforced RAC composite slab–column assemblies (RAC composite assemblies) subjected to NaCl freeze–thaw cycles (F-Cs) and dry–wet alternations (D-As) and further explores the damage mechanisms of the concrete matrix through microscopic characterization. The results show that, compared with NaCl F-Cs, NaCl D-As have a more pronounced impact on the performance degradation of RAC composite slab–column assemblies. Moreover, basalt fibers effectively mitigate the deterioration of RAC composite assemblies in chloride-rich environments, particularly under NaCl D-As, where their protective effect is more evident. At 2.5 vol% fiber content, impact toughness peaked at an 83.7% improvement after 30 D-As, while flexural toughness showed a maximum enhancement of 773.6% after 100 F-Cs. Scanning electron microscopy energy-dispersive spectroscopy (SEM-EDS) analysis revealed a marked increase in Cl content within RAC, with NaCl D-As causing more severe erosion than NaCl F-Cs. Additionally, basalt fibers significantly inhibited chloride ion penetration and associated erosion in RAC. These findings provide valuable insights into utilizing basalt fibers to enhance the durability of RAC in coastal infrastructure exposed to chloride attacks. Further research on long-term performance and fiber parameter optimization is needed to support practical implementation.

Keywords: basalt fibers; recycled aggregate concrete; dry–wet alternations; freeze–thaw cycles; flexural toughness; impact toughness

1. Introduction

Exterior walls play a crucial role in building structures, as they serve as the primary barrier against environmental elements. Therefore, selecting the appropriate materials for these walls is essential for minimizing resource consumption and reducing environmental impact. In recent years, researchers around the world have extensively explored various exterior wall systems, including fly ash bricks, lightweight foam concrete hollow blocks, and glass fiber-reinforced cement porous panels [1–4]. However, many of these systems rely heavily on mineral resources like cement and natural aggregates. The extraction and processing of these resources place significant environmental and resource pressures on rapidly developing nations such as China [5]. Moreover, manufacturing these panels typically consumes large amounts of energy and produces substantial carbon emissions, leading to less-than-ideal energy-saving and environmental performance [6,7]. Consequently, developing sustainable and low-carbon wall material solutions has become critically important and urgent.

In response to the dual challenges of resource consumption and environmental pressure posed by the above exterior wall materials, the resource recovery of construction waste offers a highly promising alternative path. For example, recycled aggregate is a favorable option [8]. Rapid urbanization has led to a marked increase in construction waste, creating enormous stockpiling pressure on the environment [9,10]. Recycled aggregates produced by crushing, washing, and grading waste concrete can effectively reduce the demand for natural aggregates and the landfill volume of construction waste, which aligns with the concept of sustainable development [11–13]. However, conventional recycled aggregate concrete often requires higher water content and typically exhibits lower compressive strength and elastic modulus compared to natural aggregate concrete [14]. These performance disadvantages mainly originate from the porous residual mortar adhered to the surface of recycled aggregates, which weakens the interfacial transition zone between aggregate and cement paste and results in the overall degradation of durability indicators, such as impermeability and frost resistance [15–18]. Therefore, recycled aggregates are currently mainly limited to the production of low-grade concrete [19].

To improve the performance of recycled aggregate concrete, researchers have developed various strengthening techniques [20]. CO_2 curing not only sequesters carbon dioxide but also reacts with the hydration products in the residual mortar on the surface of recycled aggregates to form calcium carbonate, significantly enhancing the compactness and mechanical properties of the aggregates themselves [21]. Using such strengthened recycled aggregates to produce exterior wall panels opens a new route for the high-value utilization of construction waste [22]. Nevertheless, even when using strengthened recycled aggregates, the brittleness and crack resistance of the resulting concrete materials still require further improvement. For this reason, fiber reinforcement has become an effective means of enhancement. Commonly used fiber reinforcement materials include steel fibers and polypropylene fibers [23,24]. However, steel fibers are prone to corrosion in chloride-rich environments, whereas polypropylene fibers generally exhibit low interfacial bond strength with the cement-based matrix [25,26]. By contrast, basalt fibers show superior overall performance, possessing excellent mechanical properties, high alkali resistance, and outstanding resistance to salt spray corrosion [27,28]. These characteristics render basalt fibers highly promising for applications in civil engineering materials, particularly in coastal and near-shore engineering [29,30]. Nevertheless, studies on the performance of basalt fiber-reinforced recycled aggregate concrete under chloride erosion environments remain insufficient. In particular, research on the long-term performance evolution of wall–column components made from such materials under the coupled action of mechanical loading and harsh environmental exposure in practical engineering applications is still lacking and urgently requires systematic, in-depth investigation.

Therefore, this study systematically investigates the performance evolution of basalt fiber-reinforced, CO_2-cured recycled aggregate concrete composite wall–column assemblies under chloride erosion environments by combining experimental and microstructural analyses. The experiments used concrete prism specimens to simulate columns and thin slab specimens to simulate wall panels. These specimens were reinforced with basalt fibers and prepared from CO_2-cured recycled aggregates, which were then bolted together to form an integral load-bearing system. The study focused on the performance responses of the composite slab–column assemblies after exposure to 100–300 F-Cs and 10–30 D-As in a 3% NaCl solution. Measured parameters included axial compressive strength, mass loss rate, relative dynamic modulus of elasticity, flexural toughness, and impact toughness. Additionally, SEM-EDS analysis was employed to characterize the microstructural morphology evolution and elemental distribution at the matrix interface following chloride erosion. This study provides important experimental evidence and theoretical support for the durability design and application of basalt fiber-reinforced recycled aggregate concrete in coastal engineering structures.

2. Experimental

2.1. Raw Materials

The cement used in this study was ordinary Portland cement (OPC), with its chemical composition and relevant performance indicators presented in Tables 1 and 2. The coarse aggregate used was recycled aggregate, which underwent two key pretreatment processes. Carbonation was first conducted under 8% CO_2 at 40 °C, which densified the microstructure through pore-filling precipitation of calcium carbonate ($CaCO_3$) to reduce water absorption and strengthen interfacial transition zones. Subsequently, mechanical shaping partially removed the residual surface mortar, while improving particle geometry, enhancing the physical properties of the aggregate. The processed recycled aggregate was sieved to obtain a continuously graded, carbonation-enhanced recycled aggregate with a particle size range of 9.5–26.5 mm. The fine aggregate used was river sand, with a fineness modulus of 2.87. The water-reducing agent was a high-performance polycarboxylate-based superplasticizer, providing a water reduction rate of up to 40%. The fibers used were short-cut basalt fibers (Type XK-DQXWY), with a length of 3 cm, a diameter of 6 μm, and a density of 2.635 g/cm^3.

Table 1. The chemical composition of OPC.

Chemical Composition/%									Median Particle Size D50/μm
CaO	SiO_2	Al_2O_3	Fe_2O_3	MgO	MnO	R_2O	SO_3	Loss on Ignition	
62.51	21.18	5.19	3.84	1.81	0.15	0.47	2.90	1.95	18.6

Table 2. Relevant performance indicators of OPC.

Type	Strength Grade/MPa	Density/(g·cm^{-3})	Initial Setting Time/min	Final Setting Time/min
OPC	42.5	3.01	112.3	216

2.2. Specimen Preparation

The weighed aggregates were added to an HJW-60 mixer, followed by the addition of cement and a premixed solution of water and superplasticizer. The mixture was homogenized at 45 rpm for 240 s. During mixing, basalt fibers were manually and uniformly dispersed. The superplasticizer dosage was adjusted to achieve self-compacting concrete, with a T50 cm flow time of 10 to 13 s [31]. After mixing, the concrete was poured into molds in a single lift. This casting procedure produced prism specimens (100 mm × 100 mm × 400 mm) and thin slab specimens (100 mm × 40 mm × 400 mm). The dimensions of the specimens were determined by proportionally scaling down the actual sizes of the external wall panel and column connection components in real structures. The filled molds were vibrated on a shaking table until the surface exhibited cement paste with no visible large air bubbles [32]. All specimens were covered with plastic film and cured at ambient temperature for 24 h before demolding. Finally, the demolded specimens were transferred to a standard curing room for curing [33]. Table 3 details mix proportions utilizing carbonation-modified and shaped recycled aggregate, with basalt fiber content expressed by volumetric fraction.

Table 3. Mix proportions for recycled aggregate concrete in composite slab–column assemblies (kg/m^3).

Group	Water	Cement	Sand	Recycled Aggregate	Basalt Fibers	Water Reducing Agent	Basalt Fibers Content/%
A1	175	583.31	648.50	972.95	0	0.75	0%
A2	175	583.31	648.50	972.95	13.18	1.00	0.5%
A3	175	583.31	648.50	972.95	26.35	1.25	1.0%
A4	175	583.31	648.50	972.95	39.53	1.50	1.5%
A5	175	583.31	648.50	972.95	52.70	1.75	2.0%
A6	175	583.31	648.50	972.95	65.88	2.00	2.5%

The water–cement ratio was consistently maintained at 0.30 across all groups to eliminate interference from this variable, thereby isolating the effects of basalt fibers. Previous research has established that a higher water–cement ratio reduces strength, while a lower ratio increases strength [34]. Consequently, this study employed a single water–cement ratio to evaluate the effects of basalt fiber incorporation exclusively.

As the fiber content increased from 0 to 2.5 vol%, the dosage of the water-reducing agent was progressively increased from 0.75 kg/m^3 to 2.00 kg/m^3 (Table 3, Column 7) to counteract the reduction in workability. This adjustment ensured consistent self-compacting properties (T50 = 10–13 s), allowing for a focused evaluation of fiber–matrix interactions, while controlling fresh-state variability. Increasing fiber volume is generally expected to enhance tensile strength, flexural toughness, and crack resistance but may also influence compactability and fiber dispersion.

2.3. Freeze–Thaw Cycles and Dry–Wet Alternations Test Procedure

After 24 days of curing in a standard curing room, the selected specimens were immersed in a 3% NaCl solution for 4 days. Subsequently, they were transferred to a fully automatic freeze–thaw testing chamber to undergo accelerated freeze–thaw cycles (F-Cs). During the F-Cs, the specimens were placed in rubber tubes filled with a 3% NaCl solution, with the temperature cycling between −18 °C and 8 °C. For the dry–wet alternation (D-A) test, the specimens were cured for 26 days in a standard curing room, then dried for 2 days, cooled to room temperature, and transferred to a cyclic corrosion tester for salt spray exposure. The solution used in the D-A test was a 3% NaCl solution. Each dry–wet alternation lasted 24 h, comprising 6 h of drying at 80 °C, 2 h of cooling to 25–30 °C, and 15 ± 0.5 h of salt spraying, with the solution temperature maintained at 25–30 °C. Both F-C and D-A tests were conducted in compliance with the GB/T 50082-2009 standard [35]. The total numbers of F-Cs and D-As were 300 and 30, respectively. During testing, the parameters of the specimens were evaluated after every 100 F-Cs and 10 D-As.

2.4. Measurement Methods

The test specimens were recycled aggregate concrete composite slab–column assemblies with bolted connections. The measurements focused specifically on the connection region between the slab and the column, where the strength was potentially weakest. The influence of bolted connections has been investigated in prior research [36,37]. This specific connection method was selected and held constant throughout the study. The primary objective focused on investigating the effect of incorporating basalt fibers into recycled aggregate concrete composite slab–column assemblies.

2.4.1. Dynamic Modulus of Elasticity Test

The probes of the HC-U91 ultrasonic detector were aligned with the central axis on both sides of the specimens. Measurement points were marked at 0 mm, 100 mm, and 200 mm from the center along the central axis, and ultrasonic velocity was measured [38]. The relative dynamic modulus of elasticity (RDME) of the specimens was calculated using Equation (1) [39].

$$RDME = \left(\frac{v_t}{v_1}\right)^2 \tag{1}$$

where v_t denotes ultrasonic velocity after NaCl solution exposure followed by F-Cs and D-As (m/s), and v_1 represents the initial ultrasonic velocity prior to NaCl solution exposure (m/s).

2.4.2. Axial Compressive Strength

An electro-hydraulic servo testing machine with a load capacity of 0 to 100 tons was employed to determine the axial compressive strength of specimens through graded loading. The axial load was directly measured and recorded by the calibrated load cell of the testing machine throughout the test. Testing commenced with graded force-controlled loading at 5 kN/s, where each loading stage represented 1/10 of the ultimate load and was maintained for 2 min [40]. When the load reached 80% of the ultimate load, displacement-controlled loading was applied at a rate of 0.2 mm/min until specimen failure occurred [41]. Failure was defined as the point of maximum sustained load, which is the peak axial load (Fmax) reached during the test. The axial compressive strength values were calculated as the ratio of the measured peak load (Fmax) to the original cross-sectional area perpendicular to the loading axis.

2.4.3. Flexural and Impact Toughness

Figure 1a illustrates the displacement-controlled loading test procedure. As shown, the recycled aggregate concrete composite slab–column assemblies with bolted connections were subjected to four-point bending using a 1000 kN press. The pure bending segment under loading measured 120 mm with a support span of 360 mm. Displacement-controlled loading at a rate of 0.5 mm/min was maintained until specimen failure occurred [42].

Figure 1b depicts the impact test procedure and loading method. The impact toughness of recycled aggregate concrete composite slab–column assemblies with bolted connections was determined using the drop-weight test method. This procedure was conducted by the standard proposed by the American Concrete Institute (ACI) Committee 544.2R-89 [43,44]. The composite slab–column assemblies were subjected to impact loading in a four-point loading mode. The drop hammer and drop rod masses were 2.5 kg and 5 kg, respectively. Three specimens per group were tested, with impact toughness calculated using Equation (2) [45].

$$W = (M + m)gHN \tag{2}$$

where m is drop weight mass (kg); N is total number of impact failures; M is drop rod mass (kg); H is drop height, taken as 400 mm; g is gravitational acceleration, taken as 9.8 m/s^2.

During impact testing, a 7.5 kg impact mass (2.5 kg hammer integrated with 5 kg rod) was released from a height of 400 mm onto one side of the specimen. The surface crack propagation width was monitored using a crack width detector, and the number of impacts until initial crack formation was recorded. The test was repeated until specimen failure occurred. Three repetitions were performed per test group.

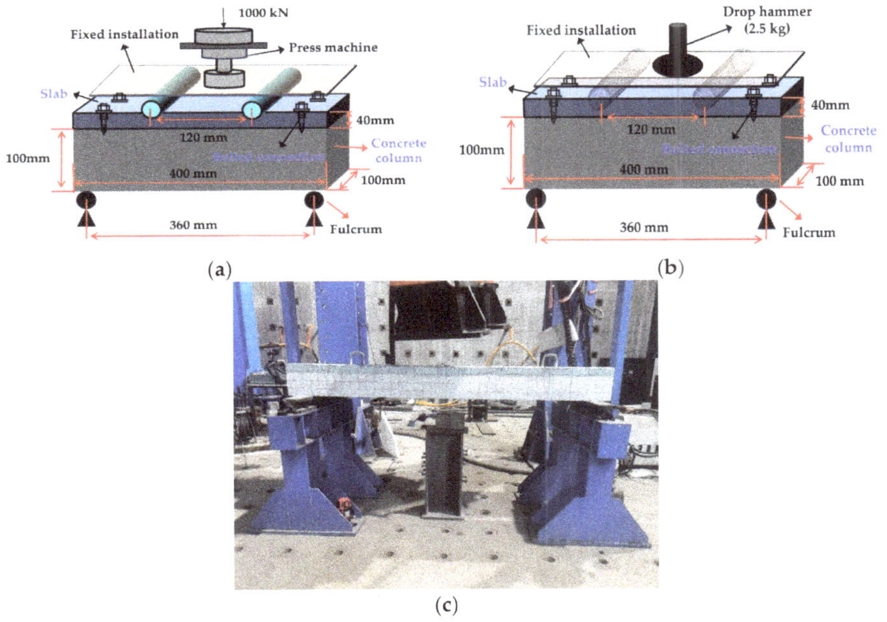

Figure 1. Test methods for composite slab–column assemblies: (**a**) displacement loading test; (**b**) impact test; (**c**) the test specimen.

The impact toughness index β for the drop-weight test is defined as the ratio of $(N_2 - N_1)$ to N_1, where N_1 is the impact count at initial crack formation, and N_2 is the impact count at failure. This index quantifies material toughness after crack initiation. The experimental procedure consisted of three sequential steps. Firstly, the bottom surface of the specimen was cleaned. Then, impact counts were recorded until a crack width detector identified the first crack with a width of 0.05 mm. Finally, the impact count at specimen failure was documented. The impact toughness index β was calculated using Equation (3) [46].

$$\beta = \frac{(N_2 - N_1)}{N_1} \tag{3}$$

2.4.4. Scanning Electron Microscopy Energy-Dispersive Spectroscopy

An SU3800 scanning electron microscope (SEM) was used to obtain micrographs of the concrete matrix within specimens and perform energy-dispersive spectroscopy (EDS) analysis. The sample preparation procedure consisted of the following steps. Firstly, a core sample was extracted from the central region of the specimen and trimmed into spherical particles 1–3 mm in diameter. These particles were then polished to a smooth finish and cleaned to remove surface dust. Subsequently, the samples were dried at 105 °C for two days in an oven. After being gold-sputtered under vacuum, the samples were analyzed using the SU3800 SEM.

The SEM characterized chloride-induced microstructural damage, including crack propagation, pore distribution, and surface morphology. The EDS analysis specifically targeted the quantitative determination of chlorine (Cl) content to evaluate the penetration depth of NaCl and the severity of corrosion. Supplementary quantitative analysis of silicon (Si) content was also performed at specific microstructural features (aggregate-paste interfaces) to assess the associated evolution.

3. Results and Discussion

3.1. NaCl Freeze–Thaw Cycles

3.1.1. Axial Compressive Strength

Figure 2 presents the axial compressive strength of recycled aggregate concrete composite slab–column assemblies with bolted connections after different numbers of NaCl F-Cs and the corresponding strength growth rate with the incorporation of basalt fibers. As shown, the axial compressive strength of the composite slab–column assemblies (control group with 0 vol% fibers) decreased by 4.4%–30.2% after exposure to 100–300 NaCl F-Cs (specifically, 4.4% at 100 F-Cs, 15.9% at 200 F-Cs, 30.2% at 300 F-Cs). This degradation can be attributed to two main factors. Firstly, fatigue damage occurs in the concrete matrix due to frost–heave stress and capillary pressure within the composite assemblies during cycling, leading to progressive deterioration [47]. Secondly, NaCl F-Cs accelerate concrete deterioration, as chloride ingress through cracks generates expansive crystalline phases that propagate cracks and reduce strength [48]. Furthermore, incorporating 0.5–2.5 vol% basalt fibers increased the axial compressive strength of the composite slab–column assemblies by 11.0%–87.1%. At 2.5 vol% fiber content, the maximum strength improvement (87.1%) occurred after 300 F-Cs, as shown in Figure 2. This improvement is primarily attributed to the crack bridging effect of the fibers during loading, which enhances concrete strength and mitigates strength loss induced by F-Cs [49]. Crucially, the degree of strength improvement and resistance to NaCl F-Cs degradation exhibited a strong positive correlation with the basalt fiber content. Higher fiber volumes (up to 2.5 vol%) consistently yielded significantly greater enhancements in both strength and freeze–thaw resistance. Consequently, the incorporation of basalt fibers enhances the NaCl freeze–thaw resistance of recycled aggregate concrete composite slab–column assemblies.

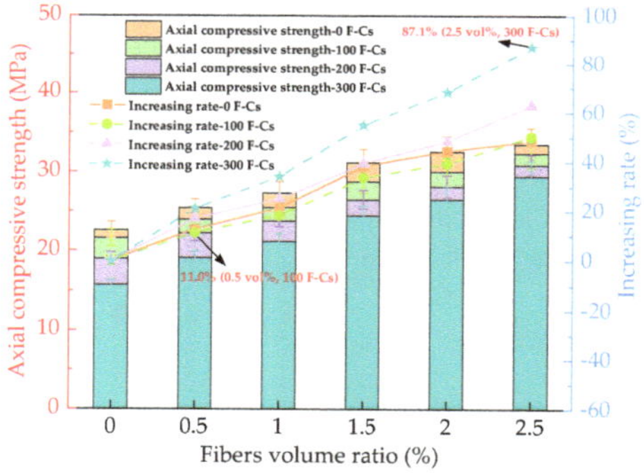

Figure 2. The axial compressive strength of recycled aggregate concrete composite slab–column assemblies after NaCl F-Cs.

3.1.2. Mass Loss Rate and Relative Dynamic Modulus of Elasticity

Figure 3a presents the mass loss rate (MLR) of recycled aggregate concrete composite slab–column assemblies with bolted connections during F-Cs in NaCl solution. As shown, the MLR of recycled aggregate concrete composite slab–column assemblies reached 0.26%–1.13% after 100–300 NaCl F-Cs. This rise in MLR is primarily attributed to frost–heave stress and crystallization stress induced in the concrete matrix by NaCl F-Cs, which lead to crack initiation, propagation, and widening in the composite slab–column assemblies [50]. Moreover, the cumulative nature of F-Cs exacerbates fatigue damage in recycled aggregate concrete composite slab–column assemblies, accelerating crack propagation and interfacial spalling, thereby elevating mass loss [51]. Furthermore, incorporating 0.5–2.5 vol% basalt fibers reduced the MLR of recycled aggregate concrete composite slab–column assemblies to 0.08%–0.21%. The minimum MLR of 0.08% was recorded at 2.5 vol% fibers after 100 F-Cs. This reduction is predominantly driven by the crack bridging effect of basalt fibers, which retard crack progression within the composite slab–column assemblies matrix, thereby mitigating mass loss [52]. These results demonstrate that the effectiveness of mass loss suppression increases significantly with increasing basalt fiber content, with higher volumes providing substantially enhanced resistance to freeze–thaw-induced deterioration.

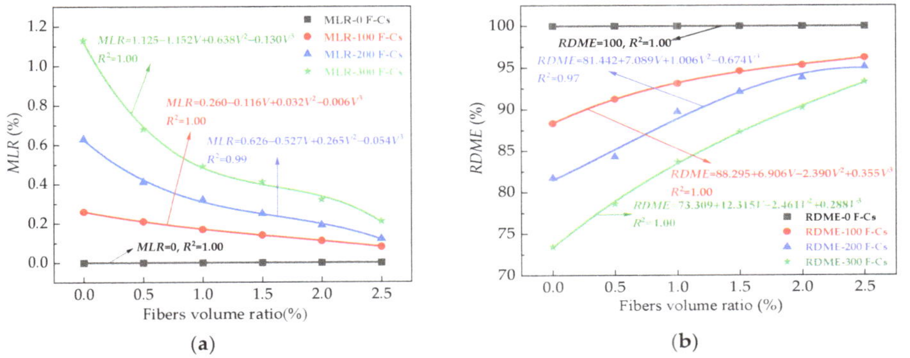

Figure 3. The MLR and RDME of recycled aggregate concrete composite slab–column assemblies after NaCl F-Cs: (**a**) MLR; (**b**) RDME.

Figure 3b displays the RDME of recycled aggregate concrete composite slab–column assemblies with bolted connections during NaCl F-Cs. The data indicate that the RDME decreased from 100% to 73.4% after 100–300 NaCl F-Cs. This phenomenon similarly stems from frost–heave stress and crystallization stress induced in the concrete matrix by NaCl F-Cs, which increase the number and width of cracks in the composite slab–column assemblies. Consequently, ultrasonic pulse propagation during RDME measurement is impeded by freeze–thaw-induced cracks, which reduce ultrasonic velocity and thereby decrease RDME [53]. As shown in the figure, incorporating 0.5–2.5 vol% basalt fibers increased the RDME of the composite slab–column assemblies by 2.6%–19.9%. The maximum improvement of 19.9% was achieved at 2.5 vol% fibers after 300 F-Cs. This enhancement primarily results from the ability of basalt fibers to delay crack development, thereby reducing the degradation of dynamic modulus. Furthermore, the magnitude of improvement exhibited a strong positive correlation with basalt fiber content, where higher volumes consistently yielded greater retention of the relative dynamic modulus.

3.1.3. Flexural Toughness Behavior Analysis

Figure 4a illustrates the flexural load-deflection curves of recycled aggregate concrete composite slab–column assemblies reinforced with basalt fibers after 100–300 NaCl F-Cs. The peak flexural load and maximum deflection decrease after these cycles. This reduction is primarily attributed to frost–heave stress and crystallization stress generated during NaCl F-Cs, which accelerate crack formation in the recycled aggregate concrete composite slab–column assemblies. As these cracks increase, they directly impair the ultimate flexural capacity and deformation tolerance of the composite slab–column assemblies [54].

Figure 4b presents flexural toughness values calculated from the area enclosed by the flexural load-deflection curves of recycled aggregate concrete composite slab–column assemblies. The flexural toughness of recycled aggregate concrete composite slab–column assemblies (control group with 0 vol% fibers) decreased by 33.9%–47.5% after 100–300 NaCl F-Cs. Specifically, the reductions of 47.5% and 33.9% were recorded after 100 F-Cs and 300 F-

Cs, respectively. This decrease primarily stems from the erosive effect of NaCl F-Cs, which exacerbate interfacial transition zone damage and crack propagation within the concrete matrix, thereby degrading flexural toughness [55]. Additionally, incorporating 0.5–2.5 vol% basalt fibers significantly enhanced the flexural toughness of recycled aggregate concrete slab stacked with concrete column. The maximum improvement in flexural toughness, reaching 773.6%, was achieved at 2.5 vol% fibers after 100 F-Cs. This improvement stems from energy dissipation mechanisms during flexural loading, primarily the pull-out of fibers and bridging of cracks, thereby increasing its flexural toughness [56]. Overall, the improvement in flexural toughness was significantly dependent on the fiber volume fraction (vol%). Generally, higher fiber content led to substantially greater improvements in flexural toughness compared to lower fiber contents.

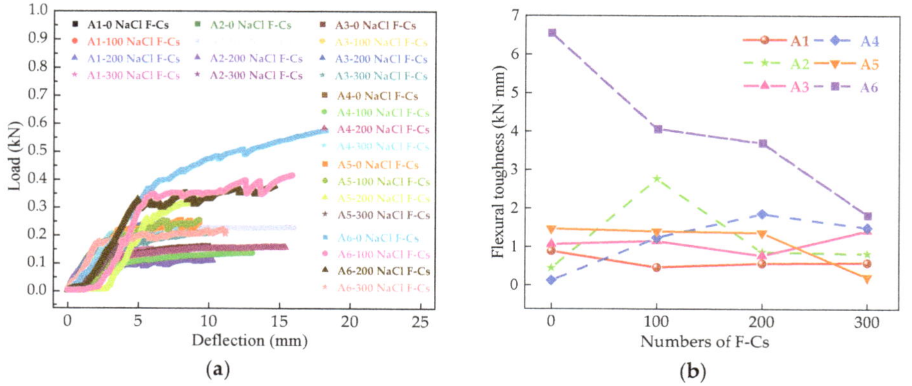

Figure 4. The load-deflection curves and flexural toughness of recycled aggregate concrete composite slab–column assemblies after NaCl F-Cs: (**a**) load-deflection curves; (**b**) flexural toughness.

3.1.4. Impact Toughness Analysis

Figures 5 and 6, respectively, depict the impact toughness of recycled aggregate concrete composite slab–column assemblies subjected to NaCl solution F-Cs, along with the number of impacts and the impact toughness index. After 100–300 NaCl F-Cs, specimens showed reduced impact toughness. Notably, the number of impacts to initial cracking, impacts to failure, and the difference in impact counts between failure and initial cracking all decreased. The impact toughness degradation rate ranged from 3.2% to 19.4% (specifically, 3.2% at 100 F-Cs, 19.4% at 300 F-Cs for the control group). In contrast, the toughness index (β, defined by Equation (3) in Section 2.4.2 as $\beta = (N_2 - N_1)/N_1$; measured using the drop-weight impact test) increased with the number of F-Cs.

Additionally, incorporating 0.5–2.5 vol% basalt fibers enhanced the impact toughness of recycled aggregate concrete composite slab–column assemblies by 6.5%–62%. The maximum improvement of 62% was achieved at 2.5 vol% fibers after 300 F-Cs. This improvement is attributed to basalt fibers effectively inhibiting crack propagation and restraining crack width development in the recycled aggregate concrete composite slab–column assemblies when subjected to impact loading, enhancing overall toughness [57]. Collectively, these findings demonstrate that the magnitude of impact toughness enhancement is strongly governed by basalt fiber content (vol%), with higher volumes yielding progressively greater improvements. Correspondingly, the number of impacts sustained by recycled aggregate concrete composite slab–column assemblies at both initial cracking and failure stages increased with a higher content of basalt fibers. This phenomenon is likely due to basalt fibers amplifying the disparity between the number of impacts required to cause failure and initial cracking. However, with further increases in the content of basalt fibers, the impact toughness index showed a downward trend. The incorporation of basalt fibers into recycled aggregate concrete composite slab–column assemblies increases the number of impacts required for initial cracking, thereby elevating the denominator in the toughness index formula and reducing its value.

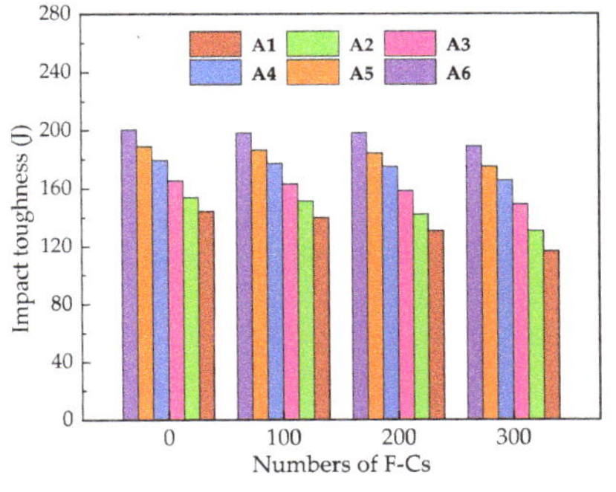

Figure 5. The impact toughness of recycled aggregate concrete composite slab–column assemblies after NaCl F-Cs.

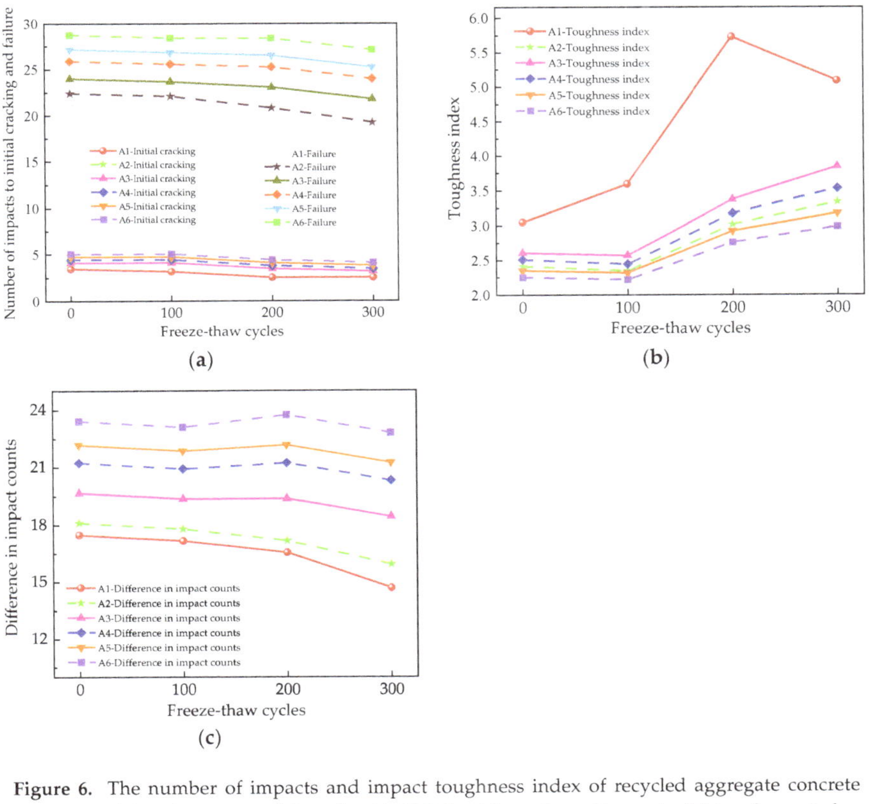

Figure 6. The number of impacts and impact toughness index of recycled aggregate concrete composite slab–column assemblies after NaCl F-Cs: (**a**) number of impacts; (**b**) toughness index; (**c**) difference in impact counts.

3.1.5. Microscopic Analysis

Figure 7a presents an SEM micrograph and EDS spectra of the concrete matrix in recycled aggregate concrete composite slab–column assembly specimens without fibers before NaCl exposure. Microstructural analysis revealed that the pre-exposure specimens exhibited a comparatively smooth and dense microstructure, with minimal micro-voids predominantly localized at the aggregate interfaces. EDS confirmed the presence of C, O, Mg, Al, Si, Ca, Mn, and Fe. Figure 7b shows EDS analysis of the recycled aggregate concrete composite slab–column assemblies without fibers after NaCl F-Cs. Compared to Figure 7a, the concrete matrix developed a roughened surface with enlarged pores after NaCl F-Cs. Additionally, the Si content (wt.%) decreased, while the Cl content increased significantly after F-Cs. This phenomenon can be explained from two perspectives. On the one hand, during chloride salt erosion, the NaCl solution permeates the concrete pores, inducing osmotic pressure and chemical corrosion within the matrix. These effects jointly modify the pore structure [58]. On the other hand, NaCl F-Cs promote the deeper penetration of free chloride ions and elevate the chloride concentration in pore solutions, resulting in a significantly increased Cl content in concrete [59].

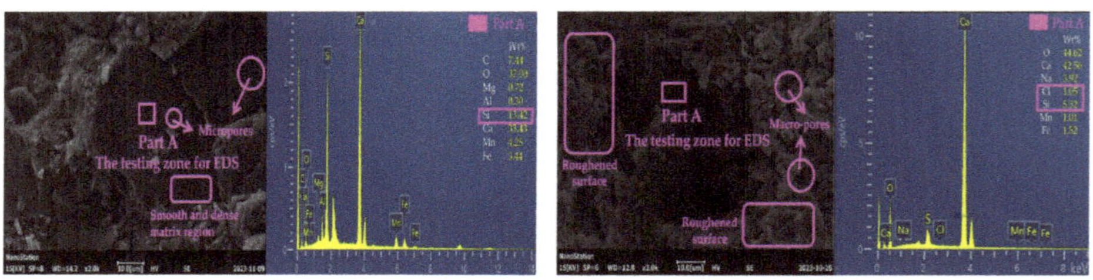

(a) (b)

Figure 7. The SEM-EDS images of recycled aggregate concrete composite slab–column assemblies: (**a**) A1-before NaCl F-Cs; (**b**) A1-after NaCl F-Cs.

The microstructural characterization in Figure 7a,b serves the following defined purpose: to establish baseline deterioration mechanisms in fiber-free control specimens under NaCl F-C exposure. Key observations, specifically surface roughening, pore enlargement, and accelerated chloride ingress, provide essential reference metrics for evaluating fiber-reinforced systems in subsequent analyses.

3.2. NaCl Dry–Wet Alternations

3.2.1. Axial Compressive Strength

Figure 8 demonstrates the axial compressive strength of recycled aggregate concrete composite slab–column assemblies with bolted connections after different numbers of NaCl D-As and the corresponding strength growth rate with basalt fiber incorporation. As shown, the axial compressive strength of the composite slab–column assemblies decreased by 8.7%–32.1% after 10–30 NaCl D-As (specifically, 8.7% at 10 D-As, 17.9% at 20 D-As, and 32.1% at 30 D-As for the control group). This degradation is primarily attributed to the cyclic precipitation and dissolution of NaCl crystals during D-As, which induces expansion and crack formation within the concrete matrix, thereby reducing the axial compressive strength [60]. Moreover, the cumulative damage from these crystallization–dissolution cycles further compromises structural strength. Conversely, the incorporation of 0.5–2.5 vol% basalt fibers resulted in a 12.4%–91.7% increase in the axial compressive strength of recycled aggregate concrete composite slab–column assemblies. The maximum improvement of 91.7% was achieved at 2.5 vol% fibers after 30 D-As, as shown in Figure 8.

A similar strength growth trend was observed in other concrete systems, where increasing basalt fiber content from 0% to 3% in sulphoaluminate cement-based reactive powder concrete induced a rising trend in compressive strength [61]. This enhancement stems from the crack bridging effect of basalt fibers, which effectively restrain microcrack propagation under load, thereby improving concrete strength and mitigating degradation induced by D-As. Importantly, the degree of strength improvement was driven by basalt fiber content, exhibiting a strong positive correlation where higher volumes (up to 2.5 vol%) yielded significantly greater enhancements. Comparative data from Figures 2 and 8 demonstrate that the axial compressive strength of basalt fiber-reinforced recycled aggregate concrete composite slab–column assemblies after NaCl D-As is marginally lower than after NaCl F-C exposure.

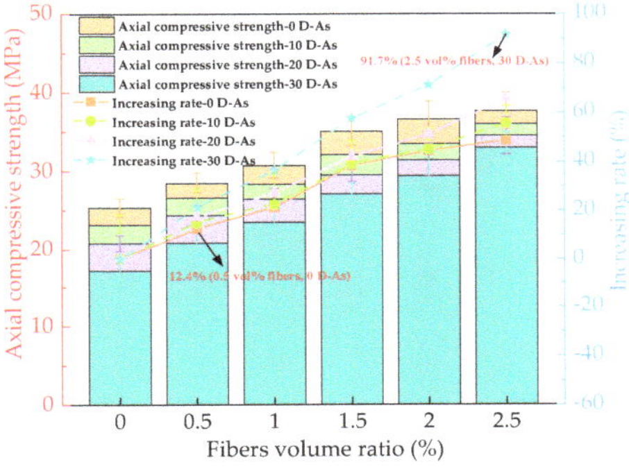

Figure 8. The axial compressive strength of recycled aggregate concrete composite slab–column assemblies after NaCl D-As.

3.2.2. Mass Loss Rate and Relative Dynamic Modulus of Elasticity

Figure 9a presents the MLR of recycled aggregate concrete composite slab–column assemblies with bolted connections during NaCl D-As. The MLR of recycled aggregate concrete composite slab–column assemblies reached 0.33%–1.19% after 10–30 NaCl D-As. This rise in MLR is primarily attributed to capillary pressure developing within the concrete matrix during NaCl D-As and fatigue effects accelerated by cyclic salt crystallization [62]. These mechanisms collectively induce progressive cracking and spalling, resulting in increased mass loss. However, incorporating 0.5–2.5 vol% basalt fibers reduced the MLR of recycled aggregate concrete composite slab–column assemblies to 0.09%–0.23%. The minimum MLR of 0.09% occurred at 2.5 vol% fibers after 10 D-As. This reduction occurs as basalt fibers restrict the development of cracks within the recycled aggregate concrete composite slab–column assemblies matrix, thereby mitigating mass loss. The effectiveness of mass loss reduction exhibited a strong dependence on basalt fiber content, with higher volumes (up to 2.5 vol%) providing progressively greater mass loss resistance. Comparative analysis of the data from Figures 3a and 9a demonstrates that NaCl D-As results in marginally higher MLR in recycled aggregate concrete composite slab–column assemblies than F-Cs.

Figure 9b depicts the RDME of recycled aggregate concrete composite slab–column assemblies with bolted connections during NaCl D-As. The RDME degraded from 100% to 72.7% after 10–30 D-As. This trend is attributed to capillary pressure induced in the concrete matrix of recycled aggregate concrete composite slab–column assemblies by NaCl D-As, which promotes crack initiation and propagation [63]. Consequently, ultrasonic wave propagation speed is reduced during RDME measurement due to D-A-induced cracks, reducing the RDME. Additionally, incorporating 0.5–2.5 vol% basalt fibers increased the

RDME of recycled aggregate concrete composite slab–column assemblies by 3.5%–20%. The maximum improvement of 20% was achieved at 2.5 vol% fibers after 30 D-As, as shown by the data. This enhancement primarily stems from the crack bridging effect of basalt fibers, which delays crack propagation in the concrete and thereby reduces the degradation of the dynamic modulus. Furthermore, the magnitude of RDME improvement correlated directly with basalt fiber content, where increased volumes consistently yielded higher dynamic modulus retention. A data comparison between Figures 9b and 3b reveals marginally greater RDME reduction under NaCl D-As than F-Cs.

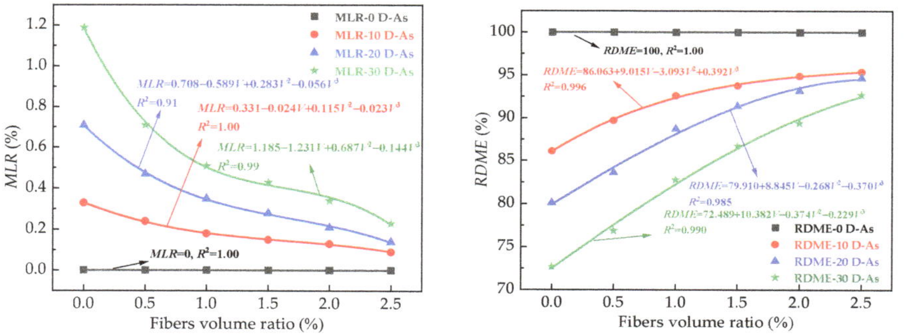

(a) (b)

Figure 9. The MLR and RDME of recycled aggregate concrete composite slab–column assemblies after NaCl D-As: (**a**) MLR; (**b**) RDME.

3.2.3. Flexural Toughness Behavior Analysis

Figure 10a presents the flexural load-deflection curves of recycled aggregate concrete composite slab–column assemblies reinforced with basalt fibers after NaCl D-As. Both the peak flexural load and maximum deflection decreased after 10–30 D-As. This reduction is primarily attributed to the fact that D-As accelerate crack formation within the recycled aggregate concrete composite slab–column assemblies matrix, consequently diminishing both flexural capacity and deflection. Moreover, comparative analysis of Figure 4a confirms that basalt fibers increase both the ultimate flexural load and maximum deflection of recycled aggregate concrete composite slab–column assemblies under F-Cs and D-As. This phenomenon is primarily attributed to the crack restricting effect of basalt fibers. The fiber reinforcement inhibits crack propagation during flexural loading, and increased fiber content further enhances this mechanism, thereby improving crack resistance, toughness, ultimate flexural load, and corresponding deflection values [64]. Collectively, these results demonstrate that incorporating basalt fibers effectively enhances the flexural toughness of recycled aggregate concrete composite slab–column assemblies.

Figure 10b displays the flexural toughness values obtained by integrating mid-span flexural load-deflection curves of recycled aggregate concrete composite slab–column assemblies with bolted connections. The figure shows that the flexural toughness of recycled aggregate concrete slab–column assemblies decreased by up to 59.7% after 10–30 NaCl DA-s. The significant deterioration in flexural toughness is attributed to the same mechanism as NaCl F-Cs. D-As equally accelerate cracking in the concrete matrix through their erosive effect, thereby degrading the flexural toughness of recycled aggregate concrete composite slab–column assemblies. Conversely, during D-As, the incorporation of 0.5–2.5 vol% basalt fibers enhanced the flexural toughness of recycled aggregate concrete composite slab–column assemblies. The maximum improvement of 254.7% was achieved at 2.5 vol% fibers after 20 D-As. This enhancement results from concrete cracking and fiber pull-out consuming fracture energy during bending, thereby increasing toughness. The magnitude of toughness enhancement exhibited strong dependence on basalt fiber content, with higher volumes (up to 2.5 vol%) generally yielding greater improvement. This content-dependent effect substantially increased resistance to deterioration-induced toughness loss.

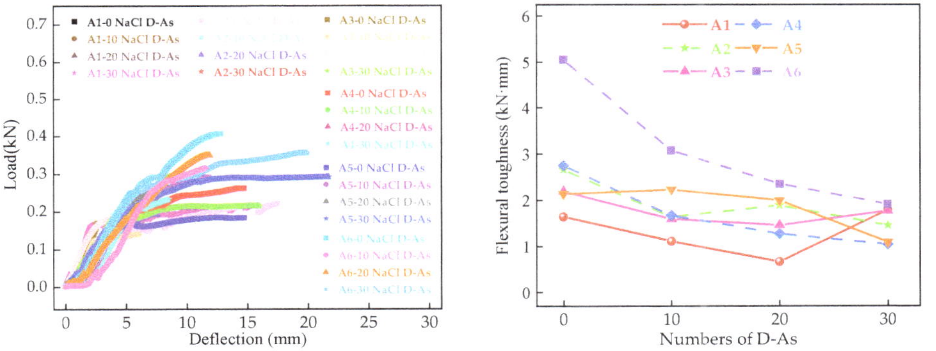

(**a**) (**b**)

Figure 10. The load-deflection curves and flexural toughness of recycled aggregate concrete composite slab–column assemblies after NaCl D-As: (**a**) load-deflection curves; (**b**) flexural toughness.

3.2.4. Impact Toughness Analysis

Figures 11 and 12, respectively, show the impact toughness of recycled aggregate concrete composite slab–column assemblies after NaCl solution D-As, along with the number of impacts and the impact toughness index. The results indicate that recycled aggregate concrete composite slab–column assemblies exhibited decreased impact toughness after 10–30 NaCl D-As. Meanwhile, the number of impacts to initial cracking, the number of impacts to failure, and the difference in impact counts between failure and initial cracking all decreased. The impact toughness degradation ranged from 3.3% to 28.3%. However, the impact toughness index (β, defined by Equation (3), as detailed in Section 3.1.4) increased with increasing D-As. Additionally, as shown in the figure, the incorporation of 0.5–2.5 vol% basalt fibers increased the impact toughness of recycled aggregate concrete composite slab–column assemblies by 5%–83.7%. The maximum improvement of 83.7% was achieved at 2.5 vol% fibers after 30 D-As. This improvement similarly originates from the effective suppression of crack initiation and propagation by basalt fibers within recycled aggregate concrete composite slab–column assemblies under impact loading, which enhances material toughness. The efficacy of impact toughness improvement was predominantly governed by basalt fiber content. Higher fiber volumes consistently improved crack initiation resistance and ultimate failure tolerance under impact loading, demonstrating content-controlled mitigation of D-A-induced deterioration. Collectively, under both NaCl F-C and D-A conditions, the incorporation of basalt fibers enhances the impact toughness of recycled aggregate concrete composite slab–column assemblies, increasing the number of impacts sustained prior to initial cracking and ultimate failure.

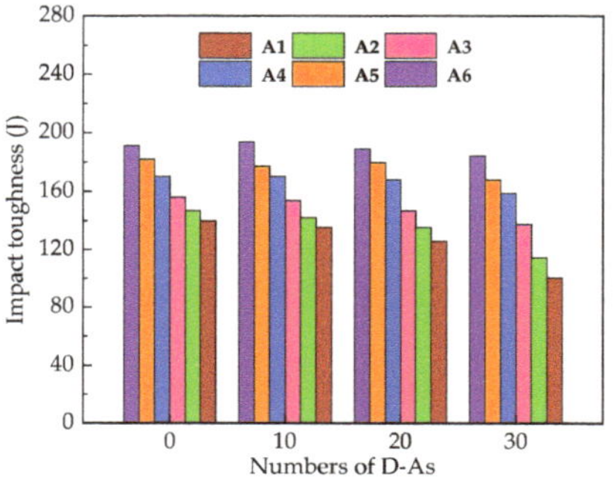

Figure 11. The impact toughness of recycled aggregate concrete composite slab–column assemblies after NaCl D-As.

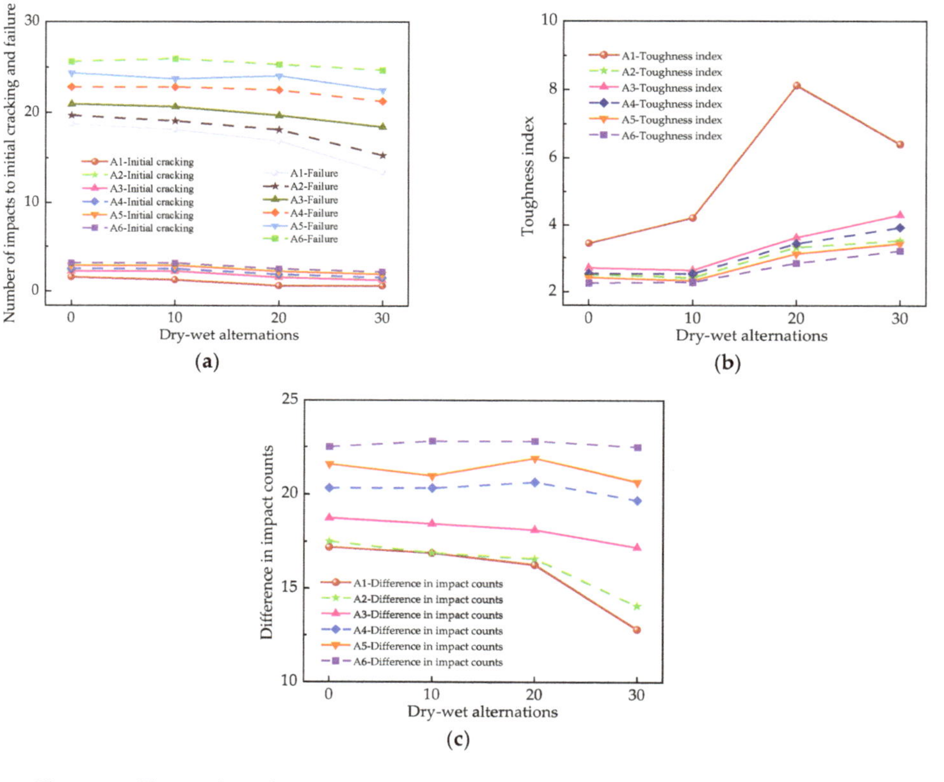

Figure 12. The number of impacts and impact toughness index of recycled aggregate concrete composite slab–column assemblies after NaCl D-As: (**a**) number of impacts; (**b**) toughness index; (**c**) difference in impact counts.

3.2.5. Microscopic Analysis

Figure 13a presents an SEM micrograph and corresponding EDS spectra of the concrete matrix in recycled aggregate concrete composite slab–column assembly specimens without fibers after NaCl D-As. The concrete microstructure exhibited a roughened surface morphology with progressively formed, interconnected crack networks, accompanied by the disintegration of the cementitious matrix around aggregates. This deterioration primarily stems from the cumulative damage effects of permeation pressure and chemical corrosion on the internal pore structure, which promotes interconnected crack formation and further aggravates degradation [65,66]. Moreover, after NaCl D-As, the Cl content (wt.%) in the concrete increased. NaCl D-As compromise the chloride erosion resistance of recycled aggregate concrete composite slab–column assemblies, facilitating continuous chloride ion penetration into concrete [67]. Additionally, comparative analysis of the Cl content disparities between Figures 13a and 7b revealed that NaCl D-As exert a significantly stronger erosive effect on recycled aggregate concrete composite slab–column assemblies than NaCl F-Cs. Figure 13b depicts the EDS results of recycled aggregate concrete composite slab–column assemblies with 2.5 vol% basalt fibers after NaCl D-As. In contrast to specimens without fibers, the 2.5 vol% basalt fibers significantly reduced Cl content in the concrete matrix. This particularly pronounced reduction in chloride content, as shown in Figure 13b, exemplifies the significant impact of fiber content on suppressing chloride ion penetration. Such an effect confirms that basalt fibers effectively enhance the chloride erosion resistance of recycled aggregate concrete composite slab–column assemblies.

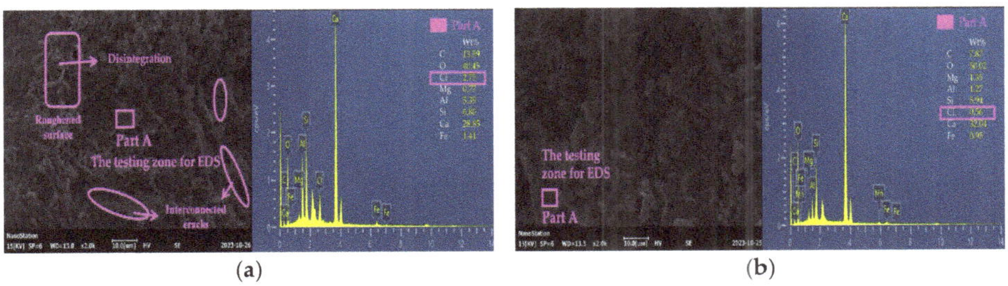

Figure 13. The SEM-EDS images of recycled aggregate concrete composite slab–column assemblies with and without fibers after NaCl D-As: (**a**) A1-after NaCl D-As; (**b**) A6-after NaCl D-As.

3.3. Comparative Analysis

The results presented in Sections 3.1 and 3.2 reveal that both NaCl F-Cs and D-As induced significant deterioration in the properties of the recycled aggregate concrete composite slab–column assemblies. Compared to F-Cs, the composite slab–column assemblies exhibited more pronounced performance degradation after 10–30 D-As. This observation is consistent with existing studies, which suggest that the detrimental effect of coupled salt and D-As on concrete properties is more severe [68,69]. The incorporation of 0.5–2.5 vol% basalt fibers improved all performance metrics of the recycled aggregate concrete composite slab–column assemblies under both F-C and D-A conditions, with the most substantial enhancement observed at 2.5 vol% fiber content. This improvement corresponds to findings from a previous study, which indicates that basalt fiber addition enhances crack resistance and toughness, thereby improving concrete durability [70]. Moreover, although the improvement in certain individual properties was more significant under F-Cs, the enhancement in overall performance was consistently more pronounced under D-As. In summary, basalt fibers effectively mitigate the performance degradation of recycled aggregate concrete composite slab–column assemblies in chloride environments, with their protective effect on overall performance being more pronounced under NaCl D-As.

4. Conclusions

This paper systematically investigated the effect of basalt fibers (0.5–2.5 vol%) on the chloride salt resistance of recycled aggregate concrete composite slab–column assemblies and explored the underlying micro-mechanism. The following conclusions are summarized below.

(1) After 100–300 NaCl F-Cs and 10–30 NaCl D-As, the axial compressive strength of recycled aggregate concrete composite slab–column assemblies decreased by 4.4%–30.2% and 8.7%–32.1%, respectively. The incorporation of 0.5–2.5 vol% basalt fibers resulted in strength increases of 11.0%–87.1% and 12.4%–91.7% relative to corroded plain assemblies under identical exposure conditions. These strength enhancements peaked at 87.1% after 300 F-Cs and 91.7% after 30 D-As with 2.5 vol% fibers.

(2) After 100–300 NaCl F-Cs and 10–30 NaCl D-As, the MLR values of recycled aggregate concrete composite slab–column assemblies were 0.26%–1.13% and 0.33%–1.19%, while the RDME decreased to 73.4% and 72.7%. Under identical exposure conditions, the incorporation of 0.5–2.5 vol% basalt fibers reduced MLR to 0.08%–0.21% and 0.09%–0.23%, while enhancing RDME by 2.6%–19.9% and 3.5%–20.0% relative to corroded plain assemblies. The minimum MLR values of 0.08% (100 F-Cs) and 0.09% (10 D-As) were achieved at 2.5 vol% fibers, while peak RDME improvements reached 19.9% after 300 F-Cs and 20.0% after 30 D-As at 2.5 vol% fibers.

(3) After 100–300 NaCl F-Cs and 10–30 NaCl D-As, the flexural toughness of recycled aggregate concrete composite slab–column assemblies was significantly reduced, with impact toughness decreasing by 3.2%–19.4% and 3.3%–28.3%, respectively. The incorporation of 0.5–2.5 vol% basalt fibers increased impact toughness by 6.5%–62.0% under NaCl F-Cs and by 5.0%–83.7% under NaCl D-As. At 2.5 vol% fibers, impact toughness increased by up to 83.7% after 30 D-As, while maximum flexural toughness improvement reached 773.6% after 100 F-Cs. In addition, NaCl D-As had a more significant effect on chloride ion erosion in recycled aggregate concrete composite slab–column assemblies compared to NaCl F-Cs. The addition of basalt fibers significantly inhibited this erosive damage.

Comprehensive analysis indicates that basalt fiber content is the key controlling factor for enhancing chloride resistance performance of recycled aggregate concrete composite slab–column assemblies. Key performance indicators, such as axial compressive strength, RDME, and toughness, exhibit an apparent positive enhancement trend with increasing basalt fiber content under NaCl erosion environment. Among these, 2.5 vol% is the optimal content. Compared to NaCl F-Cs, the protective effect of basalt fiber is more pronounced in resisting NaCl D-As erosion. This study reveals the performance evolution of basalt fiber-reinforced recycled aggregate concrete composite slab–column assemblies in chloride salt environments, providing an important theoretical basis for developing fiber-reinforced design strategies targeting high-durability recycled aggregate concrete structures in coastal salt-erosion environments.

Author Contributions: Conceptualization, F.S. and Y.W.; methodology, D.W.; software, Z.X.; validation, H.W., C.L. and Z.X.; formal analysis, X.T.; investigation, Y.W.; resources, X.T.; data curation, N.X.; writing—original draft preparation, D.W.; writing—review and editing, D.W.; visualization, H.W.; supervision, Z.X.; project administration, D.W.; funding acquisition, H.W. and D.W. All authors have read and agreed to the published version of the manuscript.

Funding: This research was funded by the 2023 Liaoning Provincial Science and Technology Programme Joint Programme Projects, grant number 2023JH2/101700001; the 2023 Liaoning Provincial Science and Technology Programme Joint Programme Projects, grant number 2023JH2/101700002; the 2024 Liaoning Provincial Department of Education Basic Research Project, grant number LJ212411779037; and the 2025 Liaoning Provincial Department of Education Basic Research Project, grant number LJ222511779001.

Institutional Review Board Statement: Not applicable.

Informed Consent Statement: Not applicable.

Data Availability Statement: The data used to support the findings of this study are available on request.

Conflicts of Interest: The authors declare no conflicts of interest.

References

1. Nguyen, M.H.; Huynh, T.P. Turning incinerator waste fly ash into interlocking concrete bricks for sustainable development. *Constr. Build. Mater.* **2022**, *321*, 126385. [CrossRef]
2. Ahmad, M.R.; Chen, B. Experimental research on the performance of lightweight concrete containing foam and expanded clay aggregate. *Compos. Part B Eng.* **2019**, *171*, 46–60. [CrossRef]
3. Song, P.; Peng, X.; Zheng, R.; Xia, J. Material properties and mechanical performances of manufactured factory-produced glass fiber-reinforced autoclaved aerated concrete panel. *Buildings* **2024**, *14*, 2895. [CrossRef]
4. Barbosa, K.; Silva, W.T.; Silva, R.; Vital, W.; Bezerra, L.M. Experimental investigation of axially loaded precast sandwich panels. *Buildings* **2023**, *13*, 1993. [CrossRef]
5. Zheng, L.; Wu, H.; Zhang, H.; Duan, H.; Wang, J.; Jiang, W.; Song, Q. Characterizing the generation and flows of construction and demolition waste in China. *Constr. Build. Mater.* **2017**, *136*, 405–413. [CrossRef]
6. Akan, A.E.; Ünal, F.; Koçyiğit, F. Investigation of energy saving potential in buildings using novel developed lightweight concrete. *Int. J. Thermophys.* **2021**, *42*, 4. [CrossRef]
7. Zhou, C.; Shi, S.Q.; Chen, Z.; Cai, L.; Smith, L. Comparative environmental life cycle assessment of fiber reinforced cement panel between kenaf and glass fibers. *J. Clean. Prod.* **2018**, *200*, 196–204. [CrossRef]
8. Zhang, M.; Liu, X.; Kong, L. Evaluation of carbon and economic benefits of producing recycled aggregates from construction and demolition waste. *J. Clean. Prod.* **2023**, *425*, 138946. [CrossRef]
9. Duan, H.; Miller, T.R.; Liu, G.; Tam, V.W. Construction debris becomes growing concern of growing cities. *Waste Manag.* **2019**, *83*, 1–5. [CrossRef]
10. Ding, Z.; Wang, N.; Zou, P.X.W. Barriers and countermeasures of construction and demolition waste recycling enterprises under circular economy. *J. Clean. Prod.* **2023**, *420*, 138235. [CrossRef]
11. Bai, G.; Zhu, C.; Liu, C.; Liu, B. An evaluation of the recycled aggregate characteristics and the recycled aggregate concrete mechanical properties. *Constr. Build. Mater.* **2020**, *240*, 117978. [CrossRef]
12. Hasheminezhad, A.; King, D.; Ceylan, H.; Kim, S. Comparative life cycle assessment of natural and recycled aggregate concrete: A review. *Sci. Total Environ.* **2024**, *950*, 175310. [CrossRef] [PubMed]
13. Hu, K.; Gillani, S.T.A.; Tao, X.; Tariq, J.; Chen, D. Eco-friendly construction: Integrating demolition waste into concrete masonry blocks for sustainable development. *Constr. Build. Mater.* **2025**, *460*, 139797. [CrossRef]
14. Wang, J.; Zhang, J.; Cao, D.; Dang, H.; Ding, B. Comparison of recycled aggregate treatment methods on the performance for recycled concrete. *Constr. Build. Mater.* **2020**, *234*, 117366. [CrossRef]
15. Neupane, R.P.; Imjai, T.; Makul, N.; Garcia, R.; Kim, B.; Chaudhary, S. Use of recycled aggregate concrete in structural members: A review focused on Southeast Asia. *J. Asian Archit. Build. Eng.* **2023**, *24*, 1197–1220. [CrossRef]
16. Thomas, B.S.; Yang, J.; Bahurudeen, A.; Chinnu, S.N.; Abdalla, J.A.; Hawileh, R.A.; Hamada, H.M. Geopolymer concrete incorporating recycled aggregates: A comprehensive review. *Clean. Mater.* **2022**, *3*, 100056. [CrossRef]
17. Abdulhaleem, K.N.; Van Mohammed, R.; Hamada, H.M.; Humada, A.M.; Majdi, A. Effect of Recycled Aggregate Concrete and Steel Fibers on the Strength of Self-Compacting Concrete. In *E3S Web of Conferences*; EDP Sciences: Paris, France, 2023; Volume 427, p. 02012.
18. Lin, D.; Wu, J.; Yan, P.; Chen, Y. Effect of residual mortar on compressive properties of modeled recycled coarse aggregate concrete. *Constr. Build. Mater.* **2023**, *402*, 132511. [CrossRef]
19. Jadon, S.; Kumar, S. Stone dust and recycled concrete aggregates in concrete construction: An efficient way of sustainable development. *Mater. Today Proc.* **2023**. [CrossRef]
20. Mistri, A.; Bhattacharyya, S.K.; Dhami, N.; Mukherjee, A.; Barai, S.V. A review on different treatment methods for enhancing the properties of recycled aggregates for sustainable construction materials. *Constr. Build. Mater.* **2020**, *233*, 117894. [CrossRef]
21. Liang, C.; Wang, S.; Cai, Z.; Yin, Y.; Gao, Y.; Guo, M.Z.; Ma, Z. Effects of CO_2 curing methods on frost resistance and mechanical properties of recycled aggregate concrete. *Case Stud. Constr. Mater.* **2024**, *20*, e02973. [CrossRef]
22. Frankovič, A.; Ducman, V.; Dolenec, S.; Panizza, M.; Tamburini, S.; Natali, M.; Bernardi, A. Up-scaling and performance assessment of façade panels produced from construction and demolition waste using alkali activation technology. *Constr. Build. Mater.* **2020**, *262*, 120475. [CrossRef]
23. Hossain, F.Z.; Pal, A.; Ahmed, K.S.; Bediwy, A.; Alam, M.S. Shear behavior of polypropylene fiber-reinforced concrete beams containing recycled aggregate and crumb rubber. *J. Clean. Prod.* **2023**, *412*, 137370. [CrossRef]

24. Zheng, Y.; Lv, X.; Hu, S.; Zhuo, J.; Wan, C.; Liu, J. Mechanical properties and durability of steel fiber reinforced concrete: A review. *J. Build. Eng.* **2024**, *82*, 108025. [CrossRef]
25. Malchiodi, B.; Pelaccia, R.; Pozzi, P.; Siligardi, C. Three sustainable polypropylene surface treatments for the compatibility optimization of PP fibers and cement matrix in fiber-reinforced concrete. *Ceram. Int.* **2023**, *49*, 24611–24619. [CrossRef]
26. Zhang, S.; Gao, D.; Zhu, H.; Chen, L.; He, Z.; Yang, L. Flexural behavior of seawater-mixed steel fiber reinforced concrete exposed to simulated marine environments. *Constr. Build. Mater.* **2023**, *373*, 130858. [CrossRef]
27. Chen, C.; Ding, Y.; Wang, X.; Bao, L. Recent advances to engineer tough basalt fiber reinforced composites: A review. *Polym. Compos.* **2024**, *45*, 12559–12574. [CrossRef]
28. Zhu, M.; Zhu, M.; Zhai, R.; Zhu, W.; He, J. Research Progress on the Surface Modification of Basalt Fibers and Composites: A Review. *Materials* **2025**, *18*, 1164. [CrossRef]
29. Al-Kharabsheh, B.N.; Arbili, M.M.; Majdi, A.; Alogla, S.M.; Hakamy, A.; Ahmad, J.; Deifalla, A.F. Basalt fiber reinforced concrete: A compressive review on durability aspects. *Materials* **2023**, *16*, 429. [CrossRef]
30. You, J.J.; Song, Q.Y.; Tan, D.; Yang, C.; Liu, Y.F. Mechanical properties and microstructure of basalt fiber-biobased-basic magnesium sulfate cement. *Cement Concr. Compos.* **2023**, *137*, 104934. [CrossRef]
31. *ASTM C1611/C1611M-21*; Standard Test Method for Slump Flow of Self-Consolidating Concrete. ASTM International: West Conshohocken, PA, USA, 2021.
32. *ASTM C192/C192M-21*; Standard Practice for Making and Curing Concrete Test Specimens in the Laboratory. ASTM International: West Conshohocken, PA, USA, 2021.
33. *GB/T 50204-2015*; Code for Acceptance of Constructional Quality of Concrete Structures. Ministry of Housing and Urban-Rural Development, People's Republic of China: Beijing, China, 2015.
34. Meng, T.; Wei, H.; Yang, X.; Zhang, B.; Zhang, Y.; Zhang, C. Effect of mixed recycled aggregate on the mechanical strength and microstructure of concrete under different water cement ratios. *Materials* **2021**, *14*, 2631. [CrossRef] [PubMed]
35. *GB/T 50082-2009*; Standard for Test Methods of Long-Term Performance and Durability of Ordinary Concrete. Ministry of Housing and Urban-Rural Development, People's Republic of China: Beijing, China, 2009.
36. He, Y.; Zhao, Y.G.; Lin, S.; Zhang, Z. Prefabricated bolted PEC beam-to-CFST column joints: Development and its seismic behavior. *Eng. Struct.* **2024**, *314*, 118442. [CrossRef]
37. Huang, A.; Qu, H.; Li, W.; Song, F.; Wang, Z. Mechanical properties of high-strength bolted connections with different bolt hole types under corrosion conditions. *Structures* **2025**, *71*, 108135. [CrossRef]
38. *GB/T 23900-2009*; Non-Destructive Testing—Practice for Measuring Ultrasonic Velocity in Materials. Standardization Administration of China, People's Republic of China: Beijing, China, 2009.
39. Wang, A.; Sun, W.; Lu, M.; Peng, X.; Wang, H.; Ji, Y. Research on the Corrosion Resistance of Reactive Powder Concrete with Straw Fibers under Chloride Environment. *Coatings* **2024**, *14*, 961. [CrossRef]
40. *GB/T 50081-2019*; Standard for Test Methods of Concrete Physical and Mechanical Properties. Ministry of Housing and Urban-Rural Development, People's Republic of China: Beijing, China, 2019.
41. Wang, H.; Wang, A.; Wang, J.; Zhang, J.; Du, T.; Shen, J. Optimizing quaternary chemical additives and steel fiber dispersion in CSA cement based HPC for rapid repair engineering. *Constr. Build. Mater.* **2025**, *489*, 142265. [CrossRef]
42. Wu, D.; Liang, Y.; Liu, H.; Peng, S. Behavior of Demountable and Replaceable Fabricated RC Beam with Bolted Connection Under Mid-Span Compression. *Buildings* **2025**, *15*, 2589. [CrossRef]
43. ACI 544.2R-89. *Measurement of Properties of Fiber Reinforced Concrete*; American Concrete Institute: Farmington Hills, MI, USA, 1999; pp. 1–12.
44. Murali, G.; Wong, L.S.; Abid, S.R. A comprehensive review of drop weight impact testing: Evaluating the pros and cons in fiber-reinforced concrete performance assessment. *J. Build. Eng.* **2024**, *94*, 109934. [CrossRef]
45. Wang, D.; Xu, Z.; Cao, Z.; Xu, N.; Li, C.; Tian, X.; Wang, H. Research on the Performance of Steel Strand-Reinforced Reactive Powder Concrete with Mixed Steel Fibers and Basalt Fibers under the Salt Dry–Wet Erosion. *Coatings* **2024**, *14*, 833. [CrossRef]
46. Cai, Z.; Ren, J.; Shen, G.; Gu, X.; Cheng, W.; Wang, H. The influence of assembly unit of fibers on the mechanical and long-term properties of reactive powder concrete. *Coatings* **2023**, *13*, 412. [CrossRef]
47. Chen, C.; Zhang, K.; Yin, Z.; Zhou, J. Deterioration performance of recycled aggregate pervious concrete under freezing–thawing cycle and chloride environment. *Buildings* **2023**, *13*, 645. [CrossRef]
48. Zhang, F.; Hu, Z.; Wei, F.; Wen, X.; Li, X.; Dai, L.; Liu, L. Study on concrete deterioration in different NaCl-Na_2SO_4 solutions and the mechanism of Cl^- diffusion. *Materials* **2021**, *14*, 5054. [CrossRef]
49. Xu, W.; Li, Z. Service life prediction of basalt fiber reinforced concrete under salt freeze–thaw cycles. *Stavební obzor—Civil Eng. J.* **2022**, *31*, 222–234. [CrossRef]
50. Sun, H.; Sun, W.; Shi, F.; Miao, L.; Wang, H. The influence of salt erosion on the mechanical performances of ultra-high-performance concrete with secondary aluminum dross. *Coatings* **2024**, *14*, 189. [CrossRef]

51. Wang, H.; Jin, K.; Zhang, A.; Zhang, L.; Han, Y.; Liu, J.; Feng, L. External erosion of sodium chloride on the degradation of self-sensing and mechanical properties of aligned stainless steel fiber reinforced reactive powder concrete. *Constr. Build. Mater.* **2021**, *287*, 123028. [CrossRef]
52. Yang, J.; Guo, Y.; Tan, J.; Shen, A.; Wu, H.; Li, Y.; Wang, L. Strength deterioration and crack dilation behavior of BFRC under dynamic fatigue loading. *Case Stud. Constr. Mater.* **2022**, *16*, e01051. [CrossRef]
53. Xiao, R.; An, B.; Wu, F.; Wang, W.; Sui, Y.; Wang, Y. Freeze–Thaw Damage Characterization of Cement-Stabilized Crushed Stone Base with Skeleton Dense Gradation. *Materials* **2024**, *17*, 1228. [CrossRef]
54. Yuan, C.; Qu, S.; Bai, W.; Guan, J.; Xie, Y. Study on the mechanical properties and mesoscopic damage mechanism of recycled aggregate concrete under different dynamic strain rates after freeze-thaw cycles. *Case Stud. Constr. Mater.* **2025**, *22*, e04182. [CrossRef]
55. Jiang, X.; Mu, S.; Liu, J. Influence of chlorides and salt concentration on salt crystallization damage of cement-based materials. *J. Build. Eng.* **2022**, *61*, 105260. [CrossRef]
56. Li, Y.; Shen, A.; Wu, H. Fractal dimension of basalt fiber reinforced concrete (BFRC) and its correlations to pore structure, strength, and shrinkage. *Materials* **2020**, *13*, 3238. [CrossRef]
57. Zhao, Y.-R.; Wang, L.; Lei, Z.-K.; Han, X.-F.; Xing, Y.-M. Experimental study on dynamic mechanical properties of the basalt fiber reinforced concrete after the freeze-thaw based on the digital image correlation method. *Constr. Build. Mater.* **2017**, *147*, 194–202. [CrossRef]
58. Da, B.; Yu, H.; Ma, H.; Wu, Z. Reinforcement corrosion research based on the linear polarization resistance method for coral aggregate seawater concrete in a marine environment. *Anti-Corros. Methods Mater.* **2018**, *65*, 458–470. [CrossRef]
59. Jiang, W.Q.; Shen, X.H.; Jin, X.; Mao, L.X.; Yang, J.; Liu, Q.F. A numerical study on chloride diffusion in freeze-thaw affected concrete. *Constr. Build. Mater.* **2018**, *179*, 553–565. [CrossRef]
60. Tang, X.; Xu, Q.; Qian, K.; Ruan, S.; Lian, S.; Zhan, S. Effects of cyclic seawater exposure on the mechanical performance and chloride penetration of calcium sulfoaluminate concrete. *Constr. Build. Mater.* **2021**, *303*, 124139. [CrossRef]
61. Huang, G.; Wang, H.; Shi, F. Coupling effect of salt freeze-thaw cycles and carbonation on the mechanical performance of quick hardening sulphoaluminate cement-based reactive powder concrete with basalt fibers. *Coatings* **2021**, *11*, 1142.
62. Chen, J.; Yang, M. Advances of environment damage mechanics in marine engineering concrete. *Mech. Eng.* **2023**, *45*, 499–512.
63. Chen, C.; Wang, L.; Liu, R.; Zhu, P.; Liu, H.; Wang, X.; Chen, Y. Chloride penetration of concrete exposed to dry-wet cycle with various dry-wet ratios and temperature. *Constr. Build. Mater.* **2023**, *400*, 132883. [CrossRef]
64. Diao, R.; Cao, Y.; Michel, M.M.; Wang, A.; Sun, L.; Yang, F. Mechanical performance study of PVA fiber-reinforced seawater and sea sand cement-based composite materials. *Sci. Rep.* **2024**, *14*, 18161. [CrossRef]
65. Liu, Y. Study on the damage characteristics of concrete structures by chloride erosion. *J. Xi'an Univ. Archit. Technol. (Nat. Sci. Ed.).* **2020**, *52*, 390–395. (In Chinese)
66. Dong, W.; Wang, J. Degradation mechanism and strength prediction of aeolian sand concrete under sulphate dry-wet cycles. *Constr. Build. Mater.* **2024**, *450*, 138617. [CrossRef]
67. Chen, C.; Wang, L.; Liu, R.; Yu, J.; Liu, H.; Wu, J. Chloride penetration of recycled fine aggregate concrete under drying–wetting cycles. *Materials* **2023**, *16*, 1306. [CrossRef]
68. Yang, W.; Huang, Y.; Tang, Z.; Xiong, X.; Li, C.; Zhong, X.; Liu, L.; Zhou, H.; Zhang, X. Impermeability performance and corrosion resistance mechanism of basalt fiber recycled concrete under the coastal tidal environment. *Constr. Build. Mater.* **2024**, *411*, 134510. [CrossRef]
69. Yan, C.; An, N.; Wang, Y.; Sun, W. Effect of Dry–Wet Cycles and Freeze–Thaw Cycles on the Antierosion Ability of Fiber–Reinforced Loess. *Adv. Mater. Sci. Eng.* **2021**, *2021*, 8834598. [CrossRef]
70. Ye, Y.; Xie, T.; Guo, T.; Ding, W. Durability of basalt fiber-reinforced aeolian sand concrete in extreme environments: Resistance to wind-sand erosion and salt freeze–thaw cycles. *Constr. Build. Mater.* **2024**, *457*, 139264. [CrossRef]

coatings

MDPI

Article

Characterizing Asphalt Binder Phase Transitions via Dynamic Mechanical Analysis: Performance Implications and SARA Fraction Correlations

Yanzhu Wang [1,*], Jincheng Wei [1], Xizhong Xu [1], Qinsheng Xu [1,*], Yue Sun [1], Na Liu [2] and Zhengchao Zhang [1]

[1] Shandong Transportation Institute, Jinan 250102, China
[2] Department of Highway Engineering, Shandong College of Highway Technician, Jinan 250104, China
* Correspondence: wangyanzhu@sdjtky.cn (Y.W.); 15966652536@163.com (Q.X.)

Abstract

Asphalt is widely used as a binder in pavement engineering. The temperature-dependent phase transition behavior of asphalt binders critically influences pavement performance. This study comprehensively evaluates phase transition characteristics to establish robust performance indicators. Dynamic mechanical analysis (DMA) was employed to characterize 30 neat asphalt binders across a broad temperature range. Phase transition temperatures and moduli were derived from complex and loss modulus curves. Correlations with saturate, aromatic, resin, and asphaltene (SARA) fractions and conventional properties (penetration, viscosity, ductility) were statistically analyzed. The results revealed significant performance variations among binders of identical penetration grades. T_g effectively differentiated low-temperature behavior, overcoming empirical limitations. High-temperature indicators (T_2, E_{20}) strongly correlated with viscosity ($R^2 > 0.96$). SARA analysis showed that saturates reduced T_g ($r = -0.566$) while asphaltenes increased E_{20} ($r = 0.804$). Multiple regression models confirm synergistic interactions among SARA fractions, although low-temperature indices exhibit a weaker dependence on composition. DMA-derived phase transition parameters provide physically meaningful performance indicators, superior to conventional metrics. Incorporating T_g and T_2/E_{20} into grading systems can enhance asphalt selection for thermal susceptibility, advancing pavement durability design.

Keywords: asphalt binder; dynamic mechanical analysis; phase transition characteristics; chemical composition; correlation analysis

1. Introduction

Asphalt binders are an engineering material commonly used as a binder in road construction. Its rheological properties are significantly affected by temperature. At elevated temperatures, asphalt binders exhibit fluid behavior, whereas they deform easily at intermediate temperatures and become solid and brittle at low temperatures. Based on these characteristics, asphalt binders can be categorized into three states: Newtonian liquid, viscoelastic, and glassy [1]. The transition of a thermodynamic system from one state to another is termed phase transition. For example, the change in asphalt binders from a glassy state to a viscoelastic state is known as glass transition, whereas the transition from the viscoelastic state to the viscous state is referred to as flow transition. The specific temperatures at which these transitions occur are designated the glass transition temperature and flow transition temperature [2,3]. Liu [1] determined the glass transition temperatures of three typical asphalt binders with penetration grades of 20/40, 40/60, and 60/80 through measurement and calculation; they were approximately −1 °C, −8 °C, and −10 °C, respectively.

The role of the phase transition properties In asphalt binders Is widely acknowledged as a critical factor affecting the performance of mixtures. At low temperatures, a compliant binder with enhanced viscoelastic relaxation capabilities exhibits superior thermal stress accommodation, effectively reducing the risk of cryogenic cracking through stress dissipation. At intermediate temperatures, an optimally balanced viscoelastic binder facilitates energy dissipation via controlled deformation mechanisms, thereby preventing stress concentration and the initiation of fatigue cracks under cyclic loading. At high temperatures, an increased stiffness combined with elastic recovery characteristics is essential for resisting cumulative plastic deformation under sustained traffic loads. Therefore, it is essential to develop an effective methodology for characterizing the phase transition performance of asphalt binders in relation to the expected performance of the mixture [4].

Dynamic mechanical analysis (DMA) has been extensively employed to evaluate the rheological properties and determine the phase transition temperature (PTT) of asphalt binders [5–9]. Asphalt binders, representative viscoelastic materials, exhibit a mechanical response consistent with the time–temperature superposition principle (TTSP). According to the TTSP, when dynamic tests are performed at small strain levels (within the linear viscoelastic range), the rheological data obtained at both elevated and reduced temperatures can be graphically correlated to high or low frequencies (which correspond to short or long loading durations), and vice versa. Therefore, a more comprehensive rheological curve in either the time or frequency domains can be constructed by shifting the rheological data from various test temperatures to a defined reference temperature. This curve is referred to as the master curve of the viscoelastic characteristic function of the asphalt material at a reference temperature. Ultimately, the glass transition temperature (T_g) of the asphalt material can be inferred from the master curve, allowing for the evaluation of the low-temperature performance of asphalt binders [10–12]. In addition to this method, temperature scanning provides a more direct method for obtaining the glass transition temperature. The mechanical response curve of the material is acquired by temperature scanning under controlled conditions, with the peak temperature of the loss modulus defined as T_g. This feature is significantly easier to identify in this way than through using the differential scanning calorimetry (DSC) method. In contrast to other techniques, the mechanical properties measured by DMA are more sensitive to changes in the microstructure [13–18]. Currently, there is a substantial body of research on the glass transition temperature of asphalt binders [19,20], yet there is comparatively limited investigation into the phase transition characteristics across a wide temperature range, such as the flow transition temperature.

The correlation between the PTT of asphalt binders and their chemical composition requires further investigation. The phase-transition characteristics of asphalt binders serve as a macroscopic representation of their microscopic molecular movements. Given the diversity and complexity of the molecular structure of asphalt binders, accurately modeling their chemical composition and its relationship with engineering properties presents significant challenges. A more pragmatic approach to elucidating the relationship between the chemical composition of asphalt binders and their engineering properties is to categorize the chemical composition based on various properties rather than attempting to determine the precise molecular structure of any specific binder [21,22]. Chromatographic separation remains the most effective and practical method for isolating asphalt binders [23–25]. Currently, asphalt binders are primarily separated into four fractions—saturates, aromatics, resins, and asphaltenes (SARA)—to facilitate comparisons regarding their chemical composition [26,27]. The relationship between the SARA fractions of asphalt binders and their physical and engineering properties has been extensively studied by numerous researchers [28–30]. It has been established that the SARA fraction content significantly affects the stiffness, viscosity, deformation behavior, and temperature sensitivity of asphalt binders. Specifically, the complex modulus of asphalt binders increases with higher asphaltene or resin content, whereas it decreases with elevated saturate or aromatic content [31–33]. Additionally, an increased asphaltene content results in a reduction in the creep stiffness modulus and m-value under low-temperature conditions [34]. With regard to phase transition characteristics, several scholars have conducted preliminary investigations into the effect of SARA components on the glass transition temperature. Wang et al. [34] demonstrated that the T_g of asphalt binders is primarily influenced by the saturates and aromatics fractions, yet it increases with higher asphaltene content. Their linear regression analysis also revealed a significant correlation between T_g and the colloidal instability index, which is derived from SARA fractions. Wieser et al. [35] indicated that saturation lowers the glass transition temperature of neat asphalt binders. For modified asphalt, the glass transition temperature increases for low saturate contents and decreases for high saturate contents. However, a comprehensive understanding of how each SARA fraction influences a broader range of phase transition temperatures and moduli—especially across a wide temperature spectrum—remains limited.

Therefore, this study acquired the phase transition characteristics of 30 types of base asphalt binders over a broad temperature range using the DMA method. The performance of the asphalt binders was evaluated based on the characteristic temperature and modulus as key indicators. An analysis was conducted to assess the correlation between the phase-transition characteristic indicators and the evaluation results of conventional indicators, demonstrating the necessity and applicability of evaluating asphalt binder phase-transition characteristics using the DMA method. Finally, a statistical analysis was performed to explore the relationship between the phase-transition characteristic indicators and the SARA fraction of asphalt binders, providing theoretical support for the evaluation of these characteristic indicators.

2. Materials and Methods

2.1. Asphalt Binder Samples

In this study, 30 types of neat asphalt binders sourced from various regions and manufacturers were evaluated. The asphalt binders were categorized by source as follows: Asphalts sourced from China Petroleum Fuel Asphalt Co., Ltd. (Qinhuangdao, Hebei, China), produced from Venezuelan crude oil via straight-run process, with penetration grades of 20/40, 40/60, and 60/80, were labeled as A1, A2, and A3, respectively. Asphalts were also provided by the Research Institute of CNPC Fuel Oil Co., Ltd. (Beijing, China), produced from Venezuelan crude oil through straight-run and blending processes, with penetration grades ranging from 20 to 90 (0.1 mm); among these, straight-run asphalts were labeled as B1, B2, B3, B4, and B5, while blended asphalts were labeled as C1, C2, C3, C4, and C5. Hard asphalts from Zhenjiang Special Asphalt Co., Ltd. (Zhenjiang, Jiangsu, China), with penetration grades ranging from 10 to 50 (0.1 mm), were labeled as D1, D2, D3, D4, D5, and D6. Straight-run asphalts from Senegal, Africa, with penetration grades ranging from 20 to 50 (0.1 mm), were labeled as E1, E2, E3, E4, and E5. Three commonly used straight-run asphalts with penetration grade 60/80, provided by Hainan Province from Foshan Gaofu China Petroleum Fuel Asphalt Co., Ltd. (Foshan, Guangdong, China), TPC Asphalt Co., Ltd. (Bangkok, Thailand), and IRPC Public Co., Ltd. (Bangkok, Thailand), were labeled as F1, F2, and F3, respectively. Three batches of straight-run asphalt with penetration grade 80/100, provided by Sinopec, were labeled as G1, G2, and G3. The fundamental physical properties and SARA fraction content are listed in Table 1.

Table 1. Physical properties and components of asphalt binders.

Asphalt Binder	Pen. (25 °C) /0.1 mm	Softening Point/°C	Ductility (15 °C) /cm	Ductility (10 °C) /cm	Dynamic Viscosity (60 °C)/Pa·s	Component Analysis/%			
						Saturates	Aromatics	Resins	Asphaltenes
A1	22	61.1	0	0	2266	7.7	38.8	37.3	16.2
A2	49	52.2	>100	10	500	10.2	40.2	35.0	14.6
A3	67	49.0	>100	>100	224	11.8	41.0	34.3	12.9
B1	30	59.0	8	0	1159	11.1	43.3	30.9	14.7
B2	37	57.2	10	0	843	9.7	41.0	34.7	14.6
B3	46	53.2	53	0	531	11.4	40.0	34.4	14.2
B4	70	45.4	>100	69	266	12.1	37.7	37.0	13.2
B5	88	44.0	>100	38	188	10.9	42.0	35.5	13.6
C1	31	59.6	8	0	1113	8.6	40.8	35.1	15.5
C2	37	57.7	12	0	808	10.4	40.8	33.8	15.0
C3	47	53.7	48	0	490	9.6	40.5	35.3	14.6
C4	71	46.1	>100	85	198	11.0	43.4	32.0	13.6
C5	90	44.7	>100	51	146	11.1	39.5	35.8	13.6
D1	11	98.9	0	0	-	10.6	29.4	32.1	27.9
D2	16	80.2	0	0	59,214	11.1	32.5	31.8	24.6
D3	16	80.4	0	0	52,809	11.1	32.5	32.2	24.2
D4	34	59.8	7	0	1533	11.7	37.2	32.9	18.2
D5	35	60.0	8	0	1377	11.6	37.6	33.4	17.4
D6	42	56.7	9	0	887	11.6	38.6	33.1	16.7
E1	27	64.3	5	0	3403	10.1	39.5	32.5	17.9
E2	47	54.6	24	8	468	11.2	41.4	30.7	16.7
E3	47	50.5	17	5	417	10.9	40.5	33.4	15.2
E4	47	50.2	37	0	443	10.0	47.0	31.7	11.3
E5	37	53.7	12	0	656	11.2	42.6	32.8	13.4
F1	69	46.9	>100	88	258	9.5	40.6	36.1	13.8
F2	71	46.7	>100	>100	194	6.2	56.1	27.3	10.4
F3	74	46.8	>100	>100	157	6.0	56.1	28.6	9.3
G1	87	43.7	>100	55	137	9.5	50.4	28.9	11.2
G2	88	43.7	>100	>100	139	9.5	49.0	30.4	11.1
G3	87	44.3	>100	>100	156	13.9	40.5	32.0	13.6

2.2. *Test and Analysis Methods*

2.2.1. Basic Performance Test

The penetration, softening point, ductility, dynamic viscosity and four fractions of asphalt binders were measured according to the Standard Test Methods of Bitumen and Bituminous Mixtures for Highway Engineering (JTG E20-2011) [36].

2.2.2. DMA Test

DMA Q800 (TA Instruments, New Castle, DE, USA) was used to investigate the phase transition characteristics of the asphalt binders. Testing was conducted using a dual-cantilever clamp with asphalt binder specimens prepared in thin sheets measuring 60 mm in length, 13 mm in width, and 3 mm in thickness (L60 mm × W13 mm × D3 mm). The dual-cantilever clamp and asphalt binder samples are illustrated in Figure 1. The temperature spectrum of the dynamic mechanical parameters of the asphalt binder was obtained using a temperature scanning mode with a fixed frequency, strain, and heating rate. The test conditions are listed in Table 2.

Table 2. Temperature scanning test setup.

Strain Amplitude	Frequency	Heating Rate	Temperature Range
25 με	10 Hz	2 °C/min	−35 °C to 35 °C

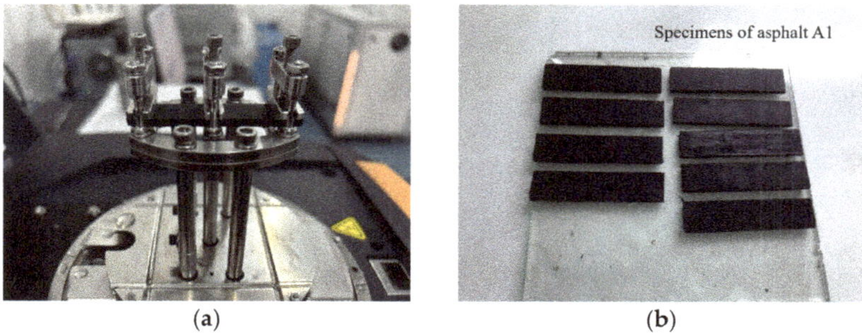

Figure 1. Double cantilever clamp (**a**) and asphalt binder specimens (**b**).

The complex modulus, storage modulus, loss modulus, and phase angle tangent value curves were derived using the dynamic mechanical test method. Figure 2 illustrates the modulus curve of the viscoelastic material over a wide temperature range. The curve representing the change in complex modulus with temperature typically exhibits an inverse S-shape. The Boltzmann function was applied to fit this curve using the function equation specified in Equation (1) [18].

$$y = \frac{A_1 - A_2}{1 + e^{(x-x_0)/dx}} + A_2 \tag{1}$$

where A_1 and A_2 are the maximum and minimum moduli, respectively, MPa; x_0 and dx are the shape parameters of the curve. x_0 is the parameter that represents the inflection point temperature of the sigmoidal curve, °C; dx is the parameter that controls the steepness (slope) of the viscoelastic states, °C.

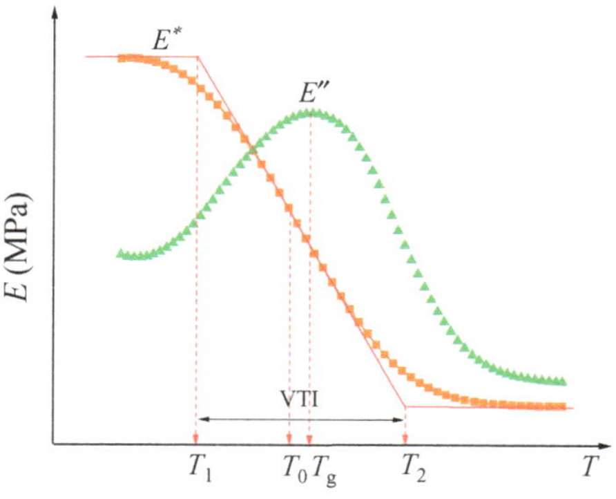

Figure 2. Temperature spectrum of the dynamic mechanical property parameters.

Utilizing the characteristics of the complex modulus test curve and the fitting parameters derived from the Boltzmann function, three PTTs can be defined from the complex modulus curve to characterize the phase transition properties of asphalt binders. These three characteristic temperatures are as follows:

- The temperatures corresponding to the midpoint (x_0, y_0) of the complex modulus curve are defined as T_0 and $T_0 = x_0$;
- The temperature at which the midpoint tangent of the complex modulus curve intersects the low-temperature asymptote is designated T_1, $T_1 = x_0 - 2dx$;
- The temperature at which the midpoint tangent of the complex modulus curve intersects the high-temperature asymptote is denoted as T_2, $T_2 = x_0 + 2dx$.

- The temperature interval between T_1 and T_2 is the viscoelastic temperature interval (VTI) of the asphalt binder. During this phase, the asphalt binder exhibits certain deformation resistance and elastic recovery capabilities. VTI = $(T_2 - T_1)$.

The loss modulus curve displays a prominent peak, commonly identified as the glass transition temperature (T_g) in polymer material studies. The loss modulus curve exhibits a degree of asymmetry. In this study, a BiGaussian function was utilized to fit the loss modulus curve, yielding the PTT T_g, $T_g = x_c$.

The BiGaussian function equation is as follows:

$$\begin{aligned} y &= y_0 + He^{-0.5(\frac{x-x_c}{w_1})^2} (x < x_c) \\ y &= y_0 + He^{-0.5(\frac{x-x_c}{w_2})^2} (x \geq x_c) \end{aligned} \tag{2}$$

where y_0 represents the offset in the data, MPa; x_c is the peak center temperature, °C; H represents the height of the peak above the baseline y_0, MPa; w_1 controls the width of the left side of the peak, °C; w_2 controls the width of the right side of the peak, °C.

In summary, the characteristic temperatures of the phase transitions (T_1, T_0, T_2, and T_g) can be determined from the curves of the complex modulus and loss modulus of the material. The characteristic moduli E_1, E_0, E_2, and E_g, corresponding to the characteristic temperatures, are obtained by substituting the characteristic temperatures into the complex modulus fitting equation. In addition to the characteristic moduli, the modulus values at specific temperatures of −20 °C and 20 °C (E_{-20} and E_{20}) can also be derived from the complex modulus fitting equation.

3. Results and Discussion

The PTTs and characteristic modulus data for the 30 asphalt binders obtained through DMA testing are presented in Table 3. The subsequent analyses and discussions were based on the data in Tables 1 and 3.

Table 3. Phase transition temperature and modulus.

Asphalt Binder	Phase Transition Temperature/°C				VTI/°C	Characteristic Modulus/MPa					
	T_1	T_0	T_2	T_g		E_1	E_0	E_2	E_g	E_{-20}	E_{20}
A1	−13.5	2.8	19.0	6.1	32.5	1910	1073	236	854	2046	210
A2	−16.4	−2.1	12.1	−0.3	28.6	1838	1043	249	911	1928	90
A3	−18.6	−4.8	9.1	−3.2	27.7	1864	1060	255	942	1905	60
B1	−14.9	0.5	15.8	3.0	30.7	1806	1023	240	856	1917	145
B2	−16.3	−1.5	13.4	0.8	29.7	1897	1077	258	914	1989	114
B3	−17.2	−2.5	12.2	−0.6	29.4	1928	1095	262	955	2004	99
B4	−17.7	−4.4	8.9	−3.7	26.6	1777	1012	248	959	1840	59
B5	−19.1	−6.0	7.1	−6.4	26.1	1641	936	232	960	1666	45
C1	−16.2	−0.2	15.9	2.8	32.0	2018	1140	263	930	2114	160
C2	−16.4	−1.2	14.1	1.6	30.5	1951	1105	259	904	2043	124
C3	−17.6	−2.6	12.3	−0.7	29.9	1993	1131	269	989	2062	104
C4	−14.2	−2.3	9.5	−1.6	23.7	1698	970	243	918	1834	58
C5	−18.5	−5.7	7.2	−5.3	25.7	1792	1019	246	992	1810	41
D1	−16.2	3.2	22.5	2.4	38.7	1887	1073	260	1117	1963	324
D2	−18.6	0.7	19.9	1.6	38.5	2069	1175	281	1118	2102	279
D3	−18.7	−0.4	17.9	0.3	36.6	1965	1116	268	1075	1997	219
D4	−19.5	−3.6	12.4	−3.4	31.8	2020	1147	275	1134	2035	115
D5	−20.2	−4.3	11.5	−3.5	31.7	1850	1051	252	999	1846	95
D6	−20.3	−4.9	10.4	−4.1	30.7	1950	1109	267	1046	1941	86
E1	−21.4	−5.3	10.9	−4.7	32.3	1844	1047	251	1013	1802	90

Table 3. *Cont.*

Asphalt Binder	Phase Transition Temperature/°C				VTI/°C	Characteristic Modulus/MPa					
	T_1	T_0	T_2	T_g		E_1	E_0	E_2	E_g	E_{-20}	E_{20}
E2	−20.1	−4.8	10.6	−3.4	30.7	1880	1067	253	969	1876	80
E3	−19.8	−4.3	11.2	−4.4	31.0	1692	963	233	971	1698	85
E4	−20.2	−4.6	10.9	−4.4	31.1	1733	986	238	971	1728	83
E5	−19.5	−3.4	12.7	−1.7	32.1	1923	1090	258	972	1937	110
F1	−18.3	−4.2	9.9	−2.2	28.1	2049	1165	281	1000	2104	77
F2	−15.4	−0.9	13.6	2.5	29.0	1916	1087	257	838	2029	113
F3	−15.6	−1.2	13.3	2.3	28.9	1899	1078	257	822	2007	109
G1	−18.0	−4.5	9.0	−4.1	27.0	1662	944	226	920	1715	51
G2	−17.9	−4.2	9.6	−3.4	27.4	1650	935	220	882	1704	50
G3	−18.3	−5.3	7.6	−5.4	25.9	1699	970	240	978	1748	50

3.1. Evaluation of Asphalt Binder Phase Transition Characteristics

3.1.1. PTTs of Asphalt Binder

(1) Evaluation of the low-temperature performance

The low-temperature performance of asphalt binders is a crucial factor affecting pavement-cracking in cold environments. As a viscoelastic material, asphalt binders transition into a rigid "glassy state" with frozen molecular segments when the temperature drops below their glass transition temperature, leading to increased modulus and brittleness. Glass transition temperature is a key parameter for material selection in practical engineering applications. In DMA, various definitions of the glass transition temperature exist. For instance, the inflection point on the complex modulus curve is identified as the glass transition temperature, aligned with T_1 and T_0. Alternatively, the peak point temperature of the loss modulus curve is considered to be the glass transition temperature corresponding to the PTT T_g. Thus, in this study, T_1, T_0, and T_g were utilized as indicators of the glass transition temperature of asphalt binder to evaluate low-temperature performance.

The significance of differences and correlations between the PTTs were assessed using Pearson correlation analysis, and the results are presented in Figure 3. The Pearson correlation analysis demonstrated the strength of the linear relationships and their statistical significance among the variables. With the exception of a smaller correlation coefficient of 0.402 between T_1 and T_2, the correlation coefficients (r) between all other variables exceeded 0.7. All were statistically significant at the 1% level ($p < 0.001$), indicating strong positive correlations. Therefore, this study evaluates the low-temperature performance of asphalt binder using T_g as the indicator; a smaller T_g value indicates better low-temperature performance.

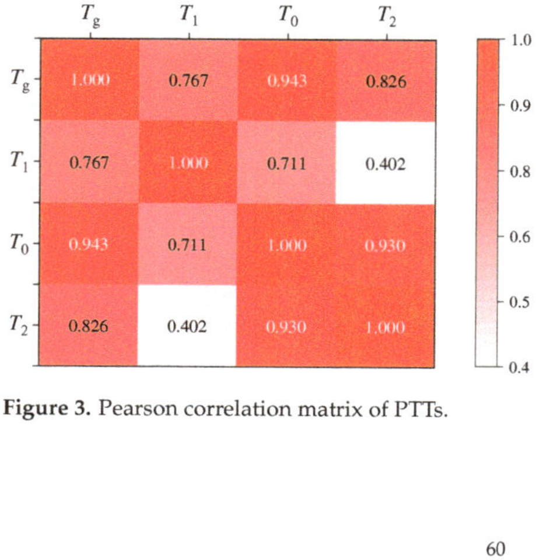

Figure 3. Pearson correlation matrix of PTTs.

The ranges of the T_g indicators for the 30 asphalt binder types studied, classified by penetration grade, are illustrated in Figure 4. The T_g spans for 20/40, 40/60, 60/80, and 80/100 penetration-grade asphalt binders are −4.7 °C to 6.1 °C, −4.4 °C to −0.3 °C, −3.7 °C to 2.5 °C, and −6.4 °C to −3.4 °C, respectively. The results indicate a significant variation in T_g values among asphalt binders of the same penetration grade, and the expected trend of lower penetration grades correlating with higher T_g values was not observed. Furthermore, the conventional penetration-grade evaluation system primarily relies on ductility values to assess the low-temperature performance. However, ductility, as an empirical indicator, fails to differentiate low-temperature performance when ductility exceeds 100 cm or when brittle fracture occurs. This highlights the limitations of the current grading framework for evaluating asphalt binder behavior under low-temperature conditions.

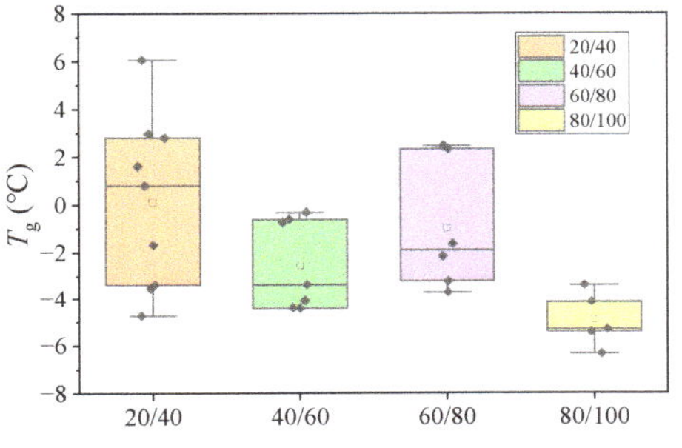

Figure 4. The range of T_g values.

The 20/40-grade asphalt binder specimens exhibited brittle fractures during ductility testing at 10 °C, resulting in zero ductility (0 cm). Furthermore, the ductility values at 15 °C showed no clear correlation with the PTT (T_g), as illustrated in Figure 5. For example:

- Asphalt binder A1 displayed a ductility of 0 cm at 15 °C, indicating the poorest low-temperature performance, consistent with its PTT evaluation.
- Asphalt binder E1, conversely, showed a ductility of 5 cm at 15 °C—lower than other 20/40-grade asphalt binders—but exhibited the lowest T_g, which contradicts the PTT assessment.
- Asphalt binders B1, C1, and D5 all recorded identical ductility values (8 cm), yet their PTTs differed by 6.5 °C.

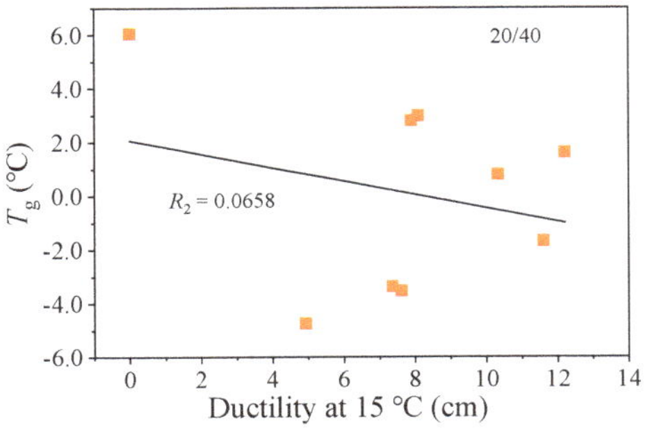

Figure 5. Relationship between T_g and ductility (15 °C) of 20/40 asphalt binders.

These results reveal a significant discrepancy between the T_g-based evaluation and ductility-based assessment for 20/40-grade asphalt binders, highlighting limitations in relying solely on ductility for characterizing low performance.

In the 40/60-grade asphalt binders, the minimum ductility at 15 °C for D6 was 9 cm, yet their T_g values were relatively low at −4.1 °C. In contrast, asphalt binder A2 demonstrates superior ductility performance, with a 15 °C ductility exceeding 100 cm, although its T_g value was the highest among the 40/60-grade asphalt binders. For the 60/80-grade asphalt binders, all 15 °C ductility values exceeded 100 cm, with three types maintaining ductility above 100 cm at 10 °C, complicating accurate assessments of low-temperature performance. However, T_g effectively distinguishes variations in the low-temperature performance of asphalt binders. Among the five 80/100-grade asphalt binders studied, the glass transition temperatures differed by approximately 3 °C, indicating relatively minor differences in low-temperature performance. A comprehensive analysis revealed that the glass transition temperature serves as an effective and scientifically valid parameter for evaluating the low-temperature properties of asphalt binders, possessing clear physical significance. This metric offers greater scientific rigor and effectiveness for assessing the low-temperature performance characteristics of asphalt materials.

(2) Evaluation of high-temperature performance

PTT (T_2), also referred to as the flow transition temperature, signifies the temperature at which an asphalt binder transitions from a viscoelastic state to a viscous flow state. An elevated T_2 indicates superior high-temperature performance of the asphalt binder. A comparison between the T_2 metric and the conventional high-temperature indicator viscosity (η) is shown in Figure 6. The analysis reveals overlapping numerical ranges of T_2 and η across asphalt binders of varying penetration grades, diverging from the expected trend of decreasing values with increased penetration. Moreover, higher viscosity does not necessarily correlate with elevated T_2 values. For instance, 20/30-grade asphalt binder E1, which demonstrated the highest viscosity in this study, possesses the lowest T_2 value within its grade, which was even lower than that of certain 40/60-grade asphalt binders. Similarly, among the 60/80-grade asphalt binders, F2 and F3 exhibit viscosity values that are lower than those of the 90-grade asphalt binders, yet their T_2 values exceed those of the 40/60-grade asphalt binders. Viscosity quantifies the resistance of asphalt binder to flow, reflecting its flow behavior at specific temperatures, whereas the flow transition temperature T_2 denotes the critical temperature at which viscosity markedly decreases. The proper interpretation and application of these two metrics offers vital guidance for material selection and engineering practices.

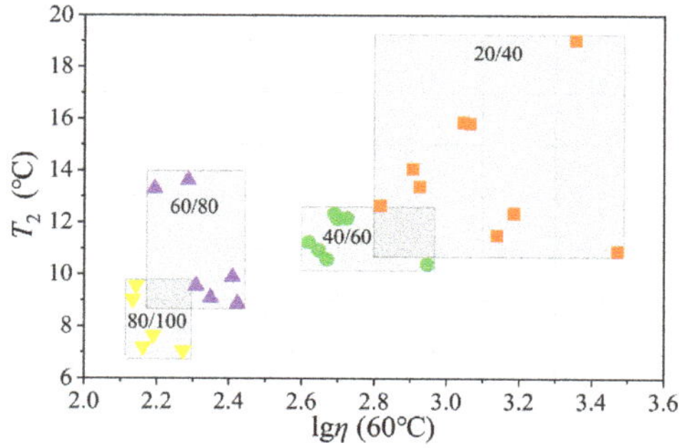

Figure 6. Schematic comparison of T_2 and viscosity.

(3) Evaluation of viscoelastic temperature interval

The viscoelastic temperature interval (VTI) between the PTTs (T_1 and T_2) is characterized by asphalt binders existing in a viscoelastic state. During this phase, the asphalt binders display a certain degree of deformation resistance and elastic recovery capability. This viscoelastic temperature range can be considered the optimal service temperature range for asphalt binders; a broader temperature range signifies enhanced performance, with improved low-temperature crack resistance and high-temperature rutting resistance. As illustrated in Figure 7, the viscoelastic state temperature interval for base asphalt binders ranges from 23 °C to 33 °C. Within the same penetration grade, variations in the VTI of asphalt binders are observed, with the 60/80-grade asphalt binders showing particularly significant dispersion. Nevertheless, the overall trend indicates that the VTI of the asphalt binders decreases as the penetration grade increases. When selecting an asphalt binder grade, it is crucial to assess its viscoelastic temperature range to ensure a wider operational temperature service range is obtained, allowing for adaptability to fluctuations in pavement temperature environments.

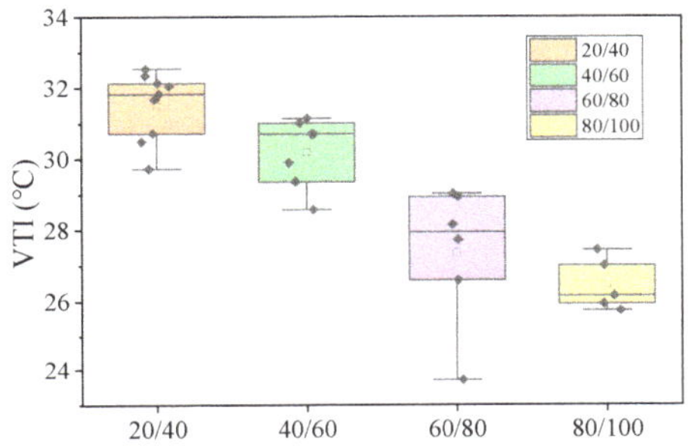

Figure 7. The range of VTI values.

3.1.2. Characteristic Modulus of Asphalt Binders

The modulus serves as a parameter reflecting the mechanical response of materials under specific conditions, wherein characteristic modulus values across varying temperature ranges reveal the macroscopic properties of the materials. The study commenced with a correlation analysis of the characteristic moduli, and the results are detailed in Figure 8. Parameters E_1, E_0, and E_2 correspond to characteristic moduli linked to PTTs derived from complex modulus curves, exhibiting correlation coefficients exceeding 0.92, indicating significant linear relationships among them. Parameters T_1, T_0, and T_g are utilized to evaluate the low-temperature performance of asphalt binders; however, the corresponding moduli E_1, E_0, and E_g demonstrate correlation coefficients below 0.37, suggesting a lack of correlation. Notably, the low-temperature modulus E_{-20} at −20 °C reveals strong correlations with E_1 and E_0 ($r > 0.93$) but shows no linear relationship with E_g ($r = 0.105$). Similarly, the high-temperature modulus, E_{20} at 20 °C, displays minimal correlation with E_2 ($r < 0.43$). This is primarily due to the fact that the characteristic moduli correspond to different temperature benchmarks, complicating the comparison of modulus variations across various asphalt materials.

Therefore, this section predominantly employed the low-temperature modulus E_{-20} (−20 °C) and the high-temperature modulus E_{20} (20 °C) as indicators to assess asphalt binders' performance under low and high temperatures. At low temperatures, the asphalt binders primarily exhibited elastic characteristics; higher modulus values signify a greater internal stress concentration and susceptibility to failure, reflecting the poor low-temperature performance. Conversely, at high temperatures, materials require adequate strength, and higher modulus values correspond to reduced deformations under load, indicating a superior high-temperature performance.

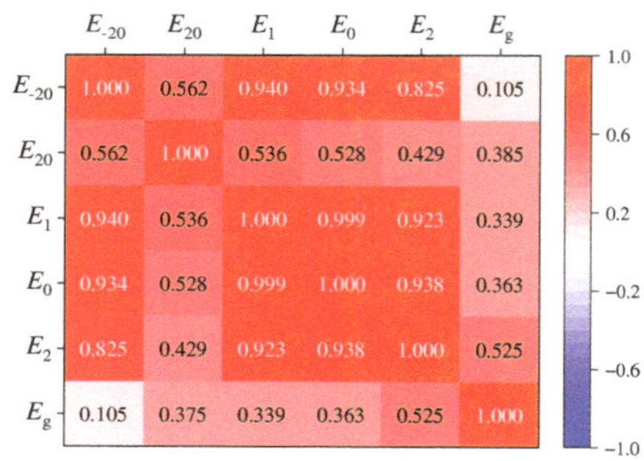

Figure 8. Pearson correlation matrix of characteristic modulus.

(1) Evaluation of the low-temperature performance

Figure 9 compares the low-temperature moduli range of the neat asphalt binders analyzed in this study. The results reveal significant variability in the low-temperature characteristic moduli within the same penetration grade, whereas the modulus values for the 20/40-, 40/60-, and 60/80-grade asphalt binders reside within similar numerical ranges. This finding indicates that there is no direct correlation between the low-temperature modulus and the penetration grade of asphalt binders (e.g., 20/40), with low-penetration-grade binders not necessarily exhibiting higher E_{-20} values than higher-grade asphalt binders (e.g., 60/80). From a modulus perspective, lower-grade asphalt binders may not have an inferior low-temperature performance. Figure 10 illustrates the correlation between the T_g of an asphalt binder and E_{-20}. While PTT and characteristic modulus show some association with low-temperature performance (R^2 = 0.5355), their correlation remains relatively weak. Practical engineering applications should therefore implement quality control by considering both PTT and modulus parameters to optimize the low-temperature performance of asphalt binders.

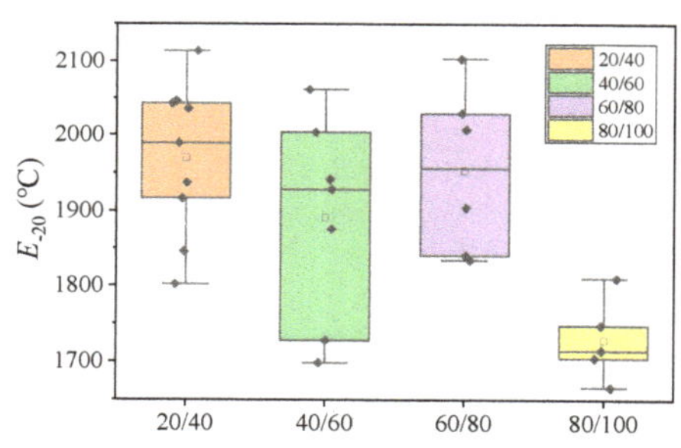

Figure 9. The range of E_{-20} values.

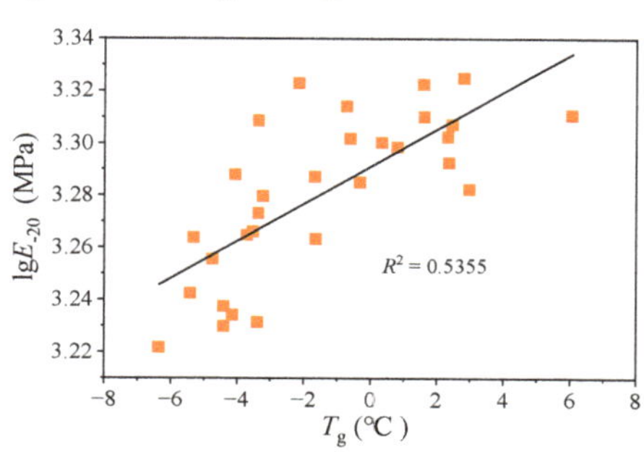

Figure 10. Relationship between T_g and E_{-20}.

(2) Evaluation of high-temperature performance

A comparison between the E_2 indicator and conventional high-temperature parameter η is shown in Figure 11. The results indicate a general correlation between the modulus values and asphalt binder penetration grades, with higher penetration grades typically exhibiting lower modulus values. However, the overlapping ranges between E_{20} and η suggest that high-viscosity asphalt binders do not necessarily correspond to elevated E_{20} values. The E_{20} indicator is closely aligned with the T_2 evaluation results presented in Figure 6. Figure 12 shows the relationship between asphalt binder's T_2 and log E_{20}, revealing a strong linear correlation (R^2 = 0.96). This confirms a solid association between PTT and the characteristic modulus in assessing the high-temperature performance of asphalt binders, with consistent evaluation results. For practical engineering applications, either PTT or modulus parameters can be selected for the quality control of asphalt binders' high-temperature performance.

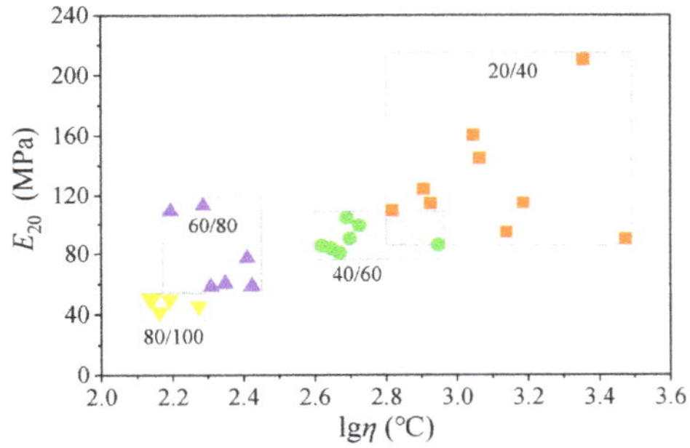

Figure 11. Schematic comparison of E_{20} and lgη.

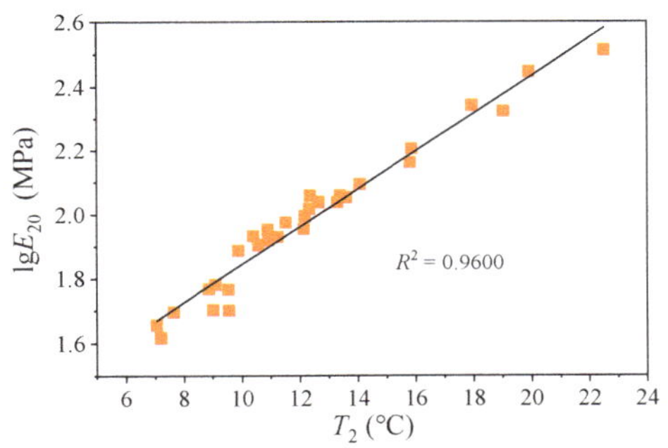

Figure 12. Relationship between T_2 and E_{20}.

3.2. The Relationship Between Phase Transition Characteristic Index and Conventional Performance Index

Under the penetration grading system, asphalt binder performance is primarily assessed through penetration, softening point, ductility, and viscosity. Typically, asphalt binders with lower penetration, a higher softening point, and greater viscosity demonstrate superior high-temperature performance. Within the same penetration grade, higher ductility correlates with improved low-temperature performance. However, evaluations based on the phase transition characteristic temperatures and modulus values reveal significant variations in the high- and low-temperature performance among asphalt binders with identical penetration grades. The penetration grading system does not limit phase transition-based evaluations; lower-grade asphalt binders may not necessarily exhibit better high-temperature performance or poorer low-temperature performance than their higher-grade counterparts. Correlation analyses were conducted for the representative low-temperature indicators (T_g and E_{-20}) and high-temperature indicators (T_2 and E_{20}), and the results are summarized in Figure 13.

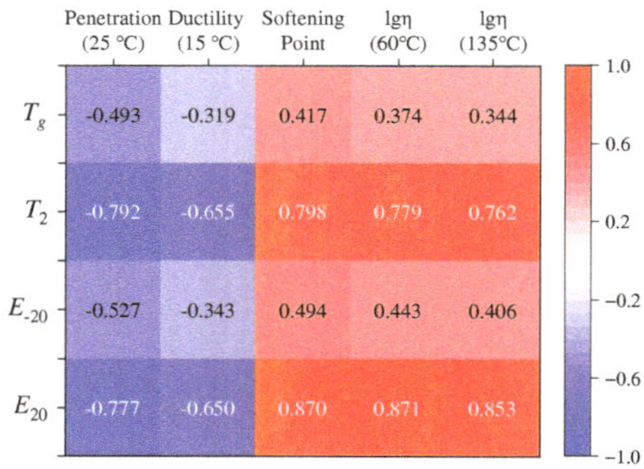

Figure 13. The Pearson correlation between phase transition characteristics and conventional indicators.

The absolute correlation coefficients between T_g/E_{-20} and conventional performance indicators are all below 0.53, indicating negligible linear correlations. This suggests inconsistencies between the phase transition characteristics derived from DMA and empirical asphalt binder performance metrics. Although ductility is a conventional low-temperature indicator, technical specifications define the ductility ranges for different penetration grades. However, ductility testing is challenging for distinguishing low-temperature performance when high-grade asphalt binders exceed the upper limit of 100 cm or when low-grade asphalt binders exhibit brittle fracture. In contrast, PTTs and characteristic moduli, grounded in clear physical significance, offer broad applicability and effectiveness for low-temperature performance assessments.

For the high-temperature performance indicators T_2 and E_{20}, significant correlations with conventional metrics (excluding ductility) are observed at the 1% significance level, with correlation coefficients of approximately 0.8, indicating strong alignment. This consistency confirms that the phase transition characteristics and empirical metrics yield coherent evaluations of high-temperature performance. Therefore, this study proposes an evaluation index of the phase transition characteristics based on DMA, which provides valuable guidance for the selection and application of asphalt materials in actual pavement engineering.

3.3. Relationship Analysis Between Phase Transition Characteristics and Asphalt Binder Components

The phase transition characteristics represent macroscopic manifestations of the microscopic molecular motion of materials. Currently, asphalt binders are primarily SARA to compare differences in chemical composition. This section focuses on analyzing the relationships between the asphalt binders' phase transition characteristic temperatures (T_g, T_2), characteristic moduli (E_{-20}, E_{20}), and their four fractions, with the aim of investigating the effects of an asphalt binder's microscopic molecular structure on dynamic mechanical response parameters.

3.3.1. Single Factor Correlation Analysis

Figure 14 presents the results of the correlation analysis of the dynamic mechanical indicators of an asphalt binder and its four-component composition. T_g shows a statistically significant negative correlation with saturates at the 0.01 significance level ($r = -0.566$). However, T_g exhibits no meaningful correlations with aromatics, resins, or asphaltenes, as indicated by p-values exceeding 0.2 and absolute correlation coefficients below 0.3. The characteristic temperature T_2 demonstrates a weak positive correlation with asphaltenes, at the 0.01 significance level ($r = 0.688$), and a very weak negative correlation with aromatics, at the 0.05 level ($r < -0.4$). No significant correlations are observed between T_2 and saturates or resins, with p-values greater than 0.1 and absolute correlation coefficients below 0.3.

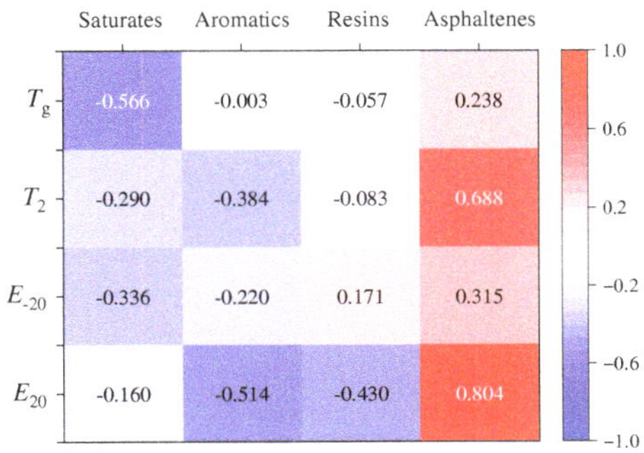

Figure 14. The Pearson correlation between the phase transition characteristics and the four fractions.

For the low-temperature characteristic modulus E_{-20}, no statistically significant correlations with any of the four components are identified at the 0.05 significance level ($|r| < 0.34$). In contrast, the high-temperature characteristic modulus E_{20} shows significant correlation with aromatics and asphaltenes at the 0.01 level. Specifically, E_{20} exhibits a weak negative correlation with aromatics ($r = -0.514$) and a strong positive linear correlation with asphaltenes ($r = 0.804$). No significant relationships are observed between E_{20} and saturates or resins.

Saturates, which are relatively light components in asphalt binders, with low molecular weight, function as softening agents. Their short molecular chains and low activation energy for segmental motion lead to a reduction in the T_g of asphalt binders with increasing saturate content, which aligns with the observed negative correlation between the saturates and T_g. Conversely, asphaltenes, which are characterized by a high molecular weight and elevated activation energy for chain movement, enhance the viscosity and high-temperature performance of asphalt binders. This is reflected in dynamic mechanical parameters, such as elevated T_2 and increased E_{20}, consistent with the results of the correlation analysis. While a higher saturate content benefits low-temperature performance and an increase in asphaltenes improves high-temperature performance, the correlation analysis indicates relatively weak associations between the phase transition characteristics and individual asphalt binder fractions.

3.3.2. Multiple Linear Regression Analysis

The macroscopic mechanical response of asphalt materials arises from synergistic interactions between their microscopic molecular components. This study employs multiple regression analysis to investigate the comprehensive relationships between the phase-transition characteristic indicators of asphalt binders (T_g, T_2, E_{-20}, E_{20}) and the content of their four fractions. The dependent variables are defined as the dynamic mechanical parameters T_g (Y1), T_2 (Y2), E_{-20} (Y3), and E_{20} (Y4), while the independent variables comprise the four fractions of saturates (X1), aromatics (X2), resins (X3), and asphaltenes (X4). Multiple regression analysis (at the 95% confidence level) was performed separately for each dependent variable to establish statistical dependencies on the compositional variables. The Data Analysis Tool in Microsoft Excel 2016 was used to analyze the sample data presented in Tables 1 and 3.

The multiple regression analysis of the dynamic mechanical indicators of asphalt binders (T_g, T_2, E_{-20}, E_{20}) and the four fractions generated the following models:

$$T_g = 214.5 - 3.52X_1 - 2.13X_2 - 1.93X_3 - 1.80X_4 \tag{3}$$

$$T_2 = 187.7 - 2.99X_1 - 1.75X_2 - 1.73X_3 - 0.98X_4 \tag{4}$$

$$E_{-20} = 14399.9 - 172.86X_1 - 128.58X_2 - 109.43X_3 - 115.03X_4 \tag{5}$$

$$E_{20} = 1592.6 - 33.48X_1 - 15.24X_2 - 15.33X_3 + 0.15X_4 \tag{6}$$

The regression equations reveal that the four fractions exhibit negative effects on the PTTs (T_g, T_2) and low-temperature modulus (E_{-20}), whereas only asphaltenes have a positive influence on the high-temperature modulus (E_{20}). These results partially contradict the findings of single-factor correlation analyses. Specifically, the sign of the coefficient for saturates in the multiple regression model for T_g is positive, whereas the Pearson correlation indicated a negative linear relationship. This apparent discrepancy is a known statistical phenomenon often arising from multicollinearity among the independent variables. The SARA fractions are compositional variables that sum to a constant, meaning they are inherently inter-correlated. The multiple regression model estimates the *unique* contribution of each fraction while holding the others constant, a condition that is difficult to achieve physically. Therefore, the multiple regression coefficients should be interpreted as indicators of synergistic and competing interactions within the asphalt's chemical system rather than as direct, isolated effects. The single-factor correlation, on the other hand, reflects the stronger, net-observable trend where an increase in saturates typically coincides with a decrease in other fractions, leading to an overall reduction in T_g. The next step was to conduct a significance test for this model, including the goodness of fit test and analysis of variance.

(1) Goodness-of-fit test

The goodness-of-fit test evaluates the extent to which a model aligns with a given set of observations. Measures of goodness-of-fit typically summarize the discrepancy between the observed values and those predicted by the model. Table 4 presents the regression statistics. It is evident that the Multiple R values for all the fitted models exceed 0.70. Multiple R serves as an indicator of how effectively a specific variable can be predicted using a linear function derived from a set of other variables, with a high Multiple R indicating a strong correlation. A Multiple R greater than 0.70 suggests a robust linear relationship between the four fractions and the dynamic mechanical index. R^2 quantifies the extent to which the observed results are replicated by the model, representing the proportion of total variation in the outcomes explained by the model. Specifically, 57.63%, 77.89%, 49.60%, and 83.70% of the T_g, T_2, E_{-20}, and E_{20} values, respectively, can be accounted for by the fitted model. The goodness-of-fit analysis reveals that the high-temperature indices T_2 and E_{20} exhibit strong alignment with the four-component regression model, whereas the low-temperature indices T_g and E_{-20} demonstrate relatively weaker alignment with the same model.

Table 4. Goodness of fit results.

	Goodness of Fit Results			
	T_g	T_2	E_{-20}	E_{20}
Multiple R	0.7591	0.8830	0.7043	0.9149
R^2	0.5763	0.7798	0.4960	0.8370
Adjusted R^2	0.5085	0.7445	0.4154	0.8109
Standard error	2.2136	1.9071	102.3	29.4
Observations	30	30	30	30

(2) ANOVA

Analysis of variance (ANOVA) was used to assess the significance of linear regression. This technique utilizes the variance of the observed data and the p value to determine the applicability of a linear regression model to the observed data.

The results of the analysis are presented in Table 5. When the significance level α is 0.05, $F_{0.05(4,25)} = 2.76$. The F values of T_g, T_2, E_{-20} and E_{20} were 8.50, 22.13, 6.15, and 32.10, respectively. All are greater than $F_{0.05(4,25)}$. The p value is the critical value of F_α at the corresponding significance level, which represents the probability of rejecting the truth; therefore, the smaller the p value, the better. In this analysis, the p value is less than 0.001, and the confidence is 99.9%. This provides strong evidence that a linear relationship exists. From the test point of F-value and p-value, T_g, T_2, E_{-20}, and E_{20} show significant multivariate fitting potential with the four components.

Table 5. ANOVA results.

		DS	SS	MS	F	p
T_g	Regression	4	166.6	41.7	8.50	1.8×10^{-4}
	Residual	25	122.5	4.9		
	Total	29	289.1			
T_2	Regression	4	321.9	80.5	22.13	6.6×10^{-8}
	Residual	25	90.9	3.6		
	Total	29	412.9			
E_{-20}	Regression	4	257,519.0	64,379.7	6.15	1.4×10^{-3}
	Residual	25	261,632.4	10,465.3		
	Total	29	519,151.4			
E_{20}	Regression	4	111,246.1	27,811.5	32.10	1.6×10^{-9}
	Residual	25	21,662.3	866.5		
	Total	29	132,908.4			

4. Conclusions

This research highlights the limitations of traditional penetration grading systems in capturing the temperature-dependent performance of asphalt binders, particularly under low-temperature conditions, where ductility fails to differentiate materials. The T_g and E_{20} values derived from DMA offer robust physics-based alternatives, demonstrating superior sensitivity to molecular dynamics and phase transitions. Key findings include the following:

- Low-temperature performance: T_g effectively distinguishes asphalt binder behavior even when ductility values exceed 100 cm or brittle fracture is observed, addressing empirical limitations.

- High-temperature performance: Strong correlations between T_2, E_{20}, and viscosity validate their utility in predicting rutting resistance; however, overlapping ranges with penetration grades suggest complementary use.
- Chemical composition: Saturates reduce T_g, enhancing low-temperature flexibility, while asphaltenes elevate E_{20}, improving high-temperature stiffness. Nonetheless, the weak correlations between individual SARA fractions and low-temperature indices highlight the necessity for a holistic compositional analysis.
- Integration into existing grading systems: The proposed DMA-based phase-transition indicators (T_g, T_2, E_{-20}, E_{20}) can be incorporated into current binder specifications as supplementary performance criteria. For instance, T_g could serve as a low-temperature performance grade supplement, especially for regions experiencing extreme cold, while T_2 and E_{20} could enhance high-temperature performance grade classification. This dual approach—combining empirical metrics with mechanistic indicators—would provide a more comprehensive evaluation framework, enabling better material selection tailored to specific climatic and loading conditions.
- Practical implications: DMA-based phase-transition indicators bridge the gap between empirical specifications and mechanistic understanding, facilitating precise material selection for diverse climatic conditions. It should be noted that the findings and models presented in this study are based on laboratory testing of neat asphalt binders under controlled conditions. The applicability to modified asphalt binders or real-world pavement environments may require further validation. Future work should explore modified asphalt binders and conduct a field validation to refine these models for broader engineering applications.

Author Contributions: Conceptualization, Y.W., J.W. and X.X.; methodology, Y.W., X.X. and Q.X.; formal analysis, Y.W. and X.X.; investigation, Q.X. and Z.Z.; resources, Y.S. and N.L.; data curation, Y.S. and Z.Z.; writing—original draft, Y.W. and J.W.; writing—review and editing, Y.W., J.W. and Q.X. All authors have read and agreed to the published version of the manuscript.

Funding: This research was funded by Shandong Provincial Natural Science Foundation, grant number: ZR2024QE425.

Institutional Review Board Statement: Not applicable.

Informed Consent Statement: Not applicable.

Data Availability Statement: Data are contained within the article.

Conflicts of Interest: The authors declare no conflicts of interest.

Abbreviations

The following abbreviations are used in this manuscript:

DMA	Dynamic Mechanical Analysis
SARA	Saturate, Aromatic, Resin, and Asphaltene
PTT	Phase Transition Temperature
TTSP	Time–Temperature Superposition Principle
DSC	Differential Scanning Calorimetry
VTI	Viscoelastic Temperature Interval

References

1. Liu, J.; Sun, Y.; Wang, W.; Chen, J. Using the viscoelastic parameters to estimate the glass transition temperature of asphalt binders. *Constr. Build. Mater.* **2017**, *153*, 908–917. [CrossRef]
2. Tan, Y.; Guo, M. Study on the phase behavior of asphalt mastic. *Constr. Build. Mater.* **2013**, *47*, 311–317. [CrossRef]
3. Guo, M. *Dynamic Mechanical Thermal Analysis of Polymers and Composites*; Chemical Industry Press: Beijing, China, 2002.

4. Elseifi, M.A.; Mohammad, L.N.; Glover, I.; Negulescu, I.; Daly, W.H.; Abadie, C. Relationship between Molecular Compositions and Rheological Properties of Neat Asphalt Binder at Low and Intermediate Temperatures. *J. Mater. Civ. Eng.* **2010**, *22*, 1288–1294. [CrossRef]
5. Kriz, P.; Stastna, J.; Zanzotto, L. Glass Transition and Phase Stability in Asphalt Binders. *Road Mater. Pavement Des.* **2008**, *9* (Suppl. 1), 37–65. [CrossRef]
6. Sun, Y.; Huang, B.; Chen, J.; Jia, X.; Ding, Y. Characterizing rheological behavior of asphalt binder over a complete range of pavement service loading frequency and temperature. *Constr. Build. Mater.* **2016**, *123*, 661–672. [CrossRef]
7. Wang, Y.; Wang, X.; Ma, Z.; Shan, L.; Zhang, C. Evaluation of the High- and Low-Temperature Performance of Asphalt Mortar Based on the DMA Method. *Materials* **2022**, *15*, 3341. [CrossRef]
8. Caro, S.; Sánchez, D.B.; Caicedo, B. Methodology to characterise non-standard asphalt materials using DMA testing: Application to natural asphalt mixtures. *Int. J. Pavement Eng.* **2014**, *16*, 1–10. [CrossRef]
9. Kang, Y.; Song, M.; Pu, L.; Liu, T. Rheological behaviors of epoxy asphalt binder in comparison of base asphalt binder and SBS modified asphalt binder. *Constr. Build. Mater.* **2015**, *76*, 343–350. [CrossRef]
10. Wang, D.; Falchetto, A.C.; Riccardi, C.; Wistuba, M.P. Investigation on the low temperature properties of asphalt binder: Glass transition temperature and modulus shift factor. *Constr. Build. Mater.* **2020**, *245*, 118351. [CrossRef]
11. Xu, J.; Fan, Z.; Lin, J.; Yang, X.; Wang, D.; Oeser, M. Predicting the low-temperature performance of asphalt binder based on rheological model. *Constr. Build. Mater.* **2021**, *302*, 124401. [CrossRef]
12. Gunay, T.; Tomkovic, T.; Hatzikiriakos, S.G. Thermorheological properties of asphalt binders. *Can. J. Chem. Eng.* **2020**, *98*, 1803–1814. [CrossRef]
13. Ayanarayanan, K.; Rasana, N.; Mishra, R.K. Chapter 6-Dynamic Mechanical Thermal Analysis of Polymer Nanocomposites. In *Thermal and Rheological Measurement Techniques for Nanomaterials Characterization*; Elsevier: Amsterdam, The Netherlands, 2017; pp. 123–157. [CrossRef]
14. Yuan, Q.; Liu, W.; Pan, Y.; Deng, D.; Liu, Z. Characterization of Cement Asphalt Mortar for Slab Track by Dynamic Mechanical Thermoanalysis. *J. Mater. Civ. Eng.* **2016**, *28*, 04015154. [CrossRef]
15. Zhang, Z.; Wang, P.; Wu, J. Dynamic Mechanical Properties of EVA Polymer-Modified Cement Paste at Early Age. *Phys. Procedia* **2012**, *25*, 305–310. [CrossRef]
16. Fu, Q.; Xie, Y.; Long, G.; Niu, D.; Song, H. Dynamic mechanical thermo-analysis of cement and asphalt mortar. *Powder Technol.* **2017**, *313*, 36–43. [CrossRef]
17. Morlat, R.; Godard, P.; Bomal, Y.; Orange, G. Dynamic mechanical thermoanalysis of latexes in cement paste. *Cem. Concr. Res.* **1999**, *29*, 847–853. [CrossRef]
18. Yang, G.; Wang, X.; Zhou, X.; Wang, Y. Experimental Study on the Phase Transition Characteristics of Asphalt Mixture for Stress Absorbing Membrane Interlayer. *Materials* **2020**, *13*, 474. [CrossRef]
19. You, L.; Spyriouni, T.; Dai, Q.; You, Z.; Khanal, A. Experimental and molecular dynamics simulation study on thermal, transport, and rheological properties of asphalt. *Constr. Build. Mater.* **2020**, *265*, 120358. [CrossRef]
20. Apostolidis, P.; Elwardany, M.; Porot, L.; Vansteenkiste, S.; Chailleux, E. Glass transitions in bituminous binders. *Mater. Struct.* **2021**, *54*, 132. [CrossRef]
21. Sultana, S.; Bhasin, A. Effect of chemical composition on rheology and mechanical properties of asphalt binder. *Constr. Build. Mater.* **2014**, *72*, 293–300. [CrossRef]
22. Oyekunle, L.O. Influence of Chemical Composition on the Physical Characteristics of Paving Asphalts. *Pet. Sci. Technol.* **2007**, *25*, 1401–1414. [CrossRef]
23. Zhang, M.; Hao, P.; Dong, S.; Li, Y.; Yuan, G. Asphalt binder micro-characterization and testing approaches: A review. *Meas. J. Int. Meas. Confed.* **2020**, *151*, 107255. [CrossRef]
24. Guo, M.; Liang, M.; Sreeram, A.; Bhasin, A.; Luo, D. Characterisation of rejuvenation of various modified asphalt binders based on simplified chromatographic techniques. *Int. J. Pavement Eng.* **2022**, *23*, 4333–4343. [CrossRef]
25. Handle, F.; Füssl, J.; Neudl, S.; Grossegger, D.; Eberhardsteiner, L.; Hofko, B.; Hospodka, M.; Blab, R.; Grothe, H. The bitumen microstructure: A fluorescent approach. *Mater. Struct.* **2016**, *49*, 167–180. [CrossRef]
26. Wei, J.; Dong, F.; Li, Y.; Zhang, Y. Relationship analysis between surface free energy and chemical composition of asphalt binder. *Constr. Build. Mater.* **2014**, *71*, 116–123. [CrossRef]
27. Wang, J.; Wang, T.; Hou, X.; Xiao, F. Modelling of rheological and chemical properties of asphalt binder considering SARA fraction. *Fuel* **2019**, *238*, 320–330. [CrossRef]
28. Weigel, S.; Stephan, D. Relationships between the chemistry and the physical properties of bitumen. *Road Mater. Pavement Des.* **2018**, *19*, 1636–1650. [CrossRef]
29. Wang, Y.; Wang, W.; Wang, L. Understanding the relationships between rheology and chemistry of asphalt binders: A review. *Constr. Build. Mater.* **2022**, *329*, 127161. [CrossRef]

30. Wang, C.; Wang, Y. Physico-chemo-rheological characterization of neat and polymer-modified asphalt binders. *Constr. Build. Mater.* **2019**, *199*, 471–482. [CrossRef]
31. Jiang, J.; Lu, F.; Guo, X.; Hao, P.; Wang, W.; Yu, J. Effect of SARA Fractions on Fatigue Properties of Hard Asphalt. *Materials* **2024**, *17*, 2713. [CrossRef] [PubMed]
32. Xu, Y.; Zhan, E.; Shan, L. Effect of SARA on Rheological Properties of Asphalt Binders. *J. Mater. Civ. Eng.* **2019**, *31*, 4019086. [CrossRef]
33. Zhang, H.; Li, Y.; Zhang, J.; Li, Z.; Chen, Z.; Pei, J.; Wang, M. Examining the rheological and adhesion performance of asphalt: Insights into the influence of SARA components. *Case Stud. Constr. Mater.* **2024**, *21*, e03427. [CrossRef]
34. Wang, T.; Wang, J.; Hou, X.; Xiao, F. Effects of SARA fractions on low temperature properties of asphalt binders. *Road Mater. Pavement Des.* **2021**, *22*, 539–556. [CrossRef]
35. Martin Wieser, A.; Schaur, S.H. Unterberger, Polymer-Bitumen Interaction: A Correlation Study with Six Different Bitumens to Investigate the Influence of SARA Fractions on the Phase Stability, Swelling, and Thermo-Rheological Properties of SBS-PmB. *Materials* **2021**, *14*, 1273. [CrossRef]
36. *JTG E20-2011;* Standard Test Methods of Bitumen and Bituminous Mixtures for Highway Engineering. China Communications Press: Beijing, China, 2011.

coatings

MDPI

Article

Analysis of the Suitability of 3D-Printed Road Surface Replicas for Laboratory Testing of Rolling Resistance

Wojciech Owczarzak *, Sławomir Sommer and Grzegorz Ronowski

Institute of Mechanics and Machine Design, Faculty of Mechanical Engineering and Ship Technology and EkoTech Center, Gdańsk University of Technology, ul. Narutowicza 11/12, 80-233 Gdańsk, Poland; slawomir.sommer@pg.edu.pl (S.S.); grzegorz.ronowski@pg.edu.pl (G.R.)
* Correspondence: wojowcza@pg.edu.pl (W.O.)

Abstract

This study investigates the influence of the method and materials used for creating road surface replicas on the evaluation of rolling resistance using the oscillatory method. While casting resin is commonly employed for this purpose, the research explores 3D printing as a viable alternative. To assess the effectiveness of the proposed approach, replicas of three road surfaces with differing rolling resistance characteristics were created using both techniques. The conventional resin-based replicas served as a reference. A range of tires—summer, winter, and all-season—were tested on the prepared samples. The results were compared to evaluate the consistency between the two replica fabrication methods and to determine the suitability of 3D-printed surfaces as substitutes for those made with casting resin.

Keywords: tire; road pavement; texture; rolling resistance; 3D printing; measurements

1. Introduction

Rolling resistance of car tires is one of the key factors influencing the interaction between the tire and the road surface. It directly affects a vehicle's energy consumption, the emission of harmful atmospheric compounds (such as NH_3, CO_2, and NO_2), and fundamental performance parameters such as top speed, acceleration, and driving range. High rolling resistance also contributes to increased tire temperatures, which accelerate aging and reduce mechanical durability [1–5].

Among the primary forces opposing vehicle motion—alongside aerodynamic drag, gradient resistance, and inertia—rolling resistance plays a particularly significant role at speeds up to approximately 80–90 km/h, when its contribution to overall energy consumption is greatest. A 10% reduction in rolling resistance can lead to a 2%–4% decrease in fuel consumption, with some studies indicating a 3%–4% reduction in CO_2 emissions under typical driving conditions [2,6].

Rolling resistance results from energy losses during the tire's cyclic deformation in the contact area with the road surface. Due to the viscoelastic hysteresis of rubber materials, more energy is required to deform the tire than is recovered when it returns to its original shape. Most of this energy loss is converted into heat, while the remainder dissipates through noise, permanent surface deformations, and tire wear. Numerous factors influence the magnitude of rolling resistance: tire construction (width, height, tread pattern, number of casing layers, belt material, and rubber compound composition), operating conditions (load, inflation pressure, temperature, and speed), and road surface characteristics (texture, stiffness, and condition) [2,7]. Rolling resistance is studied both to optimize tire design and to improve road surfaces.

Key parameters characterizing road surfaces in terms of their interaction with vehicle tires include texture, smoothness, stiffness, and the porosity of the surface layer (wearing course). Road surface texture refers to deviations in the pavement from an ideally flat plane, with characteristic wavelengths of less than 0.5 m. Depending on the wavelength, surface texture is typically classified as follows (Figure 1) [8–10]:

- Irregularities: Wavelengths greater than 500 mm.
- Megatexture: Wavelengths between 50 and 500 mm, with typical amplitudes ranging from 0.1 to 50 mm.
- Macrotexture: Wavelengths from 0.5 to 50 mm, with amplitudes between 0.1 and 20 mm.
- Microtexture: Wavelengths below 0.5 mm, with amplitudes from 0.001 to 0.5 mm.

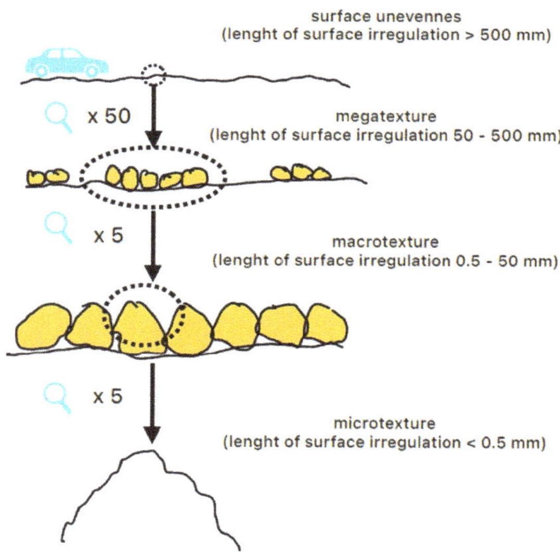

Figure 1. Individual road surface texture ranges.

The texture of the road surface has a significant impact on the local deformation of tire tread elements. Surfaces with pronounced, aggressive textures cause substantial increases in these deformations, leading to higher rolling resistance. However, when the texture is extremely pronounced, the flexibility of the tread elements (or "blocks") becomes insufficient to fully conform to the surface profile. Due to a phenomenon known as enveloping, contact between the tread and the road is limited to the highest protruding features—primarily the peaks of aggregate particles—while other areas remain out of contact.

According to the laws of physics, deformations resulting from tire–road interactions occur not only in the tire but also in the road surface. However, because the mechanical impedance of the road surface is typically much higher than that of the tire, surface deflection is minimal. For standard pavement materials, the dynamic Young's modulus ranges from 10 to 45 GPa, whereas for car tires, it is approximately 0.3 to 1 MPa [11,12]. Numerous studies have examined the complex nature of tire–road interactions [10,13].

Over the past two decades, tire manufacturers have made significant progress in reducing rolling resistance. Initially, these improvements were achieved primarily by introducing tires with large outer diameters and narrow widths (e.g., size 155/70 R19).

Today, however, low rolling resistance models are also available in more conventional sizes [6].

Reducing energy losses associated with rolling resistance has become increasingly important in light of the rapid growth of hybrid and electric vehicles. In such vehicles, energy losses during braking have been significantly reduced thanks to regenerative braking, and aerodynamic drag has been minimized through optimized vehicle design. However, their overall mass has increased—electric vehicles are typically heavier than their internal combustion counterparts due to the need for large battery packs. Since rolling resistance is proportional to the normal load on the wheels, its relative contribution to total energy losses has not decreased; in fact, it has increased.

Accurately measuring rolling resistance poses a considerable technical challenge. In modern tires, the rolling resistance force accounts for only about 1% of the wheel load. For example, in the case of a 5 kN wheel load, the rolling resistance force ranges from just 30 to 60 N. Achieving 1% measurement accuracy therefore requires detecting forces with a precision of approximately 0.3 to 0.6 N. While measurements conducted on real roads are the most representative, they are subject to numerous variables, including pavement and air temperature, speed fluctuations, surface irregularities, and surrounding traffic conditions [5,14].

These challenges are addressed through laboratory testing, where temperature, speed, load, and tire pressure can be precisely controlled. In most cases, tests are performed on steel drums using the torque method [5]. However, this approach introduces several limitations. The curvature of the drum causes tire deformations that differ from those on flat road surfaces, and the absence of realistic surface texture affects the interaction between the tread and the substrate (an example of a road–wheel facility is shown in Figure 2). As a result, the gain in measurement precision comes at the cost of real-world representativeness. Standard testing protocols are outlined in documents such as ISO 28580 (2018) and SAE J2452 (2017) [15–18]. The issue of unrealistic surface texture can be addressed by using cast replicas of actual pavements.

Figure 2. Roadwheel facility for rolling resistance measurements.

Gdańsk University of Technology operates the only laboratory in the world that uses replicas of real road surfaces made from reinforced laminates with epoxy resins, preserving the authentic texture of the original pavement (see Figure 3). Testing on these replicas yields more accurate and reliable results than tests conducted on steel surfaces, as they better simulate actual vehicle operating conditions. Measurements of rolling resistance and tire noise on these replicas correspond closely with data obtained from real roads under actual traffic conditions [19].

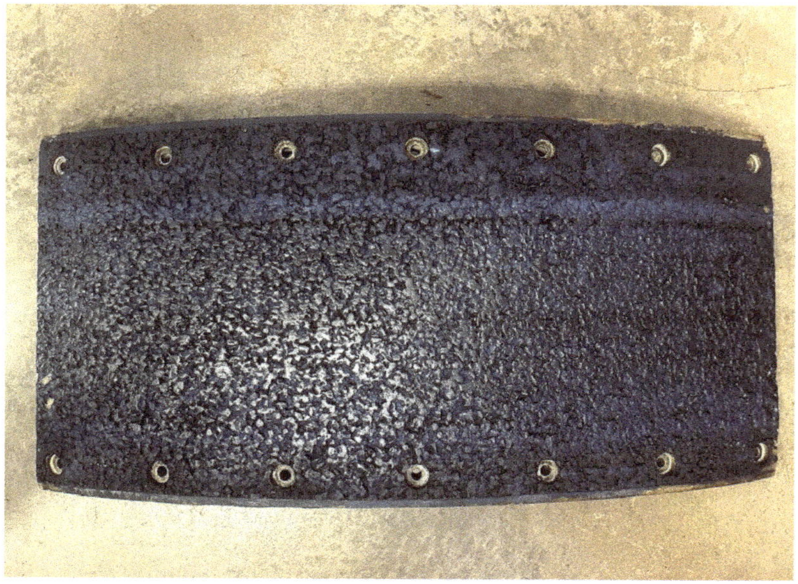

Figure 3. Replica of road surface made using the casting method.

An analysis of the available literature on simplified testing methods reveals a growing body of research focused on tire–road surface interactions [20]. One notable example is the use of high-precision pressure measurement systems, such as pressure sensitive films, to investigate the actual contact characteristics between tires and asphalt surfaces. Other studies utilize flexible force sensors embedded directly within the tire structure [21–23]. Another approach involves integrating pressure sensors into a matrix embedded in the road surface and surrounded by aggregate-simulating supports, enabling the measurement of triaxial stress distributions at the tire–pavement interface [24]. These interactions can also be studied using numerical methods [25,26].

Focusing specifically on the analysis of energy losses at the tire–road interface, researchers have explored innovative techniques such as 3D digital image correlation (3D-DIC)—a non-contact method used to estimate rolling resistance by calculating the relative velocity of primary elastic deformations. Alternatively, rolling resistance can be indirectly assessed through the pressure distribution between the tire and the road surface, obtained using specialized pressure-mapping films [26–28]. Numerical techniques also remain a valuable tool for estimating rolling resistance [29,30].

In light of these developments, alternative laboratory methods are receiving increasing attention. These methods aim to replicate realistic tire–road contact conditions while maintaining high measurement accuracy and offering simplicity and speed of testing. One such proposed approach involves the use of an oscillatory method combined with 3D printing to create test surfaces with predefined geometry and texture, thereby simulating road conditions in a controlled laboratory environment. In earlier implementations of this method, epoxy resin-based replicas were used—similar to those mentioned previously—

with the distinction that the replicas used in oscillatory testing were much smaller and flat (i.e., without curvature) [31].

However, producing such replicas using traditional casting materials presents certain challenges, primarily due to the need for an initial texture template to be placed into a mold—typically obtained through silicone rubber casting. This method inherently limits researchers to existing surface textures. In contrast, 3D printing offers the potential not only to modify existing textures but also to design entirely new surface patterns from scratch. The application of 3D printing will also contribute to a reduction in manufacturing costs, shorten production time, and minimize the generation of environmentally harmful chemical waste.

2. Materials and Methods

2.1. Oscillatory Method for Estimating Energy Losses During Tire–Road Surface Contact

To assess the impact of the manufacturing technique used for road surface replicas on measured rolling resistance values, the oscillatory method was applied. This method involves evaluating energy losses resulting from the repeated contact of a tire with a surface sample, initiated by a free-fall motion. The tire is mounted on a 1.5 m long arm, which remains level in its resting position. This motion induces oscillations within a system composed of mass, elasticity, and damping components. Replacing the wheel's rotational movement with a sliding or swinging motion simplifies both the apparatus and the measurement setup (Figure 4). This simplification enables the use of small surface samples, which are easy to prepare under laboratory conditions.

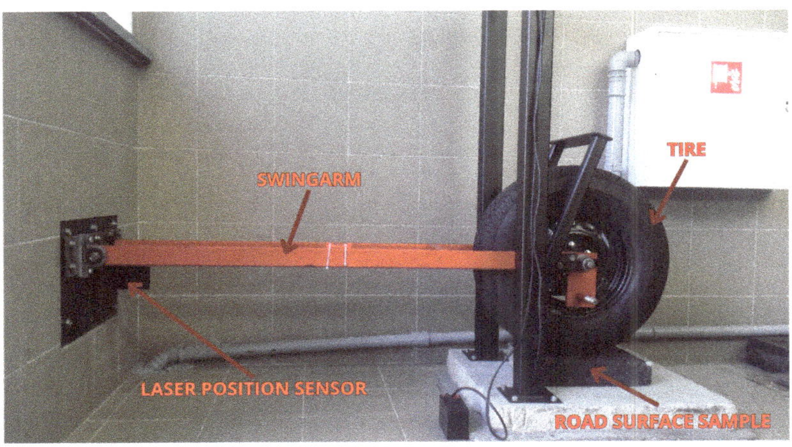

Figure 4. Oscillation method measurement stand.

During oscillation, the vibration amplitude is measured and analyzed by calculating the area between the upper and lower envelopes of the amplitude curve (Figure 5). The initial drop height and load applied to the test tire were selected based on specific assumptions. In both laboratory and real-world rolling resistance tests, the load exerted on a passenger car tire generally ranges from 2000 to over 9000 N, depending on the particular testing conditions [15]. External temperature and internal tire pressure correspond to standard conditions commonly used in ISO rolling resistance measurement procedures (25 °C, 210 kPa). Throughout the oscillation, the frequency of tire deformation increases over time, ranging from approximately 1 Hz at the start of the measurement to around 15 Hz at its conclusion. This frequency range corresponds to deformation frequencies observed in standard drum tests, which are typically conducted at speeds of about 80 km/h [15]. The method closely simulates the actual interaction between the tire and the road surface, accounting for tire structure deformation, tread behavior, and dynamic pressure distribution.

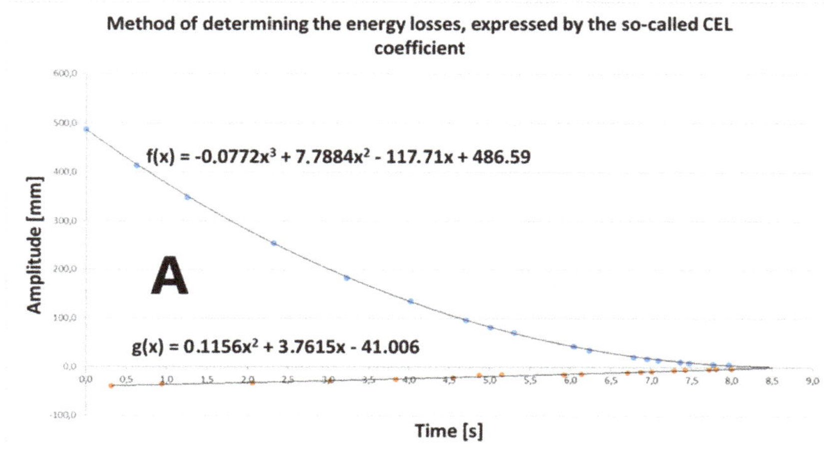

Figure 5. Method of determining energy losses by calculating the area between two envelopes of the oscillation waveform.

This technique shows a strong correlation with the torque-based method, which employs rolling test machines and complies with ISO standards. Although the oscillatory method does not provide absolute rolling resistance values, it offers a reliable comparative metric, particularly useful for evaluating prototype road surfaces. This approach enables researchers to assess surface characteristics without the need to build extensive and costly test tracks [9].

Formula for determining the CEL coefficient is as follows:

$$C_{EL} = \frac{1}{A} = \frac{1}{\int_a^b [f_x - g_x]dx} \quad (1)$$

where C_{EL}—Coefficient of Energy Losses.

2.2. *The Road Surface Replicas Used During the Tests*

The tests were conducted using three types of road surface replicas: PERS, SMA8, and APS4 (see Figures 6–8):

- PERS (Porous Elastic Road Surface) is a porous, elastic road surface designed to reduce noise generated by tire–road interaction. It is typically composed of a mixture of rubber granules (often recycled tire rubber) and a polyurethane binder. Due to its structure, PERS offers good elasticity, vibration damping capabilities, and anti-skid properties, while also allowing efficient water drainage, which enhances driving safety [32].
- SMA8 (Stone Mastic Asphalt 8) is a type of mastic asphalt featuring finer aggregate grading, with a maximum aggregate size of 8 mm. It is characterized by a high content of coarse aggregate, asphalt binder, and stabilizing additives such as cellulose fibers. SMA8 is commonly used as a wearing course in pavement structures, offering excellent rutting resistance, fatigue durability, and favorable acoustic properties. Its dense structure and high binder content ensure good impermeability and long-term

performance. Testing on SMA8 replicas is especially relevant in light of findings by Zaumanis and Haritonovs (2015), who demonstrated that SMA mixtures provide superior durability and rutting resistance under long-term, real-world traffic loads. This supports the selection of SMA8 as a reference material in laboratory studies simulating wheel–pavement interaction [33].

- APS4 is an imitation of a highly rough road surface, created as a surface treatment [34].

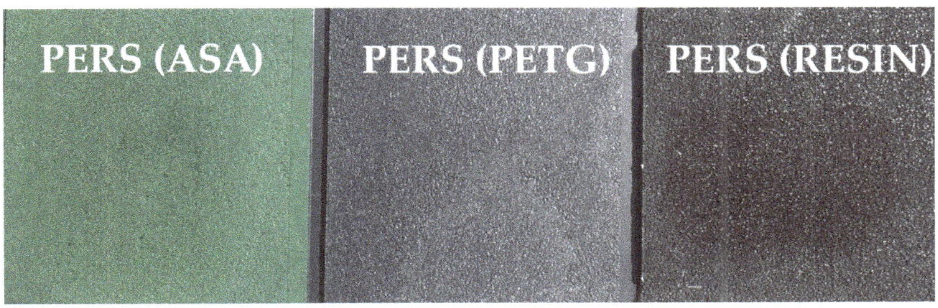

Figure 6. Road surfaces replicas (PERS).

Figure 7. Road surfaces replicas (SMA8).

Figure 8. Road surfaces replicas (APS4).

The replicas were produced using both 3D printing technology and traditional casting with epoxy resins (see Figures 6–8).

- The first and second series consisted of replicas fabricated via 3D printing. PETG (Polyethylene Terephthalate Glycol) and ASA (Acrylonitrile Styrene Acrylate) were selected due to their complementary mechanical and processing properties, which align well with the functional requirements of the test setup. The 3D printing parameters included a 5% infill density and a Tri-hexagon infill pattern. The choice of these filaments and the printing method were based on strength tests described in a previous publication by the researchers [35].

- The third series was produced using the conventional casting method with epoxy resin. This series served as the reference standard for comparison with the results obtained from the 3D-printed replicas.

2.3. The Tires Used During the Tests

Six tires of different types (summer, winter, and all-season) were used for the measurements. All tires had the same size and load index (205/55R16 and LI 91, corresponding to a maximum load capacity of 615 kg). The tests were conducted under standard conditions for the oscillatory method: an internal tire pressure of 210 kPa, an ambient temperature of 25 °C, and a load of 50 kg (including the mass of the wheel, the arm, and additional weights).

3. Results and Discussion

This section presents the results obtained from tests conducted on the individual types of road surfaces.

The results obtained for the APS4 surface replicas across all measurement series are presented in Table 1. When comparing the results for the ASA-printed surfaces to those for the reference surfaces (cast from resin), the values were on average 13.9% lower (ranging from 12% to 15.2%). For the replicas printed from PETG, the results were also lower, with an average difference of 15.9% (ranging from 13.2% to 18.1%).

Table 1. Results obtained for APS4 surface for all measurement series.

Tire	ASA [CEL]	PETG [CEL]	Resin [CEL]	ASA-Resin[%]	PETG-Resin[%]
Tire1 (winter)	0.0065	0.0064	0.0074	12.3	13.4
Tire2 (summer)	0.0072	0.0068	0.0077	7.0	11.4
Tire3 (allseason)	0.0058	0.0053	0.0062	6.6	14.5
Tire4 (summer)	0.0050	0.0049	0.0058	14.2	16.4
Tire5 (allseason)	0.0053	0.0054	0.0062	14.5	13.0
Tire6 (summer)	0.0048	0.0050	0.0056	14.3	11.4

The ranking of tires according to the decreasing oscillation coefficient is consistent across all measurement series (see Figure 9).

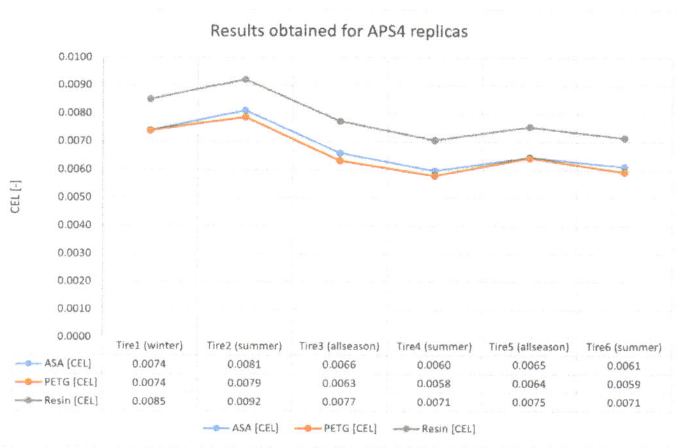

Figure 9. Results obtained for APS4 replicas.

The results obtained for the SMA8 surface replicas across all measurement series are presented in Table 2. When comparing the ASA-printed surfaces to the reference surfaces (cast from resin), the results were on average 11.5% lower (ranging from 6.6% to 14.5%). Similarly, the replicas printed from PETG showed lower results by an average of 13.4% (ranging from 11.4% to 16.4%).

Table 2. Results obtained for SMA8 surface for all measurement series.

Tire	ASA [CEL]	PETG [CEL]	Resin [CEL]	ASA-Resin[%]	PETG-Resin[%]
Tire1 (winter)	0.0065	0.0064	0.0074	12.3	13.4
Tire2 (summer)	0.0072	0.0068	0.0077	7.0	11.4
Tire3 (allseason)	0.0058	0.0053	0.0062	6.6	14.5
Tire4 (summer)	0.0050	0.0049	0.0058	14.2	16.4
Tire5 (allseason)	0.0053	0.0054	0.0062	14.5	13.0
Tire6 (summer)	0.0048	0.0050	0.0056	14.3	11.4

The ranking of tires by decreasing oscillation coefficient is identical across all measurement series (see Figure 10).

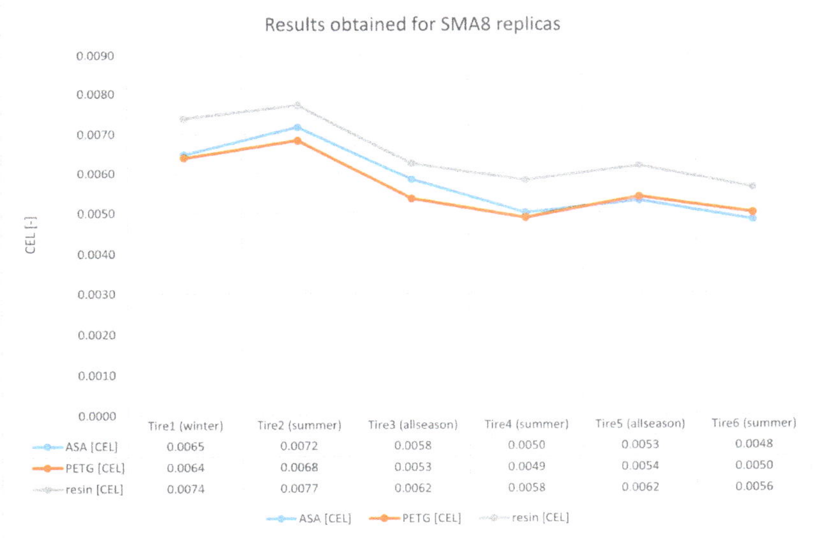

Figure 10. Results obtained for SMA8 replicas.

The results for the PERS surface across all measurement series are presented in Table 3. When comparing the ASA-printed surfaces to the reference surfaces (cast from resin), the results were, on average, 8.3% lower (ranging from 5.7% to 10.6%). For the replicas printed from PETG, the results were also lower by an average of 7.2% (ranging from 4.4% to 8.9%).

The ranking of tires by decreasing oscillation coefficient is identical for both measurement series (see Figure 11).

Table 3. Results obtained for PERS surface for both measurement series.

Tire	ASA [CEL]	PETG [CEL]	Resin [CEL]	ASA-Resin[%]	PETG-Resin[%]
Tire1 (winter)	0.0062	0.0062	0.0065	5.7	4.4
Tire2 (summer)	0.0066	0.0068	0.0072	7.8	6.4
Tire3 (allseason)	0.0056	0.0056	0.0061	8.6	8.3
Tire4 (summer)	0.0049	0.0051	0.0055	10.6	6.9
Tire5 (allseason)	0.0053	0.0052	0.0057	8.2	8.9
Tire6 (summer)	0.0049	0.0050	0.0054	8.8	8.1

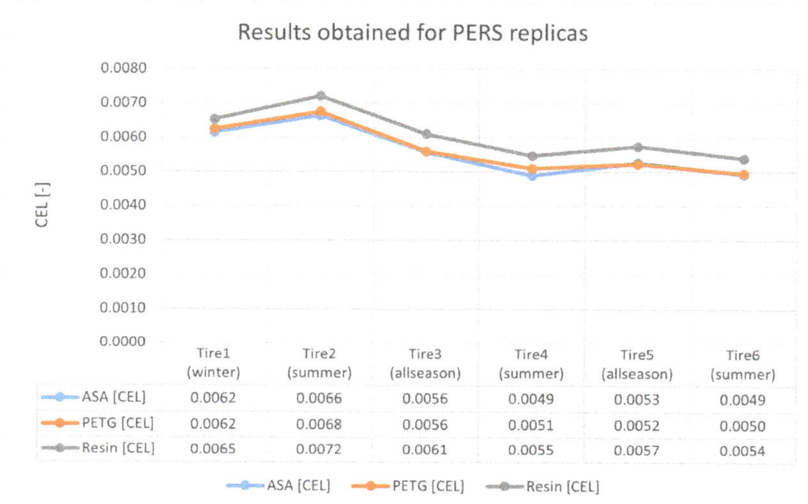

Figure 11. Results obtained for PERS replicas.

The measurement results obtained from 3D-printed replicas consistently showed lower CEL coefficient values compared to those obtained from surfaces manufactured using the traditional casting method. Specifically, replicas printed with the ASA material exhibited, on average, an 11.23% lower value, while those made with PETG showed a 12.17% reduction. To investigate the cause of this discrepancy, an additional series of measurements was conducted. In this series, SMA8 surface replicas printed with the ASA material were produced without the tri-hexagon infill base instead; only a 1 cm thick layer replicating the exact surface texture was retained (see Figure 12).

As previously mentioned, the base of the replicas was constructed using a tri-hexagon geometry with 5% infill density. From a mechanical strength perspective, such replicas fulfilled their intended function [34]. However, the low infill density may have led to excessive deflection, which in turn could have caused the tested tire to be "propelled" to a greater height than with standard replicas. This, consequently, may have resulted in the observed underestimation of the measurement values. As shown in Table 4 and in the graph (Figure 13), this modification resulted in greater consistency between the obtained results, with an average difference of 2.43% (ranging from 0.82% to 4.15%).

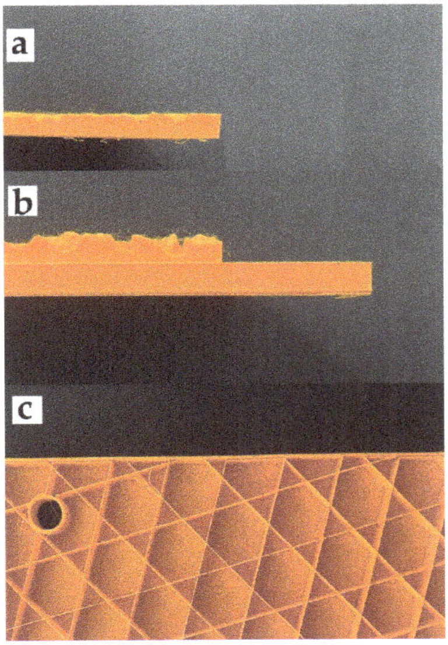

Figure 12. Replica without base (**a**); replica with base (**b**); cross-section of the base showing type and degree of filling (**c**).

Table 4. Results obtained for SMA8 surface for ASA sample with cut base and resin replica.

Tire	ASA Sample with Cut Base [CEL]	Resin [CEL]
Tire1 (winter)	0.0073	0.0074
Tire2 (summer)	0.0075	0.0077
Tire3 (allseason)	0.0062	0.0062
Tire4 (summer)	0.0059	0.0058
Tire5 (allseason)	0.0060	0.0062
Tire6 (summer)	0.0059	0.0056

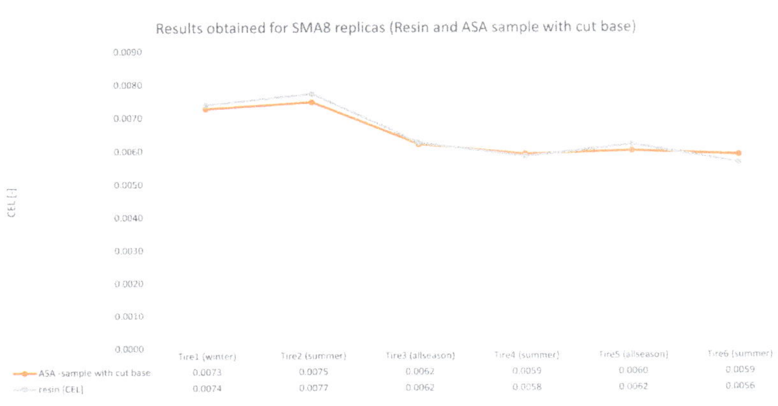

Figure 13. Results obtained for SMA8 replicas (ASA sample with cut base and resin one).

The greatest differences were observed for the APS4 surface, followed by SMA8, with the smallest differences recorded for PERS. It can be seen that the discrepancies decrease as the surface texture becomes progressively smoother. This suggests that the APS4 and SMA8 surfaces, due to their more "scratched" texture, may be more challenging to replicate accurately using 3D printing. To achieve greater printing precision, it may be beneficial to reduce the thickness of the extruded filament layer. Nevertheless, the degree of correlation remains high, and the ranking of surfaces and tires is identical across both methods.

4. Conclusions

The analysis of the obtained results led to the formulation of the following conclusions:

1. Road surface replicas produced using 3D printing exhibited lower values of the CEL (coefficient of rolling resistance) compared to replicas made using the traditional epoxy resin casting method.
2. The average reduction in CEL values was 11.23% for surfaces printed with ASA and 12.17% for those printed with PETG.
3. The greatest differences between the methods were observed for the APS4 surface, followed by SMA8, with the smallest differences noted for PERS. This suggests that rougher and more complex textures are more difficult to reproduce accurately using 3D printing.
4. Modifying the replica design—by removing the low-infill base and retaining only a thin textured layer—improved the consistency of results, reducing the average discrepancy to 2.43%.
5. Despite the absolute value differences, a high correlation was maintained between the results from 3D-printed and cast replicas, along with consistent surface and tire ranking.
6. 3D-printed replicas can be effectively used for comparative rolling resistance testing, provided that printing parameters and sample construction are optimized.

It is important to emphasize that the intended application of the printed replicas is their use as drum coverings in torque-based rolling resistance testing. Measurements obtained using the oscillatory method serve only to estimate energy losses at the tire–road interface, which directly correlate with rolling resistance. Although the oscillatory method has not yet been standardized, research is ongoing to further develop it. This method offers a practical solution for preliminary evaluations, reducing the costs associated with constructing long test tracks under real-world conditions—or, as in this case, enabling the production of full-scale printed replicas for rolling resistance testing machines. This approach allows for the ranking of road sections based on rolling resistance variability and the elimination of sections with extreme values.

Author Contributions: Conceptualization, W.O.; Methodology, W.O. and G.R.; Software, G.R.; Validation, W.O.; Formal analysis, W.O., S.S. and G.R.; Data curation, S.S.; Writing – original draft, W.O.; Visualization, S.S.; Supervision, W.O.; Funding acquisition, W.O. All authors have read and agreed to the published version of the manuscript.

Funding: Financial support of these studies from Gdańsk University of Technology by the DEC-037406 grant under the ARGENTUM TRIGGERING RESEARCH GRANTS—'Excellence Initiative—Research University' program is gratefully acknowledged.

Data Availability Statement: Data available on request.

Acknowledgments: The authors would like to acknowledge Marcin Żurek for his contribution to measurements.

Conflicts of Interest: The authors declare no conflicts of interest. The funders had no role in the design of the study; in the collection, analyses, or interpretation of data; in the writing of the manuscript; or in the decision to publish the results.

References

1. Reimpell, J.; Betzler, J. *Podwozia Samochodów Podstawy Konstrukcji*; WKŁ: Warszawa, Poland, 2001; pp. 151–154.
2. Taryma, S. *Opór Toczenia Opon Samochodowych*; Wydawnictwo PG: Gdańsk, Poland, 2007.
3. Clark, S.K. A brief history of tire rolling resistance. In Proceedings of the Symposium on Tire Rolling Resistance at 122nd Meeting of Rubber Division, American Chemical Society, Chicago, IL, USA, 5–7 October 1982.
4. Clark, S.K.; Dodge, R.N. *A Handbook for the Rolling Resistance of Pneumatic Tires*; Institute of Science and Technology, The University of Michigen: Ann Arbor, MI, USA, 1979.
5. Andersen, L.G.; Larsen, J.K.; Fraser, E.S.; Schmidt, B.; Dyre, J.C. Rolling resistance measurement and model development. *J. Transp. Eng.* **2015**, *141*, 04014075. [CrossRef]
6. 6. Ejsmont, J.; Ronowski, G.; Berge, T.; Sommer, S.; Owczarzak, W.; Szerszyńska, A. At what temperature should the tire rolling resistance be measured? *Proc. Inst. Mech. Eng. Part D J. Automob. Eng.* **2024**. [CrossRef]
7. Sun, Z.; Premarathna, W.; Anupam, K.; Kasbergen, C.; Erkens, S.M. A state-of-the-art review on rolling resistance of asphalt pavements and its environmental impact. *Constr. Build. Mater.* **2024**, *411*, 133589. [CrossRef]
8. *ISO 13473-5:2009(en)*; Characterization of Pavement Texture by Use of Surface Profiles—Part 5: Determination of Megatexture. International Organization for Standardization: Geneva, Switzerland, 2009.
9. Sandberg, U. *Influence of Road Surface Texture on Traffic Characteristics Related to Environment, Economy and Safety*; VTI Notat 53A-1997; Swedish National Road and Transport Research Institute: Linkoeping, Sweden, 1997.
10. Fathi, H.; El-Sayegh, Z.; Ren, J.; El-Gindy, M. Analysis of Tire-Road Interaction: A Literature Review. *Machines* **2024**, *12*, 812. [CrossRef]
11. Von Meier, A.; Van Blokland, G.J.; Descornet, G. The influence of texture and sound absorption on the noise of porous road surfaces. In Proceedings of the 2nd International Symposium on Road Surface Characteristics, Berlin, Germany, 23–26 June 1992.
12. Lak, M.A.; Degrande, G.; Lombaert, G. Influence of poroelastic road surfaces on ground borne vibration due to traffic. In *Deliverable D6.8*; 7FP Persuade; European Commission: Brussels, Belgium, 2013.
13. Hoever, C.; Kropp, W. A model for investigating the influence of road surface texture and tyre tread pattern on rolling resistance. *J. Sound Vib.* **2015**, *351*, 161–176. [CrossRef]
14. Ejsmont, J.; Ronowski, G.; Ydrefors, L.; Owczarzak, W.; Sommer, S.; Świeczko-Żurek, B. Comparison of Tire Rolling Resistance Measuring Methods for Different Surfaces. *Int. J. Automot. Technol.* **2024**, *25*, 965–976. [CrossRef]
15. *ISO28580*; Passenger Car, Truck and Bus Tyres—Methods of Measuring Rolling Resistance—Single Point Test and Correlation of Measurement Results, 2nd ed. International Organization for Standardization: Genève, Switzerland, 2018.
16. *J2452_201707*; Society of Automotive Engineers. Stepwise Coastdown Methodology for Measuring Tire Rolling Resistance. Society of Automotive Engineers: Warrendale, PA, USA, 2017.
17. *ISO18164*; Passenger Car, Truck, Bus and Motorcycle Tyres—Methods of Measuring Rolling Resistance, 2nd ed. International Organization for Standardization: Genève, Switzerland, 2005.
18. *J1269_201912*; Rolling Resistance Measurement Procedure for Passenger Car, Light Truck, and Highway Truck and Bus Tires. Society of Automotive Engineers: Warrendale, PA, USA, 2019.
19. Ejsmont, J.; Ronowski, G.; Sommer, S.; Owczarzak, W.; Berge, T. Replica road pavements for rolling resistance measurements. *Road Mater. Pavement Des.* **2025**, 1–11. [CrossRef]
20. Gong, Z.; Miao, Y.; Lantieri, C. Review of Research on Tire–Pavement Contact Behavior. *Coatings* **2024**, *14*, 157. [CrossRef]
21. Chen, B.; Ding, P.; Wei, G.; Xiong, C.; Wang, F.; Yu, J.; Yu, H.; Zou, Y. A Study on the Contact Characteristics of Tires–Roads Based on Pressure-Sensitive Film Technology. *Materials* **2023**, *16*, 6323. [CrossRef]
22. Chen, X.; Sathyanarayan, H.; Gong, Y.; Yi, J.; Wang, H. Dynamic Tire/Road Friction Estimation With Embedded Flexible Force Sensors. *IEEE Sens. J.* **2023**, *23*, 26608–26619. [CrossRef]
23. Yang, S.; Chen, Y.; Shi, R.; Wang, R.; Cao, Y.; Lu, J. A Survey of Intelligent Tires for Tire-Road Interaction Recognition Toward Autonomous Vehicles. *IEEE Trans. Intell. Veh.* **2022**, *7*, 520–532. [CrossRef]
24. Guan, J.; Zhou, X.; Liu, L.; Ran, M. Measurement of Tire-Pavement Contact Tri-Axial Stress Distribution Based on Sensor Array. *Coatings* **2023**, *13*, 416. [CrossRef]
25. Ge, H.; Quezada, J.C.; Le Houerou, V.; Chazallon, C. Multiscale analysis of tire and asphalt pavement interaction via coupling FEM–DEM 547 simulation. *Eng. Struct.* **2022**, *256*, 113925. [CrossRef]
26. Liu, X.; Jayme, A.; Al-Qadi, I.L. ContactGAN development–prediction of tire-pavement contact stresses using a generative and 549 transfer learning model. *Int. J. Pavement Eng.* **2023**, *24*, 2138876. [CrossRef]

27. Gao, X.; Zhuang, Y.; Liu, S. High-speed 3D digital image correlation for measuring tire rolling resistance coefficient. *Measurement* **2020**, *171*, 108830. [CrossRef]
28. Kawakami, A.; Ishigaki, T.; Shirai, Y.; Terada, M.; Kubo, K. Evaluation method of pavement surface characteristics for rolling resistance. *Road Mater. Pavement Des.* **2017**, *18* (Suppl. S2), 2–11. [CrossRef]
29. Cho, J.R.; Lee, H.W.; Jeong, W.B.; Jeong, K.M.; Kim, K.W. Numerical estimation of rolling resistance and temperature distribution of 3-D periodic patterned tire. *Int. J. Solids Struct.* **2013**, *50*, 86–96. [CrossRef]
30. Ding, Y.; Wang, H.; Qian, J.; Zhou, H. Evaluation of tire rolling resistance from tire-deformable pavement interaction modeling. *J. Transp. Eng. Part B Pavements* **2021**, *147*, 04021041. [CrossRef]
31. Ejsmont, J.; Owczarzak, W. Engineering method of tire rolling resistance evaluation. *Measurement* **2019**, *145*, 144–149. [CrossRef]
32. Ejsmont, J.; Sandberg, U.; Świeczko-Zurek, B.; Mioduszewski, P. Tyre/Road Noise Reduction by a Poroelastic Road Surface. Presented at the 43rd International Congress on Noise Control Engineering: Improving the World Through Noise Control, INTERNOISE 2014, Melbourne, Australia, 16–19 November 2014. Available online: https://urn.kb.se/resolve?urn=urn:nbn:se:vti:diva-9349 (accessed on 30 May 2025).
33. Zaumanis, M.; Haritonovs, V. Long term monitoring of full scale pavement test section with eight different asphalt wearing courses. *Mater. Struct.* **2016**, *49*, 1817–1828. [CrossRef]
34. Taryma, S.; Woźniak, R. Wpływ zużycia opony samochodowej na jej opór toczenia. *Autobusy Tech. Eksploat. Syst. Transp.* **2014**, *15*, 272–275.
35. Owczarzak, W.; Sommer, S.; Ronowski, G. Method of Forming Road Surface Replicas Using 3D Printing Technology. *Coatings* **2024**, *14*, 1455. [CrossRef]

coatings

MDPI

Article

Influence of Carbonized Garbage Fly Ash on the Performance of Foam Concrete

Di Wang [1], Zhiqiang Xu [1], Yehan Yu [2], Na Xu [1], Chuanqi Li [1], Xu Tian [1], Hui Wang [2,*], Feiting Shi [2] and Kangshuo Xia [3]

1 School of Chemical Engineering and Machinery, Liaodong University, Dandong 118000, China; wangdi@liaodongu.edu.cn (D.W.); xuzhiqiang@liaodongu.edu.cn (Z.X.); xuna@liaodongu.edu.cn (N.X.); lichuanqi@liaodongu.edu.cn (C.L.); tianxu@liaodongu.edu.cn (X.T.)

2 School of Civil Engineering and Geographic Environment, Ningbo University, Ningbo 315000, China; yuyehan0127@163.com (Y.Y.); shifeiting@ycit.cn (F.S.)

3 School of Civil Engineering, Chongqing University, Chongqing 400044, China; xiakangshuo@cqu.edu.cn

* Correspondence: huiwang123@aliyun.com

Abstract

To utilize garbage fly ash (GFA) as a resource, this research proposes a method for preparing GFA with higher reactivity through carbonation and applies it to the production of foamed concrete. The effects of CO_2-cured GFA substitution rate and foam volume on slump flow, rheological properties, mechanical strength, thermal conductivity, water absorption rate, and water resistance coefficient of foam concrete are clarified. The results show that an increase in the CO_2-cured GFA substitution rate from 0 to 100% improves the slump flow by 10.8%~34.5% and decreases the plastic viscosity by 4.8%~36.4% and yield stress by 5.6%~28.1%. The higher carbonized GFA substitution rate can prolong the initial setting time with the largest amplitude of 30.4%. In addition, increasing the CO_2-cured GFA substitution rate improves the mechanical strengths, water resistance, thermal conductivity, and solidification of heavy metals. When the CO_2-cured GFA substitution rate is 100%, the 28-day compressive strength, 28-day flexural strength, water absorption rate, water resistance coefficient, thermal conductivity, leached Zn, and leached Cr of foam concrete are 18 MPa, 3.6 MPa, 20.7%, 0.46, 0.69 $W \cdot m^{-1} \cdot K^{-1}$, 9.4×10^{-5} mg/mL, and 8.6×10^{-5} mg/mL, respectively. Moreover, more foam volume improves the fresh-mixed performance of foam concrete while reducing the mechanical strength, water resistance property, thermal conductivity, and solidification of heavy metals. It is found that the technical approach for preparing foamed concrete containing CO_2-cured GFA with 40% foam volume can achieve its large-scale use.

Keywords: foamed concrete; water absorption rate; thermal conductivity; initial setting time; mechanical strengths

1. Introduction

Garbage fly ash, steel slag, secondary aluminum dross, and other similar materials are solid wastes that can inflict substantial environmental harm if left untreated [1,2]. Among these, garbage fly ash poses the greatest threat. It contains a high concentration of heavy metal elements, leading to air pollution during windy conditions [3]. Moreover, the accumulation of garbage fly ash significantly contaminates water sources [4]. Given these factors, it is imperative to employ multiple methods to mitigate the hazards associated with garbage fly ash [5,6].

Garbage fly ash is formed by the incineration of municipal solid waste at high temperatures [7]. Although garbage fly ash contains a large amount of toxic substances, this solid waste also contains a large amount of active substances [8]. Therefore, garbage fly ash is suitable for the production of cement-based materials [9]. Several researchers have used garbage fly ash for manufacturing cement concrete [10]. It has been reported that garbage fly ash increases the flexural strength and compressive strength of cement concrete up to 11.3% and 13.2%, respectively [11]. When cement concrete is cured in a high-temperature environment, the improvements in mechanical strength and resistance to chloride ion penetration, carbonation, and sulfate corrosion are higher [12]. Liao et al. [13] have demonstrated that garbage fly ash improves the corrosion resistance of steel-reinforced concrete by increasing its electrical and polarization resistances. Although garbage fly ash has been reported to increase the mechanical strengths and durability of cement concrete, toxic substances leach from cement concrete [14,15]. Garbage fly ash contains two main heavy metal elements, Cr and Zn. In the absence of timely treatment, these elements have the potential to leach out, posing a significant threat to human health [16]. However, a promising approach involves treating garbage fly ash with carbon dioxide, which leads to the generation of $CaCO_3$ and other substances to enhance its compactness and effectively mitigate the leakage of toxic substances. The manufacturing of carbonized garbage fly ash with cement concrete can effectively solidify the elements of Cr and Zn. Solidified Cr and Zn are harmless to the environment [17].

Industrial production activities emit a large amount of CO_2 gas, which accelerates the greenhouse effect [18]. However, CO_2 is used in manufacturing certain building materials. The CO_2 curing of cement concrete can significantly improve its mechanical strength by rates up to 31.6% and also enhance its durability [19,20]. Moreover, as pointed out in previous studies, CO_2 curing of raw materials can improve the performance of cement-based materials [21]. CO_2 curing on secondary aluminum ash, furnace ash, coal ash slag, etc., can enhance the mechanical strength and long-term durability of the cement-based material. Research shows that CO_2-cured garbage fly ash can increase the compressive strength of UHPC by up to 14.2% [22]. CO_2-cured coal gangue can increase the mechanical strength of bricks by up to 222.7%. The leached Pb and Zn are reduced to less than 3.5×10^{-5} mg/mL, while the thermal conductivity is reduced by up to 32.6%, after the specimens are immersed in deionized water for six months [23]. Using CO_2-cured garbage fly ash as a partial replacement for cementitious materials can enhance the mechanical strength. At the same time, after six months of leaching, Cr and Zn are reduced to less than 8.64×10^{-5} mg/mL and 9.46×10^{-5} mg/mL. Consequently, the environmental risks associated with practical applications are rendered practically negligible [24,25]. However, little attention has been paid to the influence of CO_2-cured garbage fly ash on the properties of foam concrete.

Foam concrete is usually used for thermal insulation. However, foam concrete possesses insufficient strength. Several methods have been used to improve the mechanical properties of foam concrete [26,27]. Increasing the amount of cementitious material and reducing the porosity are often used to improve the strength of foam concrete [28]. The cost of ordinary cementitious materials is relatively high. Therefore, the use of CO_2-cured garbage fly ash to increase the mass ratio of cementitious materials presents a viable option for enhancing the mechanical properties of foam concrete [29,30].

This study investigates the rheological properties of fresh foam concrete, including slump flow, plastic viscosity, yield stress, and initial setting time, as a function of the mass ratio of CO_2-cured garbage fly ash (ranging from 0% to 100% of the total fly ash content). Meanwhile, the mechanical strengths are determined. The water absorption rate and thermal conductivity are measured. Finally, the influence of the foam volume on the corresponding performances is assessed. This research aims to develop new materials for making foam concrete while also recycling garbage fly ash and reducing excess CO_2 emissions and economic costs, thereby creating significant sustainable value.

2. Experimental Sections

2.1. Raw Materials

In this study, garbage fly ash (GFA), supplied by Zhongjielan Environmental Protection Technology Co., Ltd., Beijing, China, was utilized as a supplementary cementitious material. The garbage fly ash mainly contained SiO_2, CaO, Al_2O_3, and other minor components. Ordinary Portland cement, obtained from Fushun Cement Co., Fushun, China, served as another cementitious material, with a compressive strength of 42.5 MPa, a density of 3.1 g/cm^3, and initial/final setting times of 114 and 346 min, respectively. The aggregates used in this study were river sand (fine aggregate) with a fineness modulus of 2.35. A plant-based foaming agent, Type JD-2 (99% vegetable proteins), produced by Shanghai Fangbao Building Materials Technology Co., Ltd., Shanghai, China, was employed. This light-yellow liquid agent has a gas production capacity of 10,000 mL/g, and decomposes at 80 °C. A high-range water-reducing agent, produced by Jiangsu Subote New Materials Co., Ltd., Suzhou, China, and exhibiting a 40% water-reduction rate, was used in this study. In this study, the foam volume ranged from 0% to 60%, while the water-to-binder (w/b) ratio was maintained at 1.0. Tables 1 and 2 show the chemical compositions and the particle size distribution of the cementitious materials and river sand.

Table 1. Chemical compositions of the cementitious materials/wt.%.

Types	SiO_2	Al_2O_3	MgO	CaO	SO_3	K_2O	Na_2O	Ti_2O	CdO	CuO	ZnO	B_2O_3	Loss on Ignition
GFA	22.5	4.5	0.05	20.3	9.3	6.3	4.3	10.2	0.07	0.09	0.5	21.2	21.2
Cement	20.9	5.5	1.7	62.2	2.7	-	-	-	-	-	-	3.1	3.1
River sand	74.7	1.8	-	2.1	-	0.9	3.8	-	-	-	-	16.7	-

Table 2. Particle size distribution of the cementitious materials and river sand/wt.%.

Particle Size Types/μm	0.3	0.6	1	4	8	64	100	360
GFA	0.13	0.5	2.2	17.2	31.3	97.5	100	100
Cement	0	0.33	2.66	15.01	28.77	93.59	100	100
River sand	0	0	0	0	0	0	22.5	100

2.2. Sample Preparation and Measurement

The foaming agent was blended with water at a 1:20 ratio to prepare a foam solution, which was subsequently fed into a foaming machine to produce foam. The garbage fly ash was thoroughly blended with water and agitated for 1 min to yield a uniform fresh paste, which was subsequently combined with the foam to produce foamed garbage fly ash cement concrete. The carbonation process was performed by exposing garbage fly ash to pressurized CO_2 (atmospheric pressure of 0.4 MPa and temperature of 60 °C) for 24 h. Reactive Ca/Mg oxides were converted into stable carbonates (e.g., $CaCO_3$/$MgCO_3$), enabling simultaneous CO_2 sequestration and heavy metal immobilization. Table 3 shows the mixture design of foamed concrete per cubic meter. The proportion of GFA (carbonated GPA + uncarbonated GPA) in the cementitious materials (cement and GFA) was 33.3%. The CO_2-cured GFA accounted for 0%, 20%, 40%, 60%, 80%, and 100%, respectively, of the GFA.

Table 3. Mixture design of foamed concrete per cubic meter (kg).

Water	P·O Cement	GFA	CO_2-Cured GFA	Quartz Sand	Water-Reducing Agent
444.4	740.7	0	370.3	977.9	16.3
444.4	740.7	74.1	296.2	977.9	16.3
444.4	740.7	148.1	222.2	977.9	16.3
444.4	740.7	222.2	148.1	977.9	16.3
444.4	740.7	296.2	74.1	977.9	16.3
444.4	740.7	370.3	0	977.9	16.3

2.3. The Measurement of Workability

The flowability was tested according to the GB/T 50080-2016 standard [31]. Each set of specimens was placed in a truncated conical circular mold and compacted. The mold was gently lifted vertically, at which point the jumping table was activated 25 times at a frequency of once per second. The flowability was recorded as the average values of the change in diameter vertically in both directions. All specimens were conditioned at 20 °C prior to measuring their plastic viscosity. A Brookfield RST-SST rheometer (Beijing Boying Technology Co., Ltd., Beijing, China) was used for measurement. The plastic yield shear stress was determined using a Lamy DSR500 CP-4000 Plus machine (Bihai International Trade Co., Ltd., Shanghai, China). All specimens were placed in the rheometer, and the maximum torque was recorded.

2.4. Determination of Thermal Conductivity and Water Resistance Coefficient

The foamed garbage fly ash was cast into oiled molds to produce specimens with dimensions of $100 \times 100 \times 400$ mm^3 and $100 \times 100 \times 100$ mm^3. Three specimens were selected for each test, ensuring a standard deviation within ±10%.

After demolding, all specimens were dried in a DHG-9030 electrothermal drying oven (Suzhou Double Blue Environmental Technology Co., Ltd., Suzhou, China), with a maximum drying temperature of 400 °C, at 140 °C for 9 h. Subsequently, specimens were cured in a controlled room environment (40% relative humidity, 20 °C) for 28 days. Thermal conductivity was measured using $100 \times 100 \times 100$ mm^3 specimens with an MP-2 portable thermal conductivity tester supplied by Cymotes Technology Co., Ltd., Shenzhen, China with a measurement range of 0.001~10 W/(m·K).

These specimens were also used to evaluate the water resistance coefficient. For this purpose, the selected specimens were immersed in water until their mass stabilized. The water resistance coefficient was determined as the ratio of the compressive strength measured before water saturation to that measured after saturation.

2.5. Measurement of Mechanical Properties

Mechanical properties were evaluated in compliance with the Chinese national standard GB/T 50081-2019 [32] using an automated universal testing machine provided by Shanghai Songdun Instrument Manufacturing Co., Ltd., Shanghai, China, integrated with a humidity-controlled environmental chamber. Specimens incorporating 10%, 20%, and 30% cement by mass, combined with 30% foam agent (by volume), were tested under a controlled relative humidity of 40%. All specimens were cured for 28 days and maintained under the same environmental conditions for 6 hours prior to testing. The microstructure and crystalline phases of the hydration products were examined using an Axia-Chemi scanning electron microscope–energy spectrum analysis (SEM-EDS) (Shenzhen Shanshi Instrument Co., Ltd., Shenzhen, China). The measuring procedure is shown in Figure 1.

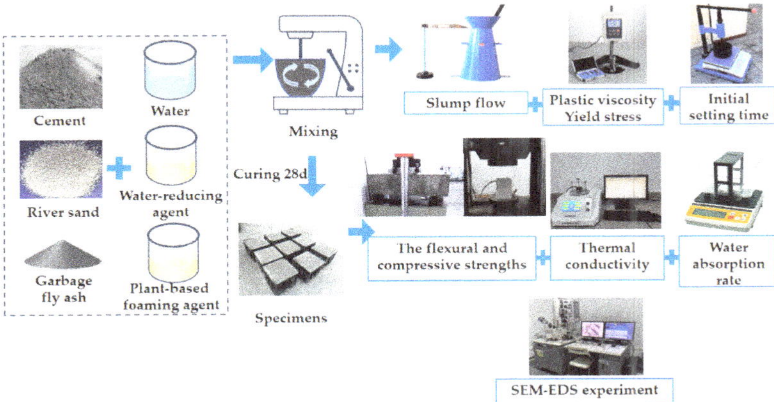

Figure 1. The measuring procedure.

3. Results and Discussions

3.1. The Influence of CO_2-Cured Garbage Fly Ash

The slump flow of fresh foamed concrete is presented in Figure 2. As shown in Figure 2, the slump flow increases in the form of cubic functions with an increased mass ratio of CO_2-cured garbage fly ash. The slump flow is increased by rates of 10.8%~34.5%. This effect is attributed to the pore refinement induced by the CO_2 curing of garbage fly ash-based materials, which promotes the release of entrapped free water and consequently enhances the slump flow of the fresh concrete [33,34]. Moreover, as depicted in Figure 2, the slump flow of fresh foamed concrete exhibits a time-dependent reduction, with the placing time by rates of 3.1%~27.5%. This reduction is primarily attributed to the evaporation of free water, which progressively diminishes the fluidity of the mixture. The goodness-of-fit (R^2 >0.9) of the fitted equations validates the accuracy of the predictive model for fluidity decay.

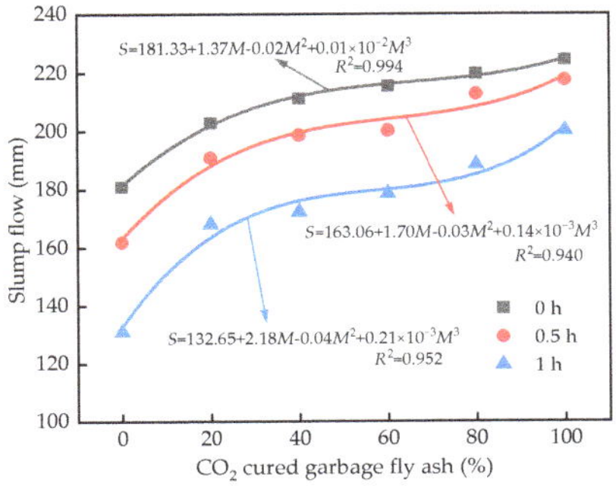

Figure 2. The slump flow of fresh foamed concrete with CO_2-cured garbage fly ash.

The plastic viscosity and the yield stress of fresh foamed concrete with CO_2-cured garbage fly ash are shown in Figure 3. The plastic viscosity and yield stress decrease quadratically with an increasing mass ratio of CO_2-cured garbage fly ash. The decreasing rates of the plastic viscosity and yield stress induced by the CO_2-cured garbage fly ash are 4.8%~36.4% and 5.6%~28.1%. The placing time demonstrates an increasing effect on the fresh foamed concrete's plastic viscosity and yield shear stress. The fresh foamed concrete's plastic viscosity and yield stress are increased by rates of 13.1%~59.0% and 11.8%~37.2%

with placing time. As reported in prior studies, the relationships between plastic viscosity, the yield shear stress, and the mass ratio of CO_2-cured garbage fly ash confirm the anti-correlation [35]. Therefore, the plastic viscosity and yield stress decrease with the mass ratio of CO_2-cured garbage fly ash, while placing time elevates both parameters.

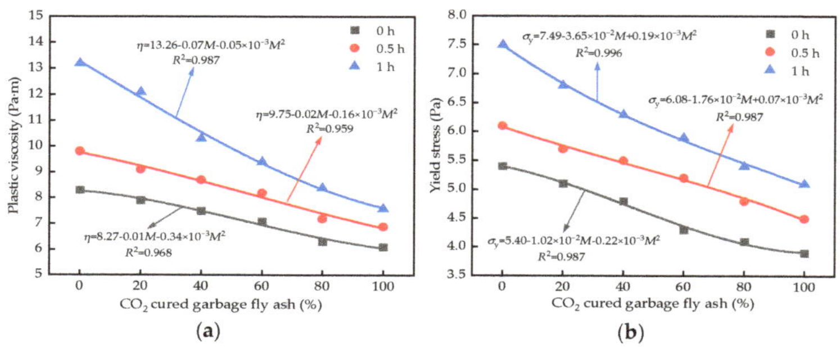

Figure 3. The plastic viscosity and yield stress of fresh foamed concrete with CO_2-cured garbage fly ash. (**a**) The plastic viscosity; (**b**) The yield stress.

The initial setting time (IST) of fresh foamed concrete with CO_2-cured garbage fly ash is shown in Figure 4. As illustrated in Figure 4, the initial setting time extends with an increasing mass ratio of CO_2-cured garbage fly ash. The relationship between the mass ratio of CO_2-cured garbage fly ash and the initial setting time fits the cubic function. This can be explained by the fact that CO_2-cured garbage fly ash can decrease the early hydration degree of cement and increase the content of free water [36]. Therefore, the initial setting time increases with the mass ratio of CO_2-cured garbage fly ash. The fitting degree of the equation is 0.996, which confirms the fitting accuracy. CO_2-cured garbage fly ash can extend the initial setting time up to 30.4%.

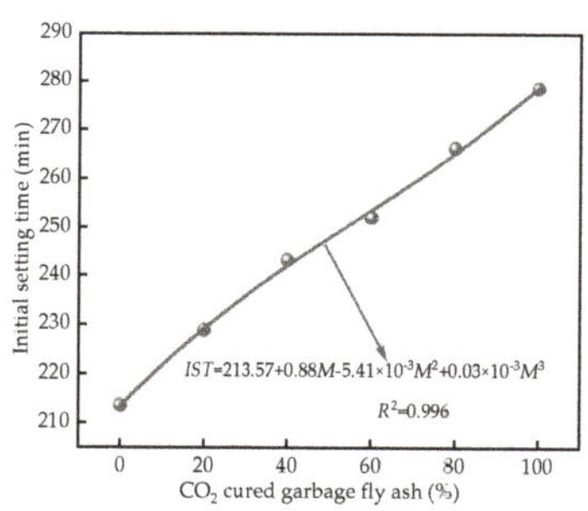

Figure 4. The initial setting time of fresh foamed concrete with CO_2-cured garbage fly ash.

Figure 5 presents the flexural and compressive strengths of foamed concrete incorporating CO_2-cured garbage fly ash. The flexural and compressive strengths increase with the mass ratio of CO_2-cured garbage fly ash, following a quadratic relationship. This can be attributed to the fact that the CO_2 curing of garbage fly ash can reduce the pores on the surface of garbage fly ash and decrease the early hydration heat of cement [37,38]. Consequently, the compactness of foamed concrete is improved and the inner cracks induced by hydration heat are reduced. Therefore, the mechanical strengths are increased by the addition of CO_2-cured garbage fly ash. The CO_2-cured garbage fly ash increases the flexural and compressive strengths by 3.4%~35.6% and 4.2%~40.8%.

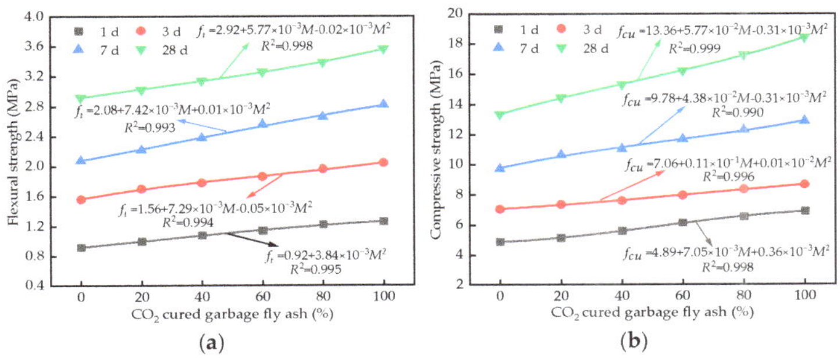

Figure 5. The mechanical strengths of foamed concrete with CO_2-cured garbage fly ash. (**a**) The flexural strength; (**b**) The compressive strength.

The water absorption rate of foamed concrete with CO_2-cured garbage fly ash is exhibited in Figure 6. The water absorption rate decreases quadratically with an increasing mass ratio of CO_2-cured garbage fly ash, exhibiting reductions ranging from 3.5% to 10.4%. This can be explained by the fact that CO_2-cured garbage fly ash can improve the compactness of foamed concrete and reduce internal cracks [39]. Therefore, the water absorption of the foamed concrete declines. The fitting degree of the equation is 0.990, ensuring the accuracy of the fitting function.

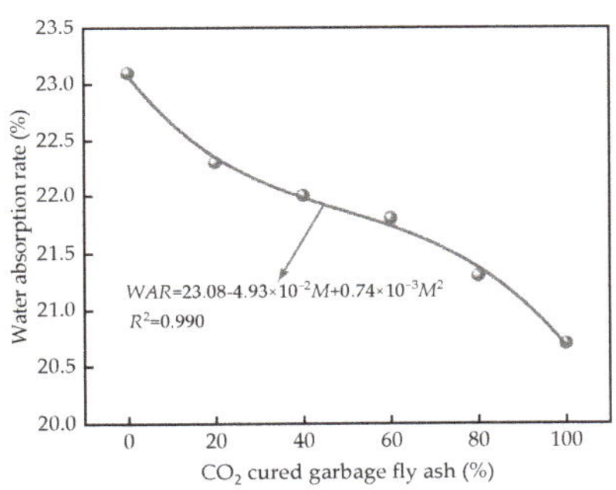

Figure 6. The water absorption of foamed concrete with CO_2-cured garbage fly ash.

The water resistance coefficient of foamed concrete with CO_2-cured garbage fly ash is shown in Figure 7. As depicted in Figure 7, the water resistance coefficient of the foamed concrete exhibits a cubic increase with the mass ratio of CO_2-cured garbage fly ash, with a fitting coefficient of 0.991, ensuring precision. The coefficient rises by 8.6%~31.4% as the garbage fly ash mass ratio increases. CO_2 curing can reduce pores in the garbage fly ash and early cracking of cement hydration, thereby enhancing the water resistance of foamed concrete.

Figure 8 illustrates the thermal conductivity of foamed concrete with CO_2-cured garbage fly ash. The thermal conductivity exhibits a linear increasing trend with the mass ratio of CO_2-cured garbage fly ash, rising by 2.5%~7.8% as the mass ratio increases. CO_2-cured garbage fly ash can improve the compactness, which increases the thermal conductivity of foamed concrete. The fitting degree of the function is 0.970, ensuring fitting accuracy. Although thermal conductivity is increased by 2.5%~7.8% through CO_2 curing of garbage fly ash, the mechanical strength is enhanced by up to 40.8%. The elevated thermal conductivity exerts minimal impacts on insulation performance, while the strength improvement enables reduced transportation damage, extended service life, decreased material loss, and improved economic efficiency.

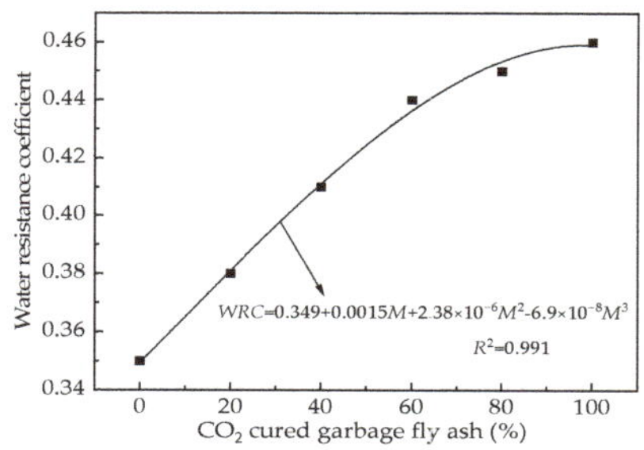

Figure 7. The water resistance coefficient of foamed concrete with CO_2-cured garbage fly ash.

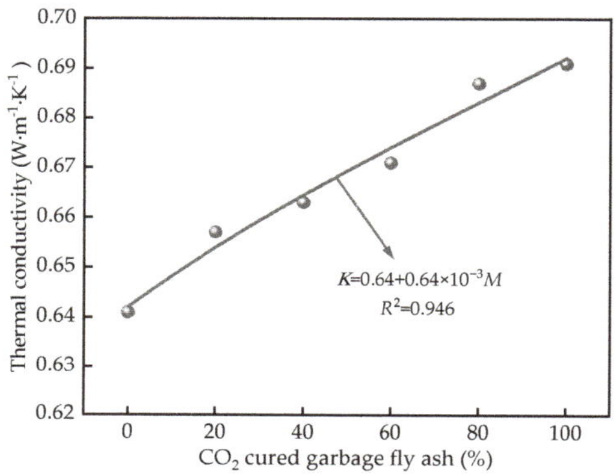

Figure 8. The thermal conductivity of foamed concrete with CO_2-cured garbage fly ash.

Figure 9 shows the leached Zn and Cr from the foamed concrete with CO_2-cured garbage fly ash. As observed in Figure 9, the leached Zn and Cr decrease in the form of a cubic function with the mass ratio of CO_2-cured garbage fly ash. The CO_2-cured garbage fly ash decreases the leached Zn and Cr by rates of 1.7%~11.7% and 2.0%~9.3%. This can be attributed to the fact that the CO_2 curing of garbage fly ash can decrease the pores on the surface of garbage fly ash, reducing the leached Zn and Cr [40,41]. Moreover, the incorporation of CO_2-cured garbage fly ash reduces the early-age hydration heat of cement, thereby enhancing the compactness of the foamed concrete. Therefore, the leached Zn and Cr are decreased by CO_2-cured garbage fly ash.

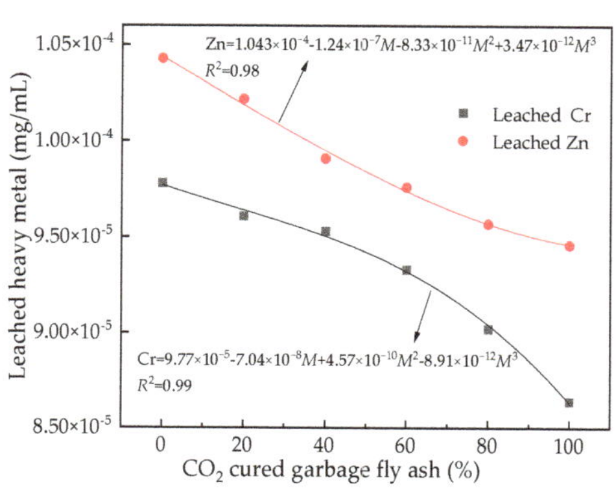

Figure 9. The leached Zn and Cr of foamed concrete with CO_2-cured garbage fly ash.

3.2. The Influence of Foam Volume

Figure 10 displays the slump flow of fresh foamed concrete with different foam volumes. As shown in Figure 10, the slump flow increases in the form of a cubic function with the increasing volume of foam, rising by 4.9%~16.7%. The increased volume of foam can increase the free water and gas, enhancing the fluidity of fresh foamed concrete [42]. The fitting degree of the fitted equation is 0.998, indicating the equation's fitting accuracy.

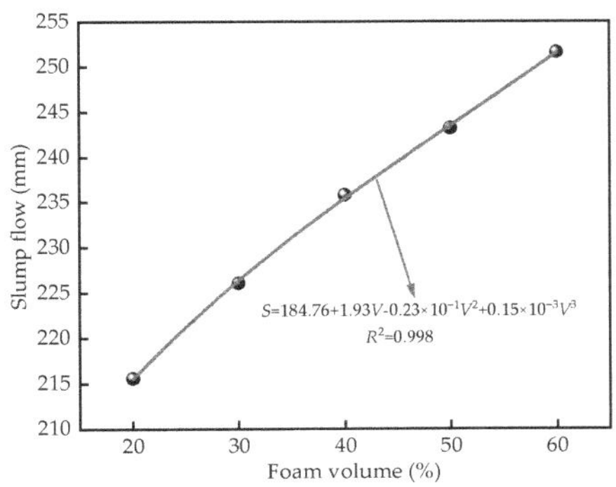

Figure 10. The slump flow of fresh foamed concrete with different foam volumes.

Figure 11 illustrates the plastic viscosity and yield stress of fresh foamed concrete with different foam volumes. As depicted in Figure 11, the plastic viscosity and yield stress decrease in the form of cubic functions. The decreasing rates of plastic viscosity and yield stress with foam volume are 4.2%~16.9% and 9.3%~32.6%, due to the fact that the plastic viscosity and yield stress of fresh cement matrix satisfy an inverse correlation with the slump flow [43]. Consequently, the increased slump flow results in reduced plastic viscosity and yield stress of fresh foamed concrete. The equations' fitting degrees are 0.998 and 0.993, respectively, ensuring high fitting precision.

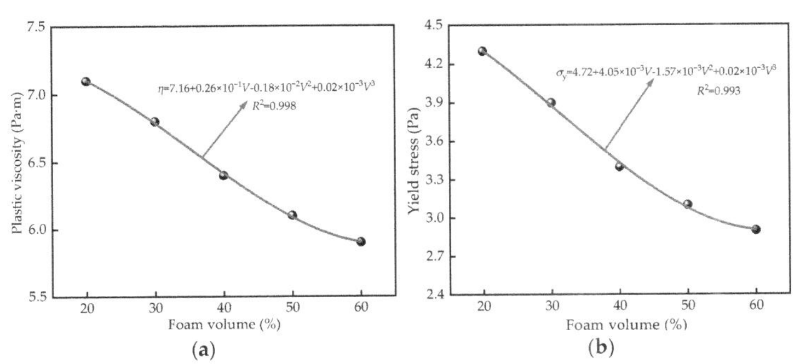

Figure 11. The plastic viscosity and yield stress of fresh foamed concrete with different foam volumes. (**a**) The plastic viscosity; (**b**) The yield stress.

The mechanical strengths of foamed concrete with different foam volumes are shown in Figure 12. The mechanical strengths decrease in the form of cubic functions with the foam volume. The flexural and compressive strengths decrease by 4.6%~22.1% and 3.1%~25.3%. This can be explained by the fact that the added foam increases the pores in the concrete, leading to a decrease in the mechanical strengths of cement concrete [44,45]. The fitting degrees of the equations are 0.958 and 0.978, ensuring the fitting precision.

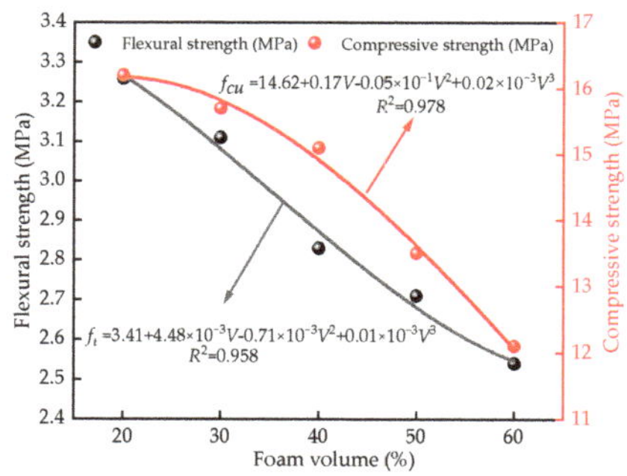

Figure 12. The mechanical strengths of foamed concrete with different foam volumes.

Figure 13 presents the water absorption rate of foamed concrete with different foam volumes. As depicted, the water absorption rate of foam concrete increases cubically with foam volume, exhibiting an increasing rate ranging from 3.2% to 12.4%. This trend is attributed to the enhanced porosity induced by higher foam volumes in foam concrete [46]. As a result, water penetrates more readily into foam concrete, resulting in an elevated water absorption rate. The fitted function's fitting degree is 0.999, which ensures the fitting precision.

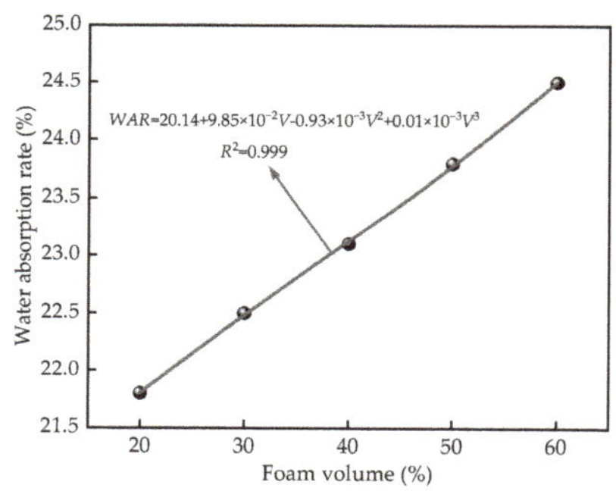

Figure 13. The water absorption of foamed concrete with different foam volumes.

The water resistance coefficient of foamed concrete with the foam volume ranging from 20% to 60% is shown in Figure 14. As shown in Figure 14, the water resistance coefficient decreases in the form of a cubic function at rates ranging from 16.2% to 43.2%. The corresponding fitting degree is 0.989, ensuring the fitting accuracy. This can be attributed to the fact that higher foam volumes increase the porosity of concrete, thereby reducing its water resistance [47].

The thermal conductivity of foamed concrete with different foam volumes is exhibited in Figure 15. As found in Figure 15, the foamed concrete's thermal conductivity displays the relationship of cubic functions with the foam volume. With the increasing foam volume, the thermal conductivity decreases by 32.0%~69.7%. The observed reduction in thermal conductivity results from increased foam volume, which expands the pore structure in foamed concrete, thereby impeding heat transfer [48]. The fitting degree of the function is 0.989, ensuring the fitting accuracy. A foam volume ratio of 40% is recommended for foamed concrete. At this ratio, the flexural and compressive strengths decrease by 13.19% and 6.79%, respectively, and the thermal conductivity decreases by up to 44.04%. When the foam content increases from 40% to 60%, these three parameters decrease by 18.51%, 8.9%, and 16.21%, respectively. This demonstrates optimal cost performance at the 40% foam volume ratio.

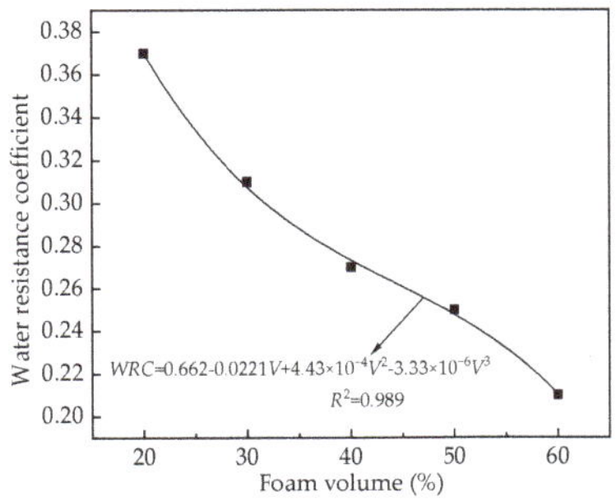

Figure 14. The water resistance coefficient of foamed concrete with different foam volumes.

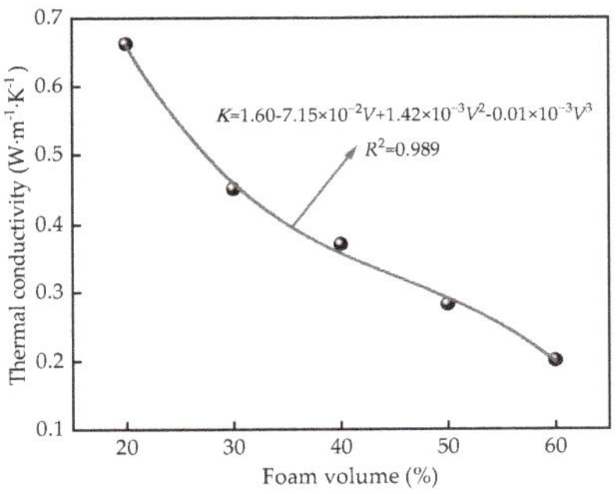

Figure 15. The thermal conductivity of foamed concrete with different foam volumes.

The leached Zn and Cr of the foamed concrete specimens with foam volumes ranging from 20% to 60% (by volume) are plotted in Figure 16. The specimens with a size of 100 mm × 100 mm × 100 mm were immersed in deionized water for 6 months. As shown, both Zn and Cr exhibit a cubic-trend increase with foam volume, with Zn leaching rising by 5.0%~27.7% and Cr leaching by 3.4%~22.8%. This trend is attributed *to* the foam, which can increase the pores in the cement concrete [49]. Moreover, the foam increases the permeability of foamed concrete, thereby increasing the leached Zn and Cr.

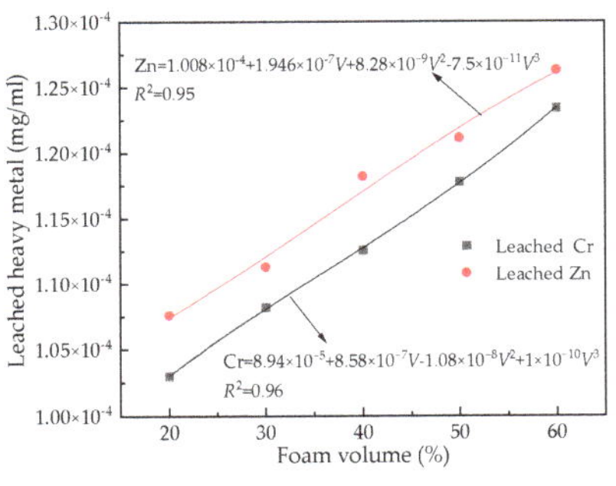

Figure 16. The leached heavy metals.

3.3. Microscopic Research Results

Figure 17 shows the scanning electron microscopy–energy spectrum analysis (SEM-EDS) of specimens with 20% un-carbonized garbage fly ash and 20% carbonized garbage fly ash. The specimens were cured under controlled conditions for 28 days. As observed in Figure 17, cracks are found in the specimen with un-carbonized garbage fly ash, while the hydration products in the specimen with carbonized garbage fly ash are more compact. The CO_2-cured garbage fly ash can decrease the early hydration cracks by reducing the early hydration heat. Additionally, the EDS results show the presence of the elements C, O, Al, Si, Ca, and Fe. The specimen with CO_2-cured garbage fly ash shows a higher C concentration than that with un-carbonized garbage fly ash.

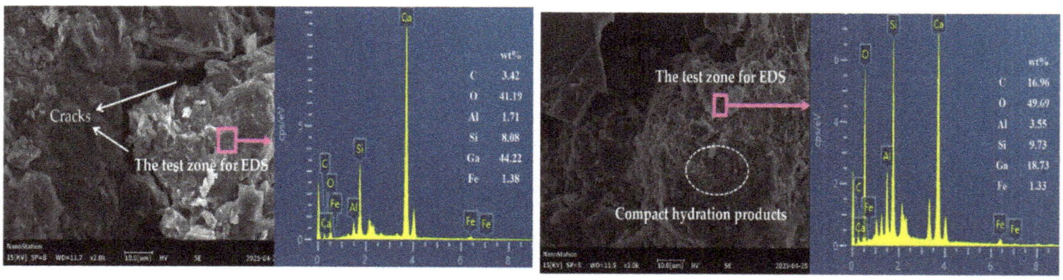

Figure 17. The SEM-EDS of foamed concrete. (**a**) With un-carbonized garbage fly ash; (**b**) With carbonized garbage fly ash.

4. Conclusions

In this study, Garbage fly ash (GFA) modified through carbonation was prepared, and foam concretes with CO_2-cured GFA were formed. The effects of CO_2-cured GFA substitution rate and foam volume on the slump flow, rheological properties, mechanical strength, thermal conductivity, water absorption rate, and water resistance coefficient of foam concrete are clarified. The main conclusions are drawn as follows:

(1) The increases in CO_2-cured GFA substitution rate and foam volume are conducive to the improvement of fresh-mixed properties. A higher CO_2-cured GFA substitution rate and foam volume increase the slump flow and initial setting time and decrease the plastic viscosity, yield stress, and loss of liquidity of foam concrete over time. When the CO_2-cured GFA substitution rate is 100%, the slump flow is increased by 34.5%, the plastic viscosity and yield stress are reduced by 59.0% and 37.2%, and the initial setting time is prolonged by 30.4%. When the foam volume is 60%, the slump flow is increased by 16.7%, and the plastic viscosity and yield stress are reduced by 16.9% and 32.6%.

(2) The increased substitution rate of CO_2-cured GFA significantly enhances the mechanical strength, demonstrating 35.6% and 40.8% improvements in flexural and compressive strength respectively, while concurrently increasing the thermal conductivity of foam concrete. However, an increase in foam volume (20%~60%) is conducive to decreasing the thermal conductivity of foam concrete, but will reduce the flexural and compressive strengths by 4.6%~22.1% and 3.1%~25.3%, respectively.

(3) A higher CO_2-cured GFA substitution rate reduces the water absorption rate and improves the water resistance coefficient, but a higher foam volume (20%~60%) increases the water absorption rate and reduces the water resistance coefficient. When the CO_2-cured GFA substitution rate is 100%, the water absorption rate is decreased by 10.4% and the water resistance coefficient is increased by 31.4%. When the foam volume is 60%, the water

absorption rate is decreased by 12.4% and the water resistance coefficient is increased by 43.2%.

(4) A higher CO_2-cured GFA substitution rate and lower foam volume are conducive to improving the solidification of heavy metals (Zn and Cr). When the CO_2-cured GFA substitution rate is 100%, the leached Zn and Cr are decreased by 11.7% and 9.3%. When the foam volume is 60%, the leached Zn and Cr are decreased by 27.7% and 22.8%.

(5) The foam concrete technology with internal addition of GFA proposed in this study significantly expands GFA resource utilization, though further research is still required in the future.

In conclusion, CO_2-cured garbage fly ash can significantly improve the strength of foam concrete and reduce the harmful substances exuded by foam concrete. Therefore, this treatment method can prepare environmentally friendly foam concrete and deal with harmful substances (garbage fly ash).

Author Contributions: Conceptualization, F.S. and Y.Y.; methodology, D.W.; software, Z.X.; validation, H.W., C.L., and Z.X.; formal analysis, X.T.; investigation, Y.Y.; resources, X.T.; data curation, N.X.; writing—original draft preparation, D.W.; writing—review and editing, D.W.; visualization, H.W.; supervision, K.X.; project administration, D.W.; funding acquisition, H.W. and D.W. All authors have read and agreed to the published version of the manuscript.

Funding: This research was funded by the 2023 Liaoning Provincial Science and Technology Programme Joint Programme Projects, grant number 2023JH2/101700001; the 2023 Liaoning Provincial Science and Technology Programme Joint Programme Projects, grant number 2023JH2/101700002; the 2024 Liaoning Provincial Department of Education Basic Research Project, grant number LJ212411779037; and the Collaborative Education Project of the Ministry of Education-Construction of Civil Engineering Materials Course Based on Job Guidance, grant number 2407134702.

Institutional Review Board Statement: Not applicable.

Informed Consent Statement: Not applicable.

Data Availability Statement: The data used to support the findings of this study are available on request.

Conflicts of Interest: The authors declare no conflicts of interest.

References

1. Li, Z.; Wu, Z.; Long, Q.; Feng, W.; Liu, S.; Yuan, X. Preparation of municipal waste incineration fly ash artificial aggregate using CO_2 curing and its properties. *J. CO2 Util.* **2025**, *93*, 103042. [CrossRef]
2. Suescum-Morales, D.; Silva, R.V.; Bravo, M.; Jiménez, J.R.; Fernández-Rodríguez, J.M.; Brito, J. Effect of incorporating municipal solid waste incinerated bottom ash in alkali-activated fly ash concrete subjected to accelerated CO_2 curing. *J. Clean. Prod.* **2022**, *370*, 133533. [CrossRef]
3. Alamri, M.; Ali, T.; Ahmed, H.; Qureshi, M.Z.; Elmagarhe, A.; Khan, M.A.; Ajwad, A.; Mahmood, M.S. Enhancing the engineering characteristics of sustainable recycled aggregate concrete using fly ash, metakaolin and silica fume. *Heliyon* **2024**, *10*, e29014. [CrossRef] [PubMed]
4. Lee, J.; Jeon, D.; Kim, S.; Lim, A.; Moon, J. Enhancement of fly ash reactivity through triethanolamine addition and CO_2 curing. *Cem. Concr. Compos.* **2025**, *157*, 105874. [CrossRef]
5. Wu, M.; Zhang, Y.; Liu, Z.; Liu, C.; She, W.; Wu, Z. Experimental study on eco-friendly one-part alkali-activated slag-fly ash-lime composites under CO_2 environment: Reaction mechanism and carbon capture capacity. *Constr. Build. Mater.* **2024**, *421*, 135779. [CrossRef]
6. Song, B.; Shi, C.; Hu, X.; Ouyang, K.; Ding, Y.; Ke, G. Effect of early CO_2 curing on the chloride transport and binding behaviors of fly ash-blended Portland cement. *Constr. Build. Mater.* **2021**, *288*, 123113. [CrossRef]
7. Chen, J.; Zhang, Z.; Shen, Y.; Li, H.; Lin, X.; Li, X.; Yan, J. A novel insight into CO_2-cured cement modified by ultrasonic carbonated waste incineration fly ash: Mechanical properties, carbon sequestration, and heavy metals immobilization. *Carbon Capture Sci. Technol.* **2025**, *14*, 100368. [CrossRef]

8. Wi, K.; Sahin, O.; Wang, K.; Lee, Y. Characterization and evaluation of cement-based systems containing solution-treated municipal solid-waste incineration fly ash. *Constr. Build. Mater.* **2024**, *416*, 135230. [CrossRef]
9. Gao, S.; Chu, H.; Shi, W.; Wang, F.; Jiang, J. Utilization of municipal solid waste incineration fly ash in ecological concrete and pavement bricks: Mechanical properties and environmental impact. *Case Stud. Constr. Mater.* **2024**, *21*, e03999. [CrossRef]
10. Fan, D.; Lu, J.; Lv, X.; Noguchi, T.; Yu, R.; Sun, C. Carbon capture and storage CO_2 foam concrete towards higher performance: Design, preparation and characteristics. *Cem. Concr. Compos.* **2025**, *157*, 105925. [CrossRef]
11. Xu, J.; Zhao, D.; Wang, S.; Chen, X.; Wu, X.; Han, Z.; Liu, Y. Quantitative analysis of pore structure's impact on early mechanical properties and durability of foam concrete: Macroscopic and mesoscopic insights. *Constr. Build. Mater.* **2025**, *459*, 139775. [CrossRef]
12. Huang, X.; Yu, H.; Wang, H.; Xie, X.; Qi, C.; Xue, F.; Zhao, X. Recycle sludge incineration ash for efficient preparation of foam concrete: Performance, microstructure, and mechanisms. *Next Sustain.* **2025**, *6*, 100125.
13. Liao, S.; Lu, W.; Chen, Y.; Shen, Y. Production of autoclaved aerated concrete by using municipal solid waste incinerator fly ash as an alternative raw material. *Case Stud. Constr. Mater.* **2024**, *21*, e03914. [CrossRef]
14. Bayraktar, O.Y.; Özel, H.B.; Benli, A.; Yılmazoğlu, M.; Türkel, İ.; Dal, B.; Şevik, H.; Kaplan, G. Sustainable foam concrete development: Enhancing durability and performance through pine cone powder and fly ash incorporation in alkali-activated geopolymers. *Constr. Build. Mater.* **2024**, *457*, 139422. [CrossRef]
15. Xue, Q.; Zhang, L.; Mei, K.; Li, X.; Wang, Y.; Cheng, X.; Fu, X. Thermal conductivity and pore structure analysis of alkali-activated foam cement with supercritical CO_2 modified slag: Feasibility evaluation for geothermal applications. *Constr. Build. Mater.* **2022**, *347*, 128506. [CrossRef]
16. Ruan, M.; Tian, Q.; Zhang, M.; Wang, C.; Xu, G.; Cai, J.; Fu, Q. Fabrication and characterization of foam ceramics from recycled waste concrete powder and waste glass powder. *Constr. Build. Mater.* **2024**, *451*, 138807. [CrossRef]
17. Messina, F.; Ferone, C.; Colangelo, F.; Roviello, G.; Cioffi, R. Alkali activated waste fly ash as sustainable composite: Influence of curing and pozzolanic admixtures on the early-age physico-mechanical properties and residual strength after exposure at elevated temperature. *Compos. Part B Eng.* **2018**, *132*, 161–169. [CrossRef]
18. Dong, K.; Chen, S.; Liu, J.; Shi, X.; Zhang, J.; Meng, J. Effect of nano- and micron-materials on the thermal properties behavior in wet environments and heat transfer mechanism of foam concrete. *J. Build. Eng.* **2025**, *107*, 112725. [CrossRef]
19. Kim, J.; Kim, T.; Park, J.; Kang, J. Durability evaluation of structural performance based on CO_2 curing conditions. *J. Build. Eng.* **2025**, *105*, 112465. [CrossRef]
20. Du, Z.; Liu, P.; Wang, P.; Jiang, S.; Shi, Y.; Sun, M.; Gao, M.; Fu, H. CSH/PCE nanoparticle as a novel foam stabilizer: Highly efficient interfacial adsorption and enhanced performance in foam concrete. *Constr. Build. Mater.* **2025**, *469*, 140509. [CrossRef]
21. Song, Q.; Zou, Y.; Zhang, P.; Xu, S.; Yang, Y.; Bao, J.; Xue, S.; Liu, J.; Gao, S.; Lin, L. Novel high-efficiency solid particle foam stabilizer: Effects of modified fly ash on foam properties and foam concrete. *Cem. Concr. Compos.* **2025**, *155*, 105818. [CrossRef]
22. Gao, S.; Zhao, S.; Yang, L.; Wang, Y.; Guan, X.; Zhang, H.; Luo, S. Synergistic effects of fly ash-cement slurry and CO_2 mineralization on coal gangue aggregate and its concrete properties. *Constr. Build. Mater.* **2025**, *465*, 140225. [CrossRef]
23. Akbulut, Z.F.; Guler, S. Structural self-compacting lightweight concrete: Effects of fly ash and basalt fibers on workability, thermal and mechanical properties under ambient conditions and high temperatures. *Constr. Build. Mater.* **2025**, *481*, 141658. [CrossRef]
24. Wang, Y.; Lu, B.; Hou, G. Preparation of a novel foam concrete for the carbonation of steel slag with low thermal conductivity and enhanced CO_2 capture. *Mater. Lett.* **2025**, *394*, 138658. [CrossRef]
25. Wu, C.; Jierula, A.; Fu, Z.; Chen, Y.; Kuang, Y.; Xia, J. Study on the mechanism of anti sulfate freeze-thaw, dry wet cycle of polypropylene fiber foam concrete. *Constr. Build. Mater.* **2025**, *463*, 140144. [CrossRef]
26. Obaidat, Y.T.; Hijazi, Y.F. Performance of bond between carbon fiber-reinforced polymer sheets and thermally shocked damaged foam concrete. *Int. J. Adhes. Adhes.* **2025**, *140*, 104042. [CrossRef]
27. Wen, C.; Li, J.; Meng, Q.; Yang, S.; Chen, Z.; Wang, X.; Li, Y.; Wang, X.; Sun, X.; Mu, L.; et al. Study of macro-/micro-properties and in-situ mineralization mechanism of CO_2 foam concrete based on solid waste-derived SAC-gypsum composite. *Constr. Build. Mater.* **2025**, *486*, 141911. [CrossRef]
28. Miah, M.J.; Pei, J.; Kim, H.; Jang, J.G. Flexural behavior, porosity, and water absorption of CO_2-cured amorphous metallic-fiber-reinforced belite-rich cement composites. *Constr. Build. Mater.* **2023**, *387*, 131668. [CrossRef]
29. Azevedo, A.G.S.; Baltazar, L.; Faria, P.; Savastano, H. Durability assessment of MgO-based fiber cement reinforced with lignocellulosic fibers and cured in a CO_2-rich atmosphere. *J. Build. Eng.* **2025**, *104*, 112327. [CrossRef]
30. Ser, M.; Hussin, A.E.; Said, A.; Kohail, M. Enhancing the sustainable production of Foamed Concrete by using different waste materials as a full sand replacement. *Constr. Build. Mater.* **2025**, *462*, 139890. [CrossRef]
31. *GB/T 50080-2016*; Standard for Test Method of Performance on Ordinary Fresh Concrete. Ministry of Housing and Urban-Rural Development of the People's Republic of China: Beijing, China, 2016.
32. *GB/T 50081-2019*; Standard for Test Methods of Concrete Physical and Mechanical Properties. Ministry of Housing and Urban-Rural Development of the People's Republic of China: Beijing, China, 2019.

33. Sun, X.; Zhong, J.; Zhang, W.; Li, G.; Cao, H.; Gao, P. Deformation and mechanical properties of foamed concrete under various components and pore structure. *J. Build. Eng.* **2024**, *94*, 109920. [CrossRef]
34. He, S.; Lei, X.; Zhang, H.; Yu, P.; Zhou, Y.; Huang, X. Durability and heavy metals long-term stability of alkali-activated sintered municipal solid waste incineration fly ash concrete in acidic environments. *Constr. Build. Mater.* **2025**, *462*, 139990. [CrossRef]
35. He, P.; Drissi, S.; Hu, X.; Liu, J.; Shi, C. Properties of CO_2-cured cement incorporating fly ash and slag subjected to further water curing. *Cem. Concr. Compos.* **2024**, *152*, 105633. [CrossRef]
36. Zhang, X.; Wu, K.; Yuan, Z.; Liu, S. Investigations into magnesium oxide carbon sequestration foamed concrete: Mechanical performance, microstructure and environmental benefits. *Constr. Build. Mater.* **2024**, *421*, 135679. [CrossRef]
37. Zhou, Z.; Xiao, J.; Ye, T.; Wang, J.; Choi, D. Influence of recycled concrete powder and CO_2 curing on the properties of thermal insulation mortars. *Constr. Build. Mater.* **2024**, *414*, 134907. [CrossRef]
38. Yılmazoğlu, M.U.; Kara, H.O.; Benli, A.; Demirkıran, A.R.; Bayraktar, O.Y.; Kaplan, G. Sustainable alkali-activated foam concrete with pumice aggregate: Effects of clinoptilolite zeolite and fly ash on strength, durability, and thermal performance. *Constr. Build. Mater.* **2025**, *464*, 140160. [CrossRef]
39. An, S.; Wang, B.; Chen, W. The cementitious properties of alkali-activated municipal solid waste incineration fly ash-phosphorus slag-secondary aluminum dross matrix composites and the mechanism of solidification of heavy metals. *Constr. Build. Mater.* **2024**, *438*, 137254.
40. Kamisetty, A.; Gandhi, I.S.R.; Kumar, A. Combined effect of fly ash and fiber on spreadability, strength and water permeability of foam concrete. *J. Build. Eng.* **2023**, *78*, 107607. [CrossRef]
41. Wang, T.; Gao, X.; Li, Y.; Liu, Y. An orthogonal experimental study on the influence of steam-curing on mechanical properties of foam concrete with fly ash. *Case Stud. Constr. Mater.* **2024**, *20*, e02665. [CrossRef]
42. Cao, X.; Jiang, T.; Shimada, H.; Sasaoka, T.; Hamanaka, A. Influence of CO_2 nano-bubble water concentration and curing time on the macroscopic and microscopic mechanical properties of cemented backfill materials. *J. Build. Eng.* **2024**, *98*, 111099. [CrossRef]
43. Liu, Y.; Li, C.; Zhai, H.; Ahmad, M.R.; Guo, D.; Dai, J. Production and performance of CO_2 modified foam concrete. *Constr. Build. Mater.* **2023**, *389*, 131671. [CrossRef]
44. Gołaszewska, M.; Gołaszewski, J.; Klemczak, B.; Koenders, E.A.B. Effect of foaming agent and admixtures on hydration in ultralight foam concrete with cement-metakaolin binders. *J. Build. Eng.* **2025**, *104*, 112365. [CrossRef]
45. Kam, E.; Yaraş, A.; Ustaoğlu, A.; Hekimoğlu, G.; Sarı, A.; Erdoğmuş, E.; Gencel, O. Properties of foam concrete containing phase change material impregnated pumice. *Constr. Build. Mater.* **2024**, *446*, 137915. [CrossRef]
46. Chen, B.; Liu, B.; Yang, L.; Zaland, S.; Ye, H. Solidification of heavy metal in municipal solid waste incineration fly ash and performance evolution of alkali-activated foam concrete. *Process Saf. Environ. Prot.* **2024**, *190 Pt A*, 850–862. [CrossRef]
47. Bersisa, A.; Moon, K.; Kim, G.M.; Cho, J.; Park, S. Microstructural characterization of CO_2-cured calcium silicate cement. *Dev. Built Environ.* **2024**, *19*, 100518. [CrossRef]
48. Wang, H.; Wen, B.; Xu, P.; Gao, G.; Zhang, L.; Niu, D. Effect of CO_2 curing on the strength and microstructure of composite waste glass concrete. *Constr. Build. Mater.* **2025**, *463*, 140042. [CrossRef]
49. Du, Z.; Liu, P.; Wang, P.; Gao, M.; Shi, Y.; Sun, M.; Jiang, S.; Cui, L.; Lv, S. Enhancing foam stability and performance in foam concrete using para-aramid nanofiber. *J. Build. Eng.* **2025**, *103*, 112139. [CrossRef]

coatings

MDPI

Article

The Properties of Self-Compacting Ultra-High Performance Concrete with Different Types of Mineral Admixtures

Lin Wang [1,*], Xiying Tian [1], Yuefan Pan [1], Dingyuan Wu [2], Shengli Xu [2], Hangyang Wang [3], Xiaolu Tian [2], Yubo Xu [1], Hong Guo [4] and Min Zou [5]

1 School of Civil and Transportation Engineering, Beijing University of Civil Engineering and Architecture, Beijing 100044, China; 2108590022101@stu.bucea.edu.cn (X.T.); panyuefan@aliyun.com (Y.P.)
2 The Fourth Engineering Co., Ltd., China Railway 16th Bureau Group, Beijing 101499, China; wudingyuan@aliyun.com (D.W.); xushengli@aliyun.com (S.X.)
3 School of Civil Engineering and Geographical Environment, Ningbo University, Ningbo 315000, China
4 Shanxi Huaxing Design and Testing Co., Ltd., China Railway 17th Bureau Group, Taiyuan 030012, China; guohong@aliyun.com
5 School of Civil and Resource Engineering, Beijing University of Science and Technology, Beijing 100083, China
* Correspondence: wanglin1@bucea.edu.cn

Abstract: This paper investigates the effects of silica fume, cenosphere, fly ash, and ground slag powder on the rheological properties and mechanical strengths of self-compacting ultra-high performance concrete (UHPC). The mass ratio of each mineral admixture varies from 0% to 15%, while the water-binder ratios are set at 0.18, 0.20, and 0.22. The slump flow and plastic viscosity of fresh UHPC are measured, and the corresponding flexural and compressive strengths of UHPC cured for 3 days and 28 days are determined. Scanning electron microscopy (SEM) and X-ray diffraction (XRD) are employed to elucidate the mechanisms underlying the observed performance changes. The results indicate that the addition of silica fume and mineral powder negatively impacts the fluidity of fresh UHPC but positively affects its plastic viscosity. Conversely, the inclusion of cenosphere and fly ash enhances the fluidity of fresh UHPC while having the opposite effect on its plastic viscosity. Increasing the water-binder ratio improves the fluidity of fresh UHPC but reduces its plastic viscosity. Mechanically, silica fume enhances the strengths of UHPC. In contrast, the cenosphere, fly ash, and mineral powder decrease the strengths of UHPC cured for 3 days but increase those cured for 28 days. UHPC containing silica fume exhibits the most compact hydration products and the lowest content of $Ca(OH)_2$.

Keywords: rheological properties; ultra-high performance concrete; mineral admixture; cenosphere; fly ash

1. Introduction

Recently, various new types of building have emerged, necessitating materials with high strength, excellent durability, and functional requirements [1,2]. It is difficult for ordinary concrete to meet the requirements of ultra-high strength, high workability, high durability, and irregular complex structures due to its high water/cement ratio, low strength, poor workability, and low compactness [3,4]. Consequently, studying ultra-high strength self-compacting concrete has become one of the development directions for addressing this issue.

Self-compacting concrete (SCC) is a high-performance concrete that can flow, fill, and compact structures without the need for vibration, relying solely on its own gravity [5,6].

SCC possesses a large amount of cementitious materials, a high sand ratio, and ultra-high fluidity, which can significantly improve the efficiency of pouring and pumping, especially for dense steel structures, irregular structures, and structures that are difficult to vibrate [7,8]. The rheological properties of self-compacting concrete (SCC) are closely associated with the effectiveness of building construction processes [9]. However, SCC characterized by excessively high fluidity often exhibits bleeding phenomena, which can compromise the material's performance. Therefore, it is of significant importance to formulate SCC that possesses high fluidity while simultaneously avoiding segregation and bleeding [10]. To achieve this goal, mineral admixtures such as silica fume, slag powder, and fly ash are commonly incorporated into SCC mixtures. These admixtures serve to enhance both the flowability and cohesion of SCC, thereby improving its self-compacting capability and effectively preventing bleeding [11]. The preparation of SCC typically requires a substantial amount of cementitious materials [12]. Notably, mineral admixtures can be utilized as cementitious components in SCC formulations. Studies have reported that ultra-fine silica fume powder, fly ash, and high-grade slag powder can increase the compressive strengths of SCC by up to 21.1%, 15.6%, and 17.8%, respectively [13]. Furthermore, ultra-fine silica fume powder and high-grade slag powder are effective in increasing the plastic viscosity and yield shear stress of fresh SCC, contributing to its improved stability and resistance to bleeding. In contrast, fly ash can enhance the fluidity of fresh SCC, facilitating its placement and compaction during construction.

Ultra-high performance concrete (UHPC) is a cement concrete with ultra-high mechanical strengths and excellent durability [14]. The macroscopic performance and microscopic mechanisms of UHPC have been reported by some researchers [15,16]. Silica fume, furnace ash, and steel slag powder have been proven to increase the flexural strength of UHPC by 17.3%, 18.1%, and 22.4%, respectively [17,18]. The corresponding compressive strengths have been increased by 25.6%, 27.3%, and 29.1% [19]. Moreover, the addition of garbage fly ash, furnace ash, and steel slag powder can increase resistance to chloride and sulfate erosion [20,21]. Furthermore, these cementitious materials are reported to improve the corrosion resistance of the inner steel bars in UHPC [22,23]. Although several studies on UHPC have been reported, the influence of various mineral admixtures on the performance of self-compacting UHPC has not been investigated [24,25]. In previous studies, ultra-high performance concrete (UHPC) often suffered from insufficient fluidity. Adding a large amount of water-reducing agent to UHPC can improve its fluidity, but it is prone to bleeding and segregation [26]. Based on these findings, mixing multiple mineral admixtures and a water-reducing agent may ensure the flowability and cohesiveness of freshly mixed self-compacting UHPC. In addition, mineral admixtures can promote the development of concrete's mechanical strengths, which are the most important performances of building structures [27,28]. However, few studies have conducted comprehensive comparative analyses on the rheological, mechanical, and microscopic properties of self-compacting UHPC after adding different combinations of multiple mineral admixtures.

In this study, the slump flow and plastic viscosity of fresh UHPC with mineral admixtures including silica fume, cenosphere, fly ash, and grade slag powder are measured. The flexural and compressive strengths are tested. Scanning electron microscopy (SEM) and X-ray diffraction (XRD) analysis are selected to reflect the mechanisms of performance changes. The mass ratios of silica fume, cenosphere, fly ash, and grade slag powder range from 0% to 15%. This research will provide a reference for the application of different mineral admixtures in UHPC and their impact on UHPC performance.

2. Materials and Methods

2.1. Raw Materials

The Dunshi P-I type 52.5 Portland cement produced by Tangshan Jidong Cement Co., Ltd., Tangshan, China was used in this study. The initial setting time and the final setting time of the Portland cement are 130 min and 185 min, respectively, with a specific surface area of 350 m^2/kg. In addition, the class F fly ash according to the standard ASTM procedures [29] and S105 grade slag powder produced by Hebei Weiran Building Materials Technology Co., Ltd., Langfang, China, silica fume offered by Shandong Boken Silica Material Co., Ltd., Rizhao, China, and cenosphere provided by Lingshou Jiagong Mineral Products Co., Ltd., Lingshou, China were used in this study. The coarse aggregate was of sizes (5–10) mm. Polycarboxylic acid, highly adaptable water-reducing agent produced by Jiangsu Subot New Material Co., Ltd., Nanjing, China, was used for adjusting the fresh UHPC's fluidity. Tables 1 and 2 show the chemical composition and the basic performance index of the raw materials.

Table 1. Chemical composition of the raw materials (%).

Types	Chemical Composition (%)							
	SiO_2	Al_2O_3	Fe_2O_3	MgO	CaO	SO_3	K_2O	Na_2O
Portland cement	21.7	4.76	3.57	2.12	63.01	1.98	2.15	0.71
Fly Ash	55.76	32.86	3.65	0.34	2.44	0.72	0.98	3.25
Grade Slag Powder	27.74	14.84	0.89	3.24	48.45	2.49	0.46	1.89
Silica fume	95.6	0.41	0.42	0.40	0.12	-	-	3.05
Cenosphere	95.3	0.44	0.41	0.41	0.11	-	-	3.06
Quartz sand	98.5	0.6	0.5	0.1	0.1	-	-	0.2

Table 2. Basic performance index of the raw materials.

Types	Fineness (45 μm%)	Density (g/cm^3)	Water Demand Ratio (%)	Specific Surface Area (m^2/kg)	Heat Loss (%)	Mobility Ratio (%)	28 d Activity Index (%)
Portland cement	6.7	3.14	25.40	350	-	-	-
Fly ash	5.5	2.21	92	560	2.9	-	-
Grade slag powder	3.47	2.85	-	450	-	86	-
Silica fume	-	2.05	110	22,000	5.10	-	125
Cenosphere	-	2.23	113	215	4.03	-	73
Quartz sand	3.4	2.65	-	430	0.5	-	-

2.2. Specimen Preparation

The specimens were prepared by the following steps. Firstly, the cement, quartz sand, and other cementitious materials were poured into the UJZ-15 mortar mixer and mixed for 30 s. Then, water mixed with the water reducer was added to the mixer, and stirring was provided for 8 min. Some fresh UHPC was used to measure the slump flow and plastic viscosity using the self-compacting concrete stability jumping table cement–mortar flowability tester, manufactured by the Yuzhan Instrument Equipment (Cangzhou) Co., Ltd., Cangzhou, China, and the Rongjida concrete rheometer, offered by Shanghai Rongjida Instrument Technology Co., Ltd., Shanghai, China. The remaining sample was poured into molds to prepare the specimens. After the samples hardened, the specimens were moved to the standard curing room (temperature of 21.1 °C, humidity of 97.8%) for standard curing. Table 3 shows the mixing proportions of UHPC per one cubic meter. The mixing proportions were obtained based on the maximum density theory obtained from prior research [30]. The mass ratio of each mineral admixture by the total mass of binder materials ranging from 0% to 15% (corresponding to the mass ranging from 0 kg to 135 kg in Table 3) is calculated from Table 3. The quartz sand including the particle sizes of 0.73 mm~1.1 mm, 0.36 mm~0.58 mm, and 0.14 mm~0.295 mm with a mass ratio of 1:1.5:1 was used as the aggregate of the UHPC.

Table 3. Mixing proportions of UHPC per one cubic meter (kg).

Water	Cement	Silica Fume	Cenosphere	Fly Ash	Grade Slag Powder	Sand	Water Reducing Agent	Water-Binder Ratio
162	900	0	0	0	0	1350	18	0.18
162	855	45	0	0	0	1350	18	0.18
162	810	90	0	0	0	1350	18	0.18
162	765	135	0	0	0	1350	18	0.18
162	855	0	45	0	0	1350	18	0.18
162	810	0	90	0	0	1350	18	0.18
162	765	0	135	0	0	1350	18	0.18
162	855	0	0	45	0	1350	18	0.18
162	810	0	0	90	0	1350	18	0.18
162	765	0	0	135	0	1350	18	0.18
162	855	0	0	0	45	1350	18	0.18
162	810	0	0	0	90	1350	18	0.18
162	765	0	0	0	135	1350	18	0.18
180	900	0	0	0	0	1350	18	0.20
180	855	45	0	0	0	1350	18	0.20
180	810	90	0	0	0	1350	18	0.20
180	765	135	0	0	0	1350	18	0.20
180	855	0	45	0	0	1350	18	0.20
180	810	0	90	0	0	1350	18	0.20
180	765	0	135	0	0	1350	18	0.20
180	855	0	0	45	0	1350	18	0.20
180	810	0	0	90	0	1350	18	0.20
180	765	0	0	135	0	1350	18	0.20
180	855	0	0	0	45	1350	18	0.20
180	810	0	0	0	90	1350	18	0.20
180	765	0	0	0	135	1350	18	0.20
198	900	0	0	0	0	1350	18	0.22
198	855	45	0	0	0	1350	18	0.22
198	810	90	0	0	0	1350	18	0.22
198	765	135	0	0	0	1350	18	0.22
198	855	0	45	0	0	1350	18	0.22
198	810	0	90	0	0	1350	18	0.22
198	765	0	135	0	0	1350	18	0.22
198	855	0	0	45	0	1350	18	0.22
198	810	0	0	90	0	1350	18	0.22
198	765	0	0	135	0	1350	18	0.22
198	855	0	0	0	45	1350	18	0.22
198	810	0	0	0	90	1350	18	0.22
198	765	0	0	0	135	1350	18	0.22

2.3. Measuring Methods

2.3.1. The Measurement of Rheological Parameters

The self-compacting concrete stability jumping table cement–mortar flowability tester, manufactured by the Yuzhan Instrument Equipment (Cangzhou) Co., Ltd., Cangzhou, China, and the Rongjida concrete rheometer, offered by Shanghai Rongjida Instrument Technology Co., Ltd., Shanghai, China, were used for measuring the slump flow and plastic viscosity of fresh UHPC. For testing the slump flow, fresh UHPC was first poured into the truncated cone mold and scraped flat with a scraper. Then, the mold was lifted vertically upwards, and at the same time, the fresh UHPC flowed onto the glass plate for 30 s. The maximum circle diameter and the corresponding perpendicular diameter were measured.

The average value of these diameters was the slump flow of the fresh UHPC. The rotor of the rheometer was placed in a beaker with a diameter of 70 mm. Then, the rotor was inserted into the fresh UHPC, and the plastic viscosity of the fresh UHPC was obtained.

2.3.2. Mechanical Strengths

Specimens with dimensions of 40 mm × 40 mm × 160 mm were used for determining the flexural and compressive strengths after curing for 3 days and 28 days. First, the specimens were moved to the flexural strength testing fixture. Then, a flexural load was applied at a loading rate of 0.05 kN/s until the specimens failed (i.e., were destroyed). The broken specimens resulting from the flexural load were used for measuring the compressive strengths. The specimens were then moved to the compressive strength testing fixture, where a compressive load was applied at a loading rate of 2.4 kN/s until the specimens failed. Six specimens were applied in the measurement of each group. In this study, the preparation and measurements of UHPC were conducted according to the Chinese standard GB/T 31387-2025 [31].

2.3.3. Experiments of Microscopic Properties

The ultra-high strength self-compacting concrete samples were soaked in anhydrous ethanol to terminate hydration. Subsequently, the samples were dried at 60 ± 5 °C. In the scanning electron microscope (SEM) experiment, a Czech TESCAN MIRA LMS microscope, with a resolution of 3.5 nm, an acceleration voltage range of 500–30,000 V, a magnification range of 18–30,000×, and a sample stage diameter of 30 mm, was used to observe the microstructure. After being sprayed with gold, the flat, rice-sized samples were placed in a vacuum environment for observation. An Ultima IV X-ray diffractometer from Bruker, Germany, was used in the XRD experiment. The samples were pulverized through a 0.075 mm sieve. The internal standard method was employed, with 10% ZnO as the marker. The scanning method was continuous slow scanning, with a working voltage of 40 kV, a current of 40 mA, a scanning speed of 0.6°/min, and a scanning range of 5–70°. After the experiment was completed, XRD patterns were obtained. The measuring process is shown in Figure 1.

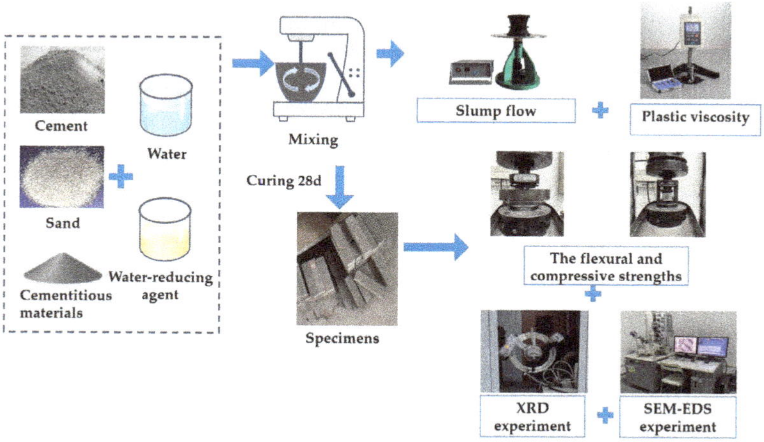

Figure 1. The measuring process.

3. Results and Discussions

3.1. The Rheological Parameters

Figure 2 illustrates the slump flow and plastic viscosity of fresh UHPC with a varying mass ratio of silica fume. The water-to-binder ratios of UHPC are 0.18, 0.20, and 0.22. As depicted in Figure 2, the slump flow of fresh UHPC decreases from 240 mm, 225 mm, and 200 mm to 220 mm, 190 mm, and 170 mm, respectively, for water-to-binder ratios of 0.22, 0.20, and 0.18. The corresponding decreasing rates are 8.3%~15%. Silica fume particles are characterized by ultra-fine, amorphous, spherical glass-like structures. The small size (0.1–0.2 μm) and large specific surface area of silica fume increase the corresponding water demand [32,33]. Concurrently, a notable decline in fluidity occurs. Meanwhile, the plastic viscosity of fresh UHPC increases from 1987.2 mPa·s, 3500 mPa·s, and 5076.3 mPa·s to 3182.0 mPa·s, 5963.8 mPa·s, and 8233.9 mPa·s, respectively, for water-to-binder ratios of 0.22, 0.20, and 0.18. The plastic viscosity's increasing rates are 60.1%~62.2%. As reported in prior research, the plastic viscosity and slump flow exhibit an anti-correlation [34,35]. Therefore, the decreased slump flow leads to an increased plastic viscosity. As observed in Figure 2, a higher water-to-binder ratio results in higher slump flow and lower plastic viscosity. This is due to the fact that a higher water-to-binder ratio provides more free water in the fresh UHPC, thereby increasing the slump flow and decreasing the corresponding plastic viscosity [36,37].

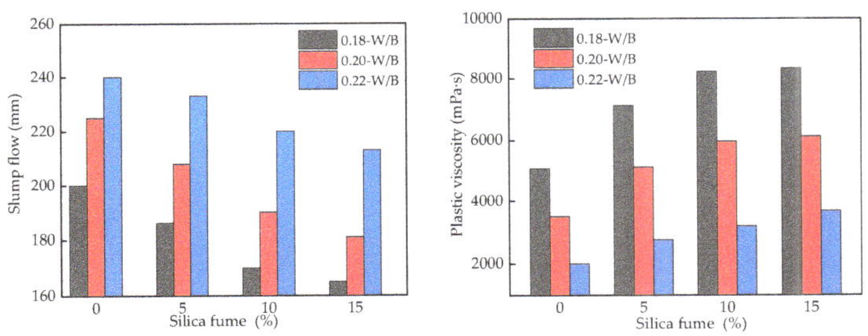

(**a**) The slump flow (**b**) The plastic viscosity

Figure 2. The rheological parameters of fresh UHPC with silica fume.

Figure 3 shows the influence of cenosphere content on the slump flow and the plastic viscosity of fresh UHPC. The water to binder ratios are 0.18, 0.20, and 0.22. In Figure 3, it can be seen that the fresh UHPC increases with the added mass ratio of cenosphere. The fresh UHPC's slump flow is increased from 200 mm, 225 mm, and 240 mm to 210 mm, 237 mm, and 248 mm, showing the increasing rates of 3.33%~5%. This can be ascribed to the fact that the microspheres consist of spherical particles which act as lubricants, thereby increasing the fresh UHPC's flowability [38,39]. Furthermore, an increase in the water-to-binder ratio also enhances the slump flow of fresh UHPC. This is due to the higher availability of free water in the mixture, which promotes greater fluidity. Meanwhile, the fresh UHPC's plastic viscosity is decreased with the increased mass ratio of cenosphere by rates of 16.6%~36.8%. This can be explained by the fact that the relationship between the fresh UHPC's slump flow and the plastic viscosity complies with the anti-correlation relationship [40,41]. Consequently, both the addition of cenospheres and an elevated water-to-binder ratio contribute to a decrease in the plastic viscosity of fresh UHPC. In Figures 2 and 3, the slump flow of each group is higher than 160 mm and the corresponding plastic viscosity is higher than 1000 mPa·s. Fresh UHPC with the slump flow and plastic viscosity has a self-compacting effect without bleeding [42].

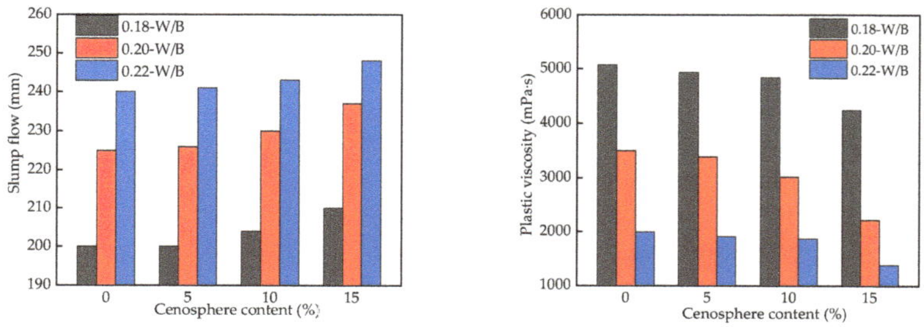

(**a**) The slump flow (**b**) The plastic viscosity

Figure 3. The rheological parameters of fresh UHPC with cenosphere.

The fresh UHPC's slump flow and plastic viscosity are shown in Figure 4. As observed in Figure 4, the slump flow increases by the decreasing rates of 1.7%~2.5% and the plastic viscosity decreases with the mass ratio of fly ash with the increasing rates of 9.7%~18.3%. The spherical shape of fly ash particles contributes to the enhanced fluidity of fresh UHPC upon the addition of fly ash [43]. However, when the fly ash content increases from 0% to 15%, the observed decrease in plastic viscosity is a result of the addition of fly ash itself. This behavior aligns with the inverse correlation between fluidity and plastic viscosity in fresh UHPC [44]. Consequently, the plastic viscosity decreases as the fly ash content increases. Additionally, as illustrated in Figure 4, an increase in the water-to-binder ratio can lead to an increase in slump flow and a decrease in plastic viscosity. This is attributed to the higher availability of free water resulting from an increased water-to-binder ratio [45].

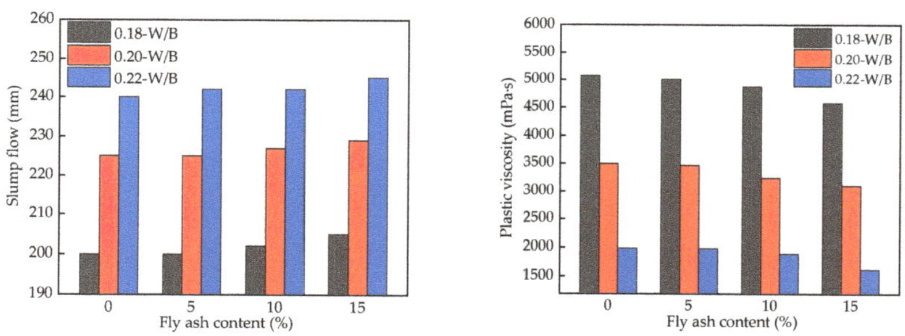

(**a**) The slump flow (**b**) The plastic viscosity

Figure 4. The rheological parameters of fresh UHPC with fly ash.

Figure 5 shows the slump flow and plastic viscosity of fresh UHPC. As illustrated in Figure 5, the slump flow decreases with the added mineral powder and the corresponding plastic viscosity is increased by adding the mineral powder. When the mass ratio of mineral powder increases from 0% to 15%, the slump flow decreases from 200 mm, 225 mm and 240 mm to 180 mm, 202 mm, and 225 mm, showing the decreasing rates of 6.25%~10%. The plastic viscosity of the samples increased from 5076.3 mPa·s, 3500 mPa·s, and 1987.2 mPa·s to 6055.8 mPa·s, 4432.8 mPa·s, and 2633.5 mPa·s, showing the increasing rates of 19.29%~32.52%. This can be attributed to the larger specific surface area of mineral powder thus absorbing more free water and decreasing the slump flow and the plastic viscosity [46]. In Figures 2–5, it can be seen that the fluidity of fresh UHPC shows in the order of UHPC with cenosphere > UHPC with fly ash > UHPC with mineral powder > UHPC with silica fume. Meanwhile, the results of the fresh UHPC's plastic viscosity is contrary to that of fluidity. This can be ascribed to the fact that the specific surface areas of the mineral admixtures exhibit in the order of cenosphere < fly ash < mineral powder < silica fume; therefore, the slump flow of fresh UHPC shows in this order.

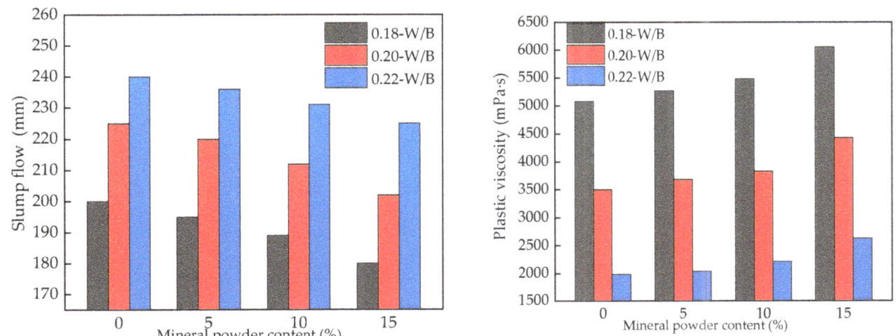

(**a**) The slump flow (**b**) The plastic viscosity

Figure 5. The rheological parameters of fresh UHPC with mineral powder.

3.2. The Mechanical Strengths

Tables A1 and A2 show the results of the mechanical strengths. The flexural and compressive strengths of UHPC with silica fume are shown in Figure 6. As depicted in Figure 6, the UHPC's flexural and compressive strengths increase with the increasing dosages of the mass ratio of silica fume. The flexural strength of the samples with the water-binder ratios of 0.18, 0.20, and 0.22 cured for 3 days increased from 12.8 MPa, 11.6 MPa, and 11.2 MPa to 13.3 MPa, 12.5 MPa, and 12.1 MPa, showing the increasing rates of 3.9%~8.0%. Meanwhile, the corresponding compressive strengths are increased from 59.2 MPa, 55.4 MPa, and 52.1 MPa to 64.3 MPa, 60.2 MPa, and 58.0 MPa with the increasing rates of 8.6%~11.3%. When the specimens are cured for 28 days, the flexural strengths increase from 16.0 MPa, 15.8 MPa, and 15.0 MPa to 17.1 MPa, 16.6 MPa, and 15.9 MPa with the increasing rates of 6%~6.9%. Meanwhile, the corresponding compressive strengths are increased from 110.1 MPa, 101.0 MPa, and 95.6 MPa to 119.5 MPa, 109.1 MPa, and 103.8 MPa, showing the increasing rates of 8.5%~8.6%. This can be ascribed to the fact that the silica fume contains the amount of nanoscale SiO_2 exceeding 90%, leading to an improvement in the corresponding secondary hydration and an increase in the corresponding mechanical strengths [47]. The mechanical strengths are increased by the decreased water/cement ratios. This can be ascribed to the fact that UHPC with lower water/cement ratio shows lower porosity [48]. As reported from prior research studies, the lower porosity leads to higher mechanical strengths. Therefore, the mechanical strengths are increased by the decreased water/cement ratios.

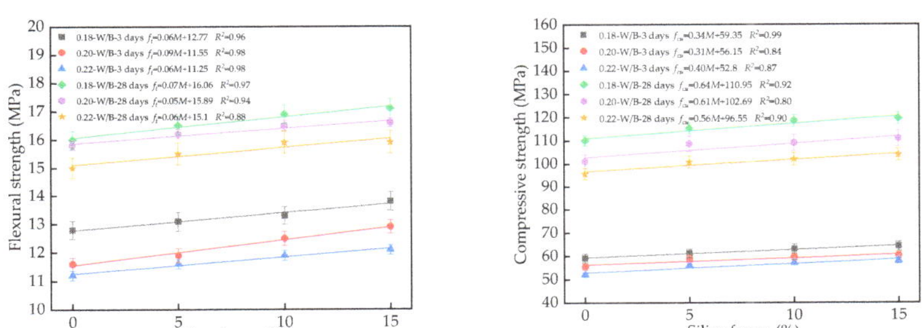

(**a**) Flexural strength (**b**) Compressive strength

Figure 6. The mechanical strengths of UHPC with silica fume.

The UHPC's mechanical strengths with cenosphere are exhibited in Figure 7. In Figure 7, it can be seen that the flexural and compressive strengths with the curing ages of 3 days decrease by rates of 5.3%~7.8% and 4.0%~8.9% with the increasing dosages of cenosphere. This can be attributed to the fact that the added cenosphere can delay the early cement hydration of the UHPC [49]. Therefore, the mechanical strengths are decreased by the increasing mass ratio of cenosphere. Meanwhile, when the specimens are cured for 28 days, the flexural and compressive strengths are increased with the increasing rates of 3.8%~5.6% and 4.3%~5.0% by the addition of cenosphere, due to the decreased cracking induced by the hydration heat [50]. Moreover, as depicted in Figure 8, the mechanical strengths are increased by higher curing age. This can be ascribed to the increased hydration degree induced by the increased curing age [51].

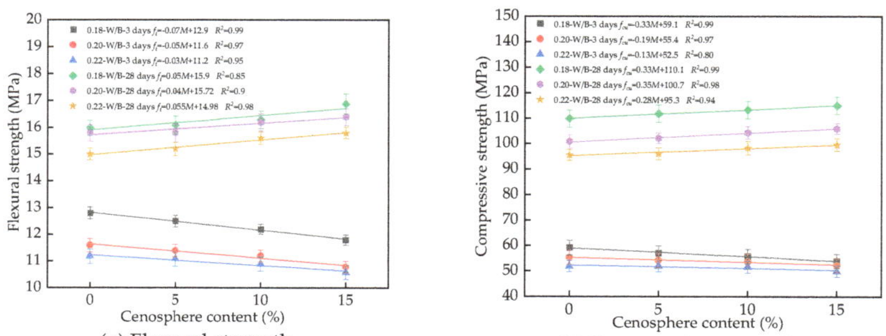

(**a**) Flexural strength (**b**) Compressive strength

Figure 7. The mechanical strengths of UHPC with cenosphere.

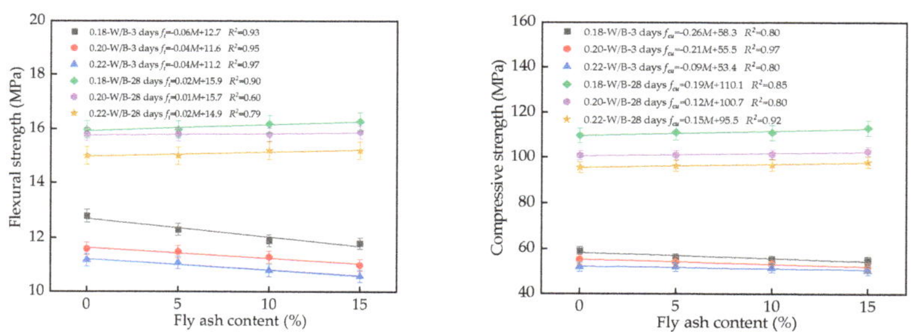

(**a**) Flexural strength (**b**) Compressive strength

Figure 8. The mechanical strengths of UHPC with fly ash.

The mechanical strengths of UHPC with fly ash are exhibited in Figure 8. It can be seen in Figure 8 that the UHPC's mechanical strengths cured for 3 days are increased by the increasing dosages of fly ash. Meanwhile, when the specimens are cured for 28 days, the addition of fly ash demonstrates a positive effect on the mechanical strengths. The decreased water/binder ratio shows an increasing effect on the mechanical strengths of the UHPC. The flexural strength of the specimens containing 15% fly ash decreased from 12.8 MPa, 11.6 MPa, and 11.2 MPa to 11.8 MPa, 11.1 MPa, and 10.6 MPa with the decreasing rates of 5.4%~7.9%. The compressive strengths of the specimens decline from 59.2 MPa, 55.4 MPa, and 52.1 MPa to 55.0 MPa, 52.0 MPa, and 50.6 MPa, showing the decreasing rates of 2.9%~7.1%. Meanwhile, the flexural strength of the specimens at the curing age of 28 days increase from 16.0 MPa, 15.8 MPa, and 15.0 MPa to 16.3 MPa, 15.9 MPa, and 15.2 MPa with the increasing rates of 1.3%~1.9%. Meanwhile, the corresponding compressive strengths increase from 110.1 MPa, 101.0 MPa, and 95.6 MPa to 113.4 MPa, 102.9 MPa, and 98.0 MPa, showing the increasing rates of 2.5%~3%. The reason for the mechanical strengths is similar to the above research results.

The mechanical strengths of UHPC with different dosages of mineral powder are shown in Figure 9. As depicted in Figure 9, the mineral powder demonstrates a negative effect on the UHPC's mechanical strengths when the curing time is 3 days. However, when the curing time is 28 days, the results are the opposite. When the curing time is 3 days, the added mineral powder can decrease the flexural strengths from 12.8 MPa, 11.6 MPa, and 11.2 MPa to 12.1 MPa, 11.4 MPa, and 10.7 MPa with the decreasing rates of 4.5%~5.5%. The compressive strengths decrease from 59.2 MPa, 55.4 MPa, and 52.1 MPa to 57.3 MPa, 54.0 MPa, and 51.4 MPa, showing the decreasing rates of 1.3%~3.2%. Meanwhile, when the specimens are cured for 28 days, the flexural strengths increase from 16.0 MPa, 15.8 MPa, and 15.0 MPa to 16.5 MPa, 16.2 MPa, and 15.6 MPa with the increasing rates of 3.1%~4%. The compressive strengths are increased from 110.1 MPa, 101.0 MPa, and 95.6 MPa to 115.1 MPa, 105.0 MPa, and 98.9 MPa, showing the increasing rates of 3.4%~4.5% by adding the mineral powder. Based on the results of the mechanical strength tests, it has been observed that the relationships between the mechanical strengths and the mass ratios of the mineral admixture adhere to linear functions. The fitting degrees (R^2 values) of these equations exceed 0.8, thereby ensuring the accuracy and reliability of the fitting results. The strengths' error bars are less than 10% of the average values, ensuring the experimental accuracy.

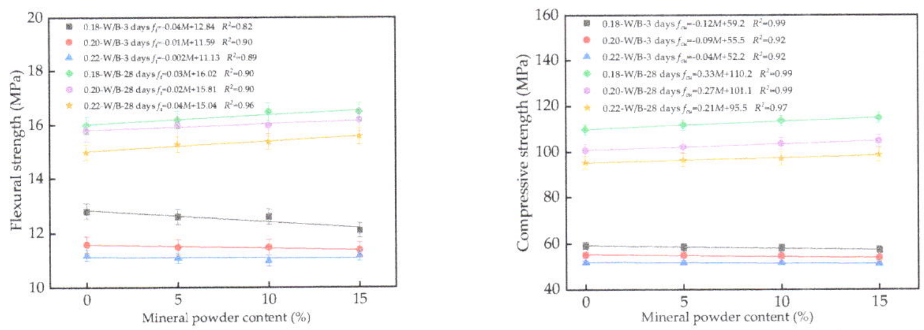

(**a**) Flexural strength (**b**) Compressive strength

Figure 9. The mechanical strengths of UHPC with mineral powder.

3.3. The Scanning Electron Microscope and X-Ray Diffraction Spectrum

Figure 10 presents the scan electron microscopy energy spectrum analysis (SEM-EDS) images of UHPC samples incorporating 10% silica fume (SF), 10% cenosphere, 10% fly ash (FA), and 10% mineral powder (MP). All specimens were cured for 28 days. From the morphology and element distribution of measuring points in Figure 10, it can be seen that hexagonal and needle-like substances, which are possibly hydroxide ($Ca(OH)_2$) and ettringite (AFt), are observed. The Ca/Si ratio contained in the hydration product enclosed by a rectangular frame ranges from 2.14 to 3.43, which are highly likely to be calcium silicate hydrate (C-S-H). As depicted in Figure 10, the hydration products identified include calcium, and calcium silicate hydrate (C-S-H). Notably, the UHPC sample containing cenosphere exhibits the highest content of $Ca(OH)_2$, whereas the UHPC sample with silica fume demonstrates the lowest $Ca(OH)_2$ content. Furthermore, the UHPC sample incorporating silica fume displays the most densely packed hydration products.

The X-ray diffraction spectrum of UHPC containing different mineral admixtures is presented in Figure 11. The samples, incorporating 10% silica fume (SF), 10% cenosphere, 10% fly ash (FA), and 10% mineral powder (MP), were Ca(OH) cured for 28 days. As shown, diffraction peaks corresponding to calcium hydroxide (C_2), tricalcium silicate (C_3S), dicalcium silicate (C_2S), calcium silicate hydrate (C-S-H), and ettringite (AFt) are observed.

The UHPC with SF exhibits the lowest intensity of $Ca(OH)_2$ diffraction peaks. This is attributed to the highest degree of cement secondary hydration in the UHPC with SF, which consumes the most $Ca(OH)_2$. These findings further confirm that UHPC with SF demonstrates the highest mechanical strengths.

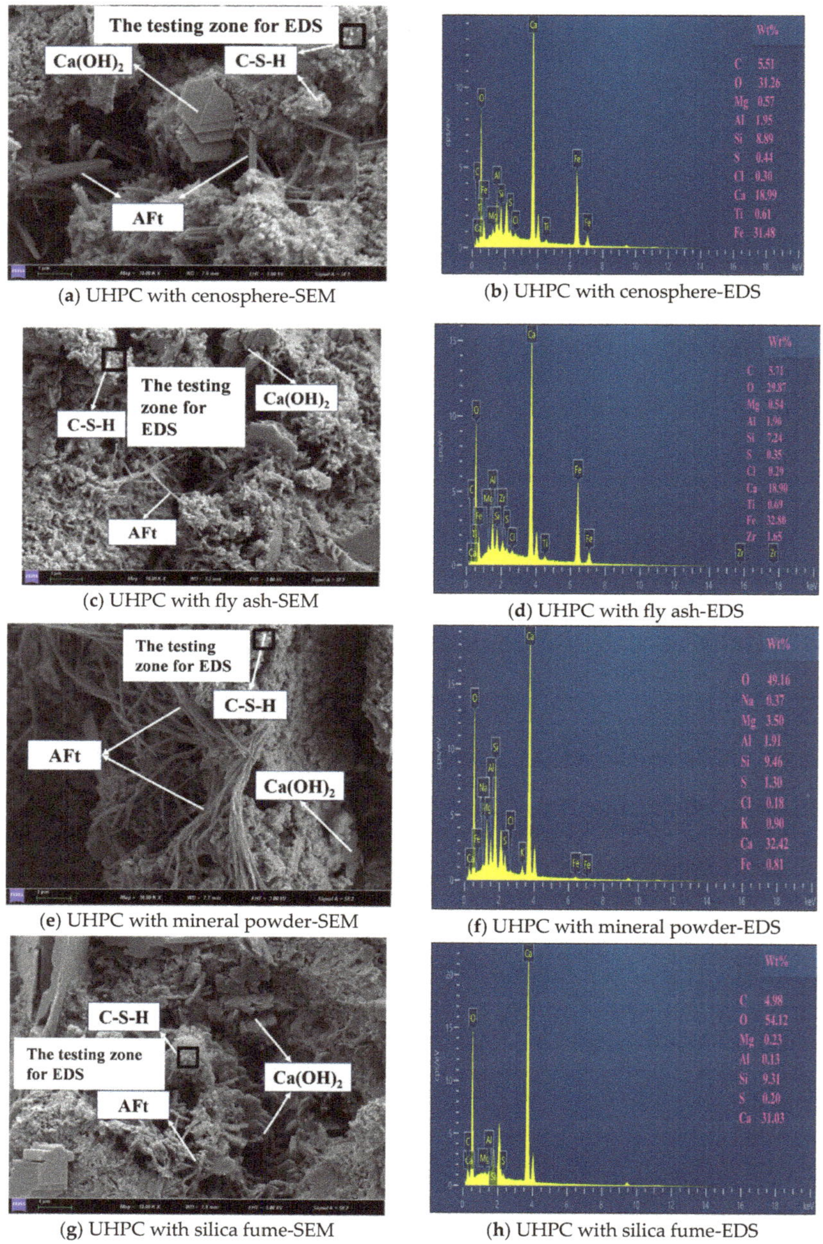

(**a**) UHPC with cenosphere-SEM (**b**) UHPC with cenosphere-EDS

(**c**) UHPC with fly ash-SEM (**d**) UHPC with fly ash-EDS

(**e**) UHPC with mineral powder-SEM (**f**) UHPC with mineral powder-EDS

(**g**) UHPC with silica fume-SEM (**h**) UHPC with silica fume-EDS

Figure 10. The scanning electron microscope photos of UHPC.

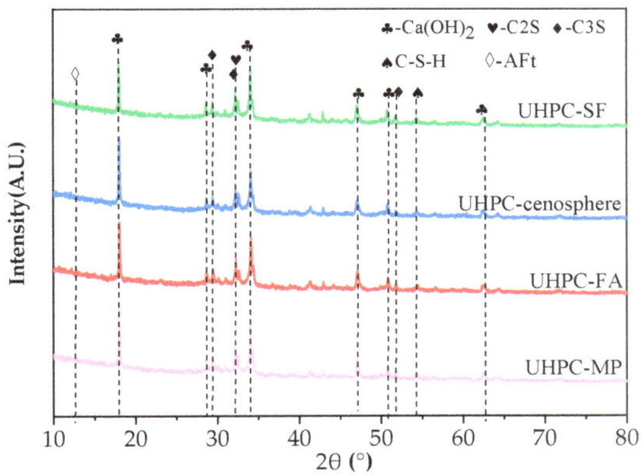

Figure 11. The X-ray diffraction spectrum of UHPC.

4. Conclusions

The influence of several mineral admixtures on the performance of self-compacting ultra-high performance concrete (UHPC) was investigated. The conclusions obtained are as follows.

The silica fume and mineral powder can decrease the fresh UHPC's slump flow by rates of 8.3%~15% and 6.25%~10%. Meanwhile, the slump flow is increased by the cenosphere and fly ash with rates of 3.33%~5% and 1.7%~2.5%. The fresh UHPC's plastic viscosity can be increased by rates of 60.1%~62.2% and 19.29%~32.52% with the added silica fume and mineral powder. Meanwhile, the cenosphere and fly ash decrease the plastic viscosity by rates of 16.6%~36.8% and 9.7%~18.3%. An increased water-to-binder (w/b) ratio enhances fluidity but negatively impacts plastic viscosity.

The silica fume can increase the flexural strength and compressive strength by rates of 3.9%~8.0% and 8.6%~11.3%. When the curing age is 3 days, the cenosphere, the fly ash, and the mineral powder decrease the flexural strengths by rates of 5.3%~7.8%, 5.4%~7.9%, and 4.5%~5.5%. The corresponding compressive strengths are decreased by rates of 4.0%~8.9%, 2.9%~7.1%, and 1.3%~3.2%. Meanwhile, when the UHPC is cured for 28 days, the cenosphere, the fly ash, and the mineral powder increase the flexural strengths by rates of 3.8%~5.6%, 1.3%~1.9%, and 3.1%~4%. The corresponding compressive strength is increased by rates of 4.3%~5.0%, 2.5%~3%, and 3.4%~4.5%.

The scanning electron microscope (SEM) images reveal that the UHPC containing SF exhibits the densest hydration products and the lowest calcium hydroxide ($Ca(OH)_2$) content.

Author Contributions: Methodology, L.W.; Validation, L.W. and Y.P.; Formal analysis, S.X. and H.W.; Investigation, H.W.; Resources, L.W., Y.P., D.W., S.X. and M.Z.; Data curation, X.T. (Xiying Tian), X.T. (Xiaolu Tian) and H.G.; Writing – original draft, L.W. and D.W.; Writing – review & editing, Y.P. and X.T. (Xiaolu Tian); Visualization, Y.X.; Supervision, Y.X. All authors have read and agreed to the published version of the manuscript.

Funding: This research was supported by Shanxi Province Key R&D Program horizontal project of China: Beijing Jianzhu University and China Railway No.16 Bureau (202102150401018, H24333).

Institutional Review Board Statement: Not applicable.

Informed Consent Statement: Not applicable.

Data Availability Statement: The original contributions presented in this study are included in the article. Further inquiries can be directed to the corresponding author.

Conflicts of Interest: Dingyuan Wu, Shengli Xu and Xiaolu Tian are employed by the company The Fourth Engineering Co., Ltd., Hong Guo is employed by the company Shanxi Huaxing Design and Testing Co., Ltd. The remaining authors declare that the research was conducted in the absence of any commercial or financial relationships that could be construed as a potential conflict of interest.

Appendix A

Table A1. The mechanical strengths of UHPC at 3 days.

W/B	Silica Fume	Cenosphere	Fly Ash	Grade Slag Powder	Flexural Strength (MPa)	Compressive Strength (MPa)
0.18	0%	0	0	0	12.8	59.2
0.18	5%	0	0	0	13.0	59.9
0.18	10%	0	0	0	13.1	60.5
0.18	15%	0	0	0	13.8	64.3
0.18	0	5%	0	0	12.5	57.1
0.18	0	10%	0	0	12.2	55.8
0.18	0	15%	0	0	11.8	53.9
0.18	0	0	5%	0	12.3	56.1
0.18	0	0	10%	0	11.9	55.3
0.18	0	0	15%	0	11.8	55.0
0.18	0	0	0	5%	12.6	58.6
0.18	0	0	0	10%	12.6	58.1
0.18	0	0	0	15%	12.1	57.3
0.20	5%	0	0	0	11.8	56.9
0.20	10%	0	0	0	12.5	59.6
0.20	15%	0	0	0	12.9	60.2
0.20	0	5%	0	0	11.4	54.3
0.20	0	10%	0	0	11.2	53.8
0.20	0	15%	0	0	10.8	52.4
0.20	0	0	5%	0	11.5	54.3
0.20	0	0	10%	0	11.3	53.6
0.20	0	0	15%	0	11.0	52.0
0.20	0	0	0	5%	11.5	55.2
0.20	0	0	0	10%	11.5	54.8
0.20	0	0	0	15%	11.4	54.0
0.22	5%	0	0	0	11.4	53.6
0.22	10%	0	0	0	11.9	57.3
0.22	15%	0	0	0	12.1	58.0
0.22	0	5%	0	0	11.1	52.1
0.22	0	10%	0	0	10.9	51.7
0.22	0	15%	0	0	10.6	50.0
0.22	0	0	5%	0	11.1	52.1
0.22	0	0	10%	0	10.8	51.7
0.22	0	0	15%	0	10.6	50.6
0.22	0	0	0	5%	11.1	52.0
0.22	0	0	0	10%	11.0	51.8
0.22	0	0	0	15%	11.2	51.4

Table A2. The mechanical strengths of UHPC at 28 days.

W/B	Silica Fume	Cenosphere	Fly Ash	Grade Slag Powder	Flexural Strength (MPa)	Compressive Strength (MPa)
0.18	0%	0	0	0	16.0	110.1
0.18	5%	0	0	0	16.2	111.4
0.18	10%	0	0	0	16.9	118.6
0.18	15%	0	0	0	17.1	119.5
0.18	0	5%	0	0	16.1	112.0
0.18	0	10%	0	0	16.3	113.5
0.18	0	15%	0	0	16.9	115.2

Table A2. *Cont.*

W/B	Silica Fume	Cenosphere	Fly Ash	Grade Slag Powder	Flexural Strength (MPa)	Compressive Strength (MPa)
0.18	0	0	5%	0	16.0	111.4
0.18	0	0	10%	0	16.2	111.3
0.18	0	0	15%	0	16.3	113.4
0.18	0	0	0	5%	16.2	112.0
0.18	0	0	0	10%	16.5	113.9
0.18	0	0	0	15%	16.5	115.1
0.20	5%	0	0	0	16.3	108.6
0.20	10%	0	0	0	16.5	109.1
0.20	15%	0	0	0	16.6	110.8
0.20	0	5%	0	0	15.8	102.2
0.20	0	10%	0	0	16.2	104.5
0.20	0	15%	0	0	16.4	106.1
0.20	0	0	5%	0	15.8	101.1
0.20	0	0	10%	0	15.8	101.5
0.20	0	0	15%	0	15.9	102.9
0.20	0	0	0	5%	16.0	102.4
0.20	0	0	0	10%	16.0	103.8
0.20	0	0	0	15%	16.2	105.0
0.22	5%	0	0	0	15.6	100.7
0.22	10%	0	0	0	15.9	102.0
0.22	15%	0	0	0	15.9	103.8
0.22	0	5%	0	0	15.2	96.1
0.22	0	10%	0	0	15.6	98.4
0.22	0	15%	0	0	15.8	99.7
0.22	0	0	5%	0	15.0	96.2
0.22	0	0	10%	0	15.2	96.6
0.22	0	0	15%	0	15.2	98.0
0.22	0	0	0	5%	15.3	96.6
0.22	0	0	0	10%	15.4	97.3
0.22	0	0	0	15%	15.6	98.9

References

1. Zong, Z.; Zhang, Q.; Liu, Y.; Guo, Z.; Lin, S.; Jiang, T. Carbonation resistance of sustainable self-compacting concrete with recycled concrete aggregate and fly ash, slag, and silica fume. *Eur. J. Environ. Civ. Eng.* **2024**, *28*, 2177–2199. [CrossRef]
2. Hao, Y.; Qin, L.; He, X.; Sun, H.; Wu, Z.; Hou, Y. Mechanical properties and constitutive model of fibre-reinforced high-performance class-F fly ash foam concrete. *Eur. J. Environ. Civ. Eng.* **2024**, *28*, 437–456. [CrossRef]
3. Hua, C.; Tang, R.; Lu, X. Radon emission characteristics and pore structure evolution of self-compacting concrete with silica fume-molybdenum tailings under different curing environments. *J. Build. Eng.* **2024**, *97*, 110769. [CrossRef]
4. Li, B.; Chen, Z.; Wang, S.; Xu, L. A Review on the Damage Behavior and Constitutive Model of Fiber Reinforced Concrete at Ambient Temperature. *Constr. Build. Mater.* **2024**, *412*, 134919. [CrossRef]
5. Wang, Y.; Qiao, P.; Sun, J.; Chen, A. Influence of Fibers on Tensile Behavior of Ultra-High Performance Concrete: A Review. *Constr. Build. Mater.* **2024**, *430*, 136432. [CrossRef]
6. Cao, Z.; Wang, K.; Peng, X.; Wang, H.; Huang, R. Influence of NaCl Solution External Erosion on Corrosion Resistance of RPC Reinforced with Straw Fiber. *Coatings* **2023**, *13*, 1308. [CrossRef]
7. Grzymski, F.; Musiał, M.; Trapko, T. Mechanical Properties of Fibre Reinforced Concrete with Recycled Fibres. *Constr. Build. Mater.* **2019**, *198*, 323–331. [CrossRef]
8. Zhagifarov, A.M.; Akhmetov, D.A.; Suleyev, D.K.; Zhumadilova, Z.O.; Begentayev, M.M.; Pukharenko, Y.V. Investigation of Hydrophysical Properties and Corrosion Resistance of Modified Self-Compacting Concretes. *Materials* **2024**, *17*, 2605. [CrossRef]
9. Djebien, R.; Hebhoub, H.; Belachia, M.; Berdoudi, S.; Kherraf, L. Incorporation of Marble Waste as Sand in Formulation of Self-Compacting Concrete. *Struct. Eng. Mech.* **2018**, *67*, 87–91.
10. Li, H.; Huang, F.; Tu, H.; Sun, D.; Wang, Z.; Yi, Z.; Yang, Z.; Xie, Y. Pumpability of Manufactured Sand Self-compacting Concrete. *J. Wuhan Univ. Technol. Mater. Sci. Ed.* **2023**, *38*, 1382–1390. [CrossRef]
11. Adhikary, S.K.; Ashish, D.K.; Sharma, H.; Patel, J.; Rudzionis, Z.; Al-Ajamee, M.; Thomas, B.S.; Khatib, J.M. Lightweight Self-compacting Concrete: A review. *Resour. Conserv. Recycl. Adv.* **2022**, *15*, 200107. [CrossRef]

12. Kumar, P.; Pasla, D.; Saravanan, T.J. Self-Compacting Lightweight Aggregate Concrete and Its Properties: A review. *Constr. Build. Mater.* **2023**, *375*, 130861. [CrossRef]
13. Song, P.; Wang, X.; Wang, Y.; Taylor, P.; Li, C.; Sun, Q.; Ma, Y. Assessment of Rheological and Toughening Behavior of Basalt Fiber Sprayed Cementitious Composites (BFSCC). *Constr. Build. Mater.* **2024**, *426*, 136169. [CrossRef]
14. ElNemr, A.; Shaltout, R. Rheological and Mechanical Characterization of Self-Compacting Concrete Using Recycled Aggregate. *Materials* **2025**, *18*, 1519. [CrossRef] [PubMed]
15. Anish, V.; Logeshwari, J. A review on ultra high-performance fibre-reinforced concrete with nanomaterials and its applications. *J. Eng. Appl. Sci.* **2024**, *71*, 1–40. [CrossRef]
16. Su, X.; Ren, Z.; Li, P. Review on physical and chemical activation strategies for ultra-high performance concrete (UHPC). *J. Cem. Concr. Compos.* **2024**, *149*, 105519. [CrossRef]
17. Ullah, R.; Qiang, Y.; Ahmad, J.; Vatin, N.I.; El-Shorbagy, M.A. Ultra-High-Performance Concrete (UHPC): A State-of-the-Art Review. *Materials* **2022**, *15*, 4131. [CrossRef]
18. He, J.; Chen, W.; Zhang, B.; Yu, J.; Liu, H. The Mechanical Properties and Damage Evolution of UHPC Reinforced with Glass Fibers and High-Performance Polypropylene Fibers. *Materials* **2021**, *14*, 2455. [CrossRef]
19. Ge, J.; Du, S.; Chen, P.; Huang, W.; Yang, R.; Gao, X.; Ji, X. Preparation and Compressive Fracture Characteristics of Fiber Foam Concrete. *J. Concr.* **2024**, *12*, 213–218.
20. Miyazaki, Y.; Nakayama, R.; Yasuo, N.; Watanabe, Y.; Shimizu, R.; Packwood, D.M.; Nishio, K.; Ando, Y.; Sekijima, M.; Hitosugi, T. Bayesian Statistics-Based Analysis of AC Impedance Spectra. *AIP Adv.* **2020**, *10*, 045231. [CrossRef]
21. Zheng, S.; Liu, Q.; Han, F.; Liu, S.; Han, T.; Yan, H. Basic Mechanical Properties of Self-Compacting Concrete Prepared with Aeolian Sand and Recycled Coarse Aggregate. *Buildings* **2024**, *14*, 2949. [CrossRef]
22. Lee, S.J.; You, I.; Kim, S.; Shin, H.O.; Yoo, D.Y. Self-Sensing Capacity of Ultra-High-Performance Fiber-Reinforced Concrete Containing Conductive Powders in Tension. *Cem. Concr. Compos.* **2022**, *125*, 104331. [CrossRef]
23. Ashteyat, A.; Obaidat, A.T.; Qerba'a, R.; Abdel-Jaber, M. Influence of basalt fiber on the rheological and mechanical properties and durability behavior of self-compacting concrete (SCC). *Fibers* **2024**, *12*, 52. [CrossRef]
24. Ahmad, J.; Zaid, O.; Aslam, F.; Shahzaib, M.; Ullah, R.; Alabduljabbar, H.; Khedher, K.M. A Study on the Mechanical Characteristics of Glass and Nylon Fiber Reinforced Peach Shell Lightweight Concrete. *Materials* **2021**, *14*, 4488. [CrossRef] [PubMed]
25. Zhang, Z.; Guo, B.; Yang, Z. Influence of fiber type on UHPC performance and its strengthening mechanism research progress. *J. Jiangxi Build. Mater.* **2024**, *5*, 13–15.
26. Zhao, W.; Kong, J.; Sun, Q. Preparation and Properties of Ultra-High Performance Concrete Incorporation Multi-Mineral Admixtures. *Mater. Res. Express* **2024**, *11*, 085311. [CrossRef]
27. Chen, Y.; Liu, P.; Sha, F.; Yin, J.; He, S.; Li, Q.; Yu, Z.; Chen, H. Study on the Mechanical and Rheological Properties of Ultra-High Performance Concrete. *J. Mater. Res. Technol.* **2022**, *17*, 111–124. [CrossRef]
28. Ju, M.; Jeong, J.-G.; Palou, M.; Park, K. Mechanical Behavior of Fine Recycled Concrete Aggregate Concrete with the Mineral Admixtures. *Materials* **2020**, *13*, 2264. [CrossRef]
29. *ASTM C1856/C1856M-17*; Standard Practice for Fabricating and Testing Specimens of Ultra-High Performance Concrete. ASTM International: West Conshohocken, PA, USA, 2017.
30. Cheng, X.; Mu, R.; Liu, X. Research progress on preparation and mechanical properties of ultra-high performance concrete. *Silic. Bull.* **2024**, *43*, 4295–4312.
31. *GB/T 31387-2025*; Ultra High Performance Concrete. Ministry of Housing and Urban Rural Development of the People's Republic of China: Beijing, China, 2025.
32. Pourjahanshahi, A.; Madani, H. Chloride diffusivity and mechanical performance of UHPC with hybrid fibers under heat treatment regime. *Mater. Today Commun.* **2021**, *26*, 102146. [CrossRef]
33. Teng, L.; Meng, W.; Khayat, K.H. Rheology Control of Ultra-High-Performance Concrete Made with Different Fiber Contents. *Cem. Concr. Res.* **2020**, *138*, 106222. [CrossRef]
34. Yoo, D.Y.; Kang, M.C.; Choi, H.J.; Shin, W.; Kim, S. Electromagnetic interference shielding of multi-cracked high-performance fiber-reinforced cement composites—Effects of matrix strength and carbon fiber. *Constr. Build. Mater.* **2020**, *261*, 119949. [CrossRef]
35. Gong, K.; Liang, Z.; Peng, X.; Wang, H. Research into Preparation and Performance of Fast-Hardening RPC Mixed with Straw. *Materials* **2023**, *16*, 5310. [CrossRef] [PubMed]
36. Liu, J.; Xie, X.; Li, L. Experimental Study on Mechanical Properties and Durability of Grafted Nano-SiO_2 Modified Rice Straw Fiber Reinforced Concrete. *Constr. Build. Mater.* **2022**, *347*, 128575. [CrossRef]
37. Tai, Y.; Yang, L.; Gao, D.; Kang, K.; Cao, Z.; Zhao, P. Properties Evolution and Deterioration Mechanism of Steel Fiber Reinforced Concrete (SFRC) under the Coupling Effect of Carbonation and Chloride Attack. *J. Build. Eng.* **2024**, *95*, 110275. [CrossRef]
38. Tu, Y.; Liu, D.; Yuan, L.; Zhang, Y. Corrosion Resistance of Concrete Strengthened with Fibre-Reinforced Polymer Sheets. *Mag. Concr. Res.* **2022**, *74*, 54–69. [CrossRef]

39. Abdel-Jaber, M.; Beale, R.; Makhoul, N. Self-compacting concrete (SCC) strength prediction via optimized duple-deep-learning model and distance ranked fire-hawk optimizer (DRFO). *Civ. Eng. Archit.* **2023**, *11*, 2447–2460. [CrossRef]
40. Guo, D.; Guo, M.; Zhou, Y.; Zhu, Z. Use of nano-silica to improve the performance of LC3-UHPC: Mechanical behavior and microstructural characteristics. *Constr. Build. Mater.* **2024**, *411*, 134280. [CrossRef]
41. Shaban, W.M.; Heniegal, A.M.; Amin, M.; Zeyad, A.M.; Agwa, I.S.; Hassan, H.H. Effect of agricultural wastes as sugar beet ash, sugarcane leaf ash, and sugarcane bagasse ash on UHPC properties. *J. Build. Eng.* **2024**, *98*, 111359. [CrossRef]
42. Khayat, K.H.; Meng, W.; Vallurupalli, K.; Teng, L. Rheological Properties of Ultra-High-Performance Concrete-An Overview. *Cem. Concr. Res.* **2019**, *124*, 105828. [CrossRef]
43. Wei, X.; Liu, W. Study on temperature crack prevention and control measures for self-compacted large volume concrete with machine-made sand. *J. China Cement* **2024**, *8*, 88–90.
44. Luo, Z.; Zhi, T.; Liu, X.; Yin, K.; Pan, H.; Feng, H.; Song, Y.; Su, Y. Effects of different nanomaterials on the early performance of ultra-high performance concrete (UHPC): C–S–H seeds and nano-silica. *Cem. Concr. Compos.* **2023**, *142*, 105211. [CrossRef]
45. Yu, R.; Liu, K.; Yin, T.; Tang, L.; Ding, M.; Shui, Z. Comparative study on the effect of steel and polyoxymethylene fibers on the characteristics of Ultra-High Performance Concrete (UHPC). *Cem. Concr. Compos.* **2022**, *127*, 104418.
46. Xie, J.; Liu, J.; Li, W.; Jin, L. Mechanical properties of magnesia fiber reinforced magnesium cement mortar composite materials. *J. Compos. Mater.* **2024**, *41*, 5492–5503.
47. Li, Y.; Mu, J.; Wang, F.; Wang, X.; Ding, Q. Research on the Effect of Apparent Density on the Rheological Properties of Lightweight and Heavyweight Concrete. *J. Build. Eng.* **2024**, *95*, 110111. [CrossRef]
48. Tariq, S.; Scott, A.; Mackechnie, J.; Shah, V. Glass powder replacement in self-compacting concrete and its effect on rheological and mechanical properties. *J. Sustain. Cem.-Based Mater.* **2022**, *11*, 240–256. [CrossRef]
49. Rajasekar, A.; Arunachalam, K.; Kottaisamy, M. Assessment of strength and durability characteristics of copper slag incorporated ultra high strength concrete. *J. Clean. Prod.* **2019**, *208*, 402–414. [CrossRef]
50. Wang, W.; Xiong, Z.; Yu, Y.; Chen, D.; Wu, C. Residual behaviour and damage assessment of UHPC-filled double-skin steel tubular columns after lateral impact. *Thin-Walled Struct.* **2024**, *205*, 112602. [CrossRef]
51. Ting, T.Z.H.; Rahman, M.E.; Lau, H.H.; Ting, M.Z.Y. Recent development and perspective of lightweight aggregates based self-compacting concrete. *Constr. Build. Mater.* **2019**, *201*, 763–777. [CrossRef]

coatings

MDPI

Article

Research on Pavement Performance of Steel Slag Asphalt Mastic and Mixtures

Jianmin Guo [1], Jincheng Wei [2,*], Feiping Xu [1], Qinsheng Xu [2,*], Liang Kang [1], Wenjuan Wu [2], Wencheng Shi [1] and Xiangpeng Yan [2]

1 Shandong Hi-Speed Company Limited, Jinan 250014, China; jtjtgkjgc@163.com (J.G.); 17865565285@163.com (F.X.); hywangyoucao@163.com (L.K.); sdtzdez@163.com (W.S.)
2 Shandong Transportation Institute, Jinan 250102, China; wuwenjuan@sdjtky.cn (W.W.); yanxiangpeng@sdjtky.cn (X.Y.)
* Correspondence: weijincheng@sdjtky.cn (J.W.); 15966652536@163.com (Q.X.)

Abstract: In order to explore the influence of steel slag on the road performance of asphalt mastic and its mixtures, steel slag powder (SSP) and steel slag aggregate (SSA) were used to replace limestone mineral powder filler (MF) and natural limestone aggregate (LA) respectively to prepare asphalt mastic and mixture. A series of standardized tests including penetration, softening point, ductility, viscosity, pull-off strength, dynamic shear rheometer (DSR), and bending beam rheometer (BBR) were carried out to evaluate the performance of asphalt mastics with SSP. Meanwhile, high- and low-temperature performance, moisture stability, volumetric stability, and fatigue resistance were evaluated by wheel tracking, uniaxial penetration strength, Hamburg, three-point bending, freeze–thaw splitting, immersed Marshall stability, water immersion expansion, and two-point bending trapezoidal beam fatigue tests. The results show that compared to the asphalt mastic with MF, enhanced high-temperature deformation resistance and reduced low-temperature cracking resistance of asphalt mastic with SSP were observed, as well as superior aging resistance. The improvements in high-temperature stability, moisture resistance, and fatigue performance were confirmed for asphalt mixtures with SSP/SSA. Additionally, compromised volumetric stability and low-temperature crack resistance were found when SSP/SSA was used in mixtures. Although asphalt mixtures with SSA exhibited 257.79%–424.60% higher expansion rate after 21-day immersion than those with LA, the 3-day immersion expansion rates complied with specification limits (<1.5% per JTG F40-2004). Critical volume expansion control measures should be adopted for full-component applications of steel slag powder/aggregates due to the hydration potential of free lime (f-CaO) and magnesium oxide (MgO) in steel slag under moisture exposure.

Keywords: steel slag powder; steel slag aggregate; asphalt mastic; asphalt mixture; performance verification

1. Introduction

Steel slag, as a highly promising road construction material, exhibits superior physical and mechanical properties unattainable by natural aggregates, positioning it as an ideal secondary resource. In recent years, extensive research has been conducted globally on its engineering applications, yielding significant advancements. Lin et al. [1] revealed through characterization studies that steel slag aggregates demonstrate an approximately three-fold higher hardness compared to natural limestone aggregates, with crushing values and abrasion loss indices being 25%–40% superior. Bessa et al. [2] conducted an evaluation of steel slag's abrasion resistance. It revealed that the superior angularity and enhanced hardness compared to conventional granite and limestone aggregates engender steel slag's exceptional wear resistance. Li et al. [3] found through 3D scanning and texture parameter analysis that the 3D texture parameters (Sq, Vmc) of steel slag aggregates were significantly better than those of conventional aggregates. Its high surface roughness and stable core support volume endow asphalt pavement with excellent anti-skid durability, with wear resistance improved by more than 20%, and the texture attenuation law is controllable. E Guangxun et al. [4] investigated the variation patterns of the skeleton void ratio in steel slag coarse aggregates. Their results revealed a multivariate nonlinear relationship between the skeleton void ratio and the proportion of different particle size fractions within the coarse aggregates. Similar patterns were observed for both basalt coarse aggregates and hybrid coarse aggregates composed of steel slag and basalt, indicating that the interstitial void characteristics of composite aggregate systems are governed by analogous gradation-dependent mechanisms. Asi M [5] et al. investigated the performance of asphalt mixtures prepared by partially replacing coarse limestone aggregates larger than 4.75 mm with steel slag aggregates at varying dosages. The results demonstrated that the asphalt content increased correspondingly with the dosage of steel slag aggregates. Notably, when 30% steel slag was incorporated, the mixture exhibited significantly enhanced skid resistance. Furthermore, at a steel slag replacement level of 75%, comprehensive improvements were observed in multiple performance characteristics of the asphalt mixture, achieving substantial optimization in overall material properties. GOLI H et al. [6] conducted a study on the road performance of steel slag asphalt mixtures with varying substitution ratios. Their findings revealed that the incorporation of steel slag aggregates led to an increase in asphalt content. Notably, the mixture achieved optimal road performance when 25% of the coarse limestone aggregates were replaced by steel slag. Chenet al. [7] found that compared with basalt, the micro-interface phase between steel slag and asphalt was more continuous and uniform. Separately, Liu et al. [8] investigated the enhancement mechanisms of skid resistance in asphalt mixtures by substituting basalt aggregates with steel slag aggregates larger than 2.36 mm. Their mechanistic analysis demonstrated that the application of steel slag aggregates endowed the mixture with enhanced surface textural characteristics, thereby significantly improving its skid resistance properties. Yan et al. [9] prepared steel slag asphalt mixtures by volumetrically replacing basalt aggregates with coarse and fine steel slag aggregates, followed by comparative analysis with basalt aggregate mixtures. Their study revealed that coarse steel slag aggregates effectively enhanced the high-temperature performance and water stability of the mixture; however, this resulted in a deterioration of low-temperature performance. Conversely, the incorporation of fine steel slag aggregates improved the low-temperature performance but adversely affected the high-temperature performance and water stability. Additionally, both types of steel slag aggregates were observed to reduce the volumetric stability of the asphalt mixtures. Shen et al. [10] found that the complex surface texture and pore characteristics of steel slag improved its adhesion to asphalt, thereby enhancing the anti-slip performance of steel slag asphalt mixtures. Wang et al. [11] investigated the effects of steel slag incorporation ratio, asphalt-to-aggregate ratio, and steel slag particle size on the expansion rate of asphalt mixtures. Their findings demonstrated that the expansion behavior could be effectively controlled under the following conditions: steel slag content not exceeding 50%, the actual asphalt-to-aggregate ratio being maintained at 1.5% above the optimum ratio, and the employment of larger-sized steel slag aggregates. Zhang et al. [12] fabricated asphalt mixtures by volumetrically replacing basalt fine aggregates with steel slag fine aggregates. Experimental results revealed that the high-temperature stability and moisture stability of the mixtures initially increased but subsequently decreased with steel slag incorporation.

Conversely, both low-temperature crack resistance and volumetric stability progressively declined as the steel slag fine aggregate content increased. Based on systematic testing, the study recommended that the substitution ratio of steel slag fine aggregates should be limited to 36% to achieve balanced performance outcomes. Dong [13] investigated the performance and fatigue damage of steel slag fine aggregates in asphalt mixtures. The study revealed that incorporating steel slag fine aggregates moderately enhances the mixture's high-temperature stability, low-temperature crack resistance, fatigue life, and skid resistance. However, it concurrently increases the water immersion-induced expansion rate. Based on a comprehensive performance evaluation, the optimal substitution ratio of steel slag fine aggregates in AC-20 asphalt mixtures is recommended as 75%, balancing mechanical enhancements with volumetric stability constraints. Kong D et al. [14] demonstrated through systematic research that steel slag powder particles exhibit richer micro-textural characteristics compared to mineral powder, with an approximately 15% higher angularity index. Additionally, parameters such as the angularity index, sphericity, and surface texture of steel slag powder were found to vary depending on the source and production processes of the steel slag aggregates, highlighting the material's inherent variability influenced by manufacturing origins and techniques. Li et al. [15] conducted a study on the partial replacement of mineral powder filler with steel slag powder in asphalt mixtures. The findings revealed that the rough surface texture and abundant silicate mineral composition of steel slag powder enhanced its compatibility with asphalt, thereby improving the adhesive bond between asphalt and aggregates. Fu et al. [16] systematically evaluated the performance of steel slag-modified asphalt mastic. A comparative analysis with conventional limestone filler-based asphalt mastic demonstrated that steel slag powder improved the viscosity and high-temperature performance of the asphalt mastic. However, this modification resulted in reduced low-temperature deformation capability and increased temperature susceptibility. The study further identified an optimal filler–bitumen ratio of 1.0, at which the steel slag asphalt mastic achieved balanced comprehensive performance characteristics. Liu [17] suggested through research that the special surface properties of steel slag powder can improve the road performance and durability of asphalt mixtures. Xiao et al. [18] demonstrated that the incorporation of steel slag powder enhances the adhesive bonding between asphalt and aggregates, whereas excessive dosage was observed to compromise the moisture stability of mixtures. Concurrently, Chen et al. [19] conducted a comprehensive rheological characterization of steel slag powder (SSP)-modified bituminous mastic by employing dynamic shear rheometry (DSR). Parallel scanning electron microscopy (SEM) analyses revealed SSP's distinct microtextural advantages over limestone filler. It is concluded that the adhesion of SSP mastics is better than that of limestone filler asphalt mastic. Li et al. [20] investigated the rheological influence of steel slag powder on asphalt mastics, revealing its capacity to improve viscosity and rutting factor. Notably, the low-temperature performance of steel slag-modified asphalt mastics was found to be comparable to that of conventional mineral filler systems. Sun et al. [21] found that steel slag micro-powder can improve the adhesion between asphalt and aggregates, enhance the water stability and high-temperature stability of asphalt mixture, and increase the rutting resistance of asphalt pavement. Furthermore, researchers have conducted in-depth investigations into the application of steel slag aggregates in asphalt mixtures, encompassing interfacial mechanisms [22–24], adhesion characteristics [25–29], and cracking behavior [30,31] at the aggregate–asphalt interface. The pursuit of resource recycling and value-added applications for steel slag materials has remained a persistent research focus globally [32–39], driven by the imperative to reconcile industrial byproduct utilization with sustainable pavement engineering requirements. The resource recycling and value-added applications of steel slag materials have remained a persistent research focus globally. However, current studies and implementations predominantly concentrate on substituting conventional stone aggregates (e.g., limestone/basalt) with steel slag in coarse or fine aggregate forms, with most applications limited to steel slag coarse aggregates in asphalt mixtures. The utilization of steel slag fine aggregates remains underdeveloped in asphalt mix design. Furthermore, research on steel slag powder filler has largely been confined to its rheological effects on asphalt mastics, while minimal attention has been given to full-component steel slag asphalt mixtures where steel slag powder replaces traditional mineral fillers. The feasibility of multi-element, multi-component cascade utilization of steel slag materials (coarse/fine aggregates and powder filler) in asphalt mixtures requires further systematic investigation to establish integrated application frameworks.

This study systematically evaluates and characterizes the performance evolution of steel slag powder-modified asphalt mastics and their corresponding mixtures through a multi-scale experimental framework. At the mastic level, investigations include conventional performance tests (penetration, softening point, ductility) and advanced rheological assessments using dynamic shear rheometry (DSR) and bending beam rheometry (BBR). For full-component mixture validation, comprehensive evaluations were conducted, encompassing high-temperature rutting resistance, low-temperature cracking resistance, moisture stability via Hamburg wheel-tracking and freeze–thaw splitting, fatigue life quantification through four-point bending tests, and volumetric stability monitoring with expansion rate measurements. The integrated experimental approach establishes critical correlations between steel slag composition, interfacial interactions, and macroscopic pavement performance, thereby providing a methodological foundation for achieving the full-component cascade utilization of steel slag materials in sustainable asphalt mixtures.

2. Materials and Experimental

2.1. Materials

2.1.1. Asphalt Binder and Aggregates

In current highway construction practices, the application of steel slag asphalt mixtures is predominantly applied to the surface course of asphalt pavements. This study adopts material specifications from an operational highway project in Shandong Province, utilizing SBS-modified asphalt binder (PG 76-22), with performance evaluation results detailed in Table 1. The technical indicators of the bitumen meet the requirements of standard JTG-F40-2004 [40].

The crushed limestone and steel slag (as shown in Figure 1) were used as aggregates, of which the technical indicators meet the requirements of standards JTG-F40-2004 [40] and GBT 25824-2010 [41], respectively, as shown in Tables 2 and 3.

(**a**) 10~15 mm (**b**) 5~10 mm (**c**) 0~5 mm

Figure 1. Steel slag aggregate.

Table 1. Test results of SBS-modified asphalt binder properties.

Test Project			Test Values	Threshold Values
Penetration (100 g, 5 s, 25 °C)/0.1 mm			53	40~60
Softening point (5 °C)/°C			83.5	≥60
Ductility (5 cm/min, 5 °C)			32	≥20
Rotational viscosity at 135 °C (Pa·s)			1.8	≤3
Flash point (°C)			326	≥230
Solubility in trichloroethylene (%)			99.92	≥99
Elastic recovery at 25 °C (%)			93	≥75
Storage stability (phase separation assessed by 48-h softening point difference) (°C)			0.3	≤2.5
Residue after thin film oven test (TFOT, 163 ± 1 °C, 5.5 ± 1 r/min, 5 h)	Mass loss (%)		0.009	±1.0
	Penetration ratio at 25 °C (%)		75	≥65
	Ductility at 5 °C (cm)		19	≥15
Original SBS-modified asphalt	Rutting factor G*/sinδ (kPa)	76 °C	1.995	≥1.0
		70 °C	3.326	
Residue after rolling thin film oven test (RTFOT, 163 ± 0.5 °C, 15 ± 0.2 r/min, 85 min)	Mass loss (%)		−0.049	±1.0
	Rutting factor G*/sinδ (kPa)	2.789	2.789	≥2.2
		4.553	4.553	
Residue after pressure-aging vessel (PAV, 90~110 °C, 2.1 MPa, 20 h)	Aging temperature (°C)		100 °C	
	Fatigue factor G*/sinδ (kPa)	31 °C	643.2	≤5000
		28 °C	990.7	
	Bending creep stiffness modulus S (MPa)	−6 °C	85.6	≤300
		−12 °C	132	
	Creep rate m-value	−6 °C	0.432	≥0.30
		−12 °C	0.369	

Table 2. Test results of coarse aggregates.

Test Project	The Size of Steel Slag Aggregate/mm		The Size of Limestone Aggregate/mm		Threshold Values
	10~15	5~10	10~15	5~10	
Apparent specific gravity	3.33	3.34	/	/	≥2.9
	/	/	2.742	2.732	≥2.6
Water absorption (%)	2.25	2.32	/	/	≤3.0
	/	/	0.69	0.54	≤2.0
Aggregate crushing value (%)	10.1	/	20.3	/	≤26
Soundness (%)	2	2	3	4	≤12
Los Angeles abrasion loss (%)	9.5	10.4	/	/	≤26
	/	/	20.2	22.3	≤28
Polished stone value (PSV)	66	/	44	/	≥42

Table 2. *Cont.*

Test Project	The Size of Steel Slag Aggregate/mm		The Size of Limestone Aggregate/mm		Threshold Values
	10~15	5~10	10~15	5~10	
Flakiness and elongation index (≥9.5 mm particles) (%)	4.5		9.9		≤12
Material finer than 0.075 mm by washing (%)	0.3	0.6	0.5	0.6	≤1
Soft particles content (%)	0.3	0.8	1.5	1.9	≤3
Adhesion to asphalt (level)	5		5		≥4
Free calcium oxide (*f*-CaO) content (%)	2.14	1.48	/	/	≤3
Expansion rate (%)	0.82		/	/	≤1.8

Table 3. Test results of fine aggregates.

Test Project	Steel Slag	Limestone	Threshold Values
Apparent specific gravity	3.314	/	≥2.9
	/	2.728	≥2.5
Water absorption (%)	2.38	1.07	/
Free calcium oxide (*f*-CaO) content (%)	1.14	/	/

2.1.2. Filler

The limestone ore powder and steel slag powder were used as the filler of the asphalt mixture. Steel slag powder is obtained by grinding. A predetermined mass of steel slag fine aggregates (0–5 mm particle size range) was weighed and subjected to mechanical grinding in a ball mill. The milled product was subsequently sieved through a series of standard sieves (2.36 mm, 1.18 mm, 0.6 mm, 0.3 mm, 0.15 mm, and 0.075 mm apertures). The residual materials from 0.15 mm and 0.075 mm standard sieves and the materials at the bottom of the sieves were mixed according to the particle size range of the mineral powder. The steel slag powder consistent with the particle size range of the mineral powder was obtained. The processed steel slag powder underwent rigorous performance characterization in compliance with Test Methods of Aggregates for Highway Engineering (JTG 3432-2024) [42], with key test results summarized in Table 4. The chemical compositions of steel slag powder and limestone mineral powder are compared in Table 5.

Table 4. Performance test results of steel slag powder/mineral powder filler.

Test Project		Limestone Mineral Powder	Steel Slag Power	Threshold Values
Apparent specific gravity		2.705	3.301	≥2.5
Water content (%)		0.42	0.63	≤1
Size range	<0.6 mm (%)	100	100	100
	<0.15 mm (%)	97.3	98.2	90~100
	<0.075 mm (%)	87.8	91.8	75~100
Appearance		No agglomeration	No agglomeration	No agglomeration
Hydrophilic coefficient (%)		0.53	0.58	<1
Specific surface area (m^2/g)		0.78	1.80	-

Table 5. Chemical composition of fillers.

Chemical Composition Content (%)	CaO	SiO_2	Fe_2O_3	MgO	MnO	P_2O_5	Al_2O_3
Steel slag power	47.61	16.74	20.67	6.25	5.14	1.85	1.74
Limestone mineral powder	73.65	16.14	0.55	2.64	0.49	3.98	2.55

2.1.3. The Preparation of Asphalt Mastic

Asphalt mastics were prepared by adding filler (limestone ore powder, steel slag powder) to asphalt. Steel slag powder replacing limestone mineral filler was added into SBS-modified asphalt. The substitution percentage of limestone ore powder filler with steel slag powder by weight was from 0 to 100%. The preparation process of asphalt mastic was as follows (Figure 2):

1. Binder Preparation: SBS-modified asphalt was homogenized in a 165 °C oven until achieving fluid state, then poured into a stainless steel mixing tank (volume 500 ± 5 mL) with a certain mass (800 g per piece) and cooled to ambient temperature.
2. Filler Treatment: Both the experimental limestone mineral filler and steel slag powder were dried in an oven at 105 °C to constant mass and cooled to room temperature prior to mixing. The two kinds of materials after drying and cooling were individually weighed according to the replacement mass ratio of the test design.
3. Mastic Fabrication:
    - The SBS-modified asphalt sample prepared in the first step was reheated to the flowing state in the oven at 165 °C.
    - Then, it was placed in the heating sleeve of the base of the asphalt high-speed mixer (the temperature of the base heating sleeve was set to 175 ± 5 °C).
    - The high-speed mixer (equipped with 20 mm four-blade inclined rotor) was initiated and allowed to stir for a certain time until the temperature of the SBS-modified asphalt in the stainless steel mixing tank was stable.
    - Then, the limestone mineral filler (MF) was added at three times, and the four-blade inclined rotor remained stirring during the addition process.
    - When adding steel slag powder as replacement for mineral filler, the steel slag powder should be added according to the adding process of the limestone mineral filler after it is stirred until no limestone ore powder float is visible to the eye in the asphalt.
    - After the steel slag powder was added, it was stirred for 90 min at the stirring rate of 800 r/min to ensure that the fillers were evenly distributed in the SBS-modified asphalt.

In order to adhere to engineering practice and truly reflect the state of asphalt mastic, the filler-to-binder ratio (F/B) selected was controlled to 1.0 according to engineering experience.

2.1.4. Mix Design of AC-13

The validation of pavement performance for asphalt mixtures represents a critical phase in the engineering application of steel slag powder filler. Building upon previous investigations, this study fabricated four AC-13 asphalt mixture variants: limestone aggregate–mineral filler (LA-MF), limestone aggregate–steel slag powder filler (LA-SSP), steel slag aggregate–mineral filler (SSA-MF), and steel slag aggregate–steel slag powder (SSA-SSP). The MF in LA-MF and SSA-MF is 100% limestone mineral powder filler, and the SSP in LA-SSP and SSA-SSP is 100% steel slag powder filler. The design of the asphalt mixture adopts the Marshall volume method according to specification JTG-F40-2004 [40].

The gradation composition of the asphalt mixtures of different aggregate types is shown in Table 6. The volumetric parameters and optimized asphalt content (OAC) for each asphalt mixture type are summarized in Table 7. As shown, all Marshall volumetric parameters of the designed mixtures comply with the prescribed specification thresholds of JTG-F40-2004 [40]. To ensure comparative validity across mixture groups, comparable percent air voids were systematically maintained during the mix design phase.

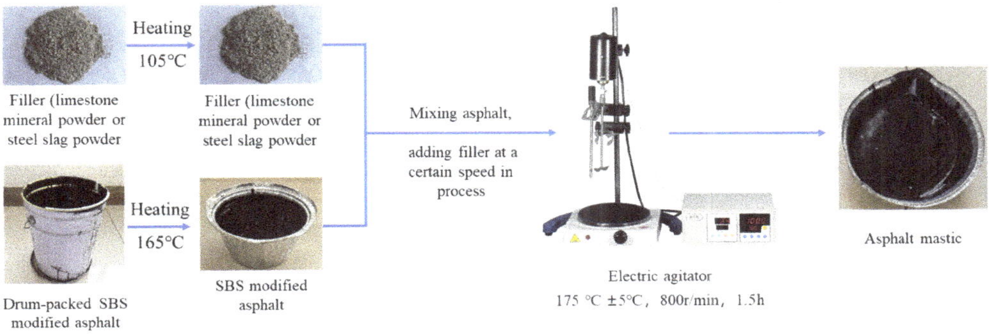

Figure 2. Asphalt mastic preparation process.

Table 6. Percent passing of the composite gradation.

Gradation Type	Sieve Size (mm)									
	16	13.2	9.5	4.75	2.36	1.18	0.6	0.3	0.15	0.075
LA-MF	100.0	96.6	74.7	44.1	31.4	23.8	15.8	10.1	7.4	5.8
LA-SSP	100.0	96.6	74.7	44.1	31.4	23.8	15.8	10.1	7.2	5.5
SSA-MF	100	96.7	74.3	43.6	30.2	23.2	15.6	10.3	7.6	5.3
SSA-SSP	100	96.7	74.3	43.6	30.2	23.2	15.6	10.2	7.2	5.4

Table 7. The volume parameters under the optimum asphalt content of different types.

Gradation Type	Optimum Asphalt Content (%)	Bulk Specific Gravity γ_f	Percent Air Void VV (%)	Percent Voids in Mineral Aggregate VMA (%)	Percent Voids Filled with Asphalt VFA (%)	Marshall Stability (kN)	Flow Value (mm)
LA-MF	4.8	2.400	4.3	14.9	71.1	14.2	2.2
LA-SSP	4.8	2.416	4.2	14.4	70.8	15.94	3.1
SSA-MF	4.6	2.762	4.4	14.0	68.6	16.12	2.9
SSA-SSP	4.5	2.782	4.5	14.0	67.9	16.3	2.6
Threshold values	/	/	4~6	14~16	65~75	≥8	1.5~4

2.2. *Test Methods*

2.2.1. Asphalt Mastic Performance Test Methods

Penetration, Softening Point, Ductility and Viscosity Index

The penetration, softening point, ductility, and viscosity are the basic properties of asphalt mastic. Here, 0, 25%, 50%, 75%, and 100% substitution percentages of mineral filler with steel slag powder by weight were designed to evaluate the effect of steel slag powder on the asphalt mastic. The tests were carried out according to the test specification (JTG E20-2011) [43].

Pull-Off Test

For evaluating the interfacial bond strength between asphalt mastic and aggregate, the pull-off test was carried out according to the standard specification ASTM D4541-22 [44]. Here, 0, 25%, 50%, 75%, and 100% substitution percentages of mineral filler with steel slag powder by weight were designed. A drawing die spindle with a diameter of 20 mm and a limestone slab with a surface polishing degree of 1000 are required for the bond strength test according to ASTM D4541-22 [44].

Dynamic Shear Rheometer (DSR)

The high-temperature rheological behaviors of asphalt mortar before and after the rolling thin film oven test (RTFOT) aging were tested individually by dynamic shear rheometry (DSR), a widely recognized method for assessing pavement performance under thermal loading. The test methods and parameters of RTFOT and DSR were carried out in accordance with the Chinese technical specification JTG E20-2011 [43]. This study employed an automated DSR system to obtain two critical viscoelastic parameters: the dynamic shear modulus (G^*) and phase angle (δ) under 5 test temperature, 58 °C, 64 °C, 70 °C, 76 °C, and 82 °C. These metrics, measured under controlled oscillatory shear conditions, serve as fundamental indicators of binder resistance to permanent deformation at elevated service temperatures.

Bending Beam Rheometer (BBR)

The low-temperature rheological characteristics of asphalt materials are commonly assessed through bending beam rheometry (BBR), a standardized methodology for characterizing thermal stress resistance in pavement engineering. In accordance with the Chinese technical specification JTG E20-2011 [43], this investigation utilized an automated BBR system to quantify two critical parameters: creep stiffness (S) and m-value (creep rate). To simulate extreme climatic conditions in cold regions, testing temperatures were established at −6 °C, −12 °C, and −18 °C. Under controlled loading conditions, data acquisition was performed at a standardized duration of 60 s to determine the S- and m-values, which serve as essential predictors of asphalt's susceptibility to thermal cracking.

2.2.2. Asphalt Mixture Performance Test Methods

A comprehensive performance evaluation protocol was implemented to characterize and compare key functional properties, including high-temperature performance, low-temperature performance, moisture stability, volumetric stability, and fatigue resistance. The test parameters of the asphalt mixture performance are presented in Table 8. The specimen dimensions, test conditions, and testing temperatures for the asphalt mixtures are all carried out in accordance with the requirements of the corresponding standard specifications.

Table 8. The test parameters of asphalt mixture performance.

Performance Parameter	Test Methods	Test Condition	Specification
High-temperature performance	Wheel-tracking test	60 °C, 1600 times of rolling	JTG E20-2011 [43]
	uniaxial penetration strength test	60 °C	JTG E20-2011 [43]
	Hamburg wheel-tracking test	50 °C, 20,000 times of rolling	AASHTO T324-2014 [45]
Low-temperature performance	Three-point bending test	−10 °C	JTG E20-2011 [43]
Moisture stability	freeze–thaw splitting test	−18 °C, freeze–thaw treatment for 16 h	JTG E20-2011 [43]
	immersed Marshall stability test	60 °C, 48 h immersion treatment	JTG E20-2011 [43]
Volumetric stability	Water immersion expansion test	60 °C, 72 h immersion treatment	JTG E20-2011 [43]
Fatigue resistance	Two-point bending trapezoidal beam fatigue test	10 °C, 25 Hz, 130 με	CSN EN 12697-24 [46]

3. Results and Discussions

3.1. Performance of Asphalt Mastic with Steel Slag Powder

3.1.1. Penetration, Softening Point, Ductility, Viscosity, and Bonding Strength

The analysis of the conventional performance metrics of asphalt mastics (Figure 3) reveals a consistent trend: as the proportion of steel slag powder (SSP) replacing mineral powder filler (MF) increases, the penetration and ductility of the mastic decrease, while the softening point and rotational viscosity exhibit progressive enhancements. Specifically, the full replacement of MF with SSP (100% substitution) results in a 10.21% reduction in penetration, 14.13% decline in ductility, and 3.24% and 75.31% increases in softening point and rotational viscosity, respectively. These findings indicate that asphalt mastics with SSP develop enhanced cohesion, rigidity, and shear resistance due to intensified interfacial interactions, albeit at the expense of reduced plastic deformation capacity and low-temperature ductility.

The mechanistic rationale lies in the complex surface texture and elevated surface roughness of SSP particles compared to conventional limestone mineral fillers. These morphological characteristics amplify larger effective adhesion area at asphalt–SSP interfaces, promoting stronger physicochemical bonding [17,24]. Such interactions restrict the mobility of asphalt molecules, thereby increasing the mastic's viscosity and thermal stability (evidenced by elevated softening points and rotational viscosity) while diminishing its capacity for plastic flow (reflected in reduced penetration and ductility).

Figure 4 demonstrates the progressive enhancement of interfacial bond strength in asphalt mastics with incremental SSP replacement of MF. Quantitative evaluations reveal a linear correlation, where bond strength increases from 1.28 MPa for MF-based mastic (0% SSP) to 1.53 MPa at full SSP replacement (100%), representing a 19.53% improvement. This trend underscores SSP's efficacy in optimizing asphalt–aggregate interfacial adhesion through enhanced physicochemical interactions. This pronounced improvement in interfacial adhesion is attributed to SSP's enhanced asphalt adsorption capacity, driven by its unique physicochemical characteristics. The rough, microporous surface texture and larger specific surface area (BET: 1.8 m^2/g vs. 0.78 m^2/g for MF) of SSP particles promote preferential adsorption of asphalt components [17], facilitating the formation of a thicker structural asphalt layer at filler–asphalt interfaces compared to conventional mineral fillers.

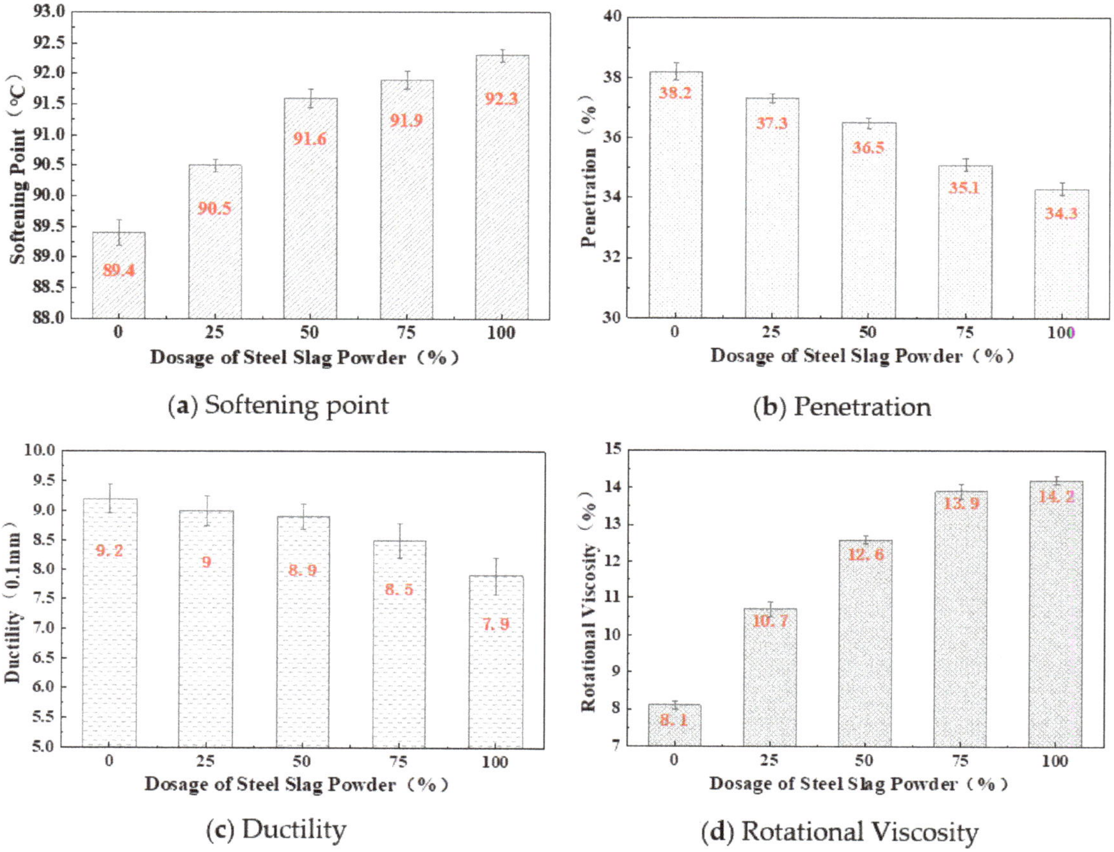

Figure 3. Soft point, penetration, ductility, and rotational viscosity for asphalt mastics.

3.1.2. High-Temperature Rheological Property

The complex shear modulus (G*), a critical rheological parameter, quantifies the resistance of asphalt binders and mastics to shear deformation under oscillatory loading, with higher G* values indicating superior deformation resistance. The analysis of the dynamic shear rheometry (DSR) temperature sweep results in Figure 5 demonstrates that all three specimens—steel slag powder-modified asphalt mastic (SSP-MAM), mineral filler-based asphalt mastic (MF-MAM), and SBS-modified asphalt (SBS-MA)—exhibit a consistent decline in G* with increasing temperature. However, at equivalent temperatures, the G* magnitudes follow the hierarchy: SSP-MAM > MF-MAM > SBS-MA. This trend substantiates that SSP-MAM possesses the highest load-deformation resistance, surpassing both conventional filler-modified and polymer-modified systems. The substitution of mineral filler with steel slag powder thereby enhances the high-temperature performance of asphalt composites through amplified viscoelastic stiffness. The incorporation of mineral filler or steel slag powder filler into asphalt induces physicochemical interactions that promote the formation of high-stiffness structural asphalt, thereby enhancing the composite's shear deformation resistance. Owing to the superior interfacial activity of steel slag powder [17]—attributed to its enhanced surface roughness, elevated specific surface area (BET: 1.8 m^2/g vs. mineral filler's 0.78 m^2/g), and alkaline composition (CaO/SiO_2 ratio > 2.5)—the asphalt mastic with SSP exhibits the highest complex shear modulus (G*) among all tested systems.

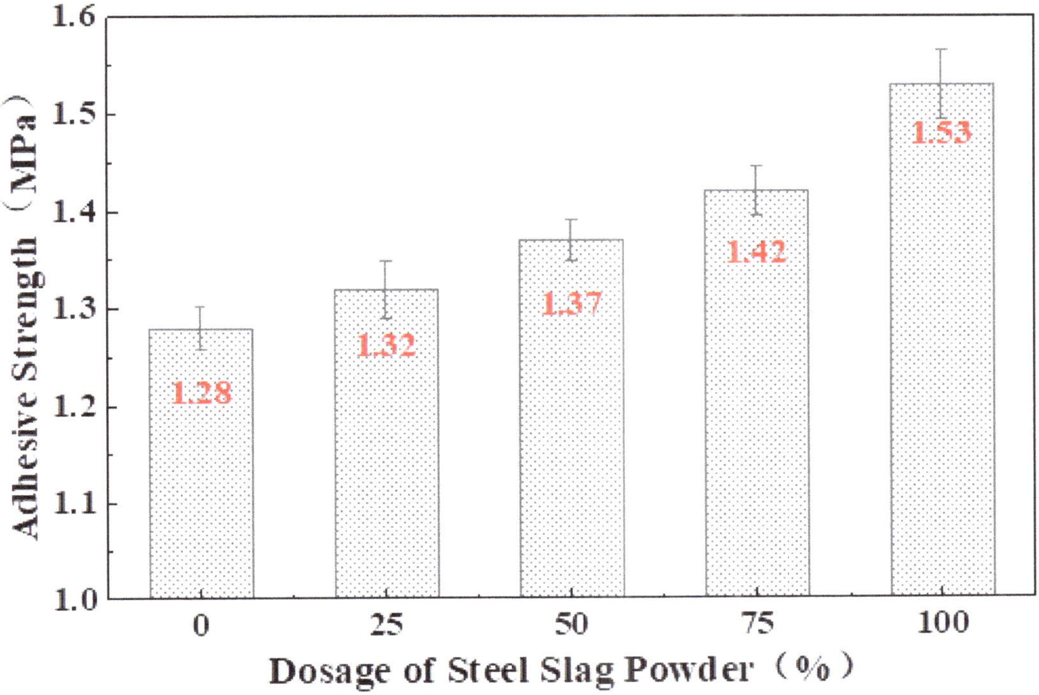

Figure 4. The interfacial bond strength of asphalt mastics.

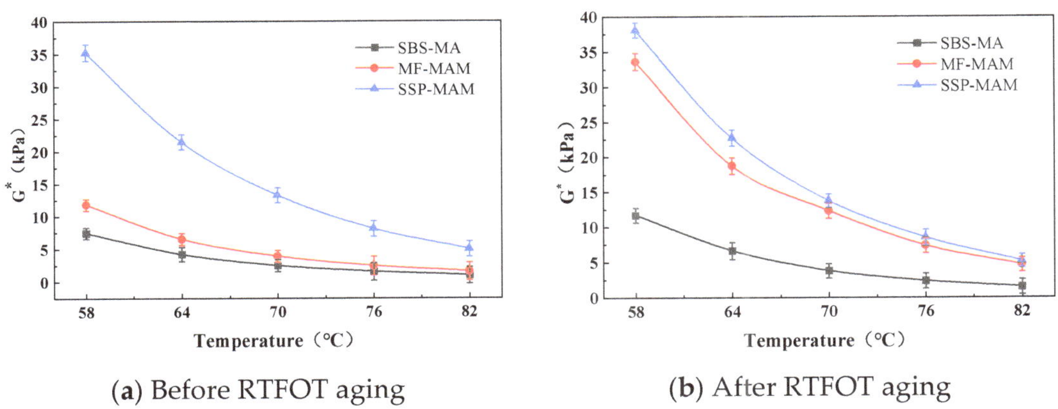

Figure 5. The complex shear modulus (G*) of SBS-modified asphalt and its asphalt mastic.

Figure 6 compares the change in complex shear modulus (ΔG*) for three material systems before and after RTFOT aging. The change in complex shear modulus (ΔG*) before and after aging is an important rheological index to evaluate its anti-aging performance. The greater the value, the more serious the hardening caused by aging and the worse the anti-aging performance [47]. The result reveals a more pronounced increase in ΔG* for the asphalt mastic with limestone powder mineral filler (MF-MAM) compared to the asphalt mastic with steel slag powder (SSP-MAM), with aged MF-MAM exhibiting significantly higher G magnitudes* than its SSP-MAM counterpart. It indicates that SSP-MAM shows superior aging resistance—a critical attribute for long-term pavement durability.

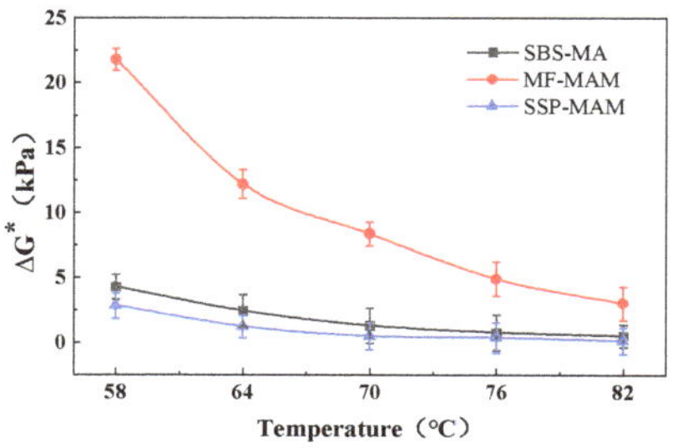

Figure 6. The change in complex shear modulus (ΔG*) before and after RTFOT aging.

The phase angle (δ), a fundamental parameter in viscoelastic material characterization, quantifies the ratio of viscous to elastic constituents within the material's mechanical response. Governed by the principles of linear viscoelasticity, δ directly reflects the material's capacity to resist deformation under cyclic loading. The smaller the δ, the more elastic components there are in the viscoelastic material, and the stronger its ability to resist deformation [48].

Figure 7 delineates the temperature-dependent evolution of the phase angle (δ) for the three material systems. While subtle disparities exist in their δ–temperature trajectories, all specimens exhibit consistent trends before and after aging. Notably, SBS-modified asphalt (SBS-MA) exhibits a monotonic increase in δ with rising temperatures, indicating a progressive dominance of viscous dissipation over elastic recovery. In contrast, both mineral filler mastic (MF-MAM) and steel slag powder mastic (SSP-MAM) demonstrate reduced δ values within the critical 58–70 °C temperature range before and after aging, reflecting the elastic components in the asphalt mastic increase while the viscous components decrease [48]. This elastic predominance enhances the mastics' recoverable deformation capacity, which is beneficial to the anti-rutting performance of asphalt mixtures. Under isothermal conditions, the phase angle hierarchy follows SSP-MAM < MF-MAM < SBS-MA, indicating that SSP-MAM possesses the highest proportion of elastic constituents and superior deformation resistance. The attenuated δ values of SSP-MAM (e.g., δ = 49.2° vs. SBS-MA's 65.0° at 64 °C) validate its enhanced improvement efficacy over conventional mastic systems with mineral filler.

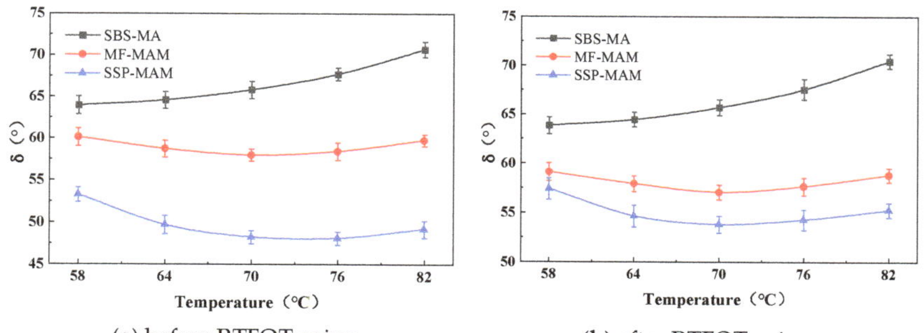

(**a**) before RTFOT aging (**b**) after RTFOT aging

Figure 7. The phase angle (δ) of SBS-modified asphalt and its asphalt mastic.

The rutting factor (G*/sinδ) serves as a critical indicator of the high-temperature rutting resistance of asphalt materials, with higher values signifying superior performance. As demonstrated in Figure 8, the temperature-dependent evolution of G*/sinδ for SBS-modified asphalt (SBS-MA) and its mastic derivatives follows a trend analogous to that of the complex shear modulus (G*): all systems exhibit a progressive decline in G*/sinδ with increasing temperature. Notably, the rate of reduction is more pronounced below 70 °C due to accelerated thermal softening of the asphalt matrix, whereas the descent plateaus above 70 °C as the binder transitions into a fully softened state. Under isothermal conditions, the G*/sinδ hierarchy is SSP-MAM > MF-MAM > SBS-MA, confirming that: Filler incorporation significantly enhances high-temperature rutting resistance by amplifying the elastic response and delaying viscous flow [49]. Steel slag powder demonstrates superior efficacy over mineral filler, attributed to its enhanced interfacial adhesion.

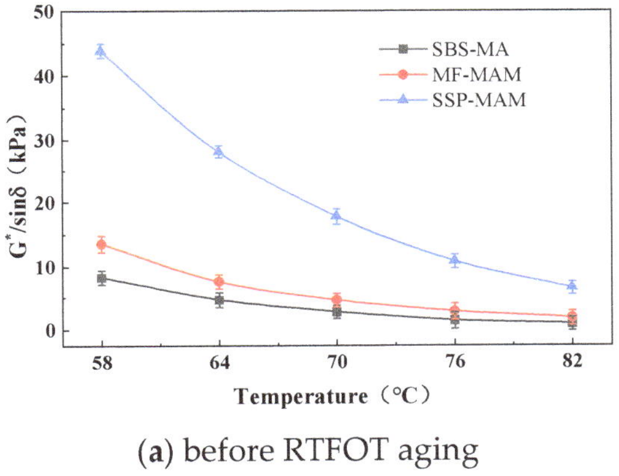

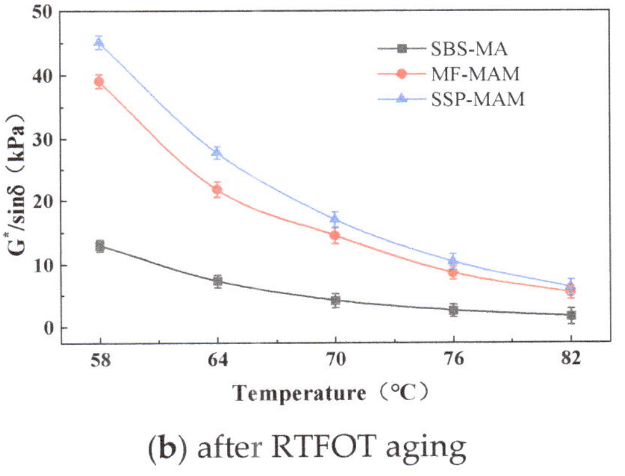

Figure 8. The rutting factor (G*/sinδ) of SBS-modified asphalt and its asphalt mastic.

3.1.3. Low-Temperature Rheological Property

The creep stiffness modulus (S) and creep rate (m-value), derived from bending beam rheometry (BBR) tests under low-temperature conditions, serve as critical indicators of asphalt and asphalt mastic low-temperature performance. These parameters quantify the material's deformation resistance and stress relaxation capacity, respectively. Specifically, lower creep stiffness (S) correlates with enhanced flexibility, reducing the risk of brittle fracture under thermal contraction. A higher m-value reflects superior stress dissipation capability. The lower the creep stiffness modulus s is and the higher the creep rate m is, the stronger the material's resistance to deformation and stress relaxation will be, and the lower the probability of low-temperature cracking caused by temperature stress will be [48].

The analysis of the low-temperature bending test results for asphalt and mastic materials in Figures 9 and 10 reveals consistent trends in the temperature-dependent evolution of creep stiffness modulus (S) and creep rate (m-value) for three types of materials. Specifically, creep stiffness (S) increases with decreasing temperature, reflecting reduced material flexibility under cryogenic conditions, while creep rate (m-value) exhibits an inverse correlation, declining as temperatures drop, indicative of attenuated stress relaxation capacity.

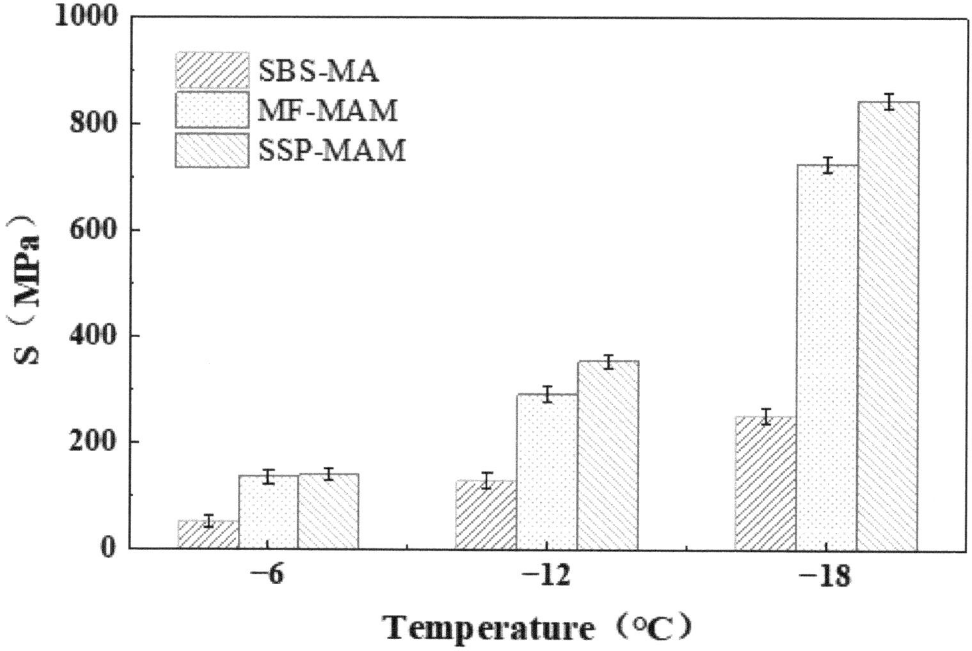

Figure 9. Creep stiffness modulus S.

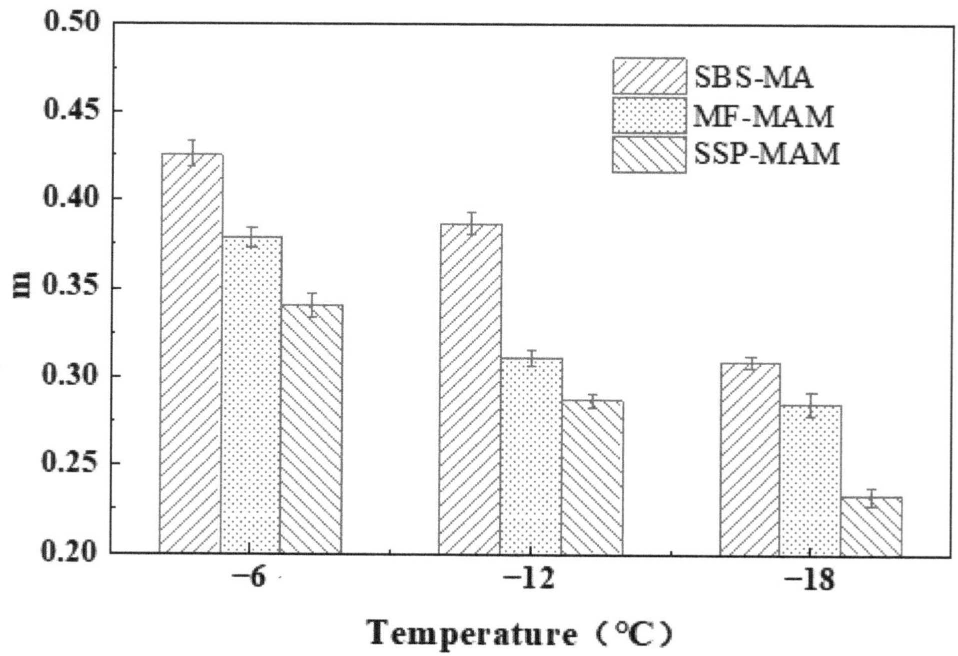

Figure 10. Creep rate m.

The hierarchical performance rankings derived from low-temperature bending tests demonstrate distinct viscoelastic behaviors among the evaluated materials: creep stiffness (S) follows SSP-MAM > MF-MAM > SBS-MA, while the creep rate (m-value) exhibits an inverse hierarchy of SBS-MA > MF-MAM > SSP-MAM. This inverse relationship demonstrates that the incorporation of mineral or steel slag fillers detrimentally impacts the low-temperature crack resistance of asphalt mastics. Notably, steel slag powder exerts a more pronounced adverse effect, attributed to its elevated rigidity (S increased by 18%–22% vs. mineral filler) and restricted stress dissipation capability (m-value reduced by 12%–15% vs. mineral filler) within the composite matrix. The addition of SSP/MF enhances the viscosity and stiffness of the asphalt mortar, reduces its low-temperature deformation resistance and stress relaxation ability, and makes it exhibit strong brittleness in low-temperature environments. Due to its own characteristics, SSP has a stronger interaction with asphalt, which leads to a greater impact of steel slag powder fillers on its low-temperature performance than MF. The elevated rigidity and constrained stress dissipation capability in SSP-MAM relative to MF-MAM highlight the critical trade-off between filler adhesion efficacy and mitigating stress accumulation capability, necessitating compensatory strategies for cold-region pavement applications.

3.2. Performance Evaluation of Steel Slag Powder Asphalt Mixture

3.2.1. High-Temperature Performance

The wheel-tracking test, a standardized method simulating tire-pavement interactions under controlled conditions (AASHTO T 324 [45]), is widely employed to evaluate the high-temperature performance of asphalt mixtures. The analysis of dynamic stability (DS) and relative deformation rate (RDR) in Figure 11 reveals the following performance hierarchy: LA-SSP mixtures exhibit an 11.11% increase in DS compared to LA-MF mixtures, SA-SSP mixtures demonstrate a 6.65% DS enhancement over SSA-MF counterparts, SSA-MF mixtures achieve 20% higher DS than LA-MF mixtures, and SSA-SSP mixtures show a 25.02% DS improvement relative to LA-SSP mixtures.

Concurrently, the RDR decreases by 14.60%, 1.86%, 16.81%, and 4.40% for these respective comparisons, confirming that steel slag powder (SSP) and steel slag aggregate (SSA) significantly enhance the mixtures' resistance to permanent deformation. This improvement stems from SSP/SSA's superior angularity, elevated hardness (LA abrasion loss < 11% vs. >20% for limestone aggregates), and enhanced interfacial bonding with asphalt, which collectively restrict shear flow and stabilize the viscoelastic matrix under repetitive loading.

Figure 12 presents the results of the uniaxial penetration strength tests for asphalt mixtures. The comparative analysis of the data reveals the following performance enhancements: LA-SSP mixtures exhibit an 18.02% increase in uniaxial penetration strength compared to LA-MF mixtures, SSA-SSP mixtures demonstrate an 8.39% improvement over SSA-MF mixtures, SSA-MF mixtures achieve a 28.83% higher uniaxial penetration strength than LA-MF mixtures, and SSA-SSP mixtures show an 18.32% enhancement relative to SA-SSP mixtures. These results conclusively validate that the incorporation of steel slag powder (SSP) and steel slag aggregate (SSA) significantly improves the high-temperature shear resistance of asphalt mixtures.

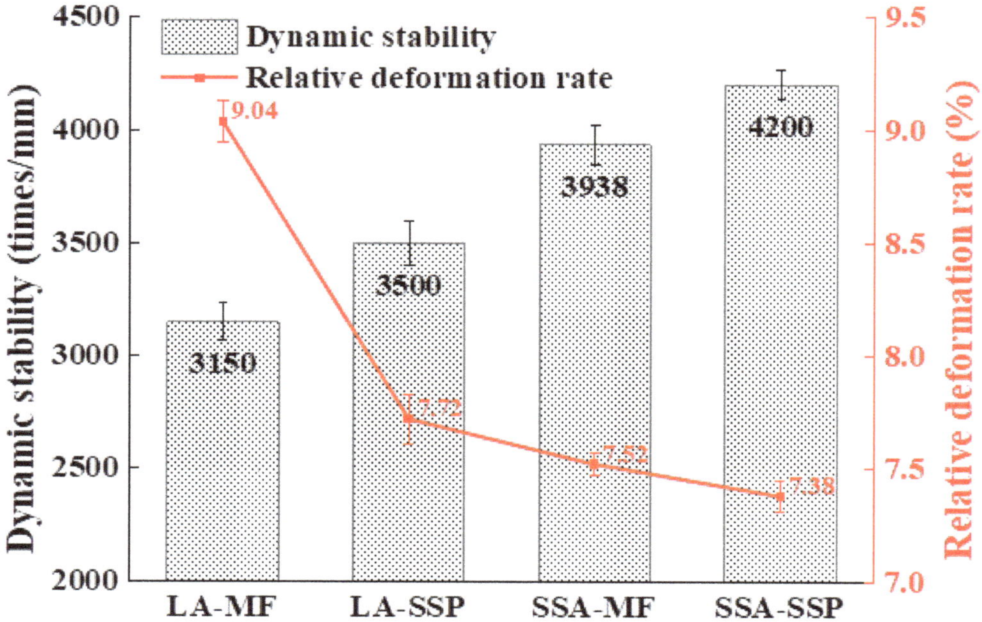

Figure 11. The wheel-tracking test results of asphalt mixtures.

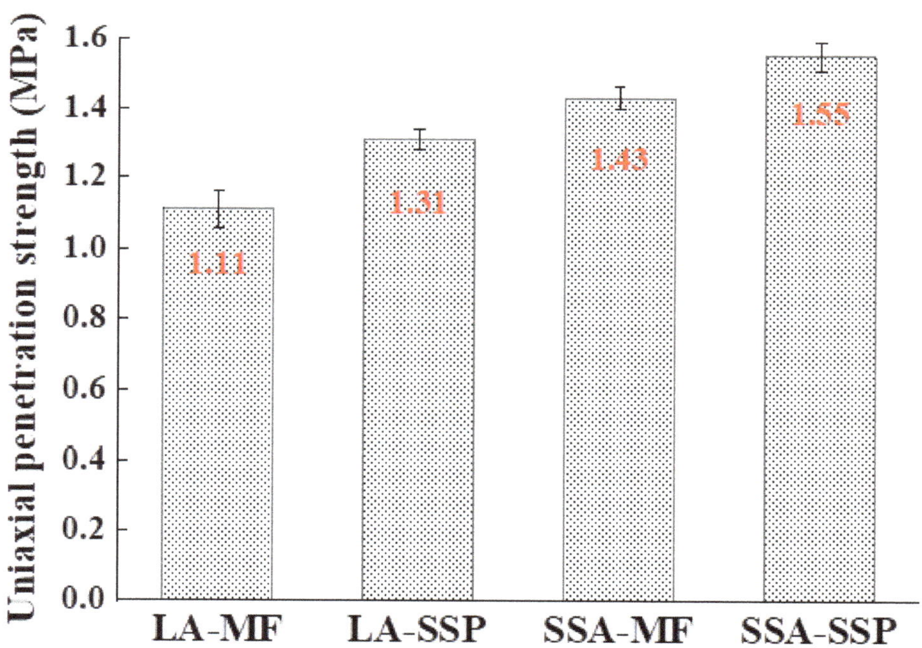

Figure 12. The uniaxial penetration strength test results of asphalt mixtures.

Figures 13 and 14 present the Hamburg wheel-tracking test (HWTT) results for asphalt mixtures under 50 °C water immersion, a method that comprehensively evaluates high-temperature performance (via cumulative deformation) and moisture susceptibility (via stripping inflection point) under coupled thermo-hydro conditions. The stripping inflection point identifies the onset of moisture-induced damage, with values below 10,000 loading cycles indicating poor moisture resistance. It can be seen that cumulative deformation reductions were observed. Notably, LA-SSP mixtures show a 20.46% decrease in cumulative deformation after 20,000 cycles compared to LA-MF. SSA-SSP mixtures exhibit a 9.62% reduction relative to SSA-MF. SSA-MF mixtures achieve a 24.83% improvement over LA-MF. SSA-SSP mixtures demonstrate a 14.58% enhancement compared to LA-SSP. All designed mixtures exhibit stripping inflection points >10,000 cycles, confirming that steel slag powder (SSP) and steel slag aggregate (SSA) significantly enhance both high-temperature rutting resistance and moisture damage resistance.

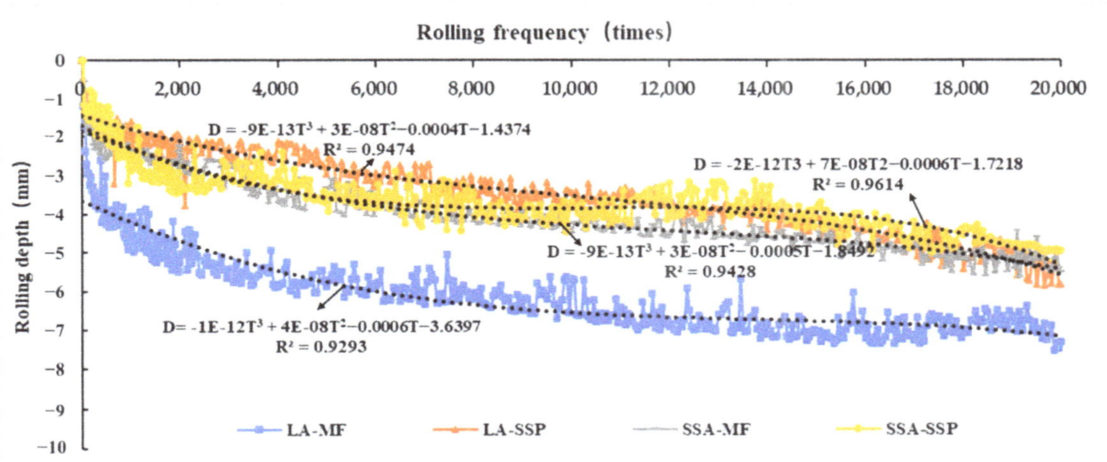

Figure 13. 20,000 rutting rolling deformation curves in Hamburg rutting test.

The application of steel slag powder and steel slag aggregates significantly enhances the high-temperature stability of asphalt mixtures, a conclusion consistent with prior findings that steel slag powder-modified asphalt mastics exhibit higher $G^*/\sin\delta$ values compared to limestone filler-modified mastics, indicating superior high-temperature performance.

The mechanistic enhancement of asphalt mixture performance through steel slag utilization is attributed to enhanced binder-aggregate adhesion resulting from the rough surface texture and high porosity of steel slag, which amplify interfacial contact area and asphalt adsorption capacity [7,17]. Meanwhile, the optimized skeletal structure driven by the angular geometry of steel slag aggregates promotes interparticle interlocking during compaction, creating a densely packed load-bearing matrix resistant to shear deformation [2,3]. The reduction in permanent strain accumulation by 20%–32% under repetitive traffic loading in the Hamburg wheel-tracking test proves shear resistance amplification. These mechanisms synergistically improve the mixture's resistance to rutting, validating steel slag's efficacy as a sustainable alternative to conventional aggregates in high-temperature pavement applications.

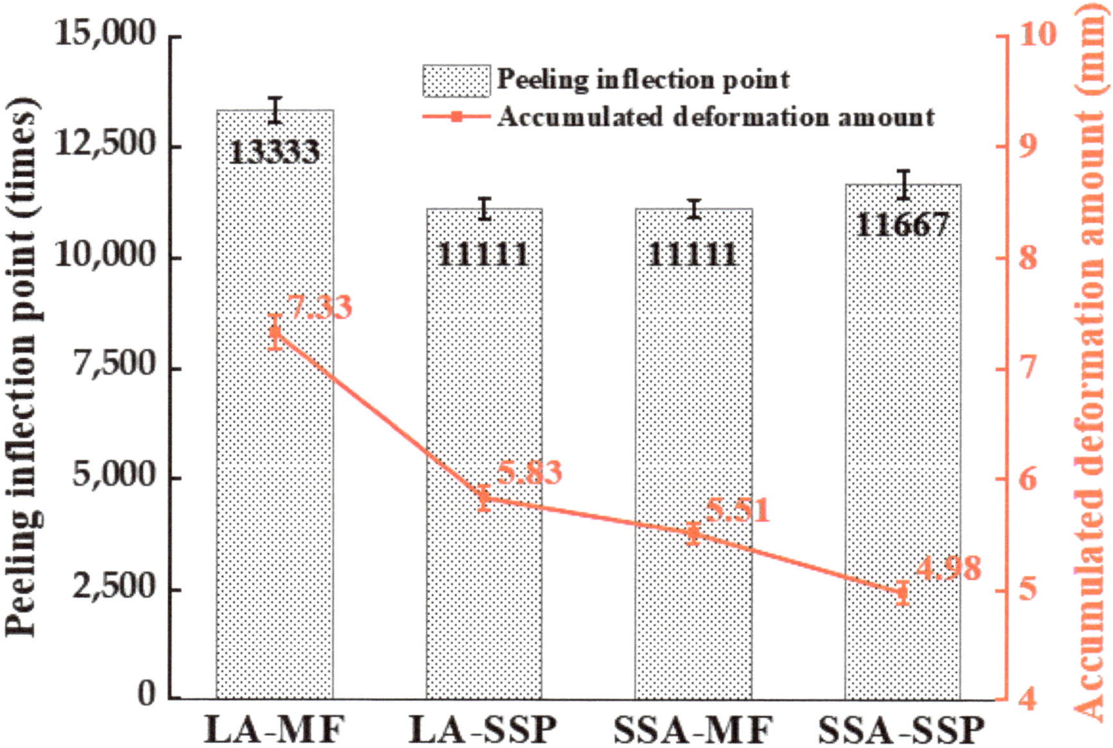

Figure 14. The cumulative deformation and spalling turning point results of Hamburg rut test.

3.2.2. Low-Temperature Performance

The low-temperature performance of asphalt mixtures, characterized by the maximum flexural tensile strain and flexural strength, was evaluated through bending beam tests at −10 °C (Figure 15). The comparative analysis of the maximum flexural tensile strain at −10 °C reveals systematic performance attenuations across the asphalt mixtures: LA-SSP mixtures exhibit a 3.54% reduction compared to LA-MF mixtures, while SSA-SSP mixtures show a 1.42% decrease relative to SSA-MF counterparts. Notably, SSA-MF mixtures demonstrate a 5.98% decline in maximum flexural strain versus LA-MF mixtures, and SSA-SSP mixtures display a further 3.92% reduction compared to LA-SSP mixtures. These results collectively highlight a progressive deterioration in low-temperature crack resistance with increasing steel slag content. The results align with prior findings on asphalt mastics with 100% SSP, which exhibited inferior stress relaxation capacity (lower m-values) at cryogenic temperatures (<−10 °C) compared to mastics with mineral filler.

The diminished low-temperature performance of steel slag-modified asphalt mixtures is mechanistically attributed to increased rigidity of mixtures with SSP/SSA, where SSP's elevated creep stiffness (S increased by 18%–22%) synergistically restricts stress relaxation, amplifying stress accumulation. From a microscopic perspective, although the surface pores of steel slag aggregates can enhance the bonding performance of the asphalt–steel slag interface, its angular and porous structure can also induce stress concentration and be prone to crack initiation [50].

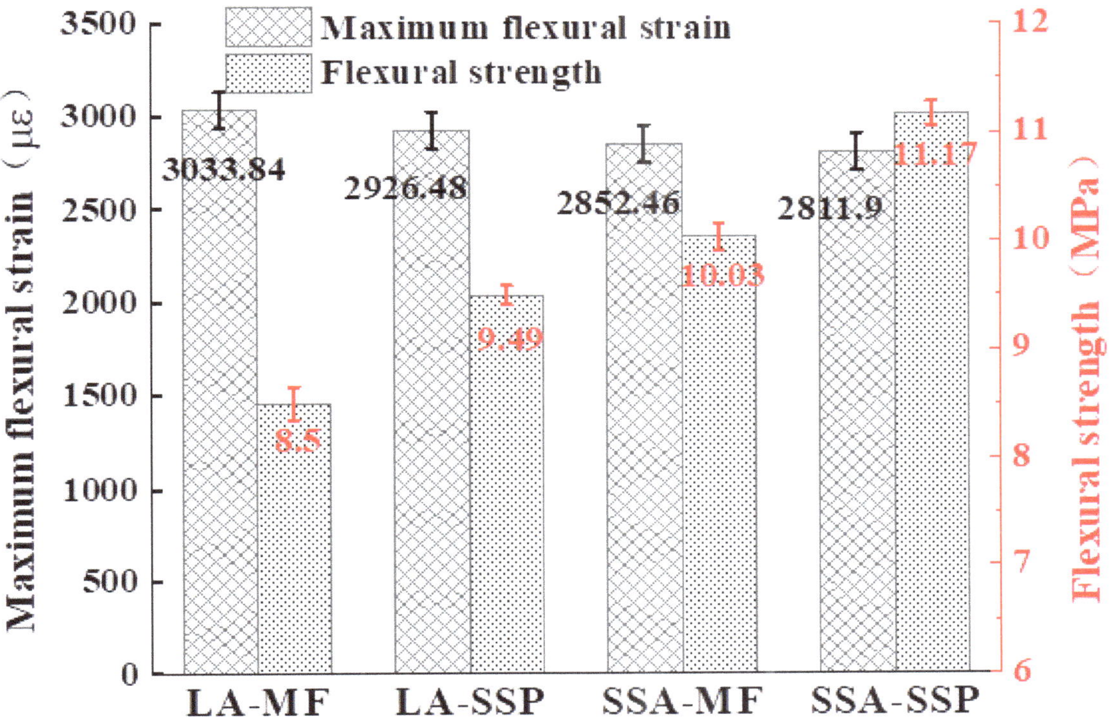

Figure 15. Low-temperature bending test results of asphalt mixtures.

3.2.3. Water Stability

The moisture resistance of asphalt mixtures is typically characterized by the freeze–thaw splitting tensile strength ratio (TSR) and immersed residual stability ratio, with higher values indicating superior resistance to moisture-induced damage. Figures 16 and 17 present the freeze–thaw splitting tensile strength ratio (TSR) and immersed residual stability results, respectively. The TSR improvements are as follows: LA-SSP mixtures exhibit a 1.79% increase compared to LA-MF mixtures, SSA-SSP mixtures demonstrate a 2.35% improvement over SSA-MF counterparts, SSA-MF mixtures achieve a 1.65% enhancement versus LA-MF mixtures, and SSA-SSP mixtures show a 2.21% gain relative to LA-SSP mixtures. Similarly, immersed residual stability enhancements include a 2.26% increase for LA-SSP over LA-MF, a 2.27% improvement for SSA-SSP compared to SSA-MF, a 2.36% gain for SSA-MF versus LA-MF, and a 4.68% rise for SSA-SSP relative to LA-SSP. While the observed improvements in TSR and residual stability are modest, they confirm that steel slag powder (SSP) and steel slag aggregate (SSA) moderately enhance the moisture resistance of asphalt mixtures. Additionally, both control and experimental groups exhibited significant increases in splitting strength and Marshall stability during testing, indirectly indicating improved deformation resistance under thermo-hydro-cyclic coupling conditions.

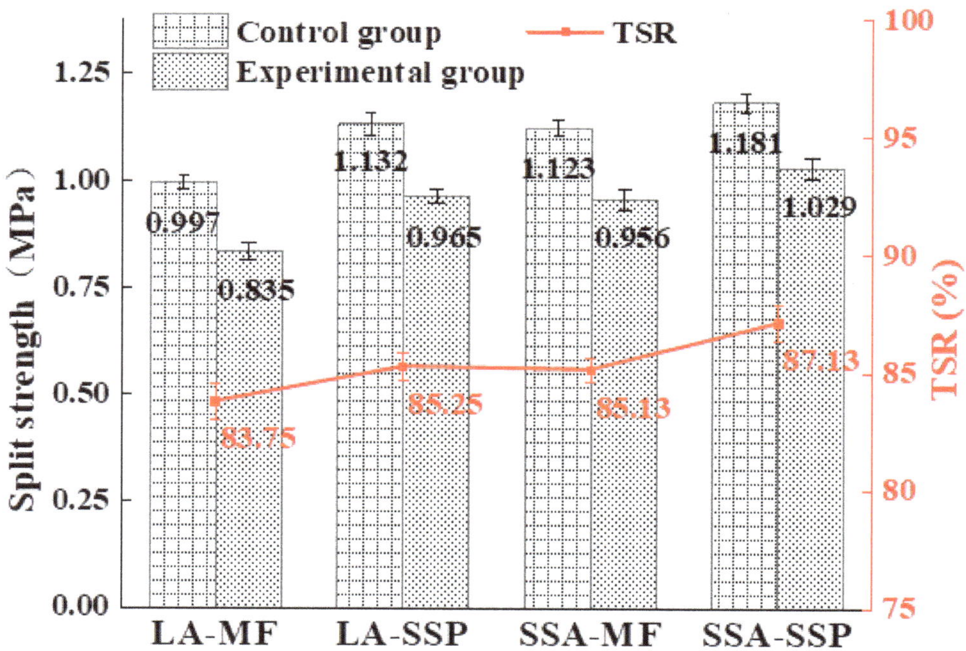

Figure 16. The freeze–thaw splitting tensile strength ratio (TSR) results of asphalt mixtures.

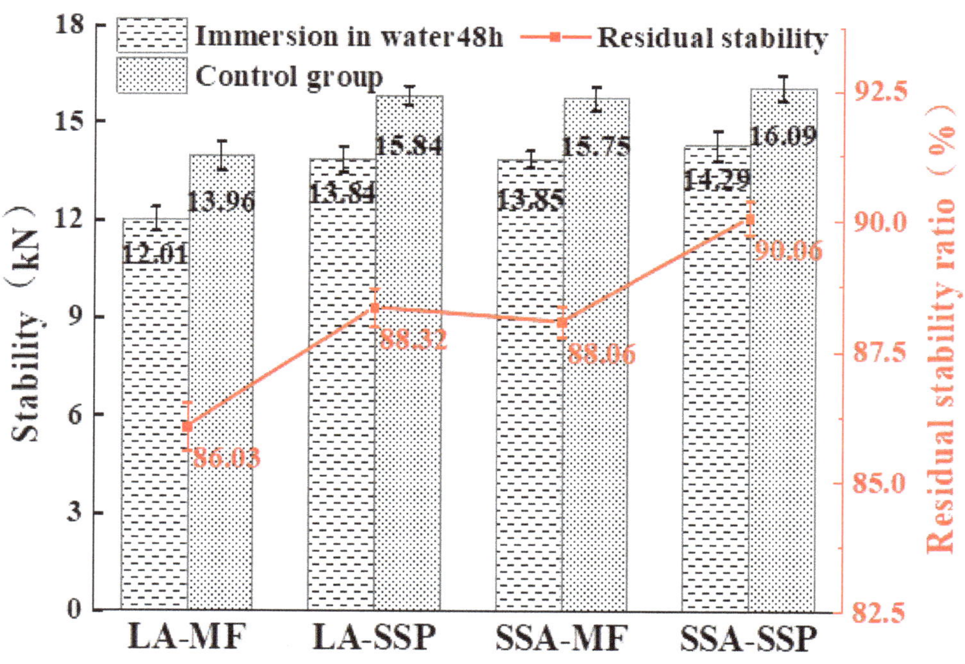

Figure 17. The immersed residual stability results of asphalt mixtures.

3.2.4. Volume Stability

The presence of unstable components such as free calcium oxide (f-CaO), magnesium oxide (MgO), and the RO phase (CaO-FeO-MnO solid solution) in steel slag powder/aggregates increases the risk of volumetric expansion in asphalt mixtures. Consequently, rigorous evaluation of volumetric stability is imperative to assess expansion characteristics during their engineering application. Figure 18 presents the long-term water immersion test results for various asphalt mixtures, revealing that the expansion rate of steel slag-modified mixtures progressively increases with prolonged immersion time. While the 3-day immersion expansion rates comply with specification limits (<1.5% per JTG F40-2004 [40]), significant differences emerge after 21 days: LA-SSP mixtures exhibit a 19.88% higher expansion rate than LA-MF mixtures, SSA-SSP mixtures show a 7.72% increase compared to SSA-MF mixtures, SSA-MF mixtures demonstrate a 424.60% surge relative to LA-MF mixtures, and SSA-SSP mixtures display a 257.79% rise versus LA-SSP mixtures.

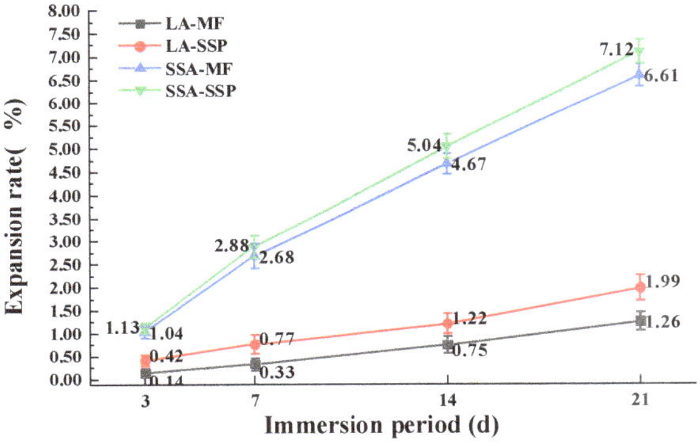

Figure 18. The long-term water immersion test results of asphalt mixtures.

These results confirm that steel slag aggregates exert a far more pronounced influence on volumetric expansion than steel slag powder. Mechanistically, steel slag aggregates—serving as the primary skeletal framework in the mixture—possess greater exposure to water, enabling sustained hydration reactions between f-CaO/MgO/RO phases and moisture. In contrast, steel slag powder, encapsulated within asphalt mastic and confined to interstitial voids, experiences limited water contact, thereby suppressing hydration-driven expansion.

3.2.5. Fatigue Performance

The fatigue resistance of asphalt mixtures represents one of the most comprehensive and stringent performance requirements in pavement engineering. To simulate real-world loading conditions, two-point bending fatigue tests were conducted on the designed mixtures at 10 °C, 25 Hz, and 130 με, with results presented in Figure 19. Analysis reveals that while the incorporation of steel slag powder (SSP) and steel slag aggregate (SSA) reduces the initial stiffness modulus of the mixtures, it slightly enhances their fatigue life: LA-SSP mixtures exhibit a 2.81% increase in fatigue life compared to LA-MF mixtures, SSA-SSP mixtures show a 0.33% improvement over SSA-MF mixtures, SSA-MF mixtures achieve a 4.09% enhancement versus LA-MF mixtures, and SSA-SSP mixtures demonstrate a 1.58% gain relative to LA-SSP mixtures.

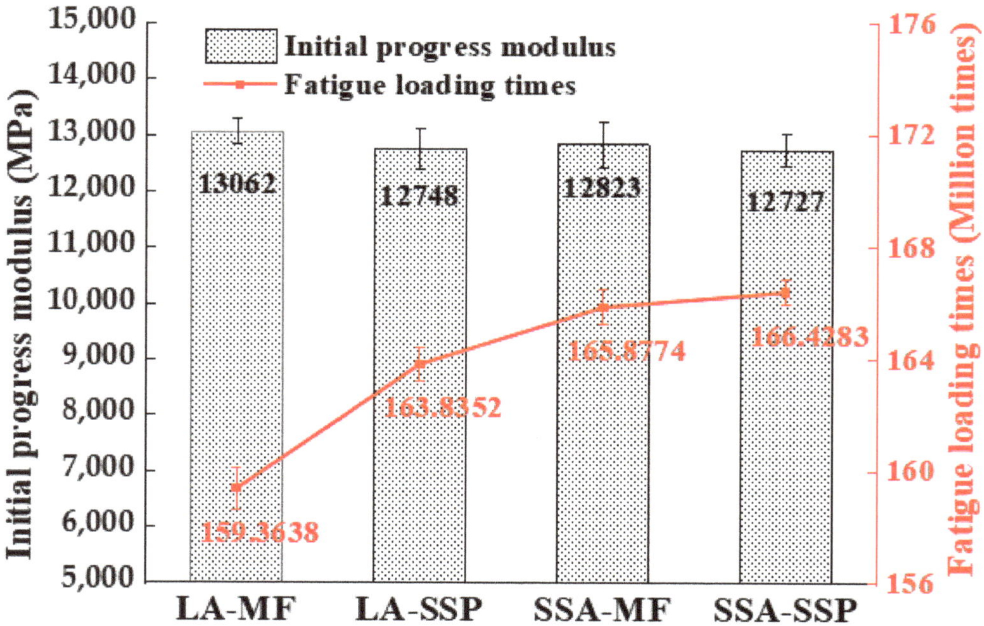

Figure 19. Two-point bending fatigue test results of asphalt mixtures.

The enhanced fatigue performance of steel slag powder (SSP)- and steel slag aggregate (SSA)-modified asphalt mixtures is attributed to their unique morphological and interfacial characteristics. SSP exhibits a complex surface texture and elevated roughness, providing a greater effective adhesion area compared to conventional mineral fillers. Concurrently, SSA's irregular porous structure facilitates the adsorption of additional asphalt mastic into voids within the coarse aggregate skeleton. Research indicates that stress concentrations in asphalt mixtures predominantly occur at mastic–coarse aggregate interfaces. The incorporation of SSP/SSA disperses these stress concentration zones, promoting the formation of distributed microcracks that gradually coalesce rather than propagating catastrophically. This microcrack dispersion mechanism reduces the overall stiffness modulus of SSP/SSA mixtures relative to limestone-mineral filler systems.

Under strain-controlled loading, the lower initial stiffness modulus of SSP/SSA mixtures reduces the applied stress required to maintain the target strain level, thereby decelerating crack propagation rates as per Paris' law ($da/dN = C(\Delta K)^n$). The diminished stress intensity factor (ΔK) at crack tips prolongs fatigue life by extending the cycles required for critical crack growth. Consequently, SSP/SSA-modified mixtures exhibit superior fatigue resistance despite their reduced stiffness, demonstrating the efficacy of steel slag in balancing stiffness–ductility trade-offs for durable pavement design.

4. Conclusions

Based on the limited testing results, the following conclusion can be drawn.

(1) The softening point, 135 °C rotational viscosity and interfacial bonding strength indicators of asphalt mastics increase as the proportion of steel slag powder (SSP) replacing mineral powder filler (MF) increases, while the penetration and ductility indicators gradually decrease with the increase of substitution dosage. When the SSP replacement percentage reaches 100%, the softening point and the rotational viscosity at 135 °C in-

crease by approximately 3.24%, 75.31% and 19.93% respectively, while the penetration and ductility decrease by approximately 10.21% and 14.13% respectively. Asphalt mastics with SSP demonstrate superior high-temperature rheological performance compared to limestone mineral filler mastic systems, albeit with compromised low-temperature rheological properties.

(2) The incorporation of steel slag materials enhances the high-temperature stability, moisture resistance, and fatigue resistance of asphalt mixtures. However, it introduces detrimental effects on low-temperature crack resistance and volumetric stability, primarily due to the hydration potential of free lime (f-CaO) and magnesium oxide (MgO) in steel slag under moisture exposure.

(3) All designed asphalt mixtures comply with the performance criteria outlined in the Technical Specifications for Highway Asphalt Pavement Construction (JTG F40-2004) [40] for Climate Zone 1-3-2 (hot summer, cold winter, humid region), achieving: high-temperature performance (dynamic stability $\geq$ 2800 cycles/mm), low-temperature flexibility (maximum flexural tensile strain $\geq$ 2500 $\mu\varepsilon$), moisture resistance (freeze–thaw splitting tensile strength ratio $\geq$ 80%, immersed residual stability ratio $\geq$ 85%), and volumetric stability (expansion rate $\leq$ 1.5%). These results validate the mixtures' conformity to critical pavement performance thresholds under the specified climatic and mechanical loading conditions.

While these mixtures satisfy conventional pavement performance requirements, full-component utilization of steel slag in engineering applications necessitates rigorous volumetric expansion control measures, such as pre-carbonation of slag or chemical stabilization, to mitigate long-term durability risks.

Author Contributions: Conceptualization, J.W.; Methodology, J.W. and Q.X.; Software, X.Y.; Validation, L.K. and W.S.; Investigation, F.X. and L.K.; Resources, J.G. and W.S.; Data curation, X.Y.; Writing—original draft, Q.X. and W.W.; Writing—review & editing, W.W. All authors have read and agreed to the published version of the manuscript.

Funding: This research was funded by the Taishan Scholars Program under grant tstp20231240.

Institutional Review Board Statement: Not applicable.

Informed Consent Statement: Not applicable.

Data Availability Statement: The data presented in this study are available in the article.

Conflicts of Interest: Author Jianmin Guo, Feiping Xu, Liang Kang, and Wencheng Shi were employed by the company Shandong Hi-Speed Company Limited. The remaining authors declare that the research was conducted in the absence of any commercial or financial relationships that could be construed as a potential conflict of interest.

References

1. Lin, D.; Chou, L.; Wang, Y.; Luo, H.-L. Performance evaluation of asphalt concrete test road partially paved with industrial waste basic oxygen furnace slag. *Constr. Build. Mater.* **2015**, *78*, 315–323. [CrossRef]
2. Bessa, L.S. Evaluation of polishing and degradation resistance of natural aggregates and steel slag using the aggregate image measurement system. *Constr. Build. Mater.* **2014**, *75*, 158–166. [CrossRef]
3. Li, J.; Yan, F.; Zhou, B. Testing and evaluation for anti-skid performance of steel slag asphalt pavement based on texture features. *Mater. Rep.* **2024**, *38* (Suppl. S1), 315–321.
4. E, G.; Cao, W.; Chen, L. Rules of voids in coarse aggregate formed by the packing of steel slag coarse aggregates and its application. *Sci. Technol. Eng.* **2024**, *24*, 8290–8296.
5. Asi, I.M.; Qasrawi, H.Y.; Shalabi, F.I. Use of steel slag aggregate in asphalt concrete mixes. *Can. J. Civ. Eng.* **2007**, *34*, 902–911. [CrossRef]
6. Goli, H.; Hesami, S.; Ameri, M. Laboratory Evaluation of Damage Behavior of Warm Mix Asphalt Containing Steel Slag Aggregates. *J. Mater. Civ. Eng.* **2017**, *29*, 04017009. [CrossRef]

7. Chen, X.; Wen, W.; Zhou, J.; Zhou, X.; Ning, Y.; Liang, Z.; Ma, Z. Research on the Interaction Capability and Microscopic Interfacial Mechanism between Asphalt-Binder and Steel Slag Aggregate-Filler. *Coatings* **2022**, *12*, 1871. [CrossRef]
8. Liu, K.; Zhu, X.; Lin, Z.; Tan, Y.; Wu, S.; Zhu, Y.; Li, Y. Study on the Improvement and Mechanism of Anti-sliding Functional Characteristics of Steel Slag SMA-13 Asphalt Mixture. *J. Wuhan Univ. Technol. (Transp. Sci. Eng.)* **2024**, *48*, 1185–1191.
9. Yan, X.; Xu, Q.; Zhang, H.; Fu, G.; Zhang, M.; Gao, H. Effect of coarse and fine steel slag aggregates on the performance of AC-13 asphalt mixture. *Shandong Jiaotong Keji* **2024**, *03*, 67–71.
10. Shen, A.; Liu, B.; Guo, Y.; Yu, P.; Yu, M. Skid resistance attenuation of steel slag asphalt mixtures on tunnel pavement. *J. Build. Mater.* **2019**, *22*, 284–291.
11. Wang, X.; Hu, L.; Cheng, Y.; Wang, P. Volume expansibility of asphalt mixture mixed with steel slag. *Sci. Technol. Eng.* **2024**, *24*, 6901–6906.
12. Zhang, X.; Song, X. Study on road performance of steel slag fine aggregate asphalt mixture. *Sichuan Build. Mater.* **2023**, *49*, 163–166.
13. Dong, R. Research on Performance and Fatigue Damage of Steel Slag Fine Aggregates in Asphalt Mixture. Master's Thesis, Kunming University of Science and Technology, Kunming, China, 2023.
14. Kong, D.; Wu, S.; Chen, M.; Zhao, M.; Shu, B. Characteristics of different types of basic oxygen furnace slag filler and its influence on properties of asphalt mastic. *Materials* **2019**, *12*, 4034. [CrossRef] [PubMed]
15. Li, J. Study on the Influence of Steel Slag Powder on the Water Stability Performance of Asphalt Concrete. *Highw. Transp. Technol. (Appl. Technol. Ed.)* **2020**, *16*, 93–95.
16. Fu, H.; Yao, L.; Ma, X.; Lv, D.; Ou, Y. Performance of asphalt mastic with steel slag filler. *J. Henan Univ. Urban Constr.* **2020**, *29*, 54–59+86.
17. Liu, H. Performance evaluation and improving mechanisms of steel slag power based asphalt mixture. *J. Shandong Univ. (Eng. Sci.)* **2023**, *53*, 32–38+48.
18. Xiao, Z.; Chen, M.; Wu, S.; Xie, J.; Kong, D.; Qiao, Z.; Niu, C. Moisture susceptibility evaluation of asphalt mixtures containing steel slag powder as filler. *Materials* **2019**, *12*, 3211. [CrossRef]
19. Chen, Y. Study on surface physical properties of steel slag powder and rheological properties of asphalt mortar at high temperature. *Fujian Archit. Constr.* **2021**, *10*, 122–127.
20. Li, C.; Chen, Z.; Wu, S.; Li, B.; Xie, J.; Xiao, Y. Effects of steel slag fillers on the rheological properties of asphalt mastic. *Constr. Build. Mater.* **2017**, *145*, 383–391. [CrossRef]
21. Sun, J.; Jiang, H. Research on characteristics of steel slag powder and its application in asphalt mixture. *J. Chongqing Jiaotong Univ. (Nat. Sci.)* **2008**, *04*, 576–579.
22. Xu, S. Research on the Preparation of Permeable Steel Slag Asphalt Mixture and Interface Mechanism. Master's thesis, Xi'an University of Architecture and Technology, Xi'an, China, 2017.
23. Zhang, C.; Li, S.; Wang, B.; Wang, Y.; Li, T.; Meng, Q. Water Stability of Steel Slag Aggregate asphalt Interface. *J. Mater. Sci. Eng.* **2024**, *42*, 99–107+152.
24. Li, S.; Zhang, C.; Ding, W.; Wang, B.; Li, T. Research on adhesion characteristics of steel slag-asphalt interface. *J. Dalian Univ. Technol.* **2022**, *62*, 254–262.
25. Hu, P.; An, W.; Chen, X. Evaluation of steel slag-asphalt adhesion based on surface energy theory. *Mod. Transp. Metall. Mater.* **2025**, *5*, 109–114.
26. Hu, P.; An, W.; Wang, L.; Chen, X. Study on the adhesion between multi-scale steel slag aggregate and asphalt interface. In Proceedings of the World Transport Conference (WTC2024) (Highway Engineering), Qingdao, China, 26 June 2024; pp. 49–54.
27. Gu, J.; Sun, S.; Zheng, B.; Guo, C.; He, L. Analysis of Adhesion Test of Steel Slag based on Microscopic Characteristics. *Transp. Energy Conserv. Environ. Prot.* **2024**, *20*, 139–144.
28. Wang, L.; Li, W.; Wang, Y.; Li, Y.; Miao, Y.; Mo, L. Effects of surface morphology of steel slag aggregates on asphalt absorption and adhesion performance. *Sci. Technol. Eng.* **2023**, *23*, 14406–14419.
29. Albayati, A.H.; Oukaili, N.K.; Mustafa, M.; Allawi, A.A.; Said, A.I.; Ibrahim, T.H. Experimental Study to Investigate the Performance-Related Properties of Modified Asphalt Concrete Using Nanomaterials Al_2O_3, SiO_2, and TiO_2. *Materials* **2024**, *17*, 4279. [CrossRef] [PubMed]
30. Feng, M.; Ma, H.; Zhang, H.; Kong, C.; Jiao, X.; Song, X. Cracking mechanism of steel slag aggregate during asphalt mixture production. *Sci. Technol. Eng.* **2024**, *24*, 14846–14853.
31. Xu, H.; Guo, P.; Chen, X.; Yao, B. Study on Re-cracking Behavior of Steel Slag Asphalt Mixture at Different Healing Efficiencies. *Highw. Eng.* **2024**, *49*, 167–174+188.
32. Qin, Y. Mix proportion design and performance verification of recycled asphalt mixture with steel slag. *Highw. Transp. Inn. Mong.* **2024**, *06*, 43–47.
33. Chen, J.; Wei, S. Engineering properties and performance of asphalt mixtures incorporating steel slag. *Constr. Build. Mater.* **2016**, *128*, 148–153. [CrossRef]

34. Liu, L.; Zhu, X.; Liu, Z.; Li, W.; Yang, C.; Huang, Y.; Liu, L. Effect of Steel Slag Content on Performance of Rubber Mixture ARAC-13. *Mater. Rep.* **2023**, *37*, 66–72.
35. Li, M.; Wu, S.; Xie, J.; Liu, T.; Wang, J. Research on Composition Design and Performance of Ultrathin Wear Layer of Asphalt Pavement. *J. Wuhan Univ. Technol. (Transp. Sci. Eng.)* **2024**, *48*, 964–968.
36. Zhuang, W.; Huang, W.; Wang, X.; Sun, J.; Luo, S. Performance of mixtures of epoxy asphalt and steel slag and its composition optimization. *Thermosetting Resin* **2024**, *39*, 7–13.
37. Wang, F.; Zhou, S.; Zheng, H.; Han, F.; Wang, H. Research on road performance of porous steel slag asphalt mixture. *New Build. Mater.* **2024**, *51*, 144–148.
38. Jiang, C. Experimental Study on the Influence of Steel Slag on the Properties of Ultra-Thin Wear Layer. Master's thesis, Yangzhou University, Yangzhou, China, 2024.
39. Wang, Z.; Ma, H.; Wu, S.; Yang, C.; Xie, J. Investigation on the Composition Design and Performance of OGFC5 Steel Slag Asphalt Mixture. *J. Wuhan Univ. Technol. (Transp. Sci. Eng.)* **2024**, *48*, 959–963+968.
40. *JTG F40-2004*; Technical Specification for Highway Asphalt Pavement Construction. People's Communications Press: Beijing, China, 2004.
41. *GBT 25824-2010*; Steel Slag for Road. Standards Press of China: Beijing, China, 2010.
42. *JTG 3432-2024*; Test Methods of Aggregates for Highway Engineering. People's Communications Press: Beijing, China, 2024.
43. *JTG E20-2011*; Highway Engineering Asphalt and Asphalt Mixture Test Regulations. People's Communications Press: Beijing, China, 2011.
44. *ASTM D4541-22*; Standard Test Method for Pull-Off Strength of Coatings Using Portable Adhesion Testers. American National Standards Institute: New York, NY, USA, 2022.
45. *AASHTO T324-2014*; Hamburg Wheel-Track Testing of Compacted Hot-Mix Asphalt (HMA). Association of State Highway and Transportation Officials: Washington, DC, USA, 2014.
46. *CSN EN 12697-24*; Bituminous Mixtures-Test Methods for Hot Mix Asphalt Part 24: Resistance to Fatigue. European Committee for Standardization: Brussels, Belgium, 2018.
47. Zhang, H.; Chen, Z.; Xu, G.; Shi, C. Physical, rheological and chemical characterization of aging behaviors of thermochromic asphalt binder. *Fuel* **2018**, *211*, 850–858. [CrossRef]
48. Zhang, W.; Wan, G.; Qiao, Y.; Zhang, J.; Xiong, Y. Study on the recovery law of rheological properties of aged asphalt with bio-oil-based recycling agents. *Mater. Rep.* **2025**, *39*, 24110036.
49. Ma, X.Y. Study on Asphalt-Filler Interaction and Prediction of Asphalt Mastic Performance. Ph.D. Thesis, Chang'an University, Xi'an, China, 2019. (In Chinese).
50. Zeng, H.; Shen, Z.; Zhi, X.; Huang, Y.; Li, T.; Chen, Y.; Xiao, Y. Investigating Mechanisms of Porous Structure of Steel Slag Affecting Low-temperature Crack Resistance of Asphalt Mixture. *J. Beijing Univ. Technol.* **2024**, *50*, 82–93.

coatings

MDPI

Article

The Performance of Sulfoaluminate Cement Mortar with Secondary Aluminum Dross

Qian Wang [1], Linchun Zhang [1,*] and Ailian Zhang [1,2]

1 School of Civil Engineering, Sichuan College of Architectural Technology, Deyang 618000, China; wangq@scac.edu.cn (Q.W.); 21b233027@hit.edu.cn (A.Z.)
2 School of Civil Engineering, Harbin Institute of Technology, Harbin 150090, China
* Correspondence: zhanglc@scac.edu.cn

Abstract: This paper endeavors to explore the impact of secondary aluminum dross (SAD) on the characteristics of sulfoaluminate cement mortar. Measurements were taken for the mortar's slump flow, plastic viscosity, initial setting time, and drying shrinkage rate (DR). Additionally, the flexural, compressive, and bonding strengths were evaluated. The leached concentrations of chromium (Cr) and zinc (Zn) in the specimens were determined. Furthermore, the carbonation depth (Dc) and chloride ion migration coefficient (CMC) were calculated. Lastly, scanning electron microscope energy spectrum analysis (SEM-EDS) and X-ray diffraction (XRD) spectrum analysis were conducted to analyze the mortar's performance. The findings revealed that the slump flow and plastic viscosity of fresh mortar exhibited negative and positive quadratic relationships, respectively, with the mass ratio of SAD. Specifically, SAD could reduce the slump flow by 1.57% to 2.72% and augment the plastic viscosity by 5.21% to 36.89%. The placement time contributed to a decrease in the slump flow of fresh mortar by up to 20.4% and an increase in plastic viscosity by up to 11.2%. The initial setting time, mechanical strength, and DR of the mortar demonstrated quadratic variations with the mass ratio of SAD. At a 15% SAD mass ratio, the mortar exhibited the highest initial setting time, mechanical strength, and DR. The inclusion of SAD could elevate the initial setting time, flexural strength, compressive strength, bonding strength, and DR of the mortar by 14.33% to 65.07%, −14.75% to 22.58%, −8.94% to 9.96%, −13.33% to 66.67%, and −13.33% to 26.67%, respectively.

Keywords: secondary aluminum dross; sulfoaluminate cement; mechanical strength; drying shrinkage rate; plastic viscosity

1. Introduction

Cement concrete structures inevitably sustain damage over a certain period of service [1,2]. Consequently, the repair of these damaged structures becomes imperative [3]. This necessitates the development of suitable repair materials tailored for concrete structures that have undergone degradation.

Asphalt and cement-based materials are commonly employed in the preparation of repair materials [4,5]. Asphalt is known for its rapid repair capabilities in damaged load-bearing areas. However, it is ill-suited for repairing building structures, due to its inadequate structural integrity. Furthermore, the application of asphalt releases toxic gases at certain concentrations, and its cost is relatively high [6,7]. Therefore, there is a pressing need to develop cement-based repair materials that are suitable for engineering applications.

Cement-based materials offer an alternative solution for the repair of concrete structures [8]. Portland cement is frequently used in the preparation of repair materials. Nonetheless, its slow setting speed renders it unsuitable for rapid repairs of concrete structures [9]. In contrast, sulfate aluminate cement is a fast-setting cement-based material that can be effectively applied to damaged concrete structures [10]. Extensive research has been conducted on the rheological, mechanical, and durability properties of sulfoaluminate cement concrete materials [11]. Studies have shown that sulfate aluminate cement paste exhibits lower fluidity and higher plastic viscosity compared to Portland cement paste [12–14]. Additionally, the mechanical strength of sulfate aluminate cement concrete cured for less than three days is higher than that of Portland cement concrete [15]. However, at later curing ages (exceeding three days), sulfate aluminate cement concrete demonstrates lower mechanical strength than Portland cement concrete [16,17]. This makes sulfoaluminate cement concrete suitable for construction projects with tight deadlines [18]. The incorporation of basalt, steel, and polypropylene fibers has been proven to enhance the flexural strength by 11.3%–21.1%, 16.7%–31.8%, and 8.6%–18.3%, respectively [19,20], while the corresponding compressive strengths increase by 5.1%–17.6%, 15.1%–28.7%, and 4.3%–14.1%, respectively [21,22]. Despite the potential of fiber-reinforced sulfoaluminate cement concrete as a repair material, there is a need to reduce its preparation cost.

Secondary aluminum dross is an industrial waste product generated during aluminum production [23]. If not properly treated, toxic elements can leach out and contaminate the environment [24]. Previous research has demonstrated that secondary aluminum dross can increase the flexural and compressive strengths by 2.1%–11.5% and 3.7%–14.3%, respectively, when specimens are cured for less than seven days [25,26]. However, when the curing age exceeds seven days, the flexural and compressive strengths decline by 1.3%–9.1% and 2.6%–19.4%, respectively. Additionally, secondary aluminum dross can reduce the drying shrinkage rate, chloride ion migration coefficient, and carbonation depth of cement concrete [27]. It has been shown to enhance the strengths of ordinary Portland cement concrete. As a fast-hardening repair cement material, sulfoaluminate cement is derived from aluminum oxide [28]. Secondary aluminum dross may provide hydraulic aluminum substances, thus potentially replacing some sulfoaluminate cement while ensuring safe disposal. However, this aspect has received limited attention.

This study investigated the slump flow and initial setting time of fresh sulfoaluminate cement mortar. It also examined the drying shrinkage rate, flexural strength, compressive strength, and bonding strength of hardened sulfoaluminate cement mortar cured for 0.25 to 28 days. The leached Zn and Cr, carbonation depth, and chloride ion migration coefficient of the sulfoaluminate cement mortar were tested. Finally, scanning electron microscope energy spectrum analysis (SEM-EDS), X-ray diffraction (XRD) spectrum analysis, and thermogravimetric curves were employed to elucidate the mechanisms underlying the performance changes in hardened sulfoaluminate cement mortar. This research provides additional raw materials and novel ideas for the preparation of sulfoaluminate cement and the treatment of secondary aluminum dross. The primary innovation of this study lies in its utilization of secondary aluminum ash to replace sulfoaluminate cement, a direction that has been scarcely explored in previous research.

2. Experimental

2.1. Raw Materials

In this study, sulfoaluminate cement (SAC) manufactured by Tangshan Polar Bear Building Materials Co., Ltd., located in Tangshan, China, was employed. The SAC exhibited a strength grade of 42.5 MPa, with an initial setting time of 21.3 minutes and a final setting time of 204.5 minutes. Quartz sand, comprising particle sizes of 0.71 mm to 1 mm, 0.35 mm to 0.59 mm, and 0.15 mm to 0.297 mm in a mass ratio of 1:1.5:1, was utilized as the aggregate for the mortar. This mass ratio of quartz sand represented the optimal grading value derived from the theory of maximum density, as per Cai's research [2]. A polycarboxylate-based water-reducer was adopted as a high-range water-reducing agent, demonstrating a water-reduction efficiency of 39.2%. Li_2SO_4 and tartaric acid were employed as the early strength agent and retarder, respectively. The purity levels of Li_2SO_4 and tartaric acid were 99.9% and 99.6%, respectively. These three additives were supplied by Yingshan New Material Technology Co., Ltd., based in Shanghai, China. Li_2SO_4, at a mass ratio of 0.15% relative to the mass of SAC, served as the early strength agent. Meanwhile, tartaric acid, at a mass ratio of 0.035%, was utilized as the retarder in this investigation. Secondary aluminum dross (SAD), provided by Dongguan Yuehui Renewable Resources Co., Ltd., located in Dongguan, China, was also incorporated in this study. The particle sizes and chemical compositions of the raw materials are detailed in Tables 1 and 2. The morphology of SAD is shown in Figure 1.

Table 1. The raw materials' particle size (%).

Types	Particle Size/μm						
	0.3	0.6	1	4	8	64	360
Quartz sand	0	0	0	0	0.059	20.89	100
SAC	0	0.34	1.91	16.24	29.85	94.86	100
SAD	0.07	0.25	0.57	1.16	3.93	25.91	87.25

Table 2. The raw materials' chemical compositions (%).

Types	SiO_2	Al_2O_3	Fe_xO_y	MgO	CaO	SO_3
Quartz sand	99.3	-	0.7	-	-	-
SAC	13.94	15.06	3.2	-	49.9	13.85
SAD	4.7	79.2	3.8	5.9	1.4	-

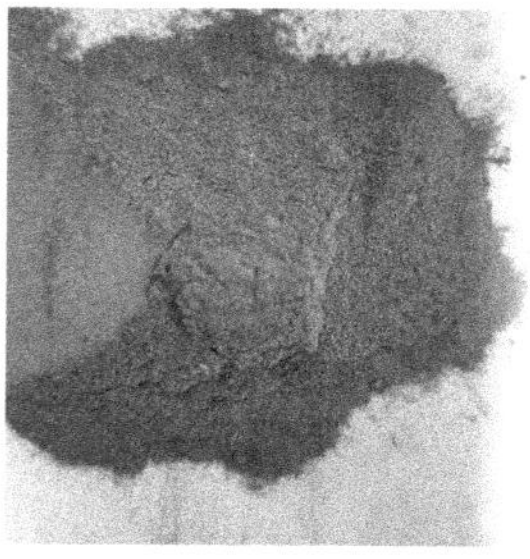

Figure 1. The morphology of SAD.

2.2. The Preparation of Mortar

The sulfoaluminate cement mortar was prepared by adhering to the following steps. Initially, the sulfoaluminate cement, standard sand, and SAD were introduced into a JJ-5 planetary cement mortar mixer provided by Shandong Huakuang High tech Development Co., Ltd., Jining, China and agitated at a mixing speed of 140 r/min for a duration of 2 minutes. Subsequently, the water-reducing agent, defoamer, and other additives were incorporated into the mixer, with the mixing speed adjusted to 285 r/min for an additional 2 minutes. Following the mixing process, a portion of the fresh mortar was poured into molds measuring 40 × 40 × 160 mm^3, 100 × 100 × 100 mm^3, and Φ100 × 50 mm^3, and cured in a room-temperature environment (47.6% relative humidity and 21.3 °C) for 4 hours until it hardened. Thereafter, the molds were removed, and the specimens were transferred to a standard curing room (96.7% relative humidity and 20.1 °C) for curing periods of 0.25 day, 1 day, 3 days, 7 days, 14 days, and 28 days. The residual sulfoaluminate cement mortar was utilized in the measurement of slump flow and plastic viscosity. The mixing proportions of the SAD sulfoaluminate cement mortar are detailed in Table 3.

Table 3. The SAD sulfoaluminate cement mortar's mixing proportions (kg/m^3).

Water	SAC	SAD	Quartz Sand	Water-Reducer	Li_2SO_4	Tartaric Acid
236.33	716.26	0	945.63	15.76	1.07	0.25
236.33	626.72	89.54	945.63	15.76	0.94	0.22
236.33	537.18	179.08	945.63	15.76	0.81	0.19
236.33	447.73	268.53	945.63	15.76	0.67	0.16
236.33	358.18	358.08	945.63	15.76	0.54	0.13

2.3. The Measuring Process

2.3.1. The Measurement of Rheological Parameters

The flowability of the cement mortar composites was assessed according to the Chinese standard GB/T 50080-2016 [29]. For each set of mixtures, the material was placed in a truncated conical circular mold and compacted. Excess mixture was carefully removed using a spatula. The mold was then gently lifted vertically, and the jumping table was activated 25 times at a rate of one activation per second. Upon completion of this process, the maximum diameter of the fresh cement mortar and the corresponding vertical section diameter were measured. The average of these diameters was recorded as the slump flow of the fresh mortar.

The plastic viscosity measurement was conducted at a temperature of 20 °C, utilizing a Brookfield RST-SST rheometer equipped with a four-blade rotor. The fresh mortar mixtures were immediately injected into the rheometer's container after preparation. The plastic viscosity was determined by fitting the shear stress and shear rate data from the ascending stage using the Bingham model.

2.3.2. The Drying Shrinkage Rate

A shrinkage rod was utilized for measuring the drying shrinkage of specimens. The length change value was recorded using a TD511-300 dial indicator, supplied by Cangzhou Tai Ding Hengye Test Instrument Co., LTD., located in Cangzhou, China, during the curing period. The drying shrinkage rate (*DSR*) was calculated using Equation (1). All specimens were securely fastened to a steel frame throughout the testing procedure.

$$DSR = \frac{L_0 - L}{L_0} \quad (1)$$

where L_0 and L denote the specimen's initial length and the length during curing, respectively.

2.3.3. The Initial Setting Time

The initial setting time of the specimens was measured using a ZKS-100A mortar setting time tester, supplied by Zhejiang Luda Mechanical Instrument Co., LTD., located in Shaoxing, China. The measurement process adhered to the Chinese standard JGJ 70-2009 [30]. All samples were placed in containers, ensuring that the surface of the mortar was 10 mm below the rim of the container. Subsequently, the tip of the penetration test needle, with a cross-sectional area of 30 mm^2, was brought into contact with the surface of the mortar, and the reading was set to zero. Finally, the mortar was vertically pressed to a depth of 25 mm within 10 seconds, and the reading was recorded. The penetration resistance value was calculated according to Equation (2). The setting time was defined as the moment when f_p reaches 0.7 MPa.

$$fp = \frac{Np}{Ap} \quad (2)$$

where f_p represents the penetration resistance (MPa), N_p denotes the static pressure exerted at a penetration depth of up to 25 mm (N), and A_p signifies the cross-sectional area of the penetration test needle (30 mm^2).

2.3.4. The Mechanical Strength

The flexural and compressive strengths of specimens cured for 0.25 day, 1 day, 3 days, 7 days, 14 days, and 28 days were tested, using specimens with a size of 40 × 40 × 160 mm^3. Firstly, the specimens were moved to the flexural strength fixture, and then a flexural load was exerted on the specimens at a 0.05 kN/s loading rate, until the specimens were destroyed. After that, the specimens were moved to the compressive strength fixture, then loading was applied with a compressive loading rate of 2.4 kN/s until the specimens were destroyed. The bonding strength was determined with the following steps. Each specimen with a size of 40 × 40 × 160 mm^3 comprised two halves. One half was made of ordinary Portland cement mortar, and the other half was made of sulfoaluminate cement mortar. The specimens were moved to the flexural strength fixture and were measured according to the same method used for determining the cement mortar's flexural strength. The Chinese standard GBT17671-2021 was used for the measurement of the mechanical strength [31].

2.3.5. Chloride Ion Permeability Coefficient of Concrete

Specimens with dimensions of Φ100 × 50 mm^3 were saturated in a vacuum saturator for 2 days. Thereafter, they were subjected to the chloride ion permeability coefficient (CMC) test using the NELD-CCM550 concrete chloride ion diffusion coefficient tester, supplied by Beijing Neerde Intelligent Technology Co., Ltd., Beijing, China. The experimental details are outlined in the Chinese standard GB/T 50082-2024 [32].

2.3.6. Carbonation Depth of Concrete

For the carbonation experiment, specimens with dimensions of 100 × 100 × 100 mm^3 were utilized. Each specimen was cured for 28 days and then placed in the CCB-70F automatic concrete carbonation test box, purchased from Zhucheng Kerun Machinery Co., Ltd. (Weifang, China). The specimens were cured under a 10% CO_2 and 90% air environment. Following carbonation, the carbonation depth was measured according to the Chinese standard GB/T 50082-2024.

2.3.7. Micro-Analysis

The SU3800 scanning electron microscope (SEM), purchased from Hitachi Scientific Instruments (Beijing) Co., Ltd., Beijing, China, was employed to obtain SEM images and energy-dispersive spectroscopy (EDS) results. Initially, samples were extracted from the core of the specimens. They were then dried in an oven at 105 °C for 2 days and coated with gold by vacuum spraying. Finally, the sprayed samples were transferred to the SU3800 for SEM and EDS measurements. The XRD experiment was conducted as follows: The powdered sample was placed in a D8 Discover X-ray diffractometer (XRD), supplied by Bruker, Berlin, Germany. The XRD experiment was performed using a working voltage of 40 kV, a current of 40 mA, a scanning speed of 0.6°/min, and a scanning range of 5° to 90°. The overall experimental process is illustrated in Figure 2.

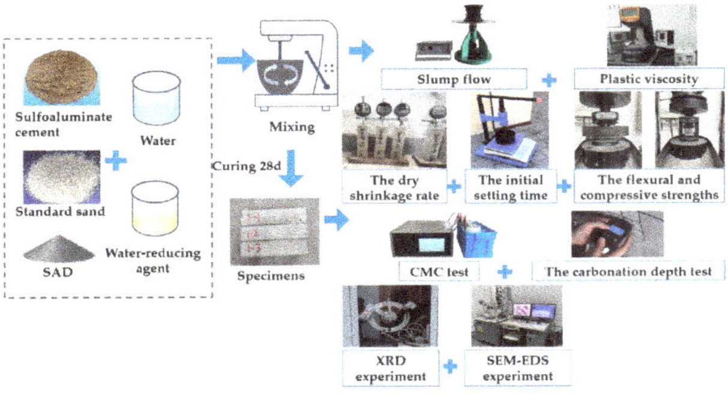

Figure 2. The overall experimental process.

3. Results and Discussion

3.1. The Rheological Parameters

The slump flow of fresh sulfoaluminate cement (SAC) mortar incorporating SAD (a specific additive) is depicted in Figure 3. As illustrated in this figure, the slump flow of fresh SAC mortar increases with an increase in the mass ratio of SAD. Previous studies [33] have highlighted that SAD exhibits a superior ball-bearing effect compared to ordinary Portland cement, thereby enhancing the fluidity of fresh SAC mortar upon its addition [34]. However, the placing time exerts a detrimental influence on the flowability of fresh SAC mortar. Free water within the fresh SAC mortar evaporates during the placing period, consequently reducing its fluidity [35]. The incremental rate of slump flow due to SAD addition and the decremental rate due to placing time are 0%–20.4% and 0%–11.2%, respectively. The error bars associated with the slump flow measurements of fresh SAD-SAC mortar are less than 10% of the slump flow values. The correlation between slump flow and the mass ratio of SAD adheres to a quadratic function.

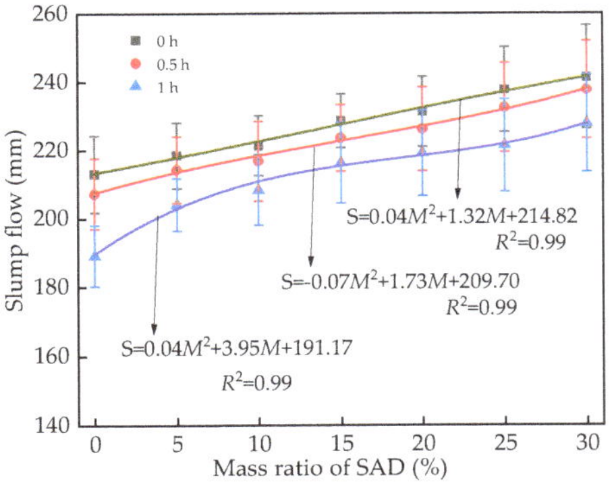

Figure 3. The slump flow of fresh SAD sulfoaluminate cement mortar.

The plastic viscosity of fresh sulfoaluminate cement (SAC) mortar with SAD (a specific additive) is illustrated in Figure 4. As depicted in this figure, the plastic viscosity of fresh SAD-SAC mortar decreases with an increase in the SAD mass ratio. Conversely, the placing time exhibits an augmenting effect on the plastic viscosity of fresh SAC mortar. Research substantiates the inverse correlation between the slump flow and the corresponding plastic viscosity of fresh SAC mortar [36]. Consequently, as the SAD content in the SAC mortar increases, its plastic viscosity decreases; conversely, when the SAD content decreases, the plastic viscosity increases. Therefore, with an increase in placing time, the slump flow decreases, while the plastic viscosity rises.

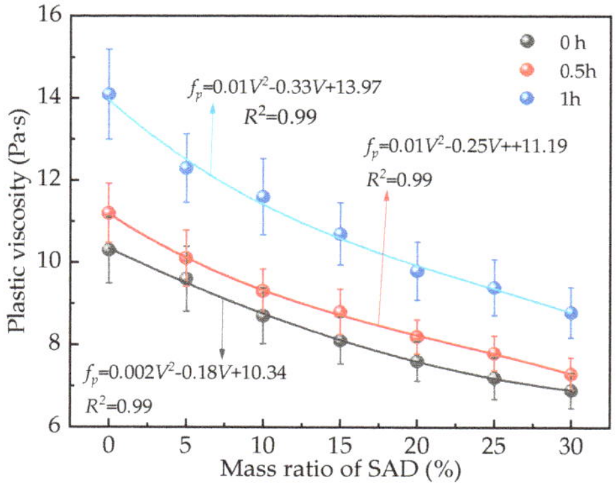

Figure 4. The plastic viscosity of fresh SAD sulfoaluminate cement mortar.

The initial setting time of SAD sulfoaluminate cement mortar is illustrated in Figure 5. As shown in the figure, the initial setting time varies as a quadratic function in accordance with the SAD mass ratio. This is due to the fact that the hydration heat generated in SAD cement-based materials is a quadratic function of the SAD mass ratio [37]. The hydration heat is closely related to the condensation time, thus explaining the quadratic function relationship between the initial setting time and the SAD mass ratio. The initial setting time of the SAD sulfoaluminate cement mortar reaches its maximum when the SAD mass ratio is 15%. The addition of SAD can increase the initial setting time of the sulfoaluminate cement mortar by 14.33% to 65.07%. The error bars for the initial setting time are less than 10% of the actual values. The SAD mass ratio of 0% to 15% has the effect of increasing the setting time of fresh SAD mortar. This can be attributed to the fact that the added SAD reduces the pozzolanic reaction of sulfoaluminate cement, thereby extending the setting time [38]. However, when the SAD mass ratio exceeds 15%, the setting time decreases with the addition of SAD. This is because SAD increases the content of hydraulic aluminum oxide, which accelerates the hydration rate of sulfoaluminate cement and shortens the setting time of the sulfoaluminate cement mortar.

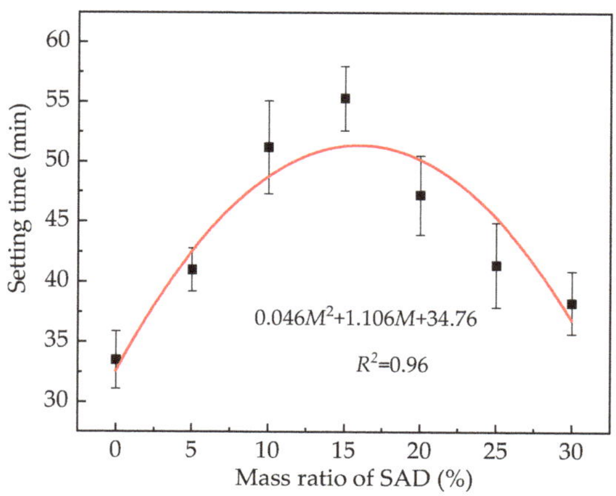

Figure 5. The setting time of fresh SAD sulfoaluminate cement mortar.

3.2. The Mechanical Strength and Drying Shrinkage Rate

The flexural strength of SAD sulfoaluminate cement mortar is presented in Figure 6. As depicted in the figure, the flexural strength of SAD sulfoaluminate cement mortar exhibits a quadratic function-based variation in relation to the SAD mass ratio. The flexural strength reaches its peak when the SAD mass ratio is 15%, with an increase of 33.6% at this ratio. When the SAD mass ratio increases from 0% to 15%, the hydration degree of sulfoaluminate cement is enhanced by the added SAD. This is because SAD promotes secondary hydration in sulfoaluminate cement, thereby increasing the mechanical strength of SAD sulfoaluminate cement mortar [39]. However, when the SAD mass ratio exceeds 15%, the mechanical strength of SAD sulfoaluminate cement mortar decreases, due to bubbles induced by SAD. The flexural strength increases with curing age, as the hydration degree is enhanced by curing time [40]. Consequently, mechanical strength improves with increased curing time. As shown in Figure 6, the flexural strength at a curing age of 0.25 days can reach 23.7% to 26.2% of the flexural strength of SAD sulfoaluminate cement mortar cured for 28 days. The error bars are less than 10% of the average flexural strength, confirming low data dispersion.

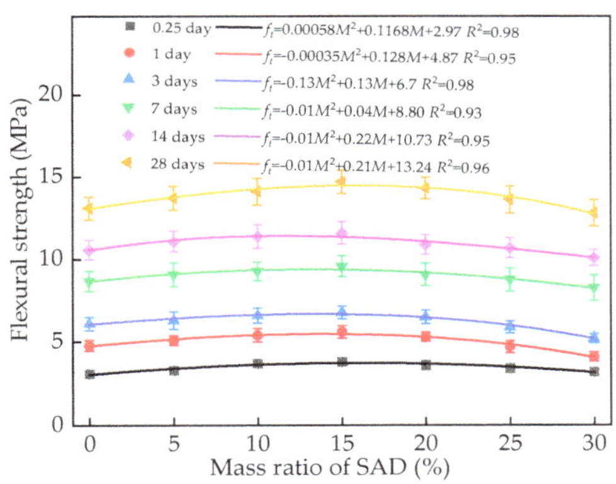

Figure 6. The flexural strength of SAD sulfoaluminate cement mortar.

Figure 7 illustrates the compressive strength of SAD sulfoaluminate cement mortar. As shown in the figure, the compressive strength initially increases and subsequently decreases with an increase in the SAD mass ratio. The relationship between the compressive strength of SAD sulfoaluminate cement mortar and the SAD mass ratio follows a quadratic function. The SAD sulfoaluminate cement mortar with 15% SAD exhibits the highest compressive strength. At this SAD mass ratio, the compressive strength increases by 11.3% to 17.1% with the addition of SAD. Furthermore, the compressive strength increases with an increase in curing age. The compressive strength of SAD sulfoaluminate cement mortar cured for 0.25 days is 9.8% to 12.7% of the compressive strength of SAD sulfoaluminate cement mortar cured for 28 days. The error bars are less than 10% of the average compressive strength, ensuring low data dispersion. The addition of SAD at a mass ratio of 0% to 15% can enhance the compressive strength of SAD sulfoaluminate cement mortar, due to the increased secondary hydration of sulfoaluminate cement [41–44]. However, when the added SAD increases from 15% to 30%, the pores induced by SAD have a negative impact on the compressive strength of SAD sulfoaluminate cement mortar [45]. The hydration degree increases with curing time; therefore, the compressive strength of SAD sulfoaluminate cement mortar increases with increased curing time.

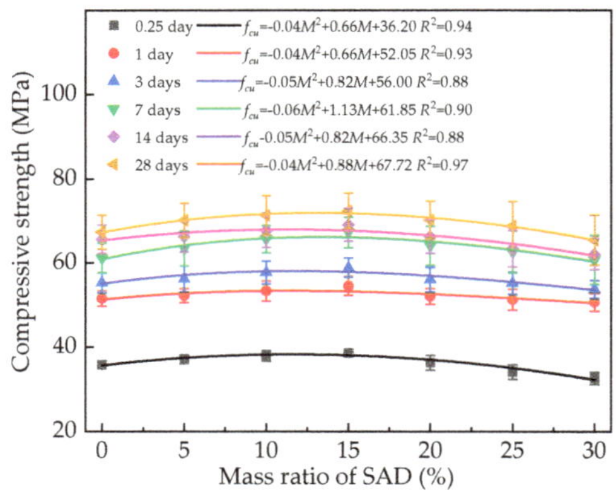

Figure 7. The compressive strength of SAD sulfoaluminate cement mortar.

The bonding strength of SAD sulfoaluminate cement mortar is depicted in Figure 8. As can be seen from the figure, the bonding strength exhibits a quadratic function relationship with the SAD mass ratio. The SAD sulfoaluminate cement mortar with 15% SAD demonstrates the highest bonding strength. At this dosage, the bonding strength of the SAD sulfoaluminate cement mortar experiences an increase in the range of −13.33% to 66.67%. The bonding strength of SAD sulfoaluminate cement mortar cured for 0.25 days is 9.8% to 12.7% of that cured for 28 days. The bonding strength accounts for 15.9% to 41.2% of the flexural strength. The error bars are less than 10% of the average compressive strength, indicating low data dispersion. The reasons for the variation in bonding strength are consistent with those for flexural and compressive strengths.

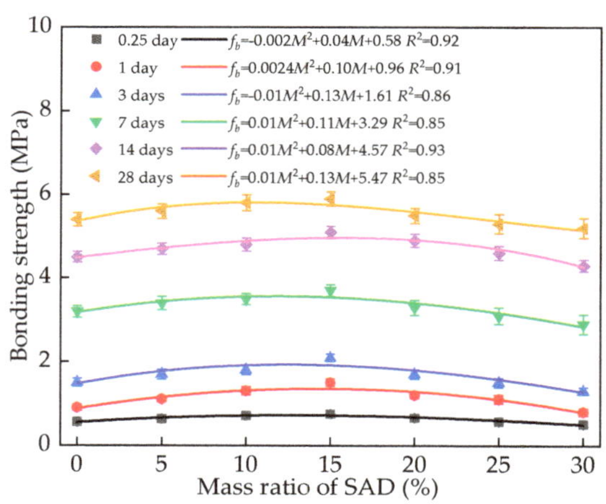

Figure 8. The bonding strength of SAD sulfoaluminate cement mortar.

The drying shrinkage rate of SAD sulfoaluminate cement mortar is presented in Figure 9. As observed in the figure, the drying shrinkage rate increases when the SAD mass ratio rises from 0% to 15%. The addition of SAD enhances the hydration degree of the SAD sulfoaluminate cement mortar, consequently increasing the drying shrinkage rate [46]. However, the drying shrinkage rate decreases when the mass ratio increases from 15% to 30%. The SAD contributes to an increase in the drying shrinkage rate of the SAD sulfoaluminate cement mortar, reaching a maximum of 10.1% to 17.2%. This phenomenon is attributed to the pores induced by the SAD, which restrict sulfoaluminate cement hydration [47]. The drying shrinkage rate of SAD sulfoaluminate cement mortar cured for 0.25 days is 18.8% to 26.9% of that cured for 28 days. The hydration degree of sulfoaluminate cement increases with curing age, leading to an increase in the drying shrinkage rate of the SAD sulfoaluminate cement mortar over time.

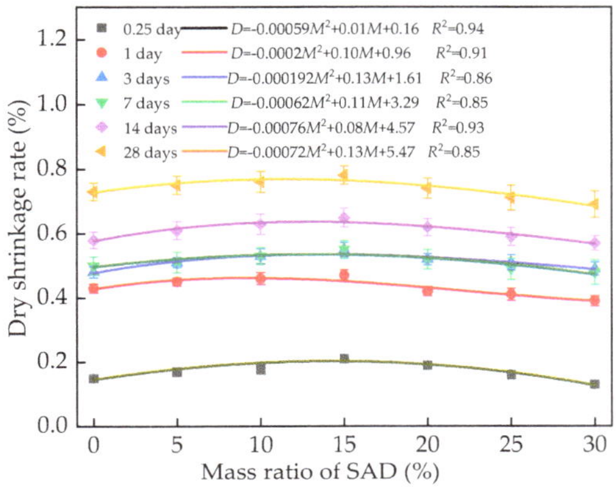

Figure 9. The drying shrinkage rate of SAD sulfoaluminate cement mortar.

3.3. The Chloride Ion Migration Coefficient and Carbonation Depth

Figure 10 illustrates the chloride ion migration coefficient (CMC) of sulfoaluminate cement mortar with SAD. As evident from the figure, the CMC follows a quadratic function pattern. The incorporation of SAD can reduce the CMC by 1.96% to 15.69%. This can be attributed to the fact that the added SAD enhances the hydration degree of the sulfoaluminate cement mortar, thereby increasing its compactness. Consequently, the CMC decreases with the addition of SAD. When the mass ratio of SAD is 15%, the CMC reaches its lowest value. The SAD increases the hydration degree, which improves the compactness of the sulfoaluminate cement mortar. However, it should be noted that SAD can also introduce gas into the sulfoaluminate cement mortar, thereby reducing its compactness. As a result, the sulfoaluminate cement mortar with 15% SAD exhibits the lowest CMC.

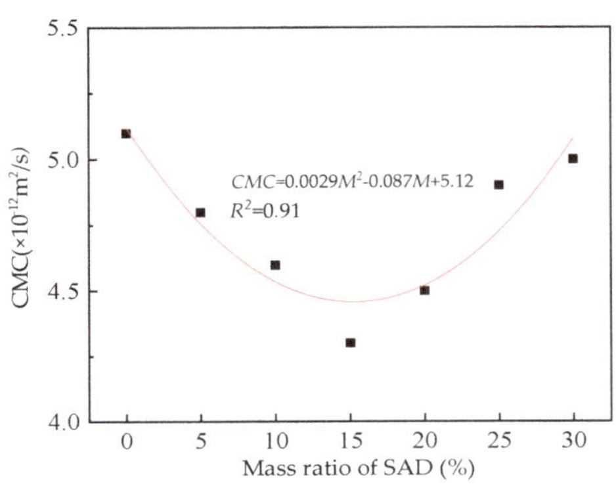

Figure 10. The CMC of SAD sulfoaluminate cement mortar.

The D_c of the SAD sulfoaluminate cement mortar is depicted in Figure 11. As observed in the figure, the Dc of the SAD sulfoaluminate cement mortar follows a quadratic function pattern. Notably, the SAD sulfoaluminate cement mortar with 15% SAD exhibits the lowest D_c. The addition of SAD can reduce the D_c by 2.1% to 18.3%. While SAD with a mass ratio below 15% negatively impacts the D_c of the SAD sulfoaluminate cement mortar, SAD with a mass ratio above 15% has a positive effect on the corresponding D_c. The SAD enhances sulfoaluminate cement hydration, thereby increasing the density of the SAD sulfoaluminate cement mortar. Consequently, the Dc decreases with the addition of SAD. However, when the mass ratio of SAD exceeds 15%, the added SAD introduces bubbles into the SAD sulfoaluminate cement mortar, leading to a decrease in its density. This, in turn, increases the migration depth of CO_2, resulting in an increase in the D_c.

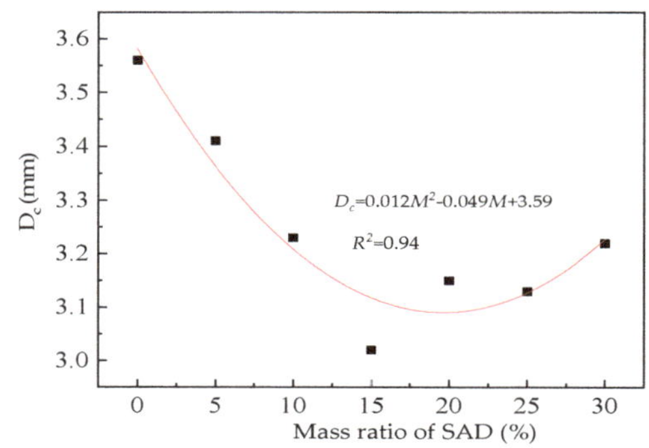

Figure 11. The D_c of SAD sulfoaluminate cement mortar.

3.4. The Leached Cr and Zn

The leached concentrations of Cr and Zn from the SAD sulfoaluminate cement mortar are illustrated in Figure 12. As is evident from the figure, the leached amounts of Cr and Zn vary in a quadratic function pattern with the SAD's mass ratio. At an SAD mass ratio of 15%, the leached Cr and Zn are at their lowest levels. At this specific dosage, the leached Cr and Zn decrease by 21.6% to 32.7% and 19.3% to 31.2%, respectively. The increasing immersion time has a pronounced effect on the mass fraction of leached Cr and Zn, causing them to increase accordingly. As the immersion time extends from 3 months to 9 months, the leached Cr and Zn increase by 161.9% to 210.3% and 130.5% to 186.3%, respectively. The error bars for the leached Cr and Zn are less than 10% of the average values, ensuring the accuracy of the experimental results.

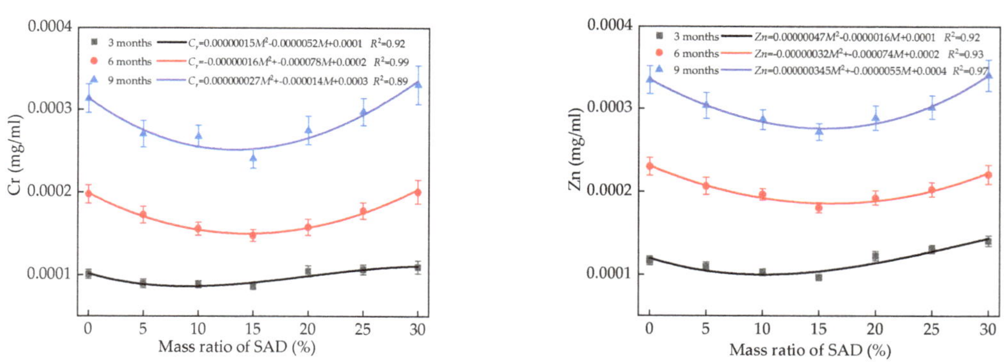

Figure 12. The leached Cr and Zn of SAD sulfoaluminate cement mortar. (**a**) The leached Cr. (**b**) The leached Zn.

3.5. Micro-Analysis Results

3.5.1. The Scanning Electron Microscope

Figure 13 presents scanning electron microscopy (SEM) images of SAD sulfoaluminate cement mortar, with SAD mass ratios of 5%, 15%, and 25%. As depicted in Figure 13, needle-shaped and flocculent hydration products are evident. When the SAD content increases from 5% to 15%, the quantity of needle-shaped hydration products and cracks diminishes, while the amount of compact hydration products rises. This phenomenon can be explained by the fact that SAD enhances the degree of hydration, thereby augmenting the production and density of hydration products. This observation corroborates that SAD sulfoaluminate cement mortar with 15% SAD exhibits superior mechanical strength. However, when the SAD mass ratio exceeds 15%, an increased number of cracks is observable (as shown in Figure 11), indicating that SAD at dosages above 15% negatively impacts the mechanical strength of SAD sulfoaluminate cement mortar.

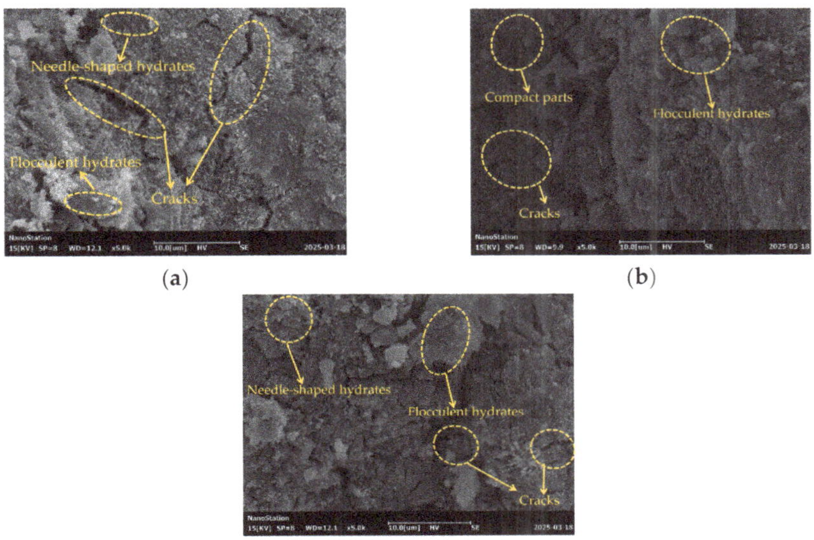

(a) (b) (c)

Figure 13. The SEM photos of the SAD sulfoaluminate cement mortar. (**a**) 5% SAD. (**b**) 15% SAD. (**c**) 25% SAD.

3.5.2. Scanning Electron Microscopy–Energy-Dispersive Spectroscopy

The SEM-EDS images are displayed in Figure 14. As illustrated in Figure 14, the elements C, O, Al, Si, S, K, and Ca are present in the SAD sulfoaluminate cement mortar. The addition of SAD can increase the content of the Al element in the SAD sulfoaluminate cement mortar. This is because the mass ratio of Al_2O_3 in SAD is higher than that in SAC. Consequently, the incorporation of SAD elevates the Al content in the SAD sulfoaluminate cement mortar.

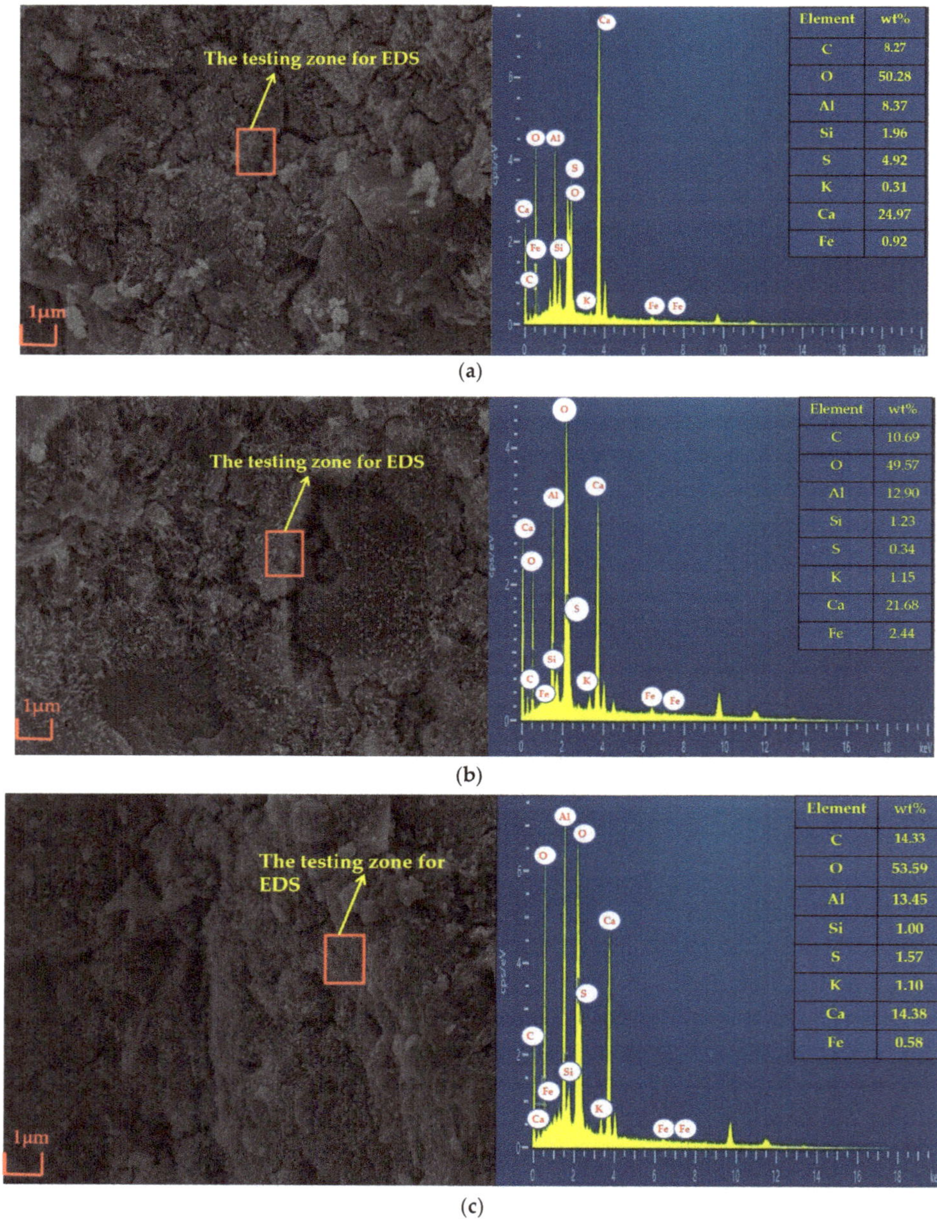

Figure 14. The SEM-EDS photos of the SAD sulfoaluminate cement mortar. (**a**) 5% SAD. (**b**) 15% SAD. (**c**) 25% SAD.

3.5.3. The X-Ray Diffraction Spectrum

Figure 15 illustrates the X-ray diffraction (XRD) spectrum of the SAD sulfoaluminate cement mortar. As shown in Figure 15, the crystals of ettringite (Aft), tetracalcium aluminoferrite (C_4A_3S), $CaSO_4{\cdot}2H_2O$, $CaSO_4$, $CaCO_3$, $Ca(OH)_2$, and $Al(OH)_3$ are present. With an increase in the amount of SAD added, the crystals of Aft, C_4A_3S, and $Al(OH)_3$ also increase.

This can be attributed to the fact that SAD contains a significant amount of nano-shaped alumina, which promotes the secondary hydration of sulfoaluminate cement, leading to the formation of more $Al(OH)_3$ [48]. This confirms that SAD can enhance the hydration degree of the sulfoaluminate cement mortar.

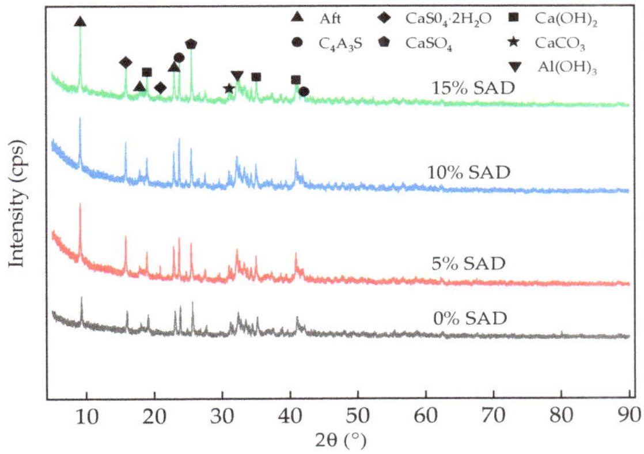

Figure 15. The XRD curves of the SAD sulfoaluminate cement mortar.

4. Conclusions

The influence of secondary aluminum dross (SAD) on the properties of sulfoaluminate cement mortar has been thoroughly investigated. The findings are summarized as follows.

The slump flow and plastic viscosity of fresh sulfoaluminate cement mortar exhibit negative and positive quadratic relationships, respectively, with the mass ratio of added SAD. Specifically, the addition of SAD results in a decrease in slump flow at rates ranging from 1.57% to 2.72%, and an increase in plastic viscosity at rates ranging from 5.21% to 36.89%. Additionally, the placing time contributes to a decrease in slump flow by up to 20.4% and an increase in plastic viscosity by up to 11.2%.

The relationships between the initial setting time, mechanical strength, and DR of the mortar, and the mass ratio of SAD, are cubic functions. When the SAD mass ratio is 15%, the sulfoaluminate cement mortar demonstrates the highest initial setting time, flexural strength, compressive strength, bonding strength, and DR. The incorporation of SAD leads to an increase in the initial setting time, flexural strength, compressive strength, bonding strength, and DR of the sulfoaluminate cement mortar at rates ranging from 14.33% to 65.07%, −14.75% to 22.58%, −8.94% to 9.96%, −13.33% to 66.67%, and −13.33% to 26.67%, respectively.

The CMC and D_c of sulfoaluminate cement mortar vary quadratically with the SAD mass ratio. The addition of SAD reduces the CMC and D_c by rates ranging from 1.96% to 15.69% and 2.1% to 18.3%, respectively. When the SAD mass ratio is 15%, the CMC and D_c of the sulfoaluminate cement mortar are at their lowest.

The leached Cr and Zn from the sulfoaluminate cement mortar show a cubic function-based relationship with the mass ratio of the SAD. When the added SAD's mass ratio is 15%, the leached Cr and Zn from the sulfoaluminate cement mortar are the highest. The SAD can decrease the leached Cr and Zn by 21.6%~32.7% and 19.3%~31.2%, respectively. Meanwhile, when the immersion time increases from 3 months to 9 months, the rates of increase in the leached Cr and Zn are 161.9%~210.3% and 130.5%~186.3%.

SAD can improve the compactness of hydration products and decrease the number and width of cracks in sulfoaluminate cement mortar. When the mass ratio of SAD is 15%, the hydration products of the sulfoaluminate cement mortar are the most compact. The element of Al in the sulfoaluminate cement mortar is increased by the added SAD. The SAD can increase the aluminum hydration products. Based on the results of SEM-EDS and XRD, the hydration degree of the sulfoaluminate cement mortar is increased by SAD.

Author Contributions: Conceptualization, Q.W.; methodology, L.Z.; validation, Q.W. and L.Z.; investigation, A.Z.; resources, Q.W.; data curation, L.Z. and A.Z.; writing—original draft, Q.W. and L.Z.; visualization, A.Z.; project administration, Q.W. All authors have read and agreed to the published version of the manuscript.

Funding: Project approved by Sichuan Higher Vocational Education Research Center (GZY23B13).

Institutional Review Board Statement: Not applicable.

Informed Consent Statement: Not applicable.

Data Availability Statement: Data are contained within the article.

Conflicts of Interest: The authors declare that there are no conflicts of interest regarding the publication of this paper.

References

1. Serelis, E.; Vaitkevicius, V. Effect of waste glass powder and liquid glass on the Physico-Chemistry of Aluminum-Based Ultra-Lightweight concrete. *Constr. Build. Mater.* **2023**, *390*, 131615. [CrossRef]
2. Cai, Z.; Ren, J.; Shen, G.; Jin, C.; Gu, X.; Cheng, W.; Wang, H. The Influence of Assembly Unit of Fibers on the Mechanical and Long-Term Properties of Reactive Powder Concrete. *Coatings* **2023**, *13*, 412. [CrossRef]
3. Wei, Y.; Gao, X.; Zhang, Q. Evaluating performance of concrete pavement joint repair using different materials to reduce reflective cracking in asphalt concrete overlay. *Road Mater. Pavement Des.* **2014**, *15*, 966–976.
4. Li, Y.; Li, X.; Tan, Y. Effect of aging on fatigue performance of cement emulsified asphalt repair material. *Constr. Build. Mater.* **2021**, *292*, 123417. [CrossRef]
5. Qiao, F.; Chau, C.; Li, Z. Property evaluation of magnesium phosphate cement mortar as patch repair material. *Constr. Build. Mater.* **2010**, *24*, 695–700. [CrossRef]
6. Wang, Y.; Ye, J.; Liu, Y.; Qiang, X.; Feng, L. Influence of freeze–thaw cycles on properties of asphalt-modified epoxy repair materials. *Constr. Build. Mater.* **2013**, *41*, 580–585. [CrossRef]
7. Li, Y.; Sun, H.; He, X.; Tan, Y. Freeze-thaw damage and creep behavior of cement asphalt composite binder. *Constr. Build. Mater.* **2020**, *245*, 118407. [CrossRef]
8. Jin, B.; Chen, L.; Chen, B. Factors assessment of a repair material for brick masonry loaded cracks using magnesium phosphate cement. *Constr. Build. Mater.* **2020**, *252*, 119098. [CrossRef]
9. Liu, S.; Yang, K.; Yu, L.; Xu, L.; Yuan, Q.; Wu, K.; Yang, Z. Integrated assessment of magnesium phosphate cement repaired concrete from the perspective of mechanical, geometric and electrochemical compatibility. *Cem. Concr. Compos.* **2025**, *157*, 105958. [CrossRef]
10. Jin, H.; Yu, H.; Fu, Z.; Shan, L.; Zhang, S.; Zhao, Y. Early performance modification of a waste-derived super-sulfated cement from arsenic-containing bio-oxidation waste via calcium aluminate and calcium sulfoaluminate cement. *Constr. Build. Mater.* **2024**, *441*, 137529.
11. Meng, Y.; Gao, D.; Yang, L.; Fang, J. A green ultra-high performance calcium sulfoaluminate cement concrete containing recycled fine aggregate: Mechanical properties, microstructure and environmental impact. *Constr. Build. Mater.* **2025**, *470*, 140641. [CrossRef]
12. Gao, D.; Che, Q.; Meng, Y.; Yang, L.; Xie, X. Properties evolution of calcium sulfoaluminate cement blended with ground granulated blast furnace slag suffered from sulfate attack. *J. Mater. Res. Technol.* **2022**, *17*, 1642–1651. [CrossRef]
13. Meng, Y.; Gao, D.; Yang, L.; Fang, J.; Li, Y.; Zhang, T. Effect of ultra-fine supplementary cementitious materials on the properties of calcium sulfoaluminate cement-based ultra-high performance concrete. *Case Stud. Constr. Mater.* **2024**, *21*, e03547. [CrossRef]
14. Al-Kheetan, M.J.; Al-Tarawneh, M.; Ghaffar, S.H.; Chougan, M.; Jweihan, Y.S.; Rahman, M.M. Resistance of hydrophobic concrete with different moisture contents to advanced freeze–thaw cycles. *Struct. Concr.* **2021**, *22*, E1050–E1061. [CrossRef]
15. Tan, B.; Okoronkwo, M.; Kumar, A.; Ma, H. Durability of calcium sulfoaluminate cement concrete. *J. Zhejiang Univ. Sci. A* **2020**, *21*, 118–128. [CrossRef]
16. Leone, M.; Blasi, G.; Colonna, D. Long-term flexural response of reinforced calcium sulfoaluminate/cement concrete beams. *Eng. Struct.* **2023**, *292*, 116585. [CrossRef]

17. Shang, C.; Wu, C.; Liu, Y.; Wang, J.; Fu, Q.; Lu, L.; Sheng, Z.; Xing, F. Impact of sulfoaluminate cement modification on corrosion resistance of reactive powder concrete subjected to ammonium sulfate-rich sewer environment. *Constr. Build. Mater.* **2024**, *411*, 134652. [CrossRef]
18. Park, S.; Lee, N.; Park, K. Early-age hydration behavior of calcium sulfoaluminate (CSA) cement/ordinary Portland cement-blended ultra-high performance concrete. *J. Build. Eng.* **2024**, *87*, 109058. [CrossRef]
19. Kumar, S.; Sharma, U. Mechanical performance of cement concrete with use of combined fibers of basalt and bamboo. *Phys. Chem. Earth Parts A/B/C* **2025**, *137*, 103799. [CrossRef]
20. Wang, J.; Zhao, C.; Li, Q.; Song, G.; Hu, Y. The synergistic effect of recycled steel fibers and rubber aggregates from waste tires on the basic properties, drying shrinkage, and pore structures of cement concrete. *Constr. Build. Mater.* **2025**, *470*, 140574. [CrossRef]
21. Katebi, A.; Asadollahfardi, G.; Salehi, A.; Ahmadi, G.; Soleymani-tushmanlo, A. Investigation of the properties of self-compacting concrete using treated rubber powder, polypropylene fibers, and wash water. *Case Stud. Constr. Mater.* **2025**, *22*, e04443. [CrossRef]
22. Wang, Y.; Zhang, J.; Du, G.; Li, Y. Synergistic effects of polypropylene fiber and basalt fiber on the mechanical properties of concrete incorporating fly ash ceramsite after freeze-thaw cycles. *J. Build. Eng.* **2024**, *91*, 109593. [CrossRef]
23. Socrates, P.; Juan, M. Influence of the secondary aluminum chip on the physical and mechanical properties of concrete. *Innov. Infrastruct. Solut.* **2023**, *8*, 45.
24. Lou, B.; Shen, H.; Liu, B.; Liu, J.; Zhang, S.G. Recycling secondary aluminum dross to make building materials: A review. *Constr. Build. Mater.* **2023**, *409*, 133989. [CrossRef]
25. Rotana, H.; Claudia, P.O. On utilization and mechanisms of waste aluminium in mitigating alkali-silica reaction (ASR) in concrete. *J. Clean. Prod.* **2019**, *212*, 864–879.
26. Zhang, S.; Wang, K.; Li, H.; Zhang, X.; Jiang, Y. Novel SCMs produced by the calcination of secondary aluminium dross with dolomite and their potential usage in cemented paste backfill. *Constr. Build. Mater.* **2023**, *365*, 130119. [CrossRef]
27. Meor Yusoff, M.; Muslimin, M.; Paulus, W. Effect of fractional precipitation on quality of nanostructured alumina produced from black aluminium dross waste. *Adv. Mater. Res.* **2011**, *173*, 24–28. [CrossRef]
28. Shen, H.; Liu, B.; Ekberg, C.; Zhang, S. Harmless disposal and resource utilization for secondary aluminum dross: A review. *Sci. Total Environ.* **2021**, *760*, 143968. [CrossRef]
29. *GB/T 50080-2016*; Standard for Test Method of Performance on Ordinary Fresh Concrete. Ministry of Housing and Urban Rural Development of the People's Republic of China: Beijing, China, 2016.
30. *JGJ/T70-2009*; Standard for Test Method of Performance on Building Mortar. Ministry of Housing and Urban Rural Development of the People's Republic of China: Beijing, China, 2009.
31. *GB/T 17671-2021*; Test Method of Cement Mortar Strength (ISO Method). State Administration for Market Regulation and National Standardization Management Committee: Beijing, China, 2021.
32. *GB/T50082-2024*; Standard for Test Methods of Long-Term Performance and Durability of Concrete. Ministry of Housing and Urban Rural Development of the People's Republic of China: Beijing, China, 2024.
33. David, O.; Opeyemi, J.; Adekunle, M.; Babatunde, F. Influence of secondary aluminum dross (SAD) on compressive strength and water absorption capacity properties of sandcrete block. *Cogent. Eng.* **2019**, *6*, 1608687.
34. Panditharadhya, B.; Sampath, V.; Shankar, A. Mechanical properties of pavement quality concrete with secondary aluminium dross as partial replacement for ordinary portland cement. *Mater. Sci. Eng.* **2018**, *431*, 032011. [CrossRef]
35. Satish, R.; Neeraja, D. Mechanical and durability aspects of concrete incorporating secondary aluminium slag. *Resour.-Effic. Technol.* **2016**, *2*, 225–232.
36. Ewais, E.; Khalil, N.; Amin, M. Utilization of aluminum sludge and aluminum slag (dross) for the manufacture of calcium aluminate cement. *Ceram. Int.* **2009**, *35*, 3381–3388. [CrossRef]
37. Elseknidy, M.H.; Salmiaton, A.; Nor Shafizah, I.; Saad, A.H. A Study on Mechanical Properties of Concrete Incorporating Aluminum Dross, Fly Ash, and Quarry Dust. *Sustainability* **2020**, *12*, 9230. [CrossRef]
38. Haneef, S.; Harish, P. An Experimental Investigation on Use of Secondary Aluminium Dross in Cement Concrete. *Eng. Technol.* **2016**, *2*, 204–207.
39. Mailar, G.; Sreedhara, B.M.; Manu, D.S.; Hiremath, P.; Jayakesh, K. Investigation of concrete produced using recycled aluminium dross for hot weather concreting conditions. *Res. Efficient Technol.* **2016**, *2*, 68–80. [CrossRef]
40. Mahinroosta, M.; Allahverdi, A. Enhanced alumina recovery from secondary aluminum dross for high purity nanostructured γ-alumina powder production: Kinetic study. *J. Environ. Manag.* **2018**, *212*, 278–291. [CrossRef]
41. Boulahbal, A.; Santamaría, L.; Azizi, A.; Boutahala, M.; Korili, S.; Gil, A. Synthesis of Cu-Al layered double hydroxides from aluminum saline slags. *Miner. Eng.* **2023**, *204*, 108413. [CrossRef]
42. Zawrah, M.; Ghanaym, E.; Sadek, H.; El Defrawy, S.; Ali, O. Synthesis, characterization and sinterability of pure and Ni-doped nano layered double hydroxides from aluminum dross. *Ceram. Int.* **2019**, *45*, 17598–17610. [CrossRef]
43. Nduka, D.; Ede, A. Mechanical and Water Absorption Properties of Normal Strength Concrete (NSC) Containing Secondary Aluminum Dross (SAD). *Int. J. Eng. Res. Afr.* **2020**, *47*, 1–13. [CrossRef]

44. Li, P.; Guo, M.; Zhang, M.; Teng, L.D. Leaching Process Investigation of Secondary Aluminum Dross: The Effect of CO_2 on Leaching Process of Salt Cake from Aluminum Remelting Process. *Metall. Mater. Trans. B* **2012**, *43*, 1220–1230. [CrossRef]
45. Kępniak, M.; Łukowski, P. Analysis of Cement Mortar with Recycled Sand. *Sustainability* **2024**, *16*, 1773. [CrossRef]
46. Dean, M. *A Practical Guide to Multi-Criteria Analysis*; University College London: London, UK, 2022.
47. Dean, M. Chapter Six-Multi-criteria analysis. *Adv. Transp. Policy Plan.* **2020**, *6*, 165–224.
48. Huang, G.; Wang, H.; Shi, F. Coupling Effect of Salt Freeze-Thaw Cycles and Carbonation on the Mechanical Performance of Quick Hardening Sulphoaluminate Cement-Based Reactive Powder Concrete with Basalt Fibers. *Coatings* **2021**, *11*, 1142. [CrossRef]

coatings

MDPI

Article

A Comparative Study on the Impact of Different Antifreeze Agents on Composite Cement at −10 °C

Bitao Zhang [1], Yongkang Du [1], Dong Liu [2] and Yanyan Hu [3,*]

[1] Shaanxi Building Materials Technology Group Co., Ltd., Xi'an 710018, China
[2] Huangling Ecological Cement Co., Ltd., Yan'an 727399, China
[3] College of Materials Science and Technology, Xi'an University of Architecture & Technology, Xi'an 710055, China
* Correspondence: huyanyan@xauat.edu.cn

Abstract: Construction in cold regions faces significant challenges due to delayed cement hydration and frost damage caused by sub-zero temperatures. This study investigated the effects of three antifreeze agents on the performance of composite cement under sub-zero temperatures. The setting time, compressive strength, freezing point, and hydration mechanisms were evaluated. The results revealed that $CaCl_2$ optimally accelerated hydration, while achieving the continuous development of compressive strengths through freezing-point depression and dense microstructure formation. $NaNO_2$ exhibited the lowest freezing point but delayed setting at high dosages, while Li_2CO_3 showed limited impact due to insufficient freezing point depression. Li_2CO_3 showed limited efficacy under continuous low temperatures but enabled strength recovery after the temperature transition from sub-zero to ambient conditions. This research provides a basis for the application of antifreeze agents in the composite cement system for construction in cold environments.

Keywords: antifreeze agents; composite cement; cold region

1. Introduction

Cement technologies for cold regions have garnered global attention, as construction under such conditions is inherently more challenging than in temperate environments due to the detrimental impacts of low ambient temperatures on both project timelines and structural quality [1–3]. During winter construction, there are two primary risks: first, extremely cold conditions delay cement hydration, leading to suppressed early compressive strength development; second, when the internal temperature of concrete falls below −5 °C, approximately 92% of capillary water freezes, damaging the microstructure, while only minimal liquid water remains for hydration [4].

To address frost heave damage, enhancing early-age cement properties, accelerating hydration, and minimizing freezable water content prior to freezing are essential [4–6]. Conventional approaches, including heating water and aggregates, employing thermal storage curing, and covering poured concrete with insulating materials, are widely implemented [7,8]. While these measures can effectively mitigate risks, the long-term investment of manpower and energy leads to excessively high construction costs and significantly increases CO_2 emissions, contributing to environmental pollution [7,8]. Therefore, there is an urgent need for a low-cost, environmentally friendly material that can be used at sub-zero temperatures, addressing the challenges posed by excessively cold weather conditions.

Portland cement, as the most widely used cement in engineering, offered advantages including substantial late-age strength growth, low price, and simple production processes.

However, sub-zero temperatures during winter construction delay the setting and hardening of Portland cement and increase CO_2 emissions [9–11]. Sulfoaluminate cement, known for its low carbon footprint, rapid hydration, and high early strength, is typically employed in cold-region projects and rapid repairs [12]. However, its high production costs and limited raw material availability restrict widespread adoption [13,14]. Studies demonstrated that blending sulfoaluminate cement with Portland cement create a composite system that combines rapid early hydration, shortening setting time, and boosting early strength at sub-zero temperatures with Portland cement's economic benefits [15,16]. Nevertheless, when the temperature decreases below −5 °C, the frost resistance of the composite cement is significantly compromised due to the freezing of 92% capillary water [6]. To address this limitation, antifreeze agents have been proposed as a viable strategy to reduce frost-induced damage and expand the applicability of cement [17–19]. Therefore, identifying appropriate antifreeze agents compatible with composite systems is of paramount importance.

Inorganic antifreeze agents have been widely utilized in winter construction due to their cost-effectiveness, accessibility, and operational simplicity. Their mechanisms of action primarily involve lowering the freezing point of pore solutions to delay ice formation and accelerating cement hydration through ionic nucleation effects, thereby enhancing early strength development [20–22]. Among these agents, chloride salts (e.g., $CaCl_2$, NaCl), nitrites (e.g., $NaNO_2$, $Ca(NO_2)_2$), and lithium carbonate (Li_2CO_3) are commonly employed, with their functional advantages and limitations varying significantly based on composition and application scenarios.

Calcium chloride ($CaCl_2$) is recognized as a highly effective antifreeze agent in cold climates. Studies demonstrate that Ca^{2+} ions improve particle dispersion and accelerate C_3S hydration, while interactions with C_3A generate insoluble chloroaluminates (e.g., Friedel's salt), promoting aluminate phase reactions [22,23]. For instance, Wang et al. [23] reported that 3%–4% $CaCl_2$ reduced the initial setting time of Portland cement by 60% and increased its 28 d compressive strength by 25% under 3 °C curing. However, the corrosion risk of Cl^- on steel reinforcement restricts its application to non-reinforced concrete structures [24,25]. Sodium nitrite ($NaNO_2$) has emerged as a prominent non-chloride antifreeze agent, which is particularly valued for its dual functionality in corrosion inhibition and hydration acceleration. Studies have demonstrated that $NaNO_2$ effectively prevents steel reinforcement corrosion by forming a protective oxide layer on the steel surface, thereby neutralizing chloride-induced electrochemical reactions [26]. In Portland cement systems, $NO_2{}^-$ ions preferentially react with C_3A to form nitrite-modified AFt (NO_2-AFt), which suppresses the conversion of AFt to $SO_4{}^{2-}$-AFm phases, thereby increasing ettringite content and early strength [11]. Compared to chloride-based agents, $NaNO_2$ exhibits superior compatibility with reinforced concrete, achieving a freezing point reduction to −15 °C at an 8% dosage without corrosion risks [11]. Lithium carbonate (Li_2CO_3) exhibits exceptional performance in sulfoaluminate cement systems. The small ionic radius and high polarization capability of Li^+ shortens the hydration induction period and promotes AFt nucleation, significantly improving early strength [27,28]. Shen et al. [28] observed that 1% Li_2CO_3 reduces the initial and final setting times of sulfoaluminate cement from 37 and 64 min to 8 and 13 min, respectively. Although the aforementioned antifreeze agents have demonstrated significant efficacy in individual cement systems, their effects in composite systems remain unclear.

In this study, three antifreeze agents ($CaCl_2$, $NaNO_2$, Li_2CO_3) are selected to evaluate their effects on the setting time, compressive strength, and freezing point of a Portland/sulfoaluminate composite cement under −10 °C curing. X-ray diffraction (XRD) and scanning electron microscopy (SEM) are employed to analyze hydration products and microstructural evolution, elucidating the mechanisms by which these agents influence composite cement performance.

2. Materials and Methods

2.1. Materials

The chemical compositions of Portland cement (P.O 42.5) and sulfoaluminate cement (52.5) are shown in Tables 1 and 2. Silica sand, with a fineness modulus of 2.3–3.0, was employed as the fine aggregate. Analytical-grade $CaCl_2$, $NaNO_2$, and Li_2CO_3 were used as antifreeze agents, with purities exceeding 96.0%, 99.0%, and 99.0%, respectively.

Table 1. Chemical compositions of Portland cement (wt. %).

SiO_2	Al_2O_3	Fe_2O_3	CaO	MgO	SO_3	Na_2O_{eq}	f-CaO	Loss
20.84	4.68	3.56	63.43	3.29	2.30	0.55	0.78	1.48

Table 2. Chemical compositions of sulfoaluminate cement (wt. %).

SiO_2	Al_2O_3	Fe_2O_3	CaO	MgO	SO_3	K_2O	P_2O_5	TiO_2	MnO_2	SrO	Cl^-
8.24	23.27	2.20	45.57	1.96	15.70	0.42	0.09	1.48	0.046	0.13	0.08

2.2. Methods

2.2.1. Setting Time

The setting time test was conducted using cement paste with a Portland/sulfoaluminate cement ratio of 9:1 and a w/c ratio of 0.3. The 9:1 ratio was selected based on prior studies, which demonstrated that exceeding a 10% sulfoaluminate cement content in the composite system significantly altered hydration mechanisms, leading to impractical setting time reductions for field applications [6]. The w/c ratio of 0.3 was optimized to balance two critical factors: freezable water mitigation and workability-structure trade-off. Antifreeze agent dosages included $CaCl_2$ (0%, 0.5%, 1%, 1.5%, 2%), $NaNO_2$ (0%, 1%, 2%, 3%, 4%), and Li_2CO_3 (0%, 0.1%, 0.2%, 0.4%, 0.6%). The cement paste was prepared according to the mix proportions. Prior to mixing, the antifreeze admixture was dissolved in water to form a homogeneous solution, which was subsequently combined with cement. The time of complete cement/water mixing was recorded as the starting time (t_1). For the initial setting test, the test needle was vertically lowered onto the paste surface, tightened for 1 to 2 s, and released. The initial setting was defined when the needle penetrated to a depth of 4 ± 1 mm from the base plate, with the corresponding time recorded as t_2. The initial setting time (Δt_1) was calculated as $t_2 - t_1$. The final setting time ($\Delta t_2 = t_3 - t_1$) was determined when the annular attachment of the final setting needle failed to leave an indentation on the specimen surface, recorded as t_3.

2.2.2. Compressive Strength

The compressive strength test was performed on mortar specimens with a Portland/sulfoaluminate cement ratio of 9:1, a w/c ratio of 0.3, and a cement/sand ratio of 1:1. The mortar preparation process, including specimen dimensions and testing procedures, was based on the Chinese standard GB/T 17671-2021 "Method for Testing the Strength of Cement Mortar" [29], as follows: the composite cement was dry-mixed for 120 s; the antifreeze solution was added and mixed for 30 s; sand was incorporated with slow mixing for 30 s, followed by fast mixing for 30 s and an additional 60 s. Mortar was cast into steel molds (40 mm × 40 mm × 160 mm), vibrated 60 times, and immediately sealed with a polyethylene film. The specimens were immediately transferred to a −10 °C curing chamber for low-temperature hydration. Two curing regimes were applied: (1) continuous −10 °C curing; (2) −10 °C curing for 7 d followed by 21 d at 20 ± 2 °C and 95

± 5% relative humidity. Compressive strength was measured at four designated ages, with six specimens tested per age to ensure statistical reliability.

2.2.3. Freezing Point

The temperature history curve test was conducted using a single mortar specimen (100 mm × 100 mm × 100 mm). A T-type thermocouple embedded in the specimen center was connected to a temperature recorder, which logged data to a computer system (Figure 1) [6]. The specimen was then transferred to a −10 °C curing chamber for monitoring. Upon test completion, the temperature history curve was plotted, and the freezing point was identified using the methodology depicted in Figure 2 [6].

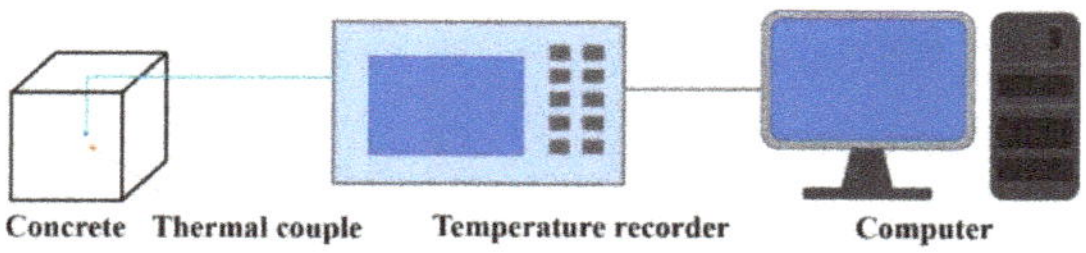

Figure 1. The schematic of the test of temperature history of cement [6].

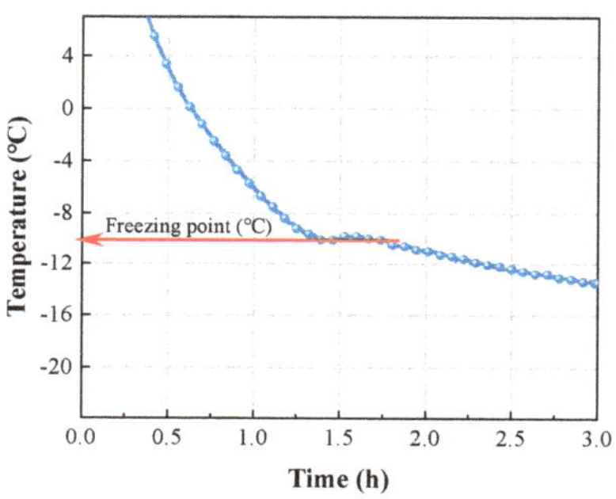

Figure 2. The schematic of selection of the freezing point [6].

2.2.4. X-Ray Diffraction

The hydration products of the composite system were analyzed using X-ray diffraction (XRD). At designated ages, core samples were extracted, ground into powder, and immersed in isopropanol to terminate hydration. The powder was then pressed into pellets and loaded into a Bruker AXS D8 X-ray diffractometer (Billerica, MA, USA) with CuKα radiation (40 kV, 40 mA). Scans were performed over 2θ angles from 5° to 90°at a rate of 10°/min.

2.2.5. Scanning Electron Microscopy

The microstructure of the composite system was analyzed using scanning electron microscopy (SEM). At designated ages, a core region was extracted from each specimen and processed into a 5 mm × 5 mm × 5 mm cube. Then, it was immersed in isopropanol for 48 h to terminate hydration, and dried for 72 h. Prior to imaging, a fresh, smooth cross-section was exposed by fracturing the cube. The sample was sputter-coated with gold prior to imaging. SEM observations were conducted using an FEI Quanta 250 instrument (Hillsboro, OR, USA).

3. Results

3.1. Setting Time

The setting time of the composite system with varying $CaCl_2$ dosages is shown in Figure 3a. Under a negative temperature, the hydration rate of the cement slows down, resulting in prolonged initial and final setting times (136 min and 188 min, respectively, for the 0% dosage). The addition of $CaCl_2$ significantly mitigated this issue, with higher dosages yielding shorter setting times. As seen in Figure 3a, adding $CaCl_2$ notably shortened both the initial and final setting time of the composite system. Additionally, a higher dosage resulted in shorter setting times. At 2.0% $CaCl_2$, initial and final setting times were reduced by 77.9% and 69.7%, respectively, compared to the control. This acceleration was attributed to $CaCl_2$ reacting with C_3A to form hydrated chloroaluminates [30]. Additionally, $CaCl_2$ reacted with CH, reducing the CH content in the system by forming calcium chlorate, which further accelerated the hydration of C_3S and C_2S [17,31,32].

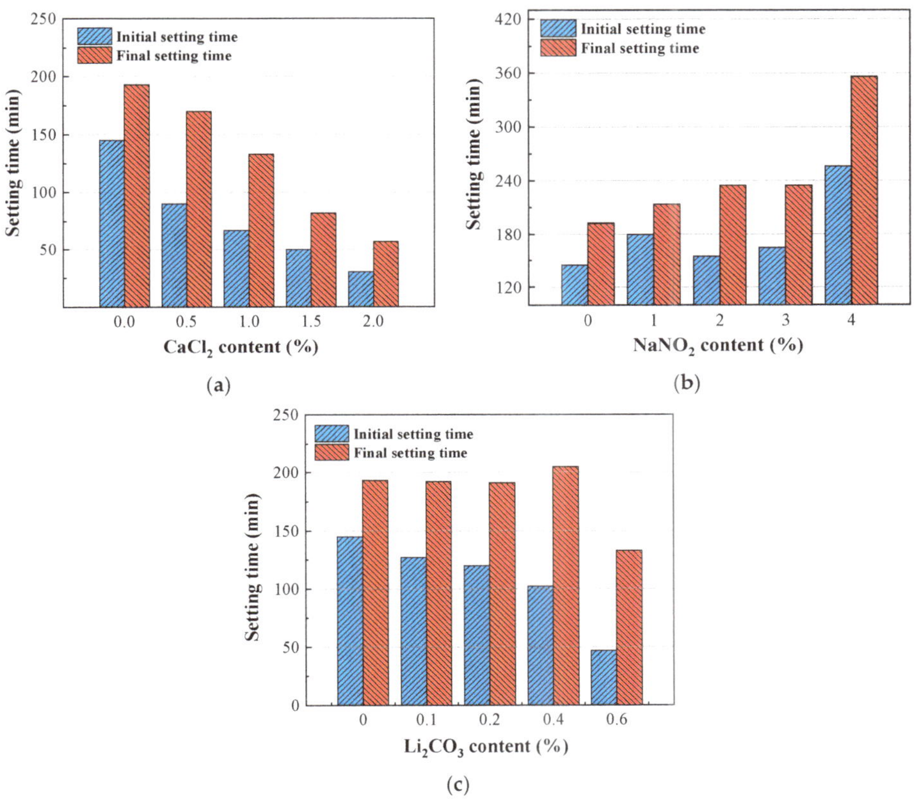

Figure 3. Effects (**a**) $CaCl_2$, (**b**) $NaNO_2$, and (**c**) Li_2CO_3 on the setting time of composite cement.

In contrast, $NaNO_2$ exhibited opposing effects, as shown in Figure 3b. While the control system (0% $NaNO_2$) showed initial and final setting times of 136 min and 188 min, $NaNO_2$ incorporation progressively delayed both. Moreover, as the $NaNO_2$ dosage increased, the setting time increased slowly at first and then rapidly when the dosage exceeded 3%. At a 4% dosage, initial and final setting times increased by 88.2% (256 min) and 89.4% (356 min), respectively, compared to the control. Thus, excessive $NaNO_2$ may delay the hydration of the composite system [33].

From Figure 3c, Li_2CO_3, similar to $CaCl_2$, reduced setting times but with lesser efficacy. At a 0.6% dosage, initial and final setting times decreased by 65.4% and 29.3%, respectively, compared to the control. The reduction was mainly due to the fact that Li^+ ions shorten the induction period, accelerating sulfoaluminate cement hydration [27]. However, its impact was weaker than $CaCl_2$, likely due to limited freezing-point depression and preferential interaction with sulfoaluminate phases rather than silicates.

3.2. Compressive Strength

When subjected to continuous negative temperature curing, as shown in Figure 4a, the incorporation of $CaCl_2$ significantly enhanced the compressive strength of the composite system at various ages. When the $CaCl_2$ content was 0%, the compressive strength of the composite system at all ages had been suppressed by the negative temperature. The compressive strength of the composite system at −3 d, −7 d, and −28 d was only 3.1 MPa, 6.2 MPa, and 4.9 MPa, respectively. As the $CaCl_2$ content increased, the compressive strength of the composite system also increased accordingly. When the $CaCl_2$ content reached 2%, the compressive strength of the composite system reached the maximum value, i.e., 22.0 MPa, 41.2 MPa, and 43.7 MPa at −3 d, −7 d, and −28 d, respectively, which had been five to eight times higher than that of the blank. $CaCl_2$ promoted the hydration of C_3S and C_3A in the composite system and accelerated the formation of hydration products. During the initial stage of continuous negative temperature curing, it had still rapidly constructed a microscopic skeleton with certain strength, thereby improving the compressive strength of the composite system.

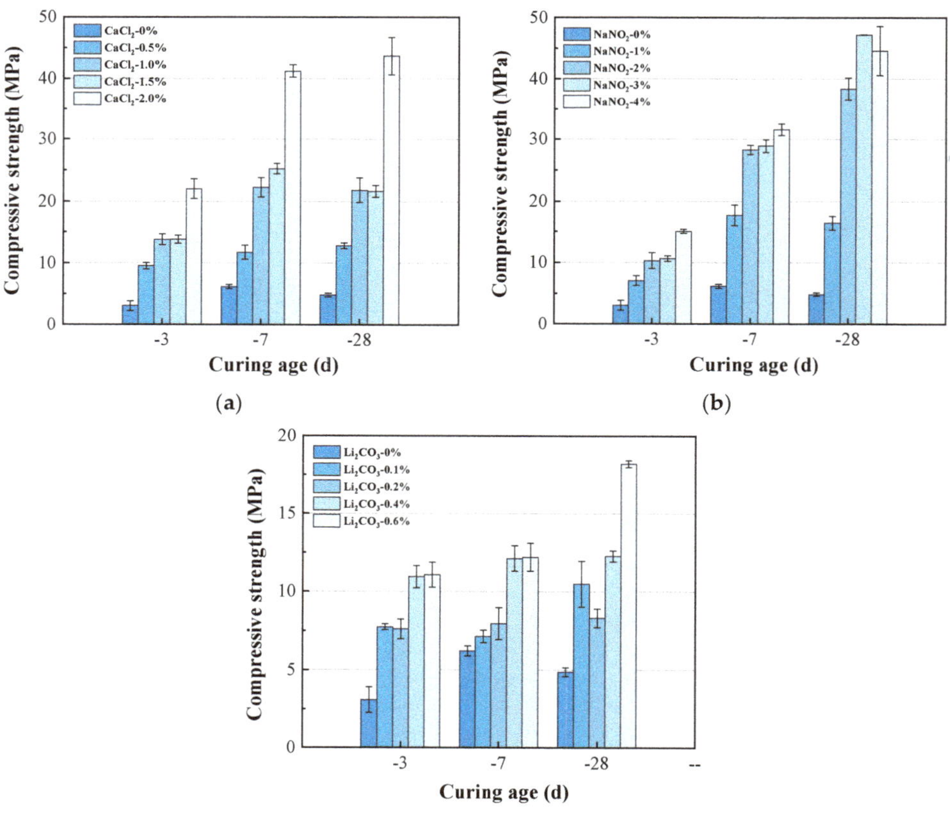

Figure 4. Effects (**a**) $CaCl_2$, (**b**) $NaNO_2$, and (**c**) Li_2CO_3 on the compressive strength of composite cement at continues curing at −10 °C.

From Figure 4b, it can be observed that the incorporation of $NaNO_2$ also improved the compressive strength of the composite system at various ages. During the early stages of negative temperature curing at −3 d and −7 d, as the $NaNO_2$ content increased from 1.0% to 4.0%, the compressive strength of the composite system at −3 d and −7 d had continued to increase. When the $NaNO_2$ content reached 4%, the compressive strength reached the maximum. During the later stage of −28 d, the compressive strength of the composite system reached the maximum when the $NaNO_2$ content was 3%, which was 47.2 MPa, more than eight times higher than that of the blank. The incorporation of $NaNO_2$ lowered the freezing point of the liquid phase, thereby reducing the content of freezable water. Under −10 °C curing, it provided a large amount of unfrozen water in the early stage to ensure the continuous hydration of the composite system.

As shown in Figure 4c, Li_2CO_3 improved the compressive strength of the composite system to a certain extent. However, as the age increased, the growth in compressive strength was not significant. The incorporation of Li_2CO_3 promoted the hydration of C_3A and $C_4A_3\bar{S}$ in the paste, thereby accelerating the formation of the hydration product AFt. This rapidly constructed a relatively dense microscopic network structure in the early stage, resisting the freezing damage caused by negative temperature curing conditions to a certain extent. However, its effect on lowering the freezing point was limited. Under continuous negative temperature curing, the free water in the system froze, making it difficult for the composite system to undergo continuous hydration.

From Figure 5a, it can be seen that when the curing condition shifted from a negative temperature to positive temperature, the compressive strength of the composite system significantly increased at −7 + 21 d compared to continuous negative temperature curing for −28 d. After the curing temperature shifted from negative to positive, some amount of frozen water melted and was reintroduced into the hydration process [19]. Under this curing condition, $CaCl_2$ also enhanced the compressive strength of the composite system. When the amount of $CaCl_2$ was 0.5%, the compressive strength at −7 + 21 d reached a maximum of 89.8 MPa, representing a 68.3% increase compared to the blank. When shifting the temperature from −10 °C to 20 °C, the previously frozen free water thawed and again became available for cement hydration. In the composite system, sulfoaluminate cement initiated rapid hydration, generating AFt to enhance strength. Concurrently, $CaCl_2$ accelerated the hydration of silicate phases, facilitating the formation of a denser microstructure, thereby achieving enhanced mechanical performance.

As shown in Figure 5b, with increasing amounts of $NaNO_2$, the compressive strength of the composite system at −7 + 21 d exhibited a trend of first increasing and then decreasing. When the amount of $NaNO_2$ was 2%, the compressive strength at −7 + 21 d reached a maximum of 64.5 MPa. However, when the amount of $NaNO_2$ was as high as 4%, the compressive strength of the composite system was only 52.4 MPa. This indicated that excessive $NaNO_2$ was detrimental to the hydration of the composite system at room temperature.

From Figure 5c, it can be observed that an appropriate amount of Li_2CO_3 significantly improved the compressive strength of the composite system compared to continuous negative temperature curing. When the amount of Li_2CO_3 was 0.2%, the compressive strength at −7 + 21 d reached a maximum of 77.0 MPa, representing a 44.4% increase compared to the control group. However, excessive Li_2CO_3 was detrimental to the development of compressive strength. When the amount of Li_2CO_3 was 0.6%, the compressive strength of the composite system was only 48.7 MPa. Compared to $NaNO_2$, Li_2CO_3 exhibited a more significant strength enhancement after the temperature transition to room temperature conditions. This contrast revealed that the freezing of free water substantially impaired Li_2CO_3's ability to promote hydration reactions.

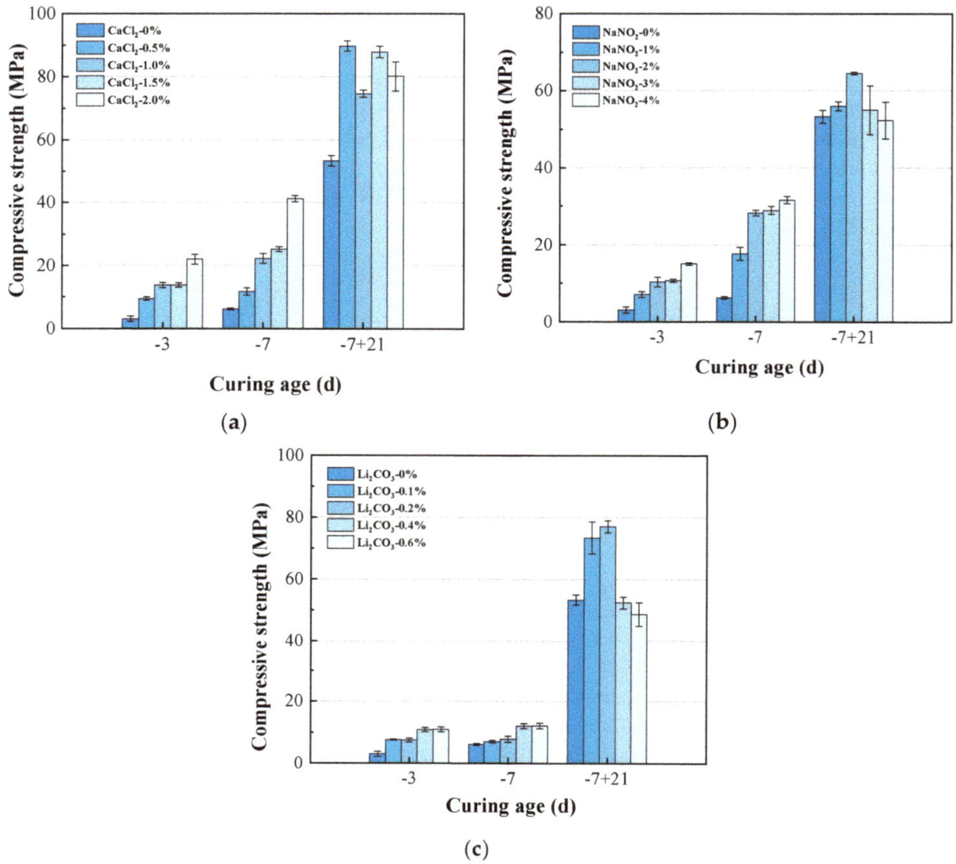

Figure 5. Effects (**a**) $CaCl_2$, (**b**) $NaNO_2$, and (**c**) Li_2CO_3 on the compressive strength of composite cement at curing condition shifted from negative temperature to positive temperature.

3.3. Freezing Point

Table 3 shows the internal liquid phase freezing point of the composite system. When without an antifreeze agent, the freezing point was −0.1 °C, indicating that the composite system was susceptible to frost damage at low temperatures of −10 °C. After adding $CaCl_2$, the freezing point of the composite system decreased significantly. As the amount of $CaCl_2$ increased, the freezing point of the composite system continued to drop. When the $CaCl_2$ content was 2%, the freezing point of the composite system decreased to −3.2 °C. As an inorganic salt, $CaCl_2$ not only lowered the freezing point of the liquid phase of the composite system but also accelerated the dissolution of C_3S and C_3A in the composite system, promoting the formation of hydration products and enabling the system to rapidly establish a microscopic network structure in the early stages [31]. As the pores inside the slurry were gradually filled with hydration products, the pore diameters decreased, and the freezing point of the composite system was reduced [32].

The effect of $NaNO_2$ in lowering the freezing point of the composite system was more significant. With increasing the $NaNO_2$ dosage, the freezing point of the composite system continued to decrease. When the $NaNO_2$ dosage was 1%, 2%, 3%, and 4%, the freezing point of the composite system decreased to −2.3 °C, −3.1 °C, −4.1 °C, and −4.9 °C, respectively. According to Raoult's Law, the addition of $NaNO_2$ effectively increased the liquid phase concentration in the composite system, and the freezing point of the liquid phase decreased as the vapor pressure inside the pores decreased [33]. Therefore, the content of unfrozen water in the composite system increased, ensuring the adequate progression of early hydration reactions. The formed hydration products filled the pores in the composite system, reducing pore diameters and further lowering the liquid phase freezing point [6]. Thus, under continuous negative temperature curing conditions, the composite system had sufficient unfrozen water available for hydration reactions, enabling continuous strength growth.

Table 3. Effects of antifreeze agent on the freezing point of composite cement.

Antifreeze Agent	Dosage	Freezing Point
Blank	/	−0.1
$CaCl_2$	0.5	−1.7
	1.0	−1.2
	1.5	−2.5
	2.0	−3.2
$NaNO_2$	1.0	−2.3
	2.0	−3.1
	3.0	−4.1
	4.0	−4.9
Li_2CO_3	0.1	−1.8
	0.2	−2
	0.4	−2.3
	0.6	−1.1

The effect of incorporating Li_2CO_3 in lowering the freezing point of the composite system was relatively insignificant. When the Li_2CO_3 dosage ranged from 0.1% to 0.6%, the freezing point of the composite system was between −1.1 °C and −2.3 °C. $CaCl_2$ and $NaNO_2$ maintained high solubility at low temperatures, and their dissociation generated a greater number of ions, enabling effective freezing point depression. In contrast, Li_2CO_3 exhibited extremely low solubility, further reduced under sub-zero conditions, with insufficient total particle concentration post-dissociation to significantly alter the freezing point. Therefore, under continuous negative temperature curing conditions, a large amount of free water froze, and Li_2CO_3 had a non-significant promoting effect on the hydration of the composite system, with minimal strength growth observed. However, when the temperature transitioned to positive, a large amount of water was released. At this point, Li_2CO_3 effectively promoted the hydration of the composite system, leading to rapid strength growth.

3.4. XRD Analysis

The effect of $CaCl_2$ on the hydration products of the composite system is shown in Figure 6a, and the diffraction peak intensities of the phases were quantitatively extracted and plotted in Figure 7a for comparative analysis. Compared to the raw materials and the blank system, the addition of $CaCl_2$ resulted in a significant decrease in the diffraction peak intensity of the clinkers C_3S and C_2S in the composite system, while the diffraction peak intensity of the hydration product AFt and CH was notably enhanced. This indicates that the incorporation of $CaCl_2$ promoted the hydration of the silicate phases in the system, thereby accelerating the formation of CH crystals. With the increase in $CaCl_2$ content, the diffraction peak intensity of gypsum in the composite system significantly decreased, while the diffraction peak intensity of the hydration product AFt further increased. Studies [11] have demonstrated that Ca^{2+} ions enhance particle dispersion through electrical double-layer effects, while their increased concentration in pore solution elevate ionic activity coefficients and reaction kinetics, thereby accelerating C_3S hydration. Concurrently, Cl^- ions chemically interact with C_3A phases to form insoluble chloroaluminates, which preferentially drive aluminate reaction pathways.

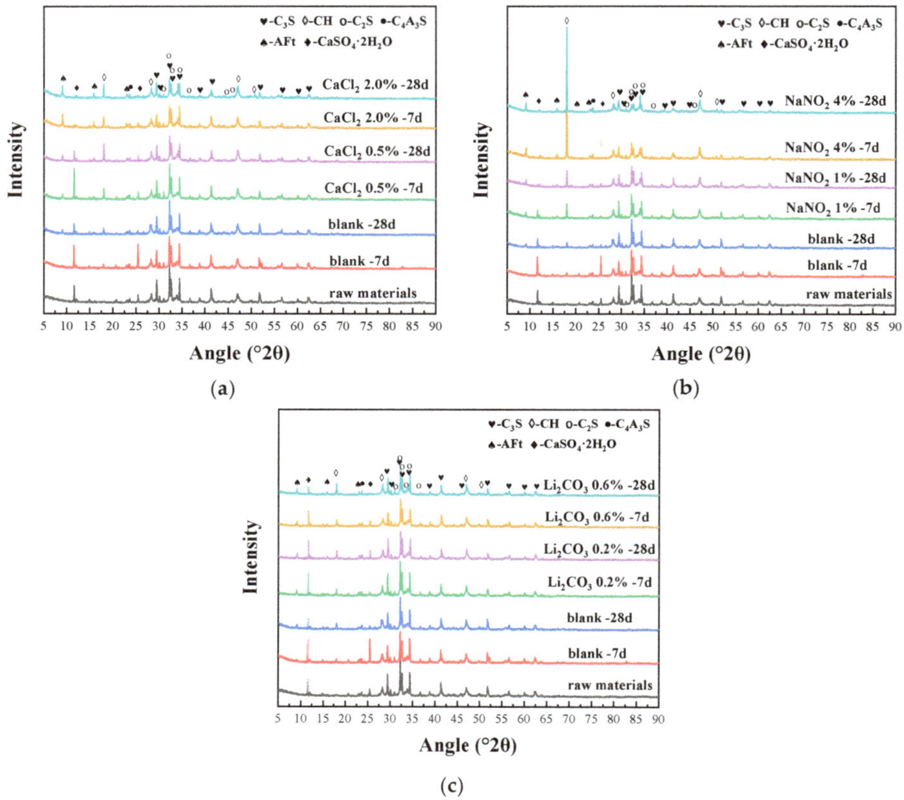

Figure 6. XRD patterns showing the effects of (**a**) $CaCl_2$, (**b**) $NaNO_2$, and (**c**) Li_2CO_3 on hydration products of composite cement.

The effect of $NaNO_2$ on the hydration products of the composite system is shown in Figure 6b, and the diffraction peak intensities of the phases were quantitatively extracted and plotted in Figure 7b for comparative analysis. Compared to the blank system, the incorporation of $NaNO_2$ decreased the diffraction peak intensity of C_3S and C_2S, while substantially increasing those of hydration products AFt and CH. When 1% $NaNO_2$ was added, the gypsum diffraction peak remained detectable after −7 d of curing but nearly disappeared by −28 d, indicating the progressive consumption of gypsum during hydration. At a higher $NaNO_2$ dosage (4%), gypsum was almost entirely consumed within −7 d of sub-zero temperature curing. The experimental results of the freezing point demonstrated that $NaNO_2$ significantly reduced the freezing point of cement paste. This phenomenon was attributed to the colligative properties of solutions, where the extent of freezing point depression correlated with the total dissolved particle concentration. Owing to its high solubility, $NaNO_2$ dissociated into abundant Na^+ and NO_2^- ions in the pore solution, resulting in an effective freezing point depression through ionic interference with ice nucleation thermodynamics. In addition, Liu et al. [11] have indicated that $NaNO_2$ promotes the rapid nucleation and growth of hydration products by providing additional nucleation sites, which also accelerate early hydration reactions.

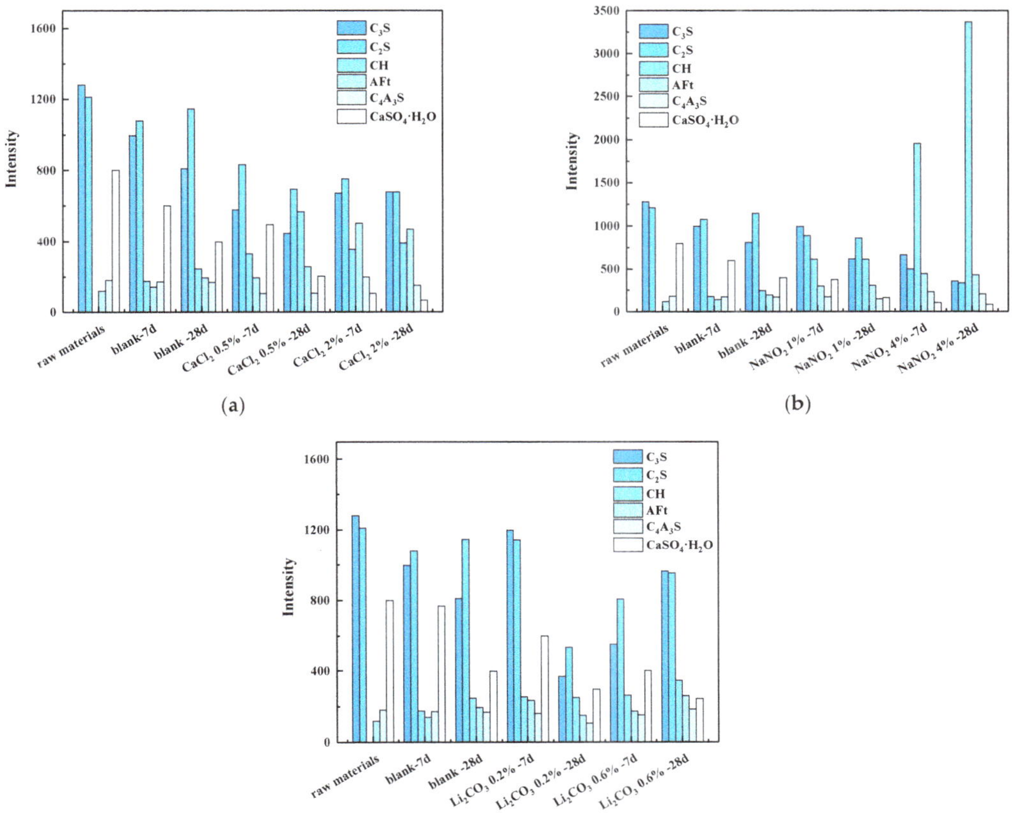

(c)

Figure 7. Effects of (**a**) $CaCl_2$, (**b**) $NaNO_2$, and (**c**) Li_2CO_3 on intensity of hydration products.

The effect of Li_2CO_3 on the hydration products of the composite system is illustrated in Figure 6c, and the diffraction peak intensities of the phases were quantitatively extracted and plotted in Figure 7c for comparative analysis. Compared to the blank system, the incorporation of Li_2CO_3 resulted in a reduction in the peak intensities of C_3S and C_2S in the composite system, while the diffraction peak of hydration products (AFt and CH crystals) gradually intensified with the increasing Li_2CO_3 content. When the Li_2CO_3 dosage increased from 0.2% to 0.6%, the diffraction peak intensity of gypsum decreased under continuous sub-zero temperature curing, whereas the peak intensity of CH crystals was progressively strengthened. Zhang et al. [27] have demonstrated that Li_2CO_3 reacts with AH_3 to produce LiOH, which increases the system's pH and enhances the dissolution of calcium sulfoaluminate phases. This process releases additional Al^{3+} and $SO_4{}^{2-}$ ions, accelerating AFt formation. Simultaneously, Li^+ ions coordinate with hydroxyl groups to stabilize $[Al(OH)_6]^{3-}$ clusters, reducing the energy barrier for AFt nucleation and promoting rapid crystallization. Therefore, the hydration of sulfoaluminate cement in the composite cement is accelerated. However, due to its limited ability to depress the freezing point, Li_2CO_3 exhibits inferior effectiveness in enhancing the strength of the composite system under sub-zero temperatures compared to $CaCl_2$ and $NaNO_2$.

3.5. SEM Analysis

The morphological effects of different admixtures on the hydration products of the composite cement at −7 d are illustrated in Figure 8. In the blank system, limited hydration products were observed on the surface of cement particles. Combined with XRD results, the hydration reaction was significantly inhibited by the sub-zero curing temperature, leaving a substantial amount of unhydrated cement particles at −7 d. With the addition of 2% $CaCl_2$, abundant gel-like hydration products covered the cement particle surfaces, accompanied by hexagonal plate-like CH crystals. Compared to the blank system, $CaCl_2$ accelerated the hydration reaction under sub-zero conditions, forming a denser microstructural network. For the system incorporating 4% $NaNO_2$, the microstructure of hydration products exhibited a high compactness, with extensive amorphous gel-like phases covering the surfaces. The addition of $NaNO_2$ depressed the freezing point of the pore solution, reduced the content of freezable water, effectively sustained hydration at sub-zero temperatures, and mitigated frost heave-induced damage. In the case of 0.6% Li_2CO_3, needle-shaped AFt crystals and amorphous gel-like hydration products were observed on cement particle surfaces. However, numerous pores were evident. Although Li_2CO_3 promoted the hydration of sulfoaluminate cement and AFt formation, its limited freezing point depression not only hindered hydration but also increased microporosity due to frost heave-induced structural damage.

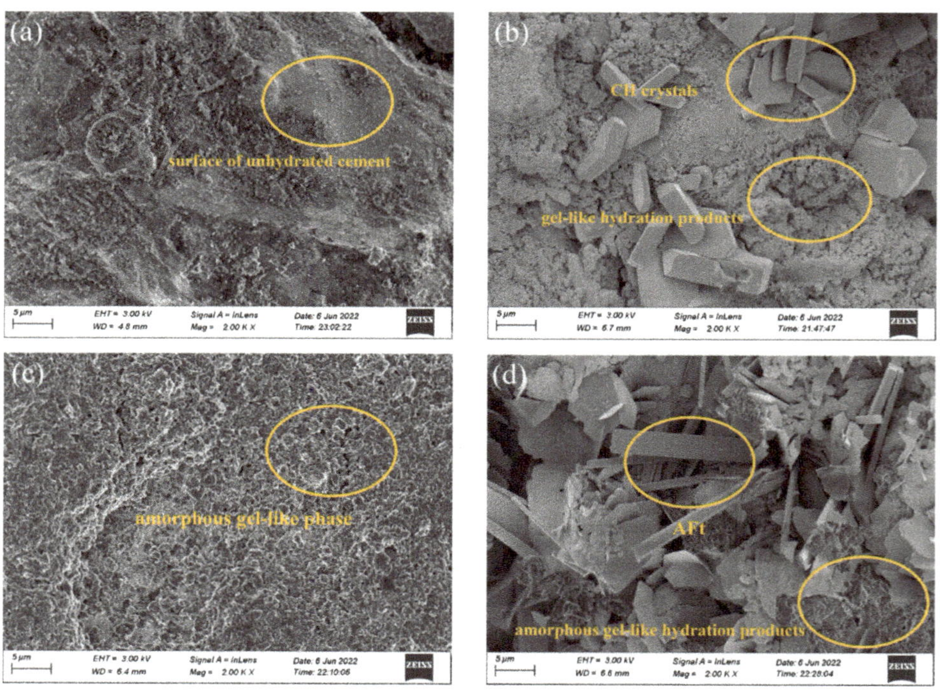

Figure 8. Effects (**a**) without addition, (**b**) $CaCl_2$, (**c**) $NaNO_2$, and (**d**) Li_2CO_3 on the morphology of composite cement.

4. Conclusions

This study investigated the effects of three antifreeze agents ($CaCl_2$, $NaNO_2$, and Li_2CO_3) on the performance of composite cement under sub-zero temperatures (−10 °C). The main conclusions are as follows:

(1) Three antifreeze agents enhanced the strength of the composite system under sub-zero conditions to different extents. Both $CaCl_2$ and $NaNO_2$ demonstrated sustained strength development under continuous sub-zero temperature curing (−10 °C), with additional strength gains observed after transitioning to room temperature. In contrast, Li_2CO_3 showed a limited early-stage improvement under persistent sub-zero conditions but achieved remarkable strength recovery following the temperature shift to room temperature.

(2) The freezing point depression efficacy in the composite system followed the order $NaNO_2 > CaCl_2 > Li_2CO_3$. Both $CaCl_2$ and $NaNO_2$ maintained substantial unfrozen water contents, enabling continuous hydration under continuous sub-zero temperature curing. In contrast, Li_2CO_3 permitted the amount of water frozen, fundamentally restricting its strength-enhancing effectiveness under sustained sub-zero conditions.

(3) The three antifreeze agents acted in the composite system through different pathways. $CaCl_2$ predominantly enhanced strength through accelerated silicate phase hydration, forming dense hydration products. $NaNO_2$ mainly provided unfrozen water to ensure the hydration of cement by lowering the freezing point, but excessive $NaNO_2$ inhibited hydration processes. Li_2CO_3 promoted the strength increase mainly by affecting the hydration rate of the sulfoaluminate phase. This only worked if unfrozen water was present, while it did not have the effect of lowering the freezing point.

(4) Practical implementation requires scenario-specific strategies: For sustained sub-zero conditions, 1.5%–2.0% $CaCl_2$ is recommended in composite cement due to its superior freezing point depression and hydration acceleration, though corrosion inhibitors must be incorporated for steel-reinforced applications. Under intermittent low-to-ambient temperature cycles, 0.2%–0.4% Li_2CO_3 offers advantages by enabling rapid strength recovery in chloride-sensitive environments. For cost-driven projects, 2%–3% $NaNO_2$ provides balanced ice suppression and strength enhancement without chloride-related durability risks. Future research should prioritize investigating the synergistic effects of hybrid antifreeze agents to advance sustainable cold-region construction practices.

Author Contributions: Conceptualization, B.Z.; methodology, B.Z.; validation, B.Z.; formal analysis, Y.D.; investigation, Y.H.; resources, D.L.; data curation, D.L.; writing—original draft preparation, D.L.; writing—review and editing, B.Z.; supervision, B.Z.; project administration, B.Z. All authors have read and agreed to the published version of the manuscript.

Funding: This research received no external funding.

Institutional Review Board Statement: Not applicable.

Informed Consent Statement: Not applicable.

Data Availability Statement: If needed, the corresponding author can be contacted to obtain the original data.

Acknowledgments: The financial support from Shaanxi Province No. S202410703094 for the current research is gratefully acknowledged.

Conflicts of Interest: Authors Bitao Zhang and Yongkang Du were employed by Shaanxi Building Materials Technology Group Co., Ltd. Author Dong Liu was employed by Huangling Ecological Cement Co., Ltd. The remaining authors declare that the research was conducted in the absence of any commercial or financial relationships that could be construed as a potential conflict of interest.

References

1. Choi, H.; Zhang, W.; Hama, Y. Method for determining early-age frost damage of concrete by using air-permeability index and influence of early-age frost damage on concrete durability. *Constr. Build. Mater.* **2017**, *153*, 630–639.
2. Elshazly, M.A.; Elakhras, A.A.; Elshami, A.A.; Ahmad, S.S.E.; Elmahdy, M.A.R. Investigating the effectiveness of a bacterial self-healing mechanism for repairing cracks in sustainable cement mortar at low temperatures. *Results Eng.* **2025**, *25*, 103907.
3. Kothari, A.; Hedlund, H.; Illikainen, M.; Cwirzen, A. Effects of sodium nitrate and OPC-GGBS concrete mix composition on phase transition of pore water at subzero temperatures. *Constr. Build. Mater.* **2022**, *327*, 126901. [CrossRef]
4. Alzaza, A.; Ohenoja, K.; Isteri, V.; Hanein, T.; Geddes, D.; Poikelispää, M.; Illikainen, M. Blending eco-efficient calcium sulfoaluminate belite ferrite cement to enhance the physico–mechanical properties of Portland cement paste cured in refrigerated and natural winter conditions. *Cem. Concr. Compos.* **2022**, *129*, 104469.
5. Liu, Y.; Sun, F.; Yu, K.; Yang, Y. Experimental and numerical research on development of synthetic heat storage form incorporating phase change materials to protect concrete in cold weather. *Renew. Energy* **2020**, *149*, 1424–1433. [CrossRef]
6. Zhang, G.; Jiang, S.; Lei, Y.; Shi, C.; Li, G.; Hanif, A. Optimizing anti-freezing agent on the properties of Portland cement-calcium sulphoaluminate cement system based on Taguchi-GRA method. *Case Stud. Constr. Mater.* **2024**, *20*, e02998.
7. Nishat, F.M.; Chakraborty, R.; Oh, H.-J.; Yoo, D.-Y.; Mohapoo, R.; Yeon, J.H. Design and performance evaluation of an electrically heated concrete panel for sustainable winter maintenance. *Dev. Built Environ.* **2023**, *16*, 100259. [CrossRef]
8. Wang, T.; Liao, C.; Qi, X.; Zhang, Y. Predicting the effect of promoting ultra-low energy buildings in hot summer and warm winter regions on CO_2 emission. *Energy Sustain. Dev.* **2025**, *85*, 101646.
9. Zhang, G.; Yang, H.; Ju, C.; Yang, Y. Novel selection of environment-friendly cementitious materials for winter construction: Alkali-activated slag/Portland cement. *J. Clean. Prod.* **2020**, *258*, 120592.
10. Li, L.; Wang, C.; Zhao, Z.; Dang, L.; He, R. Hydration behavior and micro-pore structural of Portland cement composites with crystalline nano-SiO_2 at low temperature. *J. Build. Eng.* **2024**, *98*, 111276.
11. Liu, Z.; Lou, B.; Barbieri, D.M.; Sha, A.; Ye, T.; Li, Y. Effects of pre-curing treatment and chemical accelerators on Portland cement mortars at low temperature (5 °C). *Constr. Build. Mater.* **2020**, *240*, 117893. [CrossRef]
12. Li, G.; Zhang, J.; Song, Z.; Shi, C.; Zhang, A. Improvement of workability and early strength of calcium sulphoaluminate cement at various temperature by chemical admixtures. *Constr. Build. Mater.* **2018**, *160*, 427–439. [CrossRef]
13. Isteri, V.; Ohenoja, K.; Hanein, T.; Kinoshita, H.; Kletti, H.; Rößler, C.; Tanskanen, P.; Illikainen, M.; Fabritius, T. Ferritic calcium sulfoaluminate belite cement from metallurgical industry residues and phosphogypsum: Clinker production, scale-up, and microstructural characterisation. *Cem. Concr. Res.* **2022**, *154*, 106715. [CrossRef]
14. Kim, N.; Seo, J.; Lee, H.K. Enhancement in clinker hydration degrees and later stage-ettringite stability of calcium sulfoaluminate cements by the incorporation of dolomite. *Cem. Concr. Compos.* **2025**, *155*, 105815. [CrossRef]
15. Zhang, J.; Ye, C.; Tan, H.; Liu, X. Potential application of Portland cement-sulfoaluminate cement system in precast concrete cured under ambient temperature. *Constr. Build. Mater.* **2020**, *251*, 118869. [CrossRef]
16. Chaunsali, P.; Mondal, P. Physico-chemical interaction between mineral admixtures and OPC–calcium sulfoaluminate (CSA) cements and its influence on early-age expansion. *Cem. Concr. Res.* **2016**, *80*, 10–20. [CrossRef]
17. Alzaza, A.; Ohenoja, K.; Langås, I.; Arntsen, B.; Poikelispää, M.; Illikainen, M. Low-temperature (−10 °C) curing of Portland cement paste—Synergetic effects of chloride-free antifreeze admixture, C-S-H seeds, and room-temperature pre-curing. *Cem. Concr. Compos.* **2022**, *125*, 104319. [CrossRef]
18. Khan, J.; Kumar, S.G. Influence of binary antifreeze admixtures on strength performance of concrete under cold weather conditions. *J. Build. Eng.* **2021**, *34*, 102055. [CrossRef]
19. Karagol, F.; Demirboga, R.; Khushefati, W.H. Behavior of fresh and hardened concretes with antifreeze admixtures in deep-freeze low temperatures and exterior winter conditions. *Constr. Build. Mater.* **2015**, *76*, 388–395. [CrossRef]
20. Polat, R. The effect of antifreeze additives on fresh concrete subjected to freezing and thawing cycles. *Cold Reg. Sci. Technol.* **2016**, *127*, 10–17. [CrossRef]
21. Demirboğa, R.; Karagöl, F.; Polat, R.; Kaygusuz, M.A. The effects of urea on strength development of fresh concrete under cold weather conditions. *Constr. Build. Mater.* **2014**, *64*, 114–120. [CrossRef]
22. Poirier, M.; Blotevogel, S.; Noiriel, C.; Bonnin, A.; Kaknics, J.; Olbinado, M.; Steger, L.; Patapy, C.; Cyr, M. Synchrotron X-ray micro-tomography investigation of the early hydration of blended cements: A case study on $CaCl_2$-accelerated slag-based blended cements. *Constr. Build. Mater.* **2022**, *321*, 126412.
23. Wang, H.; Tong, M. Properties and field application of the grouting material for water blocking during thawing of frozen wall of deep sand layer. *Arab. J. Geosci.* **2021**, *14*, 1–12.
24. Gómez-Luna, G.F.; Lopez-Calvo, H.Z.; Bremner, T.W.; Castro-Borges, P.; Montes-García, P. Performance of fly ash and CNI as corrosion prevention methods for steel reinforcement embedded in cracked HPC concrete exposed to a natural marine environment. *Constr. Build. Mater.* **2024**, *428*, 136260.

25. Vu, T.H.; Dang, L.C.; Kang, G.; Sirivivatnanon, V. Chloride-induced corrosion of steel reinforcement in alkali-activated slag concretes: A critical review. *Case Stud. Constr. Mater.* **2022**, *16*, e01112.
26. Jiang, M.; Liu, X.; Hang, M.; Xu, Y.; Lai, G.; Li, S. Performance and deterioration mechanism of concrete incorporated with corrosion-inhibiting admixtures under the coupling effect of composite salt and freeze-thaw cycles. *J. Build. Eng.* **2023**, *69*, 106329.
27. Zhang, Y.; Wang, Y.; Li, T.; Xiong, Z.; Sun, Y. Effects of lithium carbonate on performances of sulphoaluminate cement-based dual liquid high water material and its mechanisms. *Constr. Build. Mater.* **2018**, *161*, 374–380.
28. Shen, Y.; Zhang, W.; Wang, P.; Chen, X.; Zhu, H. Influence of lithium salt on the performance of calcium sulfoaluminate cement. *J. Therm. Anal. Calorim.* **2022**, *147*, 1–9.
29. *GB/T 17671-2021*; Test Method of Cement Mortar Strength (ISO Method). Standards Press of China: Beijing, China, 2021.
30. Ming, X.; Li, Y.; Liu, Q.; Wang, M.; Cai, Y.; Chen, B.; Li, Z. Chloride binding behaviors and early age hydration of tricalcium aluminate in chloride-containing solutions. *Cem. Concr. Compos.* **2023**, *137*, 104928.
31. Karagöl, F.; Demirboğa, R.; Kaygusuz, M.A.; Yadollahi, M.M.; Polat, R. The influence of calcium nitrate as antifreeze admixture on the compressive strength of concrete exposed to low temperatures. *Cold Reg. Sci. Technol.* **2013**, *89*, 30–35.
32. Thomas, J.J.; Allen, A.J.; Jennings, H.M. Hydration kinetics and microstructure development of normal and $CaCl_2$-accelerated tricalcium silicate pastes. *J. Phys. Chem. C* **2009**, *113*, 19836–19844.
33. Saha, O.; Boulfiza, M.; Wegner, L.D. Tracking the hydration of antifreeze treated cement paste at subfreezing temperatures using the TDR technique. *Constr. Build. Mater.* **2020**, *262*, 119915. [CrossRef]

coatings

MDPI

Article

The Influence of NaCl Internal and External Erosion on the Properties of Steel Scoria Reactive Powder Concrete

Ligai Bai [1], Haiyuan Liu [2], Jian Zhang [3,4], Youheng Zhang [1,*], Xin Zhang [2], Kangshuo Xia [5] and Feiting Shi [6]

[1] School of Architectural Engineering, North China Institute of Aerospace Engineering, Langfang 065000, China
[2] China Construction First Group Co., Ltd., 52 South West Sihuan Road, Beijing 100161, China; zhangxinzcb@cscec.com (X.Z.)
[3] School of Public Policy & Management, China University of Mining and Technology, Xuzhou 221116, China
[4] Huaxia Spatial Planning Institute (Beijing) Co. Ltd., Beijing 100176, China
[5] School of Civil Engineering, Chongqing University, Chongqing 400044, China
[6] School of Civil Engineering, Yancheng Institute of Technology, Yancheng 224051, China; shifeiting@ycit.cn
* Correspondence: zhangyh123456@aliyun.com

Abstract: This paper the flexural and compressive strengths of the reactive powder concrete (RPC) with steel scoria and quartz sand containing NaCl are investigated. Moreover, the RPC's mass, the chloride ion permeability and the carbonation depth (D_c) are determined. The mass ratios of steel scoria and the NaCl are 0%~20% and 0%~0.25% by mass of binder materials and the quartz sand respectively. The RPC specimens are exposed to the NaCl erosion environment. The scanning electron microscope-energy dispersive spectrometer (SEM-EDS) and X-ray diffraction (XRD) spectrum are acquired for analyzing the mechanism of RPC's performance. Results show that the flexural strength, the compressive strengths, the mass and the dynamic modulus of elasticity (RDME) of RPC decrease in the form of cubic function with the mass ratio of NaCl. When the mass ratio of steel scoria is 10%, the mechanical strengths and the RDME are the highest. The RPC's flexural strength, the compressive strength and the RDME decrease by rates of 4.94%~42.28%, 5.11%~48.65% and 8.72%~226.1% after NaCl erosion. Meanwhile, the corresponding mass loss rate, the chloride ion permeability, the D_c are increased by rates of 1.32%~27.63%. RPC with 10% steel scoria shows the lowest performance degradation. The SEM-EDS results show that the pores and cracks inner RPC and the Cl and Ca elements are increased by the NaCl. The Fe and Ca elements are increased by the added steel scoria. The addition of steel scoria exhibit decreasing effect and the added NaCl shows increasing effect on the Ca $(OH)_2$ crystals respectively.

Keywords: steel scoria; NaCl; mechanical strengths; chloride ion permeability; mass loss rates

1. Introduction

Steel scoria is a common solid waste during the the process of steel smelting [1]. According to statistics, the global annual production of steel scoria in 2024 is approximately 226 million to 282 million tons. Accumulated steel scoria is prone to forming dust, polluting the air and water sources [2]. Several methods have been used for processing the steel scoria. Magnetic separation method, the gravity selection method have been used for processing steel scoria by separating the substances containing irons. However, the oxides of aluminum and magnesium in steel scoria are difficult to treat [3]. Due to different processes and scales, the cost of steel scoria treatment is usually 30–50 US dollars per ton [4]. Hence, handling various elements in steel scoria reasonably is necessary. If the steel scoria can be used as raw materials for products, this industrial waste can be processed, simultaneously, the industrial resources are increased. In this article, the steel scoria is applied in the preparation of cement-based materials for the resource utilization of this solid waste.

The reuse of solid waste in cement concrete is a good method for treating the solid waste [5]. Researchers have used the garbage fly ash, secondary aluminum ash, slag ash et al. for manufacturing the cement concrete [6–8]. These solid waste at reasonable dosages can improve cement concrete's flexural strength by rates of 11.3%~26.8%, 12.6%~31.7% and 13.2%~32.6%. While, the corresponding compressive strengths can be increased by rates of 14.7%~27.3%, 17.6%~33.4% and 18.6%~33.7% [9,10]. Moreover, the resistance to chloride ion permeability, anti chloride erosion can be increased by the added solid waste [11–13]. Besides, the applications of steel scoria in cement-based materials is an excellent processing method.

The steel scoria can be used as cementitious materials or aggregates [14]. As reported in prior researchers, the steel scoria can replace some coarse aggregate in the cement concrete [15]. Liu et al. have reported that when the steel scoria replace some fine aggregate of cement concrete, the steel scoria can improve the compressive strength of cement concrete by rates of 3.1%~16.2% [16]. Moreover, the steel scoria has been proved to replace some cement in the cement concrete. When the steel scoria is applied as cementitious material, the compressive strength of the cement concrete can be improved by 2.3%~15.4% [17]. Comparing with the cement, the production of steel scoria has lower CO_2 emissions and cost. Meanwhile, the use of steel scoria in cement is cleaner than the solid waste of garbage fly ash, secondary aluminum ash and etc. [18–20]. Hence, the application of steel scoria in the cement concrete possesses a certain prospect.

Desalinated sand is a common aggregate produced in coastal cities, which can be used for manufacturing the cement concrete [21]. When the cement concrete with desalinated sand serves in seawater environment, the cement concrete is under the internal and external erosion of chloride salts [22]. However, little attention has been paid to the influence of chloride salts' internal and external erosion on the cement concrete's performance.

Reactive powder concrete (RPC) is a highly compact concrete material, showing excellent salt erosion resistance. Several kinds of solid waste including waste fly ash, secondary aluminum dross and stove ash have been reported to increase the RPC's flexural strengths by rates of 11.3%~31.4%, 12.7%~35.6% and 13.1%~36.1% after the specimens are exposed to the erosion effect of sodium chloride [23,24]. Moreover, the added solid waste can improve the impermeability and carbonation resistance. The steel scoria with desalinated sand may be applied in manufacturing the RPC, which can be served in the salt environment [25]. When the RPC with desalinated sand is served in the coastal environments, the RPC endures internal and external erosion caused by chloride salts. Hence, researches in this area are quite meaningful.

In this study, the influence of steel scoria and quartz sand on the RPC's mechanical performances is investigated. 0%~0.25% NaCl by mass ratio of quartz sand is added in the RPC, which simulates the application of desalinated sand. The RPC's mass, relative dynamic modulus of elasticity (RDME), the chloride ion permeability coefficient and the carbonation depth are measured. The NaCl freeze-thaw and dry-wet actions are exerted on the RPC specimens. The Scanning electron microscopy-energy dispersive spectroscopy (SEM-EDS) and the X-ray diffraction spectrum (XRD) curves are obtained to analyze the mechanism of RPC's properties' changes. This study will provide a new method for processing steel scoria, and more raw materials for preparing cement concrete.

Moreover, this research will offer reference for the application of steel scoria RPC in marine engineering environment.

2. Experimental

2.1. Raw Materials

Dongguan Runze Building Materials Co., Ltd., Dongguan, China, China's Ordinary Portland cement (OPC) is used in this research. The cement's initial setting time and final setting time are 106.6 min and 173.2 min respectively. The strength grade of the cement is 42.5 MPa. Henan Yunxiang New Materials Co., Ltd., Gongyi, China, China's steel scoria (SS) with the density of 2.93 × 10^3 kg/m^3, the specific surface area of 714 m^2/kg, the water content of 0.13% and PH of 7.6 is used in this study. The quartz sand (QS) with a particle size of 0.020 mm–3.350 mm, a density of 2.63 g/cm^3, and a silicon content of 99.6% provided by Changxing Huaqi New Materials Co., Ltd., Huzhou, China is utilized. The silica fume (SF), as a fine powder, with a particle size of 0.1 μm–100 μm, a density of 2.05 g/cm^3, and a silicon content of 95.5% is applied in this study. The polycarboxylate superplasticizer water reducer (WR) with a solid content of 38.5% and a water reducing rate of 28% is produced by Shandong Feike Co., Ltd., Qingdao, China. The sodium chloride analytical grade with the purity of 99.5% provided by Tianjin Dingshengxin Chemical Co., Ltd., Tianjin, China is used in this test. The particle size distribution and the chemical composition of the raw materials are shown in Tables 1 and 2. The photo of the steel scoria is shown in Figure 1.

Table 1. The raw materials' particle size (%).

Types	Particle Size/μm						
	0.3	0.6	1	4	8	64	360
P·O cement	0	0.33	2.66	15.01	28.77	93.59	100
Steel scoria	0.85	0.85	3.51	19.63	35.01	97.9	100
Silica fume	33.1	59.4	84.7	99.9	99.9	100	100
Quartz sand	0	0	0	0	0.06	20.98	100

Table 2. The raw materials' chemical compositions (%).

Types	SiO_2	Al_2O_3	Fe_xO_y	MgO	CaO	SO_3	Loss on Ignition
P·O cement	20.86	5.95	3.94	1.73	62.23	2.66	3.12
Steel scoria	34.06	14.74	4.34	9.73	35.93	0.23	–
Silica fume	44.84	32.36	4.22	–	4.63	–	–
Quartz sand	99.6	–	0.4	–	–	–	–

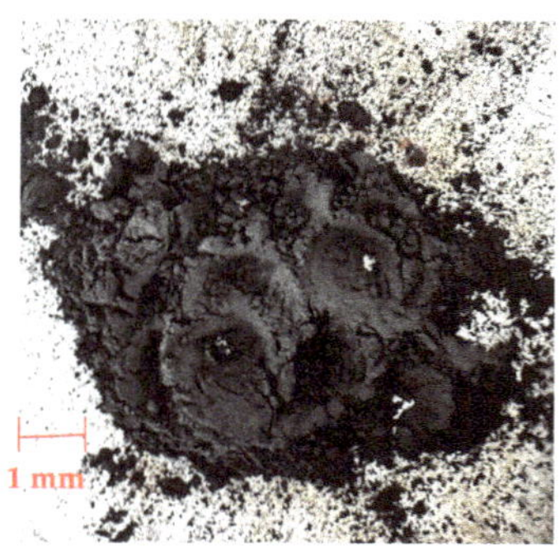

Figure 1. The steel scoria.

2.2. The Preparation of RPC

Table 3 shows the mixing proportions for preparing the RPC. According to prior research [26], when the added SS by the mass ratio of cementitious material is higher than 20%, the cement concrete reaches high expansion rate, which is disadvantage to the engineering applications. Therefore, in this study, the mass ratio of SS ranges from 0% to 20% by mass ratio of cementitious materials. 0%~0.25% NaCl by mass ratio of quartz sand is added in the RPC. The NaCl added in the RPC is used to simulate the desalinated sand. Firstly, all the powdered raw materials are poured into the Hebei Nalat Testing Machinery Co., Ltd., Cangzhou, China China's JJ-5 cement mortar mixer and mixed at the mixing speed of 140 r/min for 1 min. Then water and water reducer are added to the mortar mixer. After that another one minute's stirring with the speed of 140 r/min is provided. Finally, the last two minutes' stirring with the speed of 280 r/min is provided for manufacturing the RPC. After the fresh RPC's stirring is finished, the fresh RPC is poured to the moulds for forming the specimens. After the specimens are cured in the room environment (21.7 °C and humidity of 51.3%) for one day. Then, the specimens are demolded and moved to the standard curing environment (20.7 °C and 97.7% humidity) for 28 days.

Table 3. The UHPC's mixing proportions (kg/m^3).

Water	OPC	SS	SF	QS	WR	NaCl (%)	Mass Ratio of SS (%)
241.2	843.3	0	365.5	965.4	16.1	0	0
241.2	843.3	0	365.5	965.4	16.1	0.05	0
241.2	843.3	0	365.5	965.4	16.1	0.1	0
241.2	843.3	0	365.5	965.4	16.1	0.15	0
241.2	843.3	0	365.5	965.4	16.1	0.2	0
241.2	843.3	0	365.5	965.4	16.1	0.25	0
241.2	843.3	60.44	305.06	965.4	16.1	0	5
241.2	843.3	60.44	305.06	965.4	16.1	0.05	5
241.2	843.3	60.44	305.06	965.4	16.1	0.1	5
241.2	843.3	60.44	305.06	965.4	16.1	0.15	5
241.2	843.3	60.44	305.06	965.4	16.1	0.2	5
241.2	843.3	60.44	305.06	965.4	16.1	0.25	5
241.2	843.3	120.88	244.62	965.4	16.1	0	10
241.2	843.3	120.88	244.62	965.4	16.1	0.05	10
241.2	843.3	120.88	244.62	965.4	16.1	0.1	10
241.2	843.3	120.88	244.62	965.4	16.1	0.15	10
241.2	843.3	120.88	244.62	965.4	16.1	0.2	10
241.2	843.3	120.88	244.62	965.4	16.1	0.25	10
241.2	843.3	181.32	184.18	965.4	16.1	0	15
241.2	843.3	181.32	184.18	965.4	16.1	0.05	15
241.2	843.3	181.32	184.18	965.4	16.1	0.1	15
241.2	843.3	181.32	184.18	965.4	16.1	0.15	15
241.2	843.3	181.32	184.18	965.4	16.1	0.2	15
241.2	843.3	181.32	184.18	965.4	16.1	0.25	15
241.2	843.3	241.76	123.74	965.4	16.1	0	20
241.2	843.3	241.76	123.74	965.4	16.1	0.05	20
241.2	843.3	241.76	123.74	965.4	16.1	0.1	20
241.2	843.3	241.76	123.74	965.4	16.1	0.15	20
241.2	843.3	241.76	123.74	965.4	16.1	0.2	20
241.2	843.3	241.76	123.74	965.4	16.1	0.25	20

2.3. The Measurement of Mechanical Strengths

Specimens with sizes of 40 × 40 × 160 mm^3 are employed for the measuring the mechanical strengths cured at the standard curing environment for 28 days. Beijing Beiguang Jingyi Instrument Equipment Co., Ltd., Beijing, China China's fully automatic bending integrated testing machine is used for measuring the RPC's flexural and compressive strengths with the loading rates of 0.1 kN/s and 2.4 kN/s respectively. The experiment is carried out by the Chinese standard GBT17671-2021 [27].

2.4. The Experiment of NaCl Freeze-Thaw Cycles and Dry-Wet Alternations

Specimens are immersed in 3% NaCl solution for four days. Following the immersion, they are transferred to the TDRF-1 split type concrete rapid freeze-thaw (F-T) testing machine manufactured by Hebei Huawei Testing Instrument Co., Ltd., Cangzhou, China, for further experiments. The specimens are placed in sealed stainless steel F-T containers filled with 3% NaCl solution. The temperature in the F-T chamber varies from −15 °C to 8 °C. Firstly, the temperature decreases from room temperature to −15 °C at the decreasing rate of (10 ± 1) °C/h, the temperature of −15 °C is maintained for 3 h. Then, the temperature is increased from −15 °C to 8 °C at the increasing rate of (10 ± 1) °C/h and is maintained at this temperature for 1 h. Some specimens immersed in 3% NaCl solution for four days are moved to a concrete corrosion resistance dry-wet (D-W) cycle testing device provided by Tianjin Gangyuan Testing Instrument Factory Tianjin, China, for the NaCl D-W alternation experiments. Each D-W alternation includes a 6-h immersion in a NaCl solution, followed by wiping the specimens' surface with a cloth to remove the excess liquid and an-hour air dry. After that, the specimens are then dried in an oven at 80 °C for 15 h. When the drying is finished, the specimens are cooled in room temperature for 2 h until a D-W alternation is completed. All parameters are measured after 300 F-T cycles and 30 D-W alternations in the NaCl solution.

2.5. The Measurement of Mass Loss Rate

After being immersed in a 3% NaCl solution for four days, the surfaces of the specimens are dried with a cloth, and their weights are recorded using a Shunqiang electronic scale commercial platform scale (manufactured by Suzhou Shunqiang Electromechanical Equipment Co., Ltd., Suzhou, China). The mass loss rate (*MLR*) is computed by Equation (1).

$$MLR = \frac{m_1 - m_t}{m_1} \tag{1}$$

where m_1 is the mass after the 4-day immersion and m_t is the mass after 300 F-T cycles or 30 D-W alternations.

2.6. The Measurement of RDME

The measurements of relative dynamic modulus of elasticity (*RDME*) after NaCl freeze-thaw cycles ($RDME_f$) or after NaCl dry-wet alternations ($RDME_d$) are conducted utilizing an XUT580C digital ultrasonic flaw detector, offered by Dongguan Massive Instrument Equipment Co., Ltd., Dongguan, China. Sonar transducers are placed at the midpoint of the sample, with petroleum jelly used as a coupling agent. The determination of *RDME* is executed based on Equation (2).

$$RDME_f = (\frac{v_{ft}}{v_1})^2$$

$$RDME_d = (\frac{v_{dt}}{v_1})^2 \tag{2}$$

where v_1 is the ultrasonic velocity of the specimens after 4 days' NaCl immersion. And v_{ft} and v_{dt} are the ultrasonic velocities of the specimens after 300 NaCl F-T cycles and 30 NaCl D-W alternations respectively. Specimens with sizes of 100 × 100 × 400 mm^3 are used for the measurement of mass loss rate and *RDME*.

2.7. The Chloride Ion Migration Coefficient Measurement

The chloride ion migration coefficient (CMC) is determined using an SZ-RCM-NTB Concrete Chloride Ion Diffusion Coefficient Tester (manufactured by Shandong Sangze Instrument Co., Ltd., Jining, China). Before the measurement, all the specimens are vacuum-saturated. After that, the samples are moved to the SZ-RCM-NTB Concrete Chloride Ion Diffusion Coefficient Tester. The size of specimens is Φ100 mm × 50 mm.

2.8. The Carbonation Depth Measurement

The RPC's carbonation depth (D_c) is assessed after a 28-day exposure in a TH-2 concrete carbonation testing chamber (produced by Guangzhou Lincheng Jin Building Materials Co., Ltd., Guangzhou, China). The measurement's process is exhibited by the following steps.

The specimens are placed inside a carbonation chamber, where the CO_2 concentration is maintained at 20 ± 2%, while the temperature and humidity are controlled at 20 ± 5 °C and 70 ± 5%, respectively. After 28 days of exposure in CO_2, the specimens are removed and split across their width to facilitate the evaluation of D_c. Loose debris on the exposed surface is carefully cleared before applying a 1% phenolphthalein ethanol solution (containing 20% distilled water). The phenolphthalein solution reacts with uncarbonated areas, producing a color change that highlights the D_c. Measurements are taken 30 s after the phenolphthalein solution is applicated, spaced at 10 mm intervals along both sides of the specimen. The experiments of the CMC and the D_c are carried out according to the Chinese standard GB/T50082-2024 [28]. In this study, the coefficient of variation for all macroscopic performance test results is kept below 0.1, ensuring reliable and consistent data. Figure 2 shows the measuring process of the RPC's carbonation depth.

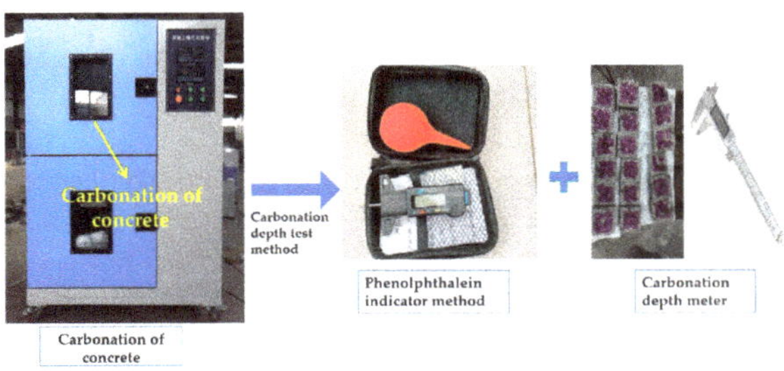

Figure 2. The measuring process of the RPC's carbonation depth.

2.9. The Microscopic Experimental Process

For the scanning electron microscopy-energy dispersive spectroscopy (SEM-EDS) analyses, samples extracted from the core of the specimen, with diameters ranging from 0.5 mm to 1 mm, are prepared. These samples are firstly dried in a vacuum oven at 105 °C for 48 h to remove any residual moisture. Following the drying process, a thin layer of gold was applied to the samples to improve the electrical conductivity and enhance image resolution. After gold coating, the samples are transferred to the SEM for observation. High-resolution images, magnified up to 5000 times, are obtained to analyze and assess the microstructure of the RPC. The SEM is determined using the Guoyi quantum Field Emission Scanning Electron Microscope SEM5000Pro (manufactured by Guoyi Quantum Technology Co., Ltd., Hefei, China). The powdered RPC sample is moved to the Haoyuan multifunctional X-ray diffractometer XRD DX-2700BH (manufactured by Dandong Haoyuan Instrument Co., Ltd., in Dandong, China) for the X-ray Diffraction (XRD) determination. Figure 3 shows the flow charts of the experimental process.

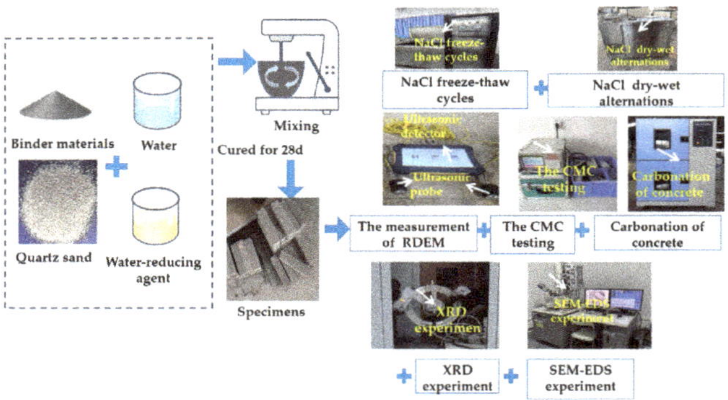

Figure 3. The flow charts of the experimental process.

3. Results and Discussions

3.1. The Mechanical Strengths

The flexural strength (f_t) and compressive strength (f_{cu}) of RPC specimens are shown in Figure 4. The fitting equations of NaCl mass ratio and the f_t and f_{cu} are shown in Table 4. It can be observed that the RPC's mechanical strengths decrease as the decreasing cubic function with the NaCl mass ratio. The added NaCl can reduce the f_t and f_{cu} by rates of 2.84%~83.86% and 2.61%~97.26%. This because the added NaCl in the RPC accelerates the corresponding alkali aggregate reaction, thus decreasing the RPC's mechanical strengths [29,30]. Moreover, the NaCl will crystallize and precipitate due to the hydration of pore water, decreasing the RPC's strength [31,32]. As found in Figure 3, the added steel scoria at mass ratio of 5% and 10% demonstrates positive effect on the RPC's mechanical strengths. While, when the mass ratios of the steel scoria are 15%~20%, the RPC's f_t and f_{cu} decrease by rates of 11.3%~21.6% and 13.4%~26.3% with the added steel scoria. The added steel scoria can improve the RPC's secondary hydration degree, thus increasing the corresponding mechanical strengths [33]. However, when the mass ratio of the steel scoria is higher than 10%, the added steel scoria can increase the alkali content in RPC. Therefore, the mechanical strengths of RPC are decreased with the steel scoria's mass ratio ranging from 15% to 20%. As obtained from Table 4, the fitting equations of NaCl mass ratio and the flexural strength (f_t) and compressive strength (f_{cu}) comply with cubic functions. Comparing the flexural and compressive strengths with RPC with traditional materials, the flexural and compressive strengths of RPC with SS are 5.6%~13.2% and 6.1%~14.3% higher [34].

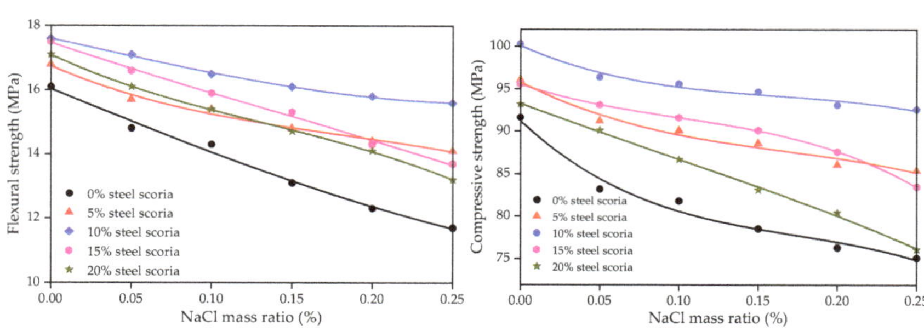

Figure 4. The mechanical strengths of RPC with NaCl and steel scoria. (**a**) The flexural strength. (**b**) The compressive strength.

Table 4. The fitting equations of NaCl mass ratio and the flexural strength (f_t) and compressive strength (f_{cu}).

Equation	Steel Scoria's Mass Ratio	a	b	c	d	R^2
$f_t = aM^3 + bM^2 + cM + d$	0%	266.37	91.05	−11.54	16.28	0.98
	5%	16.54	18.57	−15.48	16.92	0.97
	10%	6.28	17.19	−9.47	17.56	0.99
	15%	104.2	69.81	−11.54	17.33	0.99
	20%	−158.01	65.44	−21.82	17.06	0.99
$f_{cu} = aM^3 + bM^2 + cM + d$	0%	−211.42	305.85	−121.22	89.9	0.95
	5%	55.59	83.82	−62	94.93	0.94
	10%	−210	173.51	−57.31	99.47	0.94
	15%	−1129.19	337.41	−61.88	95.53	0.99
	20%	163.82	218.66	−54.59	92.88	0.99

3.2. The NaCl Freeze-Thaw Cycles and Dry-Wet Alternations

3.2.1. The Mass Loss Rate

The MLR of RPC after NaCl F-T cycles and NaCl D-W alternations are shown in Figure 5. Table 5 shows the fitting equations of NaCl mass ratio and the MLR. It can be observed from Figure 5 and Table 5, the MLR increases with the increasing mass ratio of the added NaCl in RPC. The added NaCl can increase the RPC's mass by rates of 1.72%~4.81% and 2.01%~4.92% after 300 NaCl F-T cycles and 30 NaCl D-W alternations. This can be analyzed that the inner NaCl can accelerate the alkali aggregate reaction and crystallization stress in RPC, therefore, the spalling of RPC surface has been increased [35,36]. After 300 NaCl F-T cycles and 30 NaCl D-W alternations, the MLR are increased by 1.32%~3.26% and 1.67%~3.47%. This can be explained by the fact that the NaCl F-T cycles can increase the crystallization stress and frost heave stress [37,38]. Consequently, the RPC's mass is decreased by the NaCl F-T cycles. Additionally, the NaCl D-W alternations have applied cyclic crystallization stress to RPC, leading to decreasing the RPC's mass. The addition of steel scoria at the mass ratio of 5%~10% demonstrates decreasing effect on the MLR of the RPC with the decreasing rates of 21.4%~34.6%, due to the improved secondary hydration degree by the steel scoria [39,40]. However, the MLR is increased by the added steel scoria with the mass ratio of 15%~20% with the increasing rates of 13.1%~23.6%, which is attributed to the fact that the steel slag accelerates the alkali aggregate reaction of RPC, thereby increasing the spalling of RPC's surface [41]. As shown in Table 5, the relationship between MLR and NaCl mass ratio matches a cubic function. The MLR of SS-RPC are 1.3%~13.2% and 2.1%~14.6% lower than those of RPC with traditional materials after 300 NaCl F-T cycles and 30 NaCl D-W alternations [42].

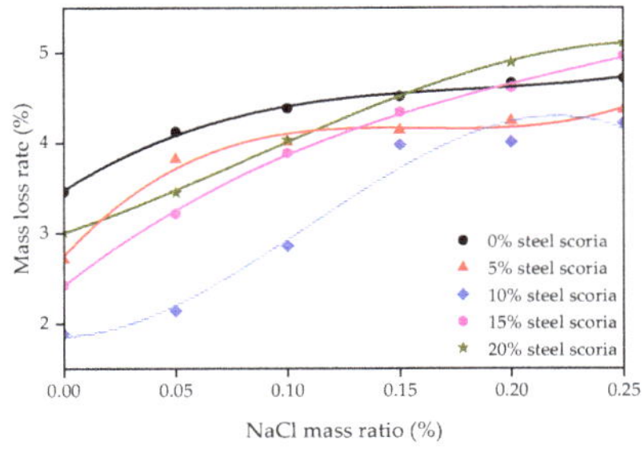

(a)

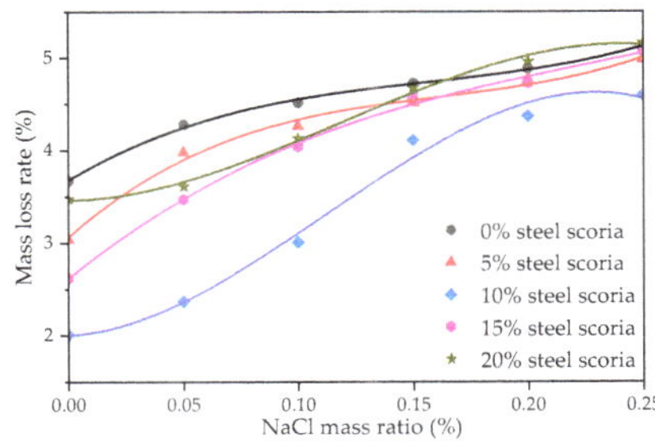

(b)

Figure 5. The mass loss rate of RPC with NaCl and steel scoria after NaCl F-T cycles and D-W alternations. (**a**) After 300 F-T cycles. (**b**) After 30 NaCl D-W alternations.

Table 5. The fitting equations of the mass loss rate of RPC with NaCl and steel scoria after NaCl F-T cycles (MLR_f) and the RPC with NaCl and steel scoria after NaCl D-W alternations (MLR_d).

Equation	Steel Scoria's Mass Ratio	a	b	c	d	R^2	Standard Deviation Values
$MLR_f = aM^3 + bM^2 + cM + d$	0%	92.23	−59.21	13.75	3.42	0.99	0.092
	5%	240.85	−127.59	22.78	2.63	0.96	0.086
	10%	−181.79	45.56	10.6	1.61	0.94	0.071
	15%	35.02	−40	17.66	2.37	0.99	0.065
	20%	−86.08	18.26	7.23	2.97	0.99	0.073
$MLR_d = aM^3 + bM^2 + cM + d$	0%	144.64	−61.73	13.51	3.64	0.99	0.058
	5%	149.15	−81.98	7.41	2.99	0.99	0.096
	10%	−228.81	69.92	−5.4	1.82	0.98	0.088
	15%	13.8	−34.83	17.11	2.53	0.99	0.079
	20%	−165.89	57.73	−2.46	3.39	0.99	0.083

3.2.2. The Relative Dynamic Modulus of Elasticity

The RDME of RPC is shown in Figure 6. Specimens are exposed to the environment of 300 NaCl F-T cycles and 30 NaCl D-W alternations. Table 6 displays the fitting equations of the RDME of RPC with NaCl and steel scoria after NaCl F-T cycles ($RDME_f$) and the RPC with NaCl and steel scoria after NaCl D-W alternations ($RDME_d$). As found in Figure 6 and Table 6, the RDME exhibits the decreased cubic function with the increased mass ratio of NaCl. This can be explained by the increased alkali aggregate reaction and crystallization stress by the inner NaCl, resulting in increasing the cracks in the RPC [43]. Consequently, the propagation speed of ultrasound in RPC decreases, hence, the RDME is decreased. Moreover, as found in Figure 6, the RDME of RPC with 10% steel scoria is the highest. At this dosage, the added steel scoria can increase the RDME by 4.23%~6.62%. This is because the hydration degree of RPC with 10% steel scoria is improved, leading to the RPC's maximum compactness [44]. Hence, the propagation speed of ultrasound in RPC is improved by the added steel scoria. However, when the steel scoria is higher than 10%, the RDME decreases by rates of 2.3%~5.62%. This can be revealed that the increased alkali aggregate reaction by the steel scoria demonstrates increased effect on the increased cracks inner RPC thus decreasing the propagation speed of ultrasound and decreasing the RDME [45]. The RDME decreases by rates of 8.72%~22.3% and 9.81%~26.1% after 300 NaCl F-T cycles and 30 NaCl D-W alternations respectively. This can be explained by the increased frost heave stress and crystallization stress, thus increasing cracks inner the RPC [46]. Therefore, the propagation speed of ultrasound is blocked and the RDME is decreased. The RDME of RPC with SS after 300 NaCl F-T cycles and 30 NaCl D-W alternations are 0%~7.8% and 0%~5.6% higher than those of RPC with traditional materials [47].

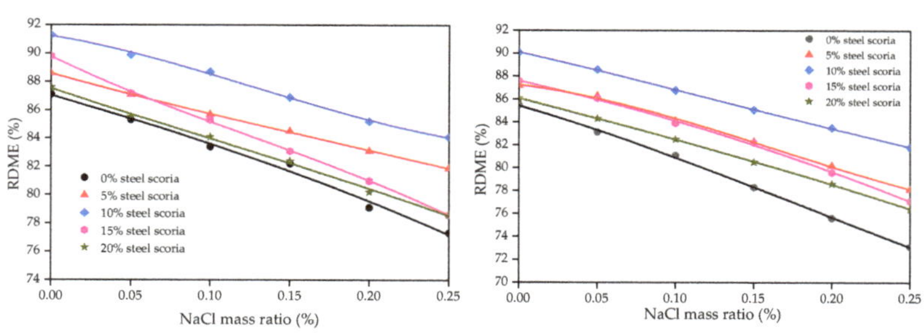

Figure 6. The RDME of RPC with NaCl and steel scoria after NaCl F-T cycles and D-W alternations. (**a**) After 300 freeze-thaw cycles. (**b**) After 30 NaCl dry-wet alternations.

Table 6. The fitting equations of the RDME of RPC with NaCl and steel scoria after NaCl F-T cycles ($RDME_f$) and the RPC with NaCl and steel scoria after NaCl D-W alternations ($RDME_d$).

Equation	Steel Scoria's Mass Ratio	a	b	c	d	R^2
$RDME_f = aM^3 + bM^2 + cM + d$	0%	457.9	184.84	−12.96	86.54	0.99
	5%	18.14	3.51	−28.46	88.66	0.99
	10%	115.37	−45.75	−11.15	91.06	0.99
	15%	−155.65	59.08	−49.06	89.65	0.99
	20%	198.28	−80.96	−26.6	87.27	0.99
$RDME_d = aM^3 + bM^2 + cM + d$	0%	−90.78	11.12	−31.43	85.23	0.99
	5%	86.72	−71.15	−7.16	87.03	0.99
	10%	−6.1	1.32	−27.46	90.04	0.99
	15%	208.11	43.29	−22.3	87.41	0.99
	20%	14.95	22.14	−33.59	86.04	0.99

3.2.3. The Chloride Ion Migration Coefficient

Figure 7 shows CMC of RPC after 300 NaCl F-T cycles and 30 NaCl D-W alternations. Table 7 shows the fitting equations of the CMC of RPC with NaCl and steel scoria after NaCl F-T cycles (CMC_f) and the RPC with NaCl and steel scoria after NaCl D-W alternations (CMC_d). As depicted that the CMC increases in the form of cubic function with the mass ratio of NaCl. This is because the added NaCl can increase the RPC's inner cracks [48]. Therefore, the migration speed of chloride ions in RPC is accelerated by the added NaCl. Moreover, the RPC with steel scoria's mass ratio of 10% shows the lowest CMC after 300 NaCl F-T cycles and 30 NaCl D-W alternations. The CMC can be decreased by rates of 5.1%~21.6% by the added steel scoria. This can be explained by the fact that the RPC with 10% steel scoria exhibits the highest compactness, resulting in RPC's lowest CMC [49]. However, when the mass ratio of steel scoria is higher than 10%, the CMC is increased by the added steel scoria, due to the increased alkali content from the steel scoria [50]. Hence, the cracks are increased by the added steel scoria and the CMC after 300 NaCl F-T cycles and 30 NaCl D-W alternations is increased. The CMC increases at rates of 4.31%~7.62% and 4.72%~8.35% after the 300 NaCl F-T cycles and 30 NaCl D-W alternations respectively. This is attributed to the fact that the NaCl F-T cycles and NaCl D-W alternations can accelerate alkali aggregate reaction and increase frost heave stress and crystallization stress thus increasing RPC's inner cracks [51]. Ultimately, the RPC's CMC is increased by the NaCl F-T cycles and NaCl D-W alternations. The CMC of RPC with SS after 300 NaCl F-T cycles and 30 NaCl D-W alternations are 0%~3.1% and 0%~4.1% lower than those of secondary aluminum ash RPC [52].

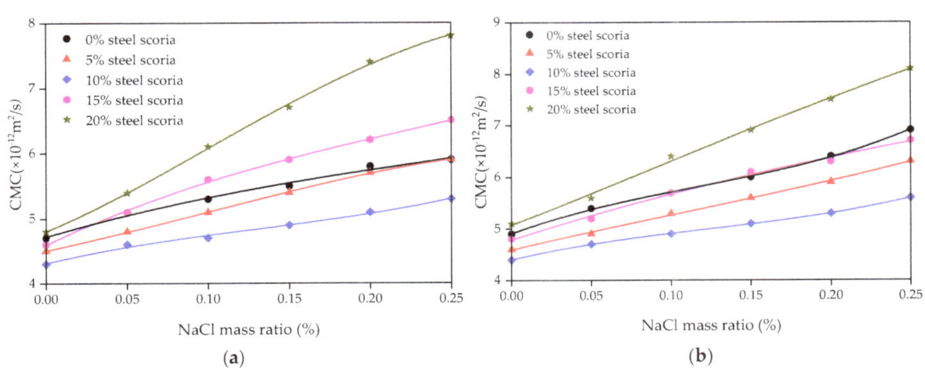

Figure 7. The CMC of RPC with NaCl and steel scoria after NaCl F-T cycles and D-W alternations. (**a**) After 300 F-T cycles. (**b**) After 30 NaCl D-W alternations.

Table 7. The fitting equations of the CMC of RPC with NaCl and steel scoria after NaCl F-T cycles (CMC_f) and the RPC with NaCl and steel scoria after NaCl D-W alternations (CMC_d).

Equation	Steel Scoria's Mass Ratio	*a*	*b*	*c*	*d*	R^2
$CMC_f = aM^3 + bM^2 + cM + d$	0%	66.15	−37.03	6.9	3.19	0.99
	5%	−15.36	7.4	8.93	2.91	0.99
	10%	−10	8.85	2.15	2.42	0.99
	15%	23.04	−10.47	4.15	2. 26	0.99
	20%	14.89	−12.27	5.71	3.16	0.99
$CMC_d = aM^3 + bM^2 + cM + d$	0%	159.89	11.12	12.47	4.93	0.99
	5%	64.22	−2.27	8.93	4.63	0.99
	10%	82.93	−31.15	7.47	4.41	0.99
	15%	96.44	−45.21	13.32	4.86	0.99
	20%	71.68	41.94	15.49	5.17	0.99

3.2.4. The Mechanical Strengths After NaCl Erosion

The f_t and f_{cu} of RPC after 300 NaCl F-T cycles and 30 NaCl D-W alternations are shown in Figure 8. Table 8 shows the fitting equations of NaCl mass ratio and the f_t and f_{cu}. As depicted in Figure 8 and Table 8, the RPC's f_t and f_{cu} decrease in the form of cubic function with the NaCl's mass ratio. After 300 NaCl F-T cycles and 30 NaCl D-W alternations, the specimens' f_t decreases by rates of 4.94%~41.30% and 5.13%~42.28%. Moreover, the f_{cu} are decreased by rates of 5.78%~43.26% and 6.92%~46.71%. This can be explained by the fact that the NaCl F-T cycles, the NaCl D-W alternations and the inner NaCl can accelerate the alkali aggregate reaction, frost heave stress, and crystallization stress of RPC, thus increasing the increase and expansion of cracks [53]. Consequently, the RPC's f_t and f_{cu} are decreased. When the mass ratio of steel scoria increases from 0% to 10%, the mechanical strengths' attenuation caused by NaCl F-T cycles and D-W alternations is reduced by rates of 5.11%~47.14% and 5.23%~48.65%. This can be ascribed to the improved cement hydration degree by the added steel scoria [54]. However, the mechanical strengths' attenuation caused by NaCl F-T cycles and D-W alternations is increased by the steel scoria at the mass ratio increasing from 10% to 20%. This is attributed to the enhancement of alkali aggregate reaction by the increased dosages of steel scoria [55]. The RPC with SS shows 0%~11.2% and 0%~13.1% higher f_t and f_{cu} than those of RPC with traditional materials after 300 NaCl F-T cycles and 30 NaCl D-W alternations [56].

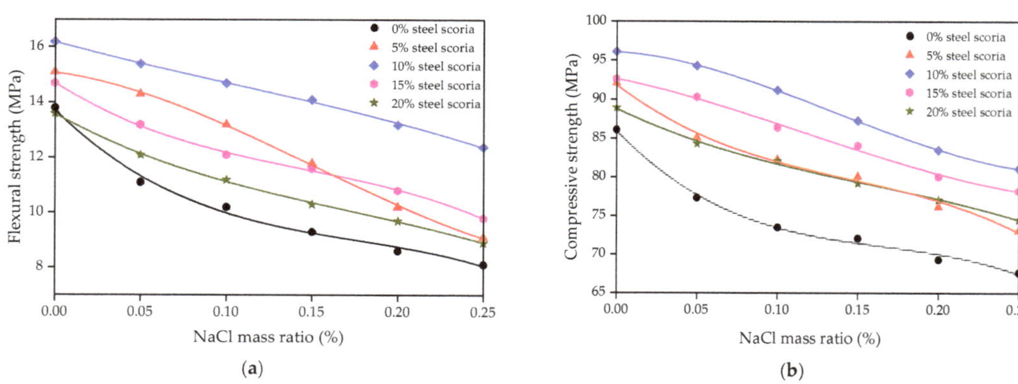

Figure 8. *Cont.*

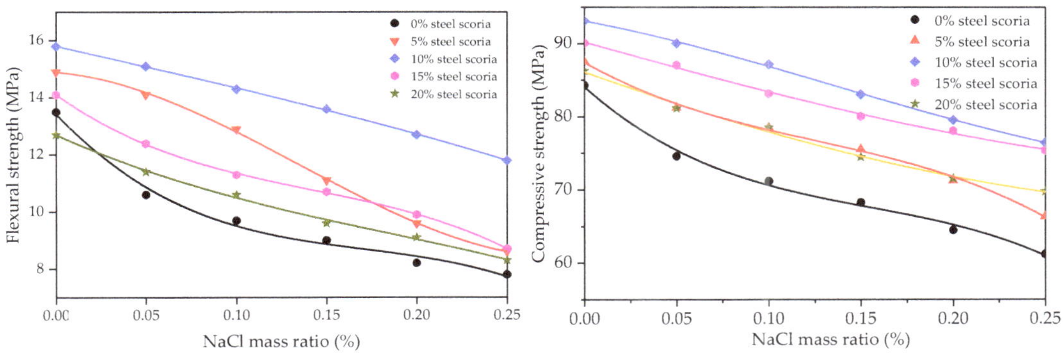

(c) (d)

Figure 8. The mechanical strengths of RPC with NaCl and steel scoria after 300 NaCl F-T cycles and 30 NaCl D-W alternations. (**a**) The flexural strength after 300 NaCl F-T cycles. (**b**) The compressive strength after 300 NaCl F-T cycles. (**c**) The flexural strength after 30 NaCl D-W alternations. (**d**) The compressive strength after 30 NaCl D-W alternations.

Table 8. The fitting equations of NaCl mass ratio and the f_t and f_{cu}.

Equation	Steel Scoria's Mass Ratio	a	b	c	d	R^2	
$f_t = aM^3 + bM^2 + cM + d$	After 300 F-T cycles	0%	−290.03	191.56	−51.04	13.47	0.98
		5%	222.96	−108.06	−4.15	14.98	0.99
		10%	−13.76	−0.98	−13.71	16.12	0.99
		15%	−328.32	149.01	−35.87	14.62	0.99
		20%	−153.21	90.59	−31.39	13.51	0.99
$f_{cu} = aM^3 + bM^2 + cM + d$		0%	−1534.59	853.28	−187.41	85.2	0.99
		5%	−945.59	457.72	−126.8	91.08	0.98
		10%	1093.76	−450.98	−8.07	95.97	0.99
		15%	29.75	3.76	−17.72	92.03	0.99
		20%	−335.21	197.66	−83.55	88.41	0.99
$f_t = aM^3 + bM^2 + cM + d$	After 30 NaCl D-W alternations	0%	−351.11	222.47	−54.65	14.06	0.99
		5%	437.57	−179.25	−0.88	14.78	0.99
		10%	14.64	15.74	−12.76	15.76	0.99
		15%	−475.45	204.43	42.74	14.06	0.99
		20%	−11.68	30.07	−23.57	12.56	0.99
$f_{cu} = aM^3 + bM^2 + cM + d$		0%	−1500.46	774.31	−186.76	83.09	0.99
		5%	−1164.58	437.48	−117.9	86.83	0.99
		10%	194.1	−88.32	−36.81	92.86	0.99
		15%	−325.1	−165.14	−52.78	89.76	0.99
		20%	−585.68	−175.29	−64.92	85.47	0.99

3.2.5. The Carbonation Depth

The D_c of RPC after 300 NaCl F-T cycles and 30 NaCl D-W alternations is shown in Figure 9. Table 9 is the fitting equations of the carbonation depth of RPC with NaCl and steel scoria after NaCl F-T cycles (D_f) and the RPC with NaCl and steel scoria after NaCl D-W alternations (D_d). As demonstrated in Figure 9, the RPC's D_c increases with the mass ratio of NaCl. This is because the inner cracks of RPC can be increased by the added NaCl [57]. The relationship between D_c and NaCl's mass ratio confirms with cubic function. Therefore, more CO_2 can react with the Ca $(OH)_2$ in the RPC, leading to increasing RPC's D_c. The added steel scoria with the mass ratio of 5%~10% exhibits decreasing effect on the D_c of RPC with the decreasing rates of 21.6%~33.1%, due to the improved compactness of RPC by the steel scoria [58]. However, when the steel scoria's mass ratio is 15%~20%, the D_c is increased by the steel scoria. Due to the increased inner cracks by the increased steel scoria's alkali aggregate reaction [59]. Consequently, more CO_2 react with the Ca $(OH)_2$ in the RPC, forming $CaCO_3$ and increasing the D_c. The D_c is decreased by rates of

11.72%~23.16% and 17.61%~27.63% after 300 NaCl F-T cycles and 30 NaCl D-W alternations. This can be explained by the increased alkali aggregate reaction due to the NaCl erosion [60]. Moreover, the NaCl F-T cycles and NaCl D-W alternations can increase the inner cracks of RPC by the increased frost heave stress and crystallization stress, leading to increasing the contact probability between CO_2 and Ca $(OH)_2$ inside RPC, forming more $CaCO_3$ and increasing the carbonation depth. After 300 NaCl F-T cycles and 30 NaCl D-W alternations, RPC with SS shows 0%~3.7% and 0%~4.1% carbonation depths lower than those of RPC with traditional materials.

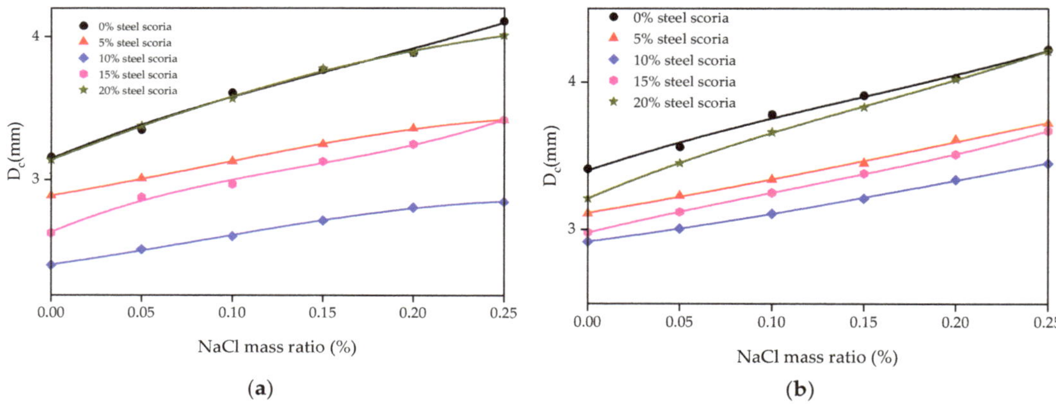

Figure 9. The D_c of RPC with NaCl and steel scoria after NaCl F-T cycles and D-W alternations. (**a**) After 300 F-T cycles. (**b**) After 30 NaCl D-W alternations.

Table 9. The fitting equations of the RPC's carbonation depth with NaCl and steel scoria after NaCl F-T cycles (D_f) and the RPC with NaCl and steel scoria after NaCl D-W alternations (D_d).

Equation	Steel Scoria's Mass Ratio	*a*	*b*	*c*	*d*	R^2
$D = aM^3 + bM^2 + cM + d$	0%	22.22	−70.79	−31.44	85.65	0.99
	5%	−23.04	15.65	8.93	4.52	0.99
	10%	97.43	−4.50	7.11	4.36	0.99
	15%	82.01	−43.83	15.42	4.63	0.99
	20%	−52.08	12.10	13.86	4.85	0.99
$D = aM^3 + bM^2 + cM + d$	0%	148.15	−58.6	−38.21	85.65	0.99
	5%	565.13	−70.79	−31.44	85.65	0.99
	10%	154.25	−59.73	−33.6	90.2	0.99
	15%	208.11	−126.46	−39.37	87.85	0.99
	20%	14.95	22.14	−38.14	86.16	0.99

3.3. *The Micro Analysis*

3.3.1. The Scanning Electron Microscope

The scanning electron microscope (SEM) photos of RPC are shown in Figure 10. In this study, specimens with the NaCl at mass ratios of 0.1% and 0.2% and the steel scoria of 10% are selected. Moreover, the specimens with 0.1% NaCl, 5% steel scoria, 10% steel scoria and 20% steel scoria are used for comparison. It can be observed that the flocculent hydrates are discovered in the SEM photos. It can be found that the increased dosages of NaCl can increase the inner cracks and pores of RPC. Moreover, the SEM microstructure of RPC with 10% steel scoria is the most compact. Therefore, the SEM results confirm that the added NaCl can decrease the mechanical strengths of the RPC and the RPC with 10% steel scoria shows the optimal mechanical strengths.

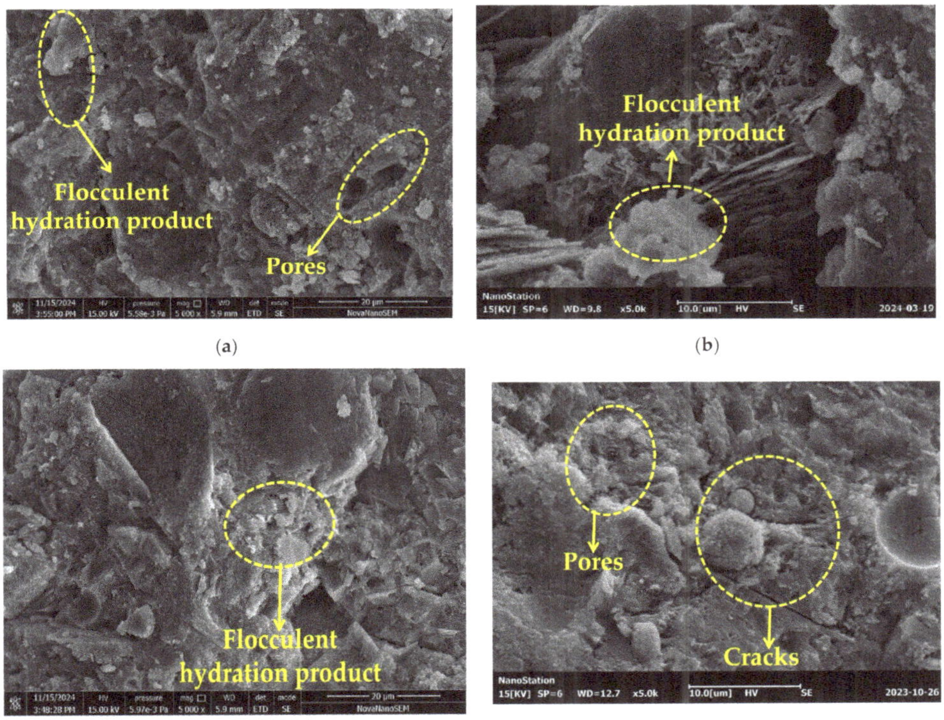

(a) (b) (c) (d)

Figure 10. The SEM photos of the RPC. (**a**) With 5% steel scoria and 0.1% NaCl. (**b**) With 10% steel scoria and 0.1% NaCl. (**c**) With 20% steel scoria and 0.1% NaCl. (**d**) With 10% steel scoria and 0.2%NaCl.

3.3.2. The Scanning Electron Microscopy-Energy Dispersive Spectroscopy

Figure 11 shows the RPC's scanning electron microscopy-energy dispersive spectroscopy (SEM-EDS). As depicted in Figure 11, the elements of C, O, Fe, Si, S, Cl, K and Ca are found. The added NaCl has increased the Cl and Ca elements in the RPC sample. The increased steel scoria demonstrates positive effect on the Fe and Ca elements.

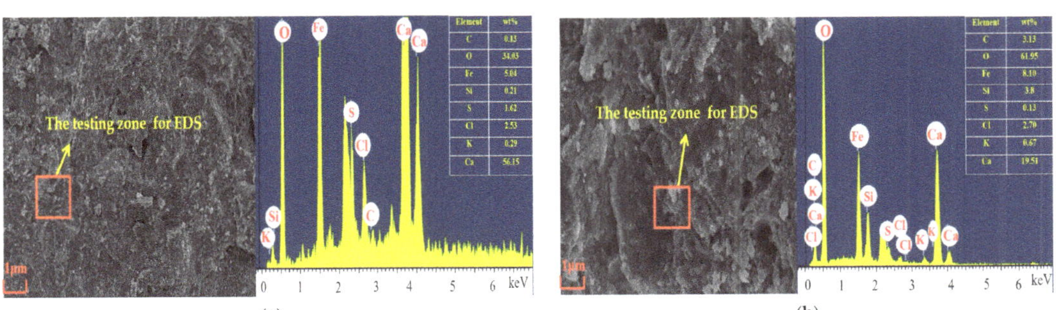

(a) (b)

Figure 11. *Cont.*

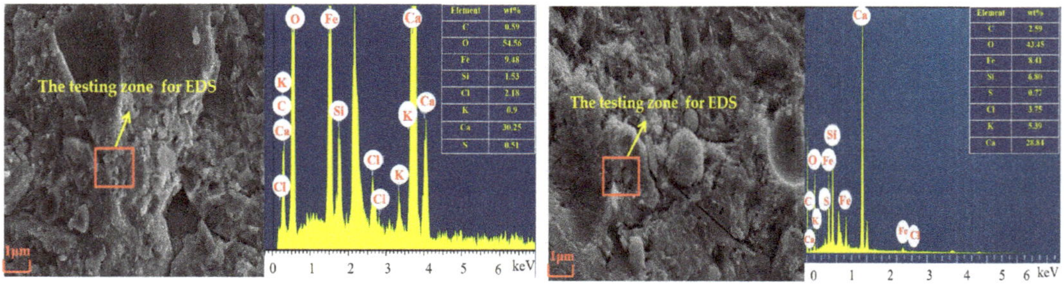

(c) (d)

Figure 11. The SEM photos of the RPC. (**a**) With 5% steel scoria and 0.1% NaCl. (**b**) With 10% steel scoria and 0.1% NaCl. (**c**) With 20% steel scoria and 0.1% NaCl. (**d**) With 10% steel scoria and 0.2%NaCl.

3.3.3. The X-Ray Diffraction Spectrum Curves

Figure 12 shows the X-ray diffraction spectrum curves of the RPC. It can be observed from the X-ray diffraction spectrum curves, the crystals of Ca $(OH)_2$, $CaCO_3$, tricalcium silicate (C_3S), dicalcium silicate (C_2S), calcium silicate hydrate (C-S-H) and FeO exist in the RPC. The increased dosages of steel scoria can decrease the amount of Ca $(OH)_2$, which indicates the fact that the added steel scoria can increase the degree of secondary hydration. The Ca $(OH)_2$ crystals are increased by the added NaCl. Therefore, the NaCl can reduce the cement's degree of secondary hydration.

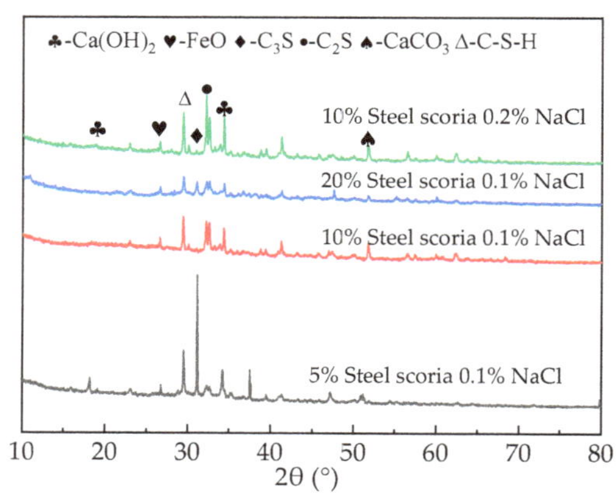

Figure 12. The X-ray diffraction spectrum curves.

4. Conclusions

The influence of NaCl internal erosion and external erosion on the RPC's properties are investigated and the conclusions are obtained as follows.

The f_t and f_{cu} of RPC show the cubic function with the NaCl's mass ratio. The added NaCl with mass ratio of 0%~0.25% can decrease the f_t and f_{cu} by rates of 2.84%~83.86% and 2.61%~97.26%. The RPC specimens with 10% steel scoria exhibit the highest f_t and f_{cu}. At this mass ratio, the addition of steel scoria can increase the f_t and f_{cu} by rates of 11.3%~21.6% and 13.4%~26.3% respectively.

The added NaCl can increase the loss of RPC's mass, RDME, f_t and f_{cu} after 300 NaCl F-T cycles by rates of 1.32%~3.26%, 8.72%~22.3%, 5.78%~43.26% and 4.94%~41.30%. Meanwhile, the mass, the RDME, the f_t and f_{cu} of RPC with 0%~0.25% NaCl are decreased by rates of 1.67%~3.47%, 9.81%~26.1%, 6.92%~46.71% and 5.23%~48.65% after 30 NaCl D-W alternations. The CMC and the D_c are increased by 4.31%~7.62% and 11.72%~23.16% after 300 NaCl F-T cycles. While, after 30 NaCl D-W alternations, the CMC and the D_c increase by rates of 4.72%~8.35% and 17.61%~27.63%.

The added steel scoria with mass ratio of 0%~10% can decrease the loss rates of mass, RDME, f_t and f_{cu} by 21.4%~34.6%, 4.23%~6.62%, 5.11%~47.14% and 5.23%~48.65%. When the added steel scoria is 5%~10%, the CMC and the D_c are decreased by rates of 5.1%~21.6% and 21.6%~33.1%.

The added NaCl can increase the pores and cracks inner RPC. RPC with 10% steel scoria shows the most compact hydration products. The elements of Cl and Ca are increased by the NaCl mass ratio. The increased dosages of steel scoria demonstrates increasing effect on the Fe and Ca elements. The Ca $(OH)_2$ crystals are decreased by the added steel scoria and increased by the added NaCl. The influence of steel scoria and other solid waste on the mechanical properties of reinforced RPC and the corrosion resistance of inner steel bars in marine engineering environments is of great significance, and comprehensive research will be conducted in the future. Electrical parameters method will be introduced for characterizing the reinforced RPC's mechanical properties and corrosion resistance.

Author Contributions: Conceptualization, L.B. and H.L.; methodology, J.Z. and Y.Z.; software, X.Z.; validation, K.X., F.S. and L.B.; formal analysis, Y.Z.; investigation, L.B.; resources, Y.Z.; data curation, L.B.; writing—original draft preparation, L.B.; writing—review and editing, Y.Z.; visualization, L.B.; supervision, H.L.; project administration, Y.Z.; funding acquisition, L.B. All authors have read and agreed to the published version of the manuscript.

Funding: This research was funded by Hebei Province Higher Education Science and Technology Research: ZC2024031.

Institutional Review Board Statement: Not applicable.

Informed Consent Statement: Not applicable.

Data Availability Statement: Data are contained within the article.

Conflicts of Interest: Author Haiyuan Liu and Xin Zhang were employed by the company China Construction First Group Co., Ltd. Author Jian Zhang was employed by the company Huaxia Spatial Planning Institute (Beijing) Co. Ltd. The remaining authors declare that the research was conducted in the absence of any commercial or financial relationships that could be construed as a potential conflict of interest.

References

1. Tyutrin, A.; Nemchinova, N.; Baranov, A.N. Analysis of Steel Slag Composition and Properties to Facilitate the Search for Rational Slag Recycling Methods. *IOP Conf. Ser. Mater. Sci. Eng.* **2020**, *969*, 012042. [CrossRef]
2. Goyal, P.K.; Mudgal, M.; Ghosh, P.K. Assessing the viability of using BOF steel slag treated with tartaric acid as coarse aggregate in concrete. *Constr. Build. Mater.* **2024**, *436*, 136912. [CrossRef]
3. Song, Q.; Zhao, H.Y.; Jia, J.W.; Zhang, F.; Wang, Z.; Lv, W.; Yang, L.; Zhang, W.; Zhang, Y.; Shu, X. Characterization of the products obtained by pyrolysis of oil sludge with steel slag in a continuous pyrolysis-magnetic separation reactor. *Fuel* **2019**, *255*, 115711. [CrossRef]
4. Liu, X.; Bao, Y. Separation and extraction of iron resources from hazardous electric arc furnace (EAF) steel slag: Aggregation of Fe-rich layers, magnetic separation, powder characterization. *Process. Saf. Environ. Prot.* **2024**, *190*, 420–428. [CrossRef]
5. Martins, A.; Carvalho, J.; Duarte, M.; Lima, G.; Pedroti, L.G.; Peixoto, R. Influence of a LAS-based modifying admixture on cement-based composites containing steel slag powder. *J. Build. Eng.* **2022**, *53*, 104517. [CrossRef]
6. Wang, F.; Du, H.; Zheng, Z.; Xu, D.; Wang, Y.; Li, N.; Ni, W.; Ren, C. The Impact of Fly Ash on the Properties of Cementitious Materials Based on Slag-Steel Slag-Gypsum Solid Waste. *Materials* **2024**, *17*, 4696. [CrossRef] [PubMed]
7. Shettima, A.U.; Hussin, M.W.; Ahmad, Y.; Mirza, J. Evaluation of iron ore tailings as replacement for fine aggregate in concrete. *Constr. Build. Mater.* **2016**, *120*, 72–79. [CrossRef]

8. Bessa, S.; Duarte, M.; Lage, G.; Mendonça, I.; Galery, R.; Lago, R.; Texeira, A.P.; Lameiras, F.; Aguilar, M.T. Characterization and Analysis of Iron Ore Tailings Sediments and Their Possible Applications in Earthen Construction. *Buildings* **2024**, *14*, 362. [CrossRef]
9. Avila-López, U.; Almanza-Robles, J.M.; Escalante-García, J. Investigation of novel waste glass and limestone binders using statistical methods. *Constr. Build. Mater.* **2015**, *82*, 296–303. [CrossRef]
10. Ewais, E.M.M.; Khalil, N.M.; Amin, M.S.; Ahmed, Y.; Barakat, M. Utilization of aluminum sludge and aluminum slag (dross) for the manufacture of calcium aluminate cement. *Ceram. Int.* **2009**, *35*, 3381–3388. [CrossRef]
11. Cai, Z.; Ren, J.; Shen, G.; Jin, C.; Gu, X.; Cheng, W.; Wang, H. The Influence of Assembly Unit of Fibers on the Mechanical and Long-Term Properties of Reactive Powder Concrete. *Coatings* **2023**, *13*, 412. [CrossRef]
12. Huang, G.; Yuan, L.; Ji, Y.; Liu, B.; Xu, Z. Cooperative action and compatibility between Portland cement and MSWI bottom ash alkali-activated double gel system materials. *Constr. Build. Mater.* **2019**, *209*, 445–453. [CrossRef]
13. Cui, L.; Wang, H. Research on the Mechanical Strengths and the Following Corrosion Resistance of Inner Steel Bars of RPC with Rice Husk Ash and Waste Fly Ash. *Coatings* **2021**, *11*, 1480. [CrossRef]
14. Li, Y.; Liu, F.; Yu, F.; Du, T. A review of the application of steel slag in concrete. *Structures* **2024**, *63*, 106352. [CrossRef]
15. Baalamurugan, J.; Kumar, V.; Padmapriya, R.; Raja, V. Recent applications of steel slag in construction industry. *Environ. Dev. Sustain.* **2024**, *26*, 2865–2896. [CrossRef]
16. Liu, X.; Zhang, C.; Yu, H.; Qian, G.; Zheng, X.; Zhou, H.; Huang, L.; Zhang, F.; Zhong, Y. Research on the Properties of Steel Slag with Different Preparation Processes. *Materials* **2024**, *17*, 1555. [CrossRef] [PubMed]
17. Weng, Y.; Liu, Y.; Liu, J. Study on mathematical model of hydration expansion of steel slag-cement composite cementitious material. *Environ. Technol.* **2021**, *42*, 2776–2783. [CrossRef]
18. Nguyen, H.T.; Nguyen, H.T.; Ahmed, S.F.; Rajamohan, N.; Yusuf, M.; Sharma, A.; Arunkumar, P.; Deepanraj, B.; Tran, H.; Al-Gheethi, A.; et al. Emerging waste-to-wealth applications of fly ash for environmental remediation: A review. *Environ. Res.* **2023**, *227*, 115800. [CrossRef] [PubMed]
19. Zhang, Y.B.; Lin, K.; Su, Z.J.; Xu, J.; Jiang, T. Self-driven hydrolysis mechanism of secondary aluminum dross (SAD) in the hydrometallurgical process without any additives. *Chem. Eng. J.* **2023**, *466*, 143141. [CrossRef]
20. Ayobami, B.; Jacques, S.; Williams, K. Data on the engineering properties of aluminum dross as a filler in asphalt. *Data Brief.* **2020**, *31*, 105934.
21. Cui, Y.; Jiang, J.; Fu, T.; Liu, S. Feasibility of using Waste Brine/Seawater and Sea Sand for the Production of Concrete: An Experimental Investigation from Mechanical Properties and Durability Perspectives. *Sustainability* **2022**, *14*, 13340. [CrossRef]
22. Kryzhanovskyi, V.; Avramidou, A.; Orlowsky, J.; Spyridis, P. Self-Compacting High-Strength Textile-Reinforced Concrete Using Sea Sand and Sea Water. *Materials* **2023**, *16*, 4934. [CrossRef] [PubMed]
23. Meshram, A.; Gautam, D.; Singh, K. Recycling of white aluminium dross: Production of potash alum. *Trans. Indian Inst. Met.* **2020**, *73*, 1239–1248. [CrossRef]
24. David, O.; Opeyemi, J.; Adekunle, M.; Babatunde, F. Influence of secondary aluminum dross (SAD) on compressive strength and water absorption capacity properties of sandcrete block. *Cogent. Eng.* **2019**, *6*, 1608687.
25. Busari, A.; Joseph, F.; Ajayi, S.; Alayande, T.; Nwachukwu, J.; Agbama, D. Index properties of aluminum dross modified pavement geo-material. *J. Phys.* **2019**, *1378*, 022102. [CrossRef]
26. Xue, G.; Wu, J.Y.; Xu, S.; Liu, J.S.; Dong, W. Study on the fracture mechanical properties of steel slag fine aggregate concrete. *Mater. Rep.* **2025**, *39*, 24090190.
27. *GB/T 17671-2021*; Test Method of Cement Mortar Strength (ISO Method). State Administration for Market Regulation and National Standardization Management Committee: Beijing, China, 2021.
28. *GB/T50082-2024*; Standard for Test Methods of Long-Term Performance and Durability of Concrete. Ministry of Housing and Urban Rural Development of the People's Republic of China: Beijing, China, 2024.
29. Ting, M.; Wong, K.; Rahman, M.; Joo, M.S. Mechanical and durability performance of marine sand and seawater concrete incorporating silicomanganese slag as coarse aggregate. *Constr. Build. Mater.* **2020**, *254*, 119195. [CrossRef]
30. Wang, L.; Yin, S.; Wang, F. Interfacial bonding performance of textile reinforced ECC and seawater sea-sand concrete under a dry-wet environment. *Constr. Build. Mater.* **2023**, *368*, 130384. [CrossRef]
31. Zhu, J.; Liu, S.; Song, L.; Qu, Z.; Wang, H. Influence of Carbon Dioxide Curing on the Corrosion Resistance of Reinforced Cement Mortar under the External Erosion of NaCl Freeze-Thaw Cycle. *Appl. Sci.* **2022**, *12*, 5061. [CrossRef]
32. Cao, Z.; Wang, K.; Peng, X.; Wang, H.; Huang, R. Influence of NaCl Solution External Erosion on Corrosion Resistance of RPC Reinforced with Straw Fiber. *Coatings* **2023**, *13*, 1308. [CrossRef]
33. Tian, E.B.; Kang, H.X.; Zhuang, Z.P. Analysis of fluidity and mechanical properties of steel slag micronized pavement concrete. *J. Sanming Coll.* **2021**, *38*, 87–92+101.
34. Sun, H.; Shi, F.; Wang, H. The Properties of Ultra-High-Performance Concrete with Assembly Unit of Secondary Aluminum Dross and Waste Fly Ash. *Coatings* **2023**, *14*, 89. [CrossRef]

35. Zhang, J.; Shu, Y.; Fang, A.; Qin, R.; Wu, Y.; Zhang, J. Effect of dry-wet cycle on dynamic mechanical characteristics of EPS cement soil under different chloride salt environments. *Case Stud. Constr. Mater.* **2023**, *18*, e02151. [CrossRef]
36. Zhang, Z.; Gong, R.; Zhang, H.; He, W. The Sustainability Performance of Reinforced Concrete Structures in Tunnel Lining Induced by Long-Term Coastal Environment. *Sustainability* **2020**, *12*, 3946. [CrossRef]
37. Mi, T.; Li, Y.; Liu, W.; Dong, Z.; Gong, Q.; Min, C.; Xing, F.; Wang, Y.; Chu, S. The effect of carbonation on chloride redistribution and corrosion of steel reinforcement. *Constr. Build. Mater.* **2023**, *363*, 129641. [CrossRef]
38. Kępniak, M.; Woyciechowski, P.; Łukowski, P.; Kuziak, J.; Kobyłka, R. The Durability of Concrete Modified by Waste Limestone Powder in the Chemically Aggressive Environment. *Materials* **2019**, *12*, 1693. [CrossRef] [PubMed]
39. Zhuang, S.Y.; Wang, Q. Inhibition mechanisms of steel slag on the early-age hydration of cement. *Cem. Concr. Res.* **2021**, *140*, 106283. [CrossRef]
40. Devi, V.S.; Gnanavel, B.K. Properties of Concrete Manufactured Using Steel Slag. *Procedia Eng.* **2014**, *97*, 95–104. [CrossRef]
41. Rui, Y.F.; Qian, C.X. CO_2-fixing steel slag on hydration characteristics of cement-based materials. *Constr. Build. Mater.* **2022**, *354*, 129193. [CrossRef]
42. Sun, H.; Sun, W.; Shi, F.; Miao, L.; Wang, H. The Influence of Salt Erosion on the Mechanical Performances of Ultra-High-Performance Concrete with Secondary Aluminum Dross. *Coatings* **2024**, *14*, 189. [CrossRef]
43. Shi, X.; Fay, L.; Peterson, M.; Yang, Z. Freeze-thaw damage and chemical change of a portland cement concrete in the presence of diluted deicers. *Mater. Struct.* **2010**, *43*, 933–946. [CrossRef]
44. Rui, Y.; Qian, C.; Zhang, X.; Ma, Z. Different carbon treatments for steel slag powder and their subsequent effects on properties of cement-based materials. *J. Clean. Prod.* **2022**, *362*, 132407. [CrossRef]
45. Nurul, H.R.; Mohammad, I.; Zaiton, A.; Seyedmojtaba, G.; Bala, M. Performance of steel slag and steel sludge in concrete. *Constr. Build. Mater.* **2016**, *104*, 16–24.
46. Zhu, J.H.; Wang, D.l.; Chai, G.G. Study on the effect of steel slag powder admixture on mechanical properties and durability of concrete. *Compr. Util. Fly Ash* **2022**, *36*, 97–101.
47. Yu, H.; Chen, B.; Zhang, Z.; Wang, H. The Influence of the Assembly Unit of CO_2-Cured Secondary Aluminum Ash and CO_2-Cured Iron Tailings on High Performance Concrete's Properties. *Coatings* **2024**, *14*, 1536. [CrossRef]
48. Hong, X.; Wang, H.; Shi, F. Influence of NaCl freeze thaw cycles and cyclic loading on the mechanical performance and permeability of sulphoaluminate cement reactive powder concrete. *Coatings* **2020**, *10*, 1227. [CrossRef]
49. Yang, C.; Liu, H.W. Research on mechanical properties and durability of steel slag coarse aggregate concrete. *Concrete* **2016**, *3*, 102–105.
50. Xing, L.L.; Shang, J.L. Study on the effect of steel slag coarse aggregate on the performance of concrete. *Concrete* **2017**, *9*, 105–107.
51. Zhu, H.; Zhang, Z.; Zhu, Y.; Tian, L. Durability of alkali-activated fly ash concrete: Chloride penetration in pastes and mortars. *Constr. Build. Mater.* **2014**, *65*, 51–59. [CrossRef]
52. Zhang, S.Y.; Wang, K.; Li, H.; Zhang, X.Q.; Jiang, Y.L. Novel SCMs produced by the calcination of secondary aluminium dross with dolomite and their potential usage in cemented paste backfill. *Constr. Build. Mater.* **2023**, *365*, 130119. [CrossRef]
53. Liang, Z.; Peng, X.; Wang, H. The Influence of Aspect Ratio of Steel Fibers on the Conductive and Mechanical Properties of Compound Cement Reactive Powder Concrete. *Coatings* **2023**, *13*, 331. [CrossRef]
54. Lai, M.H.; Zou, J.J.; Yao, B.Y.; Ho, J.C.M.; Zhuang, X.; Wang, Q. Improving mechanical behavior and microstructure of concrete by using BOF steel slag aggregate. *Constr. Build. Mater.* **2021**, *277*, 122269. [CrossRef]
55. Du, Y.; Wang, S.; Hao, W.; Shi, F.; Wang, H.; Xu, F.; Du, T. Investigations of the Mechanical Properties and Durability of Reactive Powder Concrete Containing Waste Fly Ash. *Buildings* **2022**, *12*, 560. [CrossRef]
56. Jiang, H.; Wang, K.; Wang, H. The Corrosion Resistance of Reinforced Reactive Powder Concrete with Secondary Aluminum Ash Exposed to NaCl Action. *Materials* **2023**, *16*, 5615. [CrossRef] [PubMed]
57. Ormellese, M.; Berra, M.; Bolzoni, F.; Pastore, T. Corrosion Inhibitors for Chlorides Induced Corrosion in Reinforced Concrete Structures. *Cem. Concr. Res.* **2006**, *36*, 536–547. [CrossRef]
58. Men, P.; Wang, X.; Liu, D.; Zhang, Z.; Zhang, Q.; Lu, Y. On use of polyvinylpyrrolidone to modify polyethylene fibers for improving tensile properties of high strength ECC. *Constr. Build. Mater.* **2024**, *417*, 135354. [CrossRef]
59. Wen, Y.; Sun, H.; Hu, S.D.; Xu, G.M.; Wu, X.Z.; Song, C.C.; Liu, Z.; Li, Z.J. Microstructure and life prediction model of steel slag concrete under freezing-thawing environment. *Nanotechnol. Rev.* **2021**, *10*, 1776–1788. [CrossRef]
60. Mo, L.W.; Zhang, F.; Deng, M. Mechanical performance and microstructure of the calcium carbonate binders produced by carbonating steel slag paste under CO_2 curing. *Cem. Concr. Res.* **2016**, *88*, 217–226. [CrossRef]

coatings

MDPI

Article

Consolidation Enhancement of Weathered Coal Gangue Utilized for Aggregate Filling of Cement Pavement in Mining Area

Wei Tian [1], Zike Xu [1], Kaipeng Gu [1,*], Siying Wang [2], Mingxing Huang [3] and Wei Guo [2,*]

[1] Key Laboratory for Comprehensive Energy Saving of Cold Regions Architecture, Ministry of Education, Jilin Jianzhu University, Changchun 130118, China; weitianjljz@outlook.com (W.T.); x1916456922@outlook.com (Z.X.)
[2] School of Mines, China University of Mining and Technology, Xuzhou 221113, China; ts24020256p31@cumt.edu.cn
[3] MCC Communication Construction Group Co., Ltd., Beijing 100176, China; mingxhmcc@outlook.com
* Correspondence: gukaipeng1@outlook.com (K.G.); weiguo@cumt.edu.cn (W.G.)

Abstract: The large-scale, open-air storage of coal gangue often leads to oxidation and decomposition due to natural weathering, resulting in decreased strength and instability, which limits its wider application in concrete pavement. To address these issues, this paper proposed a composite consolidation treatment for weathered coal gangue (WCG), assessing its effectiveness and enhancement mechanisms through aggregate performance tests, mixture performance tests, and microscopic visualization analyses. Results indicated that the initial and post-20 dry–wet cycle crushing values of WCG were 23.96% and 47.94%, respectively, failing to meet required standards. However, applying a composite consolidation treatment using a lithium curing agent and cement paste significantly improved WCG's robustness and stability. After 4 days of treatment, the crushing value, impact value, and Vickers hardness of WCG had reached 18.3%, 6.58%, and 113.52 kgf/mm^2, respectively, fully meeting the standards for aggregate filling in mini concrete pavements. Furthermore, tests demonstrated that the lithium curing agent induced the formation of hydrated calcium silicate and calcium aluminate on both the surface and interior of the WCG, enhancing its structural stability. Approximately 5–12 wt.% of the curing agent penetrates and encapsulates the WCG, strongly bonding and reinforcing its internal weak surfaces. These findings offer potential solutions and technical insights for the large-scale management of weathered coal gangue.

Keywords: weathered coal gangue; mining cement pavement; solidification pre-treatment; consolidation mechanism; stabilization assessment

1. Introduction

Coal gangue is an industrial waste generated during the coal production process [1], with China discharging over 700 million tons, including more than 70 million tons annually [2,3]. Throughout its accumulation, coal gangue undergoes physical and chemical reactions such as disintegration and oxidation due to natural factors, resulting in the formation of weathered coal gangue (WCG) [4–6]. The accumulation of WCG poses significant environmental challenges, including land-resource occupation and destruction, water and soil pollution, ecological degradation, and threats to ecosystem health [7,8]. During weathering and rainfall, accumulated gangue can release harmful substances like heavy metals and sulfides, leading to further pollution and adversely affecting regional ecosystems and public health [9,10]. Additionally, improper accumulation of gangue can trigger geological disasters, such as landslides and mudslides, further compromising environmental safety [11].

Efficiently utilizing WCG, converting it into usable resources, and mitigating environmental pollution have become urgent priorities [12,13]. Currently, the primary applications of gangue include the production of building materials, power generation, heat energy utilization, soil conditioners, chemical products, and mine filling [14,15]. The mineral components, particularly silicon and aluminum [16,17], in coal gangue make it suitable for use in building materials such as bricks and cement [18–20], as shown in Table 1 [15,21–23].

Table 1. Comprehensive application methods of coal gangue.

Inherent Defects of Coal Gangue	Solutions	Effect Evaluation
Coal gangue has low strength and is prone to cracking.	Zhang et al. conducted a study on surface encapsulation and solidification of coal gangue.	The compressive strength of coal gangue concrete increased by 13.0% to 32.9%.
Coal gangue accumulates in large quantities and has a low utilization value.	Coal gangue is utilized in multiple scenarios, such as the preparation of zeolite molecular sieves.	The zeolite extracted from coal gangue possesses strong ion exchange capabilities.
The accumulation of coal gangue causes environmental pollution, and its utilization rate is low.	Du et al. conducted research on the use of coal gangue in agricultural production.	The use of coal gangue effectively improves crop germination rates.

Despite considerable progress in utilizing coal gangue, many obstacles remain for large-scale applications, including low utilization rates, unscientific practices, and ongoing environmental pollution. Furthermore, many research efforts focus on fresh coal gangue, leaving the potential applications of weathered coal gangue relatively unexplored. The long-term natural weathering process contributes to WCG's fragility, low strength, and instability, limiting its applicability [24]. Currently, over 80% of coal gangue in China is weathered [11], with much of it exposed to natural conditions for more than 2 years. Thus, current research does not adequately address the challenges associated with the accumulation of weathered coal gangue, necessitating the exploration and development of more efficient and large-scale utilization pathways.

The main chemical components of WCG are SiO_2 and Al_2O_3, with the primary mineral composition comprising quartz, calcium feldspar, and kaolinite. This chemical and mineral composition is similar to that of natural crushed stone aggregates, making WCG a potential substitute for natural aggregates in concrete. Zhou et al. [25] investigated the mechanical properties of concrete with varying levels of gangue replacement (0%, 25%, 50%, 75%, and 100%) and found that the inclusion of gangue negatively impacted performance, leading to decreased mechanical properties. Similarly, Karimipour et al. [26] examined the effect of gangue content on cement concrete and discovered that replacing 25% of ordinary crushed stone with gangue reduced compressive strength and elastic modulus by 29% and 20%, respectively. The low strength of gangue in cement concrete results in significant adverse effects, including reduced overall strength, delayed early-strength development, weakened durability, compromised volume stability, and diminished elastic modulus. These drawbacks render concrete more susceptible to cracking and deformation under loads, temperature fluctuations, and humidity changes, ultimately impacting its structural integrity and service life.

To address the inherent deficiencies of weathered gangue and improve its disposal, this study selected weathered gangue from the Jilin Coal Reserve Base that had been stored for 2 years, specifically sampling from the surface to ensure the highest degree of weathering. The feasibility of utilizing weathered gangue as an aggregate substitute in mine cement pavement was explored through curing-modification technology. A systematic investigation of the physical and chemical properties of weathered gangue was conducted to identify the limitations it faces as an aggregate replacement. Subsequently, the effectiveness of curing pre-treatment on weathered gangue was evaluated through aggregate performance and stabilization tests, along with compressive strength tests to assess the effects of filling and curing treatment on the mixture's compressive properties. Finally, the curing mechanism of weathered gangue was analyzed using scanning electron microscopy (SEM), X-ray diffraction (XRD), and computed tomography (CT) techniques. The technical roadmap of this study is illustrated in Figure 1.

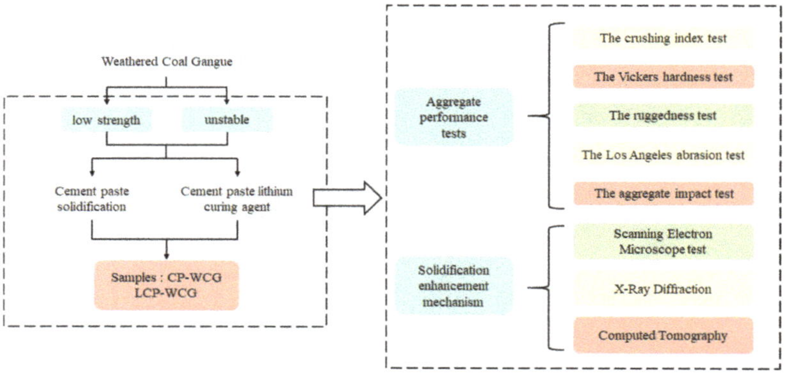

Figure 1. Technical flow chart of this paper.

2. Materials and Sample Preparation

2.1. Weathered Coal Gangue (WCG)

The weathered coal gangue examined in this study was sourced from the Jilin Coal Energy Reserve Base. The weathered coal gangue samples were originally fresh coal gangue that had been stored in the open air for 2 years. During this period, they were subjected to extreme temperatures, reaching up to 60 °C in summer and dropping to −30 °C in winter. In addition to temperature fluctuations, exposure to ultraviolet radiation and wind also played significant roles in the weathering process. As a result, the coal gangue experienced breakage, decomposition, and oxidation over the 2-year storage period. The WCG samples were taken from the surface of stockpile due to high weathering. The WCG samples are initially grayish-white in color and darkened to black upon washing mud powder, as illustrated in Figure 2.

Figure 2. The appearance of WCG.

The WCG samples exhibited certain continuous gradation characteristics after screening, as depicted in Figure 3. Notably, particles within the size range from 9.5 mm to 19.0 mm comprised the largest proportion in WCG, accounting for 24.65%.

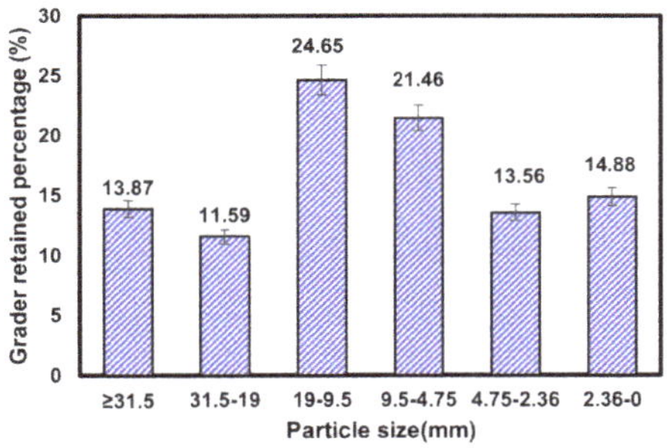

Figure 3. The gradation of WCG.

The principal constituents of WCG samples included SiO_2 Al_2O_3, and Fe_2O_3, constituting 57.24%, 26.31%, and 4.55%, respectively, as illustrated in Figure 4. Among the oxides present, SiO_2 is the primary component that contributes to the hardness of coal gangue. In contrast, the significant amounts of Al_2O_3, CaO, and MgO in weathered coal gangue are the main factors responsible for its disintegration. Notably, the Al_2O_3 content in weathered coal gangue is higher than in fresh coal gangue (ranging from 18.23% to 23.85%). Additionally, the SO_3 content in weathered coal gangue is 1.08%, which does not meet the standard requirement of <0.5%. This highlights the primary changes in oxide composition that coal gangue undergoes after weathering. Furthermore, trace elements such as Cu, Cr, As, and Zn are present in weathered coal gangue, posing potential environmental pollution risks.

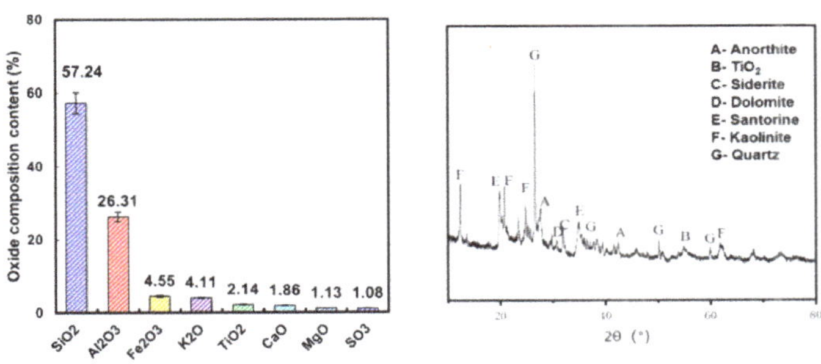

Figure 4. The element composition of WCG.

As can be seen from Figure 4, the mineral composition of WCG samples differed from that of ordinary gravel, whose main mineral components were limestone and calcium feldspar. The mineral composition of WCG was relatively complex, with the primary minerals being quartz, kaolinite, illite, calcium feldspar, TiO_2, and siderite. The high content of Al_2O_3, CaO, and MgO in WCG was identified as the main cause of collapse. Additionally, the SO_3 content in WCG was 1.08%, which was the primary oxide formed after the coal gangue underwent weathering. The Ca content in WCG was only 1.24%, while the Al and K contents were 13.41% and 3.18%, respectively. Unlike fresh coal gangue, which had a high content of anorthite, a significant portion of the anorthite in WCG had been converted into clay minerals such as kaolinite and illite after weathering. This conversion reduced its stability and strength, making the WCG prone to cracking, which was also the primary reason for the formation of weak surfaces within the material.

Additionally, relevant technical parameters of weathered coal gangue are presented in Table 2.

Table 2. The property indexes of WCG.

Test Indexes	Unit	Index Value	Test Procedure
Density	g/cm^3	2.4751	JTG T0304-2024
Moisture-absorption rate	%	4.32	JTG T0306-2024
Crushing index	%	23.96	JTG T0316-2024
Vickers hardness	kgf/mm^2	73.32	GB/T 4340.1
Ruggedness index	%	24.86	JTG T0314-2024
Los Angeles abrasion	%	28.61	JTG T0317-2024
Impact value	%	7.32	JTG T0322-2024
Specific surface area	m^2/g	54.58	JJF 1198-2023
Pore volume	cm^3/g	0.14	JJF 1198-2023

Upon examination of Table 2, it was evident that long-term weathering had caused coal gangue to deteriorate, resulting in traits such as diminished strength and instability. Notably, the crushing index surpassed the requirement of 20% for II Class aggregate in the specifications, rendering it unsuitable for direct utilization as an aggregate replacement in mining cement pavement.

2.2. Consolidation Pre-Treatment of WCG

Due to the characteristics of WCG being weak and unstable, consolidation pre-treatment was imperative to enable its utilization as a concrete aggregate substitute. Among the various curing agents available today, lithium curing agents have garnered significant attention due to their low cost and high efficiency [27]. Commonly used for the reinforcement and curing of cement-based materials, these agents primarily contain lithium silicate or other lithium salts. When applied to the surface of concrete or cement-based materials, they penetrate deeply and react with the chemical components present to form stable, insoluble compounds, thereby enhancing the material's hardness, density, and wear resistance. Lithium curing agents offer superior permeability, allowing them to penetrate more deeply into the concrete compared to sodium- or potassium-based agents [28]. Additionally, they exhibit better resistance to alkali–silica reactions and effectively reduce surface weathering and pulverization, making them an ideal choice for concrete surface treatment. This paper employed cement paste and lithium curing agents to pre-treat weathered coal gangue. The detailed procedure is outlined as follows.

(1) Cement paste pre-treatment

Step 1: A water–cement ratio of 0.75 was maintained, and the corresponding cement paste was prepared.

Step 2: Magnesium aluminum silicate (1.5% of the weight of the cement paste) was incorporated into the cement paste and stirred continuously for 15 min to prevent water precipitation.

Step 3: WCG was introduced into the mixture, stirred for 2 min, followed by a 30 min soaking period, and subsequently air-dried naturally for 30 min.

Step 4: The surface-pre-treated WCG was subjected to curing for varying durations, up to 28 days, to obtain samples designated as CP-WCG.

(2) Lithium curing agent and cement paste composite pre-treatment

Step 1: Following the procedures outlined above, the weathered coal gangue treated with cement paste was obtained.

Step 2: Lithium dilution was prepared by mixing lithium curing agent and water in a mass ratio of 1:2.

Step 3: The lithium dilution was evenly sprayed onto the surface of the weathered coal gangue treated with cement paste, ensuring that the proportion of lithium-based curing agent constituted 2 wt.% of the mixture. After uniform spraying, it was allowed to stand at room temperature for 30 min.

Step 4: The surface-pre-treated weathered coal gangue was subjected to curing for varying durations, up to 28 days, to obtain samples designated as LCP-WCG.

The lithium curing agent utilized in this study was sourced from Beijing Kaifuxin Technology Co., Ltd. (Beijing, China), with relevant technical specifications detailed in Table 3.

Table 3. The technical parameters of lithium curing agent.

Technical Parameters	Unit	Index Value
PH value	N/A	11.4
25 °C viscosity	mPa·s	14
Wear resistance	mm	0.16
Surface hardness	mm	4.5
24 h moisture absorption	Mm	1.6
VOC content	g/L	N/A
Solid content	%	43

The cement employed in this research was supplied by Jilin Yatai Cement Co., Ltd. (Changchun, China), accompanied by pertinent technical specifications presented in Table 4.

Table 4. The technical parameters of cement.

Technical Parameters		Unit	Index Value
Fineness (80 μm sieve residue)		%	2.46
Water consumption for standard consistency		%	26.92
Setting time	Initial setting	Min	183
	Final setting	Min	230
Compressive strength	3 d	MPa	23.9
	28 d	MPa	44.7
Flexural strength	3 d	MPa	5.0
	28 d	MPa	8.1

2.3. Concrete Filled with WCG Preparation

To assess the effect of weathered coal gangue-filled cement pavement, both before and after solidification, on its mechanical properties and the optimal filling particle size range, this study utilized weathered coal gangue to replace aggregates in varying proportions and prepared the corresponding cement mixtures. The concrete gradation used in this study follows the standard gradation for mining roads, specifically C20 concrete pavement ($\geq$20 MPa), as shown in Figure 5.

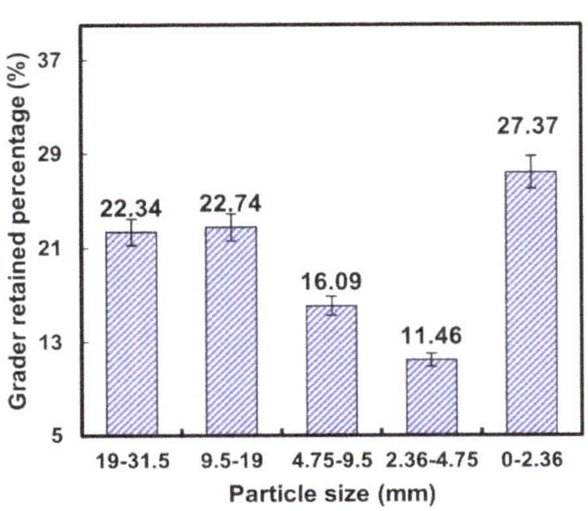

Figure 5. The gradation of cement road in mining area.

The mass ratio of cement, water, fine aggregate, and coarse aggregate was 315:120:632:1283, respectively. To determine the filling proportion of weathered coal gangue, the concrete aggregate was divided into four particle size ranges: 2.36–4.75 mm, 4.75–9.50 mm, 9.50–19.0 mm, and 19.0–31.5 mm. The weathered coal gangue was used to progressively replace the aggregate in each size range until it fully substituted the aggregate (particle size range: 2.36–31.5 mm). Corresponding concrete specimens were prepared and subjected to compressive strength tests after 28 days of curing.

3. Tests and Methods

3.1. Aggregate Performance Tests

To comprehensively analyze the fundamental properties of weathered coal gangue pre- and post-consolidation, this study employed a range of tests, including the crushing index test, ruggedness test, Los Angeles abrasion test, Vickers hardness test, and aggregate impact test. Each test serves a specific purpose.

1. The crushing index is a key characteristic that evaluates an aggregate's ability to withstand pressure without breaking or deforming. A lower crushing index indicates better performance of the aggregate. The results of the crushing index test are essential for determining the scope of application and the quality characteristics of coarse aggregates, as well as their impact on engineering materials such as concrete and roadbeds. By assessing the crushing index of coarse aggregates, the quality and stability of concrete or roadbed projects can be ensured, ultimately enhancing the durability and performance of the construction while minimizing quality issues and safety hazards. This paper employs the crushing value test to analyze the consolidation effect, adhering to specification JTG T0316-2024.

2. Vickers hardness is a standardized method used to evaluate the hardness of materials by measuring their ability to resist scratches. This is done by applying pressure to the material's surface using a specific geometric shape, typically steel balls or diamond cones, under controlled conditions. Testing the Vickers hardness of aggregates helps to determine their strength and wear resistance, providing valuable insights for the selection and optimization of aggregates. In this study, the Vickers hardness test was conducted to assess the strength and wear resistance of weathered coal gangue before and after consolidation, following the specifications outlined in GB/T 4340.1.

3. The ruggedness of aggregates refers to their ability to withstand damage from environmental factors, including climate change. A lower ruggedness value indicates better frost resistance in aggregates. Testing the durability of aggregates is essential for determining their stability across various climates and environments, which in turn influences their application range and safety. Since cement pavements in mining areas are subjected to outdoor conditions, ruggedness is a crucial property that must be assessed when selecting aggregates for use in cement pavements. The ruggedness test evaluated the resilience of weathered coal gangue against environmental stressors, in accordance with the specification JTG T0314-2024.

4. The Los Angeles abrasion rate is a crucial parameter that indicates the ability of aggregates to withstand the combined effects of impact, abrasion, and edge shear. Testing the Los Angeles wear value is significant for evaluating the durability of materials in conditions such as traffic loads and natural weathering by simulating real-world wear scenarios. A higher Los Angeles wear value typically signifies that the material possesses good wear resistance, allowing it to maintain stability and performance over time. This, in turn, enhances the durability and reliability of engineering structures, such as roads and railways. In this study, the Los Angeles abrasion test assessed the resistance of weathered coal gangue to the combined effects of impact, abrasion, and edge shear, in accordance with the specification JTG T0317-2024.

5. The aggregate impact value indicates the maximum impact load that an aggregate can withstand under specific conditions when subjected to an impact force. This measure is commonly used to evaluate the impact resistance of aggregates, reflecting the durability and stability of materials under such loads. A lower impact value signifies better properties of the aggregate. In practical scenarios, aggregates used in mine roads exhibiting better impact

values are better able to withstand pressure from traffic loads, mechanical equipment, and other external forces without breaking or sustaining damage. The aggregate impact test assessed the maximum impact load that weathered coal gangue can endure, following the guidelines outlined in the specification JTG T0322-2024.

By employing these systematic tests, this research aims to provide comprehensive insights into the performance and suitability of solidified weathered coal gangue for aggregate applications.

3.2. Stabilization Tests

Weathered coal gangue, utilized as a substitute for concrete aggregates, undergoes prolonged exposure to high temperature and moisture vapor complex mining environments, necessitating a thorough assessment of its stability in diverse environments. This study proposed a dry–wet cycle to investigate the stability of weathered coal gangue. The dry–wet cycle process involved immersing the sample in water for 12 h at room temperature and subsequently drying it at 105 °C for 12 h, constituting a single cycle. Following a maximum of 20 cycles, the sample underwent mass measurement, the crushing index test, and the Vickers hardness test to evaluate its stability under dry–wet conditions.

3.3. Compressive Strength Test

The high frequency and heavy loads of vehicles on mining roads demand high compressive strength from cement pavements. In this study, a concrete compressive strength test was conducted using the YSE-2000 hydraulic universal testing machine from Jilin Guanteng Automation Technology Co., Ltd. (Changchun, China) First, select a specimen measuring 100 mm × 100 mm × 400 mm that has been cured for 28 days. Clean the surface of the specimen, as well as the upper and lower pressure plates, then position the anti-bending test mold at the centerline between the upper and lower pressure plates. Next, place the specimen, with the loading position clearly marked, onto the anti-bending test mold. Ensure that the pressure-bearing surface corresponds to the side of the specimen that was formed and maintain an installation dimension deviation of no more than 1 mm, keeping the setup flat. Finally, select a loading speed of 0.05 MPa/s and terminate the experiment when the specimen exhibits rapid deformation under compression. The compressive strength was calculated, and the effects of weathered coal gangue filler and consolidation treatments on the compressive properties of the mixture were analyzed.

3.4. X-Ray Diffraction Test and Scanning Electron Microscope Test

XRD analysis offers insights into the crystal structure of weathered coal gangue, revealing its crystal phase composition. SEM technology, on the other hand, enables observation of the surface morphology and microstructure of weathered coal gangue, providing detailed images that help in understanding its morphological characteristics and pore structure. To analyze the alterations in chemical composition and microstructure of weathered coal gangue pre- and post-treatment, this study employed XRD and SEM tests. The SEM test utilized a Phenom Pro X desktop electron microscope manufactured by Phenom (Rotterdam, The Netherlands) in the Netherlands. For the XRD test, an Aries 600W tester produced by PANalytical (Almelo, The Netherlands) in the Netherlands was utilized, as depicted in Figure 6.

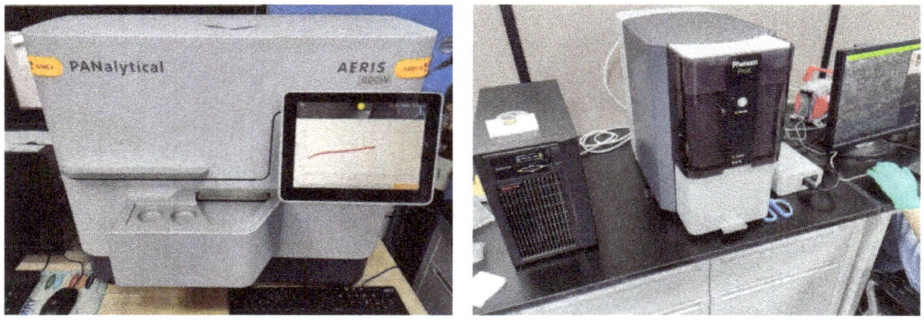

(a) XRD tester (b) SEM tester

Figure 6. XRD and SEM test.

3.5. Computed Tomography Scanning Test

After the surface consolidation of weathered coal gangue, the curing agent fills and consolidates its internal weak layers and pore structure. To effectively characterize the consolidation characterization of weathered coal gangue, this study employed a CT scanning test. The CT scanning tester utilized in this research was the Xradia 510 Versa high-resolution three-dimensional X-ray microscope manufactured by Zeiss (Oberkochen, Germany), as illustrated in Figure 7.

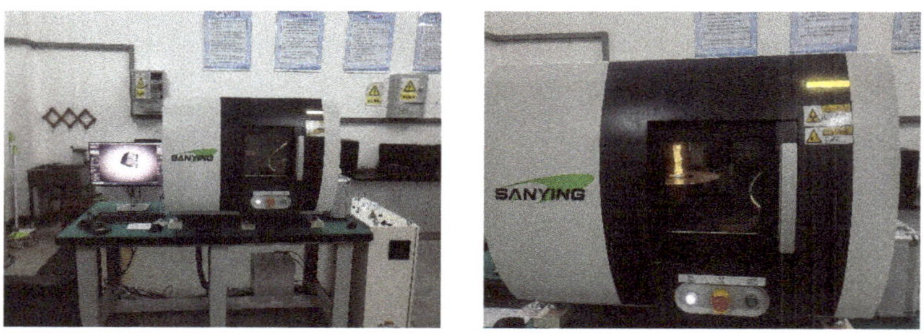

Figure 7. CT scanning test.

4. Results and Discussion

4.1. Aggregate Performance Test Results

4.1.1. Density and Moisture-Absorption-Rate Test Results

The density of the aggregate samples is presented in Figure 8.

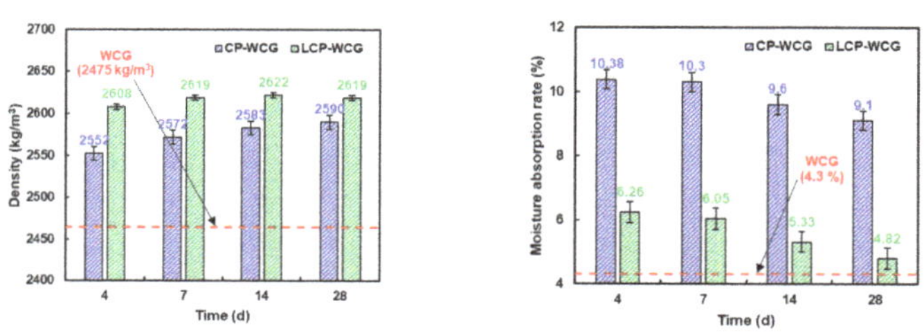

Figure 8. Density and moisture absorption of solidified WCG samples.

As illustrated in Figure 8, the density of WCG increased to varying degrees after the consolidation process, with a gradual rise observed as curing time progressed. After 28 curing days, the densities of LCP-WCG and CP-WCG reached 2619 kg/m^3 and

2590 kg/m^3, respectively, representing increases of 144 kg/m^3 and 117 kg/m^3 compared to WCG. This increase was primarily attributed to the higher density of the cement that adhered to and penetrated the WCG, resulting in a slight overall increase in density. Additionally, the moisture absorption rate of WCG increased following consolidation, with CP-WCG exhibiting a particularly significant rise. After 4 curing days, the moisture absorption rate of CP-WCG reached 10.38%, which was 6.06% higher than that of WCG (4.3%). This was due to the high moisture-absorption capacity of the cement shell formed on the surface of CP-WCG. Furthermore, the rough surface texture created by the cement slurry and WCG contributed to the increased water absorption. After 28 curing days, the moisture-absorption rate of LCP-WCG was 4.82%, while that of CP-WCG remained higher at 9.1%. The lower moisture-absorption rate of LCP-WCG could be explained by the action of the lithium-based curing agent, which catalyzed the cement-hydration reaction, forming a denser calcium silicate hydrate (C-S-H) structure that effectively limited moisture penetration into the WCG. Moreover, the lithium-based curing agent produced a compact network structure on the cement surface, mitigating the increase in moisture absorption caused by the additional cement. Whether it is WCG, CP-WCG, or LCP-WCG, the density range is between 2475 kg/m^3 and 2619 kg/m^3, which is quite comparable to that of common stone. The specification requires that the moisture-absorption rate of aggregates falls within the range of 3%–5%. After 28 days of curing, the moisture-absorption rate of LCP-WCG meets these specification requirements. It is important to note that a higher moisture-absorption rate could lead to potential moisture damage in cement pavements, emphasizing the significance of achieving an optimal moisture-absorption rate for ensuring the durability of the pavement.

4.1.2. Crushing Index Test Results

The crushing index test results for the aggregate samples are shown in Figure 9.

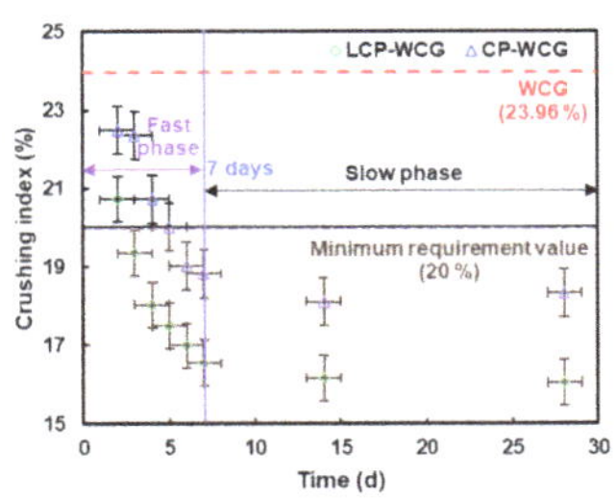

Figure 9. Crushing index of solidified WCG samples.

As indicated in Figure 9, the crushing value of WCG decreased progressively with increased curing time after consolidation pre-treatment. This demonstrated that both solidification methods effectively enhanced the compression resistance of WCG. After 28 days of curing, the crushing values of CP-WCG and LCP-WCG were 18.32% and 16.05%, respectively, representing reductions of 23.54% and 33.01% compared to WCG. In general, the crushing values of both LCP-WCG and CP-WCG showed a marked decrease within the first 7 curing days, after which the reduction slowed. This indicated that the primary solidification occurred during the initial 7-day period. Throughout the curing process, the crushing value of LCP-WCG remained consistently lower than that of CP-WCG, with the difference ranging from 1.78% to 3.02%. The smallest difference in crushing values between the two occurred on the 2nd day of curing, after which the gap increased and stabilized within the 2%–3% range. This suggested that the lithium-based curing agent began to significantly enhance the crushing value starting from the 2nd day of curing. The composite-solidification method, which combined lithium-based curing agents and cement, accelerated the hydration reaction between the cement and WCG, allowing the solidified WCG to be used more quickly for mining road applications. After 3 days of curing, the crushing value of LCP-WCG was 19.34%, while CP-WCG reached a crushing value of 19.02% after 6 days, both meeting the requirements for II-class aggregate usage. Compared to CP-WCG, LCP-WCG achieved the required crushing strength in half the curing time, making it a more efficient option for road construction in mining areas.

4.1.3. Ruggedness Test Results

The ruggedness test results for the aggregate samples are presented in Figure 10.

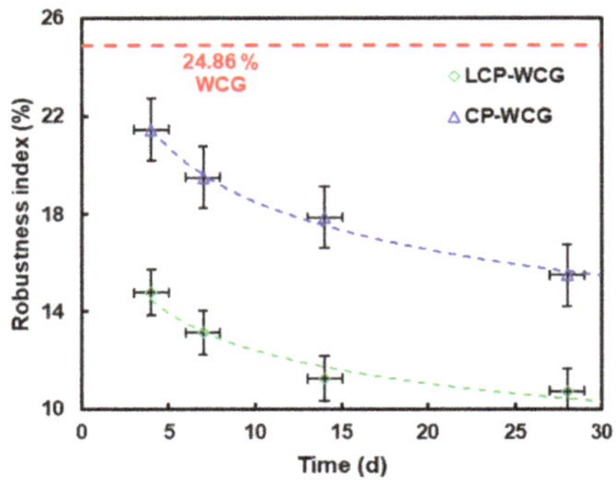

Figure 10. Robustness index of solidified WCG samples.

As shown in Figure 10, after 28 days of curing, the ruggedness index of CP-WCG was 15.48%, which represented a 37.73% reduction compared to WCG, indicating that cement curing significantly enhanced the strength of WCG. When the WCG was cured using a composite of lithium-based curing agent and cement, its robustness index decreased rapidly at first, with the rate of decrease slowing as curing time increased. After 28 days of curing, the robustness index of LCP-WCG reached 10.73%, a 56.84% reduction compared to WCG. By day 14, the robustness index of LCP-WCG was 11.24%, meeting the robustness requirements for Class III stone. Overall, the robustness index of LCP-WCG remained consistently lower than that of CP-WCG by approximately 4.75%–6.67%, indicating that the lithium-based curing agent had a more pronounced effect on the consolidation process. The enhanced performance of LCP-WCG could be attributed to the lithium-based curing agent's ability to penetrate the cement shell and react with free calcium hydroxide ($Ca(OH)_2$) in the cement. This reaction produced additional calcium silicate hydrate (C-S-H), which accelerated the hydration process, reduced the porosity of the cement shell, and improved its impermeability. Consequently, the combination of a lithium-based curing agent and cement provided superior anti-seepage properties compared to pure cement curing at the same stage of consolidation.

4.1.4. Los Angeles Abrasion Test Results

The Los Angeles abrasion test results for the aggregate samples are shown in Figure 11.

As indicated in Figure 11, the Los Angeles abrasions of LCP-WCG and CP-WCG decrease progressively with extended curing time. After 28 days of curing, the Los Angeles abrasions of LCP-WCG and CP-WCG are 19.88% and 23.81%, respectively, representing reductions of 30.51% and 16.78% compared to the 28.61% Los Angeles abrasion of WCG. Notably, after just 4 days of curing, the Los Angeles abrasion of LCP-WCG is 23.08%, which is already lower than the value of CP-WCG after 28 days of curing. This demonstrates that the composite consolidation treatment of a lithium-based curing agent and cement significantly improves the wear resistance, stability, and long-term reliability of the WCG. The lithium-based curing agent enhances the hydration reaction of the solidified cement shell around WCG, resulting in a denser and more cohesive structure. After reinforcement with the lithium-based curing agent, the cement shell tightly adheres to the WCG. Even after the Los Angeles abrasion test, the cement shell remains intact around WCG, showing only surface wear without detachment. In contrast, the cement shell of CP-WCG tends to detach under similar conditions.

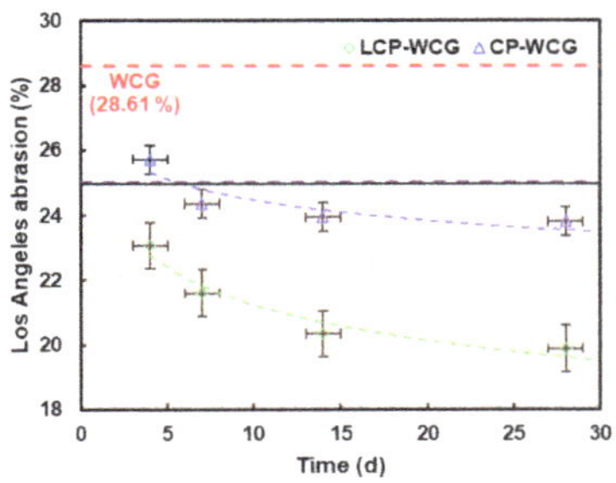

Figure 11. Los Angles abrasion of solidified WCG samples.

4.1.5. Impact Test Results

The impact test results for the aggregate samples are shown in Figure 12.

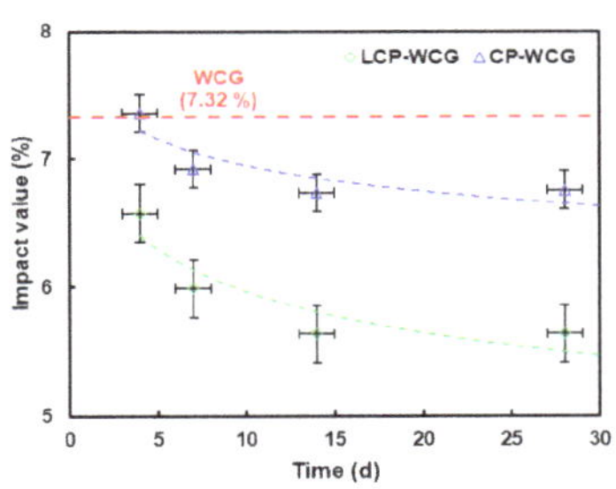

Figure 12. Impact value of solidified WCG samples.

As indicated in Figure 12, the impact value of LCP-WCG and CP-WCG decreased progressively with extended curing time. After 28 days of curing, the impact value of LCP-WCG and CP-WCG were 19.88% and 23.81%, respectively, representing reductions of 30.51% and 16.78% compared to the 28.61% impact value of WCG. Notably, after just 4 days of curing, the impact value of LCP-WCG was 23.08%, which was already lower than the value of CP-WCG after 28 days of curing. This demonstrated that the composite consolidation treatment of a lithium-based curing agent and cement significantly improved the wear resistance, stability, and long-term reliability of the WCG. The lithium-based curing agent enhanced the hydration reaction of the solidified cement shell around WCG, resulting in a denser and more cohesive structure. After reinforcement with the lithium-based curing agent, the cement shell tightly adhered to the WCG. Even after the impact-value test, the cement shell remained intact around WCG, showing only surface wear without detachment. In contrast, the cement shell of CP-WCG tended to detach under similar conditions.

4.1.6. Vickers Hardness Test

The Vickers hardness test results for the aggregate samples are depicted in Figure 13.

It can be seen from Figure 13 that the Vickers hardness of CP-WCG and LCP-WCG reached 110.49 kgf/mm^2 and 141.43 kgf/mm^2, respectively, after 28 days of curing. Compared to the 84.32 kgf/mm^2 of WCG, this represented an increase of 31.04% for CP-WCG and 67.73% for LCP-WCG. These results indicated that cement slurry solidification treatment effectively enhanced the surface strength of WCG. Notably, the Vickers hardness of LCP-WCG after just 4 days of curing was 113.52 kgf/mm^2, surpassing the hardness of CP-WCG even after 28 days of curing. Additionally, after 7 days of curing, the Vickers hardness of LCP-WCG exceeded that of limestone, reaching 127.24 kgf/mm^2. This demonstrated that the incorporation of the lithium-based curing agent significantly improved the surface hardness of WCG. The difference in Vickers hardness between LCP-WCG and CP-WCG increased progressively with curing time, although the rate of increase diminished over time. Between 2 and 4 days of curing, the Vickers hardness difference between LCP-WCG and CP-WCG grew by 14.25 kgf/mm^2, while between 14 and 28 days, the difference narrowed to 4.66 kgf/mm^2. This indicated that the lithium-based curing agent had the most pronounced effect during the first 4 days of curing. As curing time increased, the impact of the lithium-based curing agent gradually diminished due to the advanced stage of the hydration reaction.

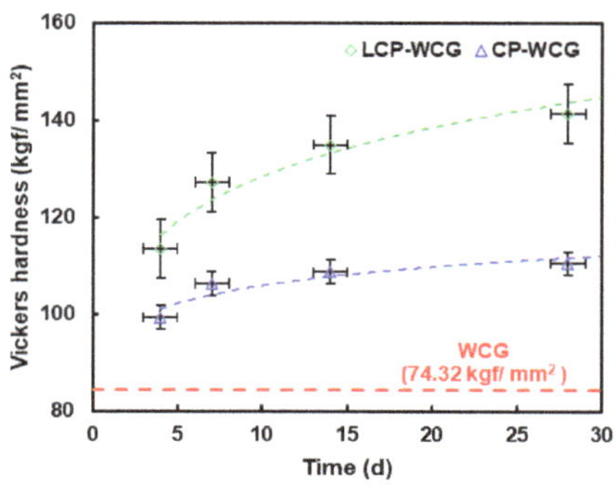

Figure 13. Vickers hardness of solidified WCG samples.

In summary, this paper presents the relevant data statistics for the three samples after 28 curing days, as shown in the following Table 5.

Table 5. The technical parameters of lithium curing agent. The crushing index, ruggedness index, Los Angeles abrasion, impact value and vickers hardness of WCG, CP-WCG and LCP-WCG.

Samples	Crushing Index	Ruggedness Index	Los Angeles Abrasion	Impact Value	Vickers Hardness
WCG	23.96%	24.86%	28.61%	7.32%	84.32 kgf/mm^2
CP-WCG	18.32%	15.48%	23.81%	6.76%	110.49 kgf/mm^2
LCP-WCG	16.05%	10.73%	19.88%	5.64%	141.43 kgf/mm^2

As shown in Table 5, both solidification methods enhance the solidity and strength of weathered coal gangue. Notably, the composite solidification method demonstrates the most significant improvements across various properties. Compared to WCG, the crushing index, ruggedness index, Los Angeles abrasion, and impact value of LCP-WCG decreased by 33.01%, 56.84%, 30.51%, and 22.95%, respectively. This indicates a significant improvement in the strength of weathered coal gangue after solidification, meeting the specification requirements for concrete pavement filling. Furthermore, the Vickers hardness of LCP-WCG is 67.73% higher than that of CP-WCG, demonstrating that solidification treatment can substantially enhance the hardness of weathered coal gangue when used in concrete pavement. This improvement helps suppress deformation and crushing under heavy loads.

4.2. Stabilization Tests Results

The mass loss rate of aggregate samples with different dry–wet cycles is shown in Figure 14.

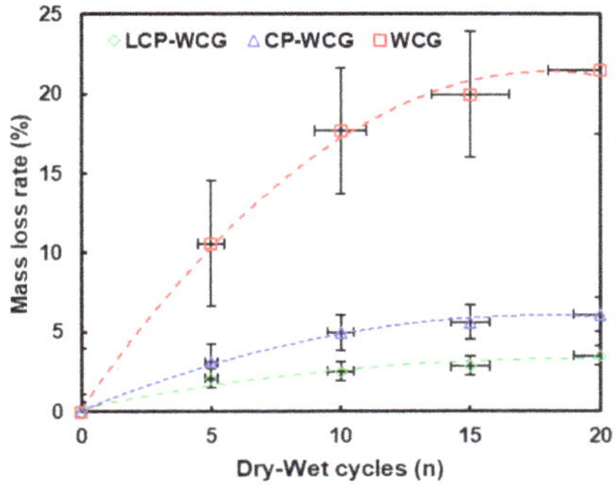

Figure 14. Vickers hardness of solidified WCG samples. The mass loss rate of aggregate samples with different dry–wet cycles.

As shown in Figure 14, the mass loss rate of the three aggregate samples gradually increased with the dry–wet cycles. WCG consistently exhibited the highest mass loss rate. After 20 dry–wet cycles, the mass loss rate of WCG reached 21.47%, significantly higher than that of CP-WCG and LCP-WCG. This high mass loss rate of WCG was primarily due to the presence of clay and other components, which were prone to softening and disintegration when exposed to moisture. In contrast, LCP-WCG consistently showed the lowest mass loss rate. After 20 dry–wet cycles, its mass loss rate was 3.52%, which was 17.95% and 2.59% lower than those of WCG and CP-WCG, respectively. This demonstrated that the lithium-based curing agent and cement paste composite consolidation process significantly enhanced the integrity of WCG, improving its stability under repeated dry–wet cycles. Notably, after 5 dry–wet cycles, the mass loss rate of LCP-WCG reached an inflection point and stabilized, while CP-WCG and WCG required approximately 10 and 15 cycles, respectively, for their mass loss rates to plateau. This indicated that the lithium-based curing agent, in combination with cement paste, effectively enhanced WCG's resistance to dry–wet cycles, improving its overall stability in complex cyclic environmental conditions.

The crushing index of aggregate samples with different dry–wet cycles is shown in Figure 15.

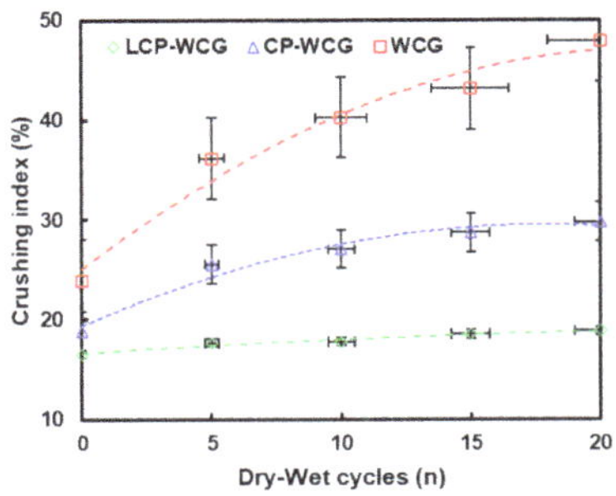

Figure 15. Crushing index of aggregate samples under dry–wet cycles.

It can be seen in Figure 15 that the crushing index of all three aggregate samples increased as the dry–wet cycles rose, indicating that changes in humidity affected the mechanical performance of the aggregates to some extent. WCG was the most sensitive to the dry–wet cycles, with its crushing index reaching 47.94% after 20 cycles—far exceeding the specified limit of 20%. This suggested that WCG, in its untreated form, was unsuitable for direct utilization in mining road fills and required pre-treatment to enhance its strength. LCP-WCG consistently exhibited the lowest crushing index and was least affected by dry–wet cycles. Over 20 cycles, the crushing index of LCP-WCG increased by only 2.29%, compared to 10.96% for CP-WCG and 23.98% for WCG. This indicated that LCP-WCG had greater resistance to the effects of dry–wet cycling and exhibited superior stability. Even after 20 cycles, the crushing index of LCP-WCG remained at 18.83%, still meeting the specified requirements. In contrast, the crushing value of CP-WCG rose to 29.77%, exceeding the acceptable limit. These results demonstrated that the compacting effect of the lithium-based curing agent on the cement shell significantly enhanced the stability of weathered coal gangue, preventing moisture vapor intrusion and reducing the softening effect caused by humidity changes.

The Vickers hardness of aggregate samples with different dry–wet cycles is shown in Figure 16.

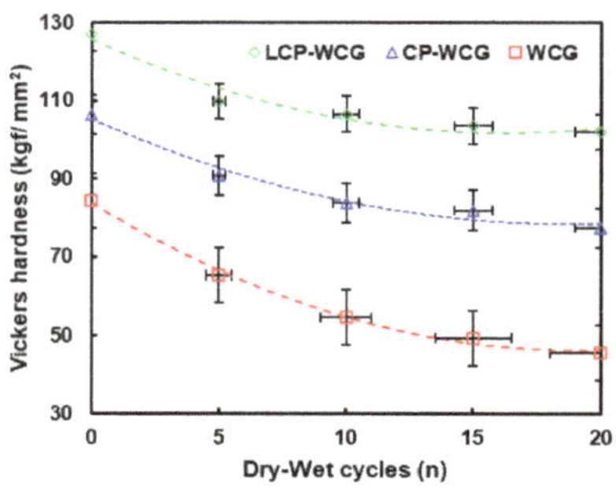

Figure 16. Vickers hardness of aggregate samples under dry–wet cycles.

As shown in Figure 16, the Vickers hardness of all three aggregate samples gradually decreased with the freeze–thaw cycles. The most significant decrease occurred in WCG, where the Vickers hardness dropped to 45.6 kgf/mm^2 after 20 dry–wet cycles, a reduction of 38.72 kgf/mm^2 from its initial value. This indicated that the loose structure of WCG was highly susceptible to moisture vapor intrusion, and under repeated cycles of high and low humidity, the material structure collapsed, leading to a rapid decrease in hardness. In contrast, the Vickers hardness of LCP-WCG after 20 dry–wet cycles was 101.93 kgf/mm^2, which was 24.55 kgf/mm^2 and 56.33 kgf/mm^2 higher than that of CP-WCG and WCG, respectively, indicating superior hardness retention. Additionally, the reduction in Vickers hardness over 20 cycles was smallest in LCP-WCG at 25.31 kgf/mm^2, compared to 28.98 kgf/mm^2 for CP-WCG and 38.71 kgf/mm^2 for WCG. This improved stability was attributed to the cement slurry and lithium-based curing agent forming a protective shell film in the pores and on the surface of the weathered coal gangue, significantly reducing water vapor intrusion and enhancing the material's overall stability.

4.3. Compressive Strength Test Results

The compressive failure characteristics of concrete samples are shown in Figure 17.

As seen in Figure 17, the surface cracks of conventional concrete specimens first occurred at the four corners and then quickly developed from the corners to the middle part of the specimen. The overall failure was rapid and brittle. Because the upper and lower surfaces of the specimen were longitudinally restricted by the press, the middle part of the concrete specimen deformed due to expansion when failure occurred. It first deformed outward, and then a small amount of detachment occurred. Finally, a large amount of peeling happened. The failure mainly occurred at the connection between the aggregate and the mortar; that is, the interface transition zone. After the failure, the concrete specimen tended to be "X" shaped as a whole.

(a) No filling (b) filled with WCG (c) filled with LCP-WCG

Figure 17. Compressive failure characteristics of concrete samples.

For the concrete filled with WCG, the ballast cracks began to appear in the middle or near the side. The cracks were perpendicular to the upper and lower surfaces. Several through cracks appeared at the same time, accompanied by the crisp sound of concrete cracking. When the damaged specimens were disassembled, obvious damage to the weathered coal gangue aggregate itself was found (indicated by the red circle in Figure 17).

The compressive failure of concrete filled with LCP-WCG was close to plastic failure, and its ballasting failure mainly occurred in the interface transition zone (ITZ) area and the interior of WCG. However, the proportion of LCP-WCG failure was much lower than that of WCG filled-in concrete specimens. This showed that the lithium-based curing agent and cement paste composite consolidation significantly improved the weak layers and weak points of weathered coal gangue, enhancing its integrity and strength.

The compressive strength of concrete samples is shown in Figure 18.

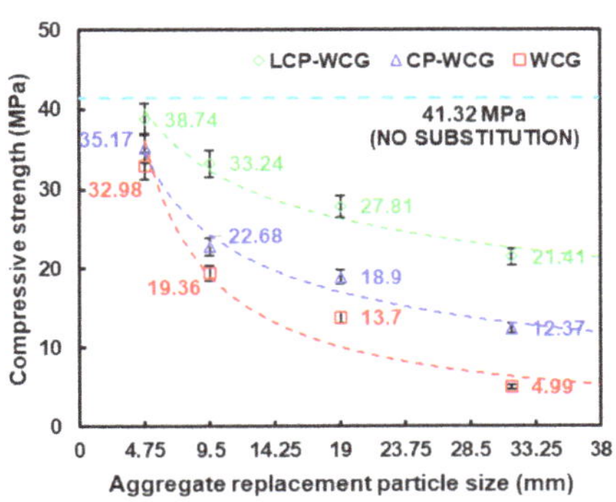

Figure 18. Compressive strength of concrete samples.

As shown in Figure 18, the compressive strength of concrete decreased gradually as the proportion of WCG increased. Notably, the compressive strength of concrete filled with WCG was most influenced by the proportion. The compressive strength of concrete filled with WCG (particle size $\leq$ 31.5 mm) was 4.99 MPa, which was 36.33 MPa lower than that of conventional concrete. This indicated that the strength of weathered coal gangue had a direct impact on the overall concrete strength. Due to its high crushing value and large porosity, weathered coal gangue was the first to fail under pressure when used in concrete, leading to stress concentration and a reduction in compressive strength. In contrast, the compressive strength of concrete filled with LCP-WCG (particle size $\leq$ 31.5 mm) was 21.41 MPa, meeting the requirements for mining roads ($\geq$20 MPa). Overall, the compressive strength of LCP-WCG-filled concrete was consistently higher than that of both CP-WCG-filled and WCG-filled concrete, showing improvements of 6.15–28.95 MPa and 8.34–36.33 MPa, respectively. This highlighted the significant effect of the composite curing mechanism involving a lithium-based curing agent and cement paste, which notably enhanced the integrity and strength of weathered coal gangue, mitigating the strength reduction typically seen in coal gangue-filled concrete.

4.4. XRD and SEM Test Results

The XRD test results of aggregate samples are shown in Figure 19.

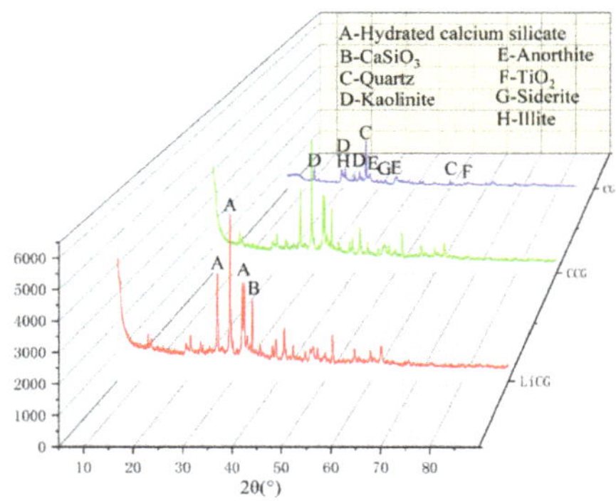

Figure 19. XRD test results of aggregate samples.

As shown in Figure 19, calcium silicate hydrate and calcium sulfonate were detected in both LCP-WCG and CP-WCG but were absent in WCG. These compounds effectively encapsulated the pore and crack structures present in WCG, contributing to enhanced density and strength. Notably, LCP-WCG exhibited a stronger calcium silicate-absorption peak, indicating that cement hydration, catalyzed by the lithium-based curing agent, produced more calcium silicate hydrate. The lithium-based curing agent penetrated the concrete and reacted with calcium hydroxide to form calcium silicate hydrate gel, which filled the voids with insoluble calcium silicate hydrate. The primary component of the lithium-based curing agent, $LiSiO_3$, enabled SiO_3^{2-} to react with calcium hydroxide, precipitating calcium silicate, while the remaining Li^+ replaced K^+ and Na^+ in the alkali-aggregate reaction, forming a non-expanding lithium silicate gel. Unlike traditional materials, this gel did not absorb water and expand, effectively preventing the expansion of cement-based materials. Moreover, experimental results demonstrated that lithium silicate gel could inhibit alkali-aggregate reactions and even repair cement-based materials already affected by such reactions. Due to the low density of lithium compared to other metals, it carried more silicate ions, enabling it to react with calcium hydroxide and generate a greater quantity of gel, further enhancing material strength and durability.

The SEM test results of aggregate samples are shown in Figure 20.

The WCG, as shown in Figure 20, exhibited a sparse, scaly, and layered structure. In low-magnification-scanning images, the surface of WCG appeared relatively rough and loose, with irregular lamellar formations and distinct long cracks. In high-magnification scans, numerous randomly distributed pores were visible on the surface of WCG, with continuous unbroken pores contributing to the formation of long cracks and weak planes. These features made the WCG prone to weakness, low strength, and instability, which were the main reasons for its susceptibility to breaking and structural degradation. After the solidification process, the WCG was coated with hydrated calcium silicate, calcium sulfonate, and some $Ca(OH)_2$, and no significant long cracks were observed. In low-magnification images, it was evident that the lithium-based curing agent catalyzed a more uniform distribution of particles on the surface of LCP-WCG, forming a distinct cement shell that effectively filled the pores. In contrast, the surface density of CP-WCG was noticeably lower than that of LCP-WCG, and it exhibited fewer granular hydrated calcium silicate and calcium sulfonate deposits of various sizes. This indicated that the degree of hydration in CP-WCG was less advanced compared to LCP-WCG. The coupling solidification process of cement and lithium-based curing agents for LCP-WCG resulted in a denser cement shell, which better resisted moisture erosion and enhanced the strength of WCG. The amount of hydrated calcium silicate and calcium sulfonate generated on the surface of LCP-WCG was significantly higher than that of CP-WCG, demonstrating the positive effect of the lithium-based curing agent on the hydration reaction at the same curing age. In high-magnification images of LCP-WCG, the weak crystalline rod-shaped points on the surface of WCG were coated with hydrated calcium silicate. The resulting calcium silicate and calcium aluminate products were interconnected to form a spiderweb-like structure, effectively protecting the WCG from environmental impacts. This filamentous spiderweb structure filled weak pores and enhanced the surface shell's density and strength, significantly improving the overall structural integrity of the WCG.

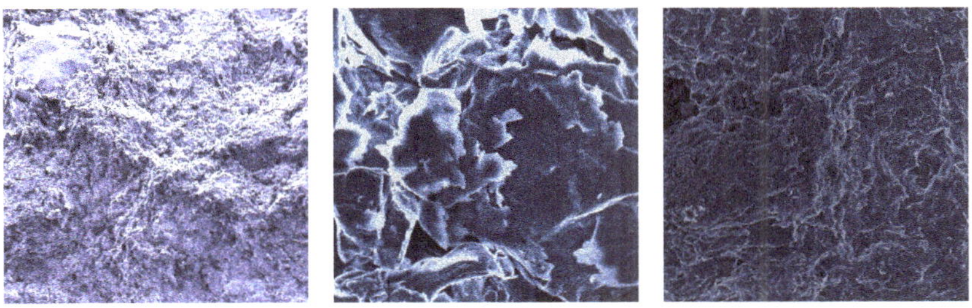

(**a**) WCG

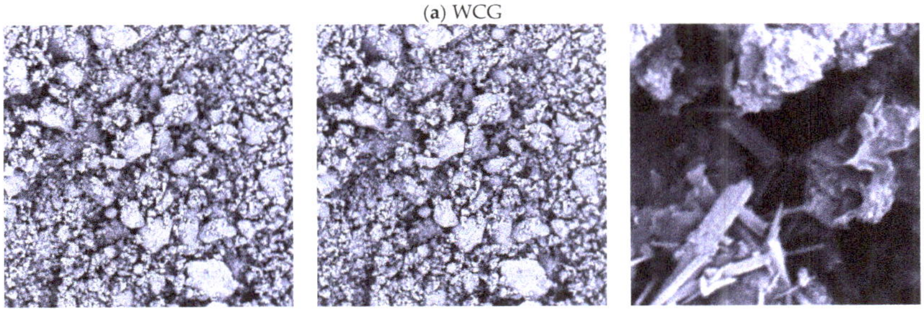

(**b**) CP-WCG

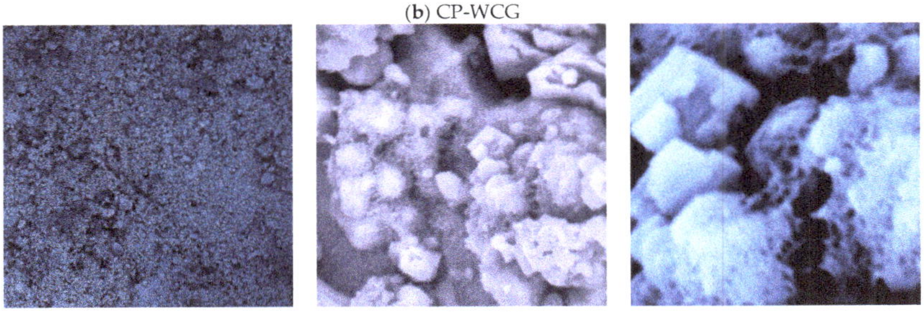

(**c**) LCP-WCG

Figure 20. SEM test results of aggregate samples.

4.5. CT Scanning Test Results

The CT scan test results of aggregate samples are shown in Figure 21.

As shown in Figure 21, the surface of WCG was very rough. Using it directly as aggregate in concrete would have required an excessive amount of cement to fill the uneven stone surfaces, leading to increased cavities within the concrete and thereby reducing its overall density and strength. Internally, WCG contained multiple weak layers, primarily carbonaceous layers and clay mineral layers, which expanded unevenly when exposed to moisture or cracked easily under pressure, ultimately causing structural failure. However, after composite curing with a lithium-based curing agent and cement, the surface roughness of the WCG became more uniform, and the particles gained a well-defined shape. This treatment significantly improved the suitability of WCG as a concrete aggregate. Statistical findings indicated that when a mixture of cement and lithium-based curing agent, comprising 5% to 12% of the total volume of the WCG, was applied, it not only coated the surface but also penetrated the internal structure. This composite curing material effectively filled deep cracks, which accounted for approximately 63% of the width of the weathered coal gangue, as well as other internal weak points. By integrating and consolidating these cracks, the integrity of the WCG was greatly enhanced, transforming it into a more solid and cohesive material. Once the lithium-based curing agent and cement worked together to solidify the WCG, the resulting surface exhibited appropriate roughness and particle shape, free from weak areas. This met the necessary standards for aggregate in cement pavement, allowing WCG to be effectively used as a concrete aggregate.

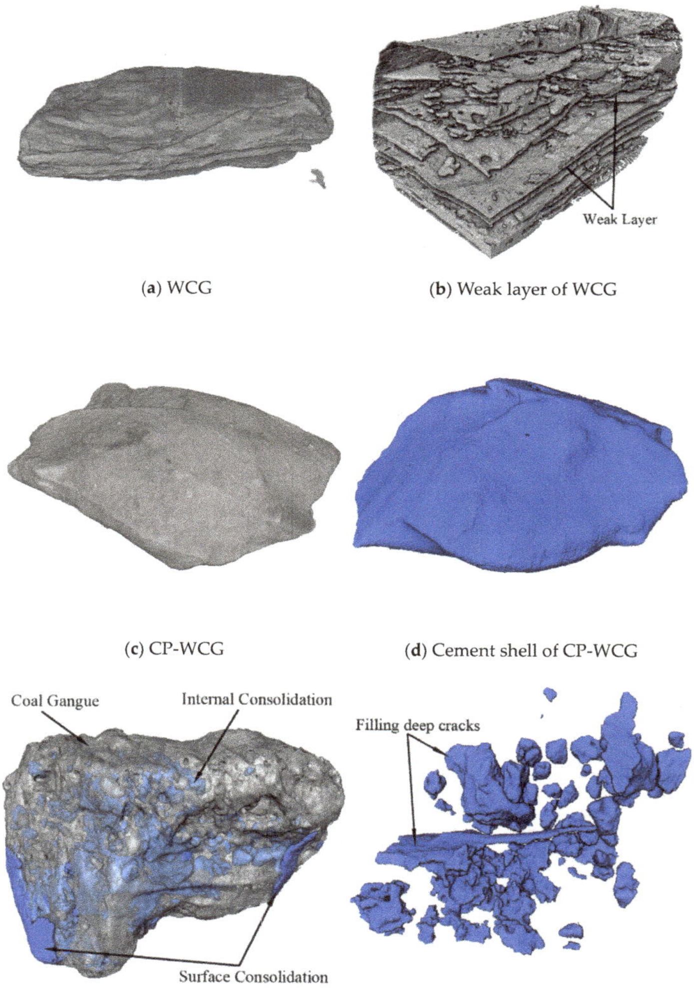

(**a**) WCG (**b**) Weak layer of WCG

(**c**) CP-WCG (**d**) Cement shell of CP-WCG

(**e**) LCP-WCG (**f**) Internal consolidation of LCP-WCG

Figure 21. CT scanning test results of aggregate samples.

5. Conclusions

To address the issues of low strength and poor stability of weathered coal gangue, this paper proposed a composite consolidation treatment using a lithium curing agent and cement paste. This method effectively enhanced the strength of weathered coal gangue and demonstrated its feasibility for use as a filler in concrete pavement. However, significant variations exist in the aggregate performance parameters of coal gangue depending on its origin region and weathering degree, which affects its potential applications in concrete pavement and warrants further systematic research. The main conclusions of this paper are as follows:

(1) The aggregate performance test results illustrated that the lithium-cement curing agent and cement paste composite consolidation method significantly improved the solidity and stability of WCG. The LCP-WCG had a crushing index, ruggedness index, Los Angeles abrasion, impact value, and Vickers hardness of 18.3%, 14.78%, 23.08%, 6.58%, and 113.52 kgf/mm^2, respectively, after 4 curing days, which met the requirements of the aggregate.

(2) Dry–wet cycles test demonstrated that the proposed composite consolidation method could improve the stabilization properties of WCG. After 20 dry–wet cycles, the mass loss rate of LCP-WCG was 3.52%, the crushing index was 18.83%, and the Vickers hardness was 101.93 kgf/mm^2, which were much better than those of CP-WCG and WCG.

(3) The compressive strength test results showed that the optimal particle size ranges of WCG, CP-WCG, and LCP-WCG that can be used to fill concrete pavement while meeting strength requirements are <4.75 mm, <9.5 mm, and <33.25 mm, respectively. This indicates that the composite curing method significantly enhances the potential for WCG utilization in concrete pavement.

(4) Microscopic tests showed that the combined action of the lithium-based curing agent and cement paste generated a large amount of calcium silicate hydrate and calcium aluminate on the surface and inside of WCG, effectively filling its weak structure, inhibiting the occurrence of alkali-aggregate reactions, and improving the integrity of weathered coal gangue.

(5) CT test results revealed that 5%–12% curing agents by mass of WCG penetrated and encapsulated the WCG, strongly bonding and reinforcing the internal weak surface.

Author Contributions: Conceptualization, W.T. and M.H.; Data curation, K.G., S.W. and M.H.; Formal analysis, S.W.; Funding acquisition, W.G.; Investigation, W.T., Z.X. and K.G.; Methodology, Z.X.; Project administration, W.T. and W.G.; Resources, M.H.; Supervision, W.G.; Writing—original draft, K.G.; Writing—review and editing, Z.X. and W.G. All authors have read and agreed to the published version of the manuscript.

Funding: This research was funded by the Key Laboratory for Comprehensive Energy Saving of Cold Regions Architecture of the Ministry of Education, Jilin Jianzhu University (Grant No. JLJZHDKF202201).

Institutional Review Board Statement: Not applicable.

Informed Consent Statement: Not applicable.

Data Availability Statement: The original contributions presented in the study are included in the article. Further inquiries can be directed to the corresponding authors.

Conflicts of Interest: Author Mingxing Huang was employed by MCC Communication Construction Group Co., Ltd. The authors declare no conflict of interest.

References

1. Yao, F.L.; Xing, C.Z.; Ke, Z. Coal Gangue in Asphalt Pavement: A Review of Applications and Performance Influence. *Case Stud. Constr. Mater.* **2024**, *20*, e03282.
2. Li, L.; Long, G.; Bai, C.; Ma, K.; Wang, M.; Zhang, S. Utilization of Coal Gangue Aggregate for Railway Roadbed Construction in Practice. *Sustainability* **2020**, *12*, 4583. [CrossRef]

3. Tan, W.F.; Wang, L.A.; Huang, C. Environmental effects of coal gangue and its utilization. *Energy Sources Part A* **2016**, *38*, 3716–3721. [CrossRef]
4. Wu, H.; Chen, C.W.; Song, W.M.; Hou, W.Q. High-capacity utilization of coal gangue as supplementary cementitious material, geopolymer, and aggregate: A review. *Constr. Build. Mater.* **2024**, *435*, 136857. [CrossRef]
5. Li, D.; Wu, D.S.; Xu, F.G.; Lai, J.H.; Shao, L. Literature overview of Chinese research in the field of better coal utilization. *J. Clean. Prod.* **2018**, *185*, 959–980. [CrossRef]
6. Vo, T.L.; Nash, W.; Del Galdo, M.; Rezania, M.; Crane, R.; Nezhad, M.M.; Ferrara, L. Coal mining wastes valorization as raw geomaterials in construction: A review with new perspectives. *J. Clean. Prod.* **2022**, *336*, 130213. [CrossRef]
7. Turhan, C.; Atalay, A.S.; Gokcen Akkurt, G. An Integrated Decision-Making Framework for Mitigating the Impact of Urban Heat Islands on Energy Consumption and Thermal Comfort of Residential Buildings. *Sustainability* **2023**, *15*, 9674. [CrossRef]
8. Lu, Y.; Qin, Y.H.; Huang, C.; Pang, X.J. Albedo of Pervious Concrete and Its Implications for Mitigating Urban Heat Island. *Sustainability* **2023**, *15*, 8222. [CrossRef]
9. Huang, Y.; Li, J.; Song, T.; Sun, Q.; Kong, G.; Wang, F. Microstructure of Coal Gangue and Precipitation of Heavy Metal Elements. *J. Spectrosc.* **2017**, *2017*, 3128549. [CrossRef]
10. Du, Z.; Hong, X.; Yang, K.; Wang, Z.; Zhang, Y.; Wang, X.; Zhang, L.; Zhu, Y. Enrichment Characteristics of Polycyclic Aromatic Hydrocarbons in Coal Gangue of the Huaibei Coalfield, China. *Environ. Technol. Innov.* **2024**, *36*, 103769. [CrossRef]
11. Li, J.Y.; Wang, J.M. Comprehensive utilization and environmental risks of coal gangue: A review. *J. Clean. Prod.* **2019**, *239*, 117946. [CrossRef]
12. Zhou, J.J.; Xiong, J.Q.; Ni, J.J.; Xie, X.F.; Liu, Y.Z. Structural changes and denitrification performance evaluation of bioretention cells with collapsible loess modified by coal gangue. *J. Clean. Prod.* **2022**, *355*, 131740. [CrossRef]
13. Zheng, Q.; Zhou, Y.; Liu, X.; Liu, M.; Liao, L.; Lv, G. Environmental Hazards and Comprehensive Utilization of Solid Waste Coal Gangue. *Prog. Nat. Sci.* **2024**, *34*, 223–239. [CrossRef]
14. Liu, C.; Xia, J.; Fan, H.; Li, W.; Zheng, G.; Ma, G.; Liang, Y. Ti Leaching Differences during Acid Leaching of Coal Gangue Based on Different Thermal Fields. *Waste Manag.* **2020**, *101*, 66–73. [CrossRef]
15. Liu, M.; Zhu, Z.; Zhang, Z.; Chu, Y.; Yuan, B.; Wei, Z. Development of Highly Porous Mullite Whisker Ceramic Membranes for Oil-in-Water Separation and Resource Utilization of Coal Gangue. *Sep. Purif. Technol.* **2020**, *237*, 116483. [CrossRef]
16. Xiao, J.; Zhang, L.; Yuan, J.; Yao, Z.; Tang, L.; Wang, Z.; Zhang, Z. Co-Utilization of Spent Pot-Lining and Coal Gangue by Hydrothermal Acid-Leaching Method to Prepare Silicon Carbide Powder. *J. Clean. Prod.* **2018**, *204*, 848–860. [CrossRef]
17. Kong, D.; Zhou, Z.; Song, S.; Feng, S.; Lian, M.; Jiang, R. Preparation of Poly Aluminum-Ferric Chloride (PAFC) Coagulant by Extracting Aluminum and Iron Ions from High Iron Content Coal Gangue. *Materials* **2022**, *15*, 3104. [CrossRef] [PubMed]
18. Dong, Z.C.; Xia, J.W.; Fan, C.; Cao, J.H. Activity of calcined coal gangue fine aggregate and its effect on the mechanical behavior of cement mortar. *Constr. Build. Mater.* **2015**, *100*, 63–69. [CrossRef]
19. Long, G.C.; Li, L.H.; Li, W.G.; Ma, K.L.; Dong, W.K.; Bai, C.N.; Zhou, J.L. Enhanced mechanical properties and durability of coal gangue reinforced cement-soil mixture for foundation treatments. *J. Clean. Prod.* **2019**, *231*, 468–482. [CrossRef]
20. Liu, H.; Bai, G.; Gu, Y.; Yan, F. The Influence of Coal Gangue Coarse Aggregate on the Mechanical Properties of Concrete Columns. *Case Stud. Constr. Mater.* **2022**, *17*, e01315. [CrossRef]
21. Zhang, S.H.; Cao, M.Y.; Zhang, K.F.; Yuan, J.; Wang, Y. Wrapped coal gangue aggregate enhancement ITZ and mechanical property of concrete suitable for large-scale industrial use. *J. Build. Eng.* **2023**, *72*, 106649. [CrossRef]
22. Wang, C.; Feng, K.; Wang, L.; Yu, Q.; Du, F.; Guo, X. Characterization of Coal Gangue and Coal Gangue-Based Sodalite and Their Adsorption Properties for Cd^{2+} Ion and Methylene Blue from Aqueous Solution. *J. Mater. Cycles Waste Manag.* **2023**, *25*, 1622–1634. [CrossRef]
23. Du, T.; Wang, D.M.; Bai, Y.J.; Zhang, Z.Z. Optimizing the formulation of coal gangue planting substrate using wastes: The sustainability of coal mine ecological restoration. *Ecol. Eng.* **2020**, *143*, 105669. [CrossRef]
24. Li, Y.Y.; Cao, Y.J.; Ruan, M.Y.; Li, R.; Bian, Q.; Hu, Z.Q. Mechanism and In Situ Prevention of Oxidation in Coal Gangue Piles: A Review Aiming to Reduce Acid Pollution. *Sustainability* **2024**, *16*, 7208. [CrossRef]
25. Zhou, M.; Dou, Y.M.; Zhang, Y.Z.; Zhang, Y.Q.; Zhang, B.Q. Effects of the variety and content of coal gangue coarse aggregate on the mechanical properties of concrete. *Constr. Build. Mater.* **2019**, *220*, 386–395. [CrossRef]
26. Karimipour, A. Effect of untreated coal waste as fine and coarse aggregates replacement on the properties of steel and polypropylene fibres reinforced concrete. *Mech. Mater.* **2020**, *150*, 103592. [CrossRef]
27. Zhou, H.J.; Qi, X.; Ma, C.; Fang, Z.F.; Lou, J.N.; Chen, H.Y.; Guo, X. Effect and mechanism of composite early-strength agents on sulfoaluminate cement-based UHPC. *Case Stud. Constr. Mater.* **2023**, *18*, e01768. [CrossRef]
28. Wang, T.Y.; Gong, J.Q.; Chen, B.; Gong, X.; Guo, W.L.; Zhang, Y.; Li, F.L. Mechanical Properties and Shrinkage of Ultrahigh-Performance Concrete Containing Lithium Carbonate and Nano-Calcium Carbonate. *Adv. Civ. Eng.* **2021**, *15*, 6646272. [CrossRef]

coatings

MDPI

Article

Evaluation Indexes of Skid Resistance of Epoxy Chip Seal Based on Texture Features at Different Scales

Kaiyi Li [1], Min Wang [2,3,*], Jie Wang [3], Hui Lv [3] and Yongzhou Deng [4]

[1] School of Materials Science and Engineering, Chongqing Jiaotong University, Chongqing 400074, China; lky19981998@163.com
[2] China Merchants Chongqing Communications Technology Research & Design Institute Co., Ltd., Chongqing 400067, China
[3] Merchants Zhixiang Road Technology (Chongqing) Co., Ltd., Chongqing 401336, China; jiewang91@163.com (J.W.); lvhui8@cmhk.com (H.L.)
[4] School of Civil Engineering, Chongqing Jiaotong University, Chongqing 400074, China; ye1800yy@163.com
* Correspondence: panda9496@163.com; Tel.: +86-18008378268

Abstract: Epoxy chip seals are widely used to improve the skid resistance of cement concrete pavements, which is significantly affected by surface texture. However, current methods for characterizing the surface texture structure are not precise, and the standard is not uniform. Therefore, a high-toughness modified epoxy resin chip seal structure was developed to conduct an indoor accelerated abrasion test. The elevation data of the epoxy chip seal under different abrasion times and different abrasion loads were obtained using laser scanning, and the Density/Sharpness combination parameters were obtained using the Hilbert–Huang transform and spectral analysis. The correlation between the texture parameters and the dynamic friction coefficient was analyzed using a stepwise multivariate quadratic polynomial method. The results showed that the texture structure of the epoxy chip seal surface is positively distributed, and the macroscopic texture occupies 49.45%, while the probability of the microscopic texture is only 4.55%. Meanwhile, the correlation coefficient between the texture parameters and the dynamic friction coefficient is 0.9307, which demonstrates that the selected texture parameters discard the texture components unrelated to skid resistance performance and reflect the distribution of the pavement texture and the skid resistance performance well.

Keywords: epoxy chip seal; coefficient of friction; surface structure; spectral analysis

1. Introduction

Currently, China's road network system is progressing at a rapid pace. Due to the high stability of cement concrete pavements, they are used in toll plazas, high-speed areas with heavy traffic, and other pavements with special needs. However, in instances of increased traffic volume, the surface layer of the road is more likely to wear out, affecting driving comfort and safety. Epoxy chip seal is not affected by the constant load and elevation of the pavement and can effectively improve the skid resistance of the pavement as well as protect the pavement structure. Under the influence of environmental factors and traffic load, however, the epoxy chip seal degrades the road surface, leading to aggregate abrasion and falling and causing anti-skid performance decline. Therefore, the long-term skid resistance of epoxy chip seal must be examined.

Researchers have investigated the material and structural construction processes to improve the performance of chip seal pavements. Lee et al. [1] compared the performance of chip seals constructed with polymer-modified emulsions (PMEs CRS-2L) and unmodified emulsions (CRS-2) and found that PMEs CRS-2L improved the durability of the sealing structure, which resulted in a longer-lasting pavement. Bekir et al. [2] evaluated the effect of aggregate surface properties on the adhesive properties of the chip seal using the HSKSC test procedure and selected pure and modified penetrating asphalt as binders.

The results showed that the modified binder has better adhesive properties. Abedini et al. [3] investigated the effect of styrene butadiene rubber latex and its microstructure on the performance of asphalt–aggregate seals at low temperatures and how changes in the asphalt–aggregate structure affect the performance of asphalt–aggregate seals at low temperatures, the results indicated that the addition of SBR and X-SBR emulsions significantly decreased the asphalt's ductility, reduced aggregate loss by fivefold, and identified the optimal SBR dosage as 4 wt.%. Changes in asphalt–aggregate bonding properties and the effect of asphalt emulsion properties on asphalt bonding properties in chip seals were investigated using the Vialit test. Zhengqi Zhang et al. [4] prepared a composite modified crushed stone sealing layer using water-based epoxy resin and styrene butadiene rubber emulsion. A series of tests showed that the structure had better wear resistance than the traditional crushed stone sealing layer and asphalt wear layer. Peide Cui et al. [5] utilized converter steel slag as an aggregate for the gravel sealing layer, and their research demonstrated that steel slag can significantly improve the interlayer bonding performance of the gravel sealing layer while also providing substantial economic benefits.

Given that epoxy chip seal pavements differ from ordinary asphalt concrete pavements, a suitable method should be selected to characterize their anti-skid performance and wear resistance. The most commonly used current texture characterization parameter is the mean profile depth (MPD). Nanxiang Zheng et al. [6] used three-dimensional texture parameters (Ssk, Spd, Spc, Vmp, and Vmc) to characterize the surface texture state of the chip seal. Jiangmiao Yu et al. [7] proposed a new type of ultra-thin asphalt overlay scheme (HUFC) for the current status quo of pavement use in South China, with the MPD to characterize the morphological state of the specimen after wear. The first International Conference on Engineering Surface Characterization and Metrology proposed various roughness parameters, and ISO 25178 [8] classified them into six categories: the average deviation from the longitudinal level of the cross-section (Ra) and the root-mean-square of the cross-section depth (Rq). However, the abrasion behavior of the texture structure related to skid resistance is a multi-scale change, and a single characterization of one factor cannot evaluate the texture structure change in the entire abrasion process with accuracy. Di Yun et al. [9] utilized the coarse aggregate ratio (CA), the coarse portion of fine aggregate (FA_c), and the fine portion of fine aggregate (FA_f) to characterize the integrity of the gradation. The mean profile depth (MPD), texture level (L_{tx}), and friction coefficient were employed to assess the degradation of pavement skid resistance. Lei et al. [10] conducted an analysis of the micro-morphology of aggregates to understand the decay law of their micro-texture. This study evaluated the influence of the nature and morphological structure of aggregates, ultimately verifying that the micro-morphology indexes Ra and Rq exhibited the strongest correlation with anti-slip performance.

Therefore, to ensure data accuracy, it is necessary to analyze the multi-dimensional surface texture and select an appropriate evaluation method.

Therefore, to correctly characterize the texture parameters of the epoxy chip seal and enhance its anti-skid performance and durability, we examine the epoxy chip seal material and its anti-skid performance based on numerical simulation and analysis of related indexes. Additionally, we determine the characterization parameters under different scales, in line with the characteristics of the distribution of the surface texture of the epoxy chip seal. We also clarify the relationship between the surface texture and the dynamic friction coefficient.

2. Raw Materials and Specimen Preparation

2.1. Cementitious Materials

In 1:3 proportion, we weighed tetramethylene ether glycol and diphenylmethane diisocyanate and added them to the reaction flask. We then added a certain amount of water-removing agent. A magnetic stirrer was used to evenly mix the mixture; the mixture was then placed in the reaction at 40 °C for two hours to obtain the polyurethane prepolymer end-isocyanate base. In 4:1 proportion, we weighed the above polyurethane prepolymer and diethylene glycol chain extender based on the proportion. We then stirred the mixture evenly and reacted it at 60 °C for 3 h to obtain the chain-expanding polyurethane prepolymer. The GPC curve and NMR spectrum of polyurethane prepolymer are shown in Figure 1.

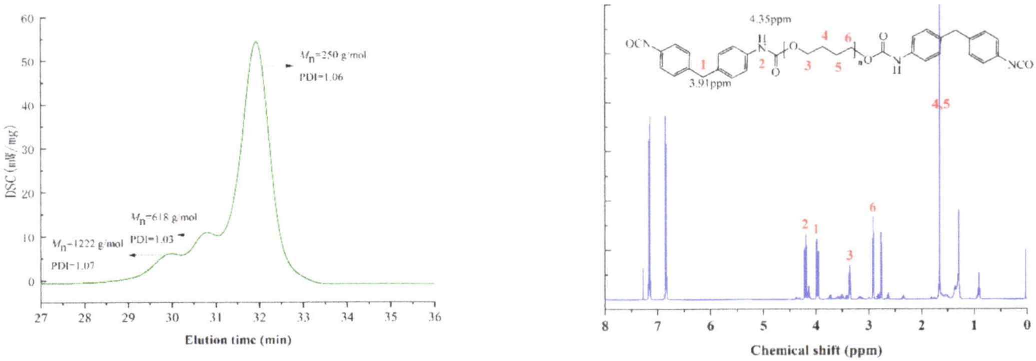

(**a**) (**b**)

Figure 1. GPC curve and NMR spectrum of polyurethane prepolymer. (**a**) GPC curve; (**b**) NMR spectrum.

As shown in Figure 1a, a polyurethane prepolymer with a molecular weight of 1222 g/mol was produced by mixing and reacting the raw materials of each component, exhibiting a molecular weight distribution width of 1.07. The small molecule polyol and diphenylmethane diisocyanate had molecular weights of 618 g/mol and 250 g/mol, respectively. Figure 1b indicates that the chemical shift at 4.35 ppm corresponds to the active hydrogen on the formed amide group, confirming the successful synthesis of the polyurethane prepolymer.

We weighed 100 parts polyurethane prepolymer and 15 parts E51 epoxy resin, mixed them proportionally, and placed them in the reaction under the protection of nitrogen at 80–90 °C for 3–5 h to obtain the modified epoxy resin. Figure 2 shows the polyurethane prepolymer and epoxy resin reaction polyurethane/epoxy resin prepolymer preparation processes.

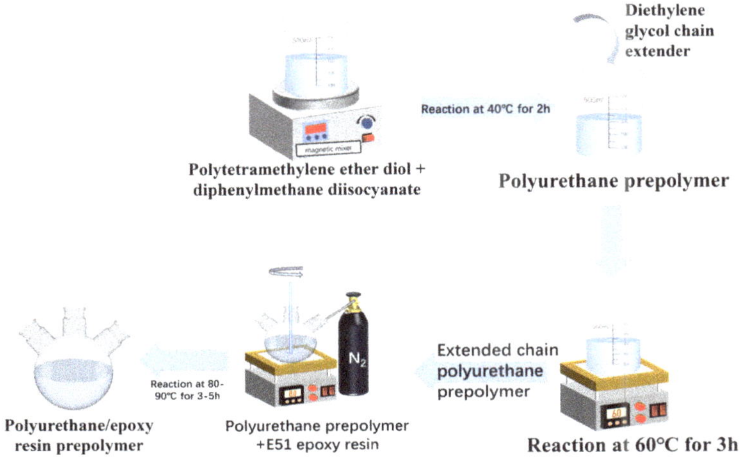

Figure 2. Polyurethane/epoxy prepolymer preparation process.

The binder to the polyurethane-modified epoxy resin (A component) and polyether curing agent (B component) were mixed in a certain ratio such that the linear structure of the epoxy resin changed into a crosslinked mesh structure of the material. Components A and B were mixed in a ratio of 10:8, and Table 1 presents its main properties. Its SEM images are shown in Figure 3.

Table 1. Main properties of high-toughness modified epoxy binder.

Pilot Projects	Unit	Test Results	Test Methods
Tensile strength (23 °C)	MPa	13.2	GB/T16777-2008 [11]
Elongation at break (23 °C)	%	66.6	
Impermeability (0.3 MPa, 30 min)	—	impermeable	

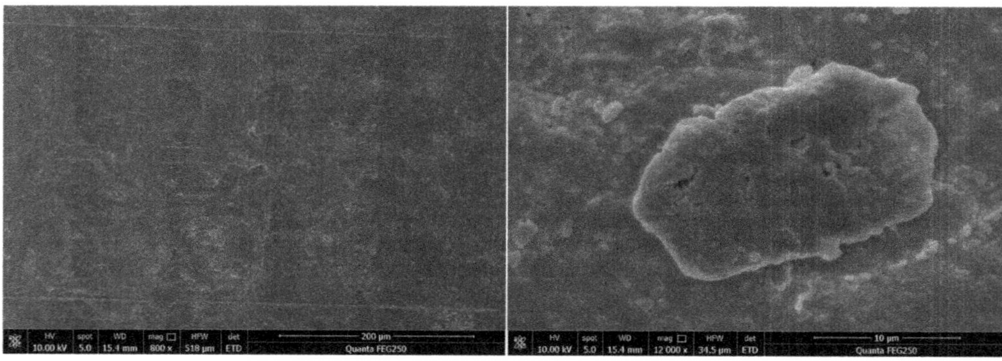

Figure 3. SEM images of modified epoxy cured compounds.

At a magnification of 12,000, uniformly distributed agglomerated particles are clearly visible, exhibiting the characteristics of a tough fracture. These fine agglomerated particles can absorb impact energy when the material is subjected to external stress. Additionally, the polyurethane prepolymer is dispersed in the continuous phase of the epoxy resin in a granular form, which enhances the cross-linking density of the epoxy resin and, consequently, increases the material's flexibility. The polyurethane prepolymer also forms a network structure within the epoxy, improving the overall mechanical properties of the material.

2.2. Aggregates

In this study, basalt aggregates with particle sizes of 3–5 mm were selected, and their performance met the requirements of the JTG 3432-2024 [12]. The primary performance indexes are listed in Table 2.

Table 2. Technical performance indexes of basalt.

Pilot Project	Test Results	Technical Requirement	Test Methods
Apparent relative density	2.967	≥2.60	JTG 3432-2024T 0304
Crushing value	15.0	≤24	JTG 3432-2024T 0316
Los Angeles abrasion losses	19.7	≤28	JTG 3432-2024T 0317
Polished value PSV	—	≥42	JTG 3432-2024T 0321

2.3. Preparation of Specimens

The steps of the accelerated abrasion specimen molding in this study are as follows:

(1) Select a C60 cement concrete slab with dimensions of 300 × 300 × 50 mm as the substrate for the epoxy chip seal and subject its surface to sandblasted roughness treatment.
(2) Adopt the developed high-tenacity epoxy resin material and scrape the modified epoxy resin binder.
(3) Prepare the epoxy chip seals with basalt aggregates of 3–5 mm particle size, respectively, as shown in Figure 4. The dosage of binder in the specimen was 1.33 kg/m^2, and the amount of chip sealing was 6 kg/m^2.
(4) Set aside at 25 °C for five hours, followed by 60 °C for 24 h. After completion of maintenance, the mold was demolded, and the unbonded aggregates were removed and cut into rectangular specimens with dimensions of 300 × 150 × 50 mm.

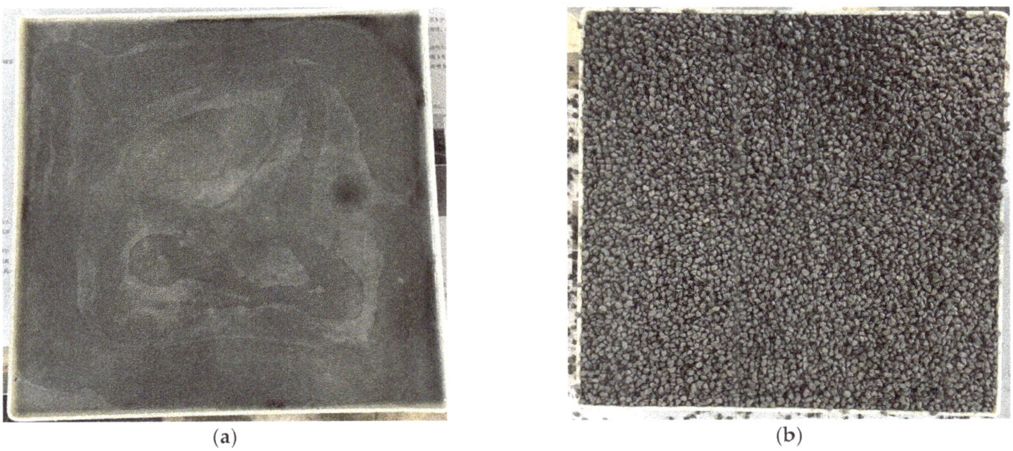

Figure 4. Test piece molding. (**a**) Spreading of cementitious materials; (**b**) spreading of aggregates.

3. Experimental Design

3.1. Accelerated Wear and Skid Resistance Tests

Accelerated abrasion tests were performed on specimens of two grain sizes using an aggregate/asphalt mixture friction characterization tester [13]. The accelerated abrasion loading end was a 120 × 100 mm rubber block abrasive, reciprocating continuous movement, and the changes in the coefficient of kinetic friction of the specimen were measured in real time during the accelerated loading abrasion.

The instrument is powered by an electric cylinder as the driving component, with its reciprocating linear motion controlled by the on–off switching of a solenoid valve. Each unidirectional movement consists of three phases: acceleration, constant velocity, and deceleration. Data are collected during the constant velocity phase, focusing on the pressure and frictional forces on the specimen. The dynamic friction coefficient is then calculated based on the relationship between the pressure and frictional force.

The single unidirectional running distance was 130 mm; the test temperature was 25 °C; the test loads were 50 kgf (490 N), 75 kgf (735 N), 100 kgf (980 N), and 150 kgf (1470 N); and the frequency was 30 times per minute.

The accelerated loading abrasion test time was 8 h, the dynamic friction coefficient was tested every 1 h, and the dynamic friction coefficient change curve was recorded. Figure 5 presents the changes in the dynamic friction coefficient.

As the number of abrasions and pressure load increase, the surface aggregates wear out, reducing the textural configuration of the specimen surface and decreasing the roughness, which decreases the coefficient of kinetic friction.

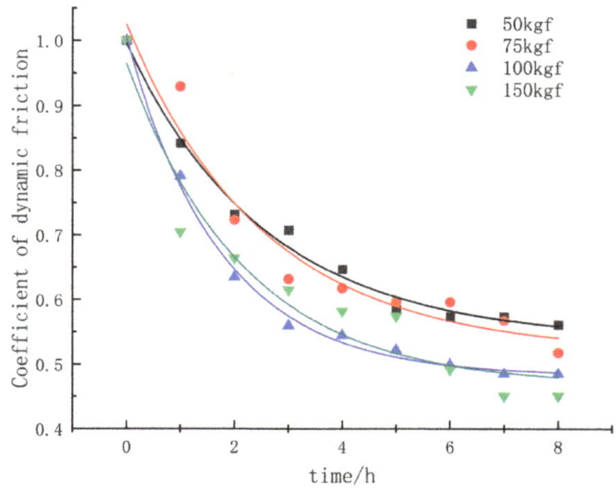

Figure 5. Variation curve of dynamic friction coefficient.

3.2. *Surface Texture Data Acquisition and Preprocessing*

(1) 3D texture acquisition

We used a laser sensor to scan and collect the surface texture of the specimen plate. The scanning data were synchronously transmitted to the control computer, and the specimen was removed every hour for scanning in the accelerated abrasion test. The texture acquisition system consisted of a laser 3D profiler, running platform, computer, and other supporting facilities.

In the laser profiler scanning, the x-axis data interval was set to 0.015 mm, the sampling frequency was set to 2500 Hz, the camera running speed was 50 mm/s, the depth of field was 12 mm, and the longitudinal sampling measurement line interval was 0.015 mm. Table 3 presents the technical parameters.

Table 3. Dynamic motion parameters of laser profiler.

Parameter	Value
z-axis height (camera depth of field)	12 mm
b-axis width	60 mm
x-axis repeatability	0.06 mm
x-axis data interval	0.02 mm
x-axis contour points	3200
z-axis repeatability	0.012 mm
Scanning speeds	2500–8000 Hz

(2) 3D Texture Preprocessing

In this study, the point cloud outliers introduced by measurement noise were removed using MATLAB R2021a and statistical filters, based on which median filtering and Gaussian filtering were sequentially performed to remove noise in the point cloud. Missing data were filled in using cubic spline interpolation [14], and planar corrections were performed using the least-squares method [15], as shown in Figure 6.

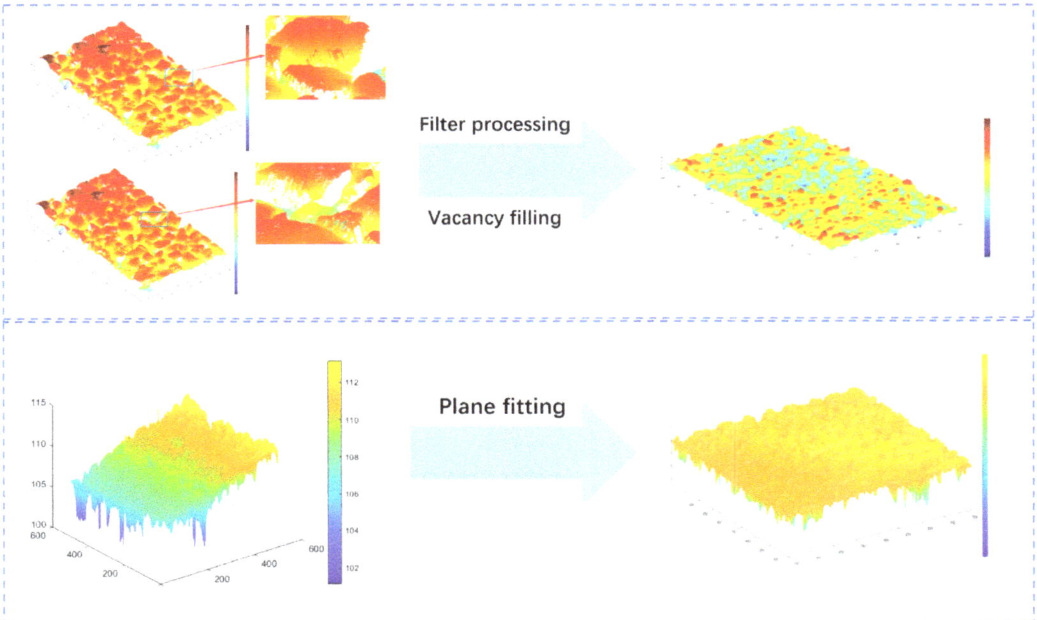

Figure 6. 3D data preprocessing.

4. Surface Texture-Based Characterization Metrics for Epoxy Chip Seals

4.1. Surface Texture Characterization of Epoxy Chip Seals Based on the Hilbert–Huang Transformation

The Hilbert–Huang transform technique is an adaptive signal time-frequency processing method that includes empirical mode decomposition (EMD) and Hilbert spectrum analysis [16]. The first step of the Hilbert–Huang transform is to decompose the complex signal into a series of intrinsic mode functions (IMF) that satisfy specific conditions using the EMD method, followed by frequency analysis of each IMF.

The empirical modal decomposition EMD is based on the direct extraction of energy associated with various intrinsic timescales, which are the most important parameters of the system. With this decomposition, each pavement texture profile was decomposed into a finite number of elementary profile functions using the following equations:

$$Z(x) = r_n(x) + \sum_{j=1}^{n} c_j(x), \tag{1}$$

where $Z(x)$ denotes the depth corresponding to the pavement contour at distance x, $r_n(x)$ denotes the residual obtained after decomposition, $C_j(x)$ denotes the jth IMF, and n denotes the total number of IMFs.

The surface texture profile was decomposed using EMD to obtain a series of IMFs. These IMFs, from high to low frequencies, were numbered from small to large [17]. The number of IMFs obtained by empirical modal decomposition is not the same for different signals, and a fixed number of IMFs must be selected to compare the surface texture profile features of the different structures. The steps of EMD are as follows:

(1) The local maxima and minima of the contour lines were connected by a cubic spline called the upper and lower envelopes.

(2) The upper and lower envelopes were averaged, and the corresponding raw data were subtracted.

(3) First, we determined whether the IMFs obtained from the decomposition satisfied the condition that the number of extremes of each IMF must be equal to the number of zero crossings or differed by at most one and that the average of the upper and lower envelopes at any point was zero. If they did so, they were classified as IMFs. However, because not every set of IMFs could satisfy the above two requirements, the iteration-stopping condition was determined using the following formula:

$$\sum_{x=0}^{T}[\frac{\left|h_{1(k-1)}(x)-h_{1k}(x)\right|^2}{{h_{1k}}^2(x)}] < SD, \quad (2)$$

where $h_{1k}(x)$ is the difference between the signal and the mean envelope after k iterations. SD is the standard deviation, which can be controlled between 0.2 and 0.3 [18].

(4) The remaining signal was used as new raw data and underwent the same process as in the previous three steps. This iterative process continued until the screening process stopped, when the residuals became monotonic functions or constants.

The original signal was divided into 10 IMFs (IMF1, IMF2, . . ., IMF10) based on the above steps, and then the IMF decomposition was distributed in descending order of frequency. The length of the scanning contour can be considered a limiting factor so that the high-frequency components are merged with the other high-frequency components during the merging of IMF components, and the low-frequency components are merged with the other low-frequency components. The low-frequency components were merged with other low-frequency components to ensure frequency consistency. Therefore, the 10 IMFs were grouped into two to obtain five BIMFs, and the Hilbert transform was performed on the five groups of BIMFs to obtain the instantaneous frequency and instantaneous amplitude of each BIMF, as shown in Figure 7.

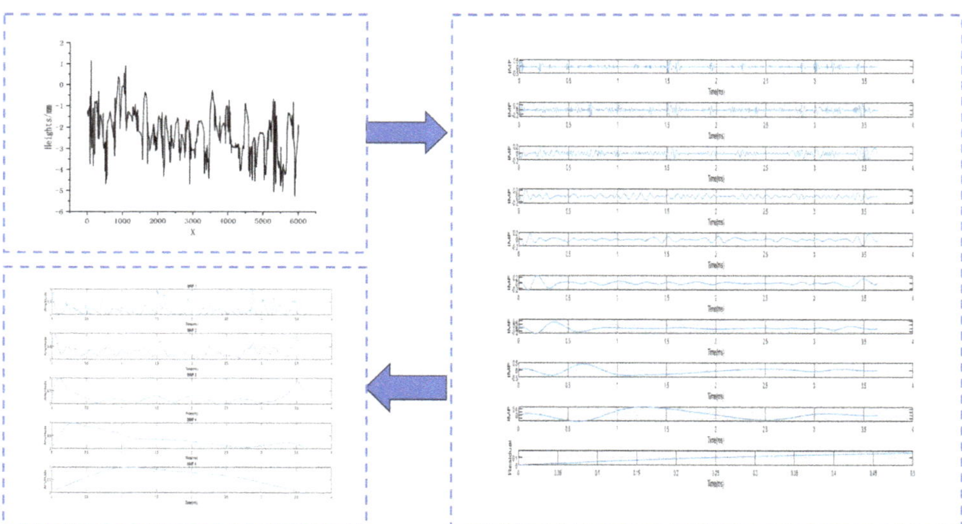

Figure 7. Hilbert–Huang transform for processing contour data.

To study the degree of undulation and constructive characteristics of each BIMF, the contour peak density (Density) and contour peak sharpness (Sharpness) metrics were introduced [19]. The contour peak density is the number of peaks per unit sampling length of the contour curve, which can be obtained using the findpeak function in MATLAB and represents the average value of the angle formed by the peaks of the contour curve, which is given by the following formula:

$$Density = \frac{Number\ of\ peaks}{l}, \quad (3)$$

$$\delta = ar\cos\left(\frac{A^2 + B^2 - C^2}{2AB}\right), \quad (4)$$

MATLAB software was used to calculate the profile peak density (Density) and profile peak sharpness (Sharpness) of the BIMF1 and BIMF2 specimens of the epoxy chip seal layered structure (3–5 mm) under different loads, and the combined density/sharpness parameter was adopted as the morphology parameter of the calculation (Table 4).

Table 4. BIMF1 and BIMF2 texture parameters.

Number of Abrasions	Density	Sharpness	D/S
0	579	3.2392	178.7478
1800	551	3.2143	171.4215
3600	546	3.1722	172.1203
5400	535	3.1653	169.0203
7200	530	3.1564	167.9128
9000	503	3.142	160.0891
10,800	488	3.1345	155.6867
12,600	486	3.1342	155.0635
14,400	485	3.1316	154.8729

4.2. Characterization of the Surface Texture Morphology of Epoxy Chip Seals Based on Spectral Analysis

The amplitude and frequency obtained using the HHT transform were characterized, but only the overall distribution level was analyzed because the surface morphology of the pavement is primarily expressed in two aspects: microtexture and macrotexture, and it is necessary to differentiate the texture by using the filtering process. The wavelength distribution of the texture of the epoxy chip seal was calculated based on mathematical statistics, as shown in Figure 8. The wavelength interval probability distribution is solved, as shown in Figure 9.

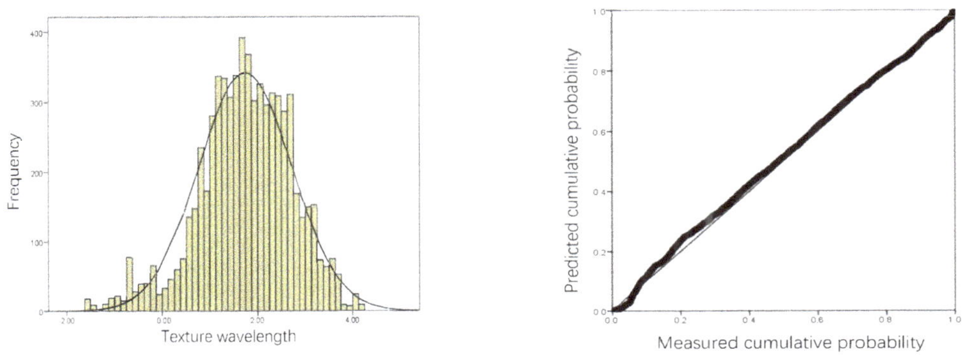

(a) (b)

Figure 8. Wavelength distribution of epoxy chip seal texture. (**a**) Texture wavelength distribution; (**b**) normal distribution tests.

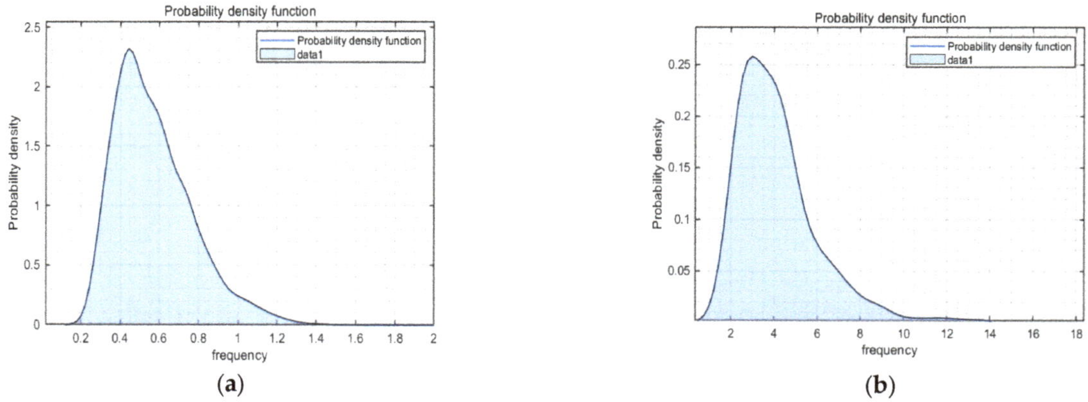

Figure 9. Plot of texture construction probability density function. (**a**) Macrotexture probability density function; (**b**) microtexture probability density function.

The figure shows that the surface texture of epoxy chip seal is normally distributed, and the macro texture occupies a larger proportion of 49.45%, while the probability of microtexture is only 4.55%. Therefore, different texture scales have different effects on the anti-skid performance, and it is necessary to categorize the texture. We characterized the microtexture and the macrotexture using discrete Fourier transform (DFT) spectral analysis [20].

Therefore, the surface texture of the crushed epoxy stone was analyzed using a DFT according to the provisions of ISO/TS13473-4 for spectral analysis. The macroscopic texture wavelength is 0.5–50 mm, and its corresponding frequency is 0.02–2 mm^{-1}; the microscopic texture wavelength is 0.05–0.5 mm, and its corresponding frequency is 2–20 mm^{-1}, and thus the band-pass filter was designed using MATLAB software to differentiate the texture of the epoxy chip seal.

In this study, the differential box analysis was used to characterize the degree of undulation of macrotexture and microtexture [21]. The idea of the process is as follows: A 3D spatial surface with a vertical projection of M × M was partitioned into a grid of s × s in the plane obtained by the projection, and $M/2 \geq s \geq 2$; therefore, r = s/M, and for the 3D spatial surface, the x-axis and y-axis denote the position of the plane, and the z-axis denotes the height. The x-y plane was partitioned into several s × s grids. On each grid was a series of s × s × h boxes, where h was the height of the box such that h = s/M.

Let the minimum value of the 3D spatial surface height in the (x, y)-th grid fall in the *k*th box, and the maximum value fall in the *t*th box.

$$n_r(i,j) = t - k + 1, \tag{5}$$

where $n_r(i, j)$ denotes the number of boxes on the surface covering the (i, j)th mesh.

The total number of boxes, $N(r)$, required to cover the entire surface is

$$N(r) = \sum_{i,j} n_r(i,j), \tag{6}$$

The fractal dimension D was then calculated according to the definition of the fractal dimension [22]:

$$D = \lim_{r \to 0} \frac{\ln N(r)}{\ln(\frac{1}{r})}, \tag{7}$$

Linear equations can be obtained by linearly fitting the data points ($\ln 1/r$, $\ln N(r)$) using the least-squares method for different r values:

$$\ln N(r) = a \ln(\frac{1}{r}) + b, \quad (8)$$

Combined with Equation (8), slope a of the line is the fractal dimension D.

The fractal dimension was calculated based on the definition, the x and y coordinates were extracted, and the number of square boxes and the number of points inside each box were calculated while traversing each scale. The proportion of cube coverage at each scale and fractal dimensions of the different contour lines were calculated.

The correlation between the texture fractal dimension and the dynamic friction coefficient was calculated to verify the validity of this distributional feature parameter. Figure 10 shows the correlation between the fractal dimension D and the dynamic friction coefficient obtained from the micro and texture calculations.

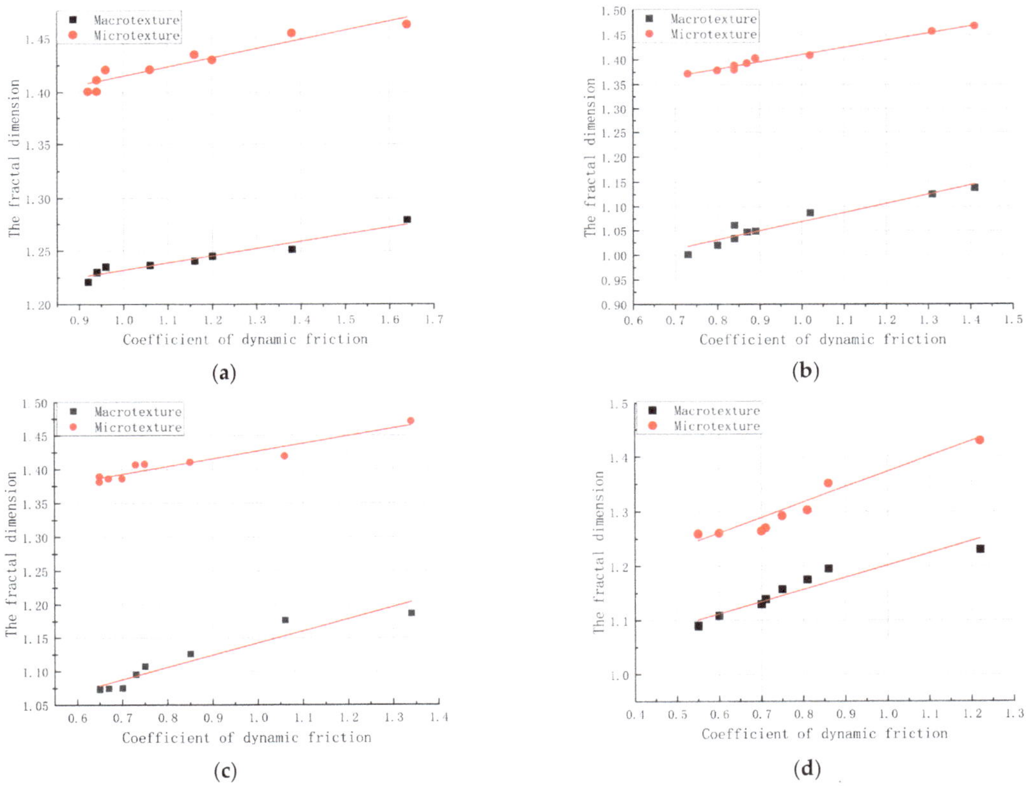

Figure 10. Texture fitting results for different abrasion loads. (**a**) 50 kgf; (**b**) 75 kgf; (**c**) 100 kgf; (**d**) 150 kgf.

As shown in Figure 9, the fitting correlations R^2 between the microtexture, macrotexture, and friction coefficient under different load abrasion cases are all above 95%; therefore, the fractal dimension D can better characterize the macrotexture and microtexture distribution characteristics. Compared with the fractal dimension of the specimen surface without filtering treatment, the fitting correlation was improved, proving that differentiating the texture according to the wavelength is more conducive to establishing a relationship between the texture characteristics and anti-skid performance. The texture configuration of the epoxy chip seal specimens was analyzed using the combined HHT + DFT method, which was required to verify the correlation between the texture feature parameters and skid resistance performance.

5. Characterization of Epoxy Chip Seal Texture Characteristic Parameters

According to previous studies, the anti-skid performance showed an exponential decay trend; therefore, it is unsuitable for linear regression. An exponential function should be selected to fit the texture parameters and friction coefficients. The data obtained from the test were analyzed using multivariate non-linear regression using the SPSS mathematical statistical analysis software, and the functional expression was established using a step-by-step regression method. The results of the step-by-step multivariate regression showed that R^2 of the model was 0.80, and there are cases in which the indicator is not fully utilized; thus, the multivariate quadratic polynomial regression model can be used.

First, the density/sharpness combination parameter (*DS*), macroscopic texture fractal dimension (*Dh*), and microscopic texture fractal dimension (*Dw*) selected in the previous section were combined into a vector indicator $M = \{DS, Dh, Dw\}$, and stepwise multivariate quadratic polynomial regression was performed based on Equation (9) [23].

$$D = MAM^T + BM^T + C, \quad (9)$$

where D is the skid resistance index and is the dynamic friction coefficient $D\mu$; M is the synthetic vector; A is the quadratic term coefficient matrix of the model, which is a symmetric matrix; B is the primary term coefficient matrix; and C is the constant term.

Tables 5–7 present the regression analysis of stepwise multivariate quadratic polynomial regression using SAS 9.4 analysis software and parameter tests. The results of the coefficient solution for the model are shown in Equations (10)–(12).

$$D\mu = MAM^T + BM^T + C, \quad (10)$$

$$M = [DS\ Dh\ \mathrm{Dw}], \quad (11)$$

$$A = \begin{bmatrix} -0.00694 & 0.10126 & 0.16754 \\ 0.10126 & 14.25601 & -30.00502 \\ 0.16754 & -30.00502 & 39.04383 \end{bmatrix}; B = \begin{bmatrix} -0.12259 & -6.50951 & -97.72689 \end{bmatrix}; C = 79.27132, \quad (12)$$

Table 5. Statistical results of the multivariate quadratic polynomial regression models.

Implicit Variable	Independent Variable	Variable Response	Value
		Standard deviation of residuals	0.0857
$D\mu$	$M = [DS, Dh, \mathrm{Dw}]$	R^2	0.9307
		Coefficient of variation	8.7915

Table 6. Parameter estimates and tests for the multivariate quadratic polynomial regression models.

Regression	DF	Discrete Sum of Squares	R^2	F	P r < F
Linearly	3	1.2609	0.7007	57.2700	<0.0001
Square	3	0.3950	0.2195	17.9400	<0.0001
Cross-cutting	3	0.0188	0.0104	0.8500	0.4837
Population	9	1.6747	0.9307	23.3500	<0.0001

Table 7. Multiple quadratic polynomial regression model factor tests.

Factor	DF	Discrete Sum of Squares	Mean Square Error	F	P r < F
DS	4	0.0214	0.0054	0.7300	0.5842
Dh	4	0.0787	0.0197	2.6800	0.0672
Dw	4	0.5016	0.1254	17.0900	<0.0001

Table 5 shows that the R^2 correlation coefficient of the model was 0.9307, indicating that the model was well fitted. Table 6 shows that the F-test p-value of the linear and squared terms of the model is lower than 0.05, and the F-test p-value of the cross term is higher than 0.05. However, the overall F-test p-value of the model is also lower than 0.05, which proves that a good quadratic polynomial relationship exists between the explanatory variables and the explanatory variables. This proves that a significant relationship exists in the regression model. Factor test results of the model showed that the F-test probability value of Dw was less than 0.0001, indicating that the overall regression coefficient was highly significant.

The correlation analysis between the texture characteristic parameters and skid resistance shows that the surface texture of the epoxy chip seal is processed by the Hilbert–Huang transform technique and spectral analysis technique. The texture components that are not related to skid resistance can be discarded by the empirical modal EMD and DFT texture differentiation. The range of textural configurations with the most significant effect on anti-slip properties during abrasion was thus obtained. This also verifies that the microtexture and macrotexture jointly determine the anti-skid performance of the pavement, and the quadratic polynomial model can well represent the relationship between the texture characteristic parameter index and the anti-skid performance index.

6. Conclusions

Based on an indoor accelerated abrasion test and surface 3D texture data, the morphological parameters of epoxy chip seals with different texture scales were investigated, and the texture structure of the epoxy chip seal was characterized using the Hilbert–Huang transform, spectral change, and calculation of the fractal dimension. The conclusions are as follows:

(1) A laser 3D profiler was used to collect the surface texture of the specimen to obtain 3D elevation data, which were preprocessed using MATLAB software through filtering, missing data filling, and planar tilt fitting to obtain the 3D reconstruction model.

(2) First, the surface texture information of the epoxy chip seal was processed using the Hilbert–Huang transform technique, the texture profile was screened using EMD, and the IMFs were composed into the base IMF. The texture structure closely related to the skid performance was obtained from BIMF, and the one unrelated to the skid performance was discarded. The texture structure range that had the most significant effect on anti-skid performance during the abrasion process was also obtained.

(3) Through spectral analysis, the surface texture parameters could be distinguished into macrotexture and microtexture; the macroscopic texture occupies 49.45% and the probability of the microscopic texture is equal to 4.55%, which can be characterized by calculating the fractal dimension through fractal theory. The correlation between the fractal dimension and the dynamic friction coefficient for both micro- and macrotextures exceeds 90%. The correlation between the fractal dimension and the dynamic coefficient of friction could be analyzed through abrasion.

(4) The R^2 correlation coefficient of the parameters (*DS*, *Dh*, and *Dw*) and the dynamic friction coefficient was 0.9307, indicating that the texture parameters *DS*, *Dh*, and *Dw* correlate well with the skid resistance and jointly affect the skid resistance of the epoxy chip seal.

Author Contributions: Conceptualization, M.W.; methodology, K.L. and J.W.; software, K.L. and H.L.; validation, K.L., M.W. and J.W.; formal analysis, K.L.; investigation, Y.D.; resources, M.W.; data curation, K.L., M.W. and J.W.; writing—original draft preparation, K.L.; writing—review and editing, K.L., M.W., H.L. and Y.D.; visualization, K.L.; supervision, M.W.; project administration, M.W.; funding acquisition, M.W. All authors have read and agreed to the published version of the manuscript.

Funding: This research was funded by the National Key Research and Development Program of China [2021YFB2601200].

Institutional Review Board Statement: Not applicable.

Informed Consent Statement: Not applicable.

Data Availability Statement: Data are contained within the article.

Acknowledgments: Thanks to China Merchants Zhixiang Road Technology (Chongqing) Co., Ltd. for providing the test platform.

Conflicts of Interest: Authors Min Wang were employed by China Merchants Chongqing Communications Technology Research & Design Institute Co., Ltd. and Merchants Zhixiang Road Technology (Chongqing) Co., Ltd. Authors Jie Wang and Hui Lv were employed by Merchants Zhixiang Road Technology (Chongqing) Co., Ltd. The remaining authors declare that the research was conducted in the absence of any commercial or financial relationships that could be construed as a potential conflict of interest.

References

1. Lee, J.; Kim, Y.R. Evaluation of Performance and Cost-Effectiveness of Polymer-Modified Chip Seals. *Transp. Res. Rec. J. Transp. Res. Board* **2010**, *2150*, 79–86. [CrossRef]
2. Aktaş, B.; Karaşahin, M.; Saltan, M.; Gürer, C.; Uz, V.E. Effect of aggregate surface properties on chip seal retention performance. *Constr. Build. Mater.* **2013**, *44*, 639–644. [CrossRef]
3. Abedini, M.; Hassani, A.; Kaymanesh, M.R.; Yousefi, A.A. Low-temperature adhesion performance of polymer-modified Bitumen emulsion in chip seals using different SBR latexes. *Pet. Sci. Technol.* **2016**, *35*, 59–65. [CrossRef]
4. Zhang, Z.; Yang, J.; Fang, Y.; Luo, Y. Design and performance of waterborne epoxy-SBR asphalt emulsion (WESE) slurry seal as under-seal coat in rigid pavement. *Constr. Build. Mater.* **2021**, *270*, 121467. [CrossRef]
5. Cui, P.; Wu, S.; Xiao, Y.; Hu, R.; Yang, T. Environmental performance and functional analysis of chip seals with recycled basic oxygen furnace slag as aggregate. *J. Hazard. Mater.* **2021**, *405*, 124441. [CrossRef] [PubMed]
6. Zheng, N.; Bi, J.; Dong, S.; Lei, J.; He, Y.; Cui, Z.; Chen, L. Testing and evaluation for long-term skid resistance of asphalt pavement composite seal using texture characteristics. *Constr. Build. Mater.* **2022**, *356*, 129241. [CrossRef]
7. Yu, J.; Chen, F.; Deng, W.; Ma, Y.; Yu, H. Design and performance of high-toughness ultra-thin friction course in south China. *Constr. Build. Mater.* **2020**, *246*, 118508. [CrossRef]
8. Marinello, F.; Pezzuolo, A. Application of ISO 25178 standard for multiscale 3D parametric assessment of surface topographies. *IOP Conf. Ser. Earth Environ. Sci.* **2019**, *275*, 012011. [CrossRef]
9. Yun, D.; Tang, C.; Gao, J.; Ran, M.; Zhou, X. Effect of asphalt mixture gradation characteristics on long-term skid resistance under high temperature and heavy load. *Constr. Build. Mater.* **2024**, *441*, 137386. [CrossRef]
10. Lei, J.; Zheng, N.; Bi, J.; Zhao, F.; Wang, Y.; Yang, J. Research on the evolution law of aggregate micro-texture during long-term wearing of asphalt pavement. *Constr. Build. Mater.* **2024**, *444*, 137846. [CrossRef]
11. *GB/T 16777-2008*; Test Methods for Building Waterproofing Coatings. National Standards of the People's Republic of China: Beijing, China, 2009.
12. *JTG 3432-2024*; Test Methods of Aggregate for Highway Engineering. Ministry of Transport of the People's Republic of China: Beijing, China, 2024.
13. Wei, T. Research on the Development of Experimental System Based on Anti-Skid Characteristics of Asphalt Pavement and Key Technology of Limestone Preference. Ph.D. Thesis, Chongqing Jiaotong University, Chongqing, China, 2017.
14. Dong, S.; Han, S.; Zhang, Q.; Han, X.; Zhang, Z.; Yao, T. Three-dimensional evaluation method for asphalt pavement texture characteristics. *Constr. Build. Mater.* **2021**, *287*, 122966. [CrossRef]
15. Wang, H.; Zhang, X.; Wang, M. Rapid texture depth detection method considering pavement deformation calibration. *Measurement* **2023**, *217*, 113024. [CrossRef]
16. de Souza, U.B.; Escola, J.P.L.; da Cunha Brito, L. A survey on Hilbert-Huang transform: Evolution, challenges and solutions. *Digit. Signal Process.* **2021**, *120*, 103292. [CrossRef]
17. Rado, Z.; Kane, M. An initial attempt to develop an empirical relation between texture and pavement friction using the HHT approach. *Wear* **2014**, *309*, 233–246. [CrossRef]

18. Liu, X.; Wu, J.; Lin, H.; Jadoon, A.; Wang, J. Research on DC bias analysis for transformer based on vibration Hilbert Huang transform and ground-state energy ratio method. *Int. J. Electr. Power Energy Syst.* **2019**, *109*, 73–82. [CrossRef]
19. Chen, C.; Luo, H.; Chen, S.; Zheng, Y.; Liu, X.; Ma, T.; Zhu, L. Exploring the relationship between pavement surface texture and friction based on Hilbert-Huang Transformation. *Tribol. Int.* **2023**, *187*, 108737. [CrossRef]
20. Su, X.; Tao, R.; Kang, X. Analysis and comparison of discrete fractional fourier transforms. *Signal Process.* **2019**, *160*, 284–298. [CrossRef]
21. Liu, C.; Zhan, Y.; Deng, Q.; Qiu, Y.; Zhang, A. An Improved Differential Box Counting Method to Measure Fractal Dimensions for Pavement Surface Skid Resistance Evaluation. *Measurement* **2021**, *178*, 109376. [CrossRef]
22. Liu, Y.U.; Chen, L.; Wang, H.; Jiang, L.; Zhang, Y.; Zhao, J.; Wang, D.; Zhao, Y.; Song, Y. An improved differential box-counting method to estimate fractal dimensions of gray-level images. *J. Vis. Commun. Image Represent.* **2014**, *25*, 1102–1111. [CrossRef]
23. Wang, Y. Study on the Relationship Between Sliding Resistance of Asphalt Pavement and Its Surface Rough Characteristics. Ph.D. Thesis, Southeast University, Nanjing, China, 2017.

coatings

MDPI

Article

Mechanical Performance Assessment of a Hot-Mix Asphalt Modified with Biochar Obtained from Pine Wood Shavings

Saieth Baudilio Chaves-Pabón [1], Hugo Alexander Rondón-Quintana [2,*] and Juan Gabriel Bastidas-Martínez [3]

1 Programa de Ingeniería Civil, Facultad de Estudios a Distancia, Universidad Militar Nueva Granada, Cajicá 111321, Colombia; saieth.chaves@unimilitar.edu.co
2 Facultad del Medio Ambiente y Recursos Naturales, Universidad Distrital Francisco José de Caldas, Bogotá 111711, Colombia
3 Facultad de Ingeniería, Universidad Piloto de Colombia, Bogotá 110311, Colombia; juan-bastidas@unipiloto.edu.co
* Correspondence: harondonq@udistrital.edu.co; Tel.: +57-71-310-8687715

Abstract: Diverse types of waste are generated during the sawmilling process. One of them is pine wood shavings (PWSs). This waste can be transformed by pyrolysis into biochar (BC) to produce a sustainable material that can serve as an asphalt binder modifier. In this study, a BC produced with PWS as biomass (BC-PWS) was used to modify the properties of an asphalt cement (AC). This type of BC has not been investigated as a modifier of ACs and asphalt mixtures. Three BC-PWS contents were used for this purpose (BC/AC = 5, 10, and 15% by weight). Conventional characterization tests such as penetration, softening point, and rotational viscosity were performed on the unmodified and modified ACs. Rheological properties were also evaluated at high and intermediate temperatures, and observations were made with a scanning electron microscope (SEM). The modified BC/AC = 10% binder was chosen to manufacture a hot-mix asphalt (HMA). Marshall, indirect tensile strength—ITS, Cantabro, resilient modulus (RM), permanent deformation, and fatigue resistance (under stress-controlled mode) tests were performed on the HMA. The Tensile Strength Ratio (*TSR*) parameter was determined from ITS tests. BC-PWS is a promising material as an AC modifier from the technical-environmental point of view, which tends to enhance the evaluated properties of AC and HMA (monotonic load, rutting, fatigue, moisture damage, and raveling resistance), without resorting to increases in asphalt content or increases in mixing and compaction temperatures.

Keywords: biochar; modified asphalt binder; hot-mix asphalt; pine wood shavings

1. Introduction

Globally, agribusiness and forestry produce approximately two hundred billion tons of lignocellulosic biomass each year [1]. Within the agroindustry, wood is an important economic pillar in multiple countries. In Colombia, the residues generated in the primary and secondary wood transformation processes can vary between 40% and 65% of the total sawn wood, and it is estimated that only between 20% and 30% of the timber species is harvestable [2–4]. The capacity of Colombian sawmills is 20,000 m^3 of sawn timber/year, which would generate approximately 12,000 m^3/year of waste in each company (mainly large and medium-sized). In total, approximately 6.7 million m^3/year of waste are produced. About 4.4 million m^3/year are recovered by the board or paper industries [4]. Some types of waste generated in the primary wood transformation processes are shavings, chips, sawdust, trimmings, selvages, and bark. These residues are obtained due to poor machine maintenance, poor operator training, and lack of optimization in cutting, among others [2]. These residues are considered a productive, environmental, and economic problem due to their slow natural degradation, transportation and disposal costs, reduction in available space in wood processing centers, and environmental risks due to fires and self-combustion or the spread of pests and diseases [4,5]. Inadequate management also causes soil, water, and air pollution. Therefore, it is essential to evaluate the capacity to use these wastes in various industrial applications, in pollution remediation and management, as a source of carbon for energy storage, in the production of other products such as pulp, paper, boards, adsorbents, activated carbon, compost, plastic wood, or in the production of bioplastics, biofuels, etc. [6]. These are all alternatives that can be adopted to promote the circular economy where the waste and by-products of a production process are converted into raw materials for recirculation, generating added and sustainable value [7–9].

On the other hand, in developed countries, total energy consumption is continuously increasing, despite energy efficiency measures [10]. This is due to population growth and industry that demands higher consumption of energy and fossil resources, etc., increasing pollutant emissions to the environment [11–13]. This is also the primary cause of waste generation, which is sometimes inappropriately incinerated or deposited in open dumps. This has a significant impact on the environment and human health. One technique that allows the reuse of waste while reducing the adverse effects on the environment is pyrolysis. This technique produces a material called biochar (BC) by thermally decomposing biomasses in an inert atmosphere and the absence of oxygen at temperatures ranging from 300 °C to 900 °C [14]. BC is seen as a cheaper substitute for coal that helps reduce greenhouse gas emissions, has the function of carbon sequestration, avoids the release of CO_2 emissions, helps absorb heavy metals, and tends to inhibit and reduce volatile organic compounds [1,15–22]. It exhibits high chemical and biological stability, high specific surface area and energy density, low thermal conductivity, and flammability [23–28]. Some uses of BC include the following: (i) soil supplement in agriculture [11,29,30]; and (ii) cement substitute in mortars and concretes in construction [27,31]. Another recent, novel, and promising use of BC identified in the literature consulted is as a modifier of asphalt binders [27]. BC is a carbonaceous, porous material, highly compatible with asphalts, which tends to increase their viscosity, stiffness, rutting, and aging resistance [32–41]. Additionally, in bio-asphalts BC produces SiO_2 particles that help improve the flow-induced crystallization ability [42,43]. In general, the use of BCs as binder modifiers is recommended for warm or temperate climates [27,41]. For low-temperature climates, asphalt mixtures made with BC-modified binders generally exhibit similar or lower performance [33,34,44].

According to [27], future studies on BCs as asphalt binder modifiers should evaluate the influence of using different types of biomasses. A waste or agro-industrial biomass that is generated in enormous quantities during the sawmilling of wood for furniture production in Colombia is pine wood shavings (PWSs). PWS biomass is used for fuel or as a source of energy and soil amendment [45]. Because of its absorbent power, it has been used as litter in broiler production. In the construction and infrastructure industry, the use of PWS has been little investigated. This biomass could be converted to BC by pyrolysis [46–49]. It is important to highlight that, in the world, multiple types of BCs with different biomasses have been used to modify asphalt binders, but BC-PWS has not been studied in asphalts, and in this sense, evaluating its use could contribute to the state of knowledge on the subject. Additionally, using BCs as binder modifiers, little research has been conducted to evaluate the performance of HMAs [27]. The understanding and design of HMAs containing BC-PWS-modified binders to meet desired performance properties pose several challenges and uncertainties for the technical-academic community, which should begin to be explored and published. Based on all the aforementioned points, in the present study, a BC produced with PWS as biomass (BC-PWS) was used to modify the properties of an asphalt cement (AC) and evaluate the mechanical performance of a hot-mix asphalt (HMA) made with this modified binder. The conventional physical properties of the modified binder with three ratios (BC/AC = 5, 10, 15%, by weight) were measured and evaluated to obtain the optimum BC/AC content. Marshall, indirect tensile strength—ITS, Cantabro, resilient modulus (RM), permanent deformation, and fatigue (stress-controlled) tests were performed on the HMA manufactured with the modified binder. To evaluate the resistance to moisture damage, the Tensile Strength Ratio (*TSR*) was obtained.

2. Materials and Methods

2.1. Materials

The AC and aggregate meet the quality requirements demanded by the Colombian standards [50,51] for manufacturing HMAs (Tables 1 and 2). Both materials come from CONCRESCOL S.A. in Bogotá.

Table 1. Properties of AC 60–70.

Test	Standard	Unit	Requirement	Result
Specific gravity	AASHTO T 228	-	-	1.024
Penetration (25 °C, 100 g, 5 s)	ASTM D-5	dmm	60–70	61.5
Softening point	ASTM D-36	°C	46–54	46.6
Absolute viscosity (60 °C)	ASTM D-4402	Pa-s	140–240	175
Ductility (25 °C, 5 cm/min)	ASTM D-113	cm	Minimum 100	122
Ignition point	ASTM D-92	°C	Minimum 232	267
Mass loss	ASTM D 2872	%	Maximum 0.80	0.17

Table 2. Aggregate properties.

Test	Standard	Requirement	Result
Specific gravity of fine aggregate	AASHTO T 84	-	2.580
Absorption of fine aggregate	AASHTO T 84	-	1.35%
Specific gravity of coarse aggregate	AASHTO T 85	-	2.623
Absorption of coarse aggregate	AASHTO T 85	-	1.77%
Fractured particles (1 side)	ASTM D5821	85% minimum	92.2%
Soundness of aggregate (magnesium sulfate)	AASHTO T 104	18.0% maximum	4.8%
Plasticity index	ASTM D4318	Not plastic	Not plastic
10% of fines (dry resistance)	DNER-ME 096	100 kN minimum	266.3 kN
10% of fines (wet resistance)	DNER-ME 096	75 kN minimum	234.7 kN
Micro-Deval	AASHTO T 327	20% maximum	7.3%
Abrasion in Los Angeles machine	AASHTO T 96	25% maximum	18.0%

The BC-PWS comes from the company Control y Controles Industriales LTDA in Bucaramanga. The biomass (PWS, Figure 1a) was transformed into BC-PWS (Figure 1b) in a furnace (pyrolysis) with a production capacity of 500 kg/h. The residence time in the furnace was 20 min and the temperature was 450 °C. The BC-PWS was crushed in a laboratory mill to obtain particles smaller than 0.075 mm (passing sieve #200). Using a scanning electron microscope (SEM) type JEOL JSM-6700F, the particles were observed (between 500 and 10,000 magnifications; Figure 2a–d), their shape was evaluated, and porosity (diameter), dimensions, and elemental chemical composition were measured. The maximum size of the amorphous particles of BC-PWS varied between 20 and 70 μm, and they presented porous surface texture (pores with average diameter varying between 250 and 800 nm) and roughness. The elemental chemical composition is shown in Table 3. In the visualizations of the PWS, two types of particles were the most common (referred to as PWS1 and PWS2), with one being more carbonaceous and the other containing biomass that did not fully carbonize and retained some elements of its base composition. The BC-PWS shows high chemical compatibility with AC as it is a carbonaceous material.

Table 3. Elemental chemical composition of BC-PWS.

Sample/ Element	C	O	Si	Ca	K	Mg	Al	Fe
PWS1	92.4 ± 5.1	6.0 ± 1.1	-	0.54 ± 0.17	0.42 ± 0.15	0.27 ± 0.11	-	-
PWS2	65.2 ± 4.7	24.3 ± 3.5	2.72 ± 0.5	1.32 ± 0.4	0.74 ± 0.2	-	1.51 ± 0.23	1.0 ± 0.15

(**a**) (**b**)

Figure 1. (**a**) PWS and (**b**) BC-PWS.

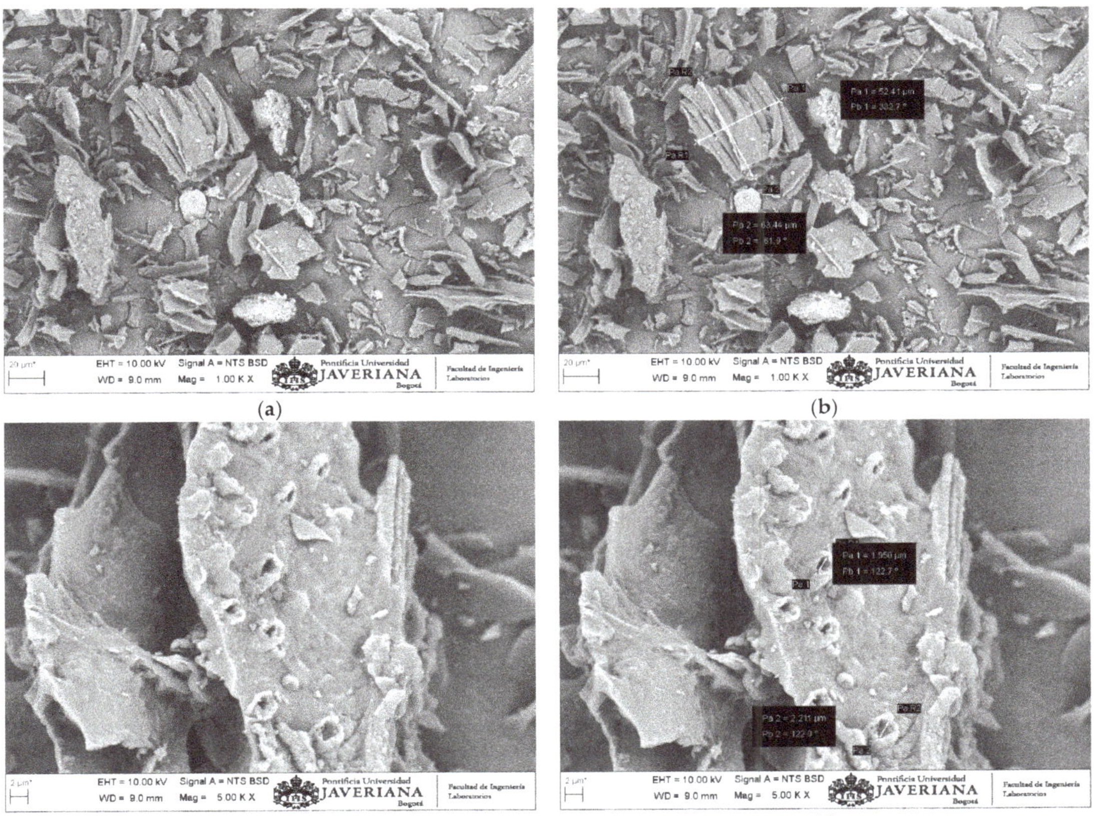

(**a**) (**b**) (**c**) (**d**)

Figure 2. SEM observations in (**a**) BC-PWS (1000× magnification); (**b**) pore measurement in BC-PWS (1000× magnification); (**c**) BC-PWS (5000× magnification); and (**d**) pore measurement in BC-PWS (5000× magnification).

2.2. Modified Asphalt and Properties

To modify the AC, three contents of BC-PWS with respect to the weight of AC were used (BC/AC = 5, 10, 15%). The AC and BC-PWS were mixed for 30 min in a high-speed shearing mixer at 150 °C and 4000 rpm. The BC-PWS contents and the modification process were chosen based on the literature review [27], and considering the physical, rheological, and chemical characterization results obtained by [41,52,53] modifying the same type of AC with two different BCs (rice husk and African palm husk as biomasses). Tests for softening point (ASTM D-36), penetration (ASTM D-5), and viscosity using a rotational viscometer (AASHTO T 316) were conducted on the unmodified control binder (BC/AC = 0%) and the modified ones (BC/AC = 5, 10, 15%). In order to determine the complex shear modulus (G^*) and phase angle (δ) to assess the performance grade (PG) at intermediate and high temperatures of the binders, a rheological characterization phase was performed using a dynamic shear rheometer (DSR), following the guidelines established by AASHTO T 315. In the DSR, the top plate oscillated at 1.59 Hz (10 rad/s) as a sine wave while the equipment measured the maximum applied stress, the resulting maximum strain, and the time lag between them. The criterion for determining the PG at elevated temperatures was $G^*/\sin\delta > 1.0$ kPa for the virgin AC, and 2.2 kPa for the short-term-aged AC after a rolling thin film oven (RTFO) procedure. For the case of PG at intermediate temperatures, the criterion was as follows: $G^*\sin\delta < 5000$ kPa for the long-term-aged AC after RTFO and the pressure aging vessel (PAV) procedure. To evaluate effects associated with the thermo-oxidative aging of the binders, the ratio between the $G^*/\sin\delta$ after RTFO and without aging, and the ratio between the $G^*\sin\delta$ after RTFO + PAV and after RTFO were calculated.

2.3. Control and Modified HMA Design

The design of the control (BC/AC = 0%) and modified (BC/AC = 10%) HMA mixes was conducted following the guidelines established by AASHTO T-245. The optimum asphalt content (OAC) was obtained according to the criteria established by [50,51] for HMA with gradation and a maximum nominal aggregate size of 19 mm (Figure 3). Twelve Marshall-type cylindrical samples per HMA type (three for each asphalt content of 4.5, 5.0, 5.5, and 6.0%) were compacted at 75 blows per face to obtain the air voids in the mix (Va), voids filled with asphalt (VFA), voids in mineral aggregates (VMA), stability (S), flow (F), and S/F ratio. S and F were measured on samples conditioned at 60 °C while applying a monotonic load in the Marshall press with a displacement rate of 50 ± 5mm/min. The compaction and mixing temperatures to make each Marshall sample (280 ± 30 cP and 170 ± 20 cP, respectively) were 145 ± 5 °C and 150 ± 3 °C, respectively. For the modified HMA, these temperatures were not changed (despite requiring an approximate increase between 5 and 10 °C) so as not to affect the environment, and age and prematurely stiffen the binder [54].

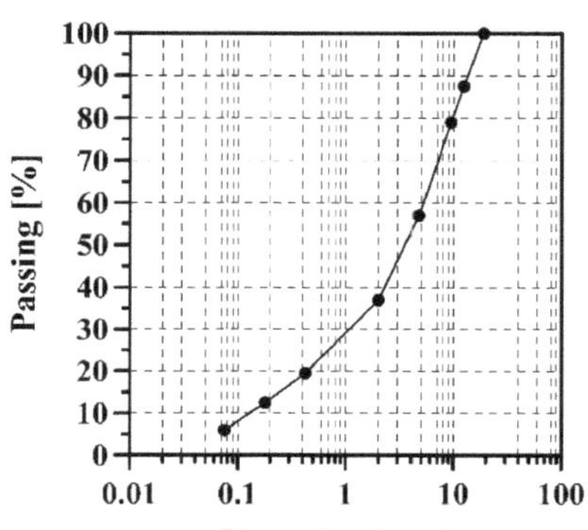

Figure 3. Particle size distribution of HMA.

2.4. Mechanical Resistance

ITS, Cantabro, resilient modulus (RM), permanent deformation under cyclic compression, and fatigue resistance tests were conducted on control (BC/AC = 0%) and modified (BC/AC = 10%) HMA.

The ITS test (AASHTO T 283) was performed on six Marshall specimens per HMA type (control and modified) manufactured with $Va = 7 \pm 0.5\%$. Three of these specimens (in a dry state) failed under monotonic loading (displacement rate of 50.8 mm/min) in the Marshall press previously conditioned at 25 °C to determine the *ITSD*. The other three failed under the same conditions (25 °C and 50.8 mm/min), but the samples were previously conditioned or immersed in water at 60 °C for one day to partially saturate them. The *ITSC* was obtained with the results of these last three samples. To evaluate the resistance to moisture damage, the *ITSC/ITSD* ratio in percentage (*TSR*) was calculated.

The Cantabro test was performed on the Los Angeles machine (without steel spheres) on three Marshall-type dry specimens per HMA type, conditioned at 20 °C. The guidelines established by Tex-245-F were followed to evaluate abrasion and raveling resistance [55,56]. The abrasion loss was measured through the parameter *CL* (Cantabro Loss) obtained with Equation (1), where *mi* is the initial mass of the sample and *mf* is the final mass of the sample after 500 revolutions in the LA machine.

$$CL = \frac{mi - mf}{mi} \cdot 100 \quad (1)$$

RM, permanent deformation, and fatigue resistance tests were conducted on a Nottingham Asphalt Tester (NAT). The RM was obtained following the guidelines established by UNE-EN 12697-26 and was calculated by averaging the results obtained on three Marshall-type samples per mix type. The repeated load was applied using frequencies of 2.5, 5.0, and 10.0 Hz and an indirect tensile "half-sine" load. The test temperatures were 10, 20, and 30 °C. To evaluate the resistance to permanent deformation, the guidelines established by UNE-EN 12697-25 were followed. The results obtained on three samples per type of mix conditioned at a temperature of 40 °C were averaged. The repeated load, whose frequency was 0.5 Hz, was square wave load type, and the magnitude of the cyclic axial stress was 100 kPa. The accumulated axial vertical displacement was measured with Linear Variable Differential Transformers (LVDT). Fatigue resistance was conducted following the guidelines established by UNE-EN 12697-24. Nine Marshall-type specimens were manufactured per type of mix to perform the test at laboratory temperature (20 °C). The type of loading was in indirect tension under stress-controlled mode (stresses—σ varying between 200 and 500 kPa) and the loading frequency was 10 Hz. The number of failure cycles (*Nf*) of each specimen was measured under total specimen rupture. The fatigue curves (Wöhler's typeσ-N_f curve) were represented mathematically by Equation (2). *a* and *b* (slope of the fatigue law) are parameters obtained by regression. To facilitate the evaluation of fatigue resistance, the cyclic stresses σ_5, σ_6, σ_7 were obtained from Equation (2) so that the HMAs would fail at $N_f = 10^5$, 10^6 and 10^7, respectively.

$$\sigma = a\left(N_f\right)^b \quad (2)$$

To evaluate whether BC-PWS produced significant changes in the measured mechanical properties, an ANOVA analysis of variance with 95% confidence (significance level of 0.05) was performed.

3. Results

3.1. Modified Asphalt Properties

The properties of the modified asphalt are shown in Figure 4 and Table 4. BC-PWS significantly increased the stiffness and rutting resistance of AC 60–70 (increased viscosity and softening point, increased PG from 64 to 70 °C with BC/AC $\geq$ 10%, and decreased penetration). Equivalent results have been reported in the reference literature employing other types of BCs as modifiers (e.g., [32–34,39–41,57–59]). Chemically speaking, the high porosity of BC-PWS could preferentially absorb maltenes (light-weight or low-molecular-weight fraction of the binder) making asphaltenes (high molecular weight) prevail, contributing to increasing the viscosity and stiffness of the binder [22,34]. In addition, the amorphous, rough, and porous structure of BC-PWS could form a stiffness zone in the binder [34]. On the other hand, BC did not generate changes in the PG of AC at intermediate temperatures (PG was maintained at 22 °C), indicating that BC-PWS does not decrease fatigue cracking resistance. On the other hand, the ratio $G^*/\sin\delta$ after RTFO and without aging had average magnitudes of 2.02, 1.64, 1.63, and 1.84 for BC/AC = 0, 5, 10, 10%, and 15%, respectively, while the ratio between $G^*\sin\delta$ after RTFO + PAV and after RTFO was 2070, 1296, 1079, and 982, respectively. The lower values in these ratios and indicators show that BC-PWS could contribute to improving the aging resistance of the base AC, since it experiences lower stiffness when subjected to thermo-oxidative processes [60]. According to [61], BC contributes to retard the thermo-oxidative and ultra-violet (UV) aging of binders. Also, some BCs contribute to providing aliphatic and aromatics to binders [43] and could modify flow-induced crystallization by improving aging resistance [42].

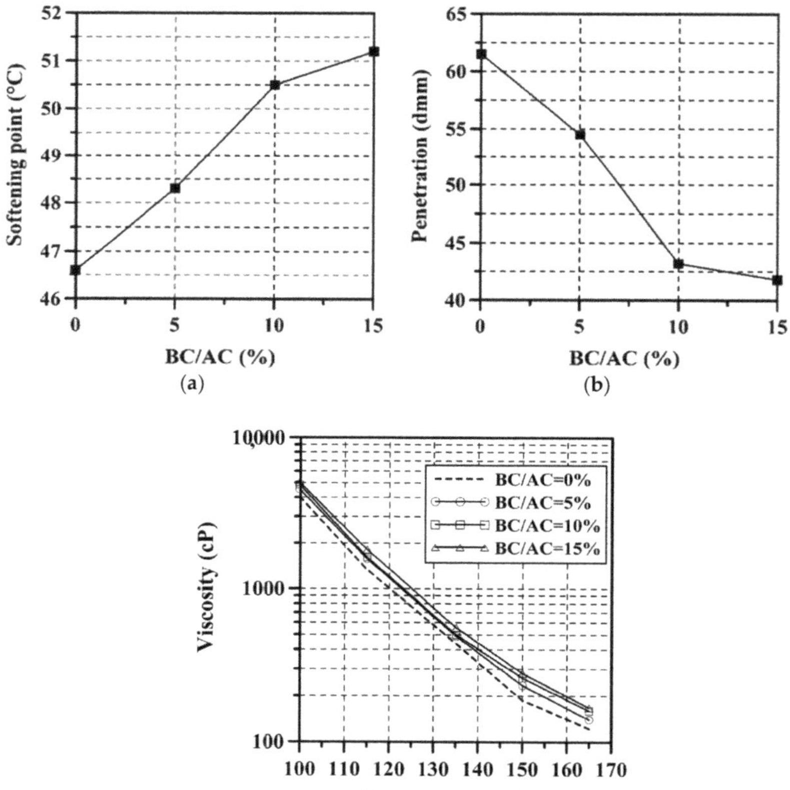

Figure 4. (**a**) Softening point, (**b**) penetration, and (**c**) rotational viscosity test results.

Table 4. Rheology test results.

Temperature (°C)	BC/AC (%)			
	0	5	10	15
	|G*|/sinδ (kPa)—Virgin asphalt binder			
58	2.42	4.69	4.90	4.90
64	1.11	2.06	2.97	2.99
70	0.55	1.22	1.51	1.56
	|G*|/sinδ (kPa)—After RTFO test			
58	4.75	7.75	8.66	9.21
64	2.26	3.32	4.39	4.84
70	1.14	2.02	2.48	3.14
	|G*| sinδ (kPa)—After RTFOT + PAV			
16	6253	6086	6151	6678
19	4618	4621	4644	4919
22	3252	3444	3643	3787

Considering the literature review and the results shown in Figure 4 and Table 4, BC/AC = 10% was chosen as the modified binder to continue the following experimental phases and manufacture the modified HMA. BC/AC = 15% was not chosen, since it makes it difficult to mix BC-PWS with AC. Additionally, lumps are formed in the binder and the magnitude of the measured parameters is similar with respect to BC/AC = 10%. SEM observations and measurements of the elemental chemical composition at more than 25 randomly chosen points were conducted on this modified binder. Images in this regard are not presented because only a dark black background color characteristic of unmodified AC was observed, which is an indicator of low lump formation and good dispersion of BC-PWS in the binder. In addition, the modified BC/AC = 10% binder showed a similar chemical composition to BC/AC = 0%, demonstrating the high compatibility that exists between BC-PWS and the binder. In some of the random spots, chemical composition like the BC-PWS was obtained mainly because it was measured in some nanometer particles of BC-PWS found in the binder.

3.2. Design of HMAs

The Marshall test results for the control (BC/AC = 0%) and modified (BC/AC = 10%) HMA mix designs are presented in Figure 5. The OAC of both HMAs is 5.5%. At either binder content, the mixtures exhibit similar volumetric composition (Figure 5a–c); i.e., BC-PWS did not generate statistically significant changes in *Va*, *VMA*, and *VFA* (Table 5; $F_T < F_{0.05} = 7.71$). The modified HMA mixture exhibits higher strength when the asphalt content is equal to or higher than the OAC. In the OAC, BC-PWS contributes to increase the strength under monotonic loading (*S* and *S/F* ratio) by approximately 15.7 and 9.1% (Figures 5d and 5f, respectively). These increases in strength were statistically significant (Table 5; $F_T > F_{0.05} = 7.71$) and are due to the increase in stiffness exhibited by the binder when BC-PWS was added (Figure 4 and Table 4). Higher *S* and *S/F* values are indirect indicators of higher resistance of the modified HMA in elevated temperature climates and higher rutting resistance.

Table 5. ANOVA analysis—Marshall test results.

AC (%)	*Va*	*VMA*	*VFA*	*S*	*F*	*S/F*
	F_T					
4.5	2.93	1.77	3.86	0.14	0.50	0.41
5.0	3.40	6.42	2.29	0.48	1.00	0.07
5.5	0.01	0.13	0.04	21.07	5.00	11.39
6.0	0.04	0.66	0.15	12.70	1.00	13.29

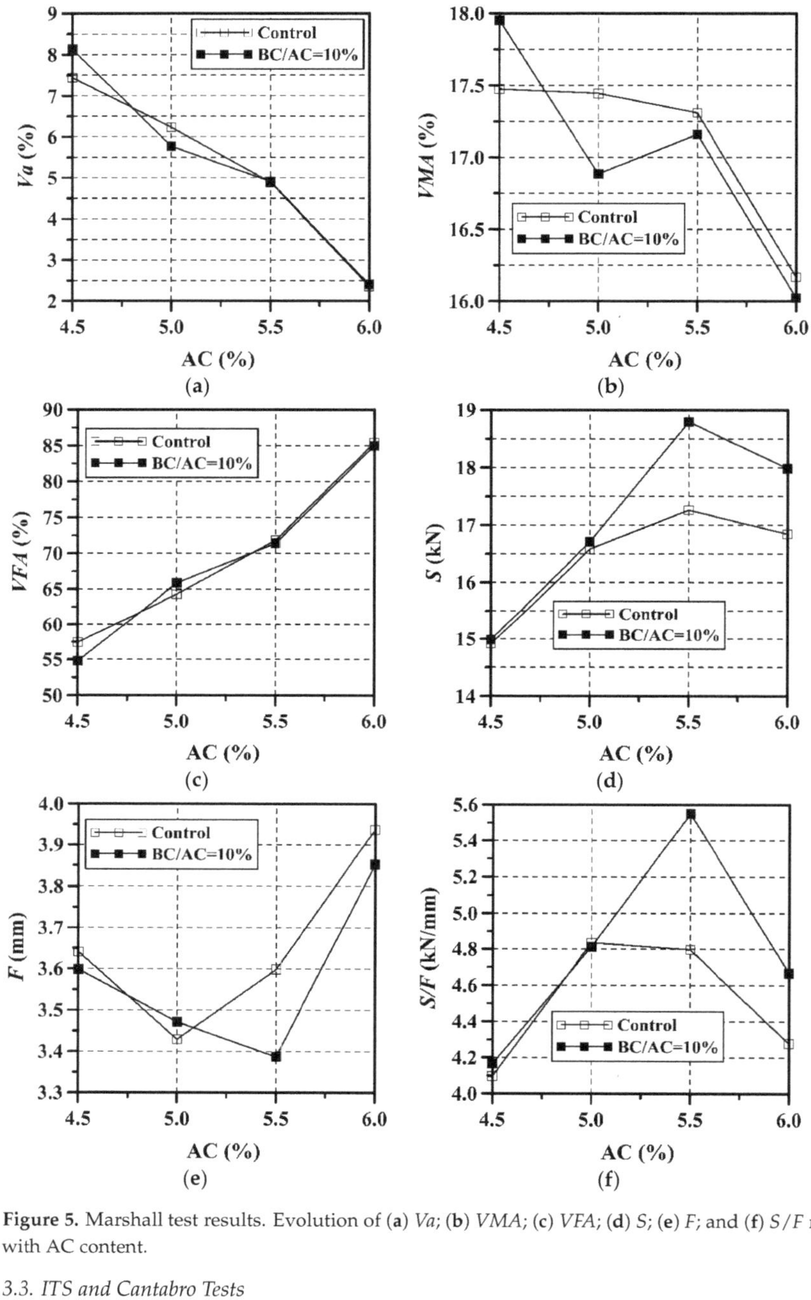

Figure 5. Marshall test results. Evolution of (**a**) *Va*; (**b**) *VMA*; (**c**) *VFA*; (**d**) *S*; (**e**) *F*; and (**f**) *S*/*F* ratio with AC content.

3.3. ITS and Cantabro Tests

The *ITSD*, *ITSC*, and *CL* parameters obtained from the ITS and Cantabro test correlate with aspects associated with the durability, adhesion, cohesion, abrasion resistance, and raveling of the mixes. On the other hand, the *TSR* parameter is used to evaluate resistance to moisture damage. The results of the ITS and Cantabro test are presented in Figures 6 and 7, respectively. The BC-PWS contributes to increasing the *ITSD* and *ITSC* parameters. The modified HMA mixture exhibits a 13.3 and 20.3% increase in these parameters over the control HMA mixture, and both increases were statistically significant ($F_T = 8.53 > F_{0.05} = 7.71$ for *ITSD* y $F_T = 60.7 > F_{0.05} = 7.71$ for *ITSC*). That is, BC-PWS helped to improve the indirect tensile strength of HMA in dry and wet states. This is consistent with the *TSR* values = 75.6% and 80.2% of the control and modified HMA, respectively. It is important to mention that the control HMA does not meet the minimum value of *TSR* = 80% recommended by [50,51]; the opposite case occurs with the modified HMA, which does meet this requirement. In summary, BC-PWS helps to improve the adhesion of the AC with the aggregates (in both dry and wet states) and contributes to increasing the resistance to moisture damage. These results are like those reported with other types of BC [27,59,62]. On the other hand, modified HMA generated a slight decrease in *CL*. However, this decrease was not statistically significant with respect to the control mixture ($F_T = 0.015 < F_{0.05} = 7.71$). In other words, the Cantabro abrasion and raveling resistance of both mixtures is similar (BC-PWS did not help to significantly decrease the *CL*).

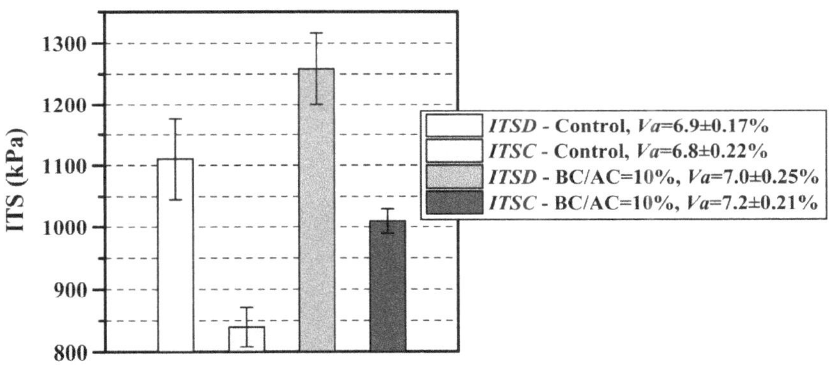

Figure 6. *ITSD* and *ITSC* parameters (ITS test).

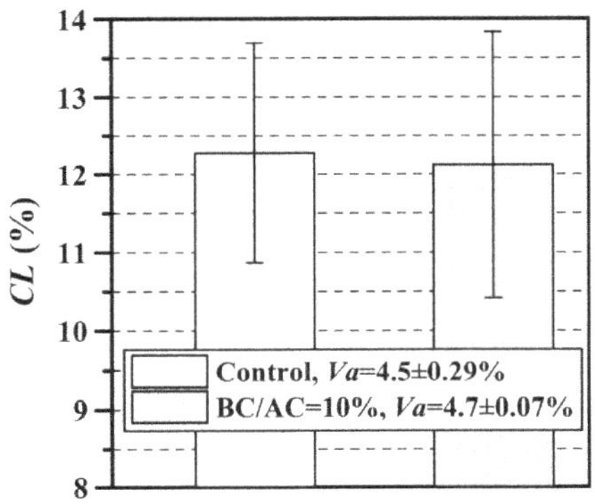

Figure 7. Cantabro loss (*CL* parameter).

3.4. RM and Permanent Deformation Tests

The RM values and permanent deformation test results of the control and modified HMA are presented in Figures 8 and 9, respectively. The BC-PWS contributes to increasing the RM of the HMA. The approximate increase in stiffness of the modified HMA over the control HMA was 17.0 ± 2.3%, 19.2 ± 1.4%, and 19.9 ± 0.8% for temperatures of 10, 20, and 30 °C, respectively. This increase in the RM of the modified mixtures was statistically significant ($F_T > F_{0.05}$; Table 6) generating an increase in the resistance to permanent deformation (Figure 9). At 3600 cycles of loading (7600 s), the control HMA showed a

0.29 mm higher displacement compared to the modified HMA, and this was statistically significant ($F_T = 8.45 > F_{0.05} = 7.71$). Additionally, the displacement rate measured between 1800 and 3600 s of loading was 48×10^{-6} and 42.8×10^{-6} mm/cycle for the control and modified HMA, respectively. That is, the BC-PWS contributes to increasing the stiffness under cyclic loading and rutting resistance of the HMA. These results agree with the increase in stiffness of AC with BC-PWS observed in softening point, penetration, viscosity, and rheology tests (Figure 4 and Table 4). They also correlate with the results found in the reference literature using other types of BC (e.g., [27,28,62–64]), with the *S/F* ratio of the Marshall test, and with the results of the ITS test (the stiffness of the mixtures tends to increase with the increase in ITS) [65]. In addition, the amorphous geometry and rough and porous surface texture of BC-PWS contributed to the strengthening of the binder microstructure, increasing the cohesion of the HMA [38,62] and of the asphalt mastic [66].

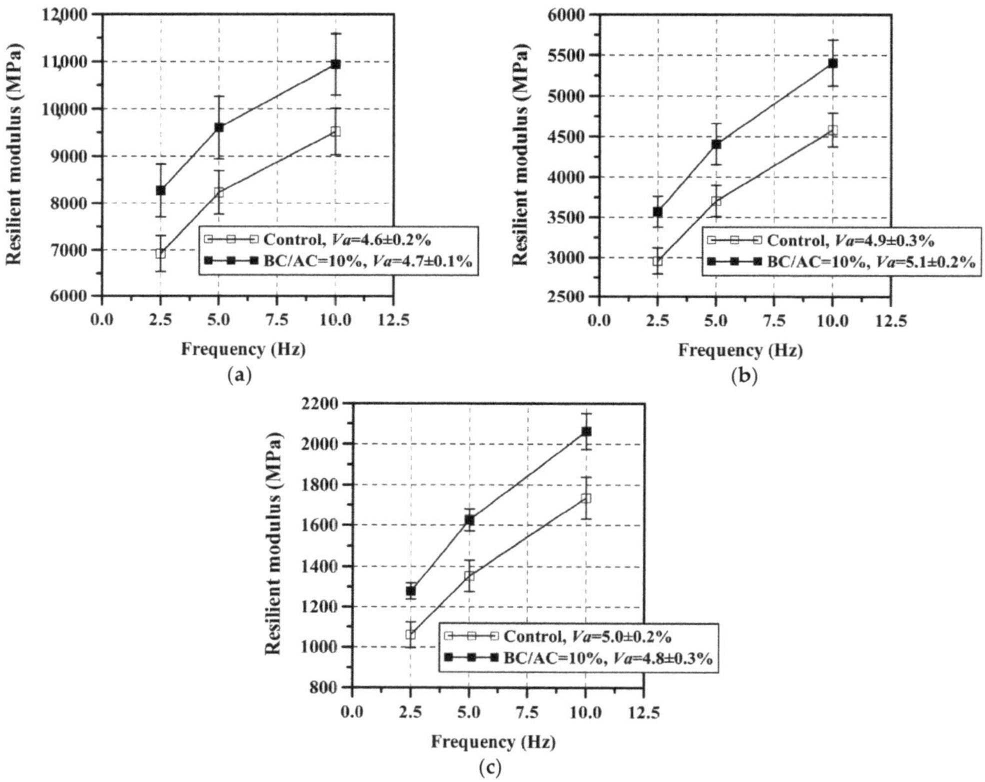

Figure 8. RM test. (**a**) 10 °C, (**b**) 20 °C, (**c**) 30 °C.

Table 6. ANOVA—Resilient modulus test.

Comparison	10 °C			20 °C			30 °C		
	2.5 Hz	5 Hz	10 Hz	2.5 Hz	5 Hz	10 Hz	2.5 Hz	5 Hz	10 Hz
Control vs. BC/AC = 10%	11.85	8.75	9.18	17.85	14.51	16.60	25.65	25.34	17.51

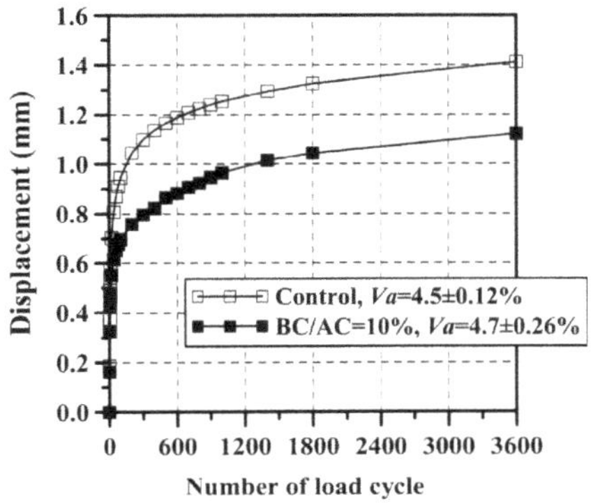

Figure 9. Permanent deformation test results.

3.5. Fatigue Test

The curves showing the fatigue results under stress-controlled mode are shown in Figure 10. The parameters a, b, σ_5, σ_6, and σ_7 obtained from these curves using Equation (2) are shown in Table 7. The higher fatigue resistance is exhibited by the modified HMA (exhibits higher values of a, σ_5, σ_6, and σ_7 and similar magnitude of b with respect to the control HMA). This is due to the higher stiffness of the BC-PWS modified binder and to the increase in the *ITSD, ITSC,* and RM parameters of the HMA due to the presence of this biochar. Generally, HMAs exhibit greater fatigue resistance under stress-controlled conditions when both ITS and RM increase [65,67–70].

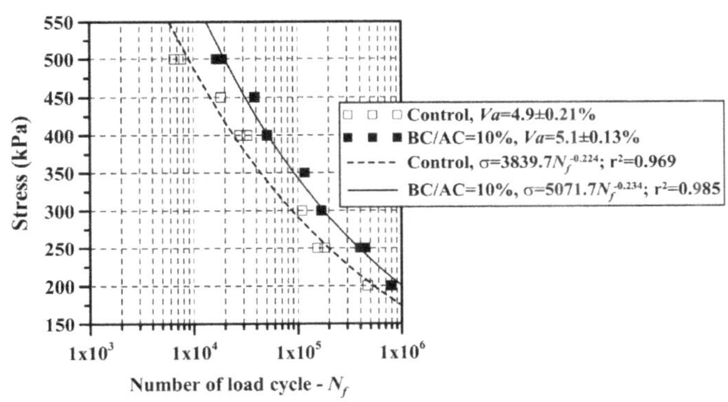

Figure 10. Fatigue test results.

Table 7. a, b, σ_5, σ_6, and σ_7 parameters.

Mixture	*VTM* (%)	R-Squared (r^2)	a	b	σ_5 (kPa)	σ_6 (kPa)	σ_7 (kPa)
Control	4.71 ± 0.11	0.969	3839.7	−0.224	291.3	173.9	103.8
BC/AC = 10%	4.80 ± 0.09	0.985	5071.7	−0.234	342.9	200.1	116.7

4. Conclusions

This study evaluated the mechanical strength of an HMA fabricated with a BC-modified AC obtained by pyrolysis using PWS as biomass. Considering the results obtained, the following conclusions can be drawn:

- Using 10% BC-PWS, the asphalt binder exhibited good performance, and the BC/AC = 10% ratio was defined as optimal.
- Concerning the unmodified AC, the modified binder exhibited higher stiffness and PG at high temperatures, maintaining the PG at intermediate temperatures. Also, the trend in the rheology test showed a modified binder that was more resistant to thermo-oxidative aging.
- BC-PWS shows high compatibility with AC, did not generate lumps in the binder, and did not hinder the mixing and compaction process of the HMA (under the same mixing and compaction temperatures, there was no loss of workability, and the volumetric composition was maintained).
- The modified HMA (BC/AC = 10%) exhibited higher stiffness under monotonic (higher S and S/F ratio) and cyclic loading (increased resilient modulus), and higher resistance to rutting. In other words, it is recommended for use in areas where elevated temperature climate prevails.
- The modified HMA showed better performance by increasing the parameters $ITSD$, $ITSC$, a, σ_5, σ_6, and σ_7 (increased indirect tensile strength and fatigue resistance under stress-controlled mode). This is an indicator of reliable performance in temperate climates and thick asphalt layers.
- The modified HMA exhibited higher resistance to moisture damage (increased TSR parameter), and similar abrasion and raveling resistance (decreased CL parameter), which are aspects associated with durability.
- All changes in the parameters measured and evaluated were statistically significant according to ANOVA analysis, except for the volumetric composition of HMA and CL parameter.
- BC-PWS is considered a promising material and a potential asphalt binder modifier. It has the potential to contribute to the circular economy. From a technical perspective, it can demonstrate reliable performance and durability in both temperate and elevated temperature climates. Additionally, it may eliminate the need to increase the OAC and the mixing and compaction temperatures.

In future research work, it is recommended to deepen the chemical characterization of the modified binder (e.g., Fourier Transform Infrared Spectroscopy—FTIR and Saturate, Aromatic, Resin, and Asphaltene—SARA tests), evaluate properties at low temperatures, evaluate the influence using different ACs, PWS, aggregates, and asphalt mixtures, and perform environmental and economic impact simulations, as well as LCA and LCCA (Life Cycle Assessment and Life Cycle Cost Analysis, respectively) modeling.

Author Contributions: Conceptualization, H.A.R.-Q. and S.B.C.-P.; methodology, H.A.R.-Q., S.B.C.-P., and J.G.B.-M.; validation, H.A.R.-Q., J.G.B.-M., and S.B.C.-P.; formal analysis, H.A.R.-Q., J.G.B.-M., and S.B.C.-P.; investigation, H.A.R.-Q., S.B.C.-P., and J.G.B.-M.; resources, H.A.R.-Q., S.B.C.-P., and J.G.B.-M.; data curation, S.B.C.-P., J.G.B.-M., and H.A.R.-Q.; writing—original draft H.A.R.-Q.; writing—review and editing, H.A.R.-Q., S.B.C.-P., and J.G.B.-M.; visualization, H.A.R.-Q.; supervision, S.B.C.-P.; project administration, S.B.C.-P.; funding acquisition, S.B.C.-P. All authors have read and agreed to the published version of the manuscript.

Funding: This research received no external funding.

Institutional Review Board Statement: Not applicable.

Informed Consent Statement: Not applicable.

Data Availability Statement: Further inquiries can be directed to the corresponding authors.

Acknowledgments: The authors express their gratitude to the institutions that supported the present research project conducted under Special Cooperation Agreement No. INV-DIS-3769 of 2023, signed between the Universidad Militar Nueva Granada, the Universidad Distrital Francisco José de Caldas, and the Universidad Piloto de Colombia. Special thanks are also expressed to the company Control y Controles Industriales LTDA for supplying the biochar (BC-PWS) and to the company CONCRESCOL S.A. for supplying the materials for the HMA (AC and aggregates).

Conflicts of Interest: The authors declare no conflicts of interest.

References

1. Zhou, H.; Shen, Y.; Zhang, N.; Liu, Z.; Bao, L.; Xia, Y. Wood fiber biomass pyrolysis solution as a potential tool for plant disease management: A review. *Heliyon* **2024**, *10*, e25509. [CrossRef] [PubMed]
2. Barua, A.; Tarek, A.; Chowdhury, A.; Mehidi, S.H.; Muhiuddin, H.M. Residue reduction and reuse in wooden furniture manufacturing industry. *Int. J. Sci. Eng. Res.* **2014**, *5*, 291–301.
3. Erhabor, T.A. Evaluation of wood wastes from sawmilling operations in Benin City, Edo State. *J. For. Res. Manag.* **2015**, *12*, 113–124.
4. López, L.T.; Vega, L.Y.; Rendón, C.D.; Tobón, S. *Aprovechamiento de Residuos Madereros: Cuantificación, Caracterización y Valorización*, 1st ed.; Corporación Universitaria Remington: Medellín, Colombia; Servicio Nacional de Aprendizaje (SENA): Medellín, Colombia, 2019; 344p.
5. Sarmiento-Oviedo, M.B. Alternativas de compostaje de aserrin de pino caribe (Pinus caribaea) en la industria maderera Refocosta S.A., municipio de Villanueva, Casanare, Colombia. *Rev. Investig. Agrar. Ambient.* **2011**, *2*, 21–32. [CrossRef]
6. Mallakpour, S.; Sirous, F.; Hussain, C.M. Sawdust, a versatile, inexpensive, readily available bio-waste: From mother earth to valuable materials for sustainable remediation technologies. *Adv. Colloid Interface Sci.* **2021**, *295*, 102492. [CrossRef] [PubMed]
7. Boloy, R.A.M.; da Cunha Reis, A.; Rios, E.M.; Martins, J.; Soares, L.O.; de Sá Machado, V.A.; de Moraes, D.R. Waste-to-energy technologies towards circular economy: A systematic literature review and bibliometric analysis. *Water Air Soil Poll.* **2021**, *232*, 306. [CrossRef]
8. Silva, U.V.; Nascimento, M.F.; Oliveira, P.R.; Panzera, T.H.; Rezende, M.O.; Silva, D.A.L.; de Moura Aquino, V.B.; Lahr, F.A.R.; Christoforo, A.L. Circular vs. linear economy of building materials: A case study for particleboards made of recycled wood and biopolymer vs. conventional particleboards. *Constr. Build. Mater.* **2021**, *285*, 122906. [CrossRef]
9. Cheah, W.Y.; Siti-Dina, R.P.; Leng, S.T.K.; Er, A.C.; Show, P.L. Circular bioeconomy in palm oil industry: Current practices and future perspectives. *Environ. Technol. Inno.* **2023**, *30*, 103050. [CrossRef]
10. Olsthoorn, D.; Haghighat, F.; Moreau, A.; Lacroix, G. Abilities and limitations of thermal mass activation for thermal comfort, peak shifting and shaving: A review. *Built. Environ.* **2017**, *118*, 113–127. [CrossRef]
11. Mensah, R.A.; Shanmugam, V.; Narayanan, S.; Razavi, S.M.J.; Ulfberg, A.; Blanksvärd, T.; Sayahi, F.; Simonsson, P.; Reinke, B.; Försth, M.; et al. Biochar-Added cementitious materials—A review on mechanical, thermal, and environmental properties. *Sustainability* **2021**, *13*, 9336. [CrossRef]
12. Legan, M.; Gotvajn, A.Ž.; Zupan, K. Potential of biochar use in building materials. *J. Environ. Manag.* **2022**, *309*, 114704. [CrossRef] [PubMed]
13. Yaro, N.S.A.; Sutanto, M.H.; Habib, N.Z.; Napiah, M.; Usman, A.; Jagaba, A.H.; Al-Sabaeei, A.M. Application and circular economy prospects of palm oil waste for eco-friendly asphalt pavement industry: A review. *J. Road Eng.* **2022**, *2*, 309–331. [CrossRef]
14. Kochanek, J.; Soo, R.M.; Martinez, C.; Dakuidreketi, A.; Mudge, A.M. Biochar for intensification of plant-related industries to meet productivity, sustainability and economic goals: A review. *Resour. Conserv. Recycl.* **2022**, *179*, 106109. [CrossRef]
15. Zhou, X.; Moghaddam, T.B.; Chen, M.; Wu, S.; Adhikari, S. Biochar removes volatile organic compounds generated from asphalt. *Sci. Total Environ.* **2020**, *745*, 141096. [CrossRef]
16. Zhou, Y.; Shen, C.; Wang, T.; Xue, Y. Inhibition effect of three types of biochar on volatile organic compounds from asphalt: Revealing chemical adsorption as the primary mechanism. *Constr. Build. Mater.* **2024**, *411*, 134322. [CrossRef]
17. Marzeddu, S.; Cappelli, A.; Ambrosio, A.; D'ecima, M.A.; Viotti, P.; Boni, M.R. A life cycle assessment of an energy-biochar chain involving a gasification plant in Italy. *Land* **2021**, *10*, 1256. [CrossRef]
18. Chen, L.; Zhou, T.; Yang, J.; Qi, J.; Zhang, L.; Liu, T.; Dai, S.; Zhao, Y.; Huang, Q.; Liu, Z.; et al. A review on the roles of biochar incorporated into cementitious materials: Mechanisms, application and perspectives. *Constr. Build. Mater.* **2023**, *409*, 134204. [CrossRef]
19. Aprianti, N.; Kismanto, A.; Supriatna, N.K.; Yarsono, S.; Nainggolan, L.M.T.; Purawiardi, R.I.; Fariza, O.; Ermada, F.J.; Zuldian, P.; Raksodewanto, A.A.; et al. Prospect and challenges of producing carbon black from oil palm biomass: A review. *Bioresour. Technol. Rep.* **2023**, *23*, 101587. [CrossRef]
20. Torres-Morales, E.; Khatiwada, D.; Xylia, M.; Johnson, F.X. Investigating biochar as a net-negative emissions strategy in Colombia: Potentials, costs, and barriers. *Curr. Res. Environ. Sus.* **2023**, *6*, 100229. [CrossRef]
21. Ge, L.; Yao, Y.; Xu, L.; Zhou, Z.; Li, J.; Zhang, X.; Liu, C.; Lv, H. Mitigation of asphalt volatile organic compounds emissions and health hazards using a TiO2-doped biochar composite: Microscopic and physiological insights. *Environ. Technol. Inno.* **2024**, *36*, 103763. [CrossRef]
22. Yaphary, Y.L.; He, M.; Lu, G.; Zou, F.; Liu, P.; Tsang, D.C.W.; Leng, Z. Experiment and multiscale molecular simulations on the Cu absorption by biochar-modified asphalt: An insight into removal capability and mechanism of heavy metals from stormwater runoff. *Chem. Eng. J.* **2023**, *462*, 142205. [CrossRef]
23. Spokas, K.A. Review of the stability of biochar in soils: Predictability of O: C molar ratios. *Carbon Manag.* **2010**, *1*, 289–303. [CrossRef]

24. Gupta, S.; Kua, H.W. Factors determining the potential of biochar as a carbon capturing and sequestering construction material: Critical review. *J. Mater. Civ. Eng.* **2017**, *29*, 04017086. [CrossRef]
25. Restuccia, L.; Ferro, G.A.; Suarez-Riera, D.; Sirico, A.; Bernardi, P.; Belletti, B.; Malcevschi, A. Mechanical characterisation of different biochar-based cement composites. *Procedia Struct. Integr.* **2020**, *25*, 226–233. [CrossRef]
26. Maljaee, H.; Madadi, R.; Paiva, H.; Tarelho, L.; Ferreira, V.M. Incorporation of biochar in cementitious materials: A roadmap of biochar selection. *Constr. Build. Mater.* **2021**, *283*, 122757. [CrossRef]
27. Rondón-Quintana, H.A.; Reyes-Lizcano, F.A.; Chaves-Pabón, S.B.; Bastidas-Martínez, J.G.; Zafra-Mejía, C.A. Use of biochar in asphalts: Review. *Sustainability* **2022**, *14*, 4745. [CrossRef]
28. He, L.; Tao, M.; Liu, Z.; Cao, Z.; Zhu, J.; Gao, J.; Van den bergh, W.; Chailleux, E.; Huang, Y.; Vasconcelos, K.; et al. Biomass valorization toward sustainable asphalt pavements: Progress and prospects. *Waste Manag.* **2023**, *165*, 159–178. [CrossRef]
29. Zakaria, M.R.; Farid, M.A.A.; Andou, Y.; Ramli, I.; Hassan, M.A. Production of biochar and activated carbon from oil palm biomass: Current status, prospects, and challenges. *Ind. Crops Prod.* **2023**, *199*, 116767. [CrossRef]
30. Zhu, H.; An, Q.; Nasir, A.S.M.; Babin, A.; Saucedo, S.L.; Vallenas, A.; Li, L.; Baldwin, S.A.; Lau, A.; Bi, X. Emerging applications of biochar: A review on techno-environmental-economic aspects. *Bioresour. Technol.* **2023**, *388*, 129745. [CrossRef]
31. Singhal, S. Biochar as a cost-effective and eco-friendly substitute for binder in concrete: A review. *Eur. J. Environ. Civ. Eng.* **2023**, *27*, 984–1009. [CrossRef]
32. Çeloğlu, M.E.; Mehmet, Y.; Kök, B.V.; Yalçin, E. Effects of various biochars on the high temperature performance of bituminous binder. In Proceedings of the 6th Euraphalt & Eurobitume Congress, Prague, Czech Republic, 1–3 June 2016. [CrossRef]
33. Gan, X.; Zhang, W. Application of biochar from crop straw in asphalt modification. *PLoS ONE* **2021**, *16*, e0247390. [CrossRef] [PubMed]
34. Ma, F.; Dai, J.; Fu, Z.; Li, C.; Wen, Y.; Jia, M.; Wang, Y.; Shi, K. Biochar for asphalt modification: A case of high-temperature properties improvement. *Sci. Total Environ.* **2022**, *804*, 150194. [CrossRef] [PubMed]
35. Wu, Y.; Cao, P.; Shi, F.; Liu, K.; Wang, X.; Leng, Z.; Tan, Z.; Zhou, C. Modeling of the complex modulus of asphalt mastic with biochar filler based on the homogenization and random aggregate distribution methods. *Adv. Mater. Sci. Eng.* **2020**, *2020*, 2317420. [CrossRef]
36. Walters, R.C.; Fini, E.H.; Abu-Lebdeh, T. Enhancing asphalt rheological behavior and aging susceptibility using biochar and Nano-clay. *Am. J. Appl. Sci.* **2014**, *7*, 66–76. [CrossRef]
37. Walters, R.C.; Begum, S.A.; Fini, E.H.; Abu-Lebdeh, T. Investigating bio-char as flow modifier and water treatment agent for sustainable pavement design. *Am. J. Appl. Sci.* **2015**, *8*, 138–146. [CrossRef]
38. Zhang, R.; Dai, Q.; You, Z.; Wang, H.; Peng, C. Rheological performance of bio-char modified asphalt with different particle sizes. *Appl. Sci.* **2018**, *8*, 1665. [CrossRef]
39. Zhang, R.; Wang, H.; Ji, J.; Wang, H. Viscoelastic properties, rutting resistance, and fatigue resistance of Waste Wood-based biochar-modified asphalt. *Coatings* **2022**, *12*, 89. [CrossRef]
40. Zhou, X.; Moghaddam, T.B.; Chen, M.; Wu, S.; Zhang, Y.; Zhang, X.; Adhikari, S.; Zhang, X. Effects of pyrolysis parameters on physicochemical properties of biochar and bio-oil and application in asphalt. *Sci. Total Environ.* **2021**, *780*, 146448. [CrossRef]
41. Rondón-Quintana, H.A.; Romero-Patiño, N.E.; Bastidas-Martínez, J.G. Performance properties of a hot-mix asphalt modified with oil palm kernel shell based biochar for road pavements. *J. Transp. Eng. B Pavements*, 2025; *accepted to publish*. [CrossRef]
42. Zhou, X.; Adhikari, S. Flow-induced crystallization of biochar in bio-asphalt under various aging conditions. *Sci. Total Environ.* **2019**, *695*, 133943. [CrossRef]
43. Zhou, X.; Zhao, G.; Wu, S.; Tighe, S.; Pickel, D.; Chen, M.; Adhikari, S.; Gao, Y. Effects of biochar on the chemical changes and phase separation of bio-asphalt under different aging conditions. *J. Clean. Prod.* **2020**, *263*, 121532. [CrossRef]
44. Dong, W.; Ma, F.; Li, C.; Fu, Z.; Huang, Y.; Liu, J. Evaluation of anti-aging performance of biochar modified asphalt binder. *Coatings* **2020**, *10*, 1037. [CrossRef]
45. Thengane, S.K.; Kung, K.S.; Gupta, A.; Ateia, M.; Sanchez, D.L.; Mahajani, S.M.; Lim, C.J.; Sokhansanj, S.; Ghoniem, A.F. Oxidative torrefaction for cleaner utilization of biomass for soil amendment. *Clean. Eng. Technol.* **2020**, *1*, 100033. [CrossRef]
46. Linhoss, J.E.; Mohammadi-Aragh, M.K.; Evans, J.D. Influence of pine and miscanthus biochar on the water activity and moisture sorption isotherms of used broiler litter. *Heliyon* **2023**, *9*, e22618. [CrossRef]
47. Maaoui, A.; Trabelsi, A.B.H.; Abdallah, A.B.; Chagtmi, R.; Lopez, G.; Cortazar, M.; Olazar, M. Assessment of pine wood biomass wastes valorization by pyrolysis with focus on fast pyrolysis biochar production. *J. Energy Inst.* **2023**, *108*, 101242. [CrossRef]
48. Sharma, A.K.; Ghodke, P.K.; Goyal, N.; Bobde, P.; Kwon, E.E.; Lin, K.A.; Chen, W. A critical review on biochar production from pine wastes, upgradation techniques, environmental sustainability, and challenges. *Bioresour. Technol.* **2023**, *387*, 129632. [CrossRef]
49. Chen, X.; Li, S.; Lin, Y.; Gao, S.; Liu, J.; Xue, Y.; Gao, X. Pyrolysis and catalytic upgrading of pine wood sawdust over modified biochar catalyst in a tandem microreactor. *J. Anal. Appl. Pyrol.* **2024**, *181*, 106607. [CrossRef]
50. INVIAS–Instituto Nacional de Vías. *Especificaciones Generales de Construcción de Carreteras*; INVIAS–Instituto Nacional de Vías: Bogotá, Colombia, 2022.
51. IDU–Instituto de Desarrollo Urbano. *Especificaciones Técnicas Generales de Materiales y Construcción, para Proyectos de Infraestructura Vial y de Espacio Público, para Bogotá D.C.*; IDU—Instituto de Desarrollo Urbano: Bogotá, Colombia, 2018.

52. Romero-Patiño, N.E. Evaluación de la Resistencia Mecánica de una Mezcla Densa en Caliente MDC-19 Fabricada con un Cemento Asfáltico (CA) Modificado con Biocarbón (BC). Master's Thesis, Universidad Distrital Francisco José de Caldas, Bogotá, Colombia, 2024.
53. Huari, R.A. Evaluación del Desempeño Físico-Químico y Reológico de un Asfalto Convencional Modificado por vía Húmeda con Biochar de Palma Africana y Cascara de Arroz. Master's Thesis, Pontificia Universidad Javeriana, Bogotá, Colombia, 2024.
54. Rincón-Estepa, J.A.; González-Salcedo, E.V.; Rondón-Quintana, H.A.; Reyes-Lizcano, F.A.; Bastidas-Martínez, J.G. Mechanical behavior of low-density polyethylenewaste modified hot mix asphalt. *Sustainability* **2022**, *14*, 4229. [CrossRef]
55. Doyle, J.D.; Howard, I.L. Characterization of dense-graded asphalt with the Cantabro test. *J. Test. Eval.* **2016**, *44*, 77–88. [CrossRef]
56. Cox, B.C.; Smith, B.T.; Howard, I.L.; James, R.S. State of knowledge for Cantabro testing of dense graded asphalt. *J. Mater. Civ. Eng.* **2017**, *29*, 04017174. [CrossRef]
57. Zhao, S.; Huang, B.; Ye, X.P.; Shu, X.; Jia, X. Utilizing bio-char as a bio-modifier for asphalt cement: A sustainable application of bio-fuel by-product. *Fuel* **2014**, *133*, 52–62. [CrossRef]
58. Kumar, A.; Choudhary, R.; Narzari, R.; Kataki, R.; Shukla, S.K. Evaluation of bio-asphalt binders modified with biochar: A pyrolysis by-product of Mesua ferrea seed cover waste. *Cogent Eng.* **2018**, *5*, 1548534. [CrossRef]
59. Yaro, N.S.A.; Sutanto, M.H.; Habib, N.Z.; Usman, A.; Tanjung, L.E.; Aliyu, I.; Jagaba, A.H. Optimizing biochar-based geopolymer composites for enhanced water resistance in asphalt mixes: An experimental, microstructural, and multi-objective analysis. *J. Eng. Appl. Sci.* **2023**, *70*, 151. [CrossRef]
60. Rondón-Quintana, H.A.; Zafra-Mejía, C.A.; Urazán-Bonells, C.F. Aging resistance evaluation of an asphalt mixture modified with zinc oxide. *Infrastructures* **2024**, *9*, 81. [CrossRef]
61. Rajib, A.; Saadeh, S.; Katawal, P.; Mobasher, B.; Fini, E.H. Enhancing biomass value chain by utilizing biochar as a free radical scavenger to delay ultraviolet aging of bituminous composites used in outdoor construction. *Resour. Conserv. Recycl.* **2021**, *168*, 105302. [CrossRef]
62. Zhao, S.; Huang, B.; Shu, X.; Ye, P. Laboratory investigation of biochar-modified asphalt mixture. *Transp. Res. Rec.* **2014**, *2445*, 56–63. [CrossRef]
63. Saadeh, S.; Al-Zubi, Y.; Zaatarah, B. *Performance Testing of Hot Mix Asphalt Containing Biochar*; Final Report 20–28; Mineta Transportation Institute: San Jose, CA, USA, 2020. [CrossRef]
64. Saadeh, S.; Al-Zubi, Y.; Katawal, P.; Zaatarah, B.; Fini, E. Biochar effects on the performance of conventional and rubberized HMA. *Road Mater. Pavement Des.* **2023**, *24*, 156–172. [CrossRef]
65. Bastidas-Martínez, J.G.; Rondón-Quintana, H.A. Assessment of asphalt binder content, temperature and loading rate in indirect tensile strength and resilient modulus tests of a hot-mix asphalt—Comparison with Marshall design method. *Constr. Build. Mater.* **2024**, *426*, 136158. [CrossRef]
66. Bastidas-Martínez, J.G.; Rondón-Quintana, H.A.; Muniz de Farias, M. Behavior of asphalt mastics containing different materials as Filler. *Can. J. Civ. Eng.* **2021**, *48*, 347–355. [CrossRef]
67. Barman, M.; Ghabchi, R.; Singh, D.; Zaman, M.; Commuri, S. An alternative analysis of indirect tensile test results for evaluating fatigue characteristics of asphalt mixes. *Constr. Build. Mater.* **2018**, *166*, 204–213. [CrossRef]
68. Bastidas-Martinez, J.G.; Camapum de Carvalho, J.; Cristhiane, L.L.; Muniz de Farias, M.; Rondón-Quintana, H.A. Effects of iron ore tailing on performance of hot-mix asphalt. *J. Mater. Civ. Eng.* **2022**, *34*, 04021405. [CrossRef]
69. Bharath, G.; Reddy, K.S.; Tandon, V.; Reddy, M.A. Aggregate gradation effect on the fatigue performance of recycled asphalt mixtures. *Road Mater. Pavement Des.* **2021**, *22*, 165–184. [CrossRef]
70. Rondón-Quintana, H.A.; Ruge-Cárdenas, J.C.; Reyes-Lizcano, F.A.; Bastidas-Martínez, J.G.; Zafra-Mejía, C.A. Mechanical resistance of hot-mix asphalt using phosphorite as filler. *J. Mater. Civ. Eng.* **2023**, *35*, 04023274. [CrossRef]

coatings

MDPI

Article

Analysis of Flexural Strength in Asphalt Concrete Specimens Using a Net-Shaped Inclusion of Waste Coffee Capsules

Juan Carlos Ruge [1,*], Juan Gabriel Bastidas-Martínez [2], Camilo E. Herrera [1], Jhan Piero Rojas [3] and Renato P. da Cunha [4]

1. Programa de Ingeniería Civil, Facultad de Ingeniería, Universidad Católica de Colombia, Bogotá 111311, Colombia; ceherrera@ucatolica.edu.co
2. Programa de Especialización en Ingeniería de Pavimentos, Facultad de Ingeniería, Universidad Católica de Colombia, Bogotá 111311, Colombia; jgbastidas@ucatolica.edu.co
3. Facultad de Ingeniería, Universidad Francisco de Paula Santander, Cúcuta 540003, Colombia; jhanpierorojas@ufps.edu.co
4. Department of Civil and Environmental Engineering, University of Brasília, Brasilia 70910-900, Brazil; rpcunha@unb.br

* Correspondence: jcruge@ucatolica.edu.co

Abstract: Asphalt mixtures can include various recycled materials, which often replace some of the coarse and fine aggregate components. In some cases, a percentage of material called mineral filler, made up of particles that pass through a No. 200-sieve, is also admitted in the preparation of the mixture. With the increasing amount of packaging used as containers for various products, many of which need to be disposed of properly and become an environmental burden in warehouses without proper reuse, there is potential for including these elements in the sustainable modification of asphalt mixtures. This research suggests reusing plastic and aluminum coffee capsules, which are difficult to recycle. While most studies crush recycled materials to sizes smaller than 0.075 mm for use in mixes, this research focuses on assembling the waste capsules into a network of cells inside specimens subjected to bending to observe the mechanical behavior of the asphalt mixture. The findings indicate that incorporating capsule networks can lead to a significant enhancement in the flexural strength of the examined beams, with an increase of up to 200%. Moreover, the deformation is reduced by an average of 66% upon the emergence of the initial crack in the specimen.

Keywords: coffee capsules; asphalt mixtures; flexural modified test

1. Introduction

Worldwide concern for the environment has made the study of the mechanical and hydraulic performance of asphalt mixes made from recycled materials a virtual necessity. This preoccupation is due to the overexploitation of virgin resources and the pollution caused by waste that is neither properly managed nor recycled for new use.

The incorporation of foreign materials into asphalt mixes aims to enhance specific properties typically associated with mechanical performance, e.g., reducing permanent deformation and improving fatigue resistance. However, additives or modifications can also enhance aspects such as durability, drainage, and aggregate–bitumen adhesion.

Plastics, widely utilized due to their cost-effectiveness, lightness, durability, and performance, have become an integral part of various applications [1–3]. Nevertheless, concerns have arisen regarding the excessive utilization of single-use plastics, which are projected to represent approximately 50% of global plastic production by 2022 [4], with only a quarter dedicated to durable plastics such as pipes, coatings, and structural applications [1].

In the realm of asphalt modification, polymers have shown potential for improving durability, particularly in terms of high-temperature stability [5], stiffness [6,7], fatigue resistance [8,9], and viscosity [10]. This modulating effect extends to compatibility and low-temperature bending. It is noteworthy that many studies emphasize the environmental benefits of these advancements, particularly in mitigating potential contamination [11–13].

The issue surrounding coffee capsules lies in the proper disposal and recycling process required by consumers. Research indicates that a substantial portion of coffee pod consumers in Spain dispose of them in regular trash, unaware that this leads to their ultimate destination in sanitary landfills, thereby imposing significant negative impacts on the environment. Moreover, the current regulations do not categorize these capsules as containers and therefore, do not indicate that they should not be discarded in the yellow container. Even attempting to separate the aluminum and organic components for disposal remains ineffective due to organic residue persisting within the capsules.

Another significant challenge in regards to coffee capsules is their composition, as they contain a small amount of aluminum along with plastic, which hinders their recyclability. This characteristic complicates the separation process, thereby leading to less efficient disposal methods, such as incineration or landfill deposition, as their ultimate fate [14,15].

Regarding environmental impact, Refs. [16,17] have conducted a comparative study of coffee capsules employing different packaging materials, including plastic, aluminum, and biodegradable options. Their findings underscore the substantial environmental impact of both plastic and aluminum production, primarily driven by the energy-intensive nature of the production and usage phases. Similarly, Marinello et al. (2021) [18] also explored the environmental impact of these capsules and highlighted the significant role of energy consumption in contributing to environmental harm, echoing the sentiments of other researchers. Their study emphasizes the urgent need to discover alternatives enabling the separation of all capsule components to enhance recyclability.

The current state of research regarding the incorporation of plastics in asphalt mixtures clearly demonstrates that plastics are commonly utilized in pulverized form or as strips to modify the mixtures themselves or the asphalt cement. However, the use of waste plastic material in reticulated form, capitalizing on the shape of the capsules, has not been explored. The initial phase of the study was proposed to assess the flexural strength of asphalt mixtures. Subsequent stages will evaluate additional tests, including dynamic modulus, fatigue, phase angle, and CBR.

The thesis presented in this study is based on using plastic coffee capsules embedded in prismatic samples of asphalt mix in a cell-like network to enhance the material's flexural strength. The study utilized a hot dense mix (MDC_19), with aggregates and bitumen from local sources that were characterized according to INVIAS standards. When applying the flexural load, the specimens were supported with a free spacing at least three times their height, with a 2% tolerance.

Background

Recently, there has been growing concern in the field of plastics recycling regarding the discovery of a new type of sedimentary rock. This rock comprises a natural parent rock fused with plastic components, and is being referred to by some scientists as "plastistones". Defined by Santos et al. (2022) [19] as rocks with plastic residues exhibiting a smooth texture and shiny appearance typical of molten plastic, plastistones are characterized by a homogeneous plastic material with a wide range of porosity, along with the presence of vesicles, flow structures, molten plastic mass, and in some cases, intercalations of lithic fragments (Figure 1).

These rocks are formed from plastic debris that has entered the environment, thus becoming part of the geological cycle of rocks and transported by environmental agents such as water and wind [20–24]. The majority of these rocks originate from microplastics that have weathered over time, resulting in residues of polyethylene (PE), polypropylene (PP), polyethylene terephthalate (PET), and polyvinyl chloride (PVC) [25–27] (Ward and Reddy, 2020; Stubbins et al., 2021; Fei et al., 2022).

Figure 1. Debris plastic melted into a natural rock.

In the domain of hot asphalt mix applications, plastic is widely employed as a modifying agent. This versatile material can exist in various forms and with several origins, impacting its incorporation into the mix. It has been observed in the form of pellets, flakes, microparticles, shreds, fibers, and even powder (refer to Figure 2 for a visual representation).

Figure 2. Waste plastics: (**a**). pellets; (**b**). shreds; (**c**). microplastic; (**d**). flakes; (**e**). fibers; (**f**). powder.

Pellets, for instance, are industrially processed small plastic spheres, with a maximum size of 0.5 mm, designed to be directly integrated into the material being modified, such as asphalt or asphalt mix (refer to Figure 2a) [28–32]. Conversely, flakes are typically derived from cuttings, e.g., from PET bottles and other containers, and are reduced to small square pieces (Figure 2b) [33,34].

In the context of microplastics, these are exceedingly small particles, measuring less than 5 mm, which may be inhaled by living organisms. In certain cases, they are artificially created through the abrasion of tires or other plastic components intended for use in hot asphalt mixes [35–38] (Lv et al., 2021; Campanale et al., 2020; Enfirn et al., 2022; Boom et al., 2023).

Shreddings, obtained through cutting, typically with scissors in a laboratory setting, can assume a slenderer form compared to that of flakes, often appearing as strips [39,40]. Notably, the use of nylon strips from army camouflage in other geomaterial applications has exhibited potential implications for their utilization in asphalt mixtures (Castilla, 2023) [41] (*vide* Figure 2d).

The application of plastic fibers in the asphalt mix serves two primary functions: as an asphalt stabilizer, mitigating the drain-down effect, and as a reinforcing additive, enhancing mechanical performance [42–47] (Tpakin, 2008; Slebi et al., 2020, Slebi et al., 2022; Guo et al., 2023; Jia et al., 2023, Alnadish et al., 2023). Furthermore, when plastic is crushed and reduced to sizes smaller than 0.075 mm, it is commonly utilized as a filler in the asphalt mix [48,49].

Refs. [50,51] provide a broader background that can furnish more details on the state-of-the-art practices in the use of plastics in asphalt mixtures and asphalt binders, particularly regarding human health, end-of-life options, legislation, and policies/concepts.

The following points underscore the rationale for utilizing coffee capsules of this kind, notwithstanding the fact that the capsules are not produced in a manner that involves cutting them into flakes, shreds, fibers, or crushing them into powder. This alternative is appealing due to its lack of need for any prior processing before employing it as a reinforcement, particularly in the case of an asphaltic concrete beam. This assertion implies that, despite being an additive, it does not undergo chemical interaction with the asphalt or the hot asphalt mix. Its usage is intended for reinforcement in the lower third, where the fibers of the specimen exhibit trajectories subjected to tension. Further elaboration on this topic will be provided subsequently.

In the process of this investigation, antecedents dating back to 1980 have been identified. For example, the utilization of plastic geogrids for reinforcing geomaterials such as pavements or soils has been documented. Komatsu et al. (1998) [52] conducted a study involving the implementation of geogrids composed of polyoxymethylene fibers, a thermoplastic material known for its high strength, toughness, and modulus stiffness. The experiments demonstrated improvements in durability, viscosity, and crack resistance when compared to the results for the control specimens that did not incorporate geogrids. The findings indicated an overall increase in stiffness in the specimens as a result of the application of geogrids.

Other studies have examined the impact of plastic geogrids on the durability of asphalt pavements, as well as the effect of initial strain and geogrid placement within the pavement layer to enhance resistance to fatigue caused by dynamic loading. A significant finding from this research is that according to Widodo et al. (2013) [53], placing a geogrid at 2/5 of the pavement layer thickness yields a better response compared to that obtained by placing the geogrid at the center of the layer.

The results of the research by Chazallon et al. (2021) [54] also concur with those from the majority of studies stating that the presence of a plastic geogrid significantly affects tensile strength and modulus of elasticity. Additionally, they discovered that despite potential damage during compaction, the geogrid's residual strength can still uphold the fatigue limit of asphalt concrete.

Numerous studies have investigated various waste products and their potential to improve the characteristics of asphalt mixtures, although these studies are not directly related to the specific focus of our investigation. For instance, research on reclaimed asphalt pavement (RAP) has demonstrated that the use of recycled mixtures does not significantly differ in regards to properties compared to those of conventional hot mix asphalt (HMA) [55,56]. Furthermore, polymer additions, particularly involving linear low-density polyethylene (LLDPE), low-density polyethylene (LDPE), and high-density polyethylene (HDPE), have gained popularity for enhancing asphalt cement performance in recent decades [57–59]. Additionally, the application of substances of biological or chemical origin, such as vegetable oils, volcanic ashes, rice husk ashes, waste soils (expansives), and blast furnace slag, has shown potential to improve the properties of asphalt mixtures or cement [60–66].

2. Materials and Methods

A total of 72 samples of asphalt mixes containing 3 cm and 4.5 cm diameter coffee capsules were prepared using a constant compaction of 75 blows. A factorial experiment was carried out in which one factor, the asphalt content, was varied by 0.5% for each mix, with the same particle-size distribution range. This was performed in order to discover the working methodology for the test, followed by the compaction procedure, and ultimately, the results for the analysis of the flexural strength were obtained.

The test molds are prepared in such a way as to ensure a constant percentage of bitumen. This bitumen percentage is defined by the previous testing of the aggregates in the mix. The laboratory tests provide a starting point for determining the final mix's working formula or optimum bitumen content, using a variable asphalt content of 5% and 5.5%. Similarly, the proportions of the mixes are formulated from the results of the granulometric analysis. MDC-19 was chosen as the hot asphalt mix because it has a finer particle size, and its aggregates can more easily penetrate the capsules.

2.1. Material Characterization

Table 1 shows the index characterization studies performed on the natural aggregate. In some cases, the local INVIAS standard was used; in other cases, the AASHTO and ASTM standards were employed.

Table 1. Index characterization of the natural aggregate.

Test	Standard	Value
Specific Gravity/Absorption (3/4″)	AASHTO T 85 [67]	2.52/1.67%
Specific Gravity/Absorption (3/8″)	AASHTO T 85 [67]	2.50/1.71%
Specific Gravity/Absorption (#4)	AASHTO T 85 [67]	
Specific Gravity/Absorption (#40)	AASHTO T 84 [68]	
Specific Gravity/Absorption (#80)	AASHTO T 84 [68]	
Specific Gravity/Absorption (#200)	AASHTO T 84 [68]	
Specific Gravity/Absorption of fines	AASHTO T 84 [68]	2.51/1.51%
Los Angeles Machine Abrasion, 500 revs.	AASHTO T 96 [69]	24.7%
Micro-Deval	AASHTO T327-05 [70]	22.4%
10% fines (dry strength)	DNER-ME 096-98 [71]	116 kN
Fractured particles, 1 face	ASTM D 5821-01 [72]	88%
Fractured particles, 2 faces	ASTM D 5821-01 [72]	81%
Plasticity index (IP)	ASTM D 4318-00 [73]	NP
Flattening index	NLT 354-91 [74]	9.6%
Elongation index	NLT 354-91 [74]	9.6%
Sand equivalent	AASHTO T 176–02 [75]	76%

The presence of larger grain sizes, such as 3/4″ and 3/8″, is correlated with an increase in absorption, as the specific gravity experiences a slight decrease. This phenomenon can be attributed to the differences in specific surface area between smaller and larger aggregates. As a result, smaller aggregates exhibit a greater overall specific surface area, thereby promoting easier adhesion of asphalt cement due to the increased available area. Moreover, the material demonstrates abrasion resistance that aligns with acceptable standards, as evidenced in AASHTO T 96 [69]. The recommended value, in accordance with the regulation, is 25% abrasion in the Los Angeles machine. Notably, there is no evidence of clay minerals or organic matter that may compromise the adhesion between the aggregate, asphalt cement, and, ultimately, the coffee pods.

Figure 3 shows the particle size distribution of virgin aggregate in the asphalt cement contents corresponding to 5 and 5.5%. It is of utmost importance to mention that although the percentage passing is the same for both samples, what is relevant is the calculation of the volumes in the asphalt, aggregates, and voids of the test specimens, since these may vary, depending on the asphalt content. It is essential to highlight that, although the coffee capsules occupy a volume in the specimen, this was calculated and occupies approximately 1% of the total volume. This value was disregarded for the estimations.

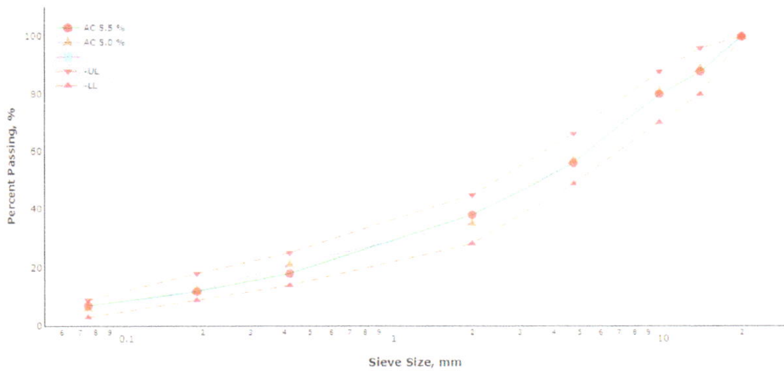

Figure 3. Particle size distribution of samples with AC 5.0 y 5.5% vs particle size range, according to INV-E-123 (2013) [76].

Likewise, the CA 60-70 asphalt cement was characterized. Table 2 shows the bitumen results that will function as an asphalt binder between aggregates. The optimum asphalt content was obtained by compacting ten Marshall samples with 75 blows per face, increasing the asphalt percentages from 4.5% to 6.0%. For mixing temperatures and subsequent compaction, the procedure outlined in ASTM D6925 [77] was followed, which adopts hot dense mix temperatures of 85 ± 15 SSF (170 cP) and 140 ± 15 SSF (280 cP).

Table 2. AC 60-70 Characterization.

Test	Standard	Unit	Material
	Original AC		
Penetration (25 °C-100 g-5 s)	ASTM D-5 [78]	0.1 mm	62.5
Penetration index	NLT 181/88 [79]	-	−0.7
Softening point	ASTM D36-95 [80]	°C	52.5
Absolute viscosity (60 °C)	ASTM D4402 [81]	Poises	1750
Specific gravity	AASHTO T 228-04 [82]	-	1.012
Viscosity 135 °C	AASHTO T-316 [83]	Pa-s	0.36
Ductility (25 °C, 5cm/min)	ASTM D-113 [84]	cm	>105
	After RTFOT		
Loss of mass	ASTM D-2872 [85]	%	0.47
Penetration (25 °C-100 g-5 s)% original	ASTM D-5 [78]	%	72
Increased softening point	ASTM D36-95 [81]	°C	5

2.2. Sample Preparation

Once the characterization of the different materials was explicit, we proceeded to prepare the samples. Since the beams will include a component not part of an ordinary asphalt mixture, this component must be carefully inserted into the sample. The coffee capsules are manually placed in the form of a net, and the capsules must be adhered to each other with silicone to ensure several points of contact between each capsule and to confirm that they can function as a joint fabric (Figures 4 and 5). It is important to mention that there is no bonding by an external agent between the cross-linked capsule layer and the asphalt mixture. The beams used two types of configurations; the first option included 52 coffee capsules, 3 cm in diameter, and the second employed 14 coffee capsules, 4.5 cm in diameter, from which the lid and the organic material (coffee) were previously removed. This procedure is imperative to ensure the penetration of the asphalt mixture into each element to avoid fewer voids in the sample.

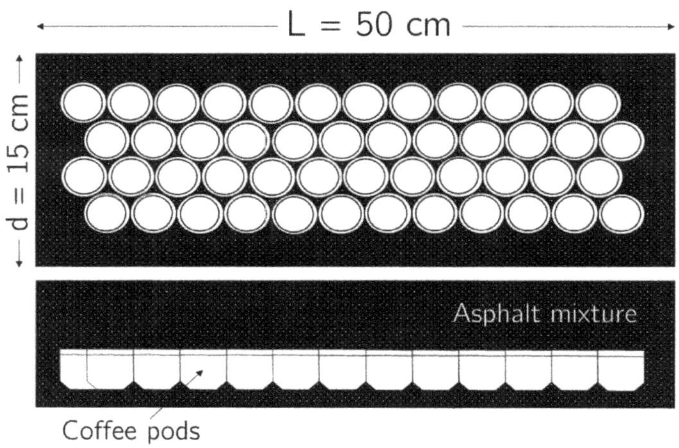

Figure 4. Configuration of 3 cm diameter coffee capsules embedded in the hot asphalt mix beam.

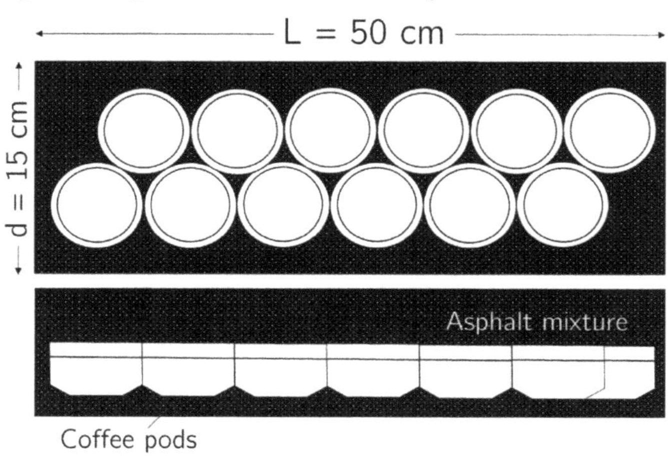

Figure 5. Configuration of 5.4 cm diameter coffee capsules embedded in the hot asphalt mix beam.

A bed of initial asphalt mix is placed under the capsule network so that the pods are not in direct contact with the mold, as shown in Figure 6. The coffee capsules are uncovered to remove the organic material and ensure they are filled with asphalt mix.

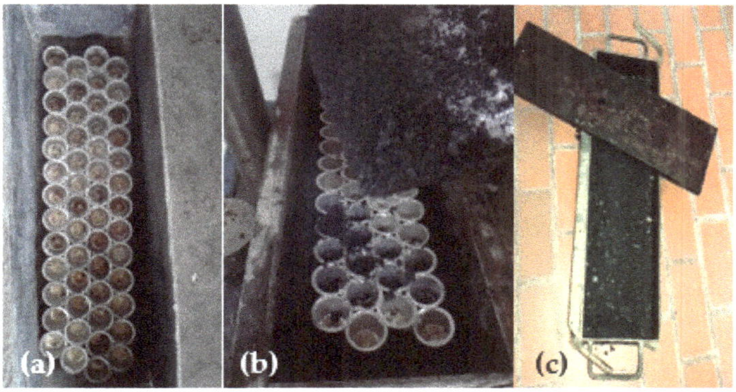

Figure 6. (**a**) Configuration of 5.4 cm diameter coffee capsules; (**b**) moment before the asphalt is poured; (**c**) after compaction.

The samples were compacted by applying a constant and uniformly distributed load, with the help of a platen on top, in three layers until a 15 cm thickness was reached, according to the configuration of each beam. The first layer, which will embed the network of coffee capsule cells, will cause some initial deformation in these elements. However, in later photographs, it can be seen that this deformation is minimal when observing a specimen after testing (Figure 7).

Figure 7. Assembling the sample for testing.

The optimum asphalt content was obtained by compacting five Marshall samples using 75 blows per face, gradually increasing the asphalt percentages from 4.5% to 6.0%. The chosen particle size distribution and typical phase volumes for AC 5.0% and 5.5% are shown in Figure 5 and Table 2. In the same way, Table 3 summarizes the Marshall test results for the optimum asphalt content of MDC-19. The optimum content obtained was 5.0% for samples with pods, and 5.5% for the non-pods option. However, it was determined to use both CA percentages for all configuration alternatives.

Table 3. Results of Marshall test MDC-19.

AC (%)	Stability-S (kN)	Flow-F (mm)	S/F	Voids (%)
4.5	11.82	3.87	3.05	6.14
5.0	12.52	3.72	3.37	4.09
5.5	11.59	3.89	2.98	2.75
6.0	11.66	4.13	2.82	1.53

The one-third loading method is used for flexure tests, using load blocks that ensure that the forces applied to the beam are perpendicular to the face of the specimen and without eccentricity. The specimens meet the requirements of INV E-402 [86]. The samples display a free distance between supports of at least three times their height, with a tolerance of 2% (Figure 7).

A total of 72 specimens were meticulously prepared and subjected to bending using a monotonic load and a fixture positioned at the top of the specimen to ensure pure bending in the beam. The deflection at the center of each specimen was precisely measured at the moment of the first crack, which coincided with the point when the monotonic load in the fixture ceased to increase.

Table 4 provides an overview of the tests conducted and presents the key characteristics of the specimens utilized. Additionally, conventional control specimens lacking a capsule network were employed to analyze the behavior and establish a reference point for comparisons.

Table 4. Test program.

Sample	AC-5.0% (with Pods)	AC-5.5% (with Pods)	AC-5.0% (no Pods)	AC-5.5% (no Pods)
MDC-19, 3.0 cm	12 specimens	12 specimens	12 specimens	12 specimens
MDC-19, 4.5 cm	12 specimens	12 specimens		

Figure 8 shows the specimens after failure, displaying the presence of the coffee capsules in the lower third of the beam. It can be seen that the failure process creates a compressibility in the capsules that does not affect the bending results. This physical expression was an expected response in the results. It is also clear that beam failure rarely occurs through any of the capsules but usually between two coffee capsules.

Figure 8. Position of the capsules in the lower third of the specimen after failure.

2.3. Load Application

The applied load should be steadily and smoothly increased at a rate between 0.9 MPa/min and 1.2 MPa/min, leading to a progressive rise in extreme fiber stress until the point of failure, as shown in Figure 9. Subsequently, three measurements of each dimension, located at the edges and center, are recorded to the nearest 1.3 mm (0.05″). These measurements are used to calculate the average width and the average height and to pinpoint the location of the fracture line within the failure area of the specimen.

Figure 9. Deformation of the beam at the end of the test.

3. Results

In this section, the flexural strength and deformation results are presented, examining the variations in asphalt content in relation to the presence of capsules in the samples. As outlined earlier, a total of 72 samples were generated, representing the three replicates of each test for 24 different combinations of asphalt and capsule content in the specimens. The average of the three replicates was then calculated, and this value was incorporated into Figures 10–14. Additionally, the specimens are evaluated in comparison to the control samples that do not contain capsules.

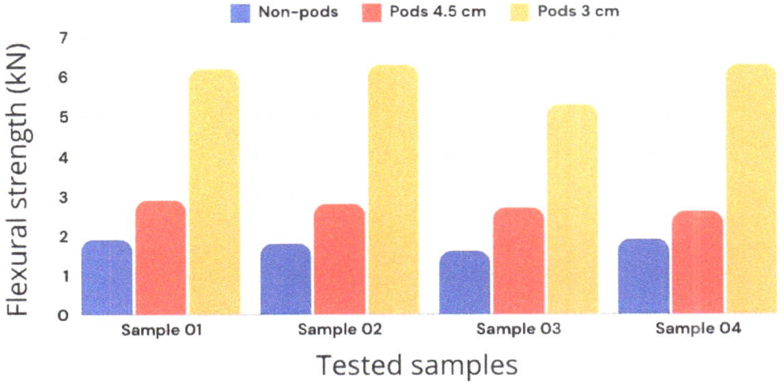

Figure 10. Behavior of the samples, without the addition and with the addition of coffee capsules, for 5.0% asphalt.

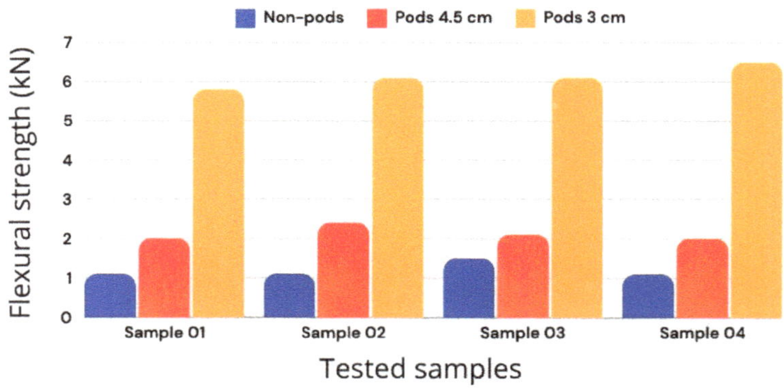

Figure 11. Behavior of the samples, without the addition and with the addition of coffee capsules, for 5.5% asphalt.

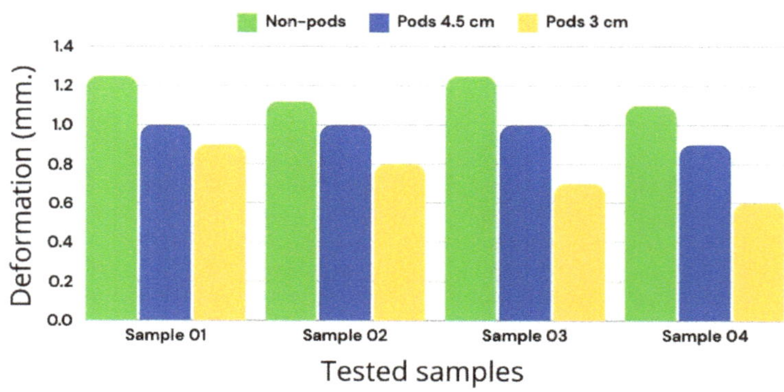

Figure 12. Deformation in the samples at the center of the beam before the appearance of the first crack (AC 5.0%).

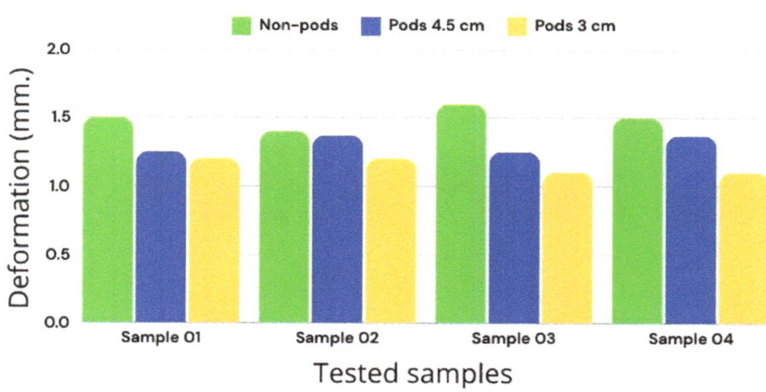

Figure 13. Deformation in the samples at the center of the beam before the appearance of the first crack (AC 5.5%).

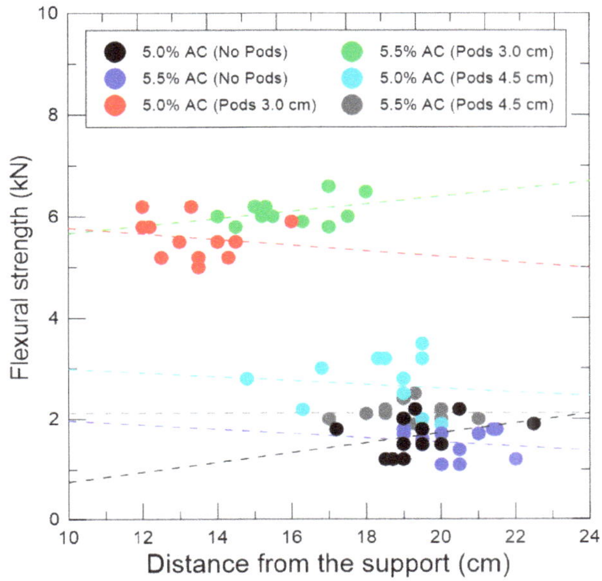

Figure 14. Distance from the support vs flexural strength.

3.1. Flexural Strength

The findings from the 5% AC specimens lacking capsules in the lower third reveal that while the loads are relatively low, they exhibit consistent stability at approximately 2 kN, indicating a lack of outliers in the results (refer to Figure 10). Conversely, in the case of specimens containing 4.5 cm diameter capsules, an increase in flexural strength to values approaching 3 kN is observed, signaling a distinct impact of the capsule network. Notably, when utilizing 3 cm diameter capsules, the value triples in comparison to that of the control sample, with average values reaching approximately 6 kN.

The specimens containing a 5.5% asphalt content (AC) exhibited a marginal decrease in flexural strength compared to those with 5.0% AC. Additionally, the specimens with 4.5 cm diameter capsules demonstrated an increase in strength compared to that of the control specimen without capsules; however, their strength decreased compared to that of the 5.0% AC specimens. In contrast, the 3 cm diameter specimens displayed enhanced incorporation of the capsules, as evidenced by increased resistance values that approached those of the specimens with 5.0% asphalt content. In essence, the findings indicate that the specimens with 5.5% AC experienced a reduction of approximately 30% in monotonic load, leading to specimen failure (Figure 11).

Upon review, it is evident that specimens featuring the 3 cm diameter pods exhibit superior performance in terms of flexural strength. This can be attributed to the smaller capsules, which yield a higher density within the same volume. As a result, the cross-linked fabric of the 3 cm specimens demonstrates greater stiffness compared to that of their 4.5 cm counterparts.

3.2. Deformations

In regards to deformations, we studied the value of the deformation at the center of the beam at the point of the first crack's appearance, which was assessed visually. The control specimens were expected to have higher deformations, and the results confirmed this expectation. For instance, as shown in Figure 12, samples without capsules deformed an average of 1.1 mm. In contrast, samples with 4.5 cm capsules showed a reduction in the deformation to about 1.0 mm. Furthermore, the deformation decreased to values close to 0.7 mm in the samples with 3 cm diameter capsules.

Regarding the findings of the specimens containing the 5.5% asphalt content, it is evident that deformations display a slight increase for both the control specimens and those with capsules of 3.0 and 4.5 cm in diameter. This particular variable does not exhibit significant variations in the percentage, contrasting with the observations related to flexural strength. While the presence of capsules demonstrates a slight enhancement, the overall improvement across all analyzed results appears to be relatively modest (Figure 13).

As in the strength analysis, the difference is not as significant for this aspect. The denser grid (3.0 cm pods) makes the sample stiffer and consequently, decreases the deformations.

3.3. First Crack

The study of cracking in asphalt mixtures is primarily focused on assessing fatigue resistance and consequently, the pavement's life cycle. Understanding the flexural strength at the onset of the initial crack is imperative for comprehending the degradation of asphalt mixtures, as it is closely linked to factors such as the quality of the asphalt cement and external elements like temperature differentials.

The asphalt binder, characterized by its coupled viscosity and elasticity, plays a pivotal role in determining the propensity for cracking within an asphalt mix. Despite its relatively low stiffness compared to other components, cracks typically manifest near the aggregate–binder interface, a phenomenon substantiated by our current investigation [87,88].

The thermal gradient, particularly the temperature differentials experienced by the mixture, also exerts considerable influence. Cracks tend to emerge on the surface layer exposed to the surroundings. Notably, the laboratory conditions in our study were controlled, minimizing the impact of this variable [89–93].

In light of the above, a comparison was made between the flexural strength and the distance from the support to the first crack, and the results are presented in Figure 14. Despite the relatively low R^2 values observed in nearly all samples, indicating a weak correlation between flexural strength and support distance, this response can be attributed to the erratic and challenging nature of first crack appearances, making measurements difficult at times. Nonetheless, discernible and consistent trends indicative of the anticipated behavior are evident. Notably, in samples exhibiting greater deformations, specifically those lacking pods, the distance from the support to the failure line is extended.

4. Discussion

It was evident from the anticipated outcomes that the incorporation of capsules, irrespective of their diameter and configuration, would enhance both flexural strength and deformation. The challenge lay in quantifying the extent of this enhancement. The inclusion of the 4.5 cm diameter capsule net exhibited no significant increase in flexural strength or reduction in deformation, likely due to its lower density, resulting in less stiffness compared to that of the 3 cm diameter capsule net, which had a higher capsule count per unit area. This configuration's higher density and greater coordination number for each capsule, as per the number of capsules in contact, contribute to its superior performance.

Such capsule network inclusions are suitable for specialized pavement structures where capsule insertion into the asphalt mix is feasible, particularly in locations with high subgrade deformations or those where reducing asphalt layer thickness is desired. This is because the capsule network enhances resistance and reduces deformation in asphalt mixtures subjected to bending. In specific contexts where the aforementioned issues are observed, it is advisable to employ these pods. This is particularly applicable in scenarios where the timely manual placement of capsules does not pose a significant constraint during road construction.

Future research should broaden its scope to encompass different asphalt and aggregate types, alternative capsule network configurations, and the use of capsules in the form of flakes or powder to modify asphalt mixes and cement.

5. Conclusions

The experimental investigation aimed at determining the flexural strength of MDC-19 incorporating a network of coffee waste capsules has been successfully accomplished. Furthermore, conventional specimens without inclusions were tested to serve as initial control specimens.

The modified test designed to evaluate the flexural strength of beam-shaped asphalt mix specimens has proven to be suitable for this study, yielding reasonable results in accordance with the proposed specimen and test configuration. In quantitative terms, the incorporation of the capsule lattice increased flexural strength by up to threefold compared to that of the samples lacking capsules.

The utilization of waste coffee capsules not only mitigates the serious environmental impact they pose but also enhances the specimens by effectively minimizing the tensile stress in the lower fibers, thereby demonstrating a favorable response under monotonic bending loads. Nonetheless, further testing is required to validate this effect in a more comprehensive manner.

Analysis of the beam deflection at the point of failure indicates lower values for specimens with waste capsule inclusions because of the stiffness it causes in the specimens. However, these values remain inconclusive, warranting additional measurements at multiple locations on the specimens to bolster the confidence in the results. It is important to note that the laboratory practices and methodology employed affirm that the results are directly correlated to the execution processes. Therefore, it can be inferred that adherence to standards and the use of calibrated equipment enhances the reliability of the obtained data.

The deformation measurements were exclusively obtained from the central region of the beam, potentially constraining the comprehensive analysis of the results. Subsequent research endeavors should address this limitation to enhance the applicability of the findings to real-world construction scenarios.

While acknowledging that the present work represents an experimental laboratory test which is not standardized for asphaltic concrete, the results are of significance and pave the way for future research endeavors in this domain.

Author Contributions: Conceptualization, J.C.R., J.G.B.-M. and C.E.H.; methodology, J.C.R., J.G.B.-M. and C.E.H.; formal analysis, J.C.R., J.G.B.-M. and R.P.d.C.; resources, J.C.R.; data curation, J.C.R.; writing—original draft preparation, J.C.R., J.G.B.-M. and J.P.R.; writing—review and editing, J.C.R., J.G.B.-M. and R.P.d.C.; visualization, J.C.R. and J.G.B.-M.; supervision, J.C.R.; project administration, J.C.R.; funding acquisition, J.C.R. and J.P.R. All authors have read and agreed to the published version of the manuscript.

Funding: This research was funded by Central Research Department of Universidad Católica de Colombia, grant number 0000636, and the APC was funded by Universidad Católica de Colombia and Universidad Francisco de Paula Santander.

Institutional Review Board Statement: Not applicable for studies not involving humans or animals.

Informed Consent Statement: Not applicable for studies not involving humans.

Data Availability Statement: Data are contained within the article.

Acknowledgments: The authors would like to thank the central research department of the Catholic University of Colombia, headed by Edwin Durán, and the research department of the Faculty of Engineering, headed by Claudia Jimenez.

Conflicts of Interest: The authors declare no conflicts of interest.

References

1. Hopewell, J.; Dvorak, R.; Kosior, E. Plastics recycling: Challenges and opportunities. *Philos. Trans. R. Soc. B Biol. Sci* **2009**, *364*, 2115–2126. [CrossRef] [PubMed]
2. Gao, G.; Zhu, Y.; Zhang, Y.; Zhang, Y. Preparation of storage stable LDPE/SBS Blend-modified Asphalt. *J. Highw. Transp. Res. Dev.* **2002**, *19*, 28–31. [CrossRef]
3. Xu, F.; Zhao, Y.; Li, K. Using Waste Plastics as Asphalt Modifier: A Review. *Materials* **2022**, *15*, 110. [CrossRef]
4. Plastics Europe. Available online: https://www.plasticseurope.org/en/resources/market-data (accessed on 10 July 2024).
5. Kalantar, Z.N.; Karim, M.R.; Mahrez, A. A review of using waste and virgin polymer in pavement. *Constr. Build. Mater.* **2012**, *33*, 55–62. [CrossRef]
6. Costa, L.M.B.; Silva, H.M.R.D.; Peralta, J.; Oliveira, J.R.M. Using waste polymers as a reliable alternative for asphalt binder modification—Performance and morphological assessment. *Constr. Build. Mater.* **2019**, *198*, 237–244. [CrossRef]
7. Punith, V.S.; Veeraragavan, A.; Amirkhanian, S.N. Evaluation of Reclaimed Polyethylene Modified Asphalt Concrete Mixtures. *Int. J. Pavement Res.* **2011**, *4*, 1–10. [CrossRef]
8. Otuoze, H.S.; Ejeh, S.P.; Amartey, Y.D.; Joel, M. Rheology and Simple Performance Test (SPT) Evaluation of High-Density Polypropylene (HDPP) Waste-Modified Bituminous Mix. *Jordan J. Civ. Eng.* **2018**, *12*, 35–44.
9. Ziari, H.; Nasiri, E.; Amini, A.; Ferdosian, O. The effect of EAF dust and waste PVC on moisture sensitivity, rutting resistance, and fatigue performance of asphalt binders and mixtures. *Constr. Build. Mater.* **2019**, *203*, 188–200. [CrossRef]
10. Hariadi, D.; Saleh, S.M.; Anwar Yamin, R.; Aprilia, S. Utilization of LDPE plastic waste on the quality of pyrolysis oil as an asphalt solvent alternative. *Therm. Sci. Eng. Prog.* **2021**, *23*, 100872. [CrossRef]
11. Costa, L.M.B.; Silva, H.M.R.D.; Oliveira, J.R.M.; Fernandes, S.R.M. Incorporation of Waste Plastic in Asphalt Binders to Improve their Performance in the Pavement. *International. J. Pavement Res. Technol.* **2013**, *6*, 457–464. [CrossRef]
12. Joohari, I.B.; Giustozzi, F. Chemical and high-temperature rheological properties of recycled plastics-polymer modified hybrid bitumen. *J. Clean. Prod.* **2020**, *276*, 123064. [CrossRef]
13. Mashaan, N.S.; Chegenizadeh, A.; Nikraz, H.; Rezagholilou, A. Investigating the engineering properties of asphalt binder modified with waste plastic polymer. *Eng. J.* **2021**, *12*, 1569–1574. [CrossRef]
14. Iftikhar, N. A Circular Economy for Coffee Capsule Waste. Master's Thesis, Faculty Technology, Policy and Management. TU Delft University of Technology, Delft, The Netherlands, 2021.
15. Pinto, S.; Gouveia, J.; Sousa, M.; Rodrigues, B.; Oliveira, J.; Pinto, C.; Baptista, A.J. Improving coffee capsules recyclability—A combined assessment of circularity and environmental performance of a novel design. *Sustain. Prod. Consum.* **2024**, *46*, 233–243. [CrossRef]
16. Li, J. Comparative Life Cycle Assessment of Single-Serve Coffee Packaging in Ontario. Master's Thesis, University of Waterloo, Waterloo, ON, Canada, 2017.
17. Li, Q.; Dai, T.; Gao, T.; Zhoing, W. Aluminum material flow analysis for production, consumption, and trade in China from 2008 to 2017. *J. Clean. Prod.* **2021**, 296. [CrossRef]
18. Marinello, S.; Balugani, E.; Gamberini, R. Coffee capsule impacts and recovery techniques: A literature review. *Packag. Technol. Sci.* **2021**, *34*, 665–682. [CrossRef]
19. Santos, F.A.; Diorio, G.R.; Guedes, C.C.F.; Fernandino, G.; Giannini, P.C.F.; Angulo, R.J.; Souza, M.C.; Ferreira, M.A.; Dos Santos, A.R. Plastic debris forms: Rock analogues emerging from marine pollution. *Mar. Pollut. Bull.* **2022**, *182*, 114031. [CrossRef]
20. Gestoso, I.; Cacabelos, E.; Ramalhosa, P.; Canning-Clode, J. Plasticrusts: A new potential threat in the Anthropocene's rocky shores. *Sci. Total Environ.* **2019**, *687*, 413–415. [CrossRef]
21. Domínguez-Hernández, C.; Villanova-Solano, C.; Sevillano-González, M.; Hernández-Sánchez, C.; González-Sálamo, J.; Ortega-Zamora, C.; Diaz-Peña, F.J.; Hernandez-Borges, J. Plastitar: A new threat for coastal environments. *Sci. Total Environ.* **2002**, *839*, 156261. [CrossRef]
22. Wang, L.; Bank, M.S.; Rinklebe, J.; Hou, D. Plastic–rock complexes as hotspots for microplastic generation. *Environ. Sci. Technol.* **2023**, *57*, 7009–7017. [CrossRef]
23. Wang, L.; Hu, Z.; Yin, H.; Bradford, S.A.; Luo, J.; Hou, D. Aging of colloidal contaminants and pathogens in the soil environment: Implications for nanoplastic and COVID-19 risk mitigation. *Soil Use Manag.* **2023**, *39*, 70–91. [CrossRef]
24. Wang, L.; Hou, D. Plastistone: An emerging type of sedimentary rock. *Earth-Sci. Rev.* **2023**, *247*, 104620. [CrossRef]
25. Ward, C.P.; Reddy, C.M. Opinion: We need better data about the environmental persistence of plastic goods. *Proc. Natl. Acad. Sci. USA* **2020**, *117*, 14618–14621. [CrossRef] [PubMed]
26. Stubbins, A.; Law, K.L.; Munoz, S.E.; Bianchi, T.S.; Zhu, L. Plastics in the Earth system. *Science* **2021**, *373*, 51–55. [CrossRef] [PubMed]
27. Fei, X.; Guo, Y.; Wang, Y.; Fang, M.; Yin, K.; He, H. The long-term fates of land-disposed plastic waste. *Nat. Rev. Earth. Environ.* **2022**, *3*, 733–735. [CrossRef]
28. Nikolaides, A.F.; Manthos, E. *Bituminous Mixtures and Pavements VII*, 1st ed.; CRC Press: Thessaloniki, Greece, 2019.
29. White, G.; Reid, G. Recycled Waste Plastic for Extending and Modifying Asphalt Binders. In Proceedings of the 8th Symposium on Pavement Surface Characteristics: SURF 2018, Brisbane, Australia, 2–4 May 2018.
30. Obaidi, H.; Gomez-Meijide, B.; Garcia, A. On-site manufacture of hot mix asphalt using pellets that can be melted by induction energy. *Powder Technol.* **2019**, *344*, 58–67. [CrossRef]

31. Zhao, Z.; Wu, S.; Xie, J.; Yang, C.; Wang, F.; Li, N.; Liu, Q.; Amirkhanian, S. Effect of direct addition of asphalt rubber pellets on mixing, performance and VOCs of asphalt mixtures. *Constr. Build. Mater.* **2024**, *411*, 134494. [CrossRef]
32. Li, M.; Zhang, M.; Rong, H.; Zhang, X.; He, L.; Han, P.; Tong, M. Transport and deposition of plastic particles in porous media during seawater intrusion and groundwater-seawater displacement processes. *Sci. Total Environ.* **2021**, *781*, 146752. [CrossRef]
33. Modarres, A.; Hamedi, H. Effect of waste plastic bottles on the stiffness and fatigue properties of modified asphalt mixes. *Mater. Des.* **2014**, *61*, 8–15. [CrossRef]
34. Almeida, A.; Capitão, S.; Estanqueiro, C.; Picado, L. Possibility of incorporating waste plastic film flakes into warm-mix asphalt as a bitumen extender. *Constr. Build. Mater.* **2021**, *291*, 123384. [CrossRef]
35. Lv, L. Challenge for the detection of microplastics in the environment. *Water Environ. Res.* **2021**, *93*, 5–15. [CrossRef]
36. Campanale, C.; Massarelli, C.; Savino, I.; Locaputo, V.; Uricchio, V.F. A detailed review study on potential effects of microplastics and additives of concern on human health. *Int. J. Environ. Res. Public Health* **2020**, *17*, 1212. [CrossRef] [PubMed]
37. Enfrin, M.; Myszka, R.; Giustozzi, F. Paving roads with recycled plastics: Microplastic pollution or eco-friendly solution. *J. Hazard. Mater.* **2022**, *437*, 129334. [CrossRef] [PubMed]
38. Boom, Y.; Xuan, D.; Enfrin, M.; Swaney, M.; Masood, H.; Pramanik, B.; Robert, D.; Giustozzi, F. Engineering properties, microplastics and emissions assessment of recycled plastic modified asphalt mixtures. *Sci. Total Environ.* **2023**, *893*, 164869. [CrossRef] [PubMed]
39. Singh, P. Use of Plastic Waste in Flexible Pavement-Green Highway. *Int. J. Eng. Res. Technol.* **2020**, *9*, 09. [CrossRef]
40. Cardoso, J.; Ferreira, A.; Almeida, A.; Santos, J. Incorporation of plastic waste into road pavements: A systematic literature review on the fatigue and rutting performances. *Constr. Build. Mater.* **2023**, *407*, 133441. [CrossRef]
41. Castilla, W. Evaluación mecánica de la grava limosa del Vichada colombiano denominada en la zona como ferrillo modificada con cemento y fibras cortas de nylon residuales de camuflado (In Spanish). Master's Thesis, Universidad Militar Nueva Granada, Bogotá, Columbia, 2023.
42. Tapkin, S. The effect of polypropylene fibers on asphalt performance. *Build. Environ.* **2008**, *43*, 1065–1071. [CrossRef]
43. Slebi, C.; Ruge, J.C.; Castro-Fresno, D. An integrated DoE–Stochastic multi criteria decision-making analysis applied for experimental evaluation of fiber reinforced porous asphalt mixtures. *Constr. Build. Mater.* **2020**, *255*, 119330. [CrossRef]
44. Slebi, C.; Lastra-González, P.; Castro-Fresno, D.; Vega-Zamanillo, A. Experimental evaluation and recyclability potential of asphalt concrete mixtures with polyacrylonitrile fibers. *Constr. Build. Mater.* **2022**, *317*, 125829. [CrossRef]
45. Jia, H.; Sheng, Y.; Guo, P.; Underwood, S.; Chen, H.; Kim, Y.; Li, Y.; Ma, Q. Effect of synthetic fibers on the mechanical performance of asphalt mixture: A review. *J. Traffic Transp. Eng.* **2023**, *10*, 331–348. [CrossRef]
46. Guo, Y.; Tataranni, P.; Sangiorgi, C. The use of fibres in asphalt mixtures: A state of the art review. *Constr. Build. Mater.* **2023**, *390*, 131754. [CrossRef]
47. Alnadish, A.M.; Singh, N.S.S.; Alawag, A.M. Applications of Synthetic, Natural, and Waste Fibers in Asphalt Mixtures: A Citation-Based Review. *Polymers* **2023**, *15*, 1004. [CrossRef]
48. You, L.; Long, Z.; You, Z.; Ge, D.; Yang, X.; Xu, F.; Hashemi, M.; Diab, A. Review of recycling waste plastics in asphalt paving materials. *J. Traffic Transp. Eng.* **2022**, *9*, 2095–7564. [CrossRef]
49. Ma, Y.; Zhou, H.; Jiang, X.; Polaczyk, P.; Xiao, R.; Zhang, M.; Huang, B. The utilization of waste plastics in asphalt pavements: A review. *Clean. Mater.* **2021**, *2*, 100031. [CrossRef]
50. Nayanathara Thathsarani Pilapitiya, P.G.C.; Sandaruwan Ratnayake, A. The world of plastic waste: A review. *Clean. Mater.* **2024**, *11*, 100220. [CrossRef]
51. Noor, A.; Atiq Ur Rehman, M. A mini-review on the use of plastic waste as a modifier of the bituminous mix for flexible pavement. *Clean. Mater.* **2022**, *4*, 100059. [CrossRef]
52. Komatsu, T.; Kikutab, H.; Tujic, Y.; Muramatsuc, E. Durability assessment of geogrid-reinforced asphalt concrete. *Geotext. Geomembr.* **1998**, *16*, 257–271. [CrossRef]
53. Widodo, S.; Subagio, B.S.; Setiadji, B.H. Geogrid as Asphalt Pavement Reinforcement. *Nat. Sci. Eng. Med.* **2013**, *4*, 4.
54. Chazallon, C.; Van Rompu, J.; Nguyen, M.L.; Hornych, P.; Mouhoubi, S.; Doligez, D.; Brissaud, L.; Gal, Y.; Godard, E. Laboratory and in Situ Damage Evaluation of Geogrid Used in Asphalt Concrete Pavement. In Proceedings of the RILEM International Symposium on Bituminous Materials. (ISBM 2020); RILEM Bookseries, 27; Springer: Cham, Switzerland, 2022. [CrossRef]
55. Barraj, F.; Elkordi, A. Investigating the effect of using unclassified fractionated reclaimed asphalt pavement materials on the properties of hot mix asphalt. *Constr. Build. Mater.* **2022**, *353*, 129099. [CrossRef]
56. Tsakoumaki, M.; Plati, C. A Critical Overview of Using Reclaimed Asphalt Pavement (RAP) in Road Pavement Construction. *Infrastructures* **2024**, *9*, 128. [CrossRef]
57. Delva, L.; Hubo, S.; Cardon, L.; Ragaert, K. On the Role of Flame Retardants in Mechanical Recycling of Solid Plastic Waste. *Waste Manag.* **2018**, *82*, 198–206. [CrossRef]
58. Dwivedi, P.; Mishra, P.K.; Mondal, M.K.; Srivastava, N. Non-Biodegradable Polymeric Waste Pyrolysis for Energy Recovery. *Heliyon* **2019**, *5*, 02198. [CrossRef]
59. Jwaida, Z.; Dulaimi, A.; Mydin, M.A.O.; Özkılıç, Y.O.; Jaya, R.P.; Ameen, A. The Use of Waste Polymers in Asphalt Mixtures: Bibliometric Analysis and Systematic Review. *J. Compos. Sci.* **2023**, *7*, 415. [CrossRef]

60. Yaro, N.S.A.; Sutanto, M.H.; Habib, N.Z.; Napiah, M.; Usman, A.; Jagaba, A.H.; Al-Sabaeei, A.M. Modeling and Optimization of Asphalt Content, Waste Palm Oil Clinker Powder and Waste Rice Straw Ash for Sustainable Asphalt Paving Employing Response Surface Methodology: A Pilot Study. *Clean. Mater.* **2023**, *8*, 100187. [CrossRef]
61. Bastidas-Martínez, J.G.; Rondón-Quintana, H.A.; Muniz de Farias, M. Behavior of asphalt mastics containing different materials as filler. *Can. J. Civ. Eng.* **2021**, *48*, 347–355. [CrossRef]
62. Franesqui, M.A.; Yepes, J.; García-González, C.; Gallego, J. Sustainable low-temperature asphalt mixtures with marginal porous volcanic aggregates and crumb rubber modified bitumen. *J. Clean. Prod.* **2019**, *207*, 44–56. [CrossRef]
63. Rondón-Quintana, H.A.; Ruge-Cárdenas, J.C.; Bastidas-Martínez, J.G.; Velandia-Castelblanco, M.Y.; Muniz de Farias, M. Use of Thermally Treated Bentonite as Filler in Hot Mix Asphalt. *J. Mater. Civ. Eng.* **2020**, *32*, 5. [CrossRef]
64. Lozano, D.A.; Molina-Gómez, F.; Ruge, J.C.; Moreno-Anselmi, L.A.; Bastidas-Martínez, J.G. Asphalts and modified dense asphalt mixtures with rubber of military boots. *DYNA* **2020**, *87*, 120–128.
65. Rondón-Quintana, H.; Ruge, J.; Bastidas-Martínez, J. Evaluation of Hot-Mix Asphalt Containing Portland Cement Treated Blast Furnace Slag. *Arch. Civ. Eng.* **2019**, *65*, 193–207. [CrossRef]
66. Rondón-Quintana, H.A.; Ruge-Cárdenas, J.C.; Reyes-Lizcano, F.; Bastidas-Martínez, J.; Zafra-Mejía, C. Mechanical Resistance of Hot-Mix Asphalt Using Phosphorite as Filler. *J. Mater. Civ. Eng.* **2023**, *35*, 9. [CrossRef]
67. *AASHTO T 85*; Standard Method of Test for Specific Gravity and Absorption of Coarse Aggregate. American Association of State Highway and Transportation Officials (AASHTO): Washington, DC, USA, 2018.
68. *AASHTO T 84*; Standard Method of Test for Specific Gravity and Absorption of Fine Aggregate. American Association of State Highway and Transportation Officials (AASHTO): Washington, DC, USA, 2000.
69. *AASHTO T 96*; Standard Method of Test for Resistance to Degradation of Small-Size Coarse Aggregate by Abrasion and Impact in the Los Angeles Machine. American Association of State Highway and Transportation Officials (AASHTO): Washington, DC, USA, 1991.
70. *AASHTO T 327*; Method of Test for Resistance of Coarse Aggregate to Degradation by Abrasion in the Micro-Deval Apparatus. American Association of State Highway and Transportation Officials (AASHTO): Washington, DC, USA, 2012.
71. *DNER-ME 096*; Agregado Graudo, Avaliacao da Resistencia Mecánica Pelo Método Dos 10% de Finos. MT–Departamento nacional de estradas de rodagem: Rio de Janeiro, Brasil, 1998.
72. *ASTM D5821*; Standard Test Method for Determining the Percentage of Fractured Particles in Coarse Aggregate. ASTM International: West Conshohocken, PA, USA, 2001.
73. *ASTM D4318*; Standard Test Methods for Liquid Limit, Plastic Limit, and Plasticity Index of Soils. ASTM International: West Conshohocken, PA, USA, 2018.
74. *NLT 354*; Índices de Lajas y Agujas de Los Áridos Para Carreteras. Normas del Laboratorio de Transporte: España, 1991.
75. *AASHTO T 176*; Plastic Fines in Graded Aggregates and Soils by Use of the Sand Equivalent Test-Final. American Association of State Highway and Transportation Officials (AASHTO): Washington, DC, USA, 2002.
76. *INV E-123-13*; Determinación de Los Tamaños de Las Partículas de Los Suelos. INVIAS: Colombia, 2013.
77. *ASTM D6925*; Standard Test Method for Preparation and Determination of the Relative Density of Asphalt Mix Specimens by Means of the Superpave Gyratory Compactor. ASTM International: West Conshohocken, PA, USA, 2015.
78. *ASTM D5*; Standard Test Method for Penetration of Bituminous Materials. ASTM International: West Conshohocken, PA, USA, 2013.
79. *NLT 181*; Índice de Penetración de Los Betunes Asfalticos. Normas del Laboratorio de Transporte: España, 1988.
80. *ASTM D36*; Standard Test Method for Softening Point of Bitumen (Ringand-Ball Apparatus). ASTM International: West Conshohocken, PA, USA, 1995.
81. *ASTM D4402*; Standard Test Method for Viscosity Determination of Asphalt at Elevated Temperatures Using a Rotational Viscometer. ASTM International: West Conshohocken, PA, USA, 2015.
82. *AASHTO T28*; Standard Method of Test for Specific Gravity of Semi-Solid Asphalt Materials. American Association of State Highway and Transportation Officials (AASHTO): Washington, DC, USA, 2004.
83. *AASHTO T316*; Standard Method of Test for Viscosity Determination of Asphalt binder Using Rotational Viscometer. American Association of State Highway and Transportation Officials (AASHTO): Washington, DC, USA, 2023.
84. *ASTM D113*; Standard Test Method for Ductility of Bituminous Materials. ASTM: West Conshohocken, PA, USA, 1991.
85. *ASTM D2872*; Standard Test Method for Effect of Heat and Air on a Moving Film of Asphalt (Rolling Thin-Film Oven Test). ASTM: West Conshohocken, PA, USA, 2012.
86. *INV E-402-13*; Elaboración y Curado de Especímenes de Concreto en el Laboratorio Para Ensayos de Compresión y Flexión. INVIAS: Colombia, 2013.
87. Roque, R.; Ruth, B.E. Mechanisms and Modeling of Surface Cracking in Asphalt Pavements. *J. Assoc. Asph. Paving Technol.* **1990**, *59*, 396–421.
88. Doll, B.; Ozer, H.; Rivera-Perez, J.; Al-Qadi, I.L.; Lambros, J. Damage Zone Development in Heterogeneous Asphalt Concrete. *Eng. Fract. Mech.* **2017**, *182*, 356–371. [CrossRef]
89. Germann, F.P.; Lytton, R.L. Methodology for Predicting the Reflection Cracking Life of Asphalt Concrete Overlays. Interim Report; Leeds, UK, 1979.

90. Mobasher, B.; Mamlouk, M.S.; Lin, H.M. Evaluation of Crack Propagation Properties of Asphalt Mixtures. *J. Transp. Eng.* **1997**, *123*, 405–413. [CrossRef]
91. Wang, Y.; Mahboub, K.C.; Hancher, D.E. Survival Analysis of Fatigue Cracking for Flexible Pavements Based on Long-Term Pavement Performance Data. *J. Transp. Eng.* **2005**, *131*, 608–616. [CrossRef]
92. Kim, M.; Buttlar, W.G.; Baek, J.; Al-Qadi, I.L. Field and Laboratory Evaluation of Fracture Resistance of Illinois Hot-Mix Asphalt Overlay Mixtures. *Transp. Res. Rec.* **2009**, *2127*, 146–154. [CrossRef]
93. Reynaldo, R.P.E.; Kim, Y.R. *Top-Down Cracking of Hot-Mix Asphalt Layers: Models for Initiation and Propagation*; National Academies Publishing: Washington, DC, USA, 2010.

coatings

MDPI

Article

A Mathematical Model for Predicting the Ultra-Early-Age Strength of Concrete

Hongfei Cao [1,2], Bing Xiao [3], Fengjiang Qin [4,*] and Qiuwei Yang [1,2,*]

[1] School of Civil and Transportation Engineering, Ningbo University of Technology, Ningbo 315211, China; caohongfei@nbut.edu.cn

[2] Engineering Research Center of Industrial Construction in Civil Engineering of Zhejiang, Ningbo University of Technology, Ningbo 315211, China

[3] Road & Bridge International Co., Ltd., Beijing 100084, China; 170057398@aliyun.com

[4] Key Laboratory of New Technology for Construction of Cities in Mountain Area, Ministry of Education, School of Civil Engineering, Chongqing University, Chongqing 400045, China

* Correspondence: qinfengjiang@cqu.edu.cn (F.Q.); yangqiuwei@nbut.edu.cn (Q.Y.)

Abstract: To accurately quantify the time-varying pattern of concrete's compressive strength, selecting an appropriate curve model is of paramount importance. Currently, widely adopted models such as polynomial, hyperbolic, and exponential models all possess limitations, particularly in terms of low fitting accuracy during the ultra-early-age stage. This paper innovatively introduces a mathematical model that utilizes a combined curve approach. This model boasts a simplified structure with only two fitting parameters. Compared to traditional models, when utilizing three or more sets of experimental data on compressive strength across different ages, the new model is capable of yielding more precise strength predictions. Due to its minimal reliance on experimental data, the new model exhibits high practicality and convenience in real-world applications. To validate its superiority, a detailed comparison between the new model and existing models was conducted based on several sets of experimental data. The results demonstrate that the new model has significant advantages in terms of mean fitting error and standard deviation, making its predictions the most reliable. For most cases, the standard deviation of the new model is reduced by approximately 30% to 80% compared to the second-best model, underscoring its exceptional stability and consistency. Additionally, the predicted long-term compressive strength values of the new model are closer to the design strength grade of the concrete. This model can also be successfully applied to predict the tensile strength of concrete during its ultra-early age. It has been demonstrated that the combined model proposed in this paper shows promising application prospects in evaluating the time-varying behavior of concrete strength.

Keywords: concrete; compressive strength; time-varying characteristics; combined model; regression analysis

1. Introduction

The assessment of concrete's compressive strength at ultra-early stages is a pivotal topic in civil engineering research. Some scholars have conducted literature reviews on the mechanical properties of concrete during its early age, including early strength. Mihashi and Leite [1] reviewed the impact of cracking in concrete during its early age, caused by factors such as insufficient strength, on the durability of the concrete. Bentz [2] provided an overview of the early strength and other properties of cement-based materials, retrospectively examining the key physical and chemical processes that occur during early stages. From the perspective of concrete engineering, Maruyama and Lura [3] outlined the early properties of concrete, with a focus on early strength, elastic modulus, and volume stability. Understanding the growth pattern of this strength during this critical period is invaluable for engineers to devise practical and efficient construction plans. Accurate strength evaluations of concrete are particularly crucial in the following scenarios: (1) dismantle the formwork as early as possible to improve the construction efficiency; (2) carry out pre-stress stretching as early as possible to avoid cracks; and (3) evaluate the negative impact of the vibration load on concrete construction. Therefore, many scholars have studied the strength evaluation model and method of concrete at an early age. Gu et al. [4] delved into the development of concrete strength by analyzing the early harmonic responses captured by embedded piezoelectric sensors. Their findings revealed a rapid surge in concrete strength within the initial few days, followed by a decline after seven days. Lee et al. [5] proposed a model to evaluate the early compressive strength according to the ultrasonic pulse velocity. It has been found that the relationship between parameters in different concrete samples is linear during initial and final setting, and parabolic after final setting. Yoon et al. [6] put forward a formula to predict the early strength of concrete based on surface wave velocity. Benaicha et al. [7] studied the strength evolution law of several kinds of concrete such as self-compacting concrete, high-performance concrete, and metal-fiber-reinforced concrete. They put forward a function to evaluate the concrete strength after pouring. Lee et al. [8] proposed a generalized rate constant model to predict the compressive strength of the early strength concrete. In order to rapidly improve the early strength of concrete, Min et al. [9] designed some new curing schemes to ensure that the early strength of concrete exceeds 10 MPa after 6 h of pouring. Chung et al. [10] put forward an evaluation model of the 28-day compressive strength of concrete. It has been shown that the slope of the strength–conductivity curve can determine the early strength grade of concrete. Voigt et al. [11] observed the microstructure change process of concrete from solidification to hardening. Demirboga et al. [12] found that the relationship between the ultrasonic velocity and compressive strength of concrete at an early age is an exponential function. Velay-Lizancos et al. [13] studied the relationship between the compressive strength and elastic modulus of early-age concrete. Jin et al. [14] predicted the compressive strength of early-age bisphenol epoxy concrete using the maturity method. Adesina et al. [15] studied the performance of additives such as calcium oxide and sodium silicate to improve the early compressive strength of slag concrete. Jiang et al. [16] studied the main factors affecting the mechanical properties of early alkali slag concrete. It was found that the compressive strength increased with the increase in binder dosage. Kaszynska [17] studied the relationship between the temperature change and early compressive strength of mass concrete during curing. Burhan et al. [18] found that polymer is more effective than silica fume in improving the workability and compressive strength of early-age concrete. Voigt et al. [19] found that the relationship between the reflection loss of ultrasonic wave and the compressive strength of concrete has nothing to do with temperature. The above research shows that the early strength of concrete increases rapidly, but there is no uniform growth law due to many factors. To anticipate the variations in concrete strength during its early stages, numerous scholars have formulated diverse mathematical models. These include the polynomial model [20–22], hyperbolic model [23–25], exponential model [26–29], and logarithmic model [30–33], each designed to capture different aspects of the strength development process. Liu [20] used a quadratic power function to fit the compressive strength data of fiber-reinforced coral concrete specimens at different ages. Ling [21] applied a polynomial model to correlate the experimental data of compressive strength for machine-made sand concrete across various ages. Jia et al. [22] established a cubic polynomial function, based on experimental data spanning from day 1 to day 28, to describe the relationship between the age and compressive strength of ordinary Portland concrete. Yang and Wang [23] determined that the hyperbolic model aptly captures the variations in the compressive strength and elastic modulus of concrete during its early stages. Wang et al. [24], after analyzing the strengths and weaknesses of the hyperbolic model, proposed an enhanced version specifically tailored to predict the compressive strength of recycled concrete. Zhao et al. [25] proposed another form of the hyperbolic model by introducing the compressive strength at the 28th day. Tang et al. [26] described the time-varying regularity of compressive strength through low-age compressive tests on concrete cube specimens using an exponential model. Other scholars [27–29] have developed variations of exponential models by introducing the strength of concrete at the 28th day. Wang and Hu [30] utilized a logarithmic model to fit the experimental data of compressive strength for shale ceramsite concrete test blocks. Zhao [31] and Li et al. [32] employed the logarithmic model to fit the experimental data pertaining to the compressive strength of high-strength concrete, leveraging this approach to forecast the subsequent strength development of the concrete. Bai and Han [33] proposed a variant form of the logarithmic model by introducing the strength data of concrete on the 28th day.

Despite the plethora of models proposed for assessing concrete strength, there remains a need to develop a novel strength–age evaluation model that can predict concrete strength with greater precision. The current evaluation models suffer from notable shortcomings, including domain gaps in the exponential and logarithmic models, the poor fitting accuracy of the hyperbolic model during the ultra-early-age stage, and potential large errors in predicting later concrete strength using the polynomial model. To address these limitations, this work introduces a combined curve model that more accurately describes the growth pattern of concrete strength. The novelty of this evaluation model lies in its fusion of two curve types, weighted to achieve higher fitting accuracy than any single curve model. The advantages of this proposed model are threefold: firstly, it requires only two unknown fitting parameters, facilitating its application with minimal experimental data; secondly, it exhibits superior fitting accuracy compared to single curve models; and thirdly, it provides more reliable predictions of later concrete strength than existing models. Several examples are employed to validate the proposed model and benchmark its performance against other models, demonstrating its superiority and reliability. The structure of this study is organized as follows. Section 2 provides a brief overview of commonly used compressive strength evaluation models, including the polynomial, hyperbolic, exponential, and logarithmic models. Section 3 elaborates on the proposed combined curve model, its parameter determination process, and error evaluation metrics. Section 4 verifies the combined model using several sets of concrete compressive strength test data. Section 5 presents a statistical analysis of the results and explores the application of the novel model for predicting concrete tensile strength. Finally, Section 6 summarizes the key findings and conclusions of this work.

2. Commonly Used Curve Models

The main task of strength evaluation is to establish the functional relationship between concrete compressive strength and time. This section reviews several commonly used curve models for strength evaluation and discusses their advantages and disadvantages. The first evaluation model is the polynomial curve (quadratic power function [20] or cubic power function [21,22]) given as follows:

$$S_t = p_2 t^2 + p_1 t + p_0 \tag{1}$$

$$S_t = p_3 t^3 + p_2 t^2 + p_1 t + p_0 \tag{2}$$

where t indicates the time from the concrete pouring, S_t represents the compressive strength of concrete at time t, p_0, p_1, p_2, and p_3 represent the unknown parameters that need to be determined by curve fitting. Obviously, the fitting accuracy will improve as the number of terms in the polynomial increases. One of the strengths of the polynomial curve model lies in its high fitting accuracy for early-age concrete compressive strength experimental data. However, its primary limitation is its inability to accurately predict the later strength of concrete. This shortcoming stems from the power function utilized in the model, which does not fully align with the actual development pattern of concrete strength, specifically failing to capture the stability of concrete strength in later stages. Additionally, the model involves three to four unknown parameters, necessitating a larger volume of experimental data for accurate curve fitting, which can be a drawback in practice.

The second evaluation model is the hyperbolic model [23,24] given as follows:

$$S_t = \frac{t}{p_0 + p_1 t} \tag{3}$$

where p_0 and p_1 represent the unknown fitting parameters. In contrast to the polynomial model, the hyperbolic model boasts a simpler structure with only two unknown parameters. This advantage translates into the ability to perform curve fitting with a reduced amount of experimental data. However, a notable disadvantage of the hyperbolic model is its limited fitting accuracy when dealing with experimental data exhibiting significant discreteness, potentially impacting the reliability of the predictions. Based on Equation (3), Zhao et al. [25] proposed another form of the hyperbolic model by introducing the compressive strength at the 28th day and modifying the power of time as follows:

$$S_t = \frac{t^{\alpha}}{p_0 + p_1 t^{\alpha}} f_{c-28} \tag{4}$$

where f_{c-28} represents the compressive strength of concrete on the 28th day and α is an adjustable coefficient. The limitations of the model presented in Equation (4) lie in two aspects: (1) the increased difficulty of curve fitting due to the introduction of an adjustable power value; (2) the need to test the concrete strength at the 28th day, which renders the model unsuitable for assessing the strength of concrete at super-early ages (before the 28th day).

The third and fourth types of models are the exponential function [26] and the logarithmic function [30], respectively, as shown below:

$$S_t = p_0 \cdot e^{-p_1/t} \tag{5}$$

$$S_t = p_0 + p_1 \cdot \ln(t) \tag{6}$$

where p_0 and p_1 represent the unknown fitting parameters. Similar to the hyperbolic model, the exponential or logarithmic model has the advantage that it contains only two unknown parameters and is easy to implement. One drawback of this model type is that its domain does not encompass the point where $t = 0$, which can limit its applicability in certain scenarios. Furthermore, similar to other models, its fitting accuracy may not be optimal when confronted with experimental data characterized by high levels of discreteness, potentially affecting the precision of the predictions it generates. By incorporating the strength of concrete at the 28th day, the literature [27–29] presents several variants of exponential models, listed sequentially as follows.

$$S_t = f_{c-28} \cdot e^{p_0 \cdot \sqrt{1 - \frac{28 - p_1}{t - p_1}}} \tag{7}$$

$$S_t = f_{c-28} \cdot f_{ult} \cdot e^{-(\frac{\tau}{t})^{\alpha}} \tag{8}$$

$$S_t = f_{c-28} \cdot e^{p_0 \cdot \sqrt{1 - \frac{28}{t}}} \tag{9}$$

The f_{ult} and τ in Equation (8) represent the final compressive strength of the concrete and a fitting parameter to be determined, respectively. Similarly, a variant of the logarithmic model that incorporates the strength data of concrete on the 28th day is presented in reference [33] as follows.

$$S_t = f_{c-28} \cdot [1 + p_0 \cdot \ln(\frac{t}{28})] \tag{10}$$

The aforementioned variants of the exponential model and logarithmic model both require the 28-day strength data of concrete, which makes them inappropriate for evaluating the compressive strength of concrete at ultra-early ages. The model shown in Equation (8) also requires the final strength of concrete, which further limits the scope of the application of this model. Furthermore, these models also suffer from issues related to the lack of definition in certain domains. For instance, Equations (8)–(10) all lack a definition for the point where $t = 0$, while Equation (7) lacks a definition for the point where $t = p_1$. Additionally, Equations (7) and (9) require that $t \geq 28$, as negative numbers cannot be squared. Furthermore, in terms of fitting accuracy, these modified forms do not exhibit significant improvements in precision. Given the limitations of the existing models, it is imperative to investigate new compressive strength–age curve models for analyzing the variation patterns of compressive strength in concrete during its ultra-early age.

3. New Combined Curve Model

The primary characteristics defining the variation law of concrete compressive strength encompass the following aspects. (1) Initial Condition: The compressive strength of concrete commences at 0 at $t = 0$, underscoring the necessity for the curve model's domain to encompass this point. (2) Bounded Growth: Within the first seven days post pouring, the strength undergoes a rapid ascent, stabilizing thereafter approximately after 28 days. This suggests that the curve model must be bounded, ensuring function values remain within a defined range as the input variable (e.g., time/age) varies, eschewing unbounded growth or excessive oscillations. (3) Increasing and Convex Functionality: The temporal evolution of concrete's compressive strength adheres to both an increasing and convex function profile, wherein the first derivative remains positive (indicating growth) and the second derivative is negative (signifying decelerating growth). (4) Multifactorial Complexity: Many factors, such as mix ratio, temperature and humidity, early shrinkage, and water evaporation, have significant impacts on the compressive strength of concrete during its early age. For example, when water evaporates too quickly or unevenly, shrinkage cracks are prone to form on the surface of the concrete. These cracks will subsequently reduce the compressive strength of the concrete. The combined influence of numerous factors often leads to a relatively large dispersion in the strength data of concrete during its early age. Consequently, a solitary curve model struggles to accurately capture the intricate age–strength relationship.

Hence, the concept of a combined model, integrating two or more curves, emerges as a potential solution to more precisely delineate the intricate strength growth pattern of concrete during its ultra-early ages. Each constituent curve within this combined model must adhere to the aforementioned prerequisites. (1) It must be an increasing function, reflecting the continuous enhancement of concrete strength over time. (2) It must exhibit convexity, indicating that the rate of strength increase diminishes as time elapses (first derivative > 0, second derivative < 0). (3) Its domain must span $t = 0$, ensuring applicability from the inception of the concrete's aging process. By amalgamating curves that fulfill these criteria, a more holistic and accurate portrayal of the complex dynamics of concrete strength development during ultra-early ages can be achieved.

To this end, a combined two-parameter model (CTM) is developed in this work as follows:

$$S_t = p_0 \cdot \frac{t}{t+1} + p_1 \cdot \frac{e^t - e^{-t}}{e^t + e^{-t}} \tag{11}$$

where p_0 and p_1 are the two unknown fitting parameters. Equation (11) is a combination of two curves, and p_0 and p_1 can be regarded as the weight coefficients of the two curves. Both of the two curves that compose this model satisfy the characteristics of being increasing functions as well as convex functions. Specifically, the first and second derivatives of these two curves are as follows.

$$\left(\frac{t}{t+1}\right)' = \frac{1}{(t+1)^2} > 0 \tag{12}$$

$$\left(\frac{t}{t+1}\right)'' = \frac{-2}{(t+1)^3} < 0 \tag{13}$$

$$\left(\frac{e^t - e^{-t}}{e^t + e^{-t}}\right)' = \frac{4e^{2t}}{(e^{2t}+1)^2} > 0 \tag{14}$$

$$\left(\frac{e^t - e^{-t}}{e^t + e^{-t}}\right)'' = \frac{8e^{2t} - 8e^{6t}}{(e^{2t}+1)^4} < 0 \tag{15}$$

In Equations (12)–(15), the superscripts $'$ and $''$ are used to denote the first-order and second-order derivatives, respectively. Furthermore, the proposed CTM is a bounded function, since $S_t = 0$ when $t = 0$ and $S_t = p_0 + p_1$ when $t \to \infty$. Through the above analysis of the CTM's characteristics, including the fact that the two curves comprising the model both exhibit the properties of being increasing functions and convex functions, it can be concluded that this newly established model aligns with the natural law governing the variation of concrete strength over time. Moreover, this novel model boasts a streamlined structure with merely two unknown parameters, allowing for curve fitting with minimal experimental data. Theoretically, just three or more sets of compressive strength test data spanning different ages suffice to accurately assess the concrete's compressive strength. Despite its concise nature with only two fitting parameters, this model surpasses the aforementioned dual or multi-parameter models in terms of fitting accuracy, offering a more precise representation of the data.

Next, the calculation method of the unknown fitting parameters is described as follows. Without loss of generality, it is assumed that n ($n \geq 3$) groups of compressive strengths at different times are measured in the concrete strength test. For these n groups of test data, the linear system of equations can be obtained from Equation (11) as follows:

$$\eta = \Omega \cdot \zeta \tag{16}$$

$$\eta = \begin{Bmatrix} S_{t_1} \\ \vdots \\ S_{t_n} \end{Bmatrix}, \zeta = \begin{Bmatrix} p_0 \\ p_1 \end{Bmatrix} \tag{17}$$

$$\Omega = \begin{bmatrix} \frac{t_1}{t_1+1} & \frac{e^{t_1}-e^{-t_1}}{e^{t_1}+e^{-t_1}} \\ \vdots & \vdots \\ \frac{t_n}{t_n+1} & \frac{e^{t_n}-e^{-t_n}}{e^{t_n}+e^{-t_n}} \end{bmatrix} \tag{18}$$

where $S_{t_1}, \cdots, S_{t_n}$ denote the measured values of compressive strength for the time $t_1, \cdots, t_n$, Ω is the coefficient matrix of the linear system. From Equation (16), the desired fitting parameters p_0 and p_1 can be obtained with the following equation:

$$\zeta_p = (\Omega^T \Omega)^{-1} \Omega^T \eta \tag{19}$$

in which the superscript "T" represents the matrix transposition operation. When the fitting parameters p_0 and p_1 are determined from ζ_p by Equation (19), the predicted compressive strength can be further calculated by the following:

$$\eta_p = \Omega \zeta_p \tag{20}$$

where:

$$\eta_p = (S^p_{t_1}, \cdots, S^p_{t_n})^T \tag{21}$$

in which $S^p_{t_1}, \cdots, S^p_{t_n}$ denote the calculated values of compressive strength for the time $t_1, \cdots, t_n$ by the proposed combined curve model. The difference between the predicted value and the measured value, which quantifies their discrepancy, is referred to as the comparative error:

$$e_{t_n} = \left| S_{t_n}^{p} - S_{t_n} \right| / S_{t_n} \tag{22}$$

where e_{t_n} denotes the comparative error for the time t_n. Ultimately, to assess the fitting accuracy of the evaluation model, the average, standard deviation, and coefficient of variation of the errors between predicted and measured values are utilized. Specifically, the average value of these errors is defined as follows:

$$\bar{e} = \frac{e_{t_1} + \cdots + e_{t_n}}{n} \tag{23}$$

where $\bar{e}$ denote the mean of the errors. The standard deviation of the errors is defined as follows:

$$\sigma = \sqrt{\frac{\sum_{i=1}^{n} (e_{t_i} - \bar{e})^2}{n-1}} \tag{24}$$

where σ denotes the standard deviation of the errors. The coefficient of variation of the errors is defined as follows.

$$c_v = \sigma / \bar{e} \tag{25}$$

As the values of the average, standard deviation, and coefficient of variation of errors decrease, so too does the discrepancy between predicted and measured outcomes, indicating a heightened fitting accuracy for the respective evaluation model. In the upcoming section, the fitting precision of the novel model will be contrasted with that of several existing models, utilizing these three statistical metrics, with the objective of emphasizing the superior performance and advantages of the new model.

4. Case Study

Examples 1–11 are utilized to compare the performance of the proposed CTM with the existing models in predicting the compressive strength of concrete. Given that the proposed model is specifically designed for assessing the strength of concrete during the ultra-early-age phase, models from the aforementioned existing ones that rely on the 28-day test data are excluded from the comparative study. Additionally, since the exponential model can be transformed into a logarithmic model by taking logarithms, the three types of existing models in the comparative study are the cubic polynomial model, hyperbolic model, and logarithmic model.

Example 1: In the literature [34], Zhao et al. produced 100 mm × 100 mm × 100 mm concrete cube specimens with a compressive strength grade of C65. The specific mix proportion of concrete is 449 kg of cement, 693 kg of sand, 1129 kg of stone, 149 kg of water, and 21.9 kg of admixture. These specimens were cured in standard environments and their compressive strength was measured at different ages of 1, 2, 3, 7, 14, and 28 days. The loading speed in the compressive strength experiment is taken as 0.8–1 Mpa/s. Four test blocks are measured for each age and the average compressive strength of these four test blocks is used as the final value for the corresponding age. The results obtained from the compression test can be found in the literature [34] or Table 1(a). Based on the test results, Equations (26)–(29) show the specific fitting equations corresponding to the polynomial model, hyperbolic model, logarithmic model, and CTM in sequence. Figure 1a shows the fitting curves corresponding to these four evaluation models more intuitively. Table 2(a) provides additional insights into the performance of various models, detailing their mean fitting error, standard deviation, coefficient of variation, and the 120-day predicted compressive strength. Notably, the proposed combined model stands out as the most accurate, with the lowest values across all these metrics. Specifically, the mean fitting error follows a descending order from logarithmic, polynomial, and hyperbolic models to the combined model. When compared to the second-best performing hyperbolic model, the new model impressively reduces the mean fitting error by 23% and the standard deviation by 52%, underscoring its substantial improvement in fitting precision. Furthermore, an examination of the predicted 120-day compressive strength values in Table 2(a) reveals that the combined model exhibits the highest reliability in forecasting the concrete's later compressive strength, thereby demonstrating its superiority in predictive capability.

$$S_t = 0.0177t^3 - 0.8546t^2 + 11.8269t + 25.4589 \tag{26}$$

$$S_t = \frac{t}{0.0160 + 0.0125t} \tag{27}$$

$$S_t = 38.507 + 12.9631 \cdot \ln(t) \tag{28}$$

$$S_t = 96.0159 \cdot \frac{t}{t+1} - 15.5774 \cdot \frac{e^t - e^{-t}}{e^t + e^{-t}} \tag{29}$$

Table 1. The experimental results of Examples 1–6.

Index		Compressive Strength (CS)						
(a) Example 1	Age (day)	1	2	3	7	14	28	
	CS (MPa)	29.7	51.3	58.9	67.3	73.6	76.1	
(b) Example 2	Age (day)	1	3	7	14	28		
	CS (MPa)	31.0	49.7	54.0	59.8	68.6		
(c) Example 3	Age (day)	1	1.5	2	2.5	3	3.5	4
	CS (MPa)	2.8	29.9	36.9	42.2	46.2	50.2	52.5
	Age (day)	5	6	7	14	28	60	90
	CS (MPa)	55.2	57.4	58.5	65.5	74.9	81	82.5
(d) Example 4	Age (day)	1	1.5	2	2.5	3	3.5	4
	CS (MPa)	2.4	20.8	28.4	36.8	40.1	42.7	45.3
	Age (day)	5	6	7	14	28	60	90
	CS (MPa)	50.3	53.4	55.6	62	69.2	76.3	78.3
(e) Example 5	Age (day)	1	2	3	5	7	11	14
	CS (MPa)	24.0	32.1	39.6	45.7	46.8	47.6	49.9
	Age (day)	28	40					
	CS (MPa)	56.8	58.4					
(f) Example 6	Age (day)	3	5	7	14	28	60	
	CS (MPa)	35.4	49.8	51.6	53.8	55.3	60.3	

Example 2: In the literature [35], Hu et al. designed C55 concrete with a mix ratio of cement 260 kg/m^3, slag powder 90 kg/m^3, limestone 90 kg/m^3, sand 693 kg/m^3, crushed stone 1131 kg/m^3, water 140 kg/m^3, and water reducer 5.28 kg/m^3. They conducted compressive strength tests on concrete and obtained compressive strength values for ages of 1, 3, 7, 14, and 28 days. The results obtained from their compression test can be found in the literature [35] or Table 1(b). Based on the test results, Equations (30)–(33) show the specific fitting equations corresponding to the four models in sequence. Figure 1b shows the fitting curves corresponding to these four evaluation models more intuitively. Table 2(b) extends the analysis by presenting the mean fitting error, standard deviation, coefficient of variation, and 120-day predicted compressive strength for a range of models. Notably, the proposed combined model emerges as the most precise, with the lowest values for all three accuracy metrics. The mean fitting error is arranged in ascending order, from the hyperbolic model, through the polynomial model, to the logarithmic model, with the combined model achieving the best performance. When benchmarked against the next-best

logarithmic model, the new model achieves a 3% reduction in mean fitting error and a significant 42% decrease in standard deviation, highlighting its substantial improvement in fitting accuracy. Additionally, an assessment of the 120-day predicted compressive strength values in Table 2(b) underscores the combined model's superior reliability in predicting the concrete's future compressive strength, reinforcing its position as the most reliable predictive tool.

$$S_t = 0.0111t^3 - 0.5257t^2 + 7.4887t + 26.6466 \tag{30}$$

$$S_t = \frac{t}{0.0249 + 0.0140t} \tag{31}$$

$$S_t = 33.6276 + 10.5328 \cdot \ln(t) \tag{32}$$

$$S_t = 72.1989 \cdot \frac{t}{t+1} - 5.4835 \cdot \frac{e^t - e^{-t}}{e^t + e^{-t}} \tag{33}$$

Table 2. The mean fitting error, standard deviation, coefficient of variation, and 120-day predicted compressive strength of Examples 1–6.

	Index	Polynomial Model	Hyperbolic Model	Logarithmic Model	CTM
(a) Example 1	Mean fitting error	0.0864	0.0487	0.1023	0.0377
	Standard deviation	0.0802	0.0676	0.0999	0.0326
	Coefficient of variation	0.9290	1.3885	0.9767	0.8657
	Predicted 120-day CS (MPa)	19,807	78.93	100.57	79.65
(b) Example 2	Mean fitting error	0.0515	0.0808	0.0413	0.0401
	Standard deviation	0.0436	0.0570	0.0436	0.0251
	Coefficient of variation	0.8476	0.7050	1.0546	0.6263
	Predicted 120-day CS (MPa)	12,599	70.57	84.05	66.12
(c) Example 3	Mean fitting error	0.8724	0.4643	0.6597	0.2286
	Standard deviation	2.8772	0.9501	2.1933	0.6462
	Coefficient of variation	3.2979	2.0460	3.3246	2.8269
	Predicted 120-day CS (MPa)	191.33	84.65	95.37	79.18
(d) Example 4	Mean fitting error	0.8697	0.4523	0.6176	0.0810
	Standard deviation	2.7353	0.9177	1.9408	0.1474
	Coefficient of variation	3.1448	2.0286	3.1422	1.8193
	Predicted 120-day CS (MPa)	194.68	80.55	91.08	74.26
(e) Example 5	Mean fitting error	0.0753	0.0574	0.0504	0.0421
	Standard deviation	0.0670	0.0492	0.0410	0.0239
	Coefficient of variation	0.8894	0.8569	0.8131	0.5690
	Predicted 120-day CS (MPa)	1625.4	59.66	70.01	56.04
(f) Example 6	Mean fitting error	0.0592	0.0486	0.0722	0.0467
	Standard deviation	0.0500	0.0362	0.0637	0.0290
	Coefficient of variation	0.8443	0.7454	0.8822	0.6211
	Predicted 120-day CS (MPa)	905.7	60.7	66.4	60.2

Example 3: In the literature [36], He designed C55 concrete with a mix ratio of cement 353 kg/m^3, fly ash 116 kg/m^3, sand 730 kg/m^3, crushed stone 1121 kg/m^3, water 142 kg/m^3, and additive agent 5.159 kg/m^3. He conducted compressive strength tests on the concrete specimens under standard curing and obtained compressive strength values for ages of 1, 1.5, 2, 2.5, 3, 3.5, 4, 5, 6, 7, 14, 28, 60, and 90 days. The results obtained from their compression test can be found in the literature [36] or Table 1(c). Based on the test results, Equations (34)–(37) show the specific fitting equations corresponding to the four models in sequence. Figure 1c shows the fitting curves corresponding to these four evaluation models more intuitively. Table 2(c) provides a comprehensive overview of the

performance metrics for various models, including their mean fitting error, standard deviation, coefficient of variation, and the 120-day predicted compressive strength. Notably, the CTM excels in all these aspects, achieving the lowest values for mean fitting error, standard deviation, and coefficient of variation. The ranking of mean fitting error, from highest to lowest, is as follows: polynomial model, hyperbolic model, logarithmic model, and the combined model. When compared to the second-best performing hyperbolic model, the CTM demonstrates a remarkable improvement, reducing the mean fitting error by 51% and the standard deviation by 32%. This underscores the significant enhancement in fitting accuracy offered by the new model. Furthermore, an analysis of the 120-day predicted compressive strength values in Table 2(c) confirms that the developed combined model possesses the highest reliability in predicting the future compressive strength of concrete, emphasizing its superiority in predictive capability.

$$S_t = 0.0005t^3 - 0.0837t^2 + 3.9424t + 29.3688 \tag{34}$$

$$S_t = \frac{t}{0.0643 + 0.0113t} \tag{35}$$

$$S_t = 25.9815 + 14.4928 \cdot \ln(t) \tag{36}$$

$$S_t = 141.2717 \cdot \frac{t}{t+1} - 60.9282 \cdot \frac{e^t - e^{-t}}{e^t + e^{-t}} \tag{37}$$

Example 4: In the literature [36], the pumped concrete was sampled at the construction site, another set of curing specimens with the site construction environment conditions was made. The compressive strength tests of field curing specimens were conducted and the results can be found in the literature [36] or Table 1(d). Equations (38)–(41) show the specific fitting equations corresponding to the four models. Figure 1d shows the fitting curves corresponding to these four evaluation models. Table 2(d) presents a detailed comparison of various models based on their mean fitting error, standard deviation, coefficient of variation, and 120-day predicted compressive strength. Notably, the CTM stands out as the most accurate, with the lowest values across all evaluation metrics. Specifically, the mean fitting error is arranged in descending order as follows: polynomial model, hyperbolic model, logarithmic model, and the combined model. When compared to the next-best performing hyperbolic model, the CTM demonstrates a substantial improvement, with a reduction of 82% in mean fitting error and an even more impressive 83% decrease in standard deviation. This remarkable performance underscores the exceptional prediction accuracy of the new model, reinforcing its superiority in forecasting the compressive strength of concrete over an extended period.

$$S_t = 0.0005t^3 - 0.0877t^2 + 4.0999t + 23.2710 \tag{38}$$

$$S_t = \frac{t}{0.0787 + 0.0118t} \tag{39}$$

$$S_t = 20.0597 + 14.8359 \cdot \ln(t) \tag{40}$$

$$S_t = 143.3004 \cdot \frac{t}{t+1} - 67.8507 \cdot \frac{e^t - e^{-t}}{e^t + e^{-t}} \tag{41}$$

Example 5: In the literature [25], Zhao designed C50 concrete with a mix ratio of cement 400 kg/m^3, fly ash 115 kg/m^3, sand 582 kg/m^3, crushed stone 1166 kg/m^3, water 170 kg/m^3, and additive agent 4.3 kg/m^3. He conducted compressive strength tests on the concrete specimens under standard curing and obtained compressive strength values for ages of 1, 2, 3, 5, 7, 11, 14, 28, and 40 days. The results obtained from their compression test can be found in the literature [25] or Table 1(e). Equations (42)–(45) present the specific fitting equations and Figure 1e shows the corresponding fitting curves. From Table 2(e), one can see that the mean fitting error, standard deviation, and coefficient of variation of the proposed combined model are all the smallest. Moreover, the 120-day compressive strength predicted by the CTM is the closest to the designed strength grade of 50 MPa.

$$S_t = 0.0019t^3 - 0.1449t^2 + 3.607t + 26.0318 \tag{42}$$

$$S_t = \frac{t}{0.0332 + 0.0165t} \tag{43}$$

$$S_t = 27.2818 + 8.9263 \cdot \ln(t) \tag{44}$$

$$S_t = 68.5525 \cdot \frac{t}{t+1} - 11.9438 \cdot \frac{e^t - e^{-t}}{e^t + e^{-t}} \tag{45}$$

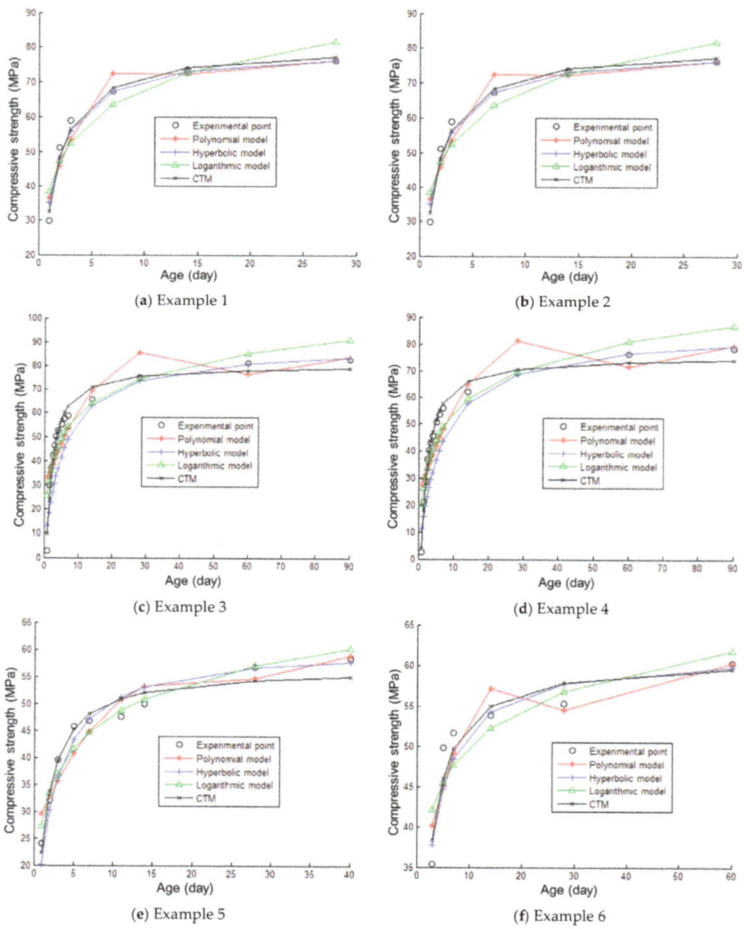

Figure 1. Fitting curves for Examples 1–6.

Example 6: In the literature [37], Chang et al. designed C55 concrete with a mix ratio of cement 486 kg/m^3, additive 7.71 kg/m^3, sand 642 kg/m^3, crushed stone 1094 kg/m^3, water 163 kg/m^3, and fly ash 65 kg/m^3. They conducted compressive strength tests on concrete and obtained compressive strength values for ages of 3, 5, 7, 14, 28, and 60 days. The results obtained from their compression test can be found in the literature [37] or Table 1(f). Equations (46)–(49) show the specific fitting equations and Figure 1f gives the corresponding fitting curves for the four evaluation models. Table 2(f) provides a comprehensive assessment of various models in terms of their mean fitting error, standard deviation, coefficient of variation, and the 120-day predicted compressive strength of concrete. Notably, the CTM emerges as the top performer, achieving the lowest values for mean fitting error, standard deviation, and coefficient of variation. This exceptional performance underscores the precision of the CTM in fitting the data. Consequently, the 120-day compressive strength predictions made by the CTM in Table 2(f) also possess the highest level of reliability, making it a trusted choice for predicting the future strength of concrete.

$$S_t = 0.0014t^3 - 0.1299t^2 + 3.4175t + 31.0187 \tag{46}$$

$$S_t = \frac{t}{0.0308 + 0.0162t} \tag{47}$$

$$S_t = 34.923 + 6.5669 \cdot \ln(t) \tag{48}$$

$$S_t = 90.393 \cdot \frac{t}{t+1} - 29.4284 \cdot \frac{e^t - e^{-t}}{e^t + e^{-t}} \tag{49}$$

Examples 7–10: The data for Examples 7–10 are all from the compressive strength tests of fly ash ceramsite concrete with different carbon fiber contents in reference [38]. These examples will further illustrate the unique advantage of the proposed model in accurately predicting the later strength of concrete based on only a small amount of experimental data. In the literature [38], Wang et al. carried out the compressive strength test of fly ash ceramsite concrete with different carbon fiber content. The specific mix proportion of concrete is 350 kg of cement, 650 kg of sand, 575 kg of stone, 140 kg of water, 135 kg of fly ash, and 5.5 kg of admixture. For these examples, the ratios of carbon fiber mass to cement mass are 0%, 0.5%, 0.8%, and 1%, sequentially. The 150 mm × 150 mm × 150 mm cube specimens were made under standard curing for compressive strength testing at different ages of 7, 14, and 28 days. The compression test results of these examples can be found in the literature [38] or Table 3. Using the data in Table 3, Equations (50)–(65) show the specific fitting equations corresponding to the polynomial model, hyperbolic model, logarithmic model, and CTM for Examples 7–10 in sequence. Figure 2 shows the fitting curves corresponding to the four models more intuitively. Table 4 extends the comparison by presenting the mean fitting error, standard deviation, coefficient of variation, and 120-day predicted compressive strength of various models. From Table 4, one can see that the mean fitting error, standard deviation, and coefficient of variation of the CTM are all the smallest. Compared to the suboptimal model, the new model shows a reduction of 41% and 69% in mean fitting error and standard deviation in Example 7, 43% and 70% in Example 8, 40% and 68% in Example 9, and 48% and 71% in Example 10, respectively. These results show that the new model significantly improves the fitting accuracy. Thus the 120-day compressive strength of concrete predicted by the CTM in Table 4 also has the highest reliability. Additionally, it is noteworthy that for Examples 7–10, the prediction results of the cubic polynomial model for the 120th day are all negative, indicating that the performance of the polynomial model is highly unstable. Despite having more fitting parameters than the two-parameter model, it fails to enhance the prediction accuracy.

$$S_t = -0.0029t^3 - 0.0313t^2 + 1.5t + 8 \tag{50}$$

$$S_t = \frac{t}{0.1834 + 0.0245t} \tag{51}$$

$$S_t = 0.6004 + 9.7382 \cdot \ln(t) \tag{52}$$

$$S_t = 154.1111 \cdot \frac{t}{t+1} - 116.1937 \cdot \frac{e^t - e^{-t}}{e^t + e^{-t}} \tag{53}$$

$$S_t = -0.001t^3 - 0.0938t^2 + 2.5t + 6 \tag{54}$$

$$S_t = \frac{t}{0.1692 + 0.022t} \tag{55}$$

$$S_t = 0.2878 + 10.8923 \cdot \ln(t) \tag{56}$$

$$S_t = 172.1369 \cdot \frac{t}{t+1} - 130.1274 \cdot \frac{e^t - e^{-t}}{e^t + e^{-t}} \tag{57}$$

$$S_t = -0.0039t^3 - 0.00t^2 + 2t + 6 \tag{58}$$

$$S_t = \frac{t}{0.165 + 0.0245t} \tag{59}$$

$$S_t = 2.1048 + 9.5218 \cdot \ln(t) \tag{60}$$

$$S_t = 150.783 \cdot \frac{t}{t+1} - 112.1832 \cdot \frac{e^t - e^{-t}}{e^t + e^{-t}} \tag{61}$$

$$S_t = -0.0039t^3 + 0.0313t^2 + 1.0t + 10 \tag{62}$$

$$S_t = \frac{t}{0.1846 + 0.0262t} \tag{63}$$

$$S_t = 1.6514 + 8.8726 \cdot \ln(t) \tag{64}$$

$$S_t = 139.1427 \cdot \frac{t}{t+1} - 103.587 \cdot \frac{e^t - e^{-t}}{e^t + e^{-t}} \tag{65}$$

Table 3. The experimental results of Examples 7–10.

Index	Age (Day)	7	14	28
(a) Example 7	CS (MPa)	18.1	29.2	31.6
(b) Example 8	CS (MPa)	19.9	32.2	35
(c) Example 9	CS (MPa)	19.2	30.1	32.4
(d) Example 10	CS (MPa)	17.8	27.3	30.1

Table 4. The mean fitting error, standard deviation, coefficient of variation, and 120-day predicted compressive strength of Examples 7–10.

	Index	Polynomial Model	Hyperbolic Model	Logarithmic Model	CTM
(a) Example 7	Mean fitting error	0.9461	0.0657	0.0751	0.0385
	Standard deviation	1.1260	0.0412	0.0271	0.0128
	Coefficient of variation	1.1902	0.6268	0.3604	0.3321
	Predicted 120-day CS (MPa)	−5324.5	38.37	47.22	36.64
(b) Example 8	Mean fitting error	0.6628	0.0652	0.0744	0.0374
	Standard deviation	0.7766	0.0408	0.0269	0.0124
	Coefficient of variation	1.1717	0.6255	0.3620	0.3325
	Predicted 120-day CS (MPa)	−2731.5	42.68	52.43	40.59
(c) Example 9	Mean fitting error	0.6626	0.0615	0.0714	0.0371
	Standard deviation	0.9323	0.0388	0.0257	0.0126
	Coefficient of variation	1.4072	0.6316	0.3595	0.3402
	Predicted 120-day CS (MPa)	−6504	38.70	47.70	37.35
(d) Example 10	Mean fitting error	0.6986	0.0508	0.0605	0.0265
	Standard deviation	0.9387	0.0320	0.0224	0.0094
	Coefficient of variation	1.3437	0.6295	0.3705	0.3549
	Predicted 120-day CS (MPa)	−6170	36.08	44.13	34.41

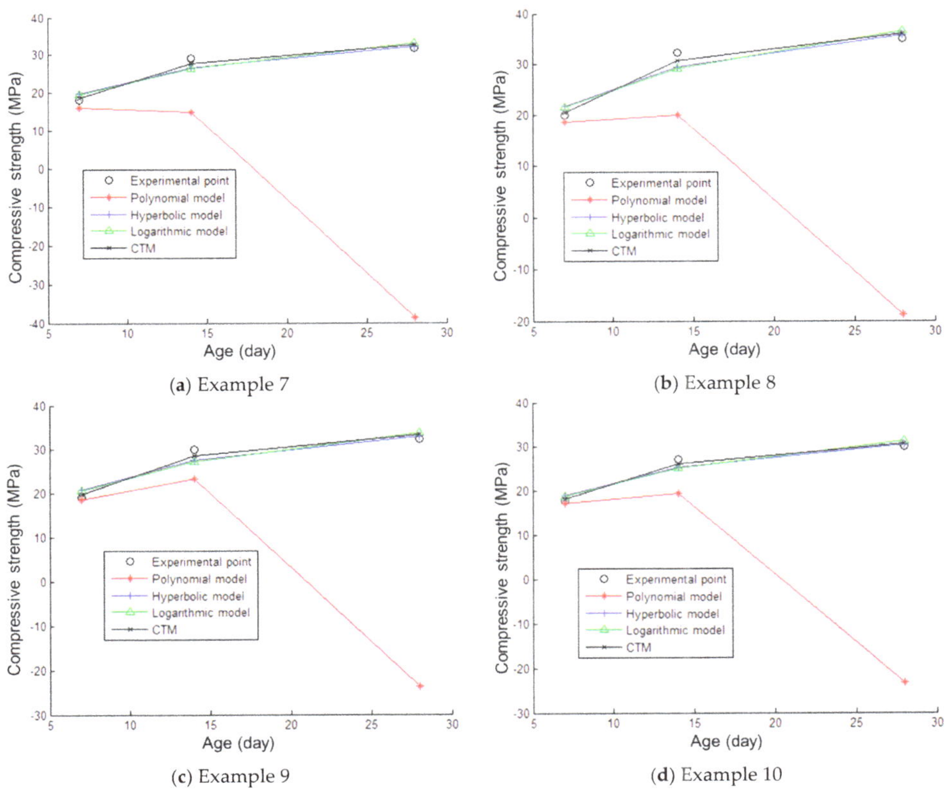

(**a**) Example 7 (**b**) Example 8 (**c**) Example 9 (**d**) Example 10

Figure 2. Fitting curves for Examples 7–10.

Example 11: An early-age compressive strength test of ultra-high strength concrete, as shown in Figure 3, was conducted by the authors to further demonstrate the proposed prediction model. The specific mix proportion of ultra-high strength concrete is 2200 kg/m^3 of powder, 158 kg/m^3 of steel fiber, and 202 kg/m^3 of water. Several test blocks with size of 100 mm × 100 mm × 100 mm were made and cured under standard conditions. The curing temperature is about (20 ± 2) °C and the relative humidity is about 95%. Three test blocks were used for each test and the average value was taken as the compressive strength for that age. The test results of compressive strength are given in Table 5.

Table 5. The test results of Example 11.

Age (Day)	0.5	1	4	7
CS (MPa)	41.2	68.5	113.9	121.8

Using the data of Table 5, Equations (66)–(69) show the specific fitting equations corresponding to the four models in sequence. Figure 4 shows the fitting curves corresponding to these models more intuitively. Table 6 summarizes the key performance metrics for various models, including the mean fitting error, standard deviation, coefficient of variation, and the 120-day predicted compressive strength. As can be seen from Table 6, while the mean and standard deviation of the fitting error of the polynomial model are both zero, the model's prediction of the concrete's later-stage strength is significantly distorted. Excluding the polynomial model, it can be found that the mean fitting error, standard deviation, and coefficient of variation of the CTM are all the smallest. Compared with the hyperbolic model, the mean fitting error and standard deviation of the new model are reduced by 38% and 76%, respectively. Combining both the fitting precision and prediction outcomes, it is reiterated that the proposed model exhibits the best performance among all the models.

$$S_t = 1.4143t^3 - 19.0548t^2 + 80.7071t + 5.4333 \quad (66)$$

$$S_t = \frac{t}{0.0079 + 0.007t} \quad (67)$$

$$S_t = 65.8094 + 31.1332 \cdot \ln(t) \quad (68)$$

$$S_t = 150.0693 \cdot \frac{t}{t+1} - 7.756 \cdot \frac{e^t - e^{-t}}{e^t + e^{-t}} \quad (69)$$

(**a**) Ultra-high-strength concrete

(**b**) Pressure experiment

Figure 3. Pressure experiment of ultra-high-strength concrete for Example 11.

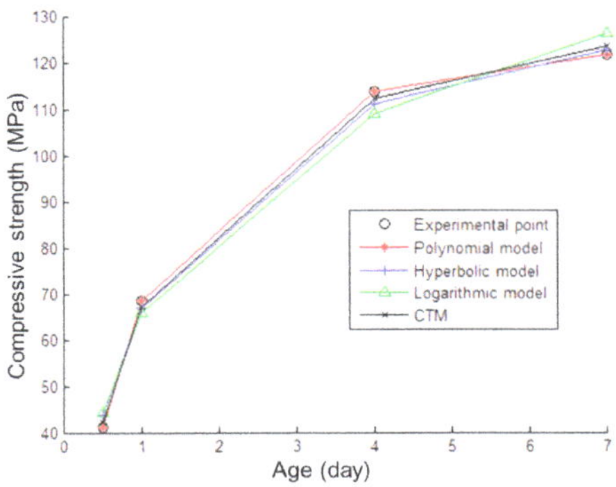

Figure 4. Fitting curves in Example 11.

Table 6. The mean fitting error, standard deviation, coefficient of variation, and 120-day predicted compressive strength of Example 11.

Example 11	Polynomial Model	Hyperbolic Model	Logarithmic Model	CTM
Mean fitting error	0.0000	0.0290	0.0485	0.0180
Standard deviation	0.0000	0.0227	0.0169	0.0055
Coefficient of variation	1.1207	0.7825	0.3485	0.3051
Predicted 120-day CS (MPa)	2179.2×10^3	141.3	214.9	141.0

5. Discussion

(1) According to Tables 2, 4 and 6, Figure 5 provides statistics on the relative values of the mean and standard deviation of fitting errors for Examples 1–11 using the four models. The relative values are calculated by taking the mean and standard deviation of CTM as 1 for each example, and then proportionally scaling the mean and standard deviation for the other models accordingly. As can be seen from Figure 5, the polynomial model is very unstable, especially for examples with little experimental data. The fitting results obtained by the polynomial model have large errors. As stated before, the predicted values of the 120-day compressive strength by the polynomial model are completely distorted. For some examples, the hyperbolic model outperforms the logarithmic model, while for others, the logarithmic model surpasses the hyperbolic model. Overall, the CTM exhibits the most stable performance and is able to achieve the most reliable prediction results.

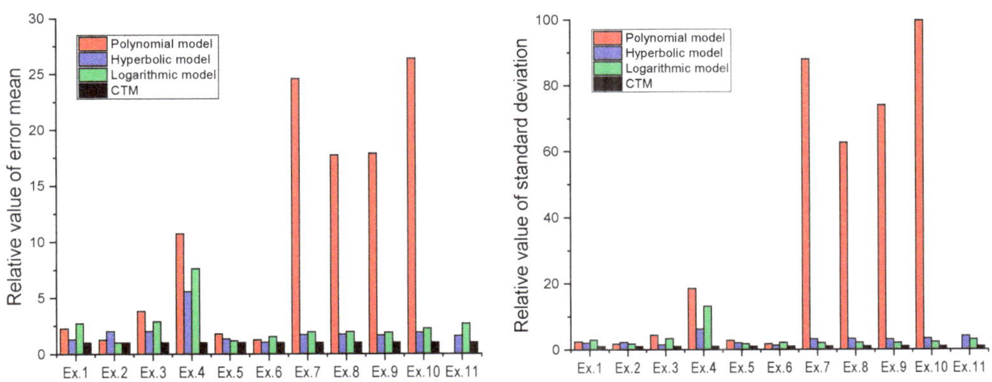

(**a**) Relative value of error mean (**b**) Relative value of standard deviation

Figure 5. Relative value of error mean and standard deviation for Examples 1–11.

(2) The proposed CTM can also be extended to predict the tensile strength of concrete. References [39,40] provide two sets of experimental data on the splitting tensile strength of concrete with strength grades of C30 and C40, respectively. Additionally, reference [41] offers two sets of experimental data, one on the splitting tensile strength and the other on the axial tensile strength, for concrete with a strength grade of C60. These four sets of experimental data are used as Examples 12–15 to illustrate the performance of the proposed model in predicting tensile strength. Figure 6 presents the experimental data points for these four examples along with the fitting curves of various models. Table 7 displays the evaluation metrics for various models, specifically focusing on the mean fitting error, standard deviation, coefficient of variation, and the predicted tensile strength values at the 120-day mark.

According to Table 7, Figure 7 provides statistics on the relative values of the mean and standard deviation of fitting errors for Examples 12–15 using the four models. The relative values are determined by setting the mean and standard deviation of the CTM to 1 for each example, and subsequently scaling the mean and standard deviation of the other models in proportion to this base. It can be observed from Figure 7 that, when considering both the mean error and standard deviation, the logarithmic model performs the worst, while the CTM performs the best. Further examination of Table 7 reveals that, despite the polynomial model having a lower mean error and standard deviation than the CTM in a very small number of cases, the polynomial model's prediction at the 120th day are completely distorted. This reinforces the conclusion that the CTM continues to demonstrate the best performance among all the models.

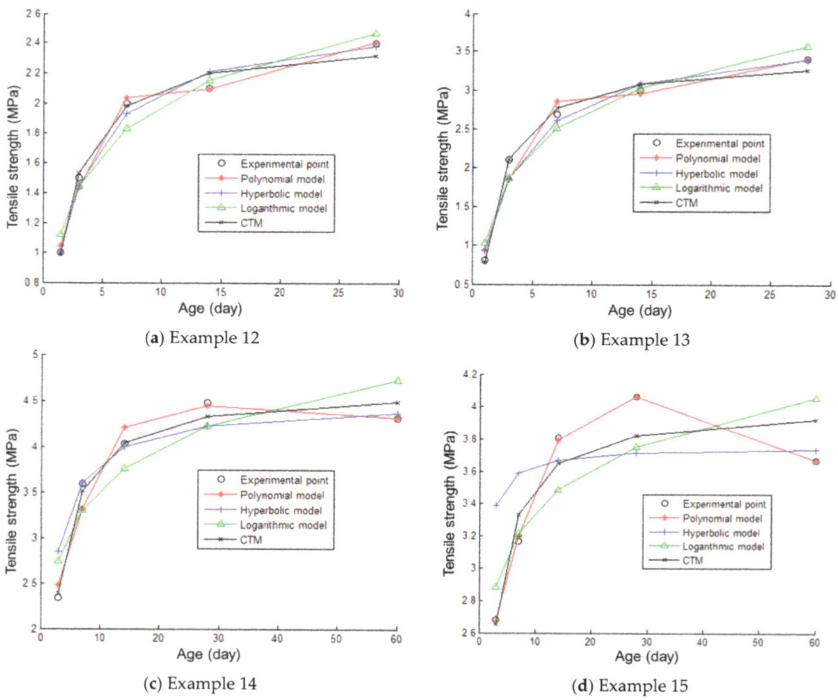

Figure 6. Fitting curves based on the tensile strength data for Examples 12–15.

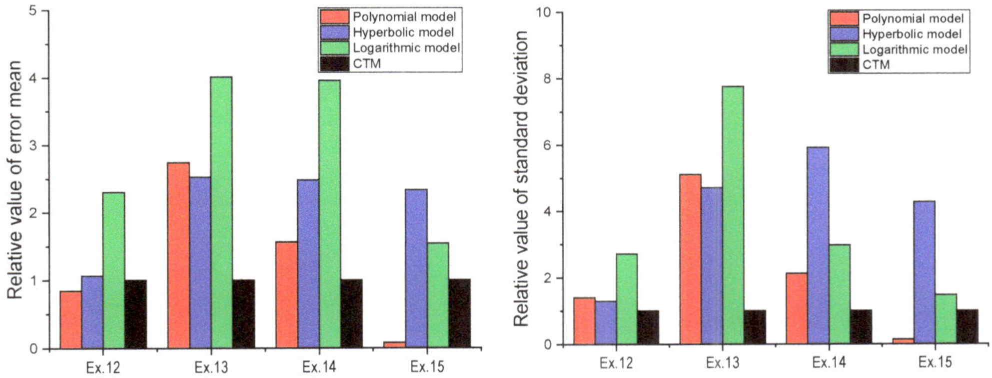

(**a**) Relative value of error mean (**b**) Relative value of standard deviation

Figure 7. Relative value of error mean and standard deviation for Examples 12–15.

Table 7. The mean fitting error, standard deviation, coefficient of variation, and 120-day predicted tensile strength of Examples 12–15.

	Index	Polynomial Model	Hyperbolic Model	Logarithmic Model	CTM
(a) Example 12	Mean fitting error	0.0218	0.0274	0.0593	0.0257
	Standard deviation	0.021	0.0194	0.0408	0.0150
	Coefficient of variation	0.9625	0.7086	0.6888	0.5830
	Predicted 120-day tensile strength (MPa)	612.7328	2.5254	3.1398	2.4094
(b) Example 13	Mean fitting error	0.0725	0.0668	0.1060	0.0264
	Standard deviation	0.0711	0.0655	0.1079	0.0139
	Coefficient of variation	0.9806	0.9794	1.0182	0.5272
	Predicted 120-day tensile strength (MPa)	1010.9	3.665	4.667	3.396
(c) Example 14	Mean fitting error	0.0374	0.0594	0.0946	0.0239
	Standard deviation	0.0326	0.0907	0.0456	0.0153
	Coefficient of variation	0.8720	1.5264	0.4826	0.6404
	Predicted 120-day tensile strength (MPa)	54.1620	4.4249	5.1819	4.5620
(d) Example 15	Mean fitting error	0.0037	0.1072	0.0709	0.0459
	Standard deviation	0.0034	0.0982	0.0339	0.0229
	Coefficient of variation	0.9035	0.9163	0.4780	0.4991
	Predicted 120-day tensile strength (MPa)	26.2071	3.7467	4.3225	3.9676

(3) According to the creation logic of the CTM presented in this work, Equation (70) offers a new model that is constructed by combining three different curves as follows.

$$S_t = p_0 \cdot \frac{t}{t+1} + p_1 \cdot \frac{t}{t+2} + p_2 \cdot \frac{e^t - e^{-t}}{e^t + e^{-t}} \quad (70)$$

Using the model of Equation (70), Table 8 presents the error analysis of Examples 7–11, encompassing the mean fitting error, standard deviation, and coefficient of variation, as well as the 120-day predictive compressive strength estimates. By comparing Table 8 with Tables 4 and 6, it can be observed that the mean and standard deviation of the model fitting errors of Equation (70) are both smaller than those of CTM. This indicates that, generally, the prediction accuracy can be improved by increasing the number of curves involved in the combination. Especially for Examples 7–10, the mean and standard deviation of the fitting errors for the model based on Equation (70) are nearly zero. This is because the number of fitting parameters in Equation (70) is exactly equal to the number of experimental points. This perfect match leads to a highly accurate fit, with minimal deviation from the experimental data. The versatility of this new model, which is composed of three curves, can be further explored in future research. This could lead to the development of more robust and reliable prediction models for a wide range of fields and applications.

Table 8. The mean fitting error, standard deviation, coefficient of variation, and 120-day predicted compressive strength obtained by Equation (70) for Examples 7–11.

Results by the Model of Equation (70)	Example 7	Example 8	Example 9	Example 10	Example 11
Mean fitting error	0.0000	0.0000	0.0000	0.0000	0.0061
Standard deviation	0.0000	0.0000	0.0000	0.0000	0.0015
Coefficient of variation	0.2415	0.2856	0.2790	0.2775	0.2529
Predicted 120-day CS (MPa)	31.3711	34.9432	32.0957	30.9492	134.2275

6. Conclusions

This work presents a novel CTM specifically designed to assess the evolving compressive strength of concrete during its ultra-early age. Comprised of two curves and featuring just two unknown parameters, this model efficiently fits experimental data, enabling the derivation of a compressive strength–age relationship for concrete with minimal data requirements. Notably, the new CTM surpasses single-curve models in terms of fitting accuracy. Key insights drawn from case studies include the following. (1) The polynomial model is limited to fitting early-age concrete data and lacks predictive capability for later stages. (2) The logarithmic model's predictions for later-age concrete strength significantly diverge from design specifications. (3) The hyperbolic model's fitting accuracy varies widely, with instances of performance inferior even to the logarithmic model. (4) The proposed CTM consistently demonstrates superior fitting accuracy across cases, underscoring its stability and versatility. Notably, its standard deviation of fitting errors is approximately 30–80% lower than the second-best model, ensuring predictions for later-age concrete strength align more closely with design grades. (5) The model's applicability extends beyond compressive strength, successfully predicting tensile strength during ultra-early ages, highlighting its unique strength in accurately characterizing concrete strength at various stages based on minimal experimental data. Collectively, these findings underscore the broad potential of this innovative combined curve model in evaluating the temporal behavior of concrete strength, opening avenues for its application in diverse fields.

Author Contributions: Methodology, Q.Y.; Software, H.C.; Investigation, F.Q.; Data curation, Q.Y.; Writing—original draft, H.C.; Visualization, F.Q.; Project administration, B.X.; Funding acquisition, B.X. All authors have read and agreed to the published version of the manuscript.

Funding: This research was supported by National Natural Science Foundation of China (No. 52078081) and Zhejiang Provincial Natural Science Foundation of China (No. LGF22E080021, No. LY24E080010).

Institutional Review Board Statement: Not applicable.

Informed Consent Statement: Not applicable.

Data Availability Statement: The data used to support the findings of this study are included in this study and are also available from the corresponding author upon request.

Conflicts of Interest: Bing Xiao was employed by the company Road & Bridge International Co., Ltd. The remaining authors declare that the research was conducted in the absence of any commercial or financial relationships that could be construed as a potential conflict of interest. The funders had no role in the collection, analyses, or interpretation of data, or in the decision to publish the results.

References

1. Mihashi, H.; Leite, J.P.B. State-of-the-art report on control of cracking in early age concrete. *J. Adv. Concr. Technol.* **2004**, *2*, 141–154. [CrossRef]
2. Bentz, D.P. A review of early-age properties of cement-based materials. *Cem. Concr. Res.* **2008**, *38*, 196–204. [CrossRef]
3. Maruyama, I.; Lura, P. Properties of early-age concrete relevant to cracking in massive concrete. *Cem. Concr. Res.* **2019**, *123*, 105770. [CrossRef]
4. Gu, H.; Song, G.; Dhonde, H.; Mo, Y.L.; Yan, S. Concrete early-age strength monitoring using embedded piezoelectric transducers. *Smart Mater. Struct.* **2006**, *15*, 1837. [CrossRef]
5. Lee, T.; Lee, J. Setting time and compressive strength prediction model of concrete by nondestructive ultrasonic pulse velocity testing at early age. *Constr. Build. Mater.* **2020**, *252*, 119027. [CrossRef]
6. Yoon, H.; Kim, Y.J.; Kim, H.S.; Kang, J.W.; Koh, H.-M. Evaluation of early-age concrete compressive strength with ultrasonic sensors. *Sensors* **2017**, *17*, 1817. [CrossRef]
7. Benaicha, M.; Burtschell, Y.; Alaoui, A.H. Prediction of compressive strength at early age of concrete—Application of maturity. *J. Build. Eng.* **2016**, *6*, 119–125. [CrossRef]
8. Lee, C.; Lee, S.; Nguyen, N. Modeling of compressive strength development of high-early-strength-concrete at different curing temperatures. *Int. J. Concr. Struct. Mater.* **2016**, *10*, 205–219. [CrossRef]
9. Min, T.-B.; Cho, I.-S.; Park, W.-J.; Choi, H.-K.; Lee, H.-S. Experimental study on the development of compressive strength of early concrete age using calcium-based hardening accelerator and high early strength cement. *Constr. Build. Mater.* **2014**, *64*, 208–214. [CrossRef]
10. Chung, K.L.; Wang, L.; Ghannam, M.; Guan, M.; Luo, J. Prediction of concrete compressive strength based on early-age effective conductivity measurement. *J. Build. Eng.* **2021**, *35*, 101998. [CrossRef]
11. Voigt, T.; Malonn, T.; Shah, S.P. Green and early age compressive strength of extruded cement mortar monitored with compression tests and ultrasonic techniques. *Cem. Concr. Res.* **2006**, *36*, 858–867. [CrossRef]
12. Demirboga, R.; Türkmen, İ.; Karakoc, M.B. Relationship between ultrasonic velocity and compressive strength for high-volume mineral-admixtured concrete. *Cem. Concr. Res.* **2004**, *34*, 2329–2336. [CrossRef]
13. Velay-Lizancos, M.; Martinez-Lage, I.; Azenha, M.; Granja, J.; Vazquez-Burgo, P. Concrete with fine and coarse recycled aggregates: E-modulus evolution, compressive strength and non-destructive testing at early ages. *Constr. Build. Mater.* **2018**, *193*, 323–331. [CrossRef]
14. Jin, N.J.; Seung, I.; Choi, Y.S.; Yeon, J. Prediction of early-age compressive strength of epoxy resin concrete using the maturity method. *Constr. Build. Mater.* **2017**, *152*, 990–998. [CrossRef]
15. Adesina, A. Influence of various additives on the early age compressive strength of sodium carbonate activated slag composites: An overview. *J. Mech. Behav. Mater.* **2020**, *29*, 106–113. [CrossRef]
16. Jiang, H.; Qi, Z.; Yilmaz, E.; Han, J.; Qiu, J.; Dong, C. Effectiveness of alkali-activated slag as alternative binder on workability and early age compressive strength of cemented paste backfills. *Constr. Build. Mater.* **2019**, *218*, 689–700. [CrossRef]
17. Kaszyńska, M. Early age properties of high-strength/high-performance concrete. *Cem. Concr. Compos.* **2002**, *24*, 253–261. [CrossRef]
18. Burhan, L.; Ghafor, K.; Mohammed, A. Enhancing the fresh and hardened properties of the early age concrete modified with powder polymers and characterized using different models. *Adv. Civ. Eng. Mater.* **2020**, *9*, 227–249. [CrossRef]
19. Voigt, T.; Sun, Z.; Shah, S.P. Comparison of ultrasonic wave reflection method and maturity method in evaluating early-age compressive strength of mortar. *Cem. Concr. Compos.* **2006**, *28*, 307–316. [CrossRef]
20. Liu, C.P. Mechanical property test of sisal fiber reinforced coral concrete. *J. Luoyang Inst. Sci. Technol. Nat. Sci. Ed.* **2019**, *29*, 1–4.
21. Ling, J.J. Compressive Strength and Nonlinear Prediction Model of High-grade Basalt Manufactured Sand Concrete. *J. Water Resour. Archit. Eng.* **2021**, *19*, 124–130.
22. Jia, M.; Han, Z.B. Analysis of Compressive Test of Silicate Concrete Based on Bridge Estimation. *J. North China Univ. Sci. Technol. Nat. Sci. Ed.* **2019**, *41*, 84–89.
23. Yang, W.J.; Wang, Y. Diachronic change model of compressive strength and elastic modulus of concrete at early age. *J. China Foreign Highw.* **2007**, *27*, 149–151.
24. Wang, W.; Zhou, A.; Feng, L.; Pan, G.; Li, J. Mathematical model for age-dependent compressive strength of recycled coarse aggregate concrete. *J. Build. Mater.* **2012**, *15*, 633–637.
25. Zhao, S.B.; Gao, Y.J.; Chen, J.H.; Liu, S.; Sun, Y.P. Application and study on time-dependent compressive strength and elastic modulus of C50 pumping concrete. *Concrete* **2015**, *303*, 98–101.
26. Tang, X.L.; Qin, Y.H.; Qu, W.J. Experimental study on time-varying regularity of compressive and bond strength of concrete at early-age. *J. Build. Struct.* **2009**, *30*, 145–150.
27. Yang, Y.; Chen, F.Y.; Zuo, T. Strength development and evaluation of high-strength concrete at early age. *China Concr. Cem. Prod.* **2004**, *3*, 1–4.

28. Mckinnie, S.B. Time Dependent Compressive Strength and Modulus of Elasticity of Florida Concrete. Ph.D. Thesis, Department of Civil and Environmental Engineering, Florida State University College of Engineering, Tallahassee, FL, USA, 2004.
29. Comite Euro-International Du Beton. *CEB-FIP Model Code 1990*; Thomas Telford Services Ltd.: London, UK, 1993; pp. 51–52, 1990p. Available online: https://vdocument.in/ceb-fip-model-code-1990pdf.html?page=1 (accessed on 23 March 2023).
30. Wang, L.; Hu, X.M. Shale ceramsite concrete early age experimental study on compressive strength. *Jiangsu Constr.* **2016**, *5*, 98–99.
31. Zhao, Z. Proportion design and strength regularity of high strength concrete. *Concrete* **2011**, *10*, 101–103.
32. Li, X.F.; Ma, W.L.; Pei, S.W.; Sun, S.N. Experimental study on compressive strength of commercial concrete at early age. *J. Zhongyuan Inst. Technol.* **2006**, *17*, 59–61.
33. Bai, R.P.; Han, K.P. Experimental study on compressive strength of polypropylene fiber-reinforced concrete at early-age. *Concrete* **2012**, *3*, 28–31.
34. Zhao, D.H.; Sun, F.L.; Shan, R.L.; Song, Y.W.; Zhao, W. Study on the dynamic mechanical properties of artificially negative temperature high-strength concrete. *Coal Eng.* **2018**, *50*, 56–60.
35. Hu, W.X.; Tao, J.F.; Zeng, X. Research on the preparation and performance of large volume concrete for marine engineering with new cement system. In Proceedings of the China Construction Technology Exchange Conference in 2023 (Part I), Beiijng, China, 26 June 2023.
36. He, X. Experimental study on the strength of C55 concrete at early ages of approach bridge of Jing-Yue Yangtze River highway bridge. *J. Guangdong Commun. Polytech.* **2012**, *11*, 15–18.
37. Chang, Z.G.; Zhang, H.X. Influence of material time-dependent effect on cantilever construction of PC box girder bridge. *J. Highw. Transp. Res. Dev.* **2019**, *36*, 67–73.
38. Wang, G.X.; Wang, Y.R.; Yang, K. Experimental study on the compression age strength of carbon fiber fly ash ceramsite concrete. *Eng. Constr.* **2014**, *46*, 4–7.
39. Wang, D.K.; Zhuang, J.P.; Cai, X.F.; Qiu, H. Experimental study on cracking tensile strength of concrete based on age degree. *Concrete* **2021**, *5*, 10–12.
40. Wang, M.M. Experimental Study on Mechanical Behavior of C40 Concrete at Early Age. Ph.D. Thesis, Beijing Jiaotong University, Beijing, China, 2014.
41. Li, X.F.; Liu, W.; Sun, S.N. Experimental research on tensile strength of premixed concrete at early ages. *Ind. Constr.* **2006**, *6*, 77–80. [CrossRef]

coatings

MDPI

Article

Fatigue-Healing Performance Analysis of Warm-Mix Rubber Asphalt Mastic Using the Simplified Viscoelastic Continuum Damage Theory

Ping Li [1], Xiao Li [2], Shangjun Yu [2], Linhao Sun [3], Jinchao Yue [2] and Ruixia Li [2,*]

1 Henan Xuxin Expressway Co., Ltd., Zhumadian 463000, China; pingli1031@163.com
2 School of Water Conservancy and Transportation, Zhengzhou University, Zhengzhou 450001, China; lixiao106118@163.com (X.L.); 15837119118@163.com (S.Y.); yuejc@zzu.edu.cn (J.Y.)
3 Zhengzhou Public Utility Investment and Development Group Co., Ltd., Zhengzhou 450001, China; sunlinhao2023@163.com
* Correspondence: ruixiali031@163.com

Abstract: As a green and low-carbon road material, warm-mix rubber asphalt (WMRA) has received extensive attention from scholars for its road performance. In the in-depth study of its properties, the fatigue characteristics of WMRA are particularly critical. However, in current studies on asphalt fatigue performance, its self-healing ability is often underestimated or neglected. Furthermore, the simplified viscoelastic continuum damage theory (S-VECD), with its accuracy, speed, and convenience, provides a powerful tool for analyzing asphalt fatigue performance. Therefore, to analyze the fatigue and self-healing performances of WMRA in practical applications, four sample materials were selected in this study: virgin asphalt mastic (VAM), rubber asphalt mastic (RAM), Sasobit rubber asphalt mastic (SRAM), and Evotherm rubber asphalt mastic (ERAM). Subsequently, the samples were subjected to a comprehensive experimental design with frequency sweep tests, linear amplitude sweep tests, and multiple intermittent loading time sweep tests under different aging conditions. The fatigue and self-healing performances of different aging degrees and different types of WMRA were evaluated based on the S-VECD theory. The results show that aging reduces the fatigue and self-healing performances of asphalt mastic to a certain extent, and at a 7% strain, the fatigue life of SRAM after long-term aging is only 30.7% of the life of the unaged sample. The greater the aging degree, the more pronounced the effect. Under different aging levels, the warm-mix agent can significantly improve the fatigue and self-healing performances of rubber asphalt mastic. After undergoing ten fatigue intermittent loading tests, the recovery rate of the complex shear modulus for the long-term aged VAM was 0.65, which is lower than that of SRAM under the same conditions, and the warm mix can further extend the fatigue life of rubber asphalt by improving the self-healing properties of the asphalt. The role of Sasobit in enhancing the fatigue and self-healing performances of rubber asphalt mastic is more significant. This study can provide a theoretical basis for the promotion and application of WMRA pavements and contribute to the sustainable development of road construction.

Keywords: warm-mix rubber asphalt mastic; fatigue-healing performance; simplified viscoelastic continuum damage theory; thermal and oxygen aging

1. Introduction

As the global economy grows and people's consumption levels rise, the automotive industry and global tire production proliferate. Relevant studies show that about 4 billion tires are discarded globally every year [1]. Traditional waste rubber disposal methods, such as landfills and incineration, cause environmental pollution and damage human health [2,3]. Grinding waste tires into rubber powder and adding them to asphalt to make rubber asphalt pavement solves the above problems and significantly improves pavement performance [4,5]. Therefore, the technical means of rubber asphalt is increasing in attention and is gradually being used in road engineering. However, the production and construction of rubber asphalt produces a lot of harmful gases due to the high temperature, which limits its popularity and application.

At present, warm-mix technology mainly includes foaming warm-mix technology, emulsified asphalt warm-mix technology, and organic viscosity reduction warm-mix technology [6–9]; among them, the organic viscosity-reducing warm mix, represented by the Sasobit warm mix and the emulsified asphalt warm mix represented by Evotherm warm mix are widely used. Suppose waste tire rubber powder is added to asphalt using warm-mix technology. In that case, it not only reduces greenhouse gas emissions during the production and construction process but also reduces energy loss [10–12]. Not only that, but with further research scholars have found that the use of warm-mix technology can improve the performance of rubber asphalt. Guo et al. [10] found that warm-mix additives improved the low-temperature crack resistance of asphalt mixtures through freeze–thaw cycle tests. Shi et al. [13] used dynamic mechanical analysis (DMA) and found that an organic viscosity-reducing temperature mix improves the high-temperature performance of warm-mix rubber asphalt. Bilema et al. [14] found that Sasobit improves the stiffness and workability of rubber asphalt by comparing the effects of three warm mixes on the physical properties of rubber asphalt.

In addition to conventional performance studies, the fatigue performance of asphalt materials has emerged as a research priority and is aimed at enhancing the service life and road performance of asphalt pavements [15,16]. Under the long-term influence of vehicle loads, fatigue cracks develop in asphalt pavements, which seriously impacts their service life. Therefore, in recent years, scholars have successively carried out related research on the fatigue performance of warm-mix rubber asphalt and its mixtures. Wang et al. [17] found that the addition of a warm mix can effectively improve the long-term fatigue performance of rubber asphalt by testing the fatigue performance of warm-mix rubber asphalt after long-term aging. Xiao et al. [18] evaluated the effects of warm-mix asphalt additives on the fatigue characteristics of the rubber asphalt mixture from the perspectives of low temperature, aging, and fatigue. They found that warm-mix agents can effectively extend the fatigue life of rubber asphalt mixture. Kumar et al. [19] compared the fatigue performance of different types of warm-mix rubber asphalt and its mixtures through various fatigue testing methods and found that the chemical warm-mix asphalt (WMA) additive improved the fatigue resistance of rubber asphalt more than the organic WMA additive. Wang et al. [20] revealed the fatigue performance enhancement mechanism of warm asphalt rubber by removing crumb rubber modifier (CRM) particles through filtration, obtaining the liquid phases of asphalt rubber (AR) and four WAR binders, and performing fatigue tests and gel permeation chromatography (GPC) tests on them.

Based on the above literature analysis, current research on the fatigue performance of WMRA pavements by scholars mainly focuses on two directions: Firstly, the use of macroscale testing to analyze the influence of the type and dosage of warm-mix additives on the fatigue performance of WMRA and its mixtures, and secondly, the employment of microscale testing to explore the mechanisms by which warm-mix additives enhance the fatigue performance of modified asphalt. However, there has been relatively little research on the impact of asphalt material's self-healing properties on the fatigue damage process. The asphalt material has a certain self-healing function (i.e., self-healing performance). Under certain conditions, the generated microcracks can be gradually recovered [21,22]. As the microcrack surface undergoes reconstruction, contact, and wetting, the stiffness of the asphalt begins to recover and the cracks gradually close. After the cracks close, the asphalt molecules begin to diffuse into each other, and eventually, the stiffness and strength of the asphalt become comparable to their original state [23–25]. This lack of research contributes to the specific differences between fatigue life measured in the laboratory and fatigue life observed under actual pavement conditions. Moreover, asphalt mastic is the basis for forming asphalt mixtures, a key factor affecting the road performance of asphalt mixtures [26–28]. In addition, the aging effect on the performance of asphalt pavement is unavoidable in practical applications. Therefore, to more accurately assess the service life of warm-mix rubber asphalt materials, it is necessary to carry out a systematic study on the fatigue performance and self-healing performance of asphalt mastic under different aging effects.

In summary, the aim of this paper is to analyze and compare the fatigue and self-healing properties of virgin asphalt mastic (VAM), rubber asphalt mastic (RAM), Sasobit rubber asphalt mastic (SRAM), and Evotherm rubber asphalt mastic (ERAM) under different degrees of aging, in order to assess the fatigue durability and self-healing capacity of warm-mixed rubber asphalt pavements in practical applications. To achieve this objective, four materials were aged in this study through short-term and long-term aging tests simulating everyday aging environments. The frequency scanning test, linear amplitude test, and multiple intermittent loading time scanning test were further carried out by a dynamic shear rheometer, and theoretical analyses were carried out by using the simplified viscoelastic continuum damage theory to provide theoretical basis and technical support for the application of warm-mix rubber asphalt pavements.

2. Materials and Methods

2.1. Materials

In this paper, 70# petroleum asphalt was selected, and the property indicators are shown in Table 1.

In this paper, a 60-mesh fine rubber powder type of rubber powder was selected, as shown in Figure 1. The dosage was 20% of the weight of asphalt. Its main technical parameters are shown in Table 2.

Figure 1. The rubber powder.

Table 1. Technical indicators of asphalt.

Technical Indicator	Test Value	Technical Requirement	Unit
Penetration (25 °C, 100 g, 5 s)	70.9	60–80	0.1 mm
Softening point	54.6	≥46	°C
Ductility (15 °C, 5 cm/min)	>100	≥100	cm

Table 2. Main technical parameters of rubber.

Technical Indicators	Test Value	Unit
Particle size	60	mesh
Moisture content	0.69	%
Relative density	1.35	-

In this paper, two warm mixes were selected: one is organic viscosity-reducing warm-mix Sasobit, as shown in Figure 2a, developed by Sasol, Johannesburg, South Africa, with a dosage of 3% of the weight of asphalt. The other is Evotherm, an emulsified asphalt warm mix developed by MeadWestvaco, Richmond, Virginia, USA, as shown in Figure 2b. In this paper, the third generation of Evotherm is used, and the dosage is 0.6% of the weight of asphalt. Due to the different modification mechanisms, the dosage of Sasobit and Evotherm is different. The dosage in this paper is the optimum dosage based on relevant studies and manufacturer recommendations [29–31]. The main basic properties of the two warm mixes are shown in Tables 3 and 4.

(**a**) (**b**)

Figure 2. The warm mix: (**a**) Sasobit and (**b**) Evotherm.

Table 3. The main basic properties of Sasobit warm mixes.

Technical Indicators	Test Value	Unit
Melting point	101	°C
Density (25 °C)	0.94	g/cm^3
Flash point	293	°C
Brinell rotational viscosity (135 °C)	5.46×10^{-3}	Pa·s

Table 4. The main basic properties of Evotherm warm mixes.

Technical Indicators	Test Value	Unit
Density (25 °C)	8.07	lbs/gal
Proportion (25 °C)	0.968	-
Viscosity (20 °C)	660–1225	mPa·s

The filler of the mastic is limestone mineral powder, as shown in Figure 3, with a density of 2.7429/cm^3, hydrophilic coefficient of 0.83, particle size of 0.075 mm through the rate of 100%, and no agglomeration phenomenon.

Figure 3. The limestone mineral powder.

The flow diagram of the research in this paper is shown in Figure 4.

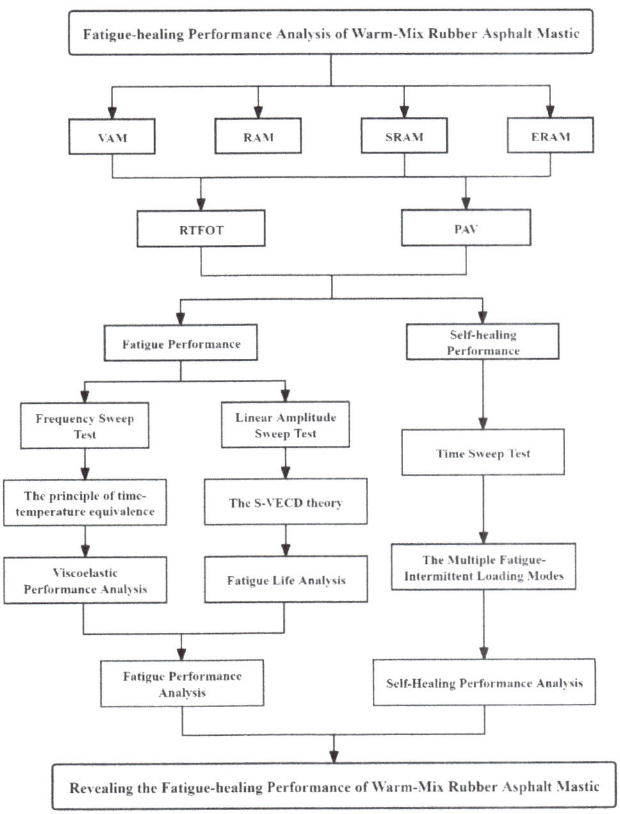

Figure 4. The flow diagram of the research.

2.2. Sample Preparation

2.2.1. Modified Asphalt

Firstly, the asphalt was heated to flow dynamically at 150 °C, and the corresponding mass of rubber powder was added to it at 170 °C. Then, it was sheared with a high-speed shear at 4500 rad/min for 60 min to produce rubber asphalt. After that, we maintained 160 °C, added the corresponding mass of warm mix to the already prepared rubber asphalt, and continued to shear at 4500 rad/min for 10 min to obtain warm-mix rubber asphalt.

2.2.2. Asphalt Mastic

We weighed the limestone mineral powder according to the powder rubber ratio of 1:1 and placed the weighed mineral powder and asphalt into the oven, heated at 150 °C for 2 h. Then, we added the mineral powder to the asphalt in batches, maintained the temperature at 150 °C, and sheared it with a high-speed shearer at a rotational speed of 3000 rad/min for 35 min to make asphalt mastic.

2.3. Methods

2.3.1. Aging Test

(1) The rolling thin film oven test (RTFOT) in JTG E20-2011 [32] was used to simulate the short-term aging of asphalt during transport, storage, and paving.

(2) To simulate the long-term aging of asphalt pavements during long-term use, the pressurized aging vessel (PAV) test from JTG E20-2011 [32] was used in this study for the asphalt after short-term aging.

The short-term aging (RTFOT) and long-term aging (PAV) test specimens are shown in Figures 5 and 6. The summary of asphalt mastic samples is shown in Table 5.

(a) (b)

Figure 5. The short-term aging test apparatus and specimens: (**a**) the rolling thin film oven and (**b**) the short-term aging specimens.

(a) (b)

Figure 6. The long-term aging test apparatus and specimens: (**a**) the pressure aging oven and (**b**) the long-term aging specimens.

Table 5. The summary of asphalt mastic samples.

Type of Asphalt Mastic	Aging Status
VAM	Unaged RTFOT PAV
RAM	Unaged RTFOT PAV
SRAM	Unaged RTFOT PAV
ERAM	Unaged RTFOT PAV

2.3.2. Frequency Sweep Test

When using the S-VECD model to characterize the fatigue properties of asphalt, it is necessary to comprehensively consider three main elements: the linear viscoelastic response, damage characteristic curve (DCC), and fatigue failure criterion (G^R) [33,34].

The dynamic shear rheometer used in this study is shown in Figure 7.

Figure 7. The dynamic shear rheometer.

Specimens were made using molds of 8 mm diameter and 2 mm height. The standard molds and asphalt mastic specimens are shown in Figure 8. To ensure the accuracy of the test data, three parallel tests were carried out for each asphalt mastic specimen, and the average of the data was taken for analysis.

Figure 8. The long-term aging test apparatus and specimens: (**a**) the standard molds and (**b**) the asphalt mastic specimens.

In this study, according to a standard testing procedure (AASHTO T 315-19 [35]), the frequency sweep test (FS) was used to obtain the parameters of the materials' linear viscoelastic properties [36]. The test temperatures were 15 °C, 25 °C, and 35 °C, and the sweep frequencies ranged from 0.1 to 100 rad/s. The test was conducted in the strain control mode with a strain level of 0.1%. The master curve was obtained by translating the frequency (time) at different test times through the principle of time-temperature equivalence.

The reduced angular frequency coordinates are used in the master curve, Equation (1):

$$\log \omega_r = \log \omega + \log \varphi_T \tag{1}$$

where ω is the angular frequency (rad/s), ω_r is the reduced angular frequency (rad/s), and φ_T is the shift factor.

2.3.3. Linear Amplitude Sweep Test

According to the standard testing procedure (AASHTO TP 101-14 [37]), the linear amplitude sweep (LAS) test was conducted using a dynamic shear rheometer [38,39]. A strain loading mode was applied at a test temperature of 25 °C, a loading frequency of 10 Hz, and a standard loading time of 5 min (300 s). During the test, the amplitude of the sinusoidal load applied to the asphalt material increased linearly from 0.1% to 30%.

Based on the simplified viscoelastic continuum damage (S-VECD) theory, the relationship between the damage modulus S of asphalt material and the corresponding virtual modulus C is established, as shown in Equation (2):

$$C(S) = \frac{\tau_P}{\gamma_P^R \cdot DMR} \tag{2}$$

where τ_P is the peak shear stress, γ_P^R is the peak virtual strain, and DMR is the ratio of the initial dynamic shear modulus to the linear viscoelastic dynamic shear modulus of the asphalt material.

The damage intensity (S) changes with time (t), and the relationship is shown in Equation (3):

$$S(t) = \sum_{i=1}^{N} \left[\frac{DMR}{2} \left(\gamma_P^R \right)^2 (C_{i-1} - C_i) \right]^{\frac{\alpha}{1+\alpha}} (t_{Ri} - t_{Ri-1})^{\frac{1}{1+\alpha}} \tag{3}$$

$$t_{Ri} = \frac{t_i}{\varphi_T} \tag{4}$$

$$\gamma_P^R = \frac{1}{G_R} \cdot \left(\left| G^* \right|_{LVE} \cdot \gamma_P \right) \tag{5}$$

where γ^R is the virtual strain; G_R is the reference modulus of the material, which is taken as 1 MPa in this study; $|G^*|_{LVE}$ is the linear viscoelastic dynamic shear modulus of the asphalt material; φ_T is the temperature shift factor of asphalt material; α is the damage evolution rate of the asphalt material, $\alpha = 1 + 1/m$; m is the slope of the master curve of the dynamic shear modulus in the linear viscoelastic range of the asphalt material; i is the loading cycle; and t_R is the curtailment time.

Fitting the imaginary modulus (C) and the damage variable (S) obtained based on the S-VECD theory, the Damage Characteristic Curve (DCC) of the asphalt material can be obtained as shown in Equation (6):

$$C(t) = 1 - K_1 \cdot S(t)^{K_2} \tag{6}$$

where K_1 and K_2 are the best fitting parameters for the DCC curve of asphalt materials.

Based on the theory of virtual strain energy and the processing of the data obtained from the LAS test, the amount of virtual strain energy stored in the asphalt specimen by the applied load can be calculated by Equation (7):

$$W_S^R = \frac{1}{2} \tau_P \gamma_P^R / DMR = \frac{1}{2} \cdot C \cdot \left(\gamma_P^R \right)^2 \tag{7}$$

where W_S^R is the stored virtual strain energy.

As the load level gradually increases, the released virtual strain energy gradually increases, and the stored virtual strain energy first increases and then decreases. A gradual increase in the released virtual strain energy indicates that the asphalt material is gradually being destroyed. When the stored virtual strain energy reaches the peak value, the maximum value of the asphalt material can store the applied load. Therefore, the peak value of stored virtual strain energy is chosen as the fatigue failure point in the LAS test.

In this paper, G^R (the average release rate of the virtual strain energy) is chosen as the fatigue failure criterion to predict the fatigue life of asphalt materials under different strains [40]. Non-standard linear amplitude tests with sweep times of 500 s and 800 s were added as a way to change the sweep time to adjust the constant strain rate for calculation.

The average released virtual strain energy for each loading cycle from the start of the test until the asphalt material reaches the point of fatigue failure is calculated and noted as $\overline{W_R^R}$. Therefore, G^R can be calculated using Equation (8):

$$G^R = \frac{\overline{W_R^R}}{N_f} = \frac{A}{(N_f)^2} \tag{8}$$

where A is the area of integration of the curve corresponding to the released virtual strain energy when the asphalt material reaches the peak value of stored virtual strain energy and N_f is the number of cyclic loading cycles of asphalt material.

Fitting G^R to the fatigue life (N_f), as shown in Equation (9):

$$G^R = a(N_f)^b \tag{9}$$

where a and b are the best fit parameters.

By combining the above relationship between G^R and N_f with the S-VECD theory and the virtual strain energy theory, the relationship between fatigue life N_f and the strain amplitude γ_P in the test can be obtained, which finally constitutes the fatigue failure criterion based on G^R, as shown in Equation (10), and thus, the fatigue life of the asphalt material can be calculated under different strains.

$$N_f = \left[\frac{X}{a} \cdot (\gamma_P)^{2+2\alpha(\frac{K_2}{Q})}\right]^{\frac{1}{b+1-\frac{K_2}{Q}}} \tag{10}$$

where the parameters X and Q are calculated by Equations (11)–(13).

$$X = \frac{1}{2} \cdot K_1 \cdot \left(\left|G^*\right|_{LVE}\right)^2 \cdot Y^{(-K_2/Q)} \cdot \frac{1}{(K_2/Q)+1} \tag{11}$$

$$Q = 1 - \alpha \cdot K_2 + \alpha \tag{12}$$

$$Y = \frac{f \cdot 2^\alpha}{Q(K_1K_2)^\alpha(|G^*|_{LVE})^{2\alpha}} \tag{13}$$

2.3.4. Time Sweep Test

This study used time sweep (TS) to conduct the self-healing research on asphalt mastic [41]. Furthermore, to more realistically simulate actual asphalt pavement loading conditions, this paper selected multiple fatigue intermittent loading modes. Ten fatigue intermittent loading tests were conducted under strain-controlled conditions, with a strain level of 1.25%, a loading frequency of 10 Hz, and a temperature of 25 °C. The loading was halted when the complex shear modulus decreased to 55% of its initial value, followed by an intermittent period of 30 min, after which the test was resumed.

As shown in Figure 9, $N_H(i)$ is the number of self-healing loading cycles increased from the previous one after the i-th interval.

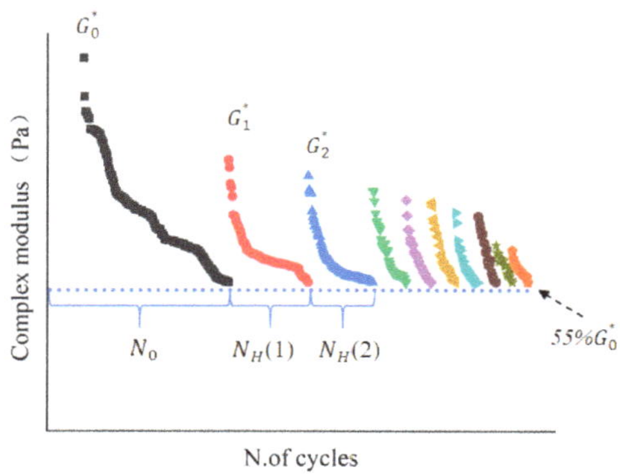

Figure 9. Relationship of complex shear modulus with the number of loadings.

$\sum N_H(i)$ is the cumulative increase in the number of self-healing loading cycles after the i-th interval, as shown in Equation (14):

$$\sum N_H(i) = \sum_{i=1}^{n} N_H(i) \tag{14}$$

As the number of intervals increases, the complex shear modulus also changes. The change in the dynamic shear modulus of the asphalt material with the increase in the number of intervals can be quantified by the following Equation (15):

$$RCM = \frac{G_i^*}{G_0^*} \tag{15}$$

where RCM is the recovery rate of the complex shear modulus, G_i^* is the complex shear modulus after the i-th healing process, and G_0^* is the initial complex shear modulus.

3. Results and Analysis

3.1. Results and Analysis of Fatigue Performance Tests

The Ishikawa cause-and-effect diagram is the evaluation of the fatigue performance of asphalt mastic based on the S-VECD theory and is shown in Figure 10.

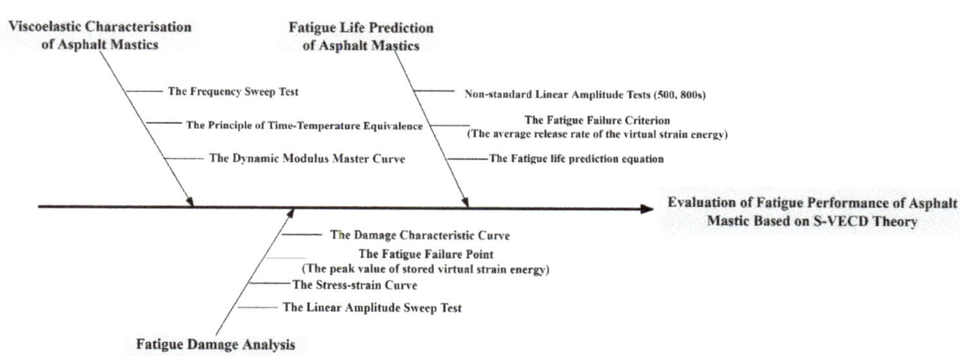

Figure 10. The Ishikawa cause-and-effect diagram: Evaluation of the fatigue performance of asphalt mastic based on the S-VECD theory.

3.1.1. Dynamic Modulus Master Curve

The viscoelastic performance of asphalt is closely related to its fatigue performance, and high-quality viscoelastic performance helps to reduce the amount of deformation of asphalt during repeated loading, which effectively reduces fatigue damage, thus enhancing its durability and service life [42].

Figure 11 shows the dynamic modulus master curves of the four asphalt mastics at different degrees of aging. As can be seen from Figure 11, the SRAM and ERAM show high dynamic shear moduli in the whole frequency range, both in the unaged, short-term aging, and long-term aging states. Moreover, the complex shear moduli of all four asphalt mastics increased to a certain extent after both the short-term and long-term aging effects.

By further comparison, SRAM has the highest complex shear modulus in the case of unaged and short-term aging. This is consistent with previous findings [29]. This may be due to the fact that Sasobit is capable of forming a network-like crystalline structure within asphalt, which enhances the asphalt's shear resistance and, as a result, elevates its complex shear modulus [29]. In contrast, ERAM has the highest complex modulus in the case of long-term aging. In actual pavement applications, asphalt will inevitably be subjected to a variety of aging effects. Therefore, when using rubber asphalt pavement, incorporating Sasobit and Evotherm warm-mix additives can enhance the complex shear modulus of the asphalt, thereby improving resistance to deformation of the asphalt pavement during its actual use.

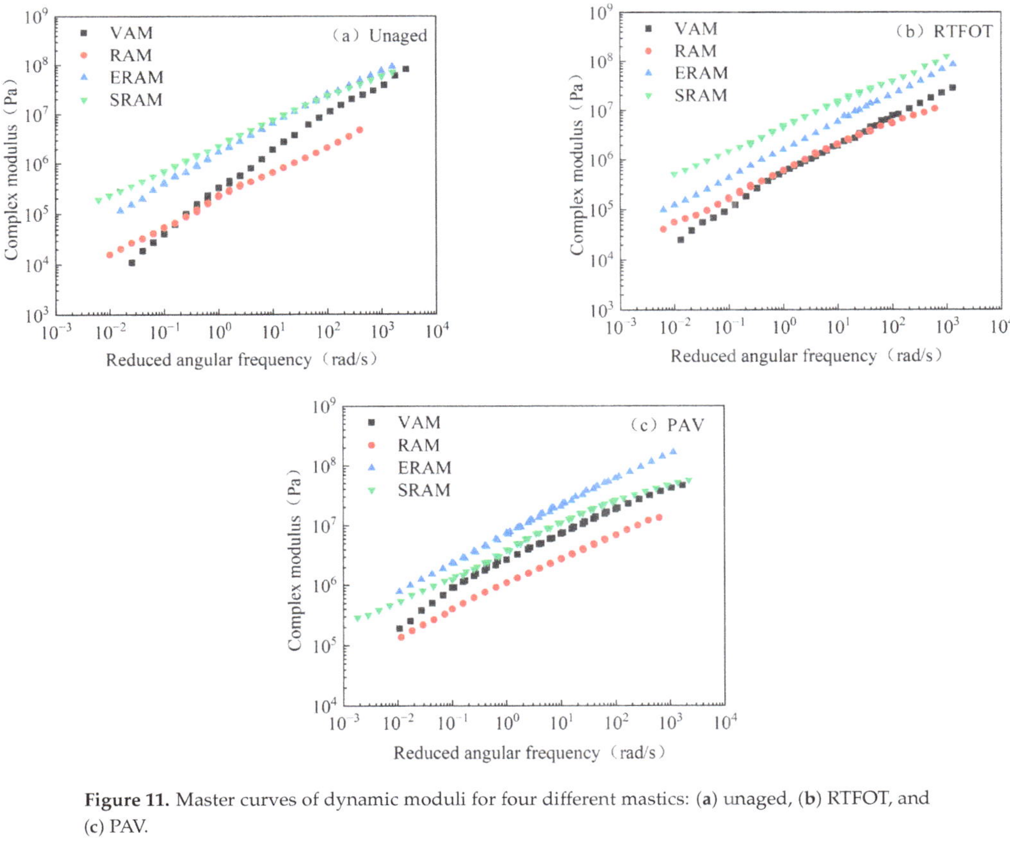

Figure 11. Master curves of dynamic moduli for four different mastics: (**a**) unaged, (**b**) RTFOT, and (**c**) PAV.

3.1.2. Stress–Strain Curves

Figure 12 presents the stress–strain curves of the four asphalt mastics under different aging degrees, with the endpoints corresponding to the shear stress and strain at the fatigue failure point. It can be seen from Figure 12 that all four asphalt mastics have peak stresses under different aging conditions, but the locations and widths have changed. The wider peak stress of asphalt mastic represents the larger shear strain it can withstand. The changes in the locations and widths of the peak stresses are due to the differences in the warm mixes and the degree of aging. Under different aging conditions, SRAM has the highest peak stress, and ERAM has the second-highest peak stress, indicating that the addition of a warm mix enhances the fatigue damage resistance of RAM to a certain extent. Comparing the stress–strain curves under different aging conditions, it was found that the Sasobit warm mix had a more noticeable effect on the fatigue resistance of rubber asphalt mastic.

Further comparison shows that the peak stresses of the same asphalt mastic have increased with the aging degree. This is because aging causes the light components of the asphalt to volatilize and the internal components to oxidize with oxygen, making the asphalt mastic harder. The deeper the aging, the more pronounced the effect.

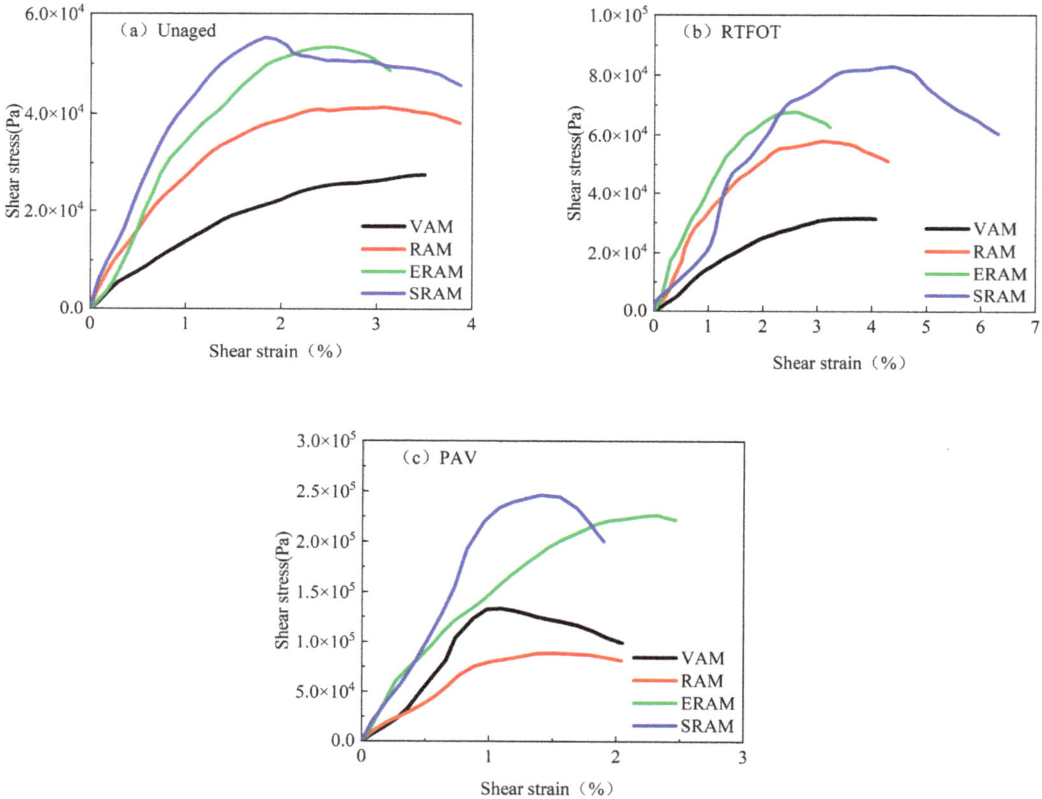

Figure 12. Stress–strain curves of four different asphalt mastics: (**a**) unaged, (**b**) RTFOT, and (**c**) PAV.

3.1.3. Damage Characterization Curves

The damage characteristic curve (DCC), i.e., the relationship between the virtual modulus C and the damage variable S, was fitted for the four asphalt mastics at different aging levels, as shown in Figure 13, where the endpoints of the curves correspond to the fatigue failure point of each asphalt mastic. As shown in Figure 13, all the DCC curves have the same trend of a decreasing virtual modulus C from 1 and increasing S from 0. This indicates that fatigue damage is produced in the asphalt mastic as the LAS test proceeds, demonstrating the damage evolution process of asphalt materials.

By comparison, the S values of SRAM and ERAM are greater than those of RAM when the asphalt mastic reaches the fatigue damage failure point, regardless of the aging action. This indicates that under the same load, the warm-mix rubber asphalt mastic can withstand greater damage, reaching fatigue damage when the time required is longer. Therefore, in practice, Sasobit and Evotherm can be added to improve the fatigue resistance of rubber asphalt pavements.

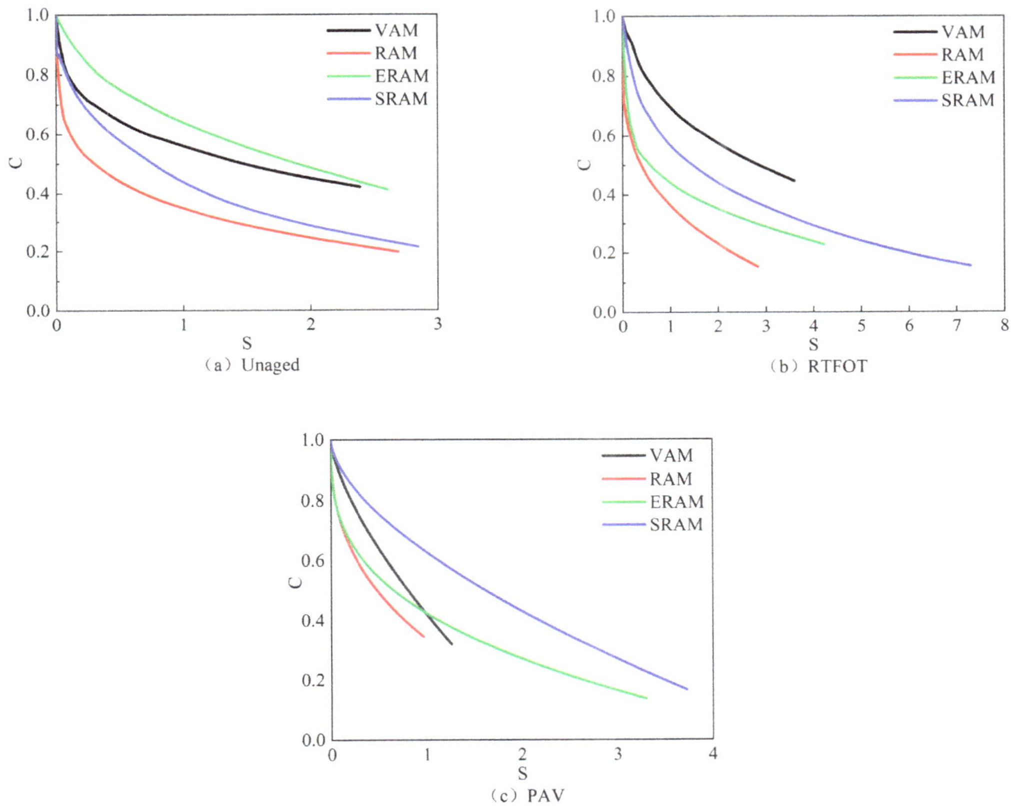

Figure 13. DCC curves of four different asphalt mastics: (**a**) unaged, (**b**) RTFOT, and (**c**) PAV.

3.1.4. Fatigue Life Analysis

Typically, each asphalt material possesses a different DCC curve, and the cumulative fatigue damage and virtual modulus analysis can only provide a qualitative understanding of the fatigue behavior of asphalt materials. To quantitatively evaluate the fatigue performance of different asphalt mastics, this paper employs Equations (9)–(13) to calculate the fatigue life of various asphalt mastics under various aging conditions and at different strain levels (4%, 7%, and 10%). The fatigue life of asphalt mastic at 7% strain is shown in Figure 14. Subsequently, a linear fitting of the strain and N_f (fatigue life) is performed using a double logarithmic coordinate system, as depicted in Figure 15, which offers a more intuitive comparison of the fatigue life performance of asphalt mastic under different strain conditions [43].

From Figure 15, it can be seen that the fatigue life of the four asphalt pastes shows a decreasing trend with an increase in the strain amplitude. In addition, regardless of the aging state and strain amplitude, the fatigue life curves of the asphalt mastics with the addition of warm-mix agents Sasobit and Evotherm are above those of RAM and VAM. This is consistent with previous findings [44]. For example, after long-term aging at 7% strain, the fatigue life of SRAM is 9.9 times the fatigue life of VAM, 2.4 times the fatigue life of RAM, and the fatigue life of ERAM is 8.7 times the fatigue life of VAM and 1.7 times the fatigue life of RAM. This shows that a warm-mix agent can significantly improve the fatigue life of asphalt mastic, and the improvement effect of Sasobit is more obvious. This is because Sasobit forms a network of crystalline structures in the asphalt. These structures make the asphalt more rigid at low and medium temperatures, which improves the deformation resistance of the asphalt [29]. On the other hand, Evotherm is able to increase the adhesion between asphalt and mineral powder, which makes the bonding interface between the two less prone to damage, thus reducing the generation and extension of fatigue cracks [45].

Comparing the fatigue life of the same asphalt mastic, it can be found that aging will reduce the fatigue life of asphalt mastic to a certain extent, and the deeper the aging level, the greater the reduction. For example, at 7% strain, the fatigue life of SRAM after long-term aging is only 30.7% of the fatigue life in the unaged state and 49.4% of the fatigue life in the short-term aging state. This shows that aging has a certain adverse effect on the fatigue performance of asphalt mastic.

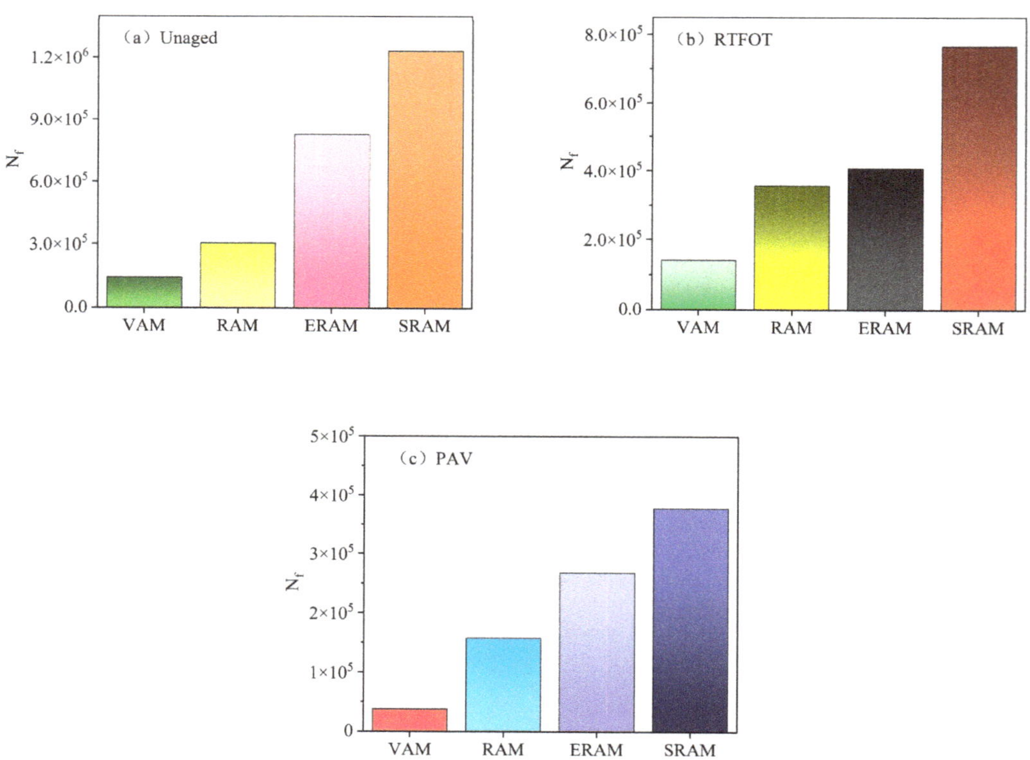

Figure 14. Fatigue life of four different asphalt mastics at 7% strain level: (**a**) unaged, (**b**) RTFOT, and (**c**) PAV.

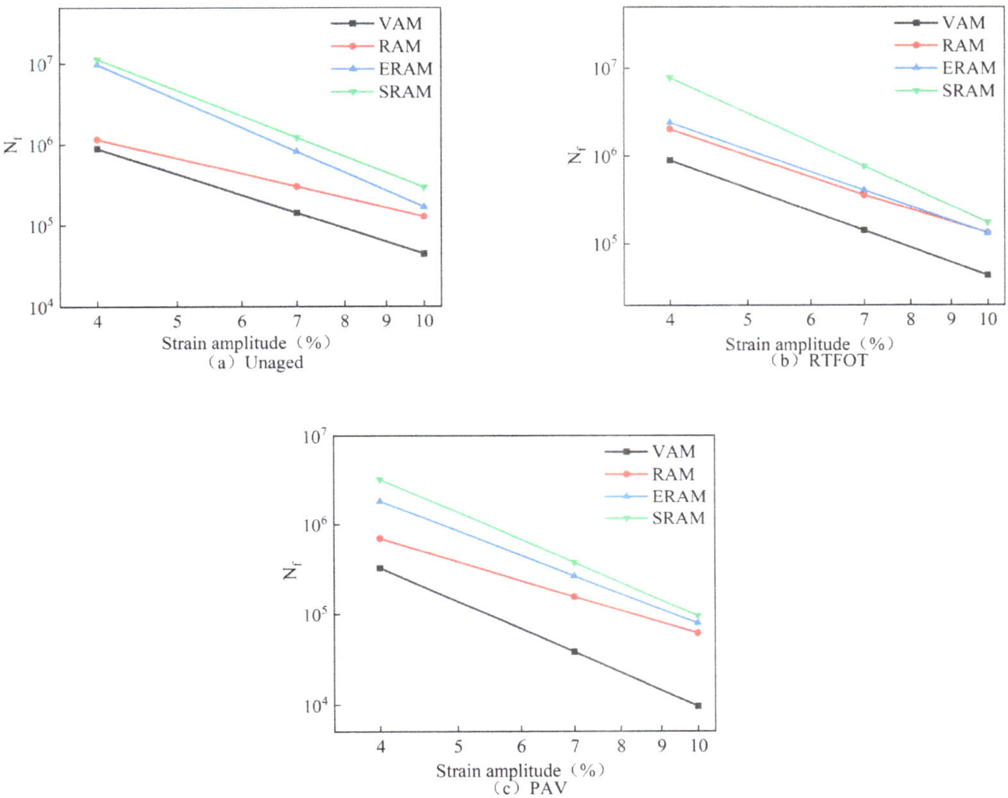

Figure 15. Fatigue life of four different asphalt mastics at different strain levels: (**a**) unaged, (**b**) RTFOT, and (**c**) PAV.

3.2. Results and Analysis of Self-Healing Performance Tests

3.2.1. Cumulative Increase in the Number of Self-Healing Loading Cycles

Based on the experimental results of the multiple intermittent test method, the values of $\sum N_H(i)$ for the four asphalt mastics at different levels of aging were obtained, as shown in Figure 16. By analyzing Figure 16, the $\sum N_H(i)$ of each asphalt mastic increases gradually with the increase in loading and the number of intervals, but its growth rate slows down gradually. This indicates that although the intervals are the same, multiple loadings gradually decrease the recovery of the asphalt mastic. The self-healing capacity of the asphalt mastic decreases continuously and tends to zero.

By further comparing the changes in $\sum N_H(i)$ of the four asphalt mastics under the effects of unaged, short-term aging, and long-term aging, it can be found that aging decreases the $\sum N_H(i)$ of each asphalt mastic. For example, after ten fatigue intermittent loads in the long-term aging state, the $\sum N_H(i)$ of the SRAM is only 53.2% of the $\sum N_H(i)$ in the unaged state and 58.8% of the $\sum N_H(i)$ in the short-term aging state. This indicates that aging decreases the self-healing ability of asphalt mastic, and the deeper the aging, the greater the decrease.

However, the $\sum N_H(i)$ of SRAM and ERAM are higher than those of VAM and RAM, regardless of the aging condition. For example, after ten fatigue intermittent loads in a long-term aging condition, the $\sum N_H(i)$ of VAM is only 43.2% of the $\sum N_H(i)$ of SRAM and, 48.2% of the $\sum N_H(i)$ of ERAM, and the $\sum N_H(i)$ of RAM is only 76.1% of the $\sum N_H(i)$ of SRAM and 84.8% of the $\sum N_H(i)$ of ERAM. This indicates that the warm mix can significantly enhance the self-healing ability of rubber asphalt mastic.

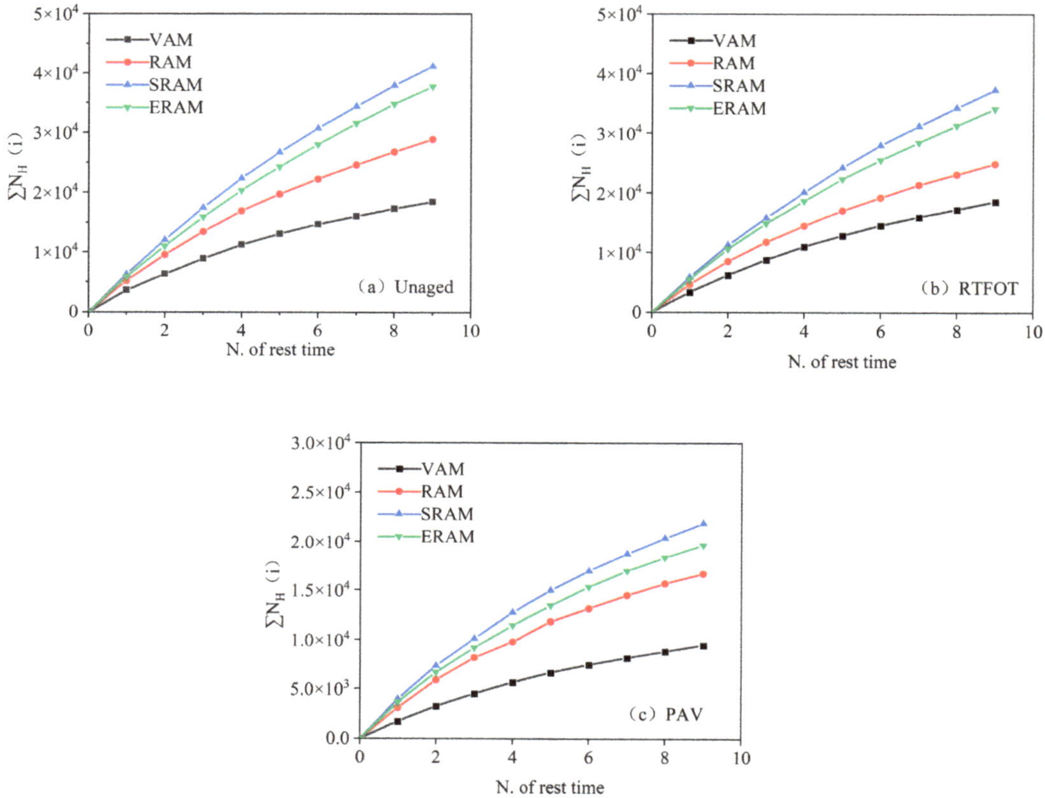

Figure 16. Relationship between $\sum N_H(i)$ and the number of rest time: (**a**) unaged, (**b**) RTFOT, and (**c**) PAV.

3.2.2. The Recovery Rate of Complex Shear Modulus

The complex shear modulus of asphalt mastic can be recovered to a certain extent after fatigue loading and the introduction of intervals. Therefore, the self-healing ability of the four asphalt mastics can be observed by the recovery rate of the complex shear modulus (RCM) under different aging degrees and multiple loading intervals. As depicted in Figure 17, the RCM values of the four asphalt mastics under the three aging conditions exhibit a decreasing trend, signifying the gradual weakening of their self-healing ability.

A further comparison can show that the RCM values of SRAM and ERAM were higher than those of VAM and RAM under different aging conditions. For example, the values of RCM for long-term aging VAM and RAM after ten fatigue intermittent loading tests are 0.65 and 0.70, respectively, which are lower than the values of RCM for SRAM and ERAM in the same state. This shows that the warm mix can significantly improve the self-healing performance of RAM, among which the improvement effect of the Sasobit warm mix is more satisfactory. The enhancement mechanisms of Sasobit and Evotherm on the self-healing performance of asphalt have not been clarified, probably because the long-chain aliphatic alkane structure of Sasobit forms reticulated crystals in asphalt, which slows down crack expansion and promotes self-healing. Evotherm reduces the viscosity of asphalt and improves the low-temperature fluidity, which is conducive to the filling and healing of microcracks. These hypotheses require further investigation, research, and validation.

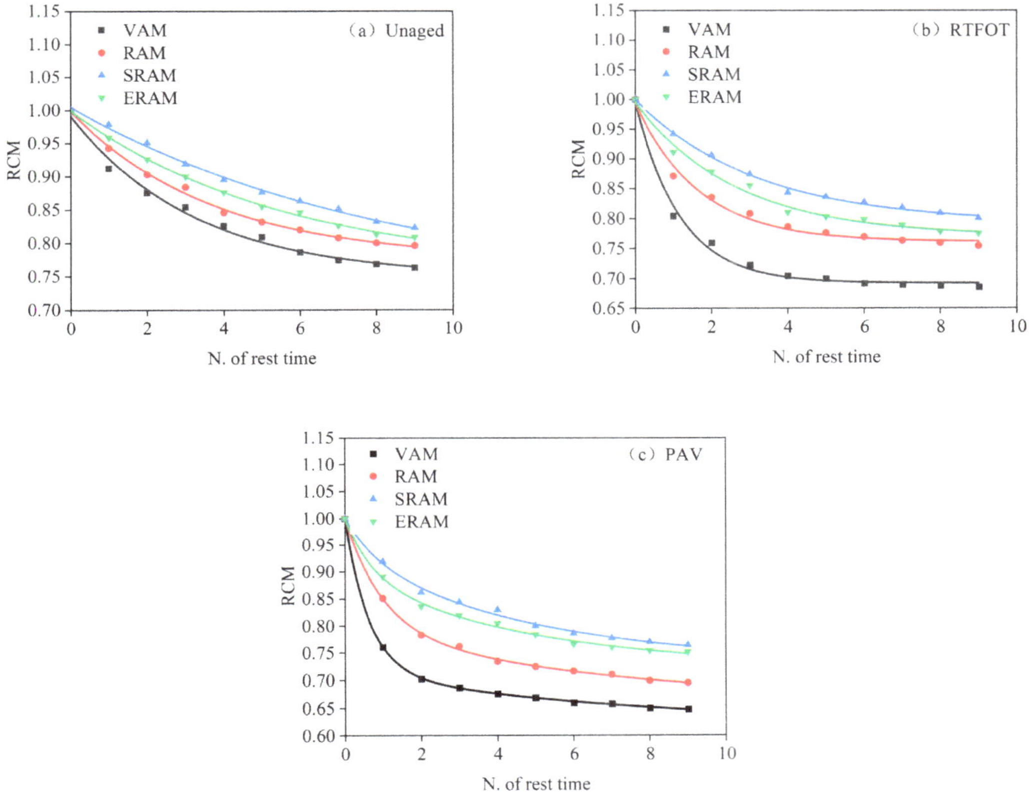

Figure 17. Relationship between RCM and the number of rest time: (**a**) unaged, (**b**) RTFOT, and (**c**) PAV.

3.3. Fatigue Life Analysis Considering the Effect of Self-Healing

From the above self-healing tests, it is evident that the introduction of the intervals increases the loading cycles of the asphalt mastic, which affects its fatigue life to a certain extent. Therefore, to more accurately predict the fatigue life of warm-mix rubber asphalt mastic, the fatigue life conversion factor μ was used in this study to calculate the fatigue life after self-healing [46]. μ is calculated as shown in (16):

$$\mu = \frac{N_0 + N_H(1)}{N_0} \quad (16)$$

where N_0 is the number of initial loading cycles, and $N_H(1)$ is the number of self-healing loading cycles increased from the previous one after the first interval. The μ values of various asphalt mastics are shown in Table 6.

Therefore, the predicted fatigue life equation considering asphalt healing is given as:

$$N_f(\text{with rest}) = \mu \times N_f'(\text{without rest}) \quad (17)$$

According to the study, the strain of asphalt was estimated to be 50 times the strain in the bulk mixture, with the strain in strong pavements greater than 4 inches thick estimated to be 500 μm, which implies a strain of 2.5% in asphalt, and the strain in weak pavements less than 4 inches thick is estimated to be 1000 μm, resulting in a strain of 5% in the asphalt [47–49]. Therefore, the fatigue life of asphalt mastics at a 5% strain was used as an example [42,43]. After calculation, the fatigue life of the four asphalt mastics under 5% strain, considering the effect of self-healing and the fatigue life without self-healing, were obtained after undergoing different aging effects, as shown in Figure 18.

Table 6. Fatigue life conversion factor before and after asphalt healing μ.

Type of Asphalt	Unaged	RTFOT	PAV
VAM	1.500896	1.469959	1.370156
RAM	1.527895	1.500586	1.449185
ERAM	1.529193	1.515128	1.474655
SRAM	1.535672	1.529622	1.491847

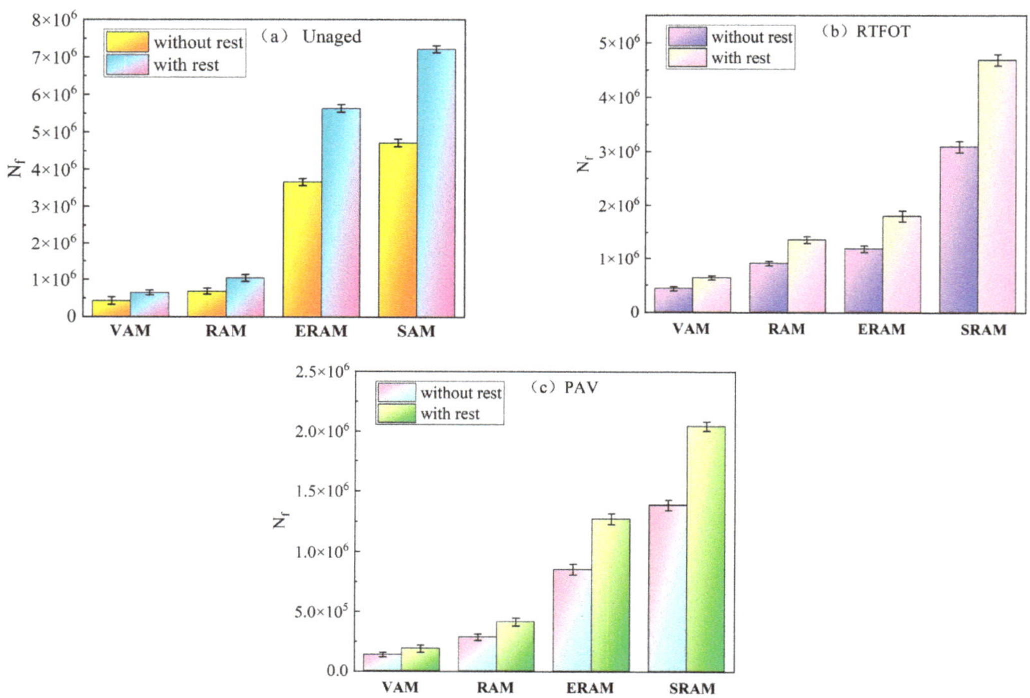

Figure 18. Comparison of fatigue life before and after asphalt mastic healing at 5% strain level: (**a**) unaged, (**b**) RTFOT, and (**c**) PAV.

From Figure 18, including the interstitial period can effectively prolong the fatigue life of asphalt mastic, thus indicating that its self-healing behavior significantly affects its fatigue life. In addition, the fatigue life growth rate of asphalt mastic decreases after short-term and long-term aging, with the largest decrease after long-term aging, which is attributed to an increase in the asphalt's hard components during long-term aging, which inhibits the healing of asphalt microcracks and leads to the deterioration of its ability to withstand the secondary loading. The fatigue life of asphalt mastic was significantly improved by both warm mixes in the unaged condition and the two aging conditions, with Sasobit providing the greatest improvement, which suggests that the Sasobit warm mix can not only positively affect the fatigue performance of asphalt but also further extend the fatigue life of asphalt by enhancing the self-healing performance of asphalt.

4. Conclusions

Through experimental tests and theoretical analyses, this paper evaluated the fatigue and self-healing performances of four asphalt mastics, VAM, RAM, SRAM, and ERAM, under simulated aging conditions, as well as the effect of self-healing behavior on fatigue performance. The following conclusions can be drawn:

(1) Both Sasobit and Evotherm can significantly enhance the fatigue performance of asphalt mastics, thereby extending their fatigue life. This is due to the fact that the former can form a solid lattice structure in the bitumen, while the latter can effectively increase the bond between the bitumen and the mineral powder. Specifically, under the long-term aging condition of 7% strain, the fatigue life of SRAM is 9.9 times as long as that of VAM and 2.4 times as long as that of RAM; the fatigue life of ERAM is 8.7 times as long as that of VAM and 1.7 times as long as that of RAM. In addition, SRAM shows better fatigue performance.

(2) The self-healing properties of four asphalt mastics under different aging states were tested and analyzed by multiple intermittent loading TS tests based on $\sum N_H(i)$ and RCM. In particular, after ten fatigue intermittent loadings in the long-term aging condition, the $\sum N_H(i)$ for VAM is only 43.2% of the $\sum N_H(i)$ for SRAM and 48.2% of the $\sum N_H(i)$ for ERAM, while the $\sum N_H(i)$ for RAM is only 76.1% of the N for SRAM and 84.8% of the $\sum N_H(i)$ for ERAM. Moreover, the values of RCM for long-term aging VAM and RAM after ten fatigue intermittent loading tests are 0.65 and 0.70, respectively, which are lower than the values of RCM for SRAM. This indicates that adding a warm-mix agent can improve the self-healing properties of asphalt mastic.

(3) The fatigue life of asphalt mastic at a 5% strain level before and after considering the effect of self-healing was calculated by the fatigue life conversion factor μ. A comparison reveals that the fatigue life of asphalt mastic is significantly prolonged after considering self-healing, which indicates that the self-healing property of asphalt mastic significantly enhances its fatigue performance.

(4) This indicates that aging has some adverse effects on fatigue performance and self-healing performance, and the deeper the aging, the more pronounced the adverse effects. This is because aging hardens the asphalt and reduces its fluidity, which affects the road's life and the asphalt's self-healing performance. At a 7% strain, the fatigue life of SRAM after long-term aging is only 30.7% of the fatigue life in the unaged state and 49.4% of the fatigue life in the short-term aging state. Moreover, after ten fatigue intermittent loads in the long-term aging state, the $\sum N_H(i)$ of the SRAM is only 53.2% of the $\sum N_H(i)$ in the unaged state and 58.8% of the $\sum N_H(i)$ in the short-term aging state.

(5) The comprehensive results of the study concluded that Sasobit has a better effect on improving the fatigue and self-healing performances of rubber asphalt mastic. Therefore, Sasobit warm mix is preferable in practical applications to improve the performance of rubber asphalt pavements. This study can provide a theoretical basis for the promotion and application of WMRA pavements and contribute to the sustainable development of road construction.

Based on the S-VECD theoretical system, this study provides an in-depth analysis of the fatigue durability and self-healing capacity of warm-mix rubber asphalt mastic under different aging conditions. This paper not only systematically examines how aging affects the fatigue properties and self-healing potential of this material but also further explores the positive role of self-healing performance on fatigue performance. This paper aims to lay a theoretical foundation for the wide application of warm-mix rubber asphalt pavements through this series of studies and to help promote the wide application of warm-mix rubber asphalt technology in the field of road construction.

Author Contributions: P.L.: Methodology, Investigation, and Writing—original draft. X.L.: Data curation. S.Y.: Data curation. L.S.: Data curation. J.Y.: Resources. R.L.: Validation. All authors have read and agreed to the published version of the manuscript.

Funding: Funding support was granted from the Projects of the Henan Provincial Department of Transportation (Grant No. 2021-2-13,2018G11) and the National Natural Science Foundation of China (Grant No. 51708513) and was greatly appreciated.

Institutional Review Board Statement: The study did not require ethical approval.

Informed Consent Statement: The study did not involve humans and we exclude this statement.

Data Availability Statement: Data are contained within the article.

Conflicts of Interest: I confirm that none of the authors of this article have a conflict of interest.

References

1. Wang, M.; Zhang, L.; Li, A.; Irfan, M.; Du, Y.; Di, W. Comparative pyrolysis behaviors of tire tread and side wall from waste tire and characterization of the resulting chars. *J. Environ. Manag.* **2019**, *232*, 364–371. [CrossRef] [PubMed]
2. Dong, R.; Zhao, M.; Tang, N. Characterization of crumb tire rubber lightly pyrolyzed in waste cooking oil and the properties of its modified bitumen. *Constr. Build. Mater.* **2019**, *195*, 10–18. [CrossRef]
3. Zheng, G.; Zhang, N.; Lv, S. Experimental Study on Dynamic Modulus of High Content Rubber Asphalt Mixture. *Buildings* **2024**, *14*, 434. [CrossRef]
4. Presti, D.L. Recycled tyre rubber modified bitumens for road asphalt mixtures: A literature review. *Constr. Build. Mater.* **2013**, *49*, 863–881. [CrossRef]
5. Su, J.; Li, P.; Zhu, G.; Wang, X.; Dong, S. Interface Interaction of Waste Rubber–Asphalt System. *Buildings* **2024**, *14*, 1868. [CrossRef]
6. Yu, H.; Leng, Z.; Zhou, Z.; Shih, K.; Xiao, F.; Gao, Z. Optimization of preparation procedure of liquid warm mix additive modified asphalt rubber. *J. Clean. Prod.* **2017**, *141*, 336–345. [CrossRef]
7. Topal, A.; Oner, J.; Sengoz, B.; Dokandari, P.A.; Özdemir, D.K. Evaluation of Rutting Performance of Warm Mix Asphalt. *Int. J. Civ. Eng.* **2017**, *15*, 705–714. [CrossRef]
8. Sharma, A.; Mir, M.S.; Farooq, M.A. Performance of WMA Additives Under Freeze–thaw Action. *Road Mater. Pavement Des.* **2019**, *20*, 1468–1480. [CrossRef]
9. Cheraghian, G.; Falchetto, A.C.; You, Z.; Chen, S.; Kim, S.Y.; Westerhoff, J.; Moon, K.H.; Wistuba, M.P. Warm Mix Asphalt Technology: An up to Date Review. *J. Clean. Prod.* **2020**, *268*, 122–128. [CrossRef]
10. Guo, J.; Chang, C.; Wang, L. Low-Temperature Crack Resistance of High-Content Rubber-Powder-Modified Asphalt Mixture under Freeze–Thaw Cycles. *Polymers* **2024**, *16*, 402. [CrossRef]
11. Gui, W.; Liang, L.; Wang, L.; Zhang, F. Cracking Resistance of Recycled Rubber Asphalt Binder Composed of Warm-Mix Additives. *Materials* **2022**, *15*, 4389. [CrossRef] [PubMed]
12. Wang, H.; Liu, X.; Apostolidis, P.; Scarpas, T. Review of warm mix rubberized asphalt concrete: Towards a sustainable paving technology. *J. Clean. Prod.* **2018**, *177*, 302–314. [CrossRef]
13. Shi, J.; Li, J.; Li, C.; Wang, T.; Liu, J.; Cao, A. Effect of Organic Viscosity-Reducing Warm-Mix Agent on the Performance of Rubber Asphalt. *Coatings* **2022**, *12*, 152. [CrossRef]
14. Bilema, M.; Wah Yuen, C.; Alharthai, M.; Hazim Al-Saffar, Z.; Oleiwi Aletba, S.R.; Md Yusoff, N.I. Influence of Warm Mix Asphalt Additives on the Physical Characteristics of Crumb Rubber Asphalt Binders. *Appl. Sci.* **2023**, *13*, 10337. [CrossRef]
15. Deng, H.B.; Ma, X.; Deng, D.Y.; Tan, L. Fatigue properties of rubber modified SMA asphalt mixture. In Proceedings of the 1st International Conference on Advances in Civil Engineering and Materials (ACEM)/1st World Symposium on Sustainable Bio-Composite Materials and Structures (SBMS), Nanjing, China, 9–11 November 2019; Volume 275.
16. Yuan, J.; Lv, S.; Peng, X.; You, L.; Borges Cabrera, M. Investigation of strength and fatigue life of rubber asphalt mixture. *Materials* **2020**, *13*, 3325. [CrossRef] [PubMed]
17. Wang, H.; Liu, X.; van de Ven, M.; Lu, G.; Erkens, S.; Skarpas, A. Fatigue performance of long-term aged crumb rubber modified bitumen containing warm-mix additives. *Constr. Build. Mater.* **2020**, *239*, 117824. [CrossRef]
18. Xiao, F.; Zhao, W.; Amirkhanian, S.N. Laboratory Investigation of Fatigue Characteristics of Rubberized Asphalt Mixtures Containing Warm Asphalt Warm Asphalt Additives at A Low Temperature. *J. Test. Eval.* **2011**, *39*, 290–295. [CrossRef]
19. Kumar, A.; Choudhary, R.; Kandhal, P.S.; Julaganti, A.; Behera, O.P.; Singh, A.; Kumar, R. Fatigue characterisation of modified asphalt binders containing warm mix asphalt additives. *Road Mater. Pavement Des.* **2020**, *21*, 519–541. [CrossRef]
20. Wang, D.; Li, D.; Yan, J.; Leng, Z.; Wu, Y.; Yu, J.; Yu, H. Rheological and chemical characteristic of warm asphalt rubber binders and their liquid phases. *Constr. Build. Mater.* **2018**, *193*, 547–556. [CrossRef]
21. Wang, H.; Yuan, M.; Wu, J.; Wan, P.; Liu, Q. Self-Healing Properties of Asphalt Concrete with Calcium Alginate Capsules Containing Different Healing Agents. *Materials* **2022**, *15*, 5555. [CrossRef]
22. Zhang, F.; Sun, Y.; Kong, L.; Cannone Falchetto, A.; Yuan, D.; Wang, W. Study on Multiple Effects of Self-Healing Properties and Thermal Characteristics of Asphalt Pavement. *Buildings* **2024**, *14*, 1313. [CrossRef]

23. Xu, H.; Xu, W.; Zheng, X.; Cao, K. A Multistage Analysis of Asphalt Binder Nanocrack Generation and Self-Healing Behavior Based on Molecular Dynamics. *Polymers* **2022**, *14*, 3581. [CrossRef] [PubMed]
24. Hasheminejad, N.; Vuye, C.; Margaritis, A.; Ribbens, B.; Jacobs, G.; Blom, J.; Van den bergh, W.; Dirckx, J.; Vanlanduit, S. Investigation of Crack Propagation and Healing of Asphalt Concrete Using Digital Image Correlation. *Appl. Sci.* **2019**, *9*, 2459. [CrossRef]
25. He, L.; Zhou, Z.; Ling, F.; Alexiadis, A.; Van den Bergh, W.; Cannone Falchetto, A.; Balieu, R.; Zhu, J.; Valentin, J.; Kowalski, K.J.; et al. A Coarse-Grained Molecular Model for Simulating Self-Healing of Bitumen. *Appl. Sci.* **2022**, *12*, 10360. [CrossRef]
26. Ma, X.; Kan, J.; Liu, S.; Tu, M.; Wang, D. Investigation of the Effect of Filler on Cohesive Bond Strength of Asphalt Mastic Using Binder Bond Strength (BBS) Test. *Coatings* **2023**, *13*, 1001. [CrossRef]
27. E, G.; Zhang, J.; Shen, Q.; Ji, P.; Wang, J.; Xiao, Y. Influence of Filler Type and Rheological Properties of Asphalt Mastic on the Asphalt Mastic–Aggregate Interaction. *Materials* **2023**, *16*, 574. [CrossRef] [PubMed]
28. Russo, F.; Veropalumbo, R.; Oreto, C.; Biancardo, S.A.; Abbondati, F.; Viscione, N. Verifying the Mechanical Performance of Cold and Hot Asphalt Mastics Containing Jet Grouting Waste as a Filler. *Coatings* **2021**, *11*, 751. [CrossRef]
29. Ge, D.; Yan, K.; You, L.; Wang, Z. Modification mechanism of asphalt modified with Sasobit and Polyphosphoric acid (PPA). *Constr. Build. Mater.* **2017**, *143*, 419–428. [CrossRef]
30. Wu, S.; Zhang, W.; Shen, S.; Li, X.; Muhunthan, B.; Mohammad, L.N. Field-aged asphalt binder performance evaluation for Evotherm warm mix asphalt: Comparisons with hot mix asphalt. *Constr. Build. Mater.* **2017**, *156*, 574–583. [CrossRef]
31. Lu, D.X.; Saleh, M. Laboratory evaluation of warm mix asphalt incorporating high RAP proportion by using evotherm and sylvaroad additives. *Constr. Build. Mater.* **2016**, *114*, 580–587. [CrossRef]
32. *JTG E20-2011*; Standard Test Methods of Bitumen and Bituminous Mixtures for Highway Engineering. Ministry of Transport of the People's Republic of China: Beijing, China, 2011.
33. Lv, S.; Ge, D.; Cao, S.; Liu, D.; Zhang, W.; Li, C.-H.; Borges Cabrera, M. Development of a Framework for Assessing Bitumen Fatigue Cracking Performance under Different Temperatures and Aging Conditions. *Buildings* **2024**, *14*, 311. [CrossRef]
34. Klug, A.; Ng, A.; Faxina, A. Application of the Viscoelastic Continuum Damage Theory to Study the Fatigue Performance of Asphalt Mixtures—A Literature Review. *Sustainability* **2022**, *14*, 4973. [CrossRef]
35. *AASHTO T 315-19*; Standard Method of Test for Determining the Rheological Properties of Asphalt Binder Using a Dynamic Shear Rheometer (DSR). AASHTO: Washington, DC, USA, 2019.
36. Huo, W.; Zhuang, Y.; Wang, Z.; Kang, X.; Wang, R. The Microscopic Mechanism and Rheological Properties of SBS-Modified Asphalt with Warm Mixing Fast-Melting. *Materials* **2023**, *16*, 5690. [CrossRef]
37. *AASHTO TP 101-14*; Standard Method of Test for Estimating Damage Tolerance of Asphalt Binders Using the Linear Amplitude Sweep. AASHTO: Washington, DC, USA, 2014.
38. Zhou, H.-Y.; Dou, H.-B.; Chen, X.-H. Rheological Properties of Graphene/Polyethylene Composite Modified Asphalt Binder. *Materials* **2021**, *14*, 3986. [CrossRef] [PubMed]
39. Zhang, Y.; Liu, X.; Apostolidis, P.; Gard, W.; van de Ven, M.; Erkens, S.; Jing, R. Chemical and Rheological Evaluation of Aged Lignin-Modified Bitumen. *Materials* **2019**, *12*, 4176. [CrossRef]
40. Yue, M.; Yue, J.; Wang, R.; Xiong, Y. Evaluating the fatigue characteristics and healing potential of asphalt binder modified with Sasobit® and polymers using linear amplitude sweep test. *Constr. Build. Mater.* **2021**, *289*, 123054. [CrossRef]
41. Luo, X.; Xie, K.; Zhao, S.; Yao, Q.; Qin, Y. Study of the Molecular Components and Rheological Properties of Asphalt after Long-Term Aging under the Action of Moisture. *Sustainability* **2022**, *14*, 9080. [CrossRef]
42. Men, B.; Li, X.; Sha, T.; Guo, F.; Tang, G.; Yue, J. Research on Fatigue Performance of Fast-Melting SBS/Epoxy Resin Composite-Modified Asphalt. *Coatings* **2024**, *14*, 789. [CrossRef]
43. Wang, H.; Du, Z.; Liu, G.; Luo, X.; Yang, C. Effect of Long-Term Aging on Fatigue and Thermal Cracking Performance of Polyphosphoric Acid and Styrene–Butadiene–Styrene-Modified Bio-Blend Bitumen. *Polymers* **2023**, *15*, 2911. [CrossRef]
44. Izadi, A.; Motamedi, M.; Alimi, R.; Nafar, M. Effect of aging conditions on the fatigue behavior of hot and warm mix asphalt. *Constr. Build. Mater.* **2018**, *188*, 119–129. [CrossRef]
45. Li, Z. Research on the Performance Experiment and Application of Evotherm Warm Mix Asphalt Mixtures. Masters's Thesis, Hebei University of Technology, Tianjin, China, 2011. [CrossRef]
46. Mannan, U.A.; Tarefder, R.A. Investigating different fatigue failure criteria of asphalt binder with the consideration of healing. *Int. J. Fatigue* **2018**, *114*, 198–205. [CrossRef]
47. Kose, S.; Guler, M.; Bahia, H.U.; Masad, E. Distribution of strains within hot-mix asphalt binders: Applying imaging and finite-element techniques. *Transp. Res. Rec.* **2000**, *1728*, 21–27. [CrossRef]
48. Bahia, H.; Tabatabaee, H.A.; Mandal, T.; Faheem, A. *Field Evaluation of Wisconsin Modified Binder Selection Guidelines—Phase II*; Technical report; University of Wisconsin-Madison: Madison, WI, USA, 2013.
49. Deef-Allah, E.; Abdelrahman, M. Thermal, chemical and rheological properties of asphalt binders extracted from field cores. *Innov. Infrastruct. Solut.* **2022**, *7*, 235. [CrossRef]

coatings

MDPI

Article

Analytical Investigation on the Dynamic Behavior of Multi-Span Continuous Beams Supported on Soil with Finite Depth

Da Li [1,*], Hang Yang [1], Jianjun Ma [1,2,*], Ju Wang [1], Chaosheng Wang [1,2] and Ying Guo [1]

1 School of Civil Engineering and Architecture, Henan University of Science and Technology, Luoyang 471023, China; yhang729@163.com (H.Y.); juwang19972022@163.com (J.W.); wchsh163@163.com (C.W.); gytha_ying@163.com (Y.G.)

2 Henan Construction Safety and Protection Engineering Technology Research Center, Luoyang 471023, China

* Correspondence: 9905639@haust.edu.cn (D.L.); majianjun@haust.edu.cn (J.M.)

Abstract: This paper investigates the influence of soil with finite depth on the vibrational behavior of a multi-span continuous beam resting on an elastic foundation. The simplified model of the Timoshenko beam supported on soil with finite depth is established, introducing the foundation displacement decay function. The numerical solution of the continuous beam's vibration response on the elastic foundation is obtained by using the transfer matrix method (TMM) and fourth-order Runge-Kutta method (RK4). Taking a two-span continuous beam as an illustrative example, the validity of the calculation theory is validated by comparing it with the outcomes obtained from the finite element method (FEM). Utilizing numerical computation and parametric analysis, the vibration response of continuous beams is evaluated in terms of its influence by various factors such as soil thickness, viscous damping coefficient of the soil, subgrade reaction coefficient, and span ratio. The findings indicate that the inertial motion of the soil with a finite depth significantly reduces the continuous beam's inherent frequency and enhances the structure's resonance effect. The rise of the subgrade response coefficient increases the system's resonant frequency while decreasing the displacement response amplitude. The ratio between the adjacent spans determines the effect of beam span vibration energy transfer to adjacent spans. In addition, compared with the span directly excited by a concentrated harmonic load, the impact of soil thickness, subgrade reaction coefficient, and viscous damping, the coefficient of the soil is more significant on the indirect influence span of a continuous beam.

Keywords: multi-span continuous beam; Winkler foundation; finite-depth soil motion; Timoshenko beam; transfer matrix method; dynamic response

1. Introduction

Continuous beam structure is typical in civil engineering, and studying its vibration characteristics has essential engineering significance. Many scholars have researched the eigenvalue of continuous beam structures. Based on the classic vibrational bending theory on beams, Hayashikawa et al. developed the analytical method for determining the eigenvalues of continuous beams [1]. Busool and Eisenberger studied the effect of tension and compression under a constant axial load on the natural frequencies of uniform multi-span beams [2]. Lee used the pseudo-spectral method to analyze the eigenvalue of a two-span Timoshenko beam [3]. Tullini et al. employ a coupling method involving finite elements and boundary integral equations to analyze the eigenvalues of Timoshenko beam under various conditions [4]. For the vibrational behavior of a multi-span beam, Seetapan analyzed the impact of span ratio on the dynamic response of vehicle and rough surface two-span beam coupled system [5]. Wang and Wei conducted a study on the acceleration response of a two-span continuous railway bridge, exploring both resonance and sub-resonance effects under the influence of moving train loads [6]. Comparing the finite element analysis and experimental results, Wang et al. conducted a study to assess the impact of beam stiffness degradation on the dynamic behavior of a multi-span continuous bridge when subjected to the dynamic interaction of traffic loads and vehicles [7]. Tamaddona studied the influence of near-source earthquakes on the impact phenomenon of a two-span continuous concrete curved bridge [8]. Lee accurately estimated the damping parameters of the two-span H-Beam by a modified continuous wavelet transform [9].

Based on the analysis of the vibrational characteristics of continuous beam structures, the effect of soil-structure interaction is also crucial in the seismic and vibration analysis of these structures [10–13]. Smith and Brown provided a comprehensive analysis of continuous beams in modern bridge construction, emphasizing the role of soil conditions on the structural integrity and dynamic response of bridges [14]. Lee and Kim conducted an in-depth vibration analysis of continuous beams on elastic foundations, highlighting how variations in soil properties can significantly alter the natural frequencies and mode shapes of the beams [15]. Zhang and Li presented several case studies on the performance of continuous beams supported by soil, offering insights into real-world applications and the practical challenges faced when dealing with varying soil conditions [16]. Miller and Thompson explored the structural behavior of continuous beams under varying soil conditions, providing valuable data on how different soil layers and their mechanical properties affect beam stability and vibration characteristics [17].

In addition, as the study of soil-structure interaction has progressed, numerous scholars have increasingly recognized the significant impact of finite-depth soil motion on the vibration characteristics of its supporting beam. Rades performed the dynamic analysis on the inertial foundation model by applying equivalent mass to the beam [18], while Jaiswal and Iyengar uncovered the vibrational properties of an infinite beam resting on an elastic foundation with finite depth under a moving concentrated force [19]. Metrikine et al. precisely substituted the 3D layer with a 1D equivalent foundation and explored the steady-state behavior of infinite beams supported by a viscoelastic layer with finite depth, considering the influence of a moving load [20]. Ma et al. derived the nonlinear motion equation of the finite depth and comprehensively analyzed the nonlinear vibration characteristics of the beam on an elastic foundation [21].

However, the studies mentioned earlier primarily focus on assessing the impact of finite-depth soil motion on the vibrational characteristics of single-span beams. In practical engineering, continuous beam structures are more common. To date, there has been limited research on the vibration response of continuous beam structures interacting with soil of finite depth. To enhance the understanding of how finite soil depth motion affects the vibration response of beams on an elastic foundation, in-depth research is needed on the impact of finite-depth soil motion on the resonance characteristics of continuous beams.

This paper investigates how the response of soil with finite depth influences the resonance characteristics of a continuous beam resting on an elastic foundation. There are six sections to this paper. In Section 2, the Timoshenko beam vibration model is introduced, considering the influence of a finite-thickness foundation's elastic response. Section 3 utilizes the solution methods [22,23] to obtain numerical solutions for the dynamic response of the continuous beam on an elastic foundation. In Section 4, the finite element method is used to verify the displacement expressions in Section 3. In Section 5, the effects of soil thickness, viscous damping coefficient of the soil, subgrade reaction coefficient, and span ratio on the vibration response of a two-span continuous beam are analyzed. Finally, the summary of the results is presented in Section 6.

2. Basic Relationships

Based on Timoshenko's theory and the elastic foundation motion theory, a continuous beam resting on the elastic foundation under harmonic excitation at an arbitrary point is investigated. In Figure 1, the cartesian coordinate system $O - xy$ is established, where $p_j(x,t) = P_j\delta(x - x_0)e^{i\Omega t}$ is the harmonic excitation acting on the jth beam at x_0.

P_j and Ω are the excitation load's amplitude and frequency, respectively. l_j is the beam's length at the jth span. H is the depth of the soil supporting the multi-span continuous beam. $w(x,t)$ signifies the displacements of the beam, $w_s(x,y,t)$ signifies the soil's displacements in the y-axis, and $w_r(x,y,t)$ is the displacement discrepancy between the beam and the foundation.

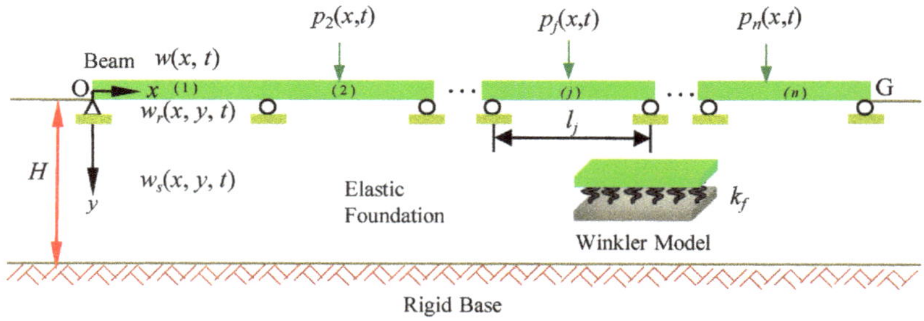

Figure 1. Continuous beam supported on soil with finite depth.

The construction of the dynamic equation is undertaken for the jth span beam resting on soil with finite depth. The axial displacement of the elastic foundation beam is neglected, and only the vertical displacement $w(x,t)$ and the rotational displacement $\theta(x,t)$ are considered. According to Timoshenko's theory [10], the equation governing the linear motion of an elastic foundation beam can be formulated.

$$KGA\left(\frac{\partial\theta(x,t)}{\partial x}-\frac{\partial^2 w(x,t)}{\partial x^2}\right)+m\frac{\partial^2 w(x,t)}{\partial t^2}+c\frac{\partial w(x,t)}{\partial t}=p(x,t)-q(x,t), \tag{1}$$

$$EI\frac{\partial^2\theta(x,t)}{\partial x^2}-KGA\left(\theta(x,t)-\frac{\partial w(x,t)}{\partial x}\right)-\gamma\frac{\partial^2\theta(x,t)}{\partial t^2}=0, \tag{2}$$

where m is the mass of the unit length beam. γ is the beam's inertia moment, c is the viscous damping coefficient per unit length of the beam, EI and KGA reflect the stiffness characteristics of bending and shear, respectively, $q(x,t)$ represents the response of the subgrade, and $p(x,t)$ represents the external stimulus or forcing function.

Assuming linear elasticity and isotropic properties for the foundation material [19], the motion equation of the elastic foundation with finite thickness is

$$\rho_s\frac{\partial^2 w_s(x,y,t)}{\partial t^2}+c_s\frac{\partial w_r(x,y,t)}{\partial t}=k_f H\frac{\partial^2 w_s(x,y,t)}{\partial y^2}, \tag{3}$$

where ρ_s represents the mass per unit depth of the foundation beneath the per unit length of the beam; c_s represents the unit thickness foundation's viscous damping factor beneath the beam per unit length; and k_f denotes the coefficient of subgrade reaction according to the Winkler foundation model. Therefore, the subgrade reaction force at the contact between the beam and the foundation is expressed as

$$q(x,t)=k_f\frac{\partial w_s(x,y,t)}{\partial y}\bigg|_{y=0}. \tag{4}$$

$\varphi(y)$ denotes the displacement decay function [24], $w_r(x,y,t)=w(x,t)[1-\varphi(y)]$, $w_s(x,y,t)=w(x,t)\varphi(y)$, where the decay function is expressed as

$$\varphi(y)=\frac{\sin\mathrm{h}[\alpha(1-\frac{y}{H})]}{\sin\mathrm{h}(\alpha)}, \tag{5}$$

In the present study, the attenuation coefficient α is set to 0.01. The boundary conditions are established to ensure the continuity of deformation among the beam and the supporting foundation, adhering to engineering experience, which translates to the conditions of $w_s(x,0,t) = w(x,t)$ and $w_s(x,H,t) = 0$.

Equation (3) is integrated over the interval $[0, H]$ and substituted Equation (4) to obtain

$$q(x,t) = \rho_s \frac{\partial^2 w(x,t)}{\partial t^2} \int_0^H \varphi(y)\mathrm{d}y + c_s \frac{\partial w(x,t)}{\partial t} \int_0^H [1 - \varphi(y)]\mathrm{d}y + \frac{k_f w(x,t)\alpha}{\sin\mathrm{h}(\alpha)} \tag{6}$$

where $\alpha/\sin\mathrm{h}(\alpha) \approx 1$.

Substitute Equation (6) into Equation (1) to obtain

$$\begin{aligned} KGA\left(\frac{\partial\theta(x,t)}{\partial x} - \frac{\partial^2 w(x,t)}{\partial x^2}\right) + (m + \rho_s \int_0^H \varphi(y)\mathrm{d}y)\frac{\partial^2 w(x,t)}{\partial t^2} + (c + c_s \int_0^H [1 - \varphi(y)]\mathrm{d}y)\frac{\partial w(x,t)}{\partial t} \\ = p_j(x,t) - k_f w(x,t). \end{aligned} \tag{7}$$

$$EI\frac{\partial^2\theta(x,t)}{\partial x^2} - KGA\left(\theta(x,t) - \frac{\partial w(x,t)}{\partial x}\right) - \gamma\frac{\partial^2\theta(x,t)}{\partial t^2} = 0, \tag{8}$$

Equations (7) and (8) are the jth span Timoshenko beam's vibration model on an elastic foundation considering of soil motion with finite thickness. The following contents are calculated and analyzed based on this model.

3. Analytical Solution of the Continuous Beam Vibration

According to the interaction dynamic model between the soil with finite depth and the jth span Timoshenko beam, this section obtains the multi-span continuous beam's natural frequencies and vibration modes by employing TMM. Then the numerical solution of the multi-span continuous beam's dynamic model excited at any position is obtained by using RK4.

The calculation process is as follows. The beam's free vibration is presumed to be a periodic motion with a constant frequency.

$$w(x,t) = W(x)e^{\mathrm{i}\omega t}, \theta(x,t) = \Psi(x)e^{\mathrm{i}\omega t}, \tag{9}$$

where $\mathrm{i} = \sqrt{-1}$, ω is the inherent frequency of the system.

Substituting Equation (9) into Equations (7) and (8), the governing equation for the free vibration state of an undamped Timoshenko beam resting on an elastic foundation, accounting for finite-depth soil motion, is derived.

$$\frac{\mathrm{d}^4 W(x)}{\mathrm{d}x^4} + \frac{(B_M + \gamma\omega^2)}{A_M}\frac{\mathrm{d}^2 W(x)}{\mathrm{d}x^2} + \frac{B_M}{A_M^2}\left(\gamma\omega^2 - KGA\right)W(x) = 0, \tag{10}$$

where $A_M = EI$, $B_M = EI(m\omega^2 + \rho_s\omega^2 \int_0^H \varphi(y)\mathrm{d}y - k_f)/KGA$.

The different general solutions can be obtained by judging from $\Delta = ((B_M - \gamma\omega^2)^2 + 4B_M KGA)/A_M^2$. Taking $\Delta < 0$ as an example to illustrate the calculation process [22], the general solution of Equation (10) is

$$W(x) = C_1 \cosh\lambda_1 x + C_2 \mathrm{sinh}\lambda_1 x + C_3 \cos\lambda_2 x + C_4 \sin\lambda_2 x, \tag{11}$$

where $\lambda_1 = \sqrt{\left(-b + \sqrt{\Delta}\right)/2}$, $\lambda_2 = \sqrt{\left(b + \sqrt{\Delta}\right)/2}$, $b = ((B_M - \gamma\omega^2)/A_M)^2$, $C_1 \sim C_4$ are undetermined coefficients.

According to the calculation principle of TMM and Equation (9), the state vector of the *j*th span's beam element, expressed in modal coordinates, can alternatively be described as:

$$\mathbf{Z}_j(x) = \left[W(x), \Psi(x), Q_y(x), M_\theta(x)\right]_j^{\mathrm{T}}, \tag{12}$$

where T stands for matrix transpose, $W(x)$, $\Psi(x)$, $Q_y(x)$ and $M_\theta(x)$ are respectively the modal coordinates of vertical displacement, angular displacement, shear force, and bending moment.

$$\begin{cases} M_\theta(x) = EI\frac{\mathrm{d}\Psi(x)}{\mathrm{d}x}, \\ \Psi(x) = \frac{\mathrm{d}W(x)}{\mathrm{d}x} - \frac{Q_y(x)}{KGA}, \\ Q_y(x) = \frac{\mathrm{d}M_\theta(x)}{\mathrm{d}x} - \gamma\omega^2\Psi(x), \\ \frac{\mathrm{d}Q_y(x)}{\mathrm{d}x} = ((m + \rho_s\int\limits_0^H \varphi(y)\mathrm{d}y)\omega^2 - k_f)W(x). \end{cases} \tag{13}$$

Equation (13) can thus be expressed in matrix form

$$\mathbf{Z}_j(x) = \mathbf{B}_j(x)\mathbf{C}^{\mathrm{T}}, \tag{14}$$

where $\mathbf{C} = [C_1, C_2, C_3, C_4]^{\mathrm{T}}$ $\mathbf{B}_j(x)$ is a matrix with four rows and four columns,

$$\mathbf{B}_j(x) = \begin{bmatrix} \cosh\lambda_1 x & \sinh\lambda_1 x & \cos\lambda_2 x & \sin\lambda_2 x \\ C_{f1}\lambda_1\sinh\lambda_1 x & C_{f1}\lambda_1\cosh\lambda_1 x & C_{f2}\lambda_2\sin\lambda_2 x & C_{f3}\lambda_2\cos\lambda_2 x \\ C_{Q1}\lambda_1\sinh\lambda_1 x & C_{Q1}\lambda_1\cosh\lambda_1 x & C_{Q2}\lambda_2\sin\lambda_2 x & C_{Q3}\lambda_2\cos\lambda_2 x \\ C_{M1}\cosh\lambda_1 x & C_{M1}\sinh\lambda_1 x & C_{M2}\cos\lambda_2 x & C_{M2}\sin\lambda_2 x \end{bmatrix}, \tag{15}$$

where $C_{f1} = \left(A_f\lambda_1^2 + B_f\right), C_{f2} = \left(A_f\lambda_2^2 - B_f\right), C_{f3} = \left(-A_f\lambda_2^2 + B_f\right), C_{Q1} = (A_Q\lambda_1^2 + B_Q)$, $C_{Q2} = (A_Q\lambda_2^2 - B_Q), C_{Q3} = (-A_Q\lambda_2^2 + B_Q), C_{M1} = (A_M\lambda_1^2 + B_M), C_{M2} = (-A_M\lambda_2^2 + B_M)$, $A_f = A_Q/(KGA), B_f = 1 - B_Q/KGA, A_Q = EIKGA/(KGA - \gamma\omega^2), B_Q = (EIk_f - EI(m + \rho_s\int\limits_0^H \varphi(y)\mathrm{d}y)\omega^2 - \gamma\omega^2 KGA)/(KGA - \gamma\omega^2)$.

Substituting the boundary condition at $x = 0$, we can get the state vector of any point on the *j*th span of the beam.

$$\mathbf{Z}_j(x) = \mathbf{B}_j(x)\mathbf{B}_j^{-1}(0)\mathbf{Z}_j(0) = \mathbf{D}_j(x)\mathbf{Z}_j(0), \tag{16}$$

where $\mathbf{D}_j(x)$ is the transfer matrix of the *j*th span of the beam.

At the multi-span continuous middle support. According to the continuity condition and the force equilibrium condition of the beam elements on both sides of the support, it can be obtained

$$\begin{cases} W_j^r(x) = W_{j+1}^l(x), \\ \Psi_j^r(x) = \Psi_{j+1}^l(x), \\ Q_{yj}^r(x) + R_{j,j+1} = Q_{yj+1}^l(x), \\ M_{zj}^r(x) = M_{zj+1}^r(x), \end{cases} \tag{17}$$

where the superscripts l, r denotes the leftmost and rightmost sides of the beam elements, respectively, and $R_{j,j+1}$ denotes the intermediate support reactions of beam elements j and $j+1$. Therefore, the state vector relation of the beams on both sides of the support can be expressed as follows

$$\mathbf{Z}_{j+1}^l(x) = \mathbf{D}_{j,j-1}\mathbf{Z}_j^r(x) = \begin{bmatrix} 1 & 0 & 0 & 0 \\ 0 & 1 & 0 & 0 \\ K_{vu} & K_{v\varphi} & 1 + K_{vq} & K_{vm} \\ 0 & 0 & 0 & 1 \end{bmatrix}\mathbf{Z}_j^r(x), \tag{18}$$

where K_{vu}, $K_{v\varphi}$, K_{vq}, and K_{vm} are the undetermined coefficient. According to the fact that the vertical displacement at $\mathbf{Z}_{j+1}^{r}(l_j)$ is zero, it can be concluded that the elements in the first row of $\mathbf{D}_{j+1}(x)\mathbf{D}_{j,j+1}$ are all zero. Thus

$$K_{vu} = -\frac{A(1,1)}{A(1,3)},\ K_{v\varphi} = -\frac{A(1,2)}{A(1,3)},\ K_{vq} = -1,\ K_{vm} = -\frac{A(1,4)}{A(1,3)}, \tag{19}$$

where $A(a_i, a_j)$ represents the element of row a_i and column a_j of the matrix $\mathbf{D}_{j+1}(x)$.

So far, the output end G of the nth span continuous beam can be calculated from the matrix input point O, that is

$$\begin{aligned} \mathbf{Z}_n(l) &= \mathbf{B}_n(l_n)\mathbf{B}_n{}^{-1}(0)\mathbf{Z}_n^{l}(0) = \mathbf{B}_n(l_n)\mathbf{B}_n{}^{-1}(0)\mathbf{D}_{n-1,n}\mathbf{Z}_{n-1}^{r}(l_{n-1}) \\ &= \mathbf{B}_n(l_n)\mathbf{B}_n{}^{-1}(0)\mathbf{D}_{n-1,n}\mathbf{B}_{n-1}(l_{n-1})\mathbf{B}_{n-1}{}^{-1}(0)\mathbf{Z}_{n-1}^{l}(0) \\ &= \mathbf{B}_n(l_n)\mathbf{B}_n{}^{-1}(0)\mathbf{D}_{n-1,n}\cdots\mathbf{B}_1(l_1)\mathbf{B}_1{}^{-1}(0)\mathbf{Z}_1(0) \\ &= \mathbf{D}_n\mathbf{D}_{n-1,n}\mathbf{D}_{n-1}\mathbf{D}_{n-2,n-1}\cdots\mathbf{D}_1\mathbf{Z}_1(0) \\ &= \mathbf{D}_n\left(\prod_{j=1}^{n-1}\mathbf{D}_{j,j+1}\mathbf{D}_j\right)\mathbf{Z}_1(0). \end{aligned} \tag{20}$$

Equation (20) is the overall transfer matrix expression of the system, where $\mathbf{D}_n\left(\prod_{j=1}^{n-1}\mathbf{D}_{j,j+1}\mathbf{D}_j\right)$ is the overall transfer matrix, which can be expressed as

$$\mathbf{D}_n\left(\prod_{j=1}^{n-1}\mathbf{D}_{j,j+1}\mathbf{D}_j\right) = \begin{bmatrix} u_{11} & u_{12} & u_{13} & u_{14} \\ u_{21} & u_{22} & u_{23} & u_{24} \\ u_{31} & u_{32} & u_{33} & u_{34} \\ u_{41} & u_{42} & u_{43} & u_{44} \end{bmatrix}. \tag{21}$$

The state vector at the boundary can be obtained according to different boundary conditions of the continuous beam at input O and output G. Taking the supported beam as an example, we can obtain

$$\begin{bmatrix} 0 \\ 0 \end{bmatrix} = \begin{bmatrix} u_{22} & u_{23} \\ u_{32} & u_{33} \end{bmatrix}\begin{bmatrix} \Psi(0) \\ Q_y(0) \end{bmatrix}. \tag{22}$$

If Equation (21) has a nontrivial solution, the coefficient matrix's determinant is zero, and the natural frequencies ω_j of each order can be obtained by solving the initial state vector $\mathbf{Z}_1(0) = [0, 1, -u_{22}/u_{23}, 0]^{\mathrm{T}}$; this can be obtained by substituting Equation (22), and the corresponding mode shapes can be obtained by coupling Equation (20).

According to the principle of modal superposition, the system's vibration response $\mathbf{v}(x,t)$ can be stated as

$$\mathbf{v}(x,t) = \sum_{k=1}^{N}\mathbf{V}^k(x)q^k(t), \tag{23}$$

where $\mathbf{v}(x,t) = [\mathbf{v}_1^{\mathrm{T}}(x,t), \mathbf{v}_2^{\mathrm{T}}(x,t), \ldots \mathbf{v}_n^{\mathrm{T}}(x,t)]^{\mathrm{T}}$,$\mathbf{v}_n = [w_n(x,t), \theta_n(x,t)]^{\mathrm{T}}$, $\mathbf{V}^k(x) = \left[\mathbf{V}_1^{k\mathrm{T}}(x), \mathbf{V}_2^{k\mathrm{T}}(x), \ldots, \mathbf{V}_n^{k\mathrm{T}}(x)\right]^{\mathrm{T}}$ is the kth mode shape of the n-span continuous beam system, $\mathbf{V}_j^k(x) = \left[W_j^k(x), \Psi_j^k(x)\right]^{\mathrm{T}}$ is the modal coordinate of the continuous beam of j-span, $q^k(x)(k = 1, 2, 3, \ldots, N)$ is the generalized coordinate, and N is that number of the truncated modal term.

Substituting Equation (23) into Equations (7) and (8), and simplifying to obtain

$$\sum_{k=1}^{N}\left(\mathbf{M}\mathbf{V}^k(x)\ddot{q}^k(t) + \mathbf{C}\mathbf{V}^k(x)\dot{q}^k(t) + \mathbf{K}\mathbf{V}^k(x)q^k(t)\right) = \sum_{j=1}^{n}\mathbf{\Phi}_j(x)P_j e^{\mathrm{i}\Omega t}, \tag{24}$$

where $\mathbf{M} = \text{diag}(\mathbf{M}_1, \mathbf{M}_2, \ldots \mathbf{M}_n)$, $\mathbf{C} = \text{diag}(\mathbf{C}_1, \mathbf{C}_2, \ldots \mathbf{C}_n)$, and $\mathbf{K} = \text{diag}(\mathbf{K}_1, \mathbf{K}_2, \ldots \mathbf{K}_n)$ represent the mass augmentation operators, the damped augmentation operators, and the stiffness augmented operators, respectively. $\mathbf{M}_j$, $\mathbf{C}_j$, and $\mathbf{K}_j$ represent the mass parameter matrices, damping parameter matrices, and stiffness parameter matrix of jth spans, P_j represents the magnitude of the force on the jth span, and $\mathbf{\Phi}_j(x)$ is a single-column matrix in which the position of the load on the jth span is represented by a Dirac function and the remaining elements are 0.

$$\mathbf{M}_j = \begin{bmatrix} m + \rho_s \int_0^H \varphi(y)\mathrm{d}y & 0 \\ 0 & \gamma \end{bmatrix}, \mathbf{C}_j = \begin{bmatrix} c + c_s \int_0^H [1 - \varphi(y)]\mathrm{d}y & 0 \\ 0 & 0 \end{bmatrix},$$
$$\mathbf{K}_j = \begin{bmatrix} k_f - KGA\frac{\partial^2}{\partial x^2} & KGA\frac{\partial}{\partial x} \\ -KGA\frac{\partial}{\partial x} & KGA - EI\frac{\partial^2}{\partial x^2} \end{bmatrix}. \tag{25}$$

Both sides of Equation (24) are multiplied by $\mathbf{V}^s(x)(s = 1, 2, 3, \ldots, N)$. According to the orthogonality of augmented eigenvectors [25], the simplification can be obtained

$$\ddot{q}^s(t) + 2\xi_s\omega_s\dot{q}^s(t) + \omega_s^2 q^s(t) = \sum_{j=1}^{n} P_j u_s e^{\mathrm{i}\Omega t} \tag{26}$$

where $2\xi_s\omega_s = \sum_{k=1}^{N} \left\langle \mathbf{C}\mathbf{V}^k(x), \mathbf{V}^s(x) \right\rangle / M_s$, $u_s = \langle \mathbf{\Phi}_j(x), \mathbf{V}^s(x) \rangle / M_s$, $K_s = \omega_s{}^2 M_s$, $M_s = \left\langle \mathbf{M}\mathbf{V}^k(x),\ \mathbf{V}^s(x) \right\rangle = \sum_{j=1}^{n} \int_0^{l_j} \mathbf{V}_j^{k\mathrm{T}}(x)\mathbf{V}_j^s(x)\mathrm{d}x$, $\langle , \rangle$ represents the inner product of two matrices.

Equation (26) can be written as $y'_i(t) = f_i(t, y_1(t), y_2(t))$, $i = 1, 2$, namely

$$\begin{cases} f_1\left(t, q^s(t), \dot{q}^s(t)\right) = \dot{q}^s(t), \\ f_2\left(t, q^s(t), \dot{q}^s(t)\right) = \sum_{j=1}^{n} P_j u_s e^{\mathrm{i}\Omega t} - 2\xi_s\omega_s\dot{q}^s(t) - \omega_s^2 q^s(t). \end{cases} \tag{27}$$

According to the calculation principle of RK4 [23], the calculation results of time t_j to $t_{j+1} = t_j + h$ are as follows.

$$q_{j+1,i}^s(t) = q_{j,i}^s(t) + \frac{h}{6}(k_{1,i} + 2k_{2,i} + 2k_{3,i} + k_{4,i})\ i = 1,\ 2, \tag{28}$$

where $k_{1,i} = f_i\left(t_j, q_j^s(t), \dot{q}_j^s(t)\right)$, $k_{2,i} = f_i\left(t_j + h/2, q_j^s(t) + hk_1/2\right)$, $k_{3,i} = f_i\left(t_j + h/2, q_j^s(t) + hk_2/2, \dot{q}_j^s(t) + hk_2/2\right)$, $k_{4,i} = f_i\left(t_j + h,\ q_j^s(t) + hk_3,\ \dot{q}_j^s(t) + hk_3\right)$, and h is the time period step.

According to the initial state $\mathbf{v}(x, 0) = 0$, $\dot{\mathbf{v}}(x, 0) = 0$ of the elastic foundation beam, the numerical solutions of the functions $q^s(t)$ and $\dot{q}^s(t)$ will be calculated step by step, According to Equation (24), the analytical solution of the system's first n-order vibration response can be obtained by superposition.

4. Numerical Tests and Model Validation

For numerical verification and analysis, based on previous studies, physical parameters required to calculate the system response are given in Table 1 [10,21], where the shear modulus is $G = E/(2(1 + \nu_0))$, the shear correction factor is $K = 10(1 + \nu_0)/(12 + 11\nu_0)$, and external excitation amplitude is $P = 65$ kN.

Table 1. Physical parameters of the beam and foundation.

Physical Meaning	Symbol	Unit	Numerical Value
The *j*th span length of a continuous beam	l_j	m	6.096
The beam's width	b	m	0.61
The beam's height	h	m	0.305
The per unit length beam's mass	m	$kg \cdot m^{-1}$	447.08
Damping factor per unit length	c	$kN \cdot s \cdot m^{-2}$	1.0
Moment of inertia of the per unit length beam	γ	$kg \cdot m$	3.466
Young's modulus of the beam	E	MPa	2.482 × 104
Poisson ratio of the beam	ν_0	——	0.25
Soil mass per unit thickness of the per unit length beam	ρ_s	$kg \cdot m^{-2}$	1037
Viscous damping coefficient of the soil per unit thickness of the per unit length beam	c_s	$kN \cdot s \cdot m^{-3}$	3.6
Modulus of subgrade reaction	k_f	MPa	16.55
Soil thickness	H	m	5

4.1. Simply Supported Beam

To ensure the accuracy of the theoretical calculations presented, a single-span beam is chosen for detailed computational and analytical examination. Based on the parameters in Table 1 and Equation (10), Table 2 compares the present methods with the first four natural frequencies in existing references [26,27]. The results show that there is no significant difference in the natural frequencies obtained between them. So, the numerical solution has been verified for accuracy and validity, demonstrating that the theory is applicable to single-span beams. Additionally, it confirms that selecting the Timoshenko beam for this study is appropriate.

Table 2. Natural frequency of the simply supported beam on an elastic foundation.

Natural Frequency (Hz)	(Timoshenko S., 1974) [10]	(Thambiratnam and Zhuge, 1996) [26]	(Friswell et al., 2007) [27]	The Present Study	
				Euler-Bernoulli	Timoshenko
ω_1	32.9063	32.9033	32.898	32.8749	32.8289
ω_2	56.8135	56.8193	56.808	56.8040	56.1037
ω_3	112.908	111.961	111.900	111.9186	108.303
ω_4			193.760	193.8085	182.7608

4.2. Two-Span Continuous Beam

To confirm the theory is also applicable to multi-span continuous beams, the theoretical calculations are contrasted with the outcomes obtained from numerical simulations. Due to the rise in the quantity of beam spans, their mutual operation forms become complicated accordingly. To not lose generality, the article selects a two-span beam for specific analysis. Based on the parameter values outlined in Table 1, the two-span continuous beam model depicted in Figure 2 is selected for consideration. The first span is the direct excitation span under load and the second is the indirect influence span without load excitation. The system damping is characterized using the Rayleigh damping coefficient. In the process of modeling, following the attenuation characteristics of the vertical displacement of the foundation, the motion effect of the soil is equivalent to the additional mass on the continuous beam. The corresponding program is written to establish the numerical simulation model by using the ANSYS software(Mechanical APDL 2023 R1), and the complete method in the finite element method (FEA) is used to analyze the harmonic response. Meanwhile, the theoretical calculations are performed using MATLAB 2024a by combining the TMM and RK4 methods. According to the outcomes derived from the theoretical calculation method and FEM, Figure 3 shows the amplitude-frequency curves in the middle of each span of the two-span continuous beam. The graphs produced by both methods are highly similar, thus verifying the applicability of this theory to continuous beams. Additionally, Figure 3 shows that the inertial motion of the soil with finite depth significantly enhances the displacement resonance characteristics of the two-span continuous beam, making the vibration performance of the continuous beam structure more complicated. Therefore, it is essential to investigate the effect of foundation response with finite-thickness on the continuous beams' dynamic response.

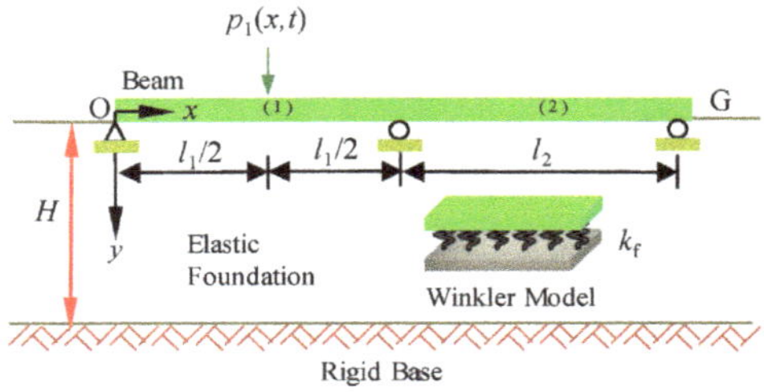

Figure 2. Two-span continuous beam model with simply supported boundary.

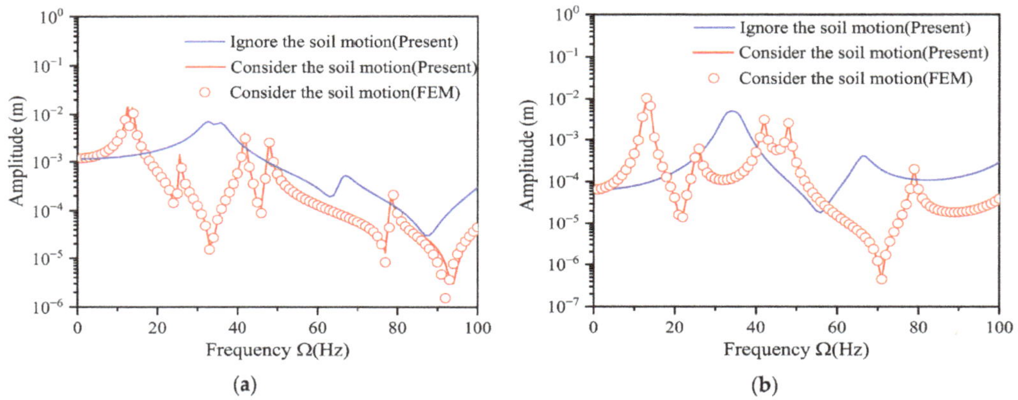

Figure 3. Amplitude-frequency curves at each span's mid-span of a two-span continuous beam: (**a**) the first span; (**b**) the second span.

5. Results and Discussion

This section focuses on the two-span continuous beam model depicted in Figure 2, utilizing the parameter values outlined in Table 1. An extensive analysis is conducted to assess the impact of various parameters, including soil thickness, viscous damping coefficient of the soil, subgrade reaction coefficient, and span ratio, on the mid-span deflection of each span of a continuous beam. The objective is to gain insights into how the finite depth foundation response impacts the resonant response of the continuous beam.

5.1. Effect of Soil Thickness

Figure 4a illustrates the relationship between soil depth and the first six natural frequencies of the two-span continuous beam. As the soil depth increases, the system's natural frequency decreases significantly, and the deeper soil motion has little effect on the change of natural frequency. This observation can be attributed to the increase in the half-wave period of the system's vibration response due to the finite depth soil's inertial motion. Figure 4b,c depict the amplitude-frequency curves of the mid-span displacement for two-span continuous beams. Figure 4b reveals that as soil thickness increases, the system's resonant frequency decreases significantly, leading to an increase in displacement resonance amplitude. On the other hand, Figure 4c shows the displacement resonance amplitudes at each span of the continuous beam are relatively close. When the excitation frequency is lower than the resonant frequency of the first-order system, the amplitude at the second span is less affected by the load excitation effect at the first span.

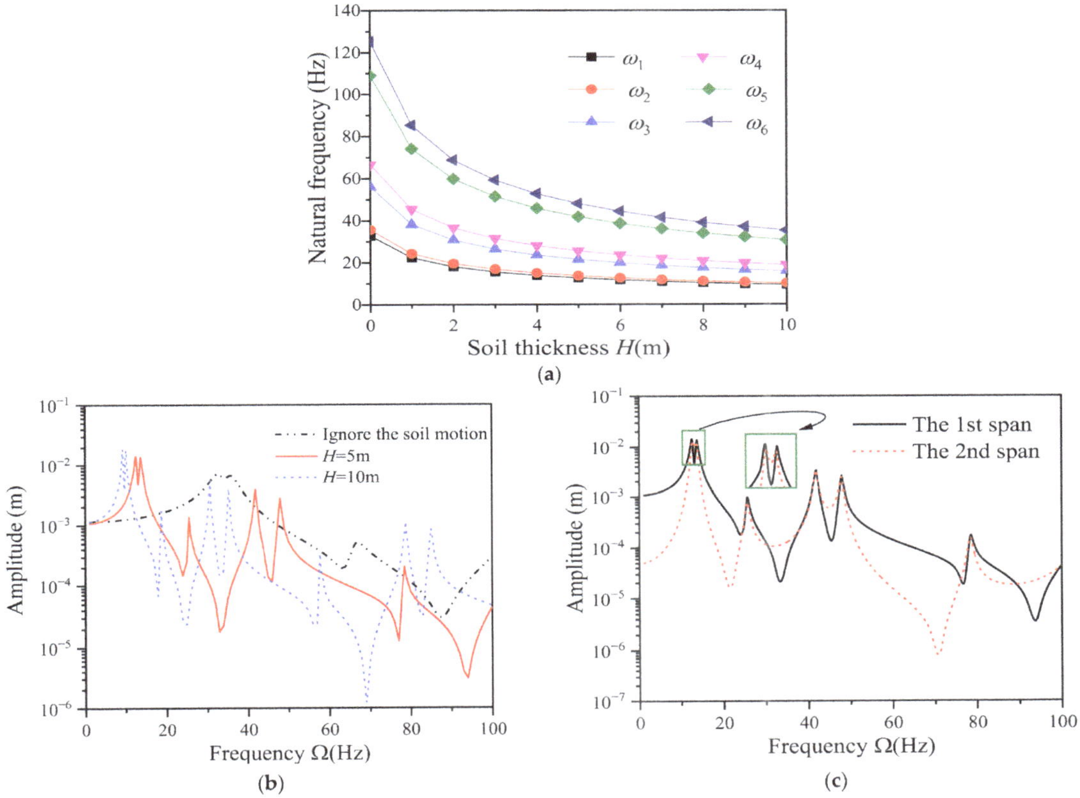

Figure 4. Effect of soil thickness on a two-span continuous beam: (**a**) natural frequency; (**b**) the first span; (**c**) H = 5 m.

Figure 5 illustrates the enhancement effect of soil depth on the first-order resonance amplitude of each span of the two-span continuous beam. w_1 represents the displacement resonance amplitude of each span when considering the fi-nite depth of the soil. As the soil depth increases, the resonance amplitude of the second span experiences a significantly greater increase compared to the first span. This empha-sizes that the influence of soil depth on the amplitude of the second span is stronger than that of the first span. This observation can be attributed to the suppression of the inertial motion of soil with finite depth by external excitation. Furthermore, it indicates that the soil depth has a lower enhancement effect on the displacement resonance amplitude of the beam span subject to direct excitation compared to the beam span influenced indirectly.

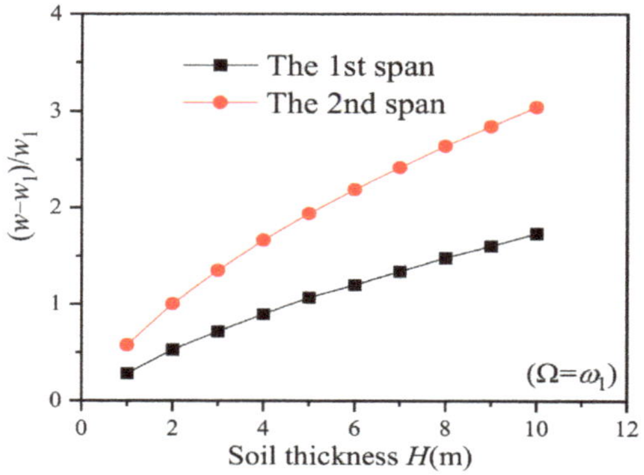

Figure 5. The enhancement effect of soil thickness on the displacement of two-span continuous beam.

5.2. Effect of the Viscous Damping Coefficient of the Soil

The viscous damping coefficient of the soil is the only variable; according to the parameters in Table 1, $c_s = 3.6\ \mathrm{kN \cdot s \cdot m^{-3}}$, $7.6\ \mathrm{kN \cdot s \cdot m^{-3}}$, and $11.6\ \mathrm{kN \cdot s \cdot m^{-3}}$ are selected for analysis. Figure 6 shows the effect of the c_s on the displacement of a two-span continuous beam. Figure 6a presents the impact of the viscous damping coefficient of the soil on the displacement response amplitude of the first span of a continuous beam. For the low damping system, as the increase of viscous damping coefficient of the soil, the displacement resonance amplitude decreases significantly, and the suppression of the non-resonance amplitude is weak. Based on the resonance amplitude w_{cs1} generated by $c_s = 3.6\ \mathrm{kN \cdot s \cdot m^{-3}}$, Figure 6b depicts the suppressing effect of the first-order resonant amplitude of each span by the viscous damping coefficient of the soil. The inhibition of the displacement resonance amplitude of the second span by c_s is significantly stronger than that of the first span, indicating that the suppression effect of the viscous damping coefficient of the soil on the indirect influence span is significantly stronger than that of the direct excitation span.

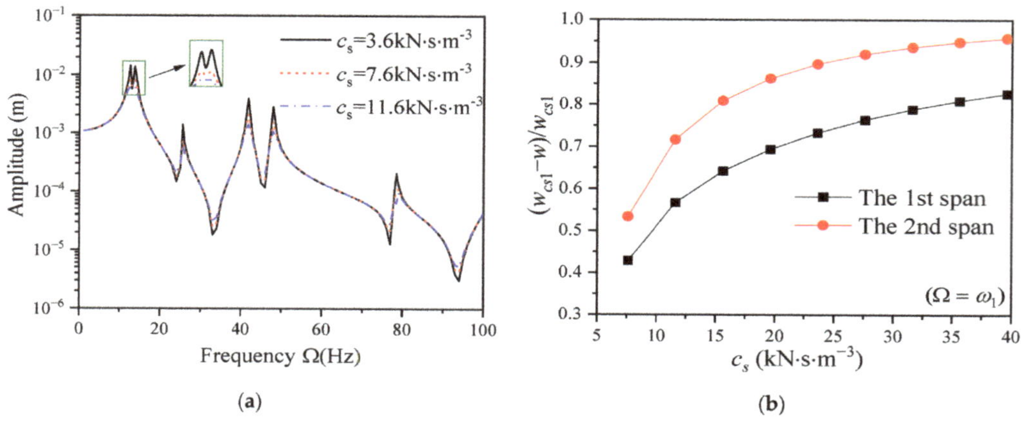

Figure 6. Impact of viscous damping coefficient of the soil on displacement of a two-span continuous beam: (**a**) the first span; (**b**) inhibition effect of each span.

5.3. Effect of Subgrade Reaction Coefficient

The impact of the subgrade reaction coefficient on the system's first six natural frequencies is graphically represented in Figure 7a. The natural frequency rises considerably as k_f increases, and compared with the influence on higher-order natural frequencies, lower-order frequencies are more significant. According to the parameter values in Table 1, k_f = 8.275 MPa, 16.55 MPa, 33.1 MPa, and 165.5 MPa are selected. Figure 7b is the impact of the foundation reaction coefficient on the amplitude-frequency response curve at the middle of the first span. As the subgrade reaction coefficient increases, the amplitude of the displacement response decreases significantly, and the resonance frequency increases. Among these changes, the first and second-order resonance characteristics exhibit the most pronounced variations.

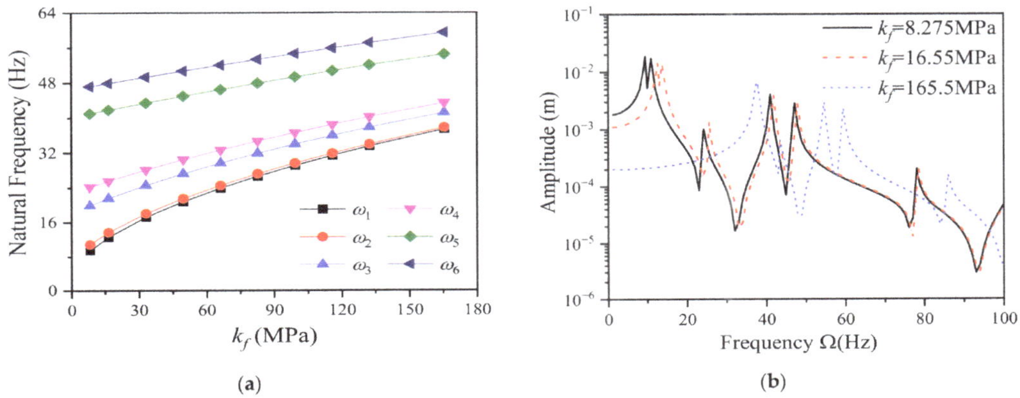

Figure 7. Impact of subgrade reaction coefficient on vibration response of a two-span continuous beam: (**a**) natural frequency; (**b**) mid-span of the first span.

The displacement resonance amplitude difference between different subgrade reaction coefficients and w_{kf1} (k_f = 8.275 MPa) is selected. Figure 8 shows the suppressing degree of the subgrade reaction coefficient on the displacement resonance amplitude ($\Omega = \omega_1$) of a two-span beam. As shown in Figure 8, the difference curve of the resonance amplitude reflects that the subgrade reaction coefficient has the same suppression trend on the resonance amplitude of each span. Compared to the first span, the subgrade reaction coefficient exerts a more pronounced influence on the second span. This indicates that the suppressive impact of the subgrade reaction coefficient on the indirectly influenced span is significantly stronger than its impact on the directly excited span. The main reason is that the direction of load excitation is opposite to that of the foundation reaction.

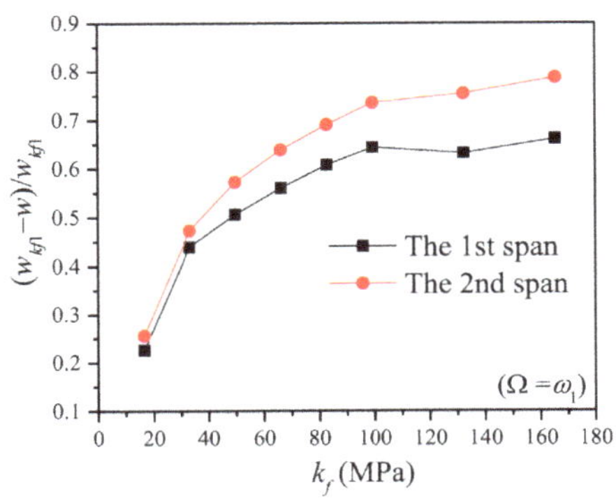

Figure 8. The suppressing impact of subgrade reaction coefficient on displacement of a two-span continuous beam.

5.4. Effect of Span Ratio

The span ratio of the first span to the second span is defined as $\varepsilon = l_1/l_2$. According to the parameter values in Table 1, select $\varepsilon = 1$ (6.096 m + 6.096 m), 0.6 (4.572 m + 7.620 m), 0.33 (3.048 m + 9.144 m). The span ratio's effect on the mode of a two-span continuous beam is shown in Figure 9. As the span ratio decreases, the mode changes significantly. The wave number of the short-span mode drops, while the wave number will increase in the long-span mode. Compared with asymmetric structures, even-order modes of symmetric structures always have the least half-wave number, while the minimum half-wavelength of odd-order modes is always the longest.

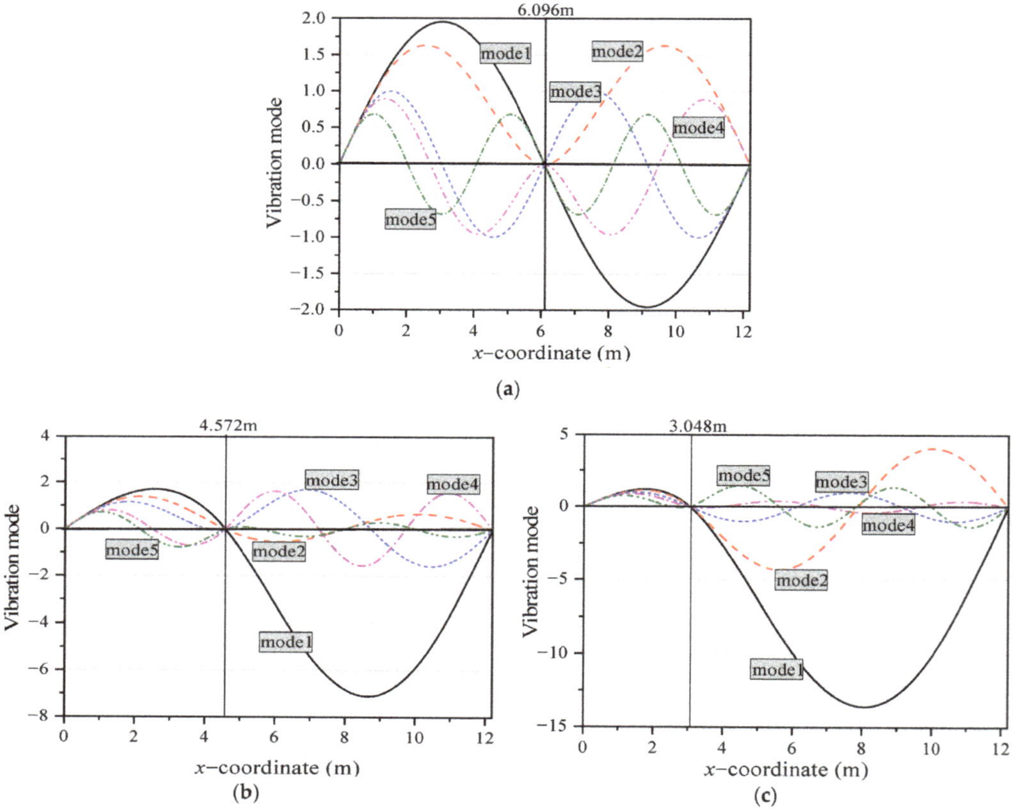

Figure 9. Modal effect of span ratio on a two-span continuous beam: (**a**) 6.096 m + 6.096 m (ε = 1.00); (**b**) 4.572 m + 7.62 m (ε = 0.60); (**c**) 3.048 m + 9.144 m (ε = 0.33).

Figure 10 further shows the effect of different span ratios on the natural frequency. As shown in Figure 10, the span ratio impacts the natural frequency of different orders differently. When ε = 1.00, compared with other span ratios, the natural frequency value of the odd-order mode is the largest, the natural frequency value of the even-order mode is the smallest. The reason is related to the half-wave number and the minimum half-wavelength of each mode in Figure 9.

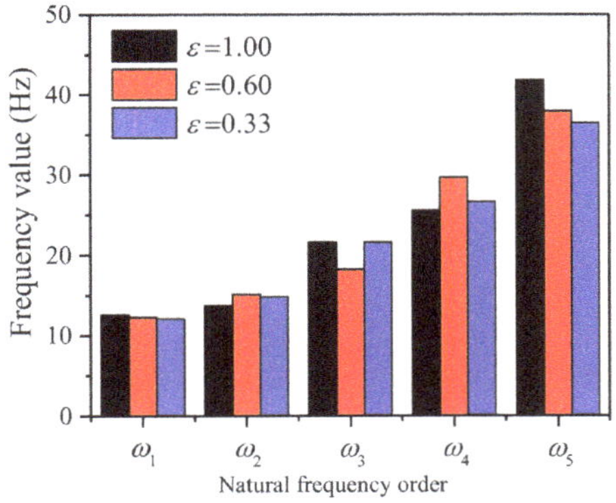

Figure 10. Effect of span ratio on the natural frequency of the two-span beam.

Figure 11 shows the amplitude-frequency curves of the two-span beams with different span ratios. When $\varepsilon \neq 1$, the structure is transformed into an asymmetric system, the difference between the continuous beam's first and second resonance rises significantly. With the decrease in span ratio, the half-wave number of the first span structure decreases, which leads to the maximum response amplitude moving to the high-order resonance. While the half-wave number of the second span structure increases, which leads to the resonance frequency of the maximum displacement amplitude decreasing compared with the first span. The amplitude-frequency curves of the second span $\varepsilon = 3.00$ and $\varepsilon = 0.33$ are almost coincident, indicating that the span ratio determines the resonance characteristics that load indirectly affect the beam span.

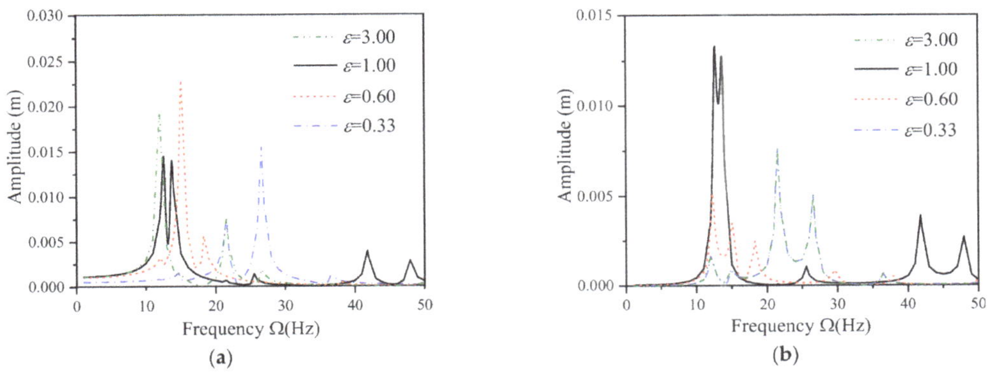

Figure 11. Effect of span ratio on response amplitude of a two-span continuous beam: (**a**) mid-span of the first span; (**b**) mid-span of the second span.

6. Conclusions

This study has conducted a detailed investigation into the impact of the finite depth foundation response on the vibration characteristics of continuous beams. By integrating the response of the finite depth foundation into a coupled model of a multi-span continuous beam resting on the elastic foundation, the study combines the Timoshenko beam theory with the motion theory of elastic foundations. The numerical solution of this coupled model is obtained using the transfer matrix method (TMM) and the fourth-order Runge-Kutta method (RK4). Taking a two-span continuous beam as a representative case, the study analyzes the influence of soil thickness, viscous damping coefficient of the soil, subgrade reaction coefficient, and span ratio on the vibration response of the continuous beam. The following conclusions can be explicitly drawn from the results and discussions presented in this research:

1. As the thickness of the soil involved in the movement increases, the continuous beam's natural frequency will decrease, and the resonance amplitude increases significantly. When comparing the dynamic responses of different beams, the enhancement impact of finite depth soil motion on the response amplitude of the indirect excitation beam section is more significant than that of the direct excitation beam section.
2. The viscous damping coefficient of the soil and the coefficient of subgrade reaction have a greater inhibition effect on the indirect influence span than on the direct excitation span. As the coefficient of the subgrade reaction increases, the system's natural frequency increases, and the resonant response amplitude decreases.
3. The adjustment of the span ratio has a significant influence on the dynamic response of the multi-span beam system. (1) Compared with other span ratios, the odd-order natural frequency is the largest and the even-order natural frequency is the smallest when $\varepsilon = 1.00$. (2) The smaller the span ratio is, the more pronounced the increasing effect of the resonant frequency is. (3) When the span ratio is determined, no matter whether the load acts on the long span or the short span, the response amplitude of the beam span indirectly affected by the load is certain.

In this paper, the model presented is limited to the dynamic response analysis of multi-span continuous beams on a linearly elastic, idealized Winkler foundation. It does not consider cases such as non-uniform beams, nonlinear soil behavior, and multiple excitation sources. Additionally, there are certain limitations in using the vertical displacement attenuation function derived from the Vlasov foundation model to incorporate finite depth soil movement into the Winkler elastic foundation model. To advance research and application in this field, it is necessary to continue modeling studies of beams on elastic foundations under complex foundation conditions. This will aim to accurately analyze the dynamic characteristics of beams on finite depth foundations.

Author Contributions: Conceptualization, D.L. and J.M.; methodology, D.L. and J.M.; software, J.W. and Y.G.; validation, D.L., H.Y. and J.W.; formal analysis, D.L. and J.M.; investigation, J.W. and C.W.; resources, D.L. and J.M.; data curation, D.L., H.Y. and J.W.; writing—original draft preparation, D.L.; writing—review and editing, D.L. and H.Y.; visualization, D.L.; supervision, D.L.; project administration, J.M.; funding acquisition, D.L. and J.M. All authors have read and agreed to the published version of the manuscript.

Funding: This article was funded by the National Natural Science Foundation of China (Grant No. 11502072), the National Key R&D Project: Research and Development of Intelligent Interconnected Equipment Network Collaborative Manufacturing/Operation and Maintenance Integration Technology and Platform (Grant No.2020YFB1712105), the National Key R&D Project: Theory and Method of Underground Space Development and Construction (Grant No. 2019YFC0605104), and the Key R&D and Promotion Special Project in Henan Province-Science and Technology Research (joint fund) Project (Grant No. 232103810082).

Informed Consent Statement: Not applicable.

Data Availability Statement: Data is contained within the article.

Conflicts of Interest: The authors declare no conflicts of interest.

References

1. Hayashikawa, T.; Watanabe, N. Free vibration analysis of continuous beams. *J. Eng. Mech.* **1985**, *111*, 639–652. [CrossRef]
2. Busool, W.; Eisenberger, M. Vibrations of axially loaded continuous beams. *Int. J. Struct. Stab. Dyn.* **2002**, *2*, 117–133. [CrossRef]
3. Lee, J. Eigenvalue analysis of double-span Timoshenko beams by pseudospectral method. *J. Mech. Sci. Technol.* **2005**, *19*, 1753–1760. [CrossRef]
4. Tullini, N.; Tralli, A.; Baraldi, D. Stability of slender beams and frames resting on 2D elastic half-space. *Arch. Appl. Mech.* **2013**, *83*, 467–482. [CrossRef]
5. Seetapan, P.; Chucheepsakul, S. Dynamic responses of a two-span beam subjected to high speed 2DOF sprung vehicles. *Int. J. Struct. Stab. Dyn.* **2020**, *6*, 413–430. [CrossRef]

6. Wang, Y.J.; Wei, Q.C.; Yau, J.D. Interaction response of train loads moving over a two-span continuous beam. *J. Struct. Stab. Dyn.* **2013**, *13*, 1350002. [CrossRef]
7. Wang, L.; Kang, X.; Jiang, P. Vibration analysis of a multi-span continuous bridge subject to complex traffic loading and vehicle dynamic interaction. *KSCE J. Civ. Eng.* **2016**, *20*, 323–332. [CrossRef]
8. Tamaddon, S.; Hosseini, M.; Vasseghi, A. Effect of non-uniform vertical excitations on vertical pounding phenomenon in continuous-deck curved box girder RC bridges subjected to near-source earthquakes. *J. Earthq. Eng.* **2022**, *26*, 5360–5383. [CrossRef]
9. Lee, S.M. Analysis of center frequency effect on damping parameters estimation using continuous wavelet transform. *KSCE J. Civ. Eng.* **2021**, *25*, 1399–1409. [CrossRef]
10. Timoshenko, S. *Vibration Problems in Engineering*; Wiley: New York, NY, USA, 1974; pp. 309–312.
11. Deng, H.Y.; Jin, X.Y.; Gu, M. A simplified method for evaluating the dynamic properties of structures considering soil–pile–structure interaction. *Int. J. Struct. Stab. Dyn.* **2018**, *18*, 1871005. [CrossRef]
12. Bao, T.; Liu, Z. Evaluation of Winkler model and Pasternak model for dynamic soil-structure interaction analysis of structures partially embedded in soils. *Int. J. Geomech.* **2020**, *20*, 04019167. [CrossRef]
13. Asadijafari, M.H.; Zarastvand, M.R.; Talebitooti, R. The effect of considering Pasternak elastic foundation on acoustic insulation of the finite doubly curved composite structures. *Compos. Struct.* **2021**, *256*, 113064. [CrossRef]
14. Smith, J.; Brown, A. Analysis of Continuous Beams in Modern Bridge Construction. *J. Struct. Eng.* **2022**, *148*, 567–580. [CrossRef]
15. Lee, C.; Kim, D. Vibration Analysis of Continuous Beams on Elastic Foundations. *Int. J. Civ. Eng.* **2021**, *19*, 345–356.
16. Zhang, X.; Li, Y. Case Studies on the Performance of Continuous Beams Supported by Soil. *Geotech. Eng. J.* **2023**, *54*, 210–225.
17. Miller, R.; Thompson, P. Structural Behavior of Continuous Beams Under Varying Soil Conditions. *Civ. Eng. Res.* **2020**, *77*, 92–108.
18. Rades, M. Dynamic analysis of an inertial foundation model. *Int. J. Solids Struct.* **1972**, *8*, 1353–1372. [CrossRef]
19. Jaiswal, O.R.; Iyengar, R.N. Dynamic response of a beam on elastic foundation of finite depth under a moving force. *Acta Mech.* **1993**, *96*, 67–83. [CrossRef]
20. Metrikine, A.V.; Popp, K. Steady-state vibrations of an elastic beam on a visco-elastic layer under moving load. *Arch. Appl. Mech.* **2000**, *70*, 399–408. [CrossRef]
21. Ma, J.J.; Liu, F.J.; Nie, M.Q.; Wang, J.B. Nonlinear free vibration of a beam on Winkler foundation with consideration of soil mass motion of finite depth. *Nonlinear Dyn.* **2018**, *92*, 429–441. [CrossRef]
22. Chen, G.L.; Zeng, X.Y.; Liu, X.B.; Rui, X.T. Transfer matrix method for the free and forced vibration analyses of multi-step Timoshenko beams coupled with rigid bodies on springs. *Appl. Math. Model.* **2020**, *87*, 152–170. [CrossRef]
23. Wang, Q.F.; Wang, L.Y.; Liu, Q.S. Seismic response of stepped frame-shear wall structures by using numerical method. *Comput. Method. Appl. Mech. Eng.* **1999**, *173*, 31–39. [CrossRef]
24. Vallabhan, C.V.G.; Das, Y.C. Parametric study of beams on elastic foundation. *J. Eng. Mech.* **1988**, *114*, 2072–2082. [CrossRef]
25. Rui, X.T.; Zhang, J.S.; Zhou, Q.B. Automatic deduction theorem of overall transfer equation of multibody system. *Adv. Mech. Eng.* **2014**, *6*, 378047. [CrossRef]
26. Thambiratnam, D.; Zhuge, Y. Free vibration analysis of beams on elastic foundation. *Comput. Struct.* **1996**, *60*, 971–980. [CrossRef]
27. Friswell, M.I.; Adhikari, S.; Lei, Y. Vibration analysis of beams with non-local foundations using the finite element method. *Int. J. Numer. Meth. Eng.* **2007**, *71*, 1365–1386. [CrossRef]

coatings

MDPI

Article

Laboratory Evaluation of Strength Performance of Full-Depth Reclamation with Portland Cement Material

Yongxiang Li [1], Chuangdan Luo [2], Kuiliang Ji [1], Haiwei Zhang [3,*] and Bowei Sun [4]

1 College of Energy and Transportation Engineering, Inner Mongolia Agricultural University, Hohhot 010018, China; lyxiang@imau.edu.cn (Y.L.); js_jkl@163.com (K.J.)
2 Henan Zhongping Jiaoke Research and Design Institute Co., Ltd., Pingdingshan 467000, China; sjylcd13937595021@163.com
3 School of Civil Engineering and Environment, Zhengzhou University of Aeronautics, Zhengzhou 450046, China
4 School of Transportation Science and Engineering, Civil Aviation University of China, Tianjin 300300, China; bwsun@cauc.edu.cn
* Correspondence: zhanghw@zua.edu.cn

Abstract: Full-depth reclamation with Portland cement (FDR-PC) represents an innovative cold recycling technology for pavements, holding significant promise due to its capacity to reuse deteriorated pavement base layers. This paper investigates the key factors influencing the strength properties of FDR-PC. The results indicate that, compared to the static compaction method, the vibratory compaction method yields cold-recycled mixtures with higher maximum dry density and unconfined compressive strength (UCS). Increasing the cement content and base-to-surface ratio, extending the curing time, and raising the curing temperature all contribute to enhancing UCS. Furthermore, increasing the base-to-surface ratio and cement content enhances both indirect tensile strength and flexural strength. An approximate linear correlation exists between indirect tensile strength and UCS, as well as between flexural strength and UCS. The strength characteristics of FDR-PC were comprehensively characterized in this study, providing effective verification of its applicability.

Keywords: full-depth reclamation with Portland cement; unconfined compressive strength; tensile strength; vibratory compaction method; base-to-surface ratio; curing temperature; regression analysis

1. Introduction

In the domain of road reconstruction, a prominent technical system known as full-depth reclamation (FDR) has emerged in recent years, employing on-site cold recovery techniques. This innovative method utilizes on-site cold recovery equipment to simultaneously crush the surface and base of the old asphalt pavement, incorporating adhesive material and water to create a recycled mixture that is subsequently applied as a sub-base, base, or underlayer in the new road structure. The bonding materials utilized in FDR encompass asphalt-based substances, such as emulsified asphalt and foamed asphalt, or inorganic bonding agents like cement, lime, etc. Considering the growing demand for the reutilization of aged pavement structures on national highways in China, full-depth reclamation with Portland cement (FDR-PC) exhibits greater potential for widespread adoption within China.

Comparative analyses demonstrate that FDR-PC offers distinct advantages over other pavement rehabilitation techniques, such as lower energy use, reduced carbon dioxide emissions, and fewer environmental impacts. The efficacy of FDR-PC hinges on understanding the mechanical properties of recycled materials, which are vital for assessing pavement performance and longevity. Key characteristics, including strength, stiffness, and durability, dictate FDR-PC's suitability for rehabilitation projects. Moreover, developing effective curing methods and strategies to minimize cracking is essential to mitigate common issues like shrinkage cracks and early fatigue fractures [1].

Studying the mechanical properties of FDR-PC is of paramount importance, particularly concerning its strength parameters. The mix design methods provided by the United States [2], Spain [3], Germany [4], Australia and New Zealand [5], South Africa [6], and Brazil [7] all regard the 7-day UCS as the primary design parameter. Moreover, tensile strength is also considered a pivotal parameter in designing pavement structures, as the bottom layer of the recycled pavements suffer the tensile stress failure.

FDR-PC is an emerging pavement recycling technology. When investigating its mechanical properties, it is prudent to draw upon established counterparts in the field. According to Chinese standard Technical Specifications for Highway Asphalt Pavement Recycling (JTG/T 5521-2019) [8], the technical framework of FDR-PC resembles that of inorganic-stabilized recycled materials. Consequently, it is viable to draw upon research concerning cement-stabilized crushed stone (CSCS) as a basis for designing experimental investigations into the mechanical properties of FDR-PC. The UCS is widely regarded as a pivotal metric for assessing the quality of CSCS [9]. Abundant research indicates that the compaction degree, cement content, curing time, and curing temperature are influential factors significantly affecting the UCS of CSCS.

Jiang et al. investigated the impact of the quasi-static compaction method (QSCM) and the vertical vibration compaction method (VVCM) on the UCS of CSCS [10] and concurrently examined the UCS under varying cement contents and curing times. The findings revealed that, irrespective of suspended dense structure specimens or skeletal dense structure specimens, VVCM demonstrates greater UCS. Through the entire curing process, the UCS growth rate of CSCS material utilizing VVCM markedly out-paced that of CSCS material employing QSCM, demonstrating VVCM's superior capability in achieving higher UCS. Furthermore, some studies [11,12] also indicate that the density of a mixture is significantly decided by the degree of compaction. The UCS increases proportionally with the augmentation of dry density, and these two factors exhibit a logarithmic–linear relationship.

The inclusion of cement enhances both the adhesion and mechanical properties of CSCS. Numerous preceding studies [13–17] have corroborated the nearly linear correlation between cement content, ranging from 3% to 7%, and UCS for a specific gravel type. Moreover, within the range of 2% to 5% (cement content), both suspended dense structures and skeleton dense structures, prepared through QSCM and VVCM, exhibit an augmentation in UCS with increasing cement content [10]. Li [18] noted a significantly positive effect on the UCS of CSCS when the cement content was below 5%. The skeleton dense structure exhibits approximately 19% higher early-stage UCS and around 10% higher late-stage UCS compared to the suspended dense structure counterparts. Sun et al. [19] revealed that, within the range of 4.5% to 4.8% cement content, the gradation type exerts a relatively minimal influence on the UCS of 7-day specimens. While the UCS of specimens decreased after 28 days of curing, the magnitude of decline gradually diminished.

The influence of curing time on UCS has been extensively documented in numerous studies [15,20]. Li's [13] research suggested an approximate logarithmic–linear relationship between UCS and curing time. The results of Lu et al. [21] indicate a progressive enhancement in the UCS of CSCS over time, reaching stabilization after approximately 90 days of curing. CSCS with shorter curing times exhibits significantly greater damage compared to specimens with longer curing times.

Regarding curing temperature, numerous studies [22,23] have addressed high-temperature curing, with the majority of findings indicating its potential to enhance initial UCS. Escalante et al. [24] conducted UCS tests on CSCS at five temperatures ranging from 10 °C to 60 °C, indicating that elevated temperatures indeed enhance initial UCS, although there is a subsequent decline in UCS under prolonged exposure to high temperatures. Similarly, Wang et al. [25] investigated the UCS of calcium sulfoaluminate cement under varied curing temperatures, observing an initial increase in UCS with rising temperatures yet a subsequent decrease within the range of 40 °C to 80 °C. Regarding low-temperature curing, Price et al. [26] found that the rate of UCS development in mixed concrete is significantly slower at low temperatures compared to room temperature. Husem et al. [27] investigated the UCS of both ordinary concrete and high-performance concrete subjected to standard curing temperatures as well as various low-temperature conditions. The findings indicated that the UCS at temperatures of 10 °C and below is inferior to the standard curing conditions. Kim et al. [28] investigated the UCS development under curing temperatures of 5 °C, 20 °C, and 40 °C. The findings suggested slight differences during the initial process, while appearing nearly indistinctive with time. Xu et al. [29] simulated local air temperature fluctuations and found that the UCS of 5% CSCS increases with prolonged curing time, with the rate of gain being lower at low temperatures compared to standard temperatures. Marzouk et al. [30] conducted experiments over three months at five temperatures ranging from −10 °C to 20 °C, revealing a proportional relationship between UCS and temperature. Furthermore, Saul [31] proposed the concept of maturity in 1951, defining it as the product of curing time and temperature. The Nurse–Saul maturity function indicates that the UCS will be approximately the same when maturity is equivalent.

Considering research on tensile strength, several studies [17,32] indicate that the ratio of flexural strength to UCS ranges between 1:10 and 1:6, with the ratio of low-strength mixtures reaching 1:6 and high-strength mixtures being 1:10, approximating a linear relationship. Additionally, studies [33–35] suggest an approximate linear relationship between UCS and indirect tensile strength. Therefore, it can be inferred that the parameters influencing UCS also influence tensile strength.

Several recent studies have investigated the mechanical properties of full-depth reclamation with Portland cement (FDR-PC). Ghanizadeh et al. [36] examined the UCS of cold-recycled mixtures composed of varying base-to-surface ratios of reclaimed asphalt pavement (RAP) (0%, 20%, and 40%) mixed with two types of soil and different cement contents (3%, 4%, and 5%). The findings revealed that increasing cement content while reducing RAP content could enhance the UCS. Gonzalo et al. [37] analyzed the relationship between UCS and tensile strength across 24 groups of FDR-PC specimens. The study found linear correlations between UCS and indirect tensile strength; indirect tensile strength and flexural strength; as well as among UCS, indirect tensile strength, and flexural strength. Ghanizadeh et al. [38] employed multivariate adaptive splines to analyze the UCS of various mixtures made of FDR-PC cold-recycled mixtures. The study concluded that cement content, RAP content, optimal moisture content, and curing time had the most significant impact on UCS.

The primary objective of this paper is to investigate the principal factors influencing the UCS, indirect tensile strength, and flexural strength of FDR-PC. Based on the research on the UCS and the tensile strength of CSCS mentioned above, and relying upon the ordinary road asphalt pavement reclamation projects in Inner Mongolia in China, this paper employed the static compaction method and the vibratory compaction method to fabricate specimens. The specimens were systematically subjected to varying cement contents (4%, 5%, and 6%), base-to-surface ratios (10:0, 8:2, and 6:4), curing temperatures (20 °C, 30 °C, 40 °C, and 50 °C), and curing times (3 days, 7 days, 28 days, and 90 days) to assess the UCS. Additionally, the indirect tensile strength and flexural strength of specimens were evaluated under different cement contents (4%, 5%, and 6%) and base-to-surface ratios (10:0, 8:2, and 6:4).

2. Materials and Specimen Preparation

2.1. Raw Materials

FDR-PC material compositions: RAP, reclaimed inorganic-binder-stabilized aggregate (RAI), cement, and water. The reclaimed materials were sourced from the 110 National Highway, a secondary road within Inner Mongolia in China. The old pavement comprises a 4 cm layer of asphalt material (AC-16) and a 20 cm base layer of cement-stabilized crushed stone (CSCS). The RAP and RAI are obtained by separately milling the asphalt layer and base layer using a milling machine.

After conducting drying treatment and screening experiments on RAP and RAI, the resulting materials were mixed in accordance with three different mass ratios to produce recycled ore, as illustrated in Figure 1. The grade curve of the blended recycled materials conforms to the specifications outlined in Highway Asphalt Pavement Recycling (JTG/T 5521-2019) [7], as depicted in Figure 2. In this study, Figure 2 was incorporated to illustrate the design of three representative recycled materials, which were formulated based on varying mass ratios of RAI and RAP. These materials exhibited distinct base-to-surface ratios, namely 10:0, 8:2, and 6:4. The selection of the aforementioned three representative base-to-surface ratios takes into account factors such as the thickness ratio between old asphalt pavement and the base layer, actual construction conditions, and economic considerations in China's national road system.

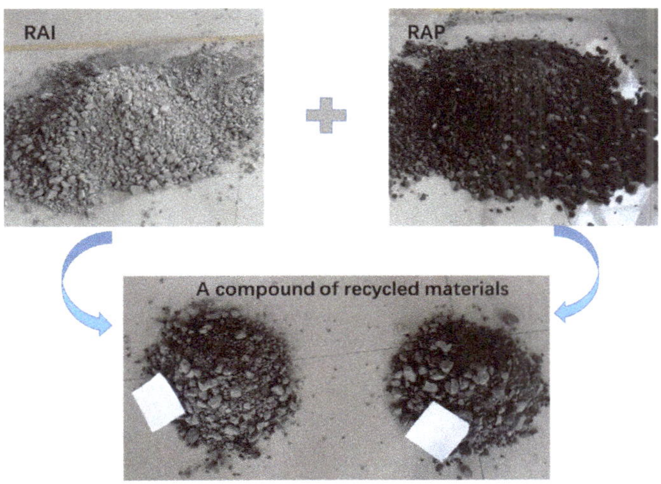

Figure 1. Reclaimed material.

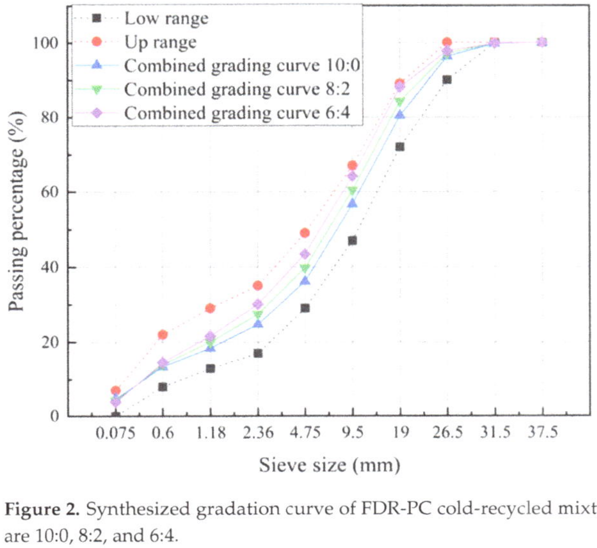

Figure 2. Synthesized gradation curve of FDR-PC cold-recycled mixture; the base-to-surface ratios are 10:0, 8:2, and 6:4.

This study employed cement of 42.5-grade ordinary Portland cement, with technical specifications detailed in Table 1. The test water requirements can be met by ordinary drinking water.

Table 1. Technical indicators of cement.

Testing Item		Technical Indicator	Testing Result
Setting Time (min)	Initial Setting	≥180	218
	Final Setting	≤600	385
Flexural Strength (MPa)	3 d	≥3.5	3.8
	28 d	≥6.5	10.6
Compressive Strength (MPa)	3 d	≥17.0	19.9
	28 d	≥42.5	45.5

2.2. Determining the Optimum Water Content and the Maximum Dry Density

Before preparing the strength test specimen, it is essential to determine the maximum dry density and optimal moisture content of th eFDR-PC mixture. Following the testing methodology delineated in the Chinese standard JTG E51-2009 [39], the heavy compaction test method (T0804-1994) and vibration compaction test method (T0842-2009) were utilized to determine the maximum dry density and optimal water content of FDR-PC. Specimens with different cement contents, consisting of 4%, 5%, and 6% of the mass of the recycled mixture, were individually prepared. By manipulating the water content, distinct curves depicting the relationship between dry density and water content for each mixture were obtained. The peaks of these curves serve as indicators to determine the optimal water content and maximum dry density, with the combined results presented in Table 2.

Table 2. The optimum water content and the maximum dry density of specimens fabricated with different cement content and base-to-surface ratio combinations.

Design Method	Cement Content (%)	Base-to-Surface Ratio	Optimal Water Content (%)	Maximum Dry Density (g/cm^3)
Heavy Compaction Test	4	10:0	6.7	2.232
	5	10:0	7.1	2.236
	6	10:0	7.3	2.238
	4	8:2	5.9	2.242
	5	8:2	6.6	2.246
	6	8:2	6.8	2.240
	4	6:4	5.8	2.246
	5	6:4	6.2	2.250
	6	6:4	6.2	2.245
Vibration Compaction Test	4	10:0	5.8	2.268
	5	10:0	6.1	2.272
	6	10:0	6.2	2.273
	4	8:2	4.7	2.282
	5	8:2	5.2	2.291
	6	8:2	5.6	2.284
	4	6:4	4.3	2.291
	5	6:4	4.6	2.297
	6	6:4	4.6	2.301

2.3. Specimen Preparation

The cylindrical specimens (with a diameter of 150 mm and a height of 150 mm) used for the UCS test were fabricated using static forming and vibration forming methods. The procedures for static forming and vibration forming followed the guidelines specified in JTG E51-2009 (T0843-2009) [39] and JTG D50-2006(Appendix A) [40], respectively. The majority of the specimens underwent standard curing in a designated room, specifically

6-day curing followed by 1-day soaking. The control conditions were as follows: the air temperature was maintained at 20 ± 2 °C and the relative humidity was kept at 95%. Another portion of the specimens was subjected to prolonged exposure to elevated temperatures and extended duration.

The cylinder specimens used in the splitting strength test have the same size as all the cylinder specimens in the UCS test, which can be formed via static pressing according to the specification JTG E51-2009 (T0843-2009) [39]. The specimens were then subjected to a standard 90-day curing period. A total of nine sets of specimens were prepared, comprising three base-to-surface ratios (10:0, 8:2, and 6:4) and three cement contents (4%, 5%, and 6%).

The flexural tensile strength test utilizes prisms measuring 100 mm × 100 mm × 400 mm as specimens, which can be prepared following the static pressure method outlined in JTG E51-2009 (T0844-2009) [39]. Subsequently, a standard curing period of 90 days was employed for the specimens prior to testing. A total of nine sets of specimens were prepared, comprising three base-to-surface ratios (10:0, 8:2, and 6:4) and three cement contents (4%, 5%, and 6%).

The weight of FDR-PC required for the preparation process of various test pieces mentioned above can be calculated and controlled using Equation (1), based on the maximum dry density and optimum water content.

$$m_0 = V \times \rho_{\max} \times \left(1 + w_{opt}\right) \times 98\% \quad (1)$$

In Equation (1), m_0 signifies the weight of an individual specimen, g; V denotes the volume relative to a single specimen, cm^3; $\rho_{\max}$ represents the maximum dry density, g/cm^3; and w_{opt} indicates the optimum water content.

3. Test Method

3.1. Unconfined Compressive Strength Test Method

The unconfined compressive strength test was conducted according to the JTG E51-2009 (T0806-1994) [39] procedure, as illustrated in Figure 3. During the test, it is imperative to maintain a constant loading rate of 1 mm/min while meticulously documenting the maximum pressure P at the precise moment of specimen failure. The UCS value for each specimen was calculated using Equation (2).

$$U_s = \frac{4P}{\pi d^2} \quad (2)$$

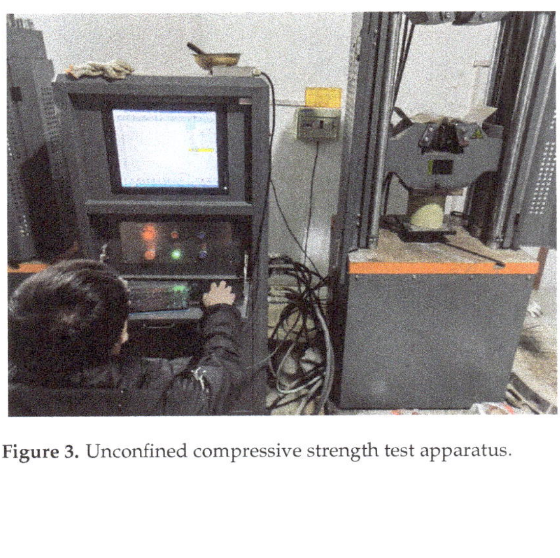

Figure 3. Unconfined compressive strength test apparatus.

In Equation (2), U_s represents the unconfined compressive strength of the specimen, MPa; P represents the maximum pressure applied to the specimen at the point of failure, N; and d represents the diameter of the specimen, mm.

3.2. Splitting Strength Test Method

The splitting strength tests of nine groups of cylinder specimens were conducted according to the JTG E51-2009 (T0806-1994) [39] procedure, as illustrated in Figure 4. During the test, the deformation rate of the specimen should be uniformly increased while maintaining a constant loading rate of 1 mm/min. The maximum pressure P at the point of specimen failure should be accurately recorded. The ITS value for each specimen was calculated using Equation (3).

$$I_s = 0.004178\frac{P}{h} \tag{3}$$

Figure 4. Splitting strength test apparatus.

In Equation (3), I_s represents the indirect tensile strength of the specimen, MPa; P represents the maximum pressure applied to the specimen at the point of failure, N; and h represents the height of the specimen, mm.

3.3. Flexural–Tensile Strength Test Method

Testing is conducted following the JTG E51-2009, as specified by T0851-2009 [39]. The specimens were placed on fixtures, with the load direction consistent with the pressure direction during specimen formation. Upper and lower pressure blocks were positioned at one-third points along the length of the specimen, as shown in Figure 5. Rapid loading was applied at a rate of 50 mm/min, with the ultimate load (P) recorded at failure. The formula for calculating flexural tensile strength is provided as Equation (4).

$$f_s = \frac{PL}{b^2h} \tag{4}$$

Figure 5. The four-point bending test apparatus.

In Equation (3), f_s represents the flexural tensile strength of the specimen, MPa. P denotes the ultimate load at failure of the specimen, N. L signifies the span, which is the distance between two supports, mm. b indicates the width of the specimen, mm. h denotes the height of the specimen, mm.

4. Results and Discussion

4.1. Influencing Factors of UCS

4.1.1. Cement Content and Base-to-Surface Ratios

The experimental results of the 7 d standard UCS of FDR-PC cold-recycled mixtures under different cement contents and base-to-surface ratios are presented in Table 3. Figure 6 illustrates a bar chart depicting the data from Table 3.

Table 3. Seven-day standard UCS of FDR-PC materials.

Forming Method	Base-to-Surface Ratio	4% Cement Content			5% Cement Content			6% Cement Content		
		$\bar{U}_s$ MPa	C_v %	$U_{s0.95}$ MPa	$\bar{U}_s$ MPa	C_v %	$U_{s0.95}$ MPa	$\bar{U}_s$ MPa	C_v %	$U_{s0.95}$ MPa
Static Compaction	10:0	5.2	4.8	4.8	5.7	6.5	5.1	6.0	2.0	5.8
	8:2	5.0	7.0	4.4	5.2	10.8	4.3	5.4	11.5	4.4
	6:4	4.6	4.1	4.3	4.7	5.1	4.3	5.0	10.8	4.1
Vibratory Compaction	10:0	6.7	7.6	5.9	6.9	4.9	6.3	7.4	5.3	6.8
	8:2	6.5	2.2	6.3	6.8	2.6	6.5	7.1	3.0	6.8
	6:4	6.1	3.8	5.7	6.3	2.1	6.1	6.5	5.4	5.9

As evident from Figure 6, the UCS under varying cement contents demonstrated an upward trend with an increase in the base-to-surface ratio. Specifically, the mixture's UCS reached its peak when the base-to-surface ratio was 10:0, followed by a slightly reduced value at a base-to-surface ratio of 8:2, and a minimum value at a base-to-surface ratio of 6:4. This phenomenon is attributed to the increase in the proportion of surface layer materials, indicating a higher content of RAP materials with relatively lower UCS. Additionally, the UCS of recycled mixtures depends on the adhesive properties of the cement, but the presence of an asphalt film on the surface of RAP materials reduces the contact between cement and RAP materials, hindering hydration reactions and resulting in a decrease in strength.

In addition, with an increase in cement content, the UCS exhibited an initial rapid increase in transitioning to a stable state. Notably, the UCS achieved through the vibratory compaction method surpasses that obtained through the static compaction method. For example, at a base-to-surface ratio of 8:2, changing the compaction method results in a sequential increase of 1.30, 1.31, and 1.31 times in UCS for cement contents of 4%, 5%, and

6%, respectively, highlighting the superior UCS exhibited by specimens fabricated through the vibratory compaction method.

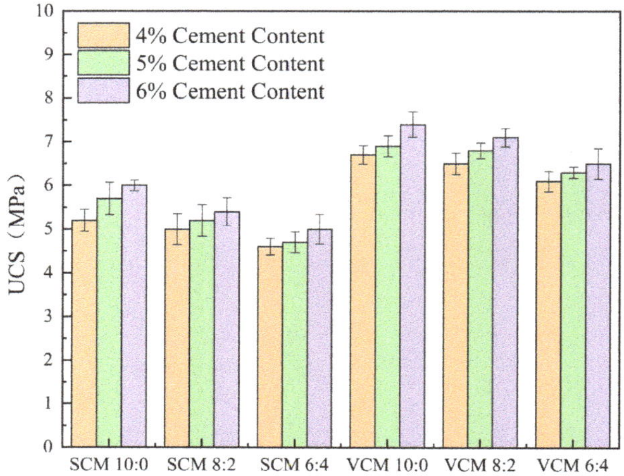

Figure 6. Comparison of UCS under different base-to-surface ratios and cement contents.

Furthermore, in accordance with the regulations, the unconfined compressive strength (UCS) of the large flow base level should range between 4.0 MPa and 6.0 MPa, while for the sub-base it should be within 3.0 MPa to 5.0 MPa. Consequently, even when employing the static compaction method, the 7-day standard UCS values of the nine types of FDR-PC cold regeneration mixtures designed in this study fall between 4.1 and 5.8 MPa, satisfying Chinese secondary and below highway specifications requirements. This demonstrates that when an adequate cement dosage is utilized and a reasonable base ratio is maintained, FDR recycled material can be effectively employed in road bases.

4.1.2. Curing Temperature

To study the influence pattern of curing temperatures on the UCS of FDR-PC cold-recycled mixtures, experiments were conducted to determine the 7 d UCS under varying curing temperature conditions. The results are presented in Table 4 and Figure 7. The selection of the curing temperature range was made considering the recommended ranges stipulated by standards and the pragmatic conditions encountered in the field. Simultaneously, it was essential to ensure that the temperature settings would not excessively impact the stability of RAP.

Table 4. Comparison in UCS under different base-to-surface ratios concerning curing temperature.

Forming Method	Base-to-Surface Ratio	20 °C			30 °C			40 °C			50 °C		
		$\bar{U}_s$ MPa	C_v %	$U_{s0.95}$ MPa	$\bar{U}_s$ MPa	C_v %	$U_{s0.95}$ MPa	$\bar{U}_s$ MPa	C_v %	$U_{s0.95}$ MPa	$\bar{U}_s$ MPa	C_v %	$U_{s0.95}$ MPa
Static Compaction	10:0	5.7	6.5	5.1	6.4	5.8	5.8	7.0	1.7	6.8	7.2	3.5	6.8
	8:2	5.2	6.9	4.6	5.7	8.1	4.9	6.2	5.2	5.7	6.4	7.3	5.6
	6:4	4.7	5.1	4.3	5.0	4.8	4.6	5.4	4.4	5.0	5.6	2.7	5.4
Vibratory Compaction	10:0	6.9	3.5	6.5	7.7	4.4	7.1	8.5	4.6	7.9	8.7	3.8	8.2
	8:2	6.8	2.6	6.5	7.5	2.4	7.2	8.2	2.6	7.9	8.4	4.0	7.8
	6:4	6.3	2.1	6.1	6.5	2.0	6.3	6.9	5.1	6.3	7.0	4.0	6.5

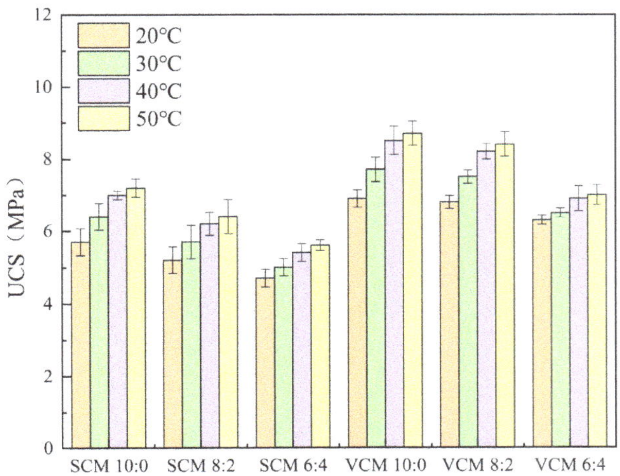

Figure 7. Comparison of UCS under different base-to-surface ratios concerning curing temperature.

According to Table 4 and Figure 7, it is evident that, whether employing the static compaction method or the vibratory compaction method, an overall increase in the curing temperature is conducive to the development of UCS in FDR-PC cold-recycled mixtures. This phenomenon arises from the accelerated formation of hydration products due to temperature rise. To clarify the pattern of UCS variation more clearly, Figure 5 presents a bar chart illustrating the data from Table 5. The graph reveals not only an increase in UCS with rising curing temperature but also a significant difference between 20 °C and 40 °C, indicating a notable enhancement in UCS within this temperature range. Beyond 30 °C, while UCS continues to rise, the rate of increase gradually diminishes. For instance, with a base-to-surface ratio of 8:2, the UCS corresponding to the static compaction method increases by factors of 1.10, 1.09, and 1.03, whereas for the vibratory compaction method the increases occur by factors of 1.11, 1.07, and 1.02, respectively. Moreover, the influence of curing temperature on UCS under different base-to-surface ratios follows a certain regularity, where a larger base-to-surface ratio experiences a more significant effect from curing temperature on its UCS, while a smaller base-to-surface ratio exhibits a lesser impact. For instance, during the curing process at 20 °C~30 °C, as the base-to-surface ratio decreases, the corresponding increases in UCS for the static compaction method are 1.13 times (10:0), 1.10 times (8:2), and 1.06 times (6:4), whereas for the vibratory compaction method the increases are 1.12 times (10:0), 1.10 times (8:2), and 1.04 times (6:4). To explain the observed characteristics, we might consider the impact of temperature on the cement hydration process, which shows larger susceptibility for the higher RAI ratio.

Table 5. Comparison in UCS of the recycled mixture under different base-to-surface ratios concerning curing time.

Forming Method	Base-to-Surface Ratio	3 d			7 d			28 d			90 d		
		$\overline{U_s}$ MPa	C_v %	$U_{s0.95}$ MPa	$\overline{U_s}$ MPa	C_v %	$U_{s0.95}$ MPa	$\overline{U_s}$ MPa	C_v %	$U_{s0.95}$ MPa	$\overline{U_s}$ MPa	C_v %	$U_{s0.95}$ MPa
Static Compaction	10:0	3.9	9.5	3.3	5.7	6.5	5.1	6.7	4.3	6.2	7.3	3.8	6.8
	8:2	3.7	9.7	3.1	5.2	6.9	4.6	6.4	5.5	5.8	6.6	5.3	6.0
	6:4	3.4	7.1	3.0	4.7	5.1	4.3	5.5	3.1	5.2	5.7	2.6	5.5
Vibratory Compaction	10:0	4.5	5.3	4.1	6.9	3.5	6.5	8.6	0.9	8.5	9.2	2.6	8.8
	8:2	4.3	4.2	4.0	6.8	2.6	6.5	8.2	3.0	7.8	8.5	3.2	8.1
	6:4	4.0	3.3	3.8	6.3	2.1	6.1	7.2	4.9	6.6	7.4	3.4	7.0

4.1.3. Curing Time

The curing time plays a crucial role in the development of UCS in cement-stabilized materials. Under normal conditions, the standard indoor curing time typically extends to 7 days. However, for FDR-PC cold-recycled mixtures, distinguished by their unique composition that includes RAP, the adequacy of a 7-day curing time to meet practical usage requirements requires further experimental verification. Moreover, few studies have focused on the evolution of UCS in relation to curing time for such materials. Consequently, this section investigates the impact of extended curing time on UCS.

Table 5 illustrates the variation in UCS with curing time under different base-to-surface ratios, with a curing temperature of 20 °C. From the table, it is evident that extending the curing time significantly enhances the UCS, with the growth rate of UCS increasing as the curing time extends. Figure 8 graphically illustrates the impact of curing time on UCS in the form of a bar chart. The graph reveals that the UCS experiences a rapid increase within the 3-day to 7-day curing time, with growth rates ranging from 1.38 to 1.58 times. Beyond 7 days, the rate of UCS growth gradually decelerates, ranging between 1.02 to 1.23 times.

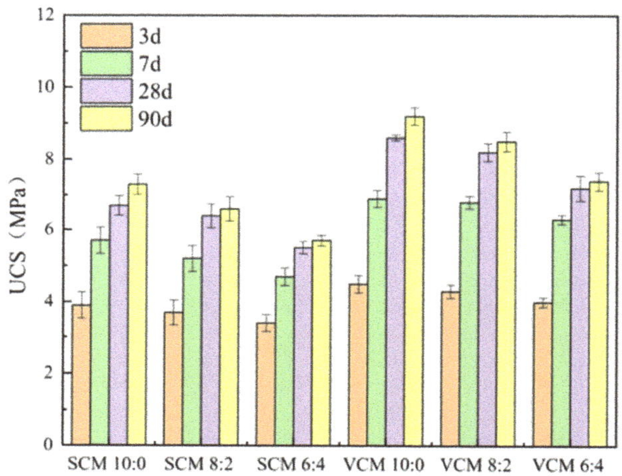

Figure 8. Comparison of UCS under different base-to-surface ratios concerning curing time.

Moreover, when the curing time surpasses 28 days, the rate of UCS growth for FDR-PC cold-recycled mixtures with base-to-surface ratios of 8:2 and 6:4 is notably less obvious compared to the counterpart with a base-to-surface ratio of 10:0. For instance, specimens fabricated using the vibratory compaction method exhibit varying rates of UCS growth at 28 days and 90 days under different base-to-surface ratio conditions, specifically 1.07 times (10:0), 1.03 times (8:2), and 1.02 times (6:4). The reason lies in the relatively lower proportion of cement-stabilized materials within the low base-to-surface mixture, resulting in fewer hydration products. Consequently, once the hydration reaction finishes, the UCS no longer exhibits significant increments. Furthermore, the aging process of RAP may induce material hardening and consequent hardness enhancement, although it is not the primary influencing factor.

4.2. The Tensile Strength of FDR-PC

4.2.1. Indirect Tensile Strength

A summary of the indirect tensile strength results for different FDR-PC cold-recycled mixtures is presented in Table 6, with a graphical representation of the representative values of indirect tensile strength illustrated in Figure 9.

Table 6. Summary of indirect tensile strength results for various FDR-PC cold-recycled mixtures.

Base-to-Surface Ratios	Cement Content 4%			Cement Content 5%			Cement Content 6%		
	$\bar{I}_s$ MPa	C_v %	$I_{s0.95}$ MPa	$\bar{I}_s$ MPa	C_v %	$I_{s0.95}$ MPa	$\bar{I}_s$ MPa	C_v %	$I_{s0.95}$ MPa
10:0	0.66	7.6	0.58	0.90	6.7	0.80	1.05	6.7	0.93
8:2	0.60	5.0	0.55	0.75	4.0	0.70	0.89	4.5	0.82
6:4	0.54	7.4	0.47	0.71	5.6	0.64	0.79	10.1	0.66

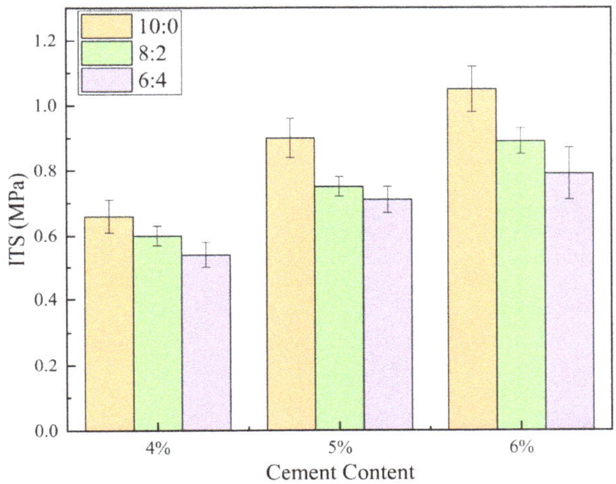

Figure 9. Comparison chart of indirect tensile strength for each mixture.

Combining Table 6 and Figure 9, it can be observed that the results of the indirect tensile strength tests exhibit relatively low dispersion, indicating that for FDR-PC cold-recycled mixtures conducting nine parallel tests for each group effectively maintains the coefficient of variation within 15%. This suggests that the quantity of specimens utilized for indirect tensile strength testing evaluation is sufficiently adequate.

For FDR-PC cold-recycled mixtures with a 4% cement content, the indirect tensile strength ranged from 0.4 MPa to 0.6 MPa across the three base-to-surface ratios. For the other two groups with different cement contents, the representative values of indirect tensile strength generally fell within the range of 0.6 to 1.0 MPa. All of these values meet the recommended indirect tensile strength values of 0.4 MPa to 0.6 MPa stipulated in JTG D50-2017 [41]. From this perspective, cautious consideration should be given to the utilization of mixtures with low cement contents and high proportions of RAP in engineering applications.

The indirect tensile strength of FDR-PC cold-recycled mixtures composed of various base-to-surface ratios increases with the rise in cement contents. Comparatively, when the base-to-surface ratio is 6:4, the variation in indirect tensile strength with cement content is more gradual. Across different cement contents, the trend of increasing indirect tensile strength with higher base-to-surface ratios is evident, indicating that the presence of RAP is detrimental to the formation of mixture strength. The fundamental reason lies in the increased proportion of reclaimed asphalt materials affecting the interconnection between cement and aggregates, hindering the smooth formation of cement hydration products. Both the cement content and base-to-surface ratio are the primary factors influencing the indirect tensile strength of mixtures. However, in comparison, the impact of cement content appears to be more significant.

4.2.2. Flexural Strength

A summary of the flexural tensile strength results for each group are provided in Table 7 and Figure 10. For FDR-PC cold-recycled mixtures, each group underwent 12 parallel tests, with the coefficient of variation overall controlled within 10%. This suggests that the quantity of specimens utilized for evaluating flexural strength tests is sufficiently effective.

Table 7. Summary of flexural strength results for various FDR-PC cold-recycled mixtures.

Base-to-Surface Ratios	Cement Content 4%			Cement Content 5%			Cement Content 6%		
	$\bar{f}_s$ MPa	C_v %	$f_{s0.95}$ MPa	$\bar{f}_s$ MPa	C_v %	$f_{s0.95}$ MPa	$\bar{f}_s$ MPa	C_v %	$f_{s0.95}$ MPa
10:0	1.25	8.8	1.07	1.62	7.4	1.42	1.98	6.1	1.78
8:2	0.97	6.2	0.87	1.44	7.6	1.26	1.71	8.2	1.48
6:4	0.89	6.7	0.79	1.25	8.8	1.07	1.48	7.4	1.30

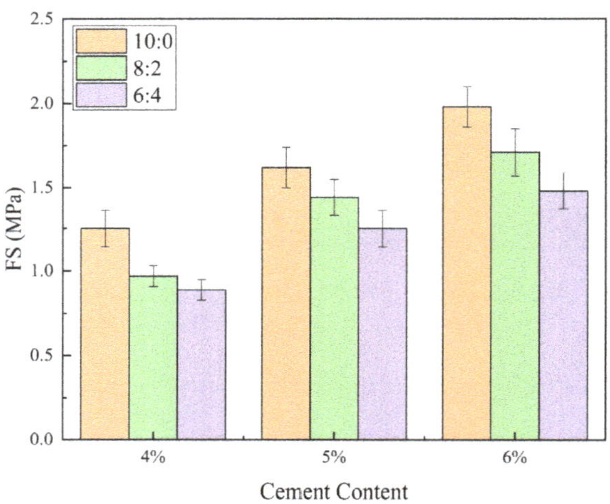

Figure 10. Comparison chart of flexural strength for each mixture.

The flexural strength of FDR-PC cold-recycled mixtures, composed of various base-to-surface ratios, increases with the rise in cement contents. Specifically, at base-to-surface ratios of 10:0, 8:2, and 6:4, the representative values of flexural strength increase by 69%, 58%, and 51%, respectively, with the increase in cement content. Moreover, at each cement content level, the flexural strength of FDR-PC cold-recycled mixtures increases with higher base-to-surface ratios. When the cement content is 4%, 5%, and 6%, the representative values of flexural strength increase by 36%, 34%, and 36%, respectively, with the increase in base-to-surface ratios. These findings indicate that both the cement content and base-to-surface ratio significantly influence the flexural strength of the cold-recycled mixtures. However, the impact of cement content appears to be more significant. Additionally, the effect of the base-to-surface ratio on flexural strength remains relatively stable and is nearly independent of the cement content.

In the test groups with 4% cement content, the flexural strength of FDR-PC cold-recycled mixtures ranges from 0.7 MPa to 1.0 MPa. These results closely align with the recommended flexural strength range of 0.6 MPa to 1.0 MPa for cement-stabilized soil materials outlined in JTG D50-2017 [41]. In the test groups with 5% and 6% cement contents, the flexural strength of FDR-PC cold-recycled mixtures ranges from 1.0 MPa to 1.8 MPa. These results also closely correspond to the recommended flexural strength range of

0.9 MPa to 2.0 MPa for cement-stabilized aggregate materials specified in the aforementioned design specifications. Therefore, from the perspective of flexural strength requirements, when utilizing FDR-PC cold-recycled mixtures for base layers, it is recommended that the cement content not be less than 5%. Particularly when the proportion of old asphalt mixtures is relatively high, a cement content of not less than 6% is recommended.

4.3. The Linear Relationship between UCS and Tensile Strength

The UCS and tensile strength values were combined for comparison on the same base-to-surface ratio, as depicted in Table 8. The 7 d UCS average values in Table 8 were obtained from the measurements of specimens outlined in Table 3. The indirect tensile strength (ITS) utilized the average value I_s from Table 6, while the flexural strength (FS) utilized the average value f_s from Table 7. Each dataset represents three sets of UCS-ITS and UCS-FS values corresponding to different cement contents under the same base-to-surface ratio. Linear regression was performed on each dataset, and the results are illustrated in Figure 11. The coefficients and R^2 of the linear regression equation in Figure 11 are shown in Table 9.

Table 8. The UCS, ITS, and FS values correspond to different cement contents under the same base-to-surface ratio.

Method	Cement Content	Base-to-Surface Ratio 10:0			8:2			6:4		
		UCS	ITS	FS	UCS	ITS	FS	UCS	ITS	UCS
Static Compaction	4%	5.2	0.66	1.25	5	0.6	0.97	4.6	0.54	0.89
	5%	5.7	0.9	1.62	5.2	0.75	1.44	4.7	0.71	1.25
	6%	6	1.05	1.98	5.4	0.89	1.71	5	0.79	1.48
Vibratory Compaction	4%	6.7	0.66	1.25	6.5	0.6	0.97	6.1	0.54	0.89
	5%	6.9	0.9	1.62	6.8	0.75	1.44	6.3	0.71	1.25
	6%	7.4	1.05	1.98	7.1	0.89	1.71	6.5	0.79	1.48

Table 9. Coefficients and R^2 of the linear regression equations for 7 d UCS-ITS and 7 d UCS-FS.

Compaction Method	Item	Base-to-Surface Ratio	$y = kx + b$ k	b	R^2
Static Compaction	UCS-ITS	10:0	0.487	−1.872	0.999
		8:2	0.725	−3.023	0.999
		6:4	0.542	−1.905	0.782
	UCS-FS	10:0	0.895	−3.425	0.982
		8:2	1.850	−8.425	0.976
		6:4	1.312	−5.045	0.843
Vibratory Compaction	UCS-ITS	10:0	0.508	−2.684	0.866
		8:2	0.483	−2.540	0.999
		6:4	0.625	−3.258	0.959
	UCS-FS	10:0	0.981	−5.249	0.939
		8:2	1.233	−7.013	0.976
		6:4	1.475	−8.086	0.984

The UCS data in Figure 11a are obtained from specimens formed via SCM. The R^2 values of the three groups of UCS-ITS are 0.782~0.999, indicating a strong linear correlation between UCS and ITS, and the R^2 values of the three groups of UCS-FS are 0.843~0.999, further confirming a highly significant linear relationship. The UCS data in Figure 11b are obtained from specimens formed via VCM. The R^2 values of the three groups of UCS-ITS are 0.866~0.999, indicating a strong linear correlation between UCS and ITS, and the R^2 values of the three groups of UCS-FS are 0.939~0.984, further confirming a highly significant linear relationship. Thus, it can be inferred that a significant linear relationship exists between 7 d UCS and tensile strength.

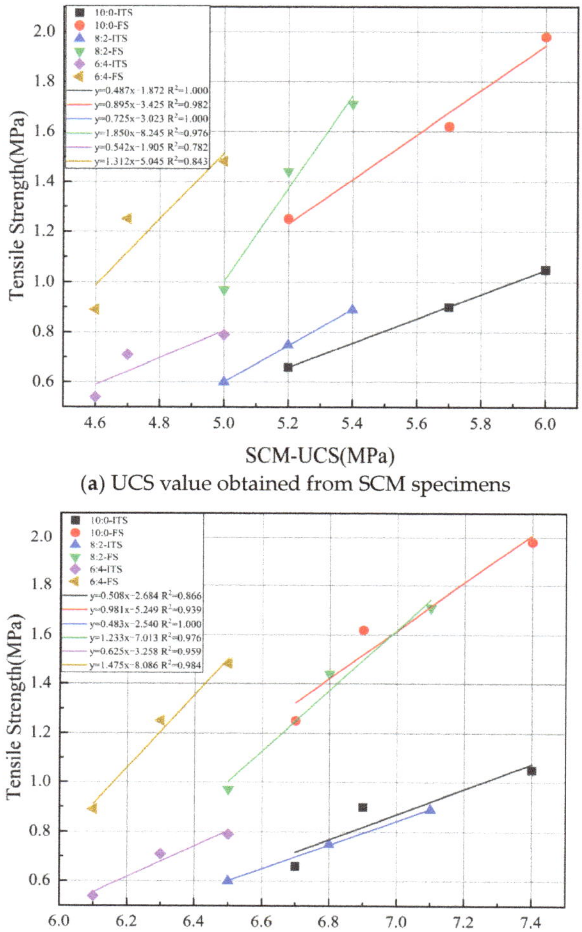

(**a**) UCS value obtained from SCM specimens

(**b**) UCS value obtained from VCM specimens

Figure 11. Linear regression for 7 d UCS-ITS and 7 d UCS-FS.

5. Conclusions

This paper investigates the strength parameters of FDR-PC, including UCS, indirect tensile strength, and flexural strength. The study examines the impact of molding methods, cement content, the base-to-surface ratio, curing temperature, and curing time on UCS. Additionally, it explores the effects of the cement content and base-to-surface ratio on indirect tensile strength and flexural strength. These factors are critical for enhancing the effectiveness of road maintenance and extending the lifespan of roads using FDR-PC technology. The significance of this study lies in providing engineers with specific data and guidance to help them select optimal material proportions and curing conditions in practice, thereby enhancing the performance and economic efficiency of road projects. The primary findings are summarized as follows:

(1) A higher cement content and base-to-surface ratio correspond to higher UCS in FDR-PC cold-recycled mixtures. The 7-day standard UCS of FDR-PC mixtures can meet the requirements of ordinary highway base strength, provided that the cement dosage exceeds 4% and the base surface ratio is no less than 6:4.
(2) Compared to static compaction methods, vibration compaction methods can achieve higher unconfined compressive strength (UCS) in FDR-PC recycled mixtures, thereby confirming the feasibility of this testing method in FDR-PC research.
(3) The UCS of the FDR-PC mixture can be enhanced by prolonging the curing duration and elevating the curing temperature, which is analogous to that observed in conventional cement stabilized materials.
(4) Both the ITS and FS exhibit a noticeable increase as the base-to-surface ratio increases, with variation rates reaching up to 30%–40%. Compared to the influence of the base-to-surface ratio, the impact of cement content is more significant. From the perspective of flexural strength requirements, when utilizing FDR-PC mixtures for base layers, it is recommended that the cement content is not less than 5%.
(5) The regression analysis revealed a strong correlation coefficient between the UCS and ITS, as well as between UCS and FS of the FDR-PC mixture, indicating a clear linear relationship among these variables. This suggests that the 7-day standard UCS can serve as a predictor for both 90-day ITS and FS.

The advantages of this study lie in its systematic investigation of multiple key factors affecting the UCS and tensile strengths of FDR-PC, providing practical reference data for engineering applications. However, the limitations of the study include a relatively limited number of samples and a focus on specific experimental conditions. Future research could validate these findings under a broader range of environmental and operational conditions to further enhance the applicability and impact of the research.

Author Contributions: Conceptualization, Y.L. and K.J.; Data curation, Y.L. and C.L.; Formal analysis, B.S.; Funding acquisition, C.L., H.Z. and B.S.; Investigation, C.L., H.Z. and B.S.; Methodology, Y.L. and H.Z.; Project administration, Y.L.; Resources, C.L.; Software, K.J. and B.S.; Visualization, B.S.; Writing—original draft, K.J. and H.Z.; Writing—review and editing, H.Z. All authors have read and agreed to the published version of the manuscript.

Funding: This research was funded by Fundamental Research Funds for the Universities (BR230145); Inner Mongolia Autonomous Region Science and Technology Plan: (2022YFHH0082); Pingdingshan Major Science and Technology Project (2021ZD07); and Fundamental Research Funds for the Central Universities (KJZ53420230066).

Institutional Review Board Statement: Not applicable.

Informed Consent Statement: Not applicable.

Data Availability Statement: The data presented in this study are available on request from the corresponding author.

Conflicts of Interest: As author Chuangdan Luo was employed by the company Henan Zhongping Jiaoke Research and Design Institute Co., Ltd. The remaining authors declare that the research was conducted in the absence of any commercial or financial relationships that could be construed as a potential conflict of interest.

References

1. Fedrigo, W.; Nunez, W.P.; Visser, A.T. A review of full-depth reclamation of pavements with Portland cement: Brazil and abroad. *Constr. Build. Mater.* **2020**, *262*, 120540. [CrossRef]
2. Portland Cement Association (PCA). *Guide to Full Depth Reclamation (FDR) with Cement*; Portland Cement Association (PCA): Skokie, IL, USA, 2005.
3. Spanish Institute of Cement and Its Applications. *Pavement In Situ Recycling with Cement: Technical Guide*; Spanish Institute of Cement and Its Applications: Madrid, Spain, 2013.
4. Wirtgen. *Wirtgen Cold Recycling Technology*; Wirtgen: Windhagen, Germany, 2012.
5. Austroads. *Mix Design for Stabilized Pavement Materials: Austroads Publication AP-T16*; Austroads: Sydney, Australia, 2002.

6. South African National Roads Agency Limited (SANRAL). *South African Pavement Engineering Manual*; South African National Roads Agency Limited (SANRAL): Pretoria, South Africa, 2014.
7. Franco, F.A.C.P. *Mechanistic-Empirical Asphalt Pavement Design Method*; Federal University of Rio de Janeiro: Rio de Janeiro, Brazil, 2007.
8. *JTG/T 5521—2019*; Technical Specifications for Highway Asphalt Pavement Recycling. Ministry of Transportation: Beijing, China, 2019.
9. Xuan, D.X.; Houben, L.J.M.; Molenaar, A.A.A.; Shui, Z.H. Mechanical properties of cement-treated aggregate material–a review. *Mater. Des.* **2012**, *33*, 496–502. [CrossRef]
10. Jiang, Y.; Fan, L. An investigation of mechanical behavior of cement-stabilized crushed rock material using different compaction methods. *Constr. Build. Mater.* **2013**, *48*, 508–515. [CrossRef]
11. Kolias, S.; Williams, R.I.T. Estimation of the modulus of elasticity of cement stabilized materials. *Geotech. Test. J.* **1984**, *7*, 26–35. [CrossRef]
12. Sherwood, P.T. *Soil Stabilization with Cement and Lime*; HMSO Press: London, UK, 1995.
13. Sherwood, P.T. *The Properties of Cement Stabilized Materials*; Road Research Laboratory: Crowthorne, UK, 1968.
14. Bell, F.G. *Engineering Treatment of Soils*; E & FN Spon: London, UK, 1993.
15. Transport, D.O.; Technical Recommendations for Highways. *Cementitious Stabilizers in Road Construction South Africa*; Technical Recommendations for Highways: Pretoria, South Africa, 1985.
16. *BS 6031*; Code of Practice for Site Investigation. Building Standards Institution: London, UK, 1981.
17. Kennedy, J. *Cement-Bound Materials for Sub-Bases and Roadbases, Material Selection and Mix Design, Construction and Control Testing*; Cement and Concrete Association: Slough, UK, 1983.
18. Li, M. Experimental Study on Influencing Factors of Strength of Cement Stabilized Macadam. *J. Highw. Transp. Res. Dev.* **2010**, *27*, 6–11.
19. Sun, Z.; Wang, T.; Xu, Z. Trial Study on Influence Factors of Cement-Stabilized Macadam Strength. *J. Build. Mater.* **2006**, *3*, 285–290.
20. Moore, R.K.; Kennedy, T.W.; Hudson, W.R. Factors affecting the tensile strength of cement-treated materials. *Highw. Res. Rec.* **1970**, *315*, 64–80.
21. Lu, S.; Zheng, J.; Zhong, W. Characteristics of Strength, Modulus and Fatigue Damage for Cement Stabilized Macadam in Curing Period. *China J. Highw. Transp.* **2015**, *28*, 9–15.
22. Burg, R.G. *The Influence of Casting and Curing Temperature on the Properties of Fresh and Hardened Concrete*; Portland Cement Association: Skokie, IL, USA, 1996.
23. *ACI Committee 517*; Accelerated Curing of Concrete at Atmospheric Pressure: State of the Art. American Concrete Institute: Farmington Hills, MI, USA, 1992.
24. Escalante-Garcıa, J.I.; Sharp, J.H. Effect of temperature on the hydration of the main clinker phases in Portland cements: Part II, blended cements. *Cem. Concr. Res.* **1998**, *28*, 1259–1274. [CrossRef]
25. Wang, P.; Li, N.; Xu, L. Hydration evolution and compressive strength of calcium sulphoaluminate cement constantly cured over the temperature range of 0 to 80 C. *Cem. Concr. Res.* **2017**, *100*, 203–213. [CrossRef]
26. Price, W.H. Factors influencing concrete strength. *J. Proc.* **1951**, *47*, 417–432. [CrossRef]
27. Husem, M.; Gozutok, S. The effects of low temperature curing on the compressive strength of ordinary and high performance concrete. *Constr. Build. Mater.* **2005**, *19*, 49–53. [CrossRef]
28. Kim, J.K.; Moon, Y.H.; Eo, S.H. Compressive strength development of concrete with different curing time and temperature. *Cem. Concr. Res.* **1998**, *28*, 1761–1773. [CrossRef]
29. Xu, Y.; Yang, G.; Zhao, H. Compressive strength gain behavior and prediction of cement-stabilized macadam at low temperature curing. *J. Adv. Transp.* **2020**, *2020*, 2469436. [CrossRef]
30. Marzouk, H.; Hussein, A. Effect of curing age on high-strength concrete at low temperatures. *J. Mater. Civ. Eng.* **1995**, *7*, 161–167. [CrossRef]
31. Saul, A.G.A. Principles underlying the steam curing of concrete at atmospheric pressure. *Mag. Concr. Res.* **1951**, *2*, 127–140. [CrossRef]
32. *Soil Stabilization in Pavement Structures: A User's Manual*; Department of Transportation, Federal Highway Administration, Office of Development, Implementation Division: Washington, DC, USA, 1979.
33. Molenaar, A.A.A. *Lecture Notes CT4860 Structural Pavement Design, Design of Flexible Pavements*; Delft University of Technology: Delft, The Netherlands, 2007; p. 165.
34. Williams, R.I.T. *The Effect of Cement Content on the Strength and Elastic Properties of Dry Lean Concrete*; Cement and Concrete Association: London, UK, 1962.
35. Kolias, S.; Williams, R.I.T. Relationships between the static and the dynamic moduli of elasticity in cement stabilised materials. *Mater. Struct.* **1980**, *13*, 99–107. [CrossRef]
36. Ghanizadeh, A.R.; Rahrovan, M.; Bafghi, K.B. The effect of cement and reclaimed asphalt pavement on the mechanical properties of stabilized base via full-depth reclamation. *Constr. Build. Mater.* **2018**, *161*, 165–174. [CrossRef]
37. Gonzalo-Orden, H.; Linares-Unamunzaga, A.; Pérez-Acebo, H.; Díaz-Minguela, J. Advances in the study of the behavior of full-depth reclamation (FDR) with cement. *Appl. Sci.* **2019**, *9*, 3055. [CrossRef]

38. Ghanizadeh, A.R.; Rahrovan, M. Modeling of unconfined compressive strength of soil-RAP blend stabilized with Portland cement using multivariate adaptive regression spline. *Front. Struct. Civ. Eng.* **2019**, *13*, 787–799. [CrossRef]
39. *JTG E51-2009*; Test Methods of Materials Stabilized with Inorganic Binders for Highway Engineering. Ministry of Transportation: Beijing, China, 2009.
40. *JTG D50-2006*; Specifications for Design of Highway Asphalt Pavement. Ministry of Transportation: Beijing, China, 2006.
41. *JTG D50-2017*; Specifications for Design of Highway Asphalt Pavement. Ministry of Transportation: Beijing, China, 2017.

coatings

Article

Study of Decay Mechanisms and Rules of Recycled Asphalt Pavement via a Full-Scale Experiment

Quanping Xia [1,2], Jiangping Gao [1], Qigong Zhang [3], Bin Xu [4], Qiang Sun [5,*], Ke Sun [6] and Zhaodi Yuan [5]

1 School of Highway, Chang'an University, Xi'an 710061, China; xiaroad@163.com (Q.X.); huangjr0124@163.com (J.G.)
2 Shandong Transportation Service Center, Jinan 250001, China
3 Dongying City Highway Development Center, Dongying 257091, China; 13589986189@163.com
4 Shandong High-Speed Group Co., Ltd., Jinan 250098, China; cyc1070@163.com
5 Shandong Transportation Institute, Jinan 250102, China; yuanzhaodi365@163.com
6 Jinan International Airport Co., Ltd., Jinan 250107, China; echo_sk@hotmail.com
* Correspondence: sunqiang@sdjtky.cn

Abstract: Under the influence of long-term vehicle loads and large attenuation degrees, asphalt pavement performance gradually decreases, which leads to failure in fulfilling the appropriate requirements and, in turn, may affect driving safety. The purpose of this paper was to study the attenuation mechanism and rule of styrene–butadiene–styrene (SBS)-modified recycled asphalt pavement, so as to determine the applicable position and rational utilization of recycled asphalt mixture. To achieve this goal, two structures were designed, and full-scale experiments were carried out. The performance of the field test road based on accelerated loading testing (ALT) was analyzed through field monitoring data. The fatigue characteristics of stone matrix asphalt-13 (SMA-13) and asphalt concrete-20 (AC-20) mixtures before and after accelerated loading were studied via the trabecular bending fatigue test and dynamic modulus test. The microscopic components in the asphalt mixtures were determined via thin-layer chromatography on chromarods with flame ionization detection (TLC-FID). The results showed that the fatigue properties of recycled asphalt mixture can meet the requirements of ordinary asphalt mixtures and meet the technical standards of asphalt pavement design. With the increase in loading times, the British pendulum number (BPN) value of the two structures tended to be stable, and the BPN of Plan 2 was six less than that of Plan 1. Under the same test conditions, the fatigue life sequence of the recycled asphalt mixture under different loading frequencies was 20 Hz > 10 Hz > 5 Hz. The contents of four components in the reclaimed asphalt mixture were similar to those in the ordinary asphalt mixture. The light component of the reclaimed asphalt mixture of SMA-13 was reduced by 11.69%, and the light component of the ordinary asphalt mixture of SMA-13 was reduced by 15.29% through the full-scale test. In summary, recycled asphalt mixture should not be applied to the upper layer of pavement but can be rationalized in the middle layer and the bottom layer of pavement.

Keywords: recycled asphalt pavement; decay mechanism; fatigue equations; full-scale experiment; rod-shaped thin-layer chromatography test

1. Introduction

In recent years, the number of roads has soared, playing an important role in global economic development [1–5]. Most of the roads built in the early stages of development have gradually entered the maintenance period. Asphalt pavement recycling technology conforms to the concept of sustainable development and green highway construction and is gradually being popularized and applied [6–9].

At present, the research on recycled asphalt mixture mainly focuses on different production processes and recycled asphalt pavement (RAP) content, and the results have shown that recycled asphalt mixture has better pavement performance [10–17]. However, since the damage of asphalt pavement is mainly fatigue damage [18,19], many road engineers are skeptical about the long-life performance of asphalt pavement regeneration technology; therefore, it is particularly important to assess the long-life performance of recycled asphalt pavement [20].

Under long-term vehicle loads and large attenuation degrees, pavement performance of asphalt pavement will be reduced, affecting driving safety. Scholars have often used full-scale experiments to simulate the long-term driving effect on asphalt pavement and study the long-life performance of asphalt mixtures [21–23]. Load, temperature, water and asphalt mixture types all lead to fatigue failure of pavement to varying degrees [24–27]. Wu et al. found that the amount of rut deformation in the pavement increased year by year during a five-year observation period, and the driving comfort was seriously affected [28]. Xu et al. analyzed the relationship between the skid resistance index and the number of accelerated loading cycles and established a logarithmic model to characterize the long-term skid resistance of asphalt pavement [29]. Park conducted a dynamic load simulation on the pavement surface profile, vehicle speed, and vehicle suspension, and their study showed that the fatigue life of the 12.5 mm + PG64-22 asphalt mixture decreased more than that of the 12.5 mm + PG58-22 asphalt mixture [30]. Although there have been a large number of experiments conducted to analyze the recycled asphalt mixture, most of the relevant research has focused on optimizing a specific aspect of its performance, such as its rut deformation or skid resistance. However, previous studies on the fatigue properties of different types of recycled asphalt mixtures under accelerated loads have not been comprehensive enough. The causes of aging of recycled asphalt mixtures have not been analyzed from the microscopic point of view, and no specific suggestions on their application location have been provided.

In this study, the variation law of pavement service performance was established through full-scale experiments and field monitoring. Trabecular bending fatigue tests of pavement structure specimens after accelerated loading were carried out and compared with those without loading, and fatigue equations for the recycled asphalt mixture under different load frequencies were obtained. Rod-shaped thin-layer chromatography (TLC-FID) was used to detect the microscopic components of the recycled asphalt mixtures of SMA-13 and AC-20 after accelerated loading, and the differences in the saturates, aromatics, resins, and asphaltenes under different conditions were analyzed.

2. Experimental Background

2.1. Sample Preparation

2.1.1. Test Road Surface Material

The test road surface materials are shown in Table 1.

Table 1. Test road material classification.

Serial Number		Material Classification
I	I-1	SMA-13 ordinary asphalt mixture
	I-2	SMA-13 ordinary asphalt mixture (after accelerated loading)
II	II-1	SMA-13 recycled asphalt mixture
	II-2	SMA-13 recycled asphalt mixture (after accelerated loading)
III	III-1	AC-20 recycled asphalt mixture
	III-2	AC-20 recycled asphalt mixture (after accelerated loading)

1. SMA-13 ordinary asphalt mixture.

According to the screening results of each hot silo, after calculating and adjusting the grading of the mineral mixture close to the target mix ratio, the weight ratio of each hot silo was determined to be the following: the 3# hot silo (10–15 mm) accounted for 43.0%, the 2# hot silo (5–10 mm) accounted for 34.0%, the 1# hot silo (0–3 mm) accounted for 13%, and the mineral powder accounted for 10%. The synthetic grading curve is displayed in Figure 1. The optimal design oil–stone ratio was determined to be 6.30%.

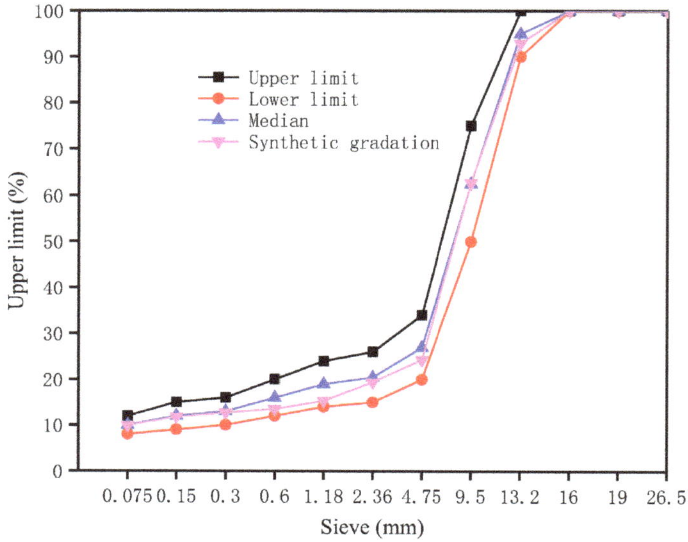

Figure 1. The synthetic grading curve of the SMA-13 ordinary asphalt mixture.

2. SMA-13 recycled asphalt mixture.

According to the screening results of each hot silo, after calculating and adjusting the grading of the mineral mixture close to the target mix ratio, the weight ratio of each hot silo was determined to be as follows: the 1#RAP silo (coarse material) accounted for 9%, the 2#RAP silo (fine material) accounted for 11%, the 3# hot silo (10–15 mm) accounted for 31.0%, the 2# hot silo (5–10 mm) accounted for 32.0%, the 1# hot silo (0–3 mm) accounted for 8.5%, and the mineral powder accounted for 8.5%. The synthetic grading curve is shown in Figure 2. The optimal design oil–stone ratio was determined to be 6.10%.

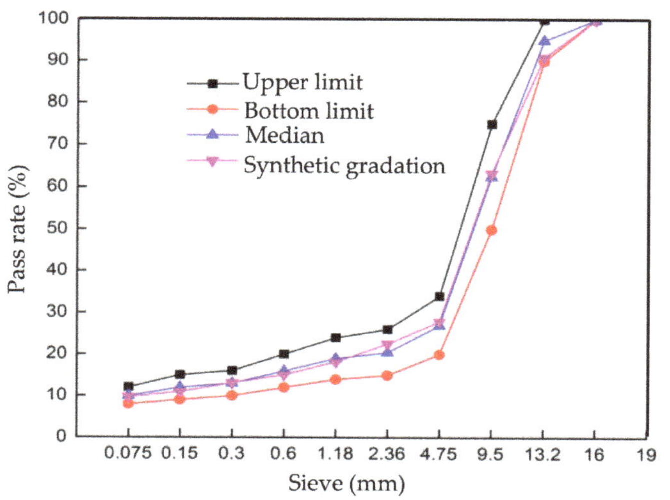

Figure 2. The synthetic grading curve of the SMA-13 recycled asphalt mixture.

3. AC-20 recycled asphalt mixture.

According to the screening results of each hot silo, after calculating and adjusting the grading of the mineral mixture close to the target mix ratio, the weight ratio of each hot silo was determined to be as follows: the 4# hot silo (19–32 mm) accounted for 18%, the 3# hot silo (11–19 mm) accounted for 21%, the 2# hot silo (6–11 mm) accounted for 8%, the 1# hot silo (0–3 mm) accounted for 10%, the 1#RAP silo accounted for 40%, and the mineral powder accounted for 3%. The regeneration agent was 5% of the old asphalt. The synthetic grade curve is displayed in Figure 3. The optimal oil–stone ratios were determined to be 4.31% and 4.50%, respectively.

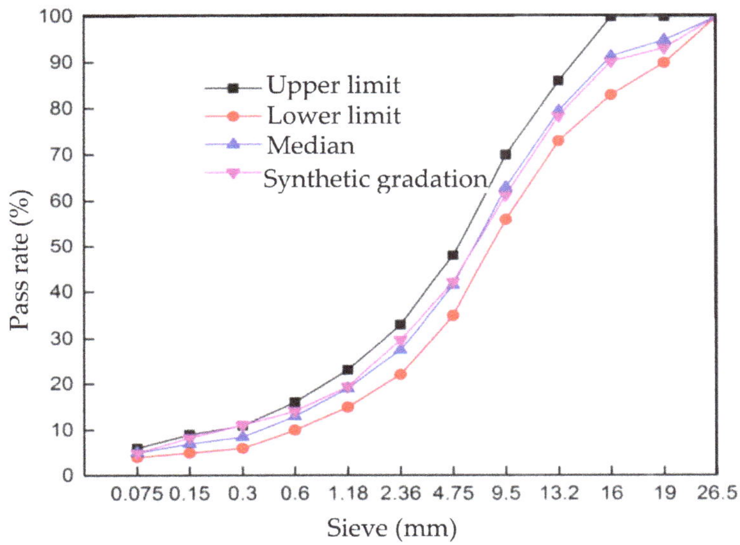

Figure 3. The synthetic grading curve of the AC-20 recycled asphalt mixture.

2.1.2. Performance Evaluation

1. Water stability.

According to the design grading and the optimal oil–stone ratios of the SMA-13 and AC-20 recycled asphalt mixtures given in Section 2.1, four groups of Marshall test pieces were made, and a Marshall test, a 48 h immersion Marshall test, and a freeze–thaw splitting test were conducted (50 compaction times of specimens) [31,32]. The water stability of the asphalt mixture was verified. The residual stability of SMA-13 was 91.8%, and its residual strength ratio was 85.5%. The residual stability and the residual strength ratio of AC-20 were 83.7% and 98.9%, respectively, both of which met the requirements of the specification.

2. Rutting resistance.

Rutting tests were carried out on the asphalt mixtures with the design grades of SMA-13 and AC-20 and the optimum oil-to-stone ratio at the temperature of 60 °C and wheel pressure of 0.7 MPa. The specimen size was 30 cm × 30 cm × 5 cm. The test results are shown in Table 2. SMA-13 met the design requirement of not less than 3000 times/mm, while AC-20 met the design requirement of not less than 1000 times/mm.

Table 2. Rutting test results.

Type of Asphalt Mixture	Oil–Stone Ratio	Dynamic Stability (times/mm)
SMA-13 recycled asphalt mixture	6.10	>6000
AC-20 recycled asphalt mixture	4.50	2896

3. Low-temperature crack resistance.

The design grading of the SMA-13 recycled asphalt mixture and the AC-20 recycled asphalt mixture and the optimal amount of asphalt were determined, and the low-temperature bending test was carried out on them. The specimens were prismatic trabeculars with a length of 250 mm ± 2.0 mm, a width of 30 mm ± 2.0 mm, and a height of 35 mm ± 2.0 mm cut using the rutting plate after wheel grinding, with a span of 200 mm ± 0.5 mm.

The low-temperature bending test of the asphalt mixture was carried out under the conditions of a −10 °C temperature and a 50 mm/min loading rate to measure the failure strength, failure strain, and failure stiffness modulus and to comprehensively evaluate the low-temperature crack resistance of the asphalt mixture [33,34]. The low-temperature bending test data for the SMA-13 recycled asphalt mixture are shown in Table 3, and the low-temperature bending test data for the AC-20 recycled asphalt mixture are shown in Table 4.

Table 3. The low-temperature bending test data for the SMA-13 recycled asphalt mixture.

Number \ Index	The Flexural Tensile Strength of the Specimen under Failure, R_B (MPa)	Maximum Flexural Tensile Strain at Failure of Specimen, $\varepsilon_B \times (10^{-6})$	The Bending Stiffness Modulus of the Specimen under Failure, S_B (MPa)
1	8.38	3371	2485
2	8.86	3898	2274
3	12.88	4387	2936
4	8.67	4222	2054
5	10.02	4064	2464
6	10.98	5232	2099
7	10.68	5218	2048
8	9.84	4806	2048
Average value	10.04	4399.7	2301
Technical requirement	-	≥2500	-

Table 4. The low-temperature bending test data for the AC-20 recycled asphalt mixture.

Number \ Index	The Flexural Tensile Strength of the Specimen under Failure, R_B (MPa)	Maximum Flexural Tensile Strain at Failure of Specimen, $\varepsilon_B \times (10^{-6})$	The Bending Stiffness Modulus of the Specimen under Failure, S_B (MPa)
1	11.28	3207	3517
2	9.70	3683	2635
3	9.32	3923	2376
4	10.11	3502	2886
5	11.16	4748	2350
6	10.77	3417	3152
7	9.24	3293	2804
8	11.30	2923	3869
Average value	10.36	3587	2949
Technical requirement	-	≥2500	-

2.2. Test Methods and Scheme

2.2.1. Full-Scale Experiment

The test road was divided into two different pavement structures: the base cement stabilized gravel, with a thickness of 360 mm, and the bottom layer is the AC-20 recycled asphalt mixture, with a thickness of 60 mm. The top layer of the SMA-13 recycled asphalt mixture and the SMA-13 ordinary asphalt mixture was 40 mm thick. The site layout of the full-scale experiment is shown in Figure 4. In order to carry out an effective comparison test, 30 m (K205 + 820–K205 + 850) of the 150 m test road was of the SMA-13 ordinary asphalt mixture, and the remaining 120 m (K205 + 700–K205 + 820) was of the SMA-13 recycled asphalt mixture, as shown in Table 5 and Figure 5.

Figure 4. Site layout of the full-scale experiment.

Table 5. Test section pavement structure combination.

Plan	The Upper Layer (Thickness of 40 mm)	The Bottom Layer (Thickness of 60 mm)
Plan 1	SMA-13 ordinary asphalt mixture	AC-20 recycled asphalt mixture; RAP content is 40%
Plan 2	SMA-13 recycled asphalt mixture; RAP content is 20%	AC-20 recycled asphalt mixture; RAP content is 40%

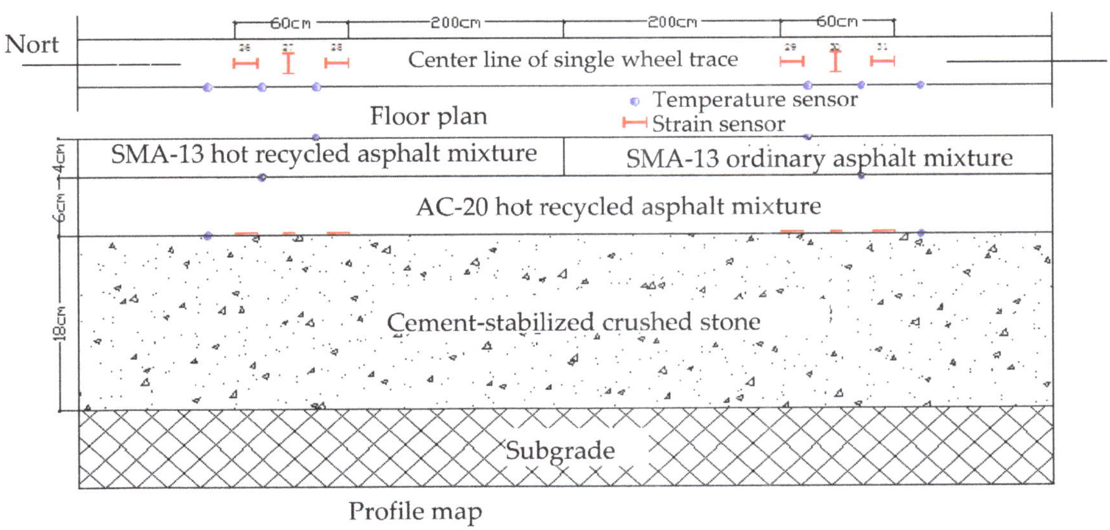

Figure 5. Test pavement structure drawing.

ALT large acceleration loading equipment was used. The maximum external dimensions, length × width × height, were 26,342 mm × 4220 mm × 7934 mm. The maximum gradient was 6%, the maximum transverse was 3%, the loading length was 10 m, the running speed was 24.6 km/h, and the shaft load was 150 kN.

2.2.2. Field Monitoring Method

A resistance strain gauge in the shape of an "I" was buried in the test section. The field layout diagram of the sensor in the field test road and the location and wiring of the strain gauge are shown in Figure 6. The strain gauge was installed using the method of core hole superposition excavation. The depth was about 60 mm. At the edge of the core hole, the slit length was about 200 mm, made using a slit cutter.

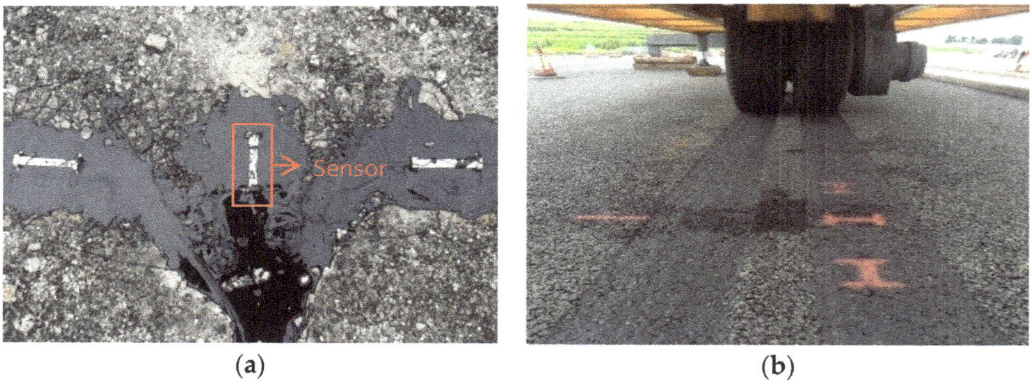

Figure 6. Sensor mounting position. (**a**) Field layout of sensors on the field test road. (**b**) Location and wiring of the strain gauge.

The more commonly used embedded thermocouple temperature sensor was adopted, which has a large measuring range and high accuracy up to 0.002 °C. The temperature sensors P1 and P2 were embedded, and P3 was excavated. The difference between the measured data from the three temperature sensors was within 0.1 °C, and the average value of the three data points was taken as the final measurement. In order to facilitate data reading, the strip and display were transformed into the plug-and-read display, as shown in the figure, and the temperature was recorded once every hour. In the field test section, the DI-510-32 data acquisition system (see Figure 7) could be used to extract sample dynamic data from the system.

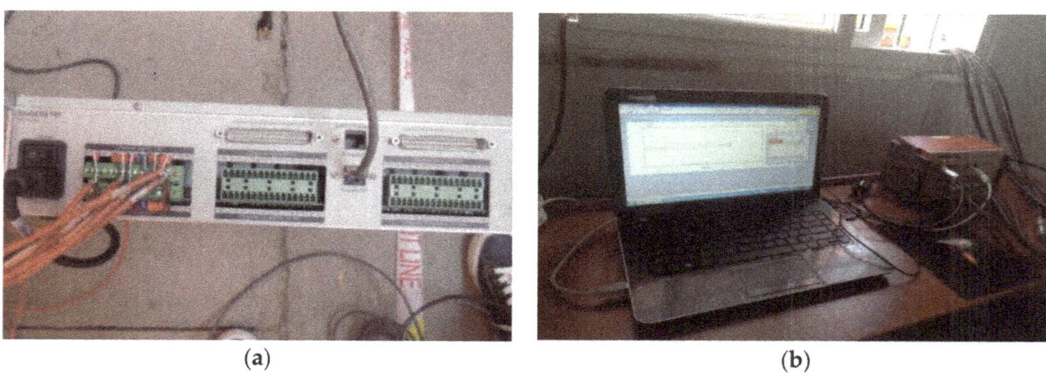

Figure 7. Data acquisition system. (**a**) Strain data acquisition. (**b**) Sensor indoor detection.

1. Pavement temperature monitoring.

In order to continuously and automatically detect the temperature field of the accelerated loading site, an automatic temperature field acquisition system was added, as shown in Figure 8. The locations of the temperature field data collection were set at 1/2 thickness of the water stabilization base, and 1/2 thickness of the upper layer and bottom layer, the road surface, and in the air. According to field testing experience, the collection interval was set to 30 min.

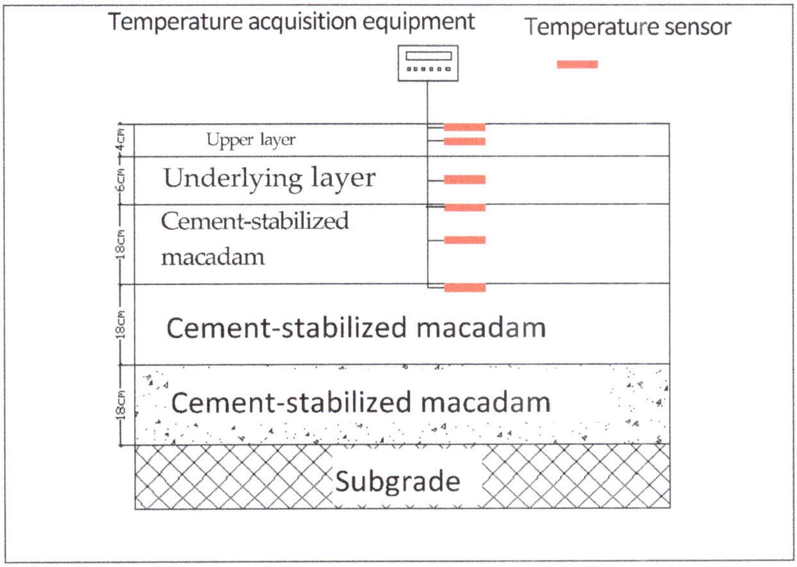

Figure 8. Layout diagram of an automatic pavement temperature field acquisition system.

2. Skid resistance monitoring.

In the full-scale experiment, the pendulum tribometer was used to monitor the BPN, and the data from the test road were collected every 5000 loading times. During the test, the friction coefficient was detected at a fixed point, and 4 detection points (a total of 8 measurement points) were set for the 2 wheel tracks. Figure 9 shows the layout of the fixed monitoring points.

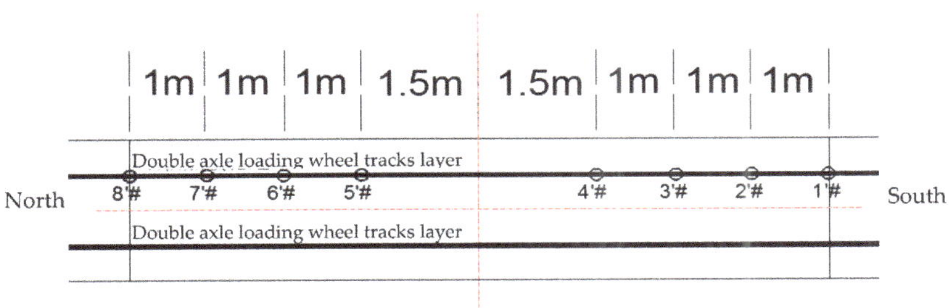

Figure 9. Schematic diagram of the fixed monitoring points.

3. Rut monitoring.

This full-scale experiment road adopted a 3 m straightedge to collect rutting data every 5000 loading times. According to previous experience, rutting appears more obvious under a high-temperature environment and heavy load, and it serves as the main control index of pavement fatigue failure.

4. Monitoring of tensile and compressive strain in the asphalt layer bottom.

The tensile and compressive strain data from the bottom layer were collected every 5000 loading times in this loading acceleration test. According to previous experimental observations, the whole-day temperature difference in coastal cities is large. As such, the bottom strain was greatly affected by the bottom temperature of the pavement. Therefore, the strain data were collected once an hour, and the dynamic change in strain value under loading times at the same temperature was compared.

2.2.3. Trabecular Bending Fatigue Test

After the completion of the field ALT full-scale experiment, the field test road was sampled. The representativeness of the field core sample was strong, and it was of great significance to test and analyze the asphalt mixture fatigue performance of the field core sample. The specimen size was 500 mm × 400 mm × 100 mm. The specimens are shown in Figure 10.

Figure 10. The specimen diagram. (**a**) Field sample diagram. (**b**) Field rectangular sample. (**c**) Trabecular cutting specimen of SMA-13. (**d**) Trabecular cutting specimen of AC-20.

Under room temperature, an asphalt mixture exhibits significant viscoelastic properties [9]. With the increase in the number of loads, the specimen's flexural residual deformation gradually increases, the stiffness (or modulus) of the material gradually decreases, micro-cracks continuously develop, and, finally, there is a complete fracture. As this expression is simple and clear, and the test data are stable, the complete fracture of the specimen is usually taken as the fatigue failure standard under the control of stress.

A trabecular bending fatigue test, using the MTS-810 material testing machine, was carried out according to the specification requirements and test scheme design. The trabecular bending fatigue test is displayed in Figure 11. The specific test steps are as follows:

(1) The standard specimen is stored in a YDDL 40-440 digital display high- and low-temperature refrigerator (15 °C) for two hours.
(2) The MTS environment chamber regulates the constant temperature (15 °C).
(3) The specimen is placed on the mold.
(4) The initial contact load is set via MTS. The specimen is in contact with the load device.
(5) The load is applied according to the design requirements.
(6) The test data are collected according to the design requirements.

Figure 11. MTS environmental chamber, loading die, and MTS load test.

This test adopts the stress control loading mode and three-point loading [35]. The loading waveform and frequencies in this paper were continuous semi-positive vector loads of 5 Hz, 10 Hz, and 20 Hz, and the stress ratios were 0.2, 0.3, 0.4, 0.5, and 0.6. The test temperature was 15 °C.

2.2.4. Dynamic Modulus Test

For the core samples taken on site, a core sample with a diameter of φ100 mm was drilled using a core machine; then, after cutting and polishing, the height of the specimen met the test requirements, as shown in Figure 12a.

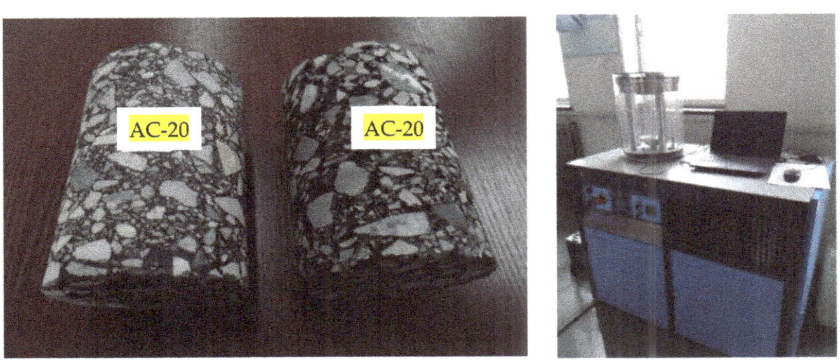

(**a**) (**b**)

Figure 12. Data acquisition system. (**a**) Specimen. (**b**) AMPT dynamic modulus testing machine.

The dynamic modulus test was carried out according to the test methods of "Standard Method of Test for Determining Dynamic Modulus of Hot Mix Asphalt (HMA)" AASHTO TP62 and "Standard Test Methods of Bitumen and Bituminous Mixtures for Highway Engineering" JTG E20-2011 (T0738-2011) [36,37]. The instrument used in the dynamic compressive resilience modulus test system was the asphalt mixture simple performance test (AMPT). The AMPT dynamic modulus testing machine is shown in Figure 12b. The four temperatures of 6 °C, 21.1 °C, 37 °C, and 54 °C were adopted in the test process, and the following loading frequencies of 25 Hz, 20 Hz, 10 Hz, 5 Hz, 2 Hz, 1 Hz, 0.5 Hz, 0.2 Hz, and 0.1 Hz were adopted at each test temperature. This test adopted the constant strain control mode, and the sinusoidal load was applied to the specimen. The control range of strain varied from 75 μm to 125 μm centimeters.

2.2.5. Rod-Shaped Thin-Layer Chromatography Test

This experiment employed TLC-FID to test the microscopic components of the recycled material, which involved the analysis of the saturate, aromatic, resin, and asphaltene components' differences under different conditions [38]. The rod-shaped thin-layer chromatography instrument in shown in Figure 13.

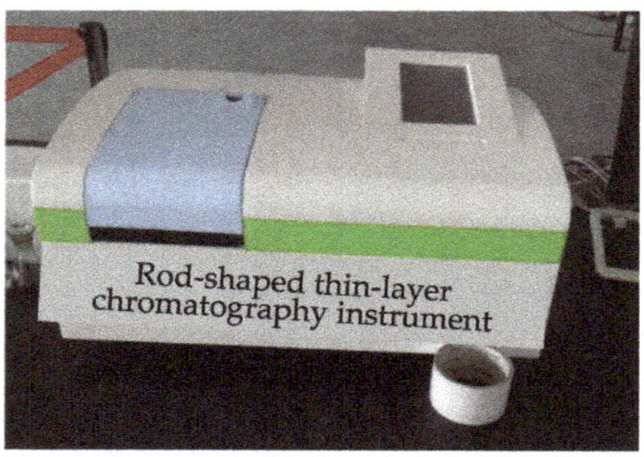

Figure 13. Rod-shaped thin-layer chromatography instrument.

The test process of the TLC-FID analysis of the four components of asphalt is shown in Figure 14, and the test conditions are shown in Table 6.

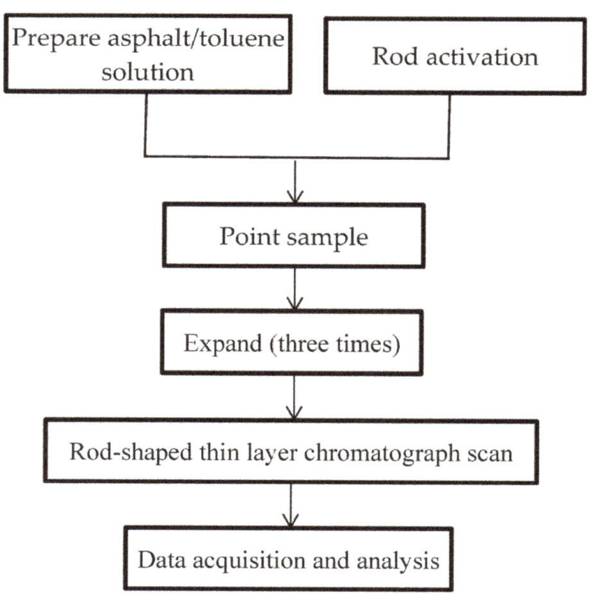

Figure 14. The test process of TLC-FID analysis of four components of asphalt.

Table 6. Test conditions.

Test Condition	Parameter
Air flow	2.0 L/min
Hydrogen flow rate	160 mL/min
Test rate	30 s/scan
Activation rate	50 s/scan
Rod load	100–150 μg
Sampling quantity	1 μL

3. Results and Discussion

3.1. Study on the Service Performance of a Field Test Road Based on ALT

3.1.1. Temperature Variation in Asphalt Pavement

The relation curve between pavement loading times and temperature is shown in Figure 15. Due to the long period of the full-scale experiment, the temperature during the test changed with the ambient temperature. It can be seen that the temperature measured in Figure 15 is divided into two stages. In the first stage, the average temperature of the upper layer of the pavement was 57.9 °C, the average temperature of the bottom layer of the pavement was 36.2 °C, and the total number of loading times was 120,000. In the second stage, the average temperature of the upper layer of the pavement was 36.8 °C, the average temperature of the bottom layer of the pavement was 26.3 °C, and the total number of loading times was 180,000. With the increase in loading times, the temperature of the pavement decreased. The temperature of the upper layer of the pavement was 25% lower than that of the bottom layer of the pavement, indicating that the upper layer of the pavement was more affected by ambient temperature and loading.

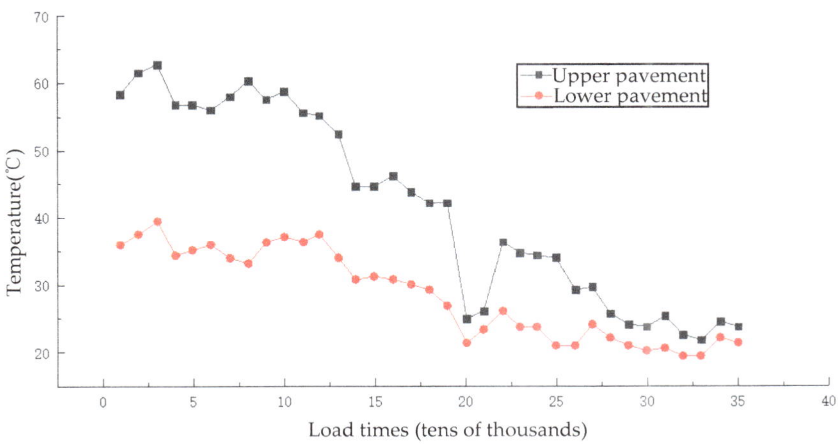

Figure 15. Temperature change curve during loading times.

3.1.2. Skid Resistance Variation in Asphalt Pavement

The results of a systematic analysis of the change results of the friction coefficient after different rolling times are shown in Figure 16. After loading the two pavement structures, their swing values rapidly decreased in the early stage, but with the gradual increase in loading times, the two pavement structures had different changes. When the loading times of Plan 1 reached 100,000, the BPN reached 44 and then slowly decreased. When the loading times of Plan 2 reached 140,000, the BPN dropped to 38, six less than that of Plan 1. Under the same external conditions, the BPN of Plan 1 was larger.

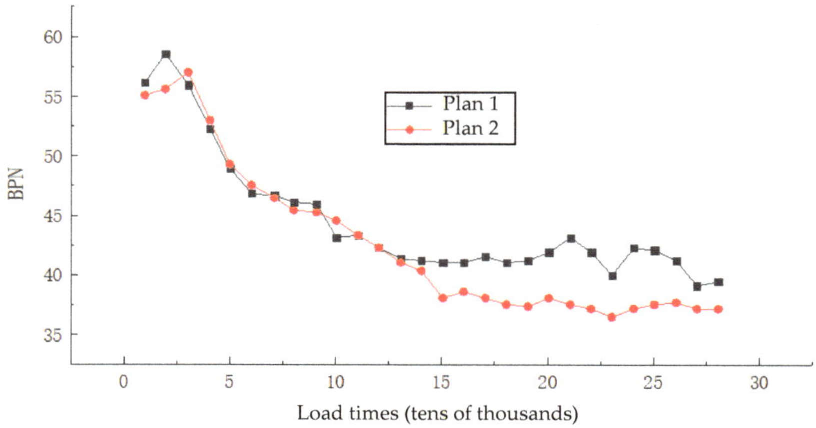

Figure 16. The friction coefficient of the road varies with the loading times.

The decay law of the pavement friction coefficient is in exponential form; hence, this study conducted regression on the decay law of the friction coefficient of Plan 1 and Plan 2 in logarithmic form, and the results are shown in Figure 17.

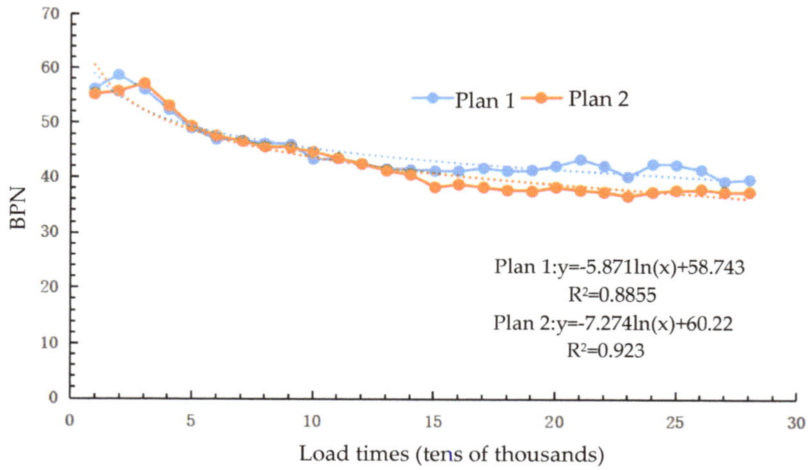

Figure 17. The fitting relationship between the road swing value and the number of loading times.

At first, the BPN values of Plan 1 and Plan 2 were similar. However, starting from about 150,000 times, Plan 1 exhibited a stronger anti-slip effect. The pavement material had a decisive influence on the surface shape of the pavement, and the angularity of the aggregate in the RAP material was reduced, which reduced the overall friction coefficient of the pavement.

3.1.3. Rutting Variation in Asphalt Pavement

The rutting development of pavement under different loading times was sorted and analyzed, as shown in Figure 18. There was a rapid development of rutting in the two types of road structure in the early stage. When the number of loadings reached 80,000, it changed towards slow development. The system entered the high-temperature stage at this point, and the development remained slow. The final rutting data for Plan 1 and

Plan 2 were 6 mm and 5.6 mm, respectively. Since the bottom layer of the two pavement structures on the test road was of the AC-20 recycled asphalt mixture with a RAP content of 25%, this indicates that the formation of rutting was due to the secondary compaction of the overall pavement structure and involved the small permanent plastic deformation of each pavement structural layer under the action of the load. In addition, the content of light oil in the aged asphalt in the hot asphalt pavement was reduced, which led to the increase in viscosity and the enhancement of compressive and shear resistance; therefore, the rut depth was less than that of ordinary recycled pavement.

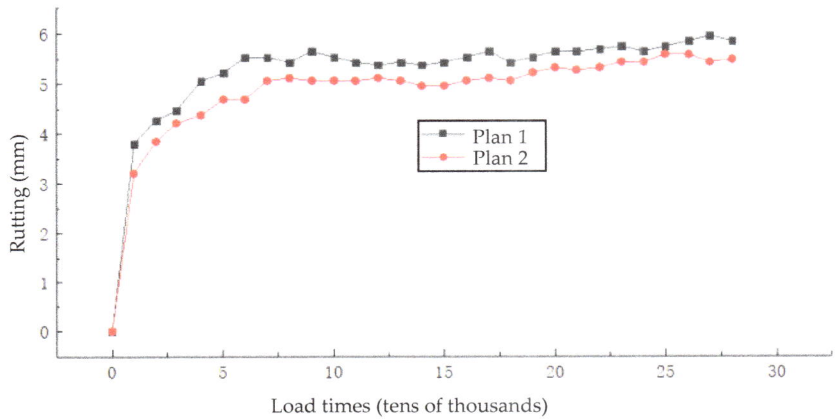

Figure 18. The variation in road rutting with loading times.

3.1.4. Strain Variation in Asphalt Pavement (Bottom Layer)

The strain corresponding to different loading times was monitored in real time, and the correlation curve results between the strain and loading times were obtained through analysis. These results are shown in Figure 19, where 2# and 5# are transverse sensors for Plan 1 and Plan 2, respectively, and 3# and 6# are longitudinal sensors for Plan 1 and Plan 2, respectively.

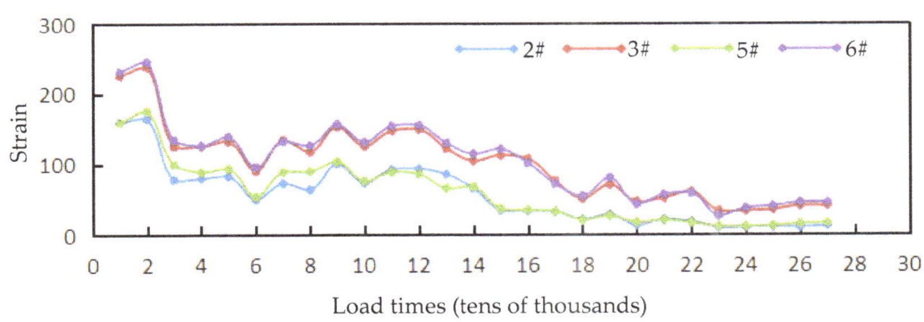

Figure 19. Relationship between the bottom strain and loading times.

Plan 1 and Plan 2 basically displayed the same strain law. When the loading times reached between 120,000 and 300,000, the environment entered the low-temperature stage. Therefore, the bottom strain turned to the sharply decreasing stage after 120,000 times, and the strain in this stage was affected more by temperature than by the loading times. These results indicated that the macroscopic mechanical responses and road service performances of the two pavement structures under load were basically the same, reflecting the good road performance of the recycled asphalt mixture.

3.2. Analysis of the Load–Deflection Relationship

3.2.1. Analysis of the Load–Deflection Relationship

The load and deflection data were analyzed, and the recycled asphalt mixture and ordinary asphalt mixture were compared and analyzed under different load frequencies and stress ratios. The specific relationship diagrams are shown in Figures 20–24.

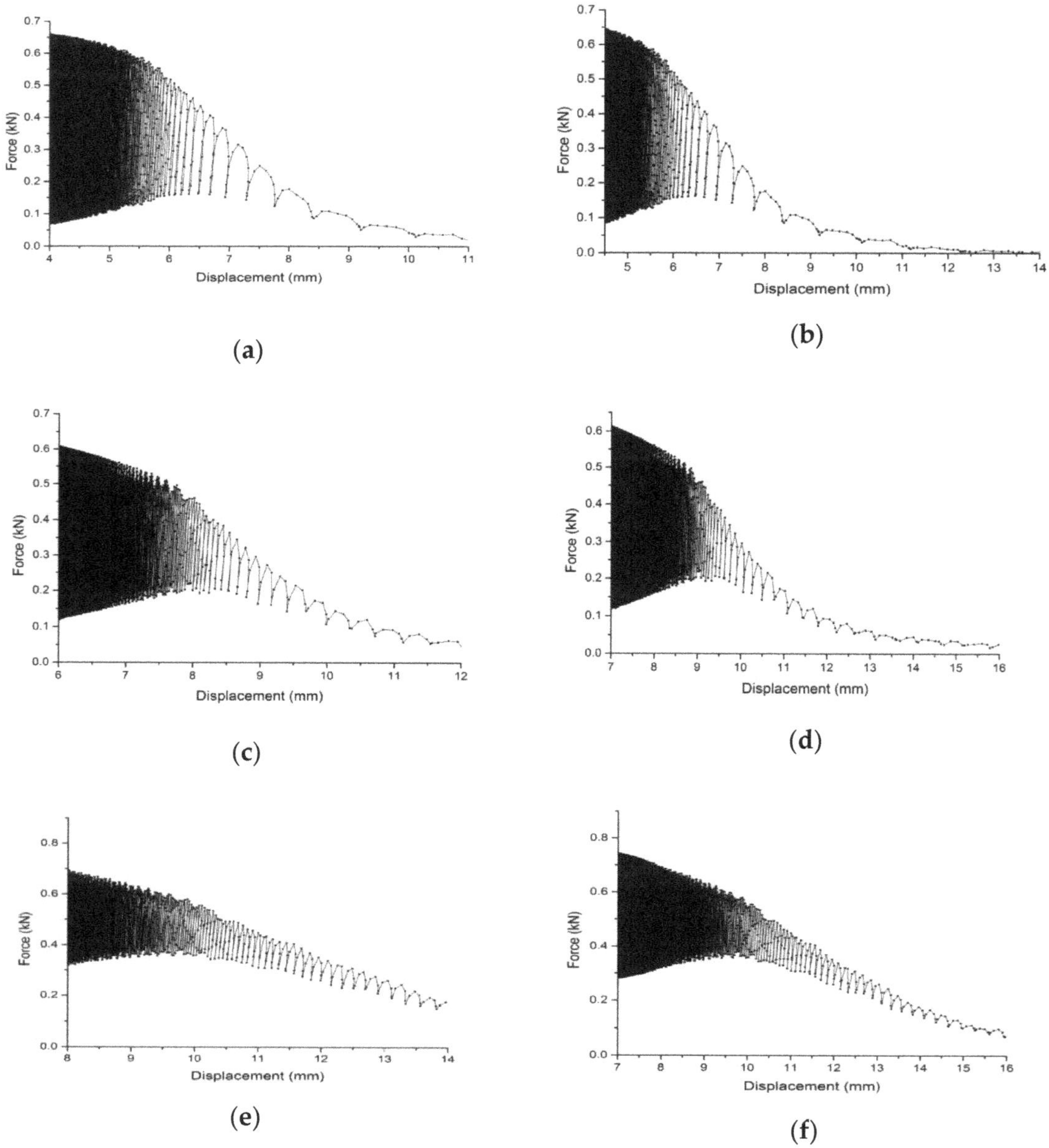

Figure 20. The relationship between load and deflection at different loading frequencies when the stress ratio was 0.2. (**a**) I-1 (5 Hz); (**b**) II-1 (5 Hz); (**c**) I-1 (10 Hz); (**d**) II-1 (10 Hz); (**e**) I-1 (20 Hz); and (**f**) II-1 (20 Hz).

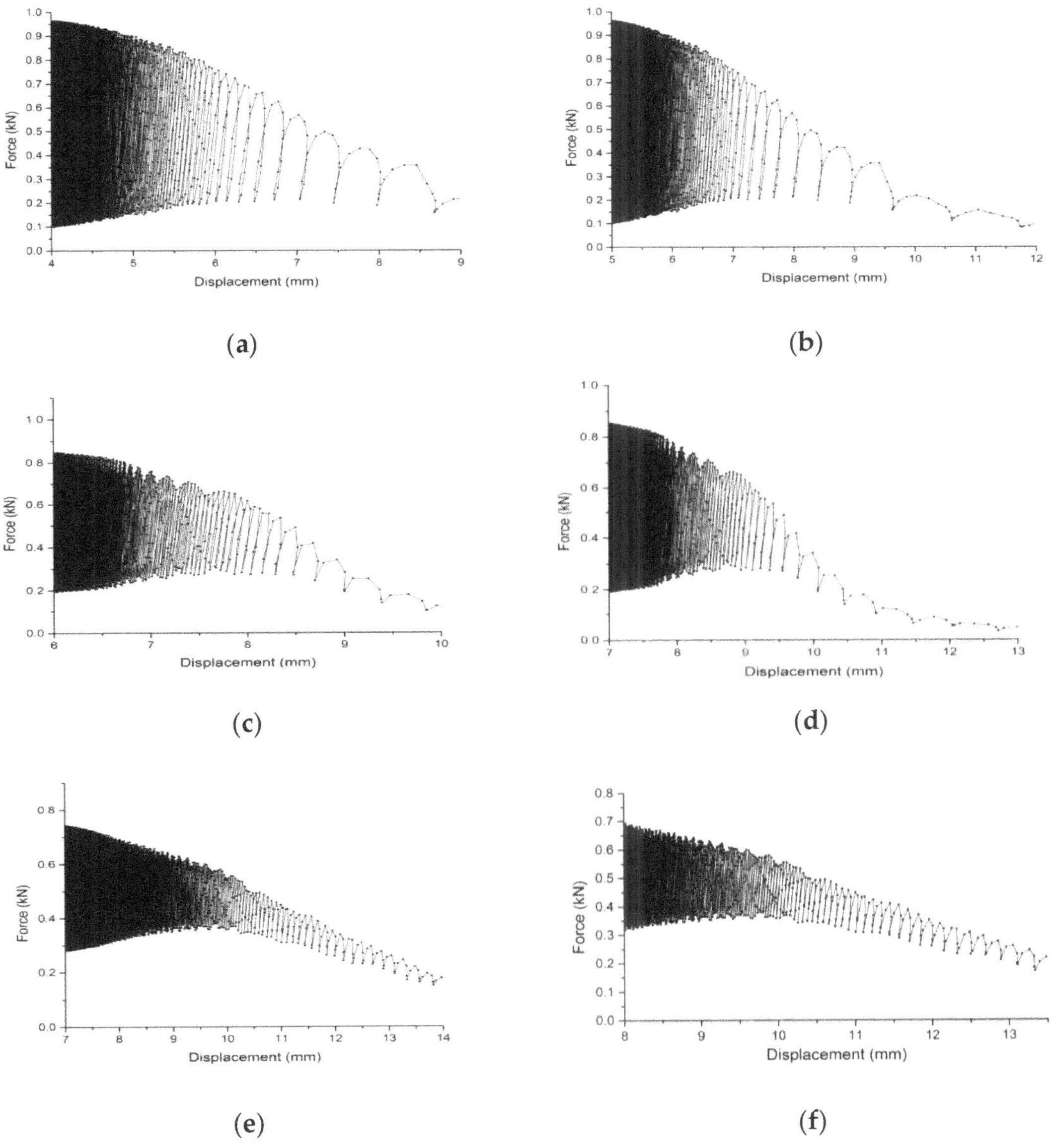

Figure 21. The relationship between load and deflection at different loading frequencies when the stress ratio was 0.3. (**a**) I-1 (5 Hz); (**b**) II-1 (5 Hz); (**c**) I-1 (10 Hz); (**d**) II-1 (10 Hz); (**e**) I-1 (20 Hz); and (**f**) II-1 (20 Hz).

The following can be seen from Figures 20–24:

(1) When the stress ratio was 0.2, 0.3, 0.4, 0.5, and 0.6, the failure forms of the ordinary asphalt mixture and the recycled asphalt mixture were basically the same under different loading frequencies. That is, in the process of repeated loading and unloading, the specimens' flexural deformation gradually accumulated until fractures occurred.

(2) Under the same stress level and loading frequency, the curve relationships between load and deflection for the ordinary asphalt mixture and the recycled asphalt mixture were similar (basically consistent). The results showed that the fatigue performance of the recycled asphalt mixture can meet the requirements of ordinary asphalt mixture and meet the technical standards of asphalt pavement design.

(3) The relationships between load and fatigue deformation for the ordinary asphalt mixture and the recycled asphalt mixture were different under different stress ratios and different loading frequencies. When the stress level was small (stress ratios of 0.2 and 0.3), the cumulative fatigue failure deformation of the ordinary asphalt mixture and the recycled asphalt mixture increased with the increase in loading frequency.

(4) When the stress level was large (stress ratios of 0.4, 0.5, and 0.6), the cumulative fatigue failure deformation of the ordinary asphalt mixture and the recycled asphalt mixture decreased with the increase in loading frequency. The reason for this was that when the stress ratio was small, the flexural deformation of the trabecular specimen caused by the applied load was very small, far less than the bending and tensile failure limit deformation. Under high loading frequencies, the cumulative deformation of deflection increased faster, and the cumulative deformation of fatigue failure was larger. When the stress ratio was large, the load exerted great influence on the bending deformation of the trabecular specimen. The greater the loading frequency, the faster the time of action on the asphalt mixture viscoelastic specimen. The specimen was mostly elastomer, and the stiffness modulus was large. When the fatigue failure was reached, the cumulative deformation caused via fatigue failure was relatively small. However, under a certain stress level, due to the larger stiffness modulus, the flexural deformation will be relatively small, and fatigue cracking will not easily occur. Therefore, the higher the loading frequency of an asphalt mixture specimen, the greater the fatigue life.

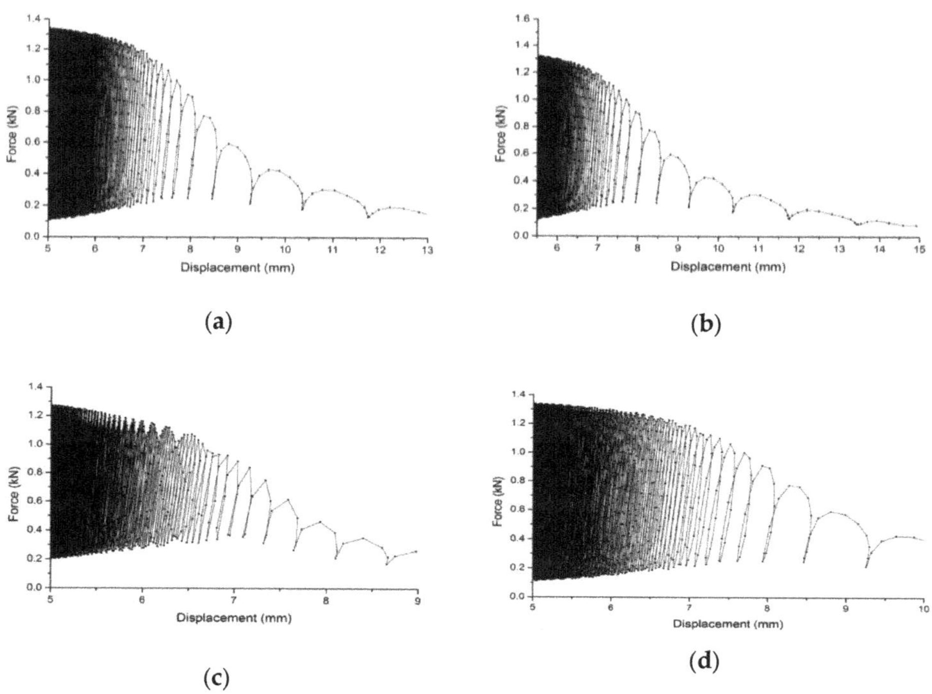

Figure 22. *Cont.*

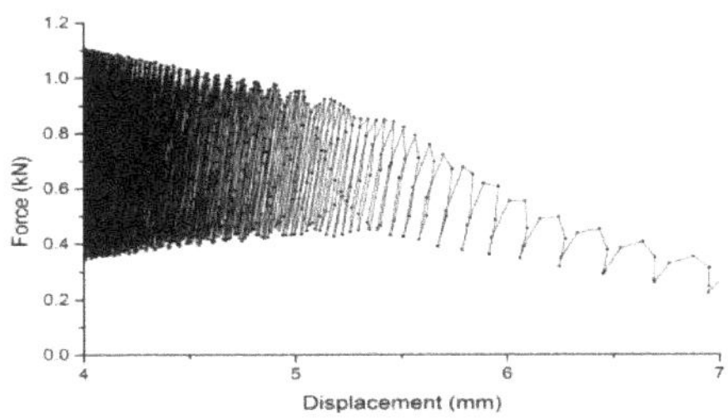

(e)

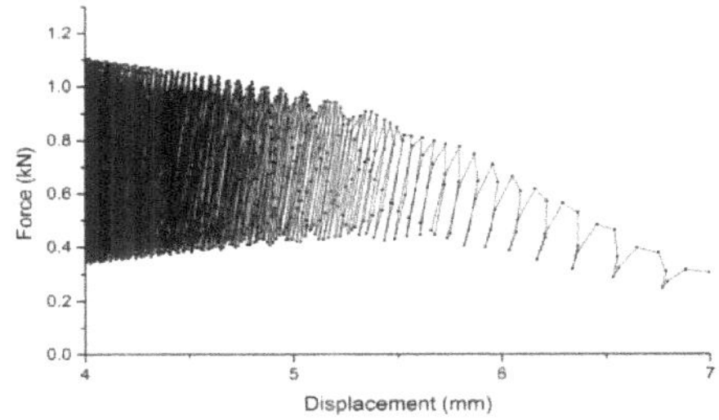

(f)

Figure 22. The relationship between load and deflection at different loading frequencies when the stress ratio was 0.4. (**a**) I-1 (5 Hz); (**b**) II-1 (5 Hz); (**c**) I-1 (10 Hz); (**d**) II-1 (10 Hz); (**e**) I-1 (20 Hz); and (**f**) II-1 (20 Hz).

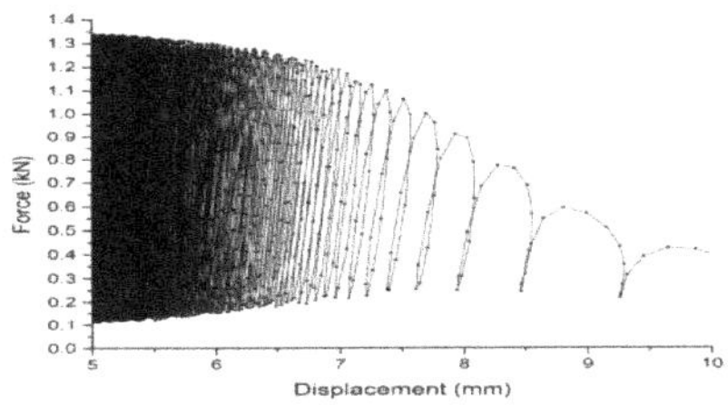

(a)

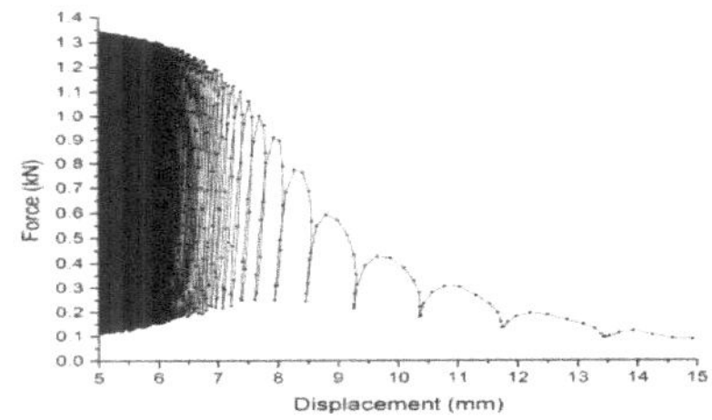

(b)

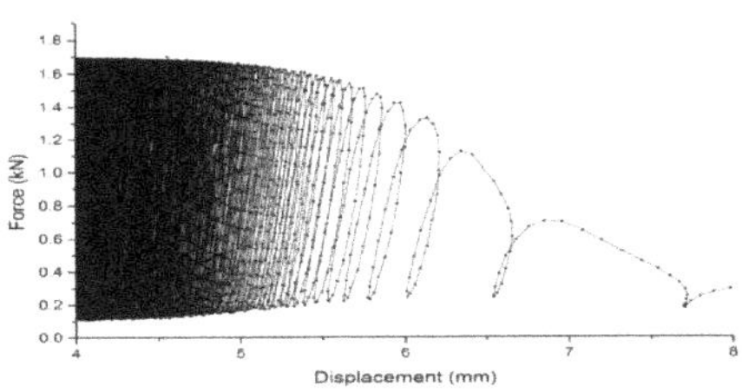

(c)

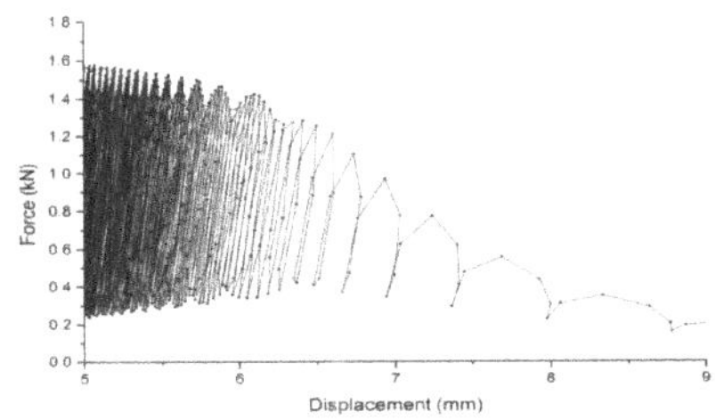

(d)

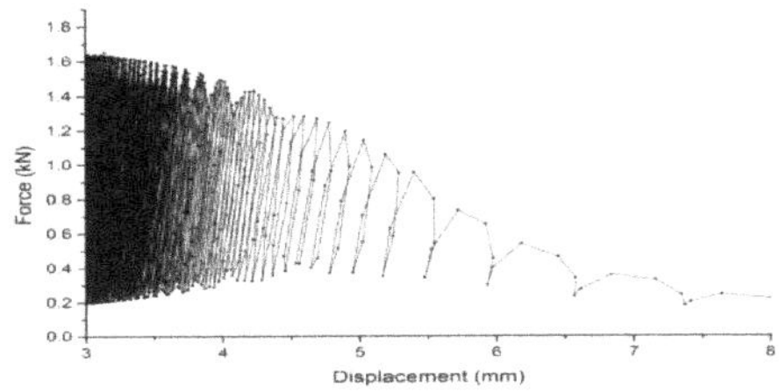

(e)

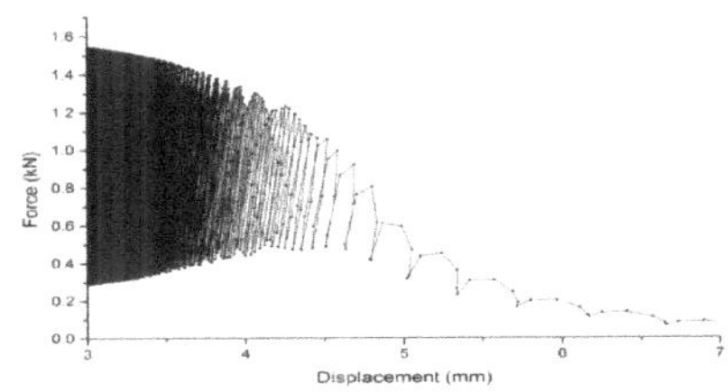

(f)

Figure 23. The relationship between load and deflection at different loading frequencies when the stress ratio was 0.5. (**a**) I-1 (5 Hz); (**b**) II-1 (5 Hz); (**c**) I-1 (10 Hz); (**d**) II-1 (10 Hz); (**e**) I-1 (20 Hz); and (**f**) II-1 (20 Hz).

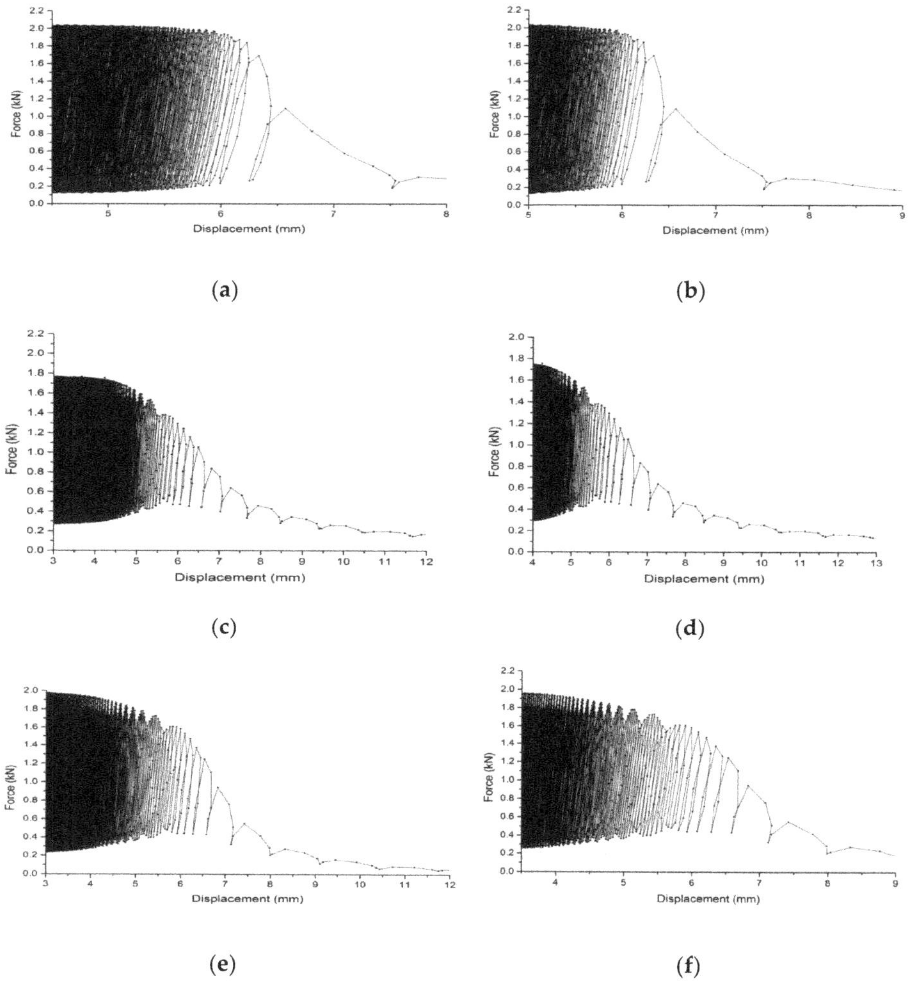

Figure 24. The relationship between load and deflection at different loading frequencies when the stress ratio was 0.6. (**a**) I-1 (5 Hz); (**b**) II-1 (5 Hz); (**c**) I-1 (10 Hz); (**d**) II-1 (10 Hz); (**e**) I-1 (20 Hz); (**f**) II-1 (20 Hz).

3.2.2. Three-Point Bending Fatigue Life Analysis

Figure 25 shows the three-point bending fatigue life data analysis, according to different load frequencies, and the different stress ratio comparative analysis.

Four stress levels of 0.3, 0.4, 0.5, and 0.6 were selected to carry out bending fatigue tests on the SMA-13 ordinary asphalt mixture, SMA-13 recycled asphalt mixture, and AC-20 recycled asphalt mixture at 5 Hz, 10 Hz, and 20 Hz, respectively. The following can be seen from Figure 25:

(1) Under the same stress ratio, the fatigue life of II-1 and III-1 increased with the increase in loading frequency. When the stress ratio was 0.3, the fatigue life difference between the asphalt mixtures with loading frequencies of 5 Hz and 10 Hz was not obvious, but the fatigue life of the asphalt mixture with a loading frequency of 20 Hz was 3.8 times than that of the 5 Hz mixture and 3.1 times than that of the 10 Hz mixture.

This indicates that the influence of higher loading frequency on the fatigue life of an asphalt mixture is more obvious when other test conditions are the same. Under the same test conditions, the order of fatigue life of the recycled asphalt mixtures under different loading frequencies was 20 Hz > 10 Hz > 5 Hz.

(2) Under most of the same test conditions, the fatigue life of I-1 and II-1 was similar. This indicates that the recycled asphalt mixture met the requirements for long-term use of asphalt pavement. Although the aging of the RAP material in the recycled asphalt mixture has adverse effects on the fatigue life, the addition of a regenerant can better restore the pavement performance of the aged asphalt, thus improving the fatigue life of the recycled asphalt mixture.

(3) Under the same loading frequency, the fatigue life significantly decayed with the increase in stress ratio, which indicates that the stress ratio has a major effect on the fatigue life of an asphalt mixture. When the loading frequency was 20 Hz, with the increase in stress ratio from 0.3 to 0.6, the fatigue life of I-1 decreased by 66.3%, 85.3%, and 94.8%, and the fatigue life of II-1 decreased by 65.5%, 85.8%, and 95.1%. As the stress ratio gradually increased, the applied load was closer to the ultimate stress of bending and tensile failure of the specimen. With the increase in load times, the residual deformation of the specimen gradually increased, the stiffness (or modulus) of the material gradually decreased, the micro-cracks continuously developed, and, finally, the complete fracture occurred.

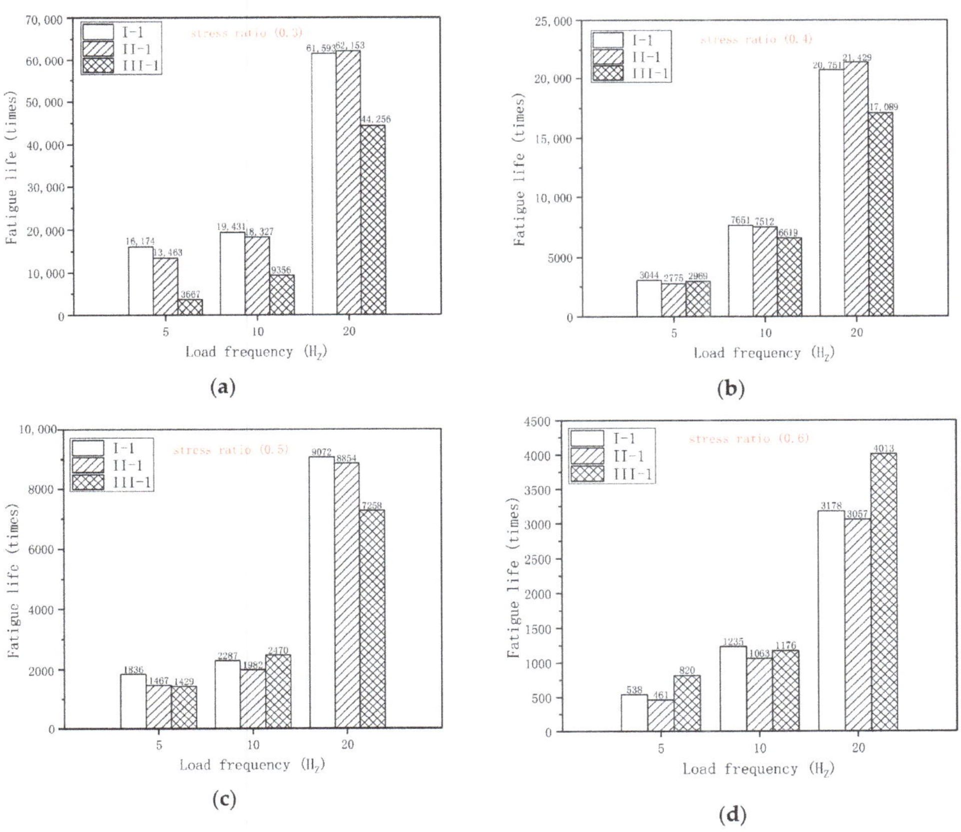

Figure 25. Relationship between load frequency and fatigue life under different stress ratios. (**a**) Stress ratio of 0.3. (**b**) Stress ratio of 0.4. (**c**) Stress ratio of 0.5. (**d**) Stress ratio of 0.6.

3.2.3. Fatigue Equation under Stress Control Mode

The test load level represents the load borne by the real road surface to some extent. According to the analysis of relevant data at home and abroad, five stress levels of 0.2, 0.3, 0.4, 0.5, and 0.6 were selected to replace the ordinary load and heavy load of different traffic axles. This experiment simulated different vehicle speeds on the road under the three loading frequencies of 5 Hz, 10 Hz, and 20 Hz. In accordance with a large number of experimental studies, it can be verified from the figure that the stress ratio and fatigue life showed a good linear relationship on the single logarithmic curve, which can be expressed via Formula (1):

$$\lg N_f = \mathrm{k} + n \times S_i \quad (1)$$

where N_f is the number of times the load is applied; that is, the fatigue life; S_i is the stress ratio; k; and n is the regression constant.

At 15 °C, the stress ratio and fatigue life of the SMA-13 recycled asphalt mixture under the three loading frequencies of 5 Hz, 10 Hz, and 20 Hz showed linear regression, and the single logarithmic fatigue equation was obtained, as shown in Figures 26 and 27.

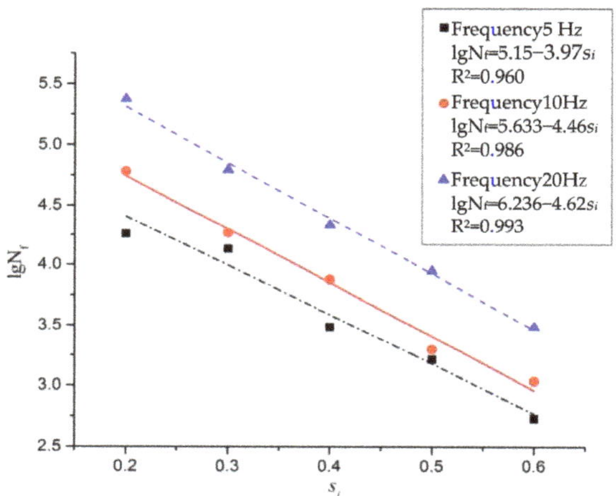

Figure 26. Stress ratio–fatigue curve of II-1.

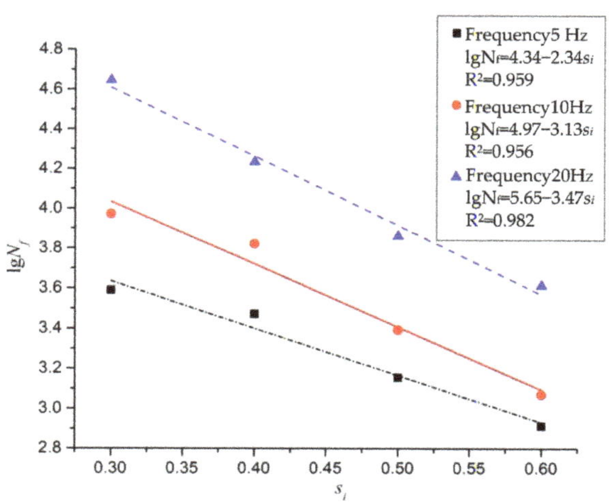

Figure 27. Stress ratio–fatigue curve of III-1.

As can be seen from Figures 26 and 27, the fatigue equations under different loading frequencies were basically parallel. The stress level had a significant effect on the fatigue life. With the increase in stress level, the fatigue life of the recycled asphalt mixture gradually decreased. The n value in the fatigue equation for the SMA-13 recycled asphalt mixture ranged from 3.97 to 4.62, and the variation range was relatively small. It was shown that the loading frequency had little difference in the sensitivity of the fatigue life and stress level of the recycled asphalt mixture specimens. The n value in the fatigue equation for the AC-20 recycled asphalt mixture ranged from 2.34 to 3.47, and the variation range was relatively large. This indicates that the loading frequency was more sensitive to the fatigue life and stress level of the specimen. The intercept (k) of the fatigue equation for the SMA-13 recycled asphalt mixture ranged from 5.15 to 6.236, and that of the AC-20 recycled asphalt mixture ranged from 4.34 to 5.65, with a relatively large variation range. This indicates that there were two kinds of grading regeneration under three kinds of load frequency differences between the fatigue performances of the recycled asphalt mixture.

For the indoor small bending fatigue test, the loading frequency of the test load was selected considering the actual vehicle load action time and determined according to Vander Poel's Formula (2):

$$t = \frac{1}{2\pi f} \tag{2}$$

where t is the loading time and f is the loading frequency.

Therefore, the larger the loading frequency of the trabecular specimen, the shorter the actual vehicle loading time, and the faster the vehicle running speed. The asphalt mixture at room temperature was viscoelastic. As the loading frequency increased, the action time on the asphalt mixture viscoelastic specimens was faster, and the specimens were mostly elastomer with a larger stiffness modulus. At a certain stress level, the bending and tensile deformation was relatively small, and it was not easy to produce fatigue cracking. Therefore, the higher the loading frequency of the recycled asphalt mixture specimen, the greater the fatigue life.

3.3. Dynamic Modulus Analysis of the Asphalt Mixture Based on the Test Road

Dynamic modulus tests were carried out on the ordinary SBS-modified asphalt mixture and the SBS-modified recycled asphalt mixture before and after wheel loading, and the test results are shown in Figures 28–30.

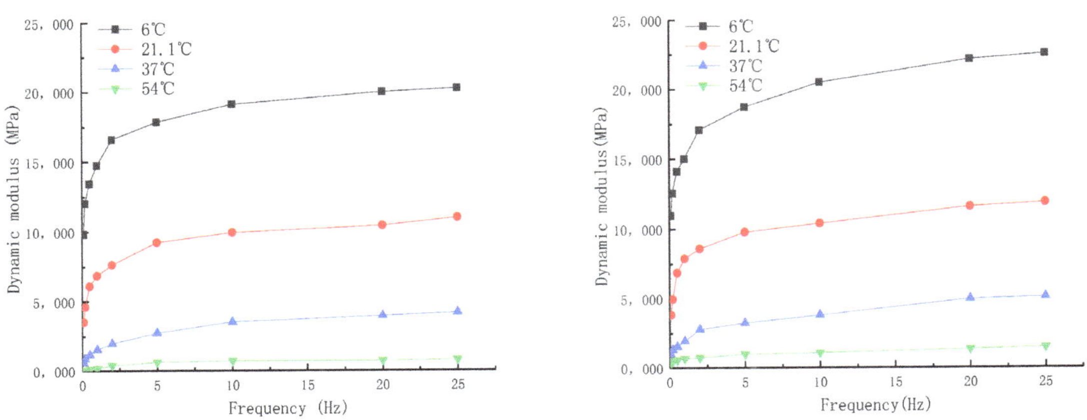

(a) (b)

Figure 28. Dynamic modulus of the ordinary SMA-13 asphalt mixture. (**a**) The dynamic modulus of I-1. (**b**) The dynamic modulus of I-2.

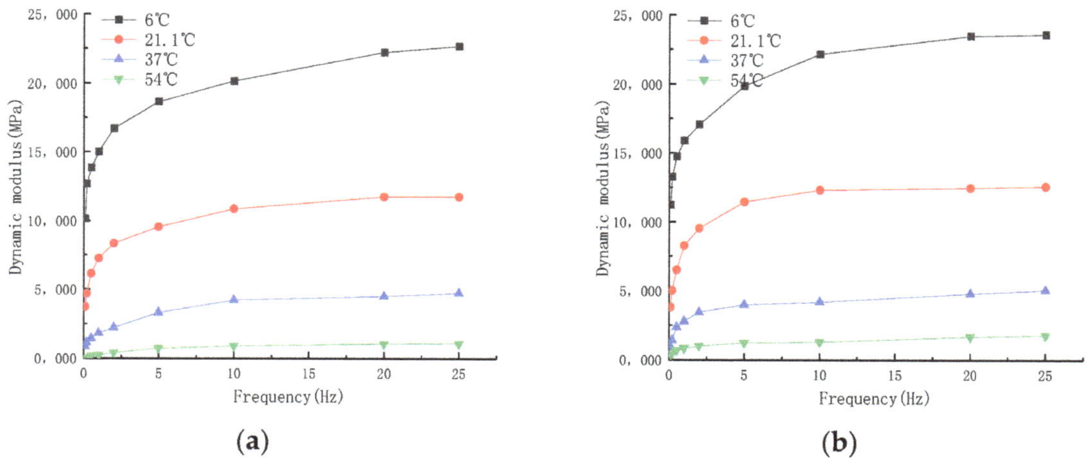

Figure 29. Dynamic modulus of the SMA-13 recycled asphalt mixture. (**a**) The dynamic modulus of II-1. (**b**) The dynamic modulus of II-2.

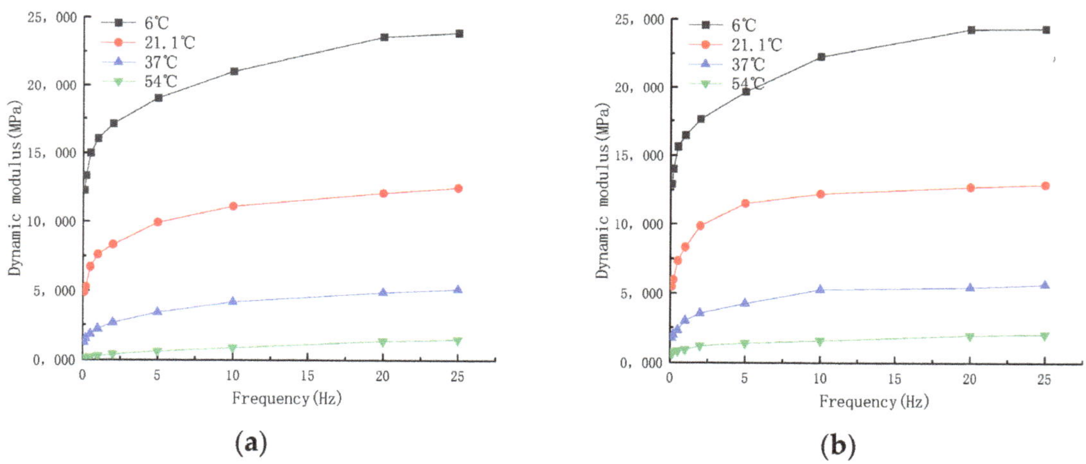

Figure 30. Dynamic modulus of the AC-20 recycled asphalt mixture. (**a**) The dynamic modulus of III-1. (**b**) The dynamic modulus of III-2.

(1) The dynamic modulus of the asphalt mixture material increased with the increase in load frequency under the same temperature condition, but with the continuous increase in temperature, the dynamic modulus gradually decreased the growth rate until it approached a quantitative value. The dynamic modulus of the asphalt mixture rapidly increased before 3 Hz and then tended to gradually slow down. Therefore, the dynamic modulus of the asphalt mixture will not become infinitely higher due to the increasing frequency, as when the loading frequency continues to increase, the viscosity of the asphalt mixture will gradually decrease, and its elasticity will gradually increase.

(2) Based on the ALT full-scale experiment, it was found that under the same temperature conditions, the dynamic elastic modulus of the asphalt mixture before the equipment wheel load was greater than that after the wheel load. The reason for this is that after long-term wheel loading, the fine aggregate will be compacted, the coarse aggregate

will be crushed, and the asphalt will age, which will lead to the structural stiffness and dynamic modulus of the mixed material being reduced.

(3) The modulus of the SMA-13 recycled asphalt mixture was smaller than that of the AC-20 recycled asphalt mixture. Structurally speaking, SMA-13 is characterized by discontinuous gradation, with more coarse aggregates and less fine aggregates, and its mixture belongs to the skeleton compact structure. AC is characterized by continuous gradation, a continuous amount of each file material, and step-by-step gap filling, and its mixed material structure belongs to the suspension dense type. This is the cause of the smaller modulus of the former and the larger modulus of the latter.

(4) The dynamic modulus of the asphalt mixture increased at temperature and rapidly decreased at different loading frequencies. When the test temperature was different, the dynamic modulus of the asphalt mixture increased with the increase in loading frequency. At the same frequency, the dynamic modulus decreased with the increase in temperature, which was due to the gradual decrease in the resilience modulus of asphalt as a binder. It can be seen that the basic mechanical properties of a viscoelastic asphalt mixture can be accurately reflected through dynamic modulus testing.

(5) The modulus of the recycled asphalt mixture was slightly larger than that of the ordinary asphalt mixture, which was because the recycled asphalt mixture was mixed with 20% RAP material, and the elastic modulus became larger due to the existence of the RAP material. Under different temperatures and frequencies, the dynamic modulus of the SMA-13 ordinary asphalt mixture and SMA-13 recycled asphalt mixture after accelerated loading were close to each other, which is consistent with the residual fatigue life rule of the two given in Section 3.2, and the same was true for the AC-20 recycled asphalt mixture.

3.4. Analysis of Asphalt Secondary Aging Components

The four-component test data of asphalt after secondary aging were compared and analyzed, as shown in Figures 31 and 32.

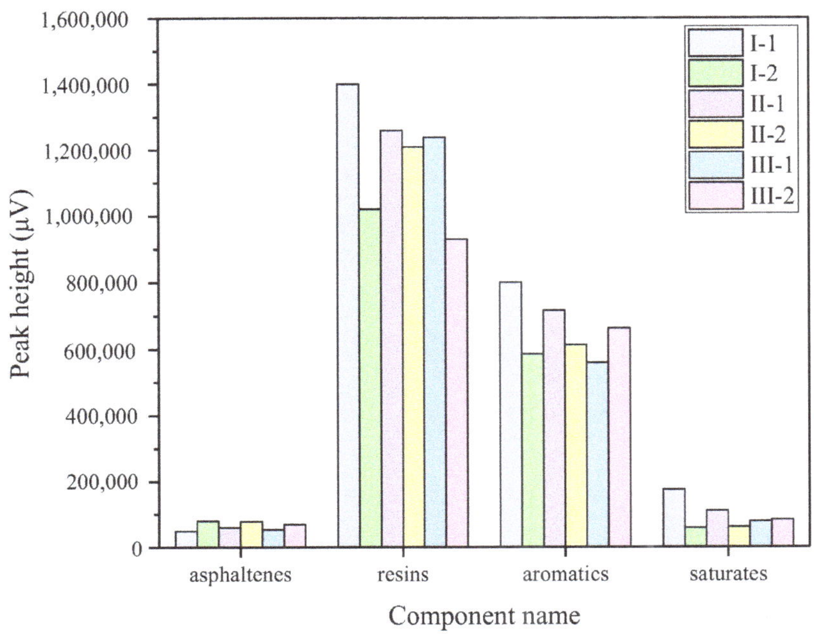

Figure 31. Peak heights of the four components of different asphalt mixtures.

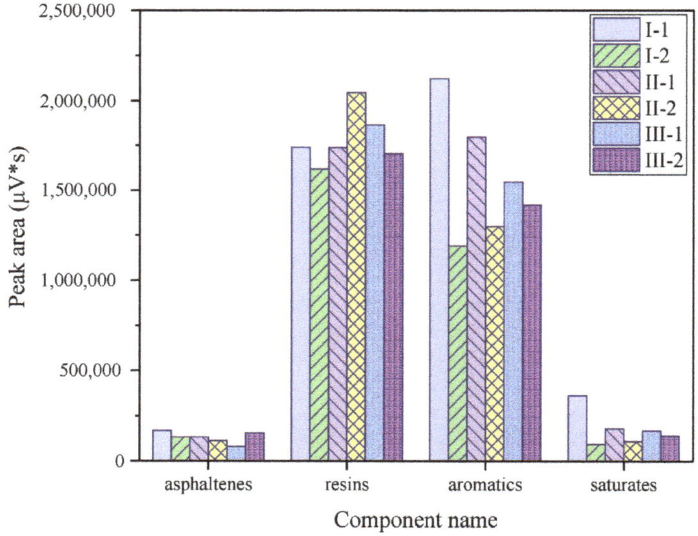

Figure 32. Peak areas of the four components of different asphalt mixtures.

The contents of each component (saturates, aromatics, resins, and asphaltenes) in the sample, W_i, were calculated according to Formulas (3) and (4):

$$W_i = (A_i/A) \times 100\% \quad (3)$$

$$A = \sum A_i \quad (4)$$

where W_i is the content of a component in the standard sample, A_i is the corrected peak area of a component in the sample, and A is the total peak area of each component in the standard sample.

The specific test data statistics according to Formulas (3) and (4) are shown in Figure 33.

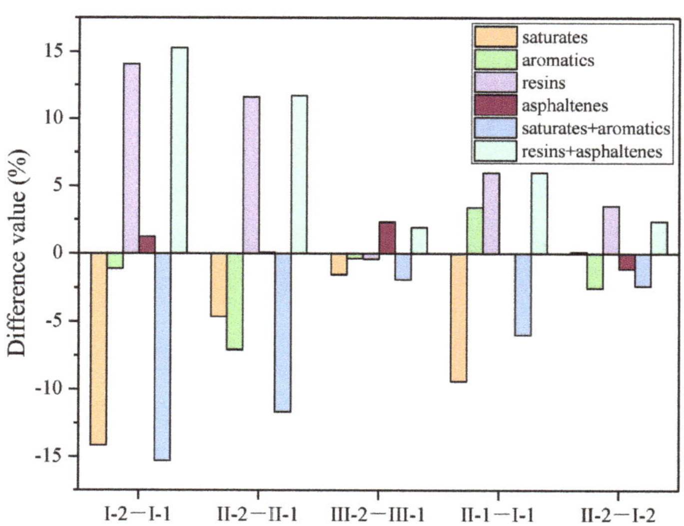

Figure 33. Contents of four components in different types of asphalt mixtures.

(1) Under accelerated loading, different loads under the same thermal oxygen, light, and water conditions will have different influences on the pavement over a period of time. According to the experimental data, the saturates + aromatics content of I-2 was 58.74%, and the saturates + aromatics content of I-1 was 74.03%, with I-2 < I-1. The resins + asphaltenes content of I-2 was 41.26%, and the resins + asphaltenes content of I-1 was 25.97%. The saturates and aromatics of asphalt following the accelerated loading section were reduced, and the lighter components were converted more into resins and asphaltenes. Under the coupling action of load, hot oxygen, light, and water, the aging of asphalt was accelerated, and the aging degree was normal, without serious aging phenomena.
(2) The saturates + aromatics content of II-2 was 56.35%, and the content of saturates + aromatics of II-1 was 11.69% more than II-2. The contents of resins + asphaltenes under unaccelerated loading and accelerated loading were 31.96% and 43.65%, respectively. Under the coupling action of running load, hot oxygen, light, and water, certain aging conditions were generated, but no serious aging phenomena occurred.
(3) The content of saturates + aromatics in II-1 was 5.99% less than that in I-1, and the content of saturates + aromatics in II-2 was 2.39% less than that in I-2. Based on the content of the four components, there were no excessive aging phenomena in the recycled asphalt mixture. After the fatigue test, the contents of each component in the recycled asphalt mixture after aging were basically the same as that of the ordinary asphalt mixture. This may be because the recycled asphalt mixture contains RAP, which has undergone aging and has stable content of light components. The old and new asphalt were combined and the four components were recombined, and the content of light components of I-1 was 5.99% different from that of I-1. After accelerated loading, the light component content of II-2 was 2.39% different from that of I-2, and the light components of II-2 and II-1 were only 11.69% lower. Therefore, the recycled asphalt mixture exhibited better fatigue resistance under the same aging conditions. The road performance and durability of the recycled asphalt mixture met the application requirements.
(4) Distinct from the accelerated loading of the SMA-13 ordinary asphalt mixture and SMA-13 recycled asphalt mixture on the upper layer, the different components of the AC-20 recycled asphalt mixture on the bottom layer displayed little difference under the two conditions of accelerated loading and unaccelerated loading. The saturates + aromatics content of III-2 was 62.59%, while the saturates + aromatics content of III-1 was 64.51% (basically the same). The resins + asphaltenes content of III-2 was not different from that of III-1. This indicated that the aging degree of the accelerated loading section and the unaccelerated loading section in the following layer were basically the same, with no obvious change, and that the road performance and durability met the use requirements. After accelerated loading, the light component of the SMA-13 thermal reclaimed asphalt mixture decreased by 11.39%, and that of the AC-20 thermal reclaimed asphalt mixture by 1.92%, mainly because the full-scale experiment directly acted on the upper layer, and the influence of the thermal oxygen, light, and water conditions also mainly acted on the upper layer. However, the AC-20 thermal reclaimed asphalt mixture was located at the bottom layer of the pavement structure, which was less affected and did not experience accelerated asphalt aging.

4. Conclusions

In this study, a series of tests were carried out to evaluate the decay mechanisms and rules of recycled asphalt pavement using a full-scale experiment. The following conclusions can be drawn from the tests:

(1) As the number of loading times increases, the BPN value of the two structures tended to be stable, and the BPN of Plan 2 was six lower than that of Plan 1. On the basis of the full-scale experiment, the recycled asphalt mixture and ordinary asphalt mixture exhibited good road performance. However, the ordinary asphalt mixture pavement

had better skid resistance. As the bottom layer, each component of the AC-20 recycled asphalt mixture showed little difference after accelerated loading and unaccelerated loading and remained consistent. This indicated that accelerated loading aging mainly affects the upper layer of pavements. In summary, recycled asphalt mixtures should not be applied to the upper layer of pavements but can be rationalized in the middle layer and the bottom layer of pavements.

(2) Under the same stress level and loading frequency, the curve relationships between load and deflection for the SMA-13 and AC-20 ordinary asphalt mixtures and the recycled asphalt mixture were similar (basically consistent). The results showed that the fatigue performance of a recycled asphalt mixture can meet the requirements of an ordinary asphalt mixture and meet the technical standards of asphalt pavement design. The macroscopic mechanical response and road service performance of the two pavement structures under load were basically the same, which reflects the good road performance of the blended recycled asphalt mixture.

(3) When the stress ratio was 0.3, the fatigue life difference between the asphalt mixtures with loading frequencies of 5 Hz and 10 Hz was not obvious, but the fatigue life of the asphalt mixture with a loading frequency of 20 Hz was 3.8 times than that of the 5 Hz mixture and 3.1 times than that of the 10 Hz mixture. Under the same test conditions, the order of fatigue life of the recycled asphalt mixtures under different loading frequencies was 20 Hz > 10 Hz > 5 Hz.

(4) The stress ratio and fatigue life showed a good linear relationship on the single logarithm curve, and the fatigue equations for the recycled asphalt mixtures under different load frequencies were obtained; that is, $\lg N_f = k + n \times S_i$. The n value of AC-20 ranged from 2.34 to 3.47, and the variation range was relatively large, indicating that the loading frequency was more sensitive to the fatigue life and stress level of the specimens.

(5) The saturates and aromatics of asphalt in the accelerated loading section were decreased, while the light components were transformed more into resins and asphaltenes. Following the full-scale experiment, the light component of the SMA-13 recycled asphalt mixture decreased by 11.69%, while that of the SMA-13 ordinary asphalt mixture decreased by 15.29%, indicating that the coupling effect of load, thermal oxygen, light, and water accelerated the aging of the asphalt, but no serious aging phenomena occurred. The recycled asphalt mixture exhibited better fatigue resistance under the same aging conditions.

(6) The pendulum tribometer used in this study could not fully simulate tire contact with the road, as real vehicle tires are much rougher and can provide better grip on the road surface. It is well known that BPN measurements are more influenced by non-slip micro-texture components than by macroscopic texture. This problem may lead to slight changes in the BPN value, especially when the sample surface void is blocked by dust or the road is wet in rainy days, and the BPN cannot be accurately detected. In addition, temperature, load, light, water, and other factors have a great impact on pavement aging. Therefore, it is necessary to monitor the road temperature, load, and other parameters and comprehensively analyze the decay mechanisms and laws of recycled asphalt pavement from a multi-directional perspective, so as to improve the utilization rate of recycled materials to make full use of resources and reduce waste.

Author Contributions: Conceptualization, J.G.; methodology, J.G. and Q.X.; design of analysis, Q.X. and Q.Z.; data curation, Q.Z., Q.S. and K.S.; validation, B.X.; writing—original draft, Q.X., Q.S. and Z.Y.; writing—review and editing, J.G. and Q.X. All authors have read and agreed to the published version of the manuscript.

Funding: This research was funded by the 2023 Shandong Provincial Transportation Science and technology plan, (Contract number: 2023B26).

Institutional Review Board Statement: Not applicable.

Informed Consent Statement: Not applicable.

Data Availability Statement: All of the data used to support the results of this study can be obtained from the corresponding authors according to the requirements.

Acknowledgments: The authors would like to thank the Shandong Transportation Service Center and the Shandong Transportation Institute for their financial support and technical services for this research work.

Conflicts of Interest: The authors declare no conflict of interest.

References

1. Ruiz, A.; Guevara, J. Sustainable Decision-Making in Road Development: Analysis of Road Preservation Policies. *Sustainability* **2020**, *12*, 872. [CrossRef]
2. Roach, P.E.; Ugursal, I. Development and technoeconomic evaluation of pathways to new construction net-zero energy houses in Nova Scotia. *Trans. Can. Soc. Mech. Eng.* **2020**, *45*, 221–232. [CrossRef]
3. Perumal, L.; New, M.; Jonas, M.; Liu, W. The impact of roads on sub-Saharan African ecosystems: A systematic review. *Environ. Res. Lett.* **2021**, *16*, 113001. [CrossRef]
4. Sidle, R.C. Dark Clouds over the Silk Road: Challenges Facing Mountain Environments in Central Asia. *Sustainability* **2020**, *12*, 9467. [CrossRef]
5. Suprayoga, G.B.; Witte, P.; Spit, T. Coping with strategic ambiguity in planning sustainable road development: Balancing economic and environmental interests in two highway projects in Indonesia. *Impact Assess. Proj. Apprais.* **2020**, *38*, 233–244. [CrossRef]
6. Zhao, H.; Su, J.; Ma, S.; Su, C.; Wang, X.; Li, Z.; Wei, J.; Cui, S. Study on Cold Recycled Asphalt Mixtures with Emulsified/Foamed Asphalt in the Laboratory and On-Site. *Coatings* **2022**, *12*, 1009. [CrossRef]
7. Li, J.; Zhu, L.; Yu, M.; Zuo, S.; Cui, X.; Liu, P. Long-Term Performance of Recycled Asphalt Pavement with Recycled Engine Oil Bottom Based on Accelerated Loading Test. *Coatings* **2022**, *12*, 522. [CrossRef]
8. Li, M.; Tang, W.; Yu, X.; Zhan, H.; Wang, X.; Wang, Z. Laboratory investigation on blending process of reclaimed asphalt mixture. *Constr. Build. Mater.* **2022**, *325*, 126793. [CrossRef]
9. Li, X.; Wang, Q.; Liu, S. Research on Application of Warm Mix Asphalt Technology in Hot In-Place Recycled Engineering. *Appl. Mech. Mater.* **2013**, *361-363*, 1655–1658. [CrossRef]
10. Reza, I.; Luis, L.; Alan, C. Integrated performance evaluation of asphalt mixtures with very high reclaimed asphalt pavement (RAP) content. *Constr. Build. Mater.* **2022**, *347*, 129607. [CrossRef]
11. Yin, P.; Pan, B. Effect of RAP content on fatigue performance of hot-mixed recycled asphalt mixture. *Constr. Build. Mater.* **2022**, *328*, 127077. [CrossRef]
12. Marcandré, B.; Sébastien, L.; Kevin, B.; Alan, C. Laboratory Study of the Effects of the Mixer Type and Mixing Time on the Volumetric Properties and Performance of a HMA with 30 Percent Reclaimed Asphalt Pavement. *Materials* **2023**, *16*, 1300. [CrossRef]
13. Li, F.; Zhu, H.; Yan, J.; Liu, Q.; Zhang, Z. Experimental Study of Hot Recycled Asphalt Mixtures with High Percentages of Reclaimed Asphalt Pavement and Different Recycling Agents. *J. Test. Eval.* **2014**, *42*, 1183–1190.
14. Gao, J.; Yao, Y.; Huang, J.; Yang, J.; Song, L.; Xu, J.; Lu, X. Effect of Hot Mixing Duration on Blending, Performance, and Environmental Impact of Central Plant Recycled Asphalt Mixture. *Buildings* **2022**, *12*, 1057. [CrossRef]
15. Prosperi, E.; Bocci, E.; Bocci, M. Effect of Bitumen Production Process and Mix Heating Temperature on the Rheological Properties of Hot Recycled Mix Asphalt. *Sustainability* **2022**, *14*, 9677. [CrossRef]
16. Bocci, E.; Bocci, M.; Prosperi, E. Influence of the Hot-Mix Asphalt Production Temperature on the Effectiveness of the Reclaimed Asphalt Rejuvenation Process. *Infrastructures* **2022**, *8*, 8. [CrossRef]
17. Cong, P.; Zhang, Y.; Liu, N. Investigation of the properties of asphalt mixtures incorporating reclaimed SBS modified asphalt pavement. *Constr. Build. Mater.* **2016**, *113*, 334–340. [CrossRef]
18. Homsi, F.; Bodin, D.; Yotte, S.; Breysse, D.; Balay, J.M. Fatigue life modelling of asphalt pavements under multiple-axle loadings. *Road Mater. Pavement Des.* **2012**, *13*, 749–768. [CrossRef]
19. Dong, C.; Liu, W.; Zhou, L.; Zhang, R.; Kan, Q.; Leng, W. Evaluation on Fatigue Life of Expressway Asphalt Pavement Based on Tire-Pavement-Subgrade Coupling Model. *Am. J. Civ. Eng.* **2017**, *5*, 400–407.
20. Ziari, H.; Aliha, M.; Mojaradi, B.; Jebalbarezi, S. Investigating the effects of loading, mechanical properties and layers geometry on fatigue life of asphalt pavements. *Fatigue Fract. Eng. Mater. Struct.* **2019**, *42*, 1563–1577. [CrossRef]
21. Yu, Y.; Tang, S.; Xu, G.; Fan, Y.; Wu, Y.; Wang, H.; Yang, J. Investigations on the long-term skid resistance of epoxy asphalt mixture based on accelerated loading test. *Constr. Build. Mater.* **2023**, *365*, 130150. [CrossRef]
22. Hoha, D.; Boboc, V.; Boboc, A. The Behaviour of a Flexible Road Pavement under Accelerated Loading Test. *Bull. Transilv. Univ. Brasov. Eng. Sciences. Ser. I* **2018**, *11*, 257–266.
23. Cao, W.; Mohammad, L.; Barghabany, P. Use of indirect tension test and viscoelastic continuum damage theory for fatigue characterization of asphalt mixtures. *Constr. Build. Mater.* **2018**, *187*, 38–49. [CrossRef]

24. Luo, X.; Hu, S.; Zhou, F.; Crockford, W.; Karki, P. Simple Asphalt Mixture Shear Rutting Test and Mechanical Analysis. *J. Mater. Civ. Eng.* **2022**, *34*, 04022220. [CrossRef]
25. Ibrahim, O.; Yared, D.; Björn, B. Hierarchical approach for fatigue cracking performance evaluation in asphalt pavements. *Front. Struct. Civ. Eng.* **2017**, *11*, 257–269.
26. Mensching, D.; Daniel, J.; Underwood, B. Exploring indicators for fatigue cracking in hot mix asphalt pavements using simplified-viscoelastic continuum damage theory. *Road Mater. Pavement Des.* **2018**, *19*, 536–545. [CrossRef]
27. Fu, Y. Research on the Fatigue Damage of the Asphalt Pavement. *Appl. Mech. Mater.* **2012**, *256–259*, 1776–1779.
28. Wu, Y.; Zhou, X.; Wang, X.; Shan, L. Long-Term Service Performance of Hard-Grade Asphalt Concrete Base Pavement Based on Accelerated Loading Test of Full-Scale Structure. *Sustainability* **2022**, *14*, 9712. [CrossRef]
29. Xu, X.; Yu, Y.; Yang, J.; Wu, C. Long-Term Skid Resistance Evaluation of GAC-16 Based on Accelerated Pavement Testing. *Adv. Mater. Sci. Eng.* **2020**, *2020*, 1–9. [CrossRef]
30. Park, D. Evaluation of Predicted Pavement Fatigue Life Based on Surface Profiles and Asphalt Mixture Types. *KSCE J. Civ. Eng.* **2010**, *14*, 191–196. [CrossRef]
31. Xiao, Z.; Chen, M.; Wu, S.; Xie, J.; Kong, D.; Qiao, Z.; Niu, C. Moisture Susceptibility Evaluation of Asphalt Mixtures Containing Steel Slag Powder as Filler. *Materials* **2019**, *12*, 3211. [CrossRef] [PubMed]
32. Syyed, A.; Hunain, A.; Ahsan, W.; Muhammad, H.; Salman, H.; Junaid, M.; Abdullah, N.; Muhammad, A. Saving Energy in the Transportation Sector: An Analysis of Modified Bitumen Application Based on Marshall Test. *Energies* **2018**, *11*, 3025. [CrossRef]
33. Lv, S. Study on Low Temperature Performance Evaluation Index of Asphalt Mixture Based on Bending Test at Low Temperature. *Appl. Mech. Mater.* **2012**, *1931*, 427–430. [CrossRef]
34. Li, Z.; Yu, W.; Ying, M. The Research of Low Temperature Performance of Asphalt Mixture. *Adv. Mater. Res.* **2010**, *168–170*, 2507–2512.
35. Kim, M.; Mohammad, L.; Jordan, T.; Cooper, S. Fatigue performance of asphalt mixture containing recycled materials and warm-mix technologies under accelerated loading and four point bending beam test. *J. Clean. Prod.* **2018**, *192*, 656–664. [CrossRef]
36. *AASHTO TP 62*; Standard Method of Test for Determining Dynamic Modulus of Hot Mix Asphalt (HMA). American Association of State Highway and Transportation Officials: Washington, DC, USA, 2007.
37. *JTG E20-2011*; Standard Test Methods of Bitumen and Bituminous Mixtures for Highway Engineering. Ministry of Transport of the People's Republic of China: Beijing, China, 2011.
38. Nie, Y.; Zhang, Y.; Yu, J.; Kuang, D.; Zhang, X. Research on Aging Mechanism and Recycling Mechanism Based on Asphalt Four Components Analysis. *Appl. Mech. Mater.* **2012**, *1975*, 1659–1664. [CrossRef]

coatings

Article

Research on Flexural and Freeze–Thaw Properties of Polypropylene-Fiber-Reinforced Pavement Concrete Containing Waterborne Epoxy

Li Wang [1,*], Bing Zhang [2], Hongliang Zhang [3], Bendong Zhao [4], Biao Wang [5], Quanman Zhao [1] and Min Sun [1]

[1] School of Transportation Engineering, Shandong Jianzhu University, Jinan 250101, China
[2] Shaanxi Road and Bridge Engineering Testing Co., Ltd., Xi'an 710000, China
[3] Key Laboratory for Special Area Highway Engineering of Ministry of Education, School of Highway, Chang'an University, Xi'an 710064, China
[4] CCCC First Highway Consultants Co., Ltd., Xi'an 710000, China
[5] Shaanxi Road and Bridge Group Co., Ltd., Xi'an 710065, China
* Correspondence: wangli20@sdjzu.edu.cn

Abstract: In order to further enhance the reinforcing effectiveness of polypropylene (PP) fibers on pavement concrete, waterborne epoxy (WBE) was introduced in this research and its effect on the flexural properties and freeze–thaw resistance of PP-fiber-reinforced concrete was evaluated. Compressive-strength tests, flexural-strength tests, three-point bending tests, freeze–thaw cycling tests and a scanning electron microscopic observation were carried out to analyze mainly the influence of WBE on the flexural properties and freeze–thaw resistance of PP-fiber-reinforced concrete. WBE contents of 0, 5%, 10%, 15% and 20% by weight of the cement were employed. The experimental results indicated that WBE was beneficial to improving the flexural properties of PP-fiber-reinforced concrete. With increasing content of WBE, the flexural strength and the peak load showed significant increases. Although a slight degradation in the abovementioned flexural parameters was observed when the WBE content was above 15%, the deflection at the peak, the fracture energy and the fracture toughness still showed an upward trend. In addition, the freeze–thaw resistance of PP-fiber-reinforced concrete was improved remarkably with the increasing addition of WBE content, leading to smaller mass loss and higher residual flexural strength. Moreover, microstructural images revealed that with the addition of WBE, the PP fiber/concrete interfacial bonding was effectively improved, and the concrete matrix tended to be denser as well, which provided higher resistance for crack initiation and propagation. In consideration of maximally improving the flexural properties of PP-fiber-reinforced pavement concrete, and while ensuring the compressive strength and meeting the freeze–thaw requirements, it was recommended that the content of WBE in PP-fiber-reinforced concrete should be 15%.

Keywords: concrete pavement; polypropylene fibers; waterborne epoxy; flexural properties; freeze–thaw resistance; microstructure

1. Introduction

Currently, improving the sustainability of construction and pavement projects is still one of the major challenges confronting the construction and infrastructure industry. Cementitious materials, however, as the most widely used construction materials, intrinsically possesses the well-known disadvantages of high brittleness, low tensile strength and low fracture toughness, which weakens its long-term durability and lowers the service life of construction projects, thereby hindering the achievement of sustainability. Concrete pavement is often utilized for structures such as highways, airport runways and bridge decks, which are directly subjected to repeated vehicle loads and environmental actions. Under these repeated loads, concrete pavement will suffer from the gradual deterioration of pavement performance, leading to the occurrence of various diseases, such as surface cracks, joint cracking and even fracture diseases during the service period and accordingly a reduced driving comfort and safety [1,2]. To guarantee the performance of concrete pavement and greatly postpone the occurrence of the diseases mentioned above, one of the most effective measures is to improve the flexural properties of the pavement concrete used.

As is well known, the incorporation of various kinds of fibers is deemed as a practical method to largely enhance the engineer properties of cementitious materials, such as flexural strength, ductility, fracture toughness and resistance to fatigue [3,4]. Polypropylene (PP) fibers, as one of the commonly used synthetic fibers, have been widely applied in cementitious materials in order to improve the toughness and deformability of them and restrain the appearance of cracks and the subsequent crack propagation, so as to prolong the service life of engineering projects due to their low cost, high tensile strength, excellent toughness and good corrosion resistance to chemicals [5–7]. Qin et al. indicated that compared to plain concrete, the compressive strength of PP-fiber-reinforced concrete showed a slight increase, and the fracture process was significantly lower [8]. Akça et al. and Das et al. discovered that the flexural tensile strength and splitting tensile strength of concrete were both enhanced through the introduction of PP fibers [9,10]. Investigations by Li et al. revealed that with the addition of 0.9 kg/m^3 of PP fibers, the flexural strength of concrete notably improved; thus, the formation and propagation of microcracks were reduced [11]. Wang et al. performed a series of laboratory tests and found that with the PP fibers, the flexural strength of concrete was remarkably enhanced [12]. Islam and Gupta found that except for the increase in tensile strength, the crack width and propagation of plastic shrinkage cracks could also be reduced by incorporating PP fibers into concrete [13]. In addition, Del Savio et al. investigated the relationship between toughness and PP macrofiber volume and found that the toughness of the concrete was enhanced with increases in the PP macrofiber volume [14]. To obtain these excellent performances of PP-fiber-reinforced concrete (PFRC), the key factor is that the PP fibers should be tightly bonded to their surrounding concrete matrix so as to ensure a consistent load transfer from the concrete matrix to PP fibers [15,16].

However, owing to the hydrophobicity and surface smoothness of PP fibers, the fiber–matrix connection zones may be weaker due to the generation of voids in these zones, which may lead to a reduction in PFRC flexural performance. With an increasing PP fiber content, this reduction can be even greater because of the formation of fiber balling to trap a considerable amount of air. Various methods have been adopted to enhance the interfacial bonding and reduce the porosity between PP fibers and the cementitious matrix; among those methods, it is reported that the addition of polymers can be an effective method [15–18]. Our previous studies also found that the interfacial bonding between PP fibers and their surrounding matrix was improved through the addition of appropriate content of waterborne epoxy (WBE) [19]. Generally, the incorporation of polymers in PFRC is to ensure the toughening effect of PP fibers, so as to effectively reduce cracking in the PFRC and further improve the fracture toughness. Therefore, from the viewpoint of enhancing the actual engineering application, it is essential to study the impact of the polymers on the flexural properties (mainly the fracture properties) of PFRC to obtain an in-depth understanding of polymer-modified PFRC. Presently, the study's focus is mainly on the flexural properties of concrete containing PP fibers, and it is certainly demonstrated that the addition of PP fibers plays an active role in restricting the cracking in concrete [20,21]. However, there is limited research information concerning the impact of polymers on the fracture properties of PFRC, which may impose restrictions on the efficient application of polymers in PFRC.

Moreover, because concrete pavement is directly exposed to the changeable climate environment, the effect of temperature change on the performance of pavement concrete should be attached importance to as well, especially when the concrete pavement is built in an area characterized by extreme low temperatures. Previous studies conducted by other researchers have revealed that the durability properties of cementitious materials (not limited to the freeze–thaw resistance) were improved through the incorporation of PP fibers. Chen et al. investigated the impact of PP fibers on the frost resistance of airport pavement concrete and discovered that the addition of 0.1% PP fiber could significantly increase the frost resistance grade of the concrete [22]. Nam et al. found that after the freeze–thaw cycles, the mass loss of concrete was greatly reduced by adding 1.5% PP fiber by volume [23]. Richardson et al. pointed out that PP fiber was capable of improving the frost resistance of concrete and reducing the water absorption [24]. Zhang and Li found that the durability properties of concrete composites were remarkably improved with the inclusion of PP fibers, and the freeze–thaw resistance increased when the PP fiber volume percentage was below 0.08% [25]. In addition, studies were also conducted to investigate the influence of additive materials, mainly the supplementary cementitious materials, on the frost resistance of PFRC. Karahan and Atiş investigated the durability of PFRC and found that with the addition of fly ash, the frost resistance of PFRC was significantly improved [26]. Similar research was carried out by Abadel and Alghamdi to investigate fly ash and ground blast furnace slag on the freeze–thaw resistance of geopolymer mortar and found that the two additives had positive effects [27]. However, studies about the impact of polymers on the freeze–thaw resistance of PFRC are limited.

The utilization of polymers in PFRC should be a promising practice to improve its performance. Currently, although the influence of different types of polymers on the mechanical strength of PFRC have been reported in the literature, there is little research available on fracture properties and freeze–thaw resistance of PFRC-containing polymers, whereas this knowledge is of significant importance in promoting the rational application of polymers in PFRC. In this research, WBE was employed to modify PFRC as it is a kind of environmentally friendly polymer and has also been proved to be beneficial to improving the mechanical and durability properties of PFRC [19]. Furthermore, the current research regarding the effect of WBE on the fracture properties and freeze–thaw resistance of PFRC is limited. Therefore, the primary purpose of this paper is to investigate the flexural and fracture properties of PFRC containing different levels of WBE and to study the freeze–thaw resistance of this kind of concrete composite as well to obtain a comprehensive understanding of material behavior of PFRC containing WBE. In addition, microstructural analysis was conducted using a scanning electron microscopy (SEM) to explain the enhancement mechanism on the introduction of WBE into the PFRC. The results of this study can provide a reference for the efficient utilization of WBE in PFRC, while laying the foundation for the coupled application of PP fibers and WBE in concrete pavement.

2. Materials and Methods

2.1. Materials

In this study, raw materials involved are listed as follows: (1) ordinary Portland cement 42.5, produced by Yangchun cement Co., Ltd. in Shandong Province, Zhucheng, China. The initial setting time and final setting time are 160 min and 240 min, respectively. Its chemical composition is displayed in Table 1; (2) PP filament fiber produced by Ningxiang building materials Co., Ltd. in Hunan Province, Changsha, China. It has a length of 19 mm, with a diameter of about 50 μm. The density of the PP fiber used in this study is 0.91 g/cm^3, whereas its tensile strength and elastic modulus are more than 486 MPa and 4.8 GPa, respectively. Figure 1 presents the image of the PP fiber used; (3) waterborne epoxy produced by Dongyang coating Co., Ltd. in Liaoning Province, Shenyang, China. Its primary properties and its matched curing agent are shown in Table 2 (provided by the manufactures); (4) natural river sand with a fineness modulus of 2.7; (5) crushed limestone with a continuous size from 0.5 to 2 cm.

Table 1. Chemical composition of cement (wt%).

Oxides	CaO	SiO_2	Al_2O_3	Fe_2O_3	MgO	SO_3	Na_2O	K_2O	LOI
Cement	58.34	20.87	5.83	3.64	1.06	3.61	1.36	0.94	4.35

Figure 1. Polypropylene fibers.

Table 2. Properties of WBE and its matched hardener.

Items	Exterior	Solid Content (%)	pH	Viscosity (mPa·s)	Epoxy Equivalent (g/eq)
WBE	Milky white	50 ± 3	6–8	1218	192.3
Hardener	Light yellow	50 ± 1	9.5–10.5	6000	-

2.2. Specimen Preparation

To satisfy the requirements of flexural strength and the workability, the mix proportions of PFRCs in this study were designed according to the JTG/T F30-2014 technical guidelines [28]. A total of five types of concrete specimens were fabricated with a constant water cement ratio of 0.4, as present in Table 3. The PP fibers content for all specimens was 0.1% by volume fraction of concrete, and this content was chosen according to JTG/T F30-2014 technical guidelines [28] with an aim to significantly improve the early-age cracking resistance of the concrete. Additionally, on the basis of our previous studies, PFRCs containing 0.1% PP fibers by volume of concrete exhibited superior performance in terms of workability and mechanical strength [19]. On the other hand, studies by other researchers also showed that the addition of 0.1% PP fibers by volume of concrete provided the best mechanical and durability properties [24]. Furthermore, five levels of WBE of 0, 5%, 10%, 15% and 20% by weight of cement were added to PFRC to investigate the impact of WBE on its fracture and freeze–thaw properties. It should be mentioned that the content of WBE was in solid content; thus, the water contained in WBE should be considered in the total water.

Table 3. Mixture proportions (unit: kg/m^3).

Mixture Type	Cement	PP Fibers	Sand	Coarse Aggregate	Water	WEB		
						Total	Solid	Water
PFRC(control)	400	0.91	690	1170	160	-	-	-
PFRC-E5					140	40	20	20
PFRC-E10					120	80	40	40
PFRC-E15					100	120	60	60
PFRC-E20					80	160	80	80

According to the mixture proportions, concrete specimens were prepared in the following steps. First, cement, PP fibers, and fine and coarse aggregates were blended together in the concrete mixer for 3 min. Then, the mixing water, in which the water-reducing agent was dissolved in advance, was added to the dry mixtures and stirred for another 3 min to ensure that the concrete mixtures evenly distributed. After that, WBE was blended with its matched curing agent and then this epoxy mixture was subsequently added to the concrete mixtures and stirred for 3 min. After the mixing process, the fresh mixtures were immediately poured into plastic molds of different sizes and vibrated on a vibrating table to obtain a desirable compactness. The concrete specimens were cured at ambient temperature and demolded after 24 h, then cured in the curing room at a temperature of 20 ± 2 °C and a relative humidity of 95% until a specific age. Figure 2 presents the preparation procedures of the concrete specimens.

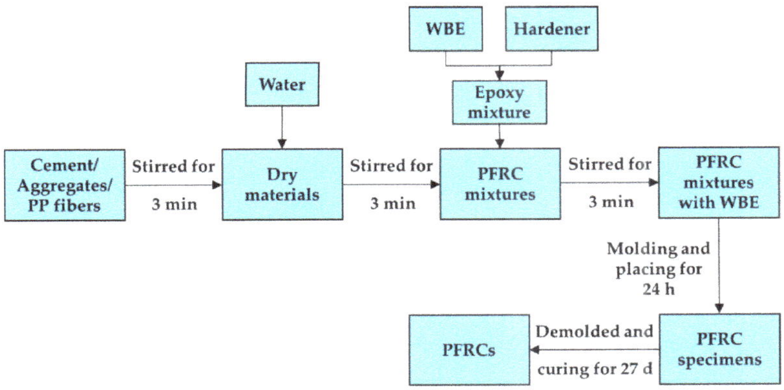

Figure 2. Preparation procedures of concrete specimens.

2.3. Experimental Methods

In this experiment, the mechanical and fracture properties, together with freeze–thaw resistance of PFRC containing different dosages of WBE were investigated. Cube specimens with a side length of 100 mm and beam specimens with a dimension of 100 × 100 × 400 mm^3 were employed.

2.3.1. Mechanical Properties

A compressive-strength test, which conformed to the JTG E30-2005 standard test method [29], was conducted on cube specimens after 28 day through the use of a WDW-300 universal testing machine manufactured by Shanghai Hualong Test instrument Co., Ltd., Shanghai, China. Three cube specimens were tested, and the average compressive strength was determined for each type of concrete mixture.

A flexural-strength test was performed on beam specimens in accordance with the JTG E30-2005 standard test method [29], and for each type of concrete, three beam specimens were tested and the average value was recorded.

2.3.2. Fracture Properties

A three-point bending test was conducted according to the JCI-S-002-2003 standard [30] to evaluate the fracture properties of PFRC containing different levels of WBE. The centrally notched beam specimens, with a 30 mm deep sawed notch at the midspan, were used to investigate the fracture properties of concrete at 28 day of curing. The effective span and the ligament length of the specimen were 300 mm and 70 mm, respectively. A WDW-300 universal testing machine was employed, and the load was applied at a displacement rate of 0.02 mm/min. During the test, two linear variable differential transformers (LVDTs) were installed at both sides of the beam to measure the deflections. Meanwhile, at the notch, a clip gauge was attached to the bottom of the beam to measure the crack width. Figure 3 provides the schematic diagram of the test set-up for notched-beam specimens. The fracture energy (G_f) and the critical stress intensity factor (K_{IC}) were determined from the three-point bending tests. Three notched-beam specimens were tested, and the average value was recorded for each type of concrete mixture.

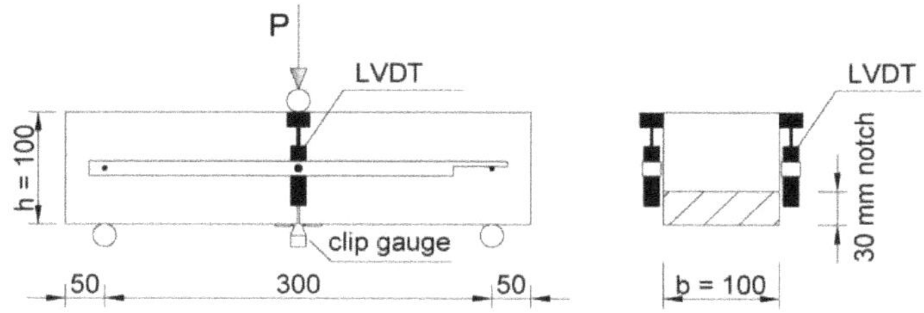

Figure 3. The schematic diagram of the three-point bending test.

2.3.3. Freeze–Thaw Resistance

The freeze–thaw resistance test was conducted on PFRC containing different dosages of WBE according to JTG E30-2005 standard test method [29]. Beam specimens were used, and after 28 day of curing, they were subjected to freeze–thaw cycles. Each freeze–thaw cycle consisted of a 2 h freeze at -18 ± 2 °C and 2 h thaw at 5 ± 2 °C. After 0, 25, 50, 75 and 100 cycles, the mass loss and flexural strength of the concrete specimens were tested. For each type of concrete, three beam specimens were examined and the average value was recorded.

2.3.4. SEM

Microstructural images of the concrete mixtures were obtained using SEM (VEGA *II* XMU, TESCAN CO., LTD., Brno, Czech Republic). The specimens used for SEM test were obtained from fractured specimens after 28 day. The specimens were immersed in anhydrous ethanol, and before the observation, they were dried and sprayed with gold to make them conductive.

3. Results and Discussion

3.1. Mechanical Properties

3.1.1. Compressive Strength

The compressive strength of PFRC containing different contents of WBE is presented in Figure 4. It is well known that the compressive strength of concrete is strongly linked to the porosity, as well as the quantity of the cement hydrates [31]. According to the results, it can be found that the compressive strength was gradually improved with the increase in WBE content up to 10%. This phenomenon can be attributed to the filling effect of epoxy particles to fill the voids and defects in PFRC and decrease the porosity of PFRC, thus the compressive strength showed an upward trend compared with the control PFRC. This result is in harmony with findings of other researchers indicating that superior compactability of the polymer-modified concrete was achieved to obtain a denser microstructure and, accordingly, a higher compressive strength compared to plain concrete, as the "ball bearing" effect of polymer particles enhances the workability of polymer-modified concrete [32,33]. In addition, WBE may act as a binder to bind the concrete components tightly together due to its high viscosity after polymerization [34]. In general, with a lower quantity of added WBE, the components in PFRC may be bonded together to obtain a denser structure, which improves the compressive strength of PFRC. However, the compressive strength showed a decreased tendency when the content of WBE exceeded 10%. This tendency was mainly because the formation of polymer film is a gradual process and the higher incorporation of polymers may result in the formation of a thicker polymer film to adhere to the surfaces of cement particles, which could depress the diffusion of ions and, accordingly, retard the hydration of cement, thus leading to a reduced quantity of hydration products, whereas the quantity of hydration products is closely related to the compressive strength of concrete [15,35]. On the other hand, with the large addition of WBE content, polymer phases may be regarded as defects in concrete because epoxy film has a lower mechanical capacity compared to concrete [34,36]. Therefore, the compressive strength of PFRC-E15 and PFRC-E20 exhibited a gradual decrease, leading to a rather lower mean compressive strength of PFRC-E20 by about 2.6% compared with the control PFRC, whereas the compressive strength of PFRC-E15 was still slightly higher compared to the control PFRC.

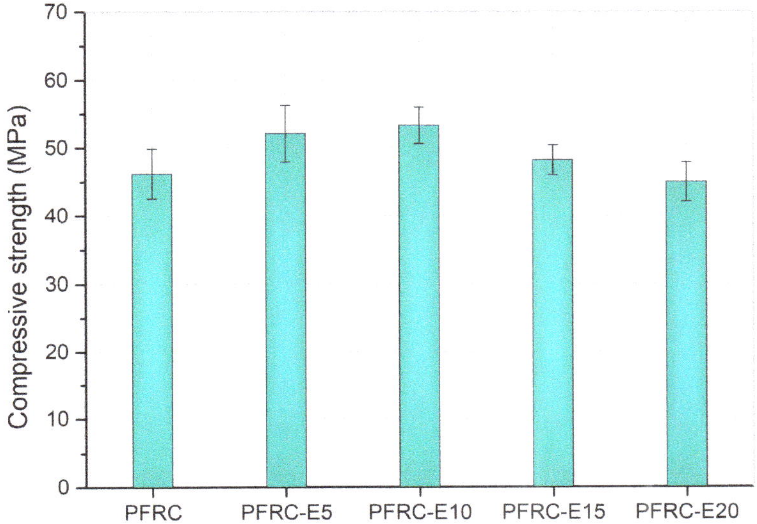

Figure 4. Compressive strength of PFRC with different contents of WBE.

3.1.2. Flexural Strength

The flexural strength of PFRC with different levels of WBE is displayed in Figure 5. It can be observed from the results that with the increased addition of WBE, the flexural strength of PFRC was improved obviously, and this result is in harmony with the investigations of other researchers [36–38]. Compared to the control PFRC, the flexural strength of PFRC with 5% WBE increased by about 16%. This improvement may probably be attributed to the pore-filling effect of epoxy particles, which can effectively fill the internal porosity in concrete and reduce the porosity between PP fibers and their surrounding concrete matrix and thus improve the bond between PP fibers and concrete matrix and, accordingly, enhance the reinforcing effect of PP fibers [37]. Well-bonded discrete fibers can arrest microcracks in concrete; hence, the crack resistance of concrete can be enhanced, and eventually the flexural strength can be improved. On the other hand, as the dosage of WBE continued to increase, PFRC-E15 exhibited the highest flexural strength, with an increase of approximately 40% compared to the control PFRC. This notable increase in flexural strength could have resulted from the formation of a fully developed, coherent polymer film, which could effectively dissipate part of the energy under the flexural loads [34,38]. Additionally, due to the active groups in epoxy chains, the epoxy film and hydrates may connect with each other to form a strong organic–inorganic cross-linking structure to withstand higher flexural loads, thereby postponing the appearance and propagation of cracks [34,38]. However, a slight reduction in flexural strength was observed in PFRC-E20 compared to PFRC-E15. This may be due to the fact that the over-percolation of polymer film may negatively affect the cement hydration and destroy the continuity of hydration products, which is consistent with the findings of other researchers indicating that there is a threshold polymer-to-cement ratio, beyond which the properties of concrete could not be improved [36]. Moreover, with a higher content of WBE in concrete, epoxy phase is likely to mutually reunite and form larger colloidal particles. These solidified particles are relatively soft compared to plain concrete and may be regarded as localized weak zones in concrete to decrease the strength [34].

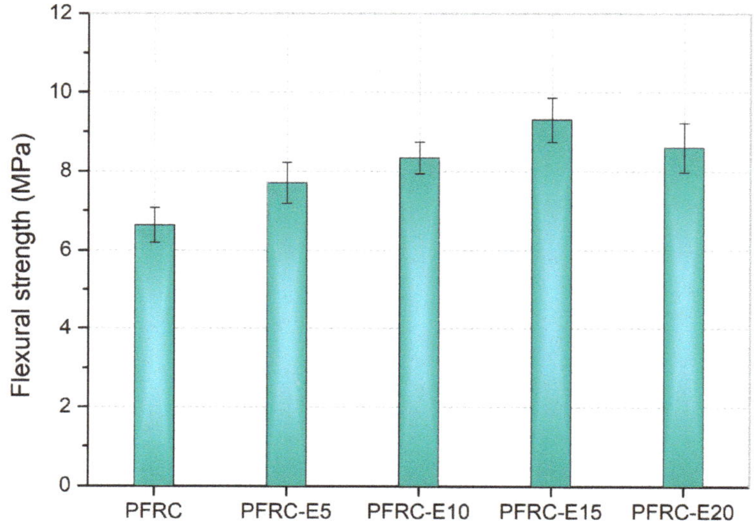

Figure 5. Flexural strength of PFRC with different contents of WBE.

3.1.3. Flexural Strength/Compressive Strength Ratio

To some extent, the ratio of flexural strength to compressive strength is a significant index that represents the toughness of the concrete; the higher the strength ratio, the larger the toughness of the concrete [36,39]. The flexural strength-to-compressive strength ratios of PFRC containing different levels of WBE are shown in Table 4. According to Table 4, the strength ratios of PFRCs increased with an increasing content of WBE until the WBE content exceeded 15%, indicating that the toughness of PFRC-E15 was the highest. When the WBE content was 20%, a subtle reduction was observed.

Table 4. Ratio of flexural strength to compressive strength.

Mixture Type	PFRC	PFRC-E5	PFRC-E10	PFRC-E15	PFRC-E20
Flexural strength/compressive strength	0.143	0.148	0.157	0.193	0.191

3.2. Fracture Properties

3.2.1. Load versus Deflection Response

The load versus deflection curves were obtained through the three-point bending test for PFRC with different dosages of WBE, as shown in Figure 6. These curves were plotted according to a method suggested by other researchers [40,41]. The deflection value was assumed to increase in 0.002 mm increments, and the load data were calculated on the basis of a linear interpolation of the measured data, which is to say that the load value was the mean value at the same deflection value.

It has long been established that the flexural and fracture properties of PFRC are influenced by three parameters, which are the properties of the matrix, the bonding properties of the interfaces between the fibers and their surrounding matrix and the properties of fibers. Among the three parameters, the properties of the matrix and the bonding properties of the interfaces between the fibers and matrix play a significant part [42]. From Figure 6, it can be observed that the load deflection curve of the control PFRC reached the peak load and then dropped suddenly. This brittle behavior likely resulted from the poor adhesion between PP fibers and their surrounding matrix. On the other hand, as the addition of WBE increased up to a 15% content, the peak loads, and deflections at the peak gradually increased. In contrast, for a WBE content beyond 15%, no further increase in the peak load was observed. Meanwhile, it is noteworthy that the slopes in the pre-peak region were increasingly smaller, and the curves descended relatively gently in the post-peak region with the increase in WBE content, which indicated that the deformability and ductility of PFRC with incorporated WBE were higher compared to PFRC without WBE. In particular, it was evident that the load-carrying capacity of PFRCs in the post-peak region showed an upward trend with the increase in WBE. This is because with the addition of WBE, the total porosity of the PFRC can be reduced, and the adhesion between PP fibers and their surrounding matrix can be enhanced as well through the filling effect of WBE to fill in the voids in concrete. As a result, the strength of matrix and the reinforcing effect of PP fibers can both be enhanced to obtain a higher load-carrying capacity. In addition, the formation of highly viscous and ductile epoxy films can reduce the defects and microcracks in concrete and improve the interfacial bonding properties so as to improve the ability to withstand external loads. Therefore, with the incorporation of WBE, the dense microstructure of the matrix and the improved interfacial bonding between PP fibers and the matrix yield a higher matrix strength and higher debonding resistance of PP fibers than for PFRC without WBE, eventually leading to higher peak load and higher deflection at the peak. Furthermore, with high addition of WBE, the fully developed epoxy film and hydrates may also interpenetrate to form a strong reinforcement so as to withstand higher flexural loads, which is in agreement with the findings of other researchers [34,38]. Ultimately, even if cracking occurs in the concrete matrix during the test, both the PP fibers and the epoxy film across the crack surfaces can still bear the loads and maintain a relatively higher postcracking strength. When the incorporation of WBE reaches 20%, the reduction in the peak load may be because the hydrates lose the continuity, but the deformability of PFRC-E20 was still the largest.

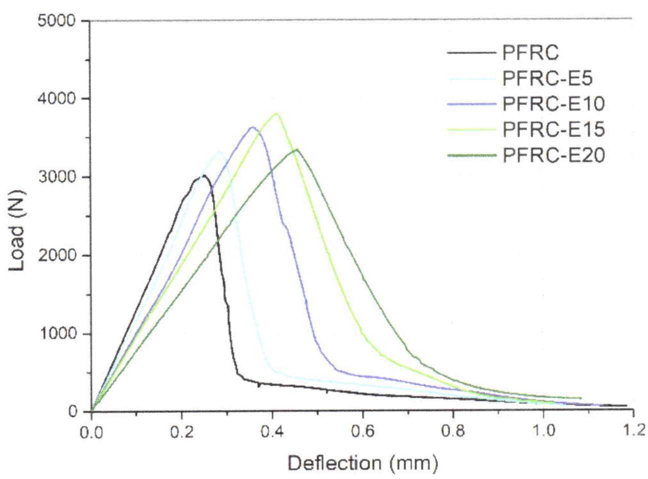

Figure 6. Load versus deflection curves of PFRC with different contents of WBE.

3.2.2. Fracture Energy

Fracture energy represents the energy required to create a crack of unit area in the material; thus, higher fracture energy indicates a higher capacity for cracking resistance.

Figure 7 presents the fracture energy of PFRC with different levels of WBE. The fracture energy (G_f) was calculated according to the following equation [43,44]:

$$G_f = \frac{W_0 + mg\delta}{A} \quad (1)$$

where W_0, m, g, δ and A denote the area under the load versus deflection curve (N·m), the mass of the concrete (kg), the gravity acceleration (N/kg), the deflection at the midspan of the beam (m) and the ligament area (m^2), respectively.

According to Equation (1), the fracture energy depends on the area under the load versus deflection curve to a large extent. It can be found from Figure 7 that the inclusion of WBE had a pronounced effect on the fracture energy of PFRC, and increasing the WBE content significantly increased the fracture energy of PFRC until the WBE content reached 15%, beyond which the improvement in fracture energy was subtle. Compared to the control PFRC, the improvements in fracture energy in PFRC-E5, PFRC-E10, PFRC-E15 and PFRC-E20 were 20.5%, 64.3%, 88.9% and 91.5%, respectively. The reason may be that with the addition of WBE, the interfacial bonding between PP fibers and their surrounding matrix can be strongly enhanced because the porosity of the interfaces can be reduced due to the pore-filling effect of WBE; therefore, more energy will be consumed for cracks to propagate and, accordingly, the propagation rate of the cracks will be reduced because fibers may cross the fracture surfaces to carry and transfer the load to the matrix [44]. In addition, the concrete matrix can be strengthened due to the addition of WBE to reduce the porosity in concrete, which also increases the energy consumption for the formation of cracks and crack propagation. Moreover, with a high dosage of WBE, the three-dimensional interpenetrating structure of epoxy film and cement hydrates could act as strong macro-reinforcements, and much more energy will be consumed for crack propagation because this interpenetrating organic–inorganic structure can retard the extension of tiny cracks and decrease the rigidity of concrete. Both the improved bonding between PP fibers and the concrete matrix and the interpenetrating organic–inorganic structure benefit from increasing the energy consumption for crack propagation because cracks in concrete can be bridged with PP fibers and the interpenetrating structure; thus, the coalescence of cracks will be postponed. As a result, PFRC-E20 exhibited the highest fracture energy among the five concretes. As presented in Figure 8, a regression analysis was performed to visually illustrate the relationship between fracture energy (G_f) and WBE content by weight of cement (W_f), and the fitting degree was 0.913 in the proposed relation.

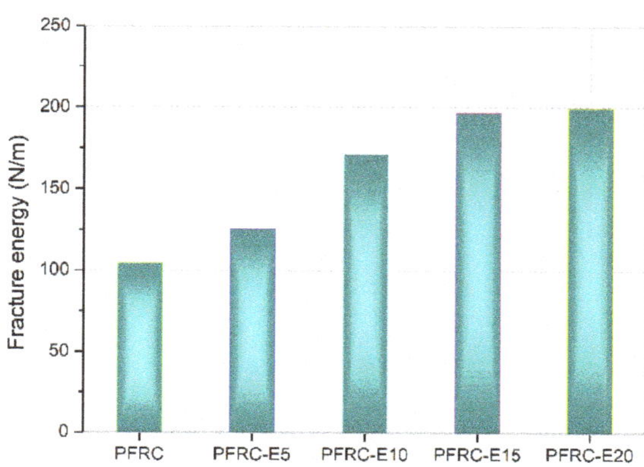

Figure 7. G_f of PFRC with different contents of WBE.

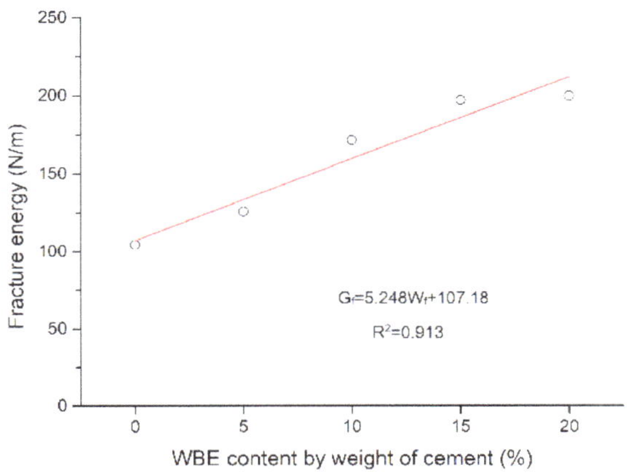

Figure 8. Proposed relation between G_f and W_f.

3.2.3. Fracture Toughness

The fracture toughness is determined using the critical stress intensity factor (K_{IC}) by using the following equation [45,46]:

$$K_{IC} = \frac{PL}{bh^{1.5}} F\left(\frac{a}{h}\right) \tag{2}$$

$$F\left(\frac{a}{h}\right) = \left[2.9\left(\frac{a}{h}\right)^{0.5} - 4.6\left(\frac{a}{h}\right)^{1.5} + 21.8\left(\frac{a}{h}\right)^{2.5} - 37.6\left(\frac{a}{h}\right)^{3.5} + 38.7\left(\frac{a}{h}\right)^{4.5}\right], \tag{3}$$

where P, L, b, h and a stand for the peak load (N), the clear span of the specimen (mm), the width of the specimen (mm), the height of the specimen (mm) and the effective crack length (mm), respectively.

As is well known, the critical stress intensity factor is a material property index at which the crack in the concrete begins to propagate; therefore, if the concrete material has a large critical stress intensity factor, it may present ductile fracture, and its ability to resist brittle fracture will be high [40,44]. The values of the critical stress intensity factors for PFRC with different dosages of WBE are presented in Figure 9, and it can be evidently observed from the results that with an increasing WBE content, the critical stress intensity factor showed an upward tendency compared to PFRC. The critical stress intensity factors of PFRC-E5, PFRC-E10, PFRC-E15 and PFRC-E20 were 22.3%, 43.5%, 62% and 71.7%, respectively, which were higher than that of the control PFRC, indicating that the ductility of PFRC was enhanced through the addition of WBE. This was because the inclusion of PP fibers probably creates some defects in the concrete, mainly in the interfaces of the PP fibers and their surrounding matrix, leading to a lower load-carrying capacity of the matrix, as well as a limited improvement in ductility via the PP fibers. However, with the addition of WBE, these defects may be improved through the physical filling of WBE; therefore, the reinforcing effect of the PP fibers was enhanced, and the ductility of PFRC was improved accordingly. In addition, the ductility of PFRC can be further improved when a high quantity of WBE is incorporated because the polymer film can be interlinked with hydrates to make the concrete much more ductile and to reduce the rigidity of the concrete [34,38]. Compared with the control PFRC, the higher capacity of PFRCs with WBE for resisting the initiation of cracks may be attributed to the dense microstructure of the concrete matrix, the effective toughening effect brought about by PP fibers and the interpenetrating organic–inorganic structure, all of which had the beneficial effect of inhibiting the appearance of cracks. In the present study, the relationship between the critical stress intensity factor and the WBE content was obtained. The relationship proposed in Figure 10 had a fitting degree of 0.976, showing that there was a strong correlation between the critical stress intensity factor (K_{IC}) and WBE content by weight of cement (W_f).

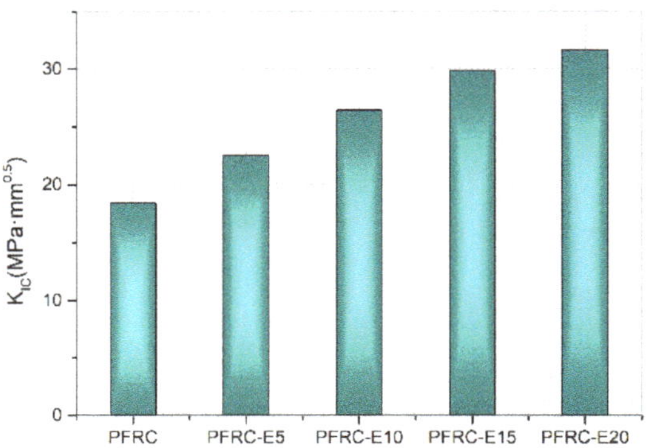

Figure 9. K_{IC} of PFRC with different contents of WBE.

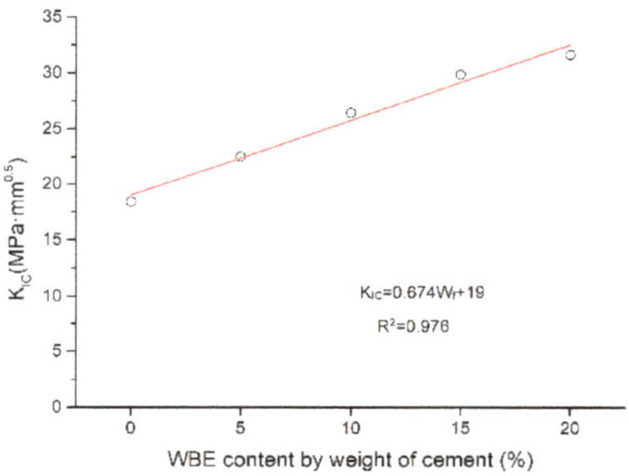

Figure 10. Proposed relation between K_{IC} and W_f.

3.3. Freeze–Thaw Resistance

In order to promote the application of PFRC with WBE in concrete pavement, the freeze–thaw resistance of PFRC with different dosages of WBE was evaluated in terms of mass loss rate and residual flexural strength.

3.3.1. Mass Loss Rate

Figure 11 exhibits the mass loss rate of PFRC with different contents of WBE subjected to 0, 25, 50, 75 and 100 freeze–thaw cycles. As shown in Figure 11, the mass loss rate of all the concrete mixtures exhibited an increase with the increase in freeze–thaw cycles. This is because the repeated freezing and thawing causes the internal water in the concrete to continuously freeze and melt. During this process, static pressure and penetration pressure are created in the surrounding concrete when the volume of water expands due to freezing. As a consequence, cracks may be induced when these pressures finally exceed the tensile strength of the concrete. With increasing freezing and thawing cycles, this damage will be more severe, and the concrete stripping off will arise [45]. Therefore, the freeze–thaw resistance of concrete could be improved by reducing the ingress of water.

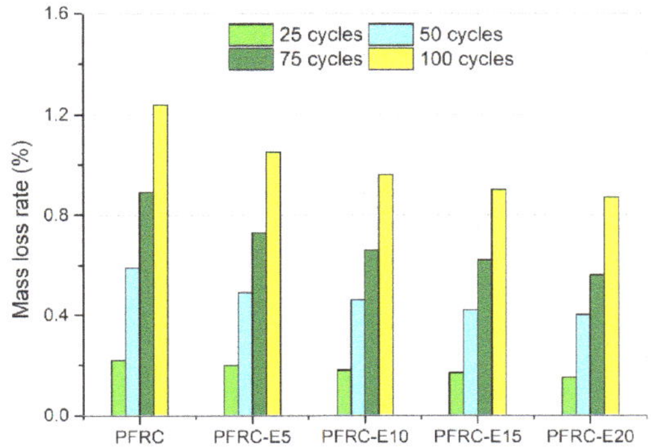

Figure 11. Mass loss rate of PFRC with different contents of WBE.

It can be found that among the five PFRCs, the control PFRC had the largest mass loss rate of approximately 1.2% after 100 cycles. The reason why the control PFRC had the largest mass loss rate may lie in the fact that, with the addition of PP fibers, the defects in PFRC might increase because the adhesion between PP fibers and their surrounding concrete matrix may be weak, leading to a relatively higher porosity in the PP fiber/matrix interfaces. Additionally, the inclusion of PP fibers may negatively affect the workability of concrete and result in a reduction in compactness. Therefore, stripping off of the control PFRC will be larger under the repeated freeze–thaw cycles. After the addition of WBE, the mass loss rate showed a decreased tendency. The mass loss rates of PFRC-E5, PFRC-E10, PFRC-E15 and PFRC-E20 after 100 cycles were 15.3%, 22.6%, 27.4% and 29.8% lower than that of the control PFRC, respectively. This phenomenon may be attributed to the fact that epoxy particles are capable of filling the voids and blocking the capillaries in PFRC to lower the porosity of PFRC and make PFRC denser; thus, water ingress can be reduced to improve the freeze–thaw resistance [26]. In addition, as the adhesion between PP fibers and their surrounding concrete matrix is improved, the PP fibers can assist in reducing the freeze–thaw damage because a large amount of discrete fibers may intertwine together to maintain the integrity of the concrete and effectively restrain cracking, bridging the cracks so as to reduce the peeling off of the concrete. Furthermore, due to the low modulus of the PP fibers compared to the surrounding concrete, PP fibers may yield before concrete under the hydrostatic pressure, thereby relieving the pressure and improving the freeze–thaw resistance [26]. Furthermore, the workability of PFRC can be improved through the "ball-bearing" effect of WBE to increase the compactness of PFRC, which also contributes to its freeze–thaw resistance [33,46]. On the other side, the impermeability and integrity of PFRC can be strengthened through the formation of the continuously interpenetrating structure of epoxy film and hydrates; therefore, the damage to freeze–thaw cycles can be reduced as well.

3.3.2. Residual Flexural Strength

The residual flexural strength of PFRC with different contents of WBE after 0, 25, 50, 75 and 100 freeze–thaw cycles is exhibited in Figure 12, and the residual flexural strength is determined as the ratio of flexural strength subjected to the specific freeze–thaw cycles to the flexural strength without freeze–thaw cycles.

As shown in Figure 12, with the increasing freeze–thaw cycles, the residual flexural strength of all the PFRCs exhibited a gradual decline. This is because that with the increase in freeze–thaw cycles, the damage in PFRC mixtures will be much more severe since more cracks will be formed, the width of cracks will be bigger and the stripping off of the concrete will be larger, leading to a loosely distributed microstructure and a smaller load bearing area to withstand lower flexural loads. For the control PFRC, the flexural strength was reduced by about 37.8% after 100 cycles, whereas for PFRCs with 5%, 10%,

15% and 20% WBE, the reductions in flexural strength were 32.5%, 29.2%, 26.1% and 24.9%, respectively. It can be obviously observed that the control PFRC had the lowest residual flexural strength after 100 freeze–thaw cycles, and this phenomenon is probably due to the relatively higher porosity in the PP fiber/matrix interfaces. Furthermore, under the freeze–thaw actions, damages may initiate from these interfaces. Moreover, the repeated freeze–thaw cycles may further impair the bonding in these interfaces, leading to rather higher porosity in these interfaces and even the debonding of PP fibers; therefore, the flexural strength of the control PFRC showed a notable decline. On the other hand, the incorporation of WBE can fill the pores in PFRC to make it denser, so as to improve the adhesion between PP fibers and matrix, inhibiting the ingress of water as well, thereby improving the freeze–thaw resistance of PFRC and increasing the residual flexural strength compared to the control PFRC. Furthermore, the interpenetrating structure of epoxy film and hydrates can significantly improve the flexural strength of PFRC and hold the integrity of PFRC under the action of freeze–thaw cycles, because the epoxy film can hinder the penetration of water.

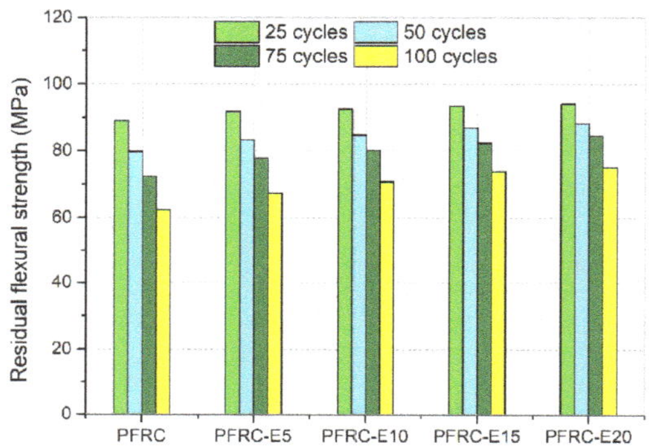

Figure 12. Residual flexural strength of PFRC with different contents of WBE.

3.4. SEM Analysis

According to the results of above experiments, it could be concluded that the addition of WBE improved the mechanical properties and freeze–thaw resistance of PFRC and the improvement was more significant with the increase in WBE content. However, the incorporation of 20% WBE decreased the compressive strength and the flexural property of PFRC. Therefore, from the viewpoint of engineering application, PFRC with 15% WBE was suggested as the optimum content, which synthetically considered the requirements of strength, toughness and durability. Figures 13 and 14 present the SEM images of the control PFRC and PFRC-E15 to reveal the reinforcing mechanism of the WBE in PFRC.

It can be seen from Figure 13a that PP fibers were scattered in the fractured surface of the control PFRC. In addition, there were some holes where PP fibers were pulled out under the external loads. Figure 13b,c show the intact surface of a PP fiber that remained smooth after fracture and the loose bonding between the PP fiber and the matrix. The three images indicated that the adhesion between PP fibers and the surrounding matrix was relatively weak, and some fibers were easily pulled out from the matrix under the loads; therefore, the bridging effect provided by PP fibers was not effective. This was consistent with our previous study in which the interfacial bonding between PP fibers and their surrounding matrix was relatively weak [19]. Meanwhile, Figure 13d presents the microstructural image of a concrete matrix in the control PFRC. It can be seen from Figure 13d that the cement hydrates, such as calcium silicate hydrate, calcium hydroxide and ettringite, were distributed in the concrete, while the structure was not very dense and there were some pores and cracks in the concrete. Therefore, the formation of the cracks and crack propagation may easily occur in the control PFRC due to the poor adhesion between PP fibers/concrete matrix and the less-dense microstructure of the matrix.

Figure 14 displays the microstructural images of PFRC-E15. In Figure 14a,b, it can be observed that the PP fibers were distorted, and some scratch marks were found on their surfaces. This degradation of PP fibers, which could have resulted from the friction during the loading process, confirmed the improved bonding between PP fibers and their surrounding matrix. In addition, some materials, probably hydrates and epoxy films still attached to the surfaces of PP fibers after fracture, which might be attributed to the cement–epoxy reaction to form the interpenetrating organic–inorganic structure [34], can also be observed in Figure 14a,c. The existence of the attachments also indicated the improved bonding between PP fibers and their surrounding matrix. This strong bonding could be attributed to the physical effect of WBE, filling the holes and cracks in concrete, and the chemical effect of WBE, absorbing calcium ions to form strong cross-linking structures [34]. With higher interfacial bonding, better flexural properties for PFRC can be achieved. The reason may be that when the fibers are tightly bonded with the surrounding concrete matrix, fibers can consume partial energy owing to the increased pull-out resistance; therefore, more energy will be required for the formation of cracks and the subsequent crack propagation. Moreover, Figure 14d presents the microstructural image of the matrix, from which it can be seen that the microstructure was dense, and the epoxy film wrapped on the surfaces of hydrates and bonded tightly with the hydrates. Therefore, additional energy will be needed for crack initiation and propagation in this matrix, because epoxy film has the larger deformability in contrast to concrete. This phenomenon also implied that with the combination of concrete and WBE as the matrix, PFRC had the superior performance compared to the only concrete as the matrix.

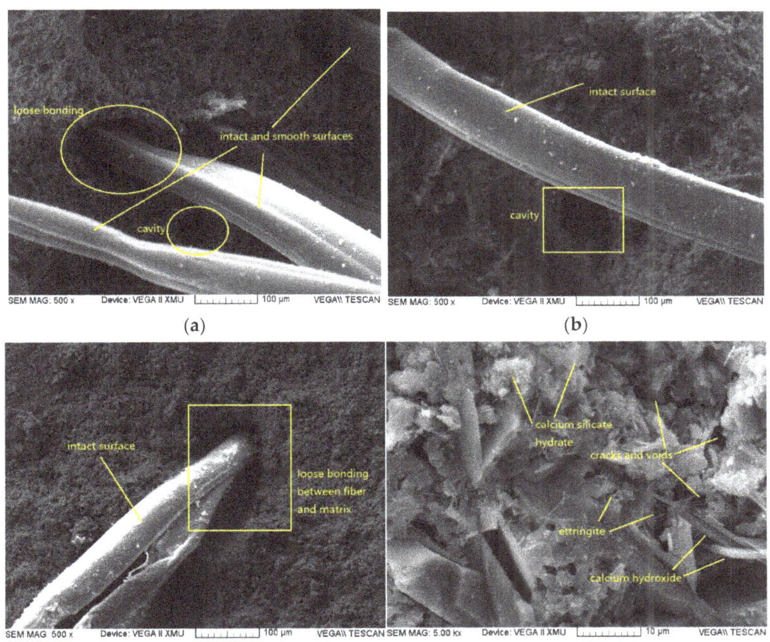

Figure 13. Microstructural images of PFRC by SEM. (**a**–**c**) PP fiber surfaces and interfacial bonding between PP fibers and the matrix, (**d**) the matrix.

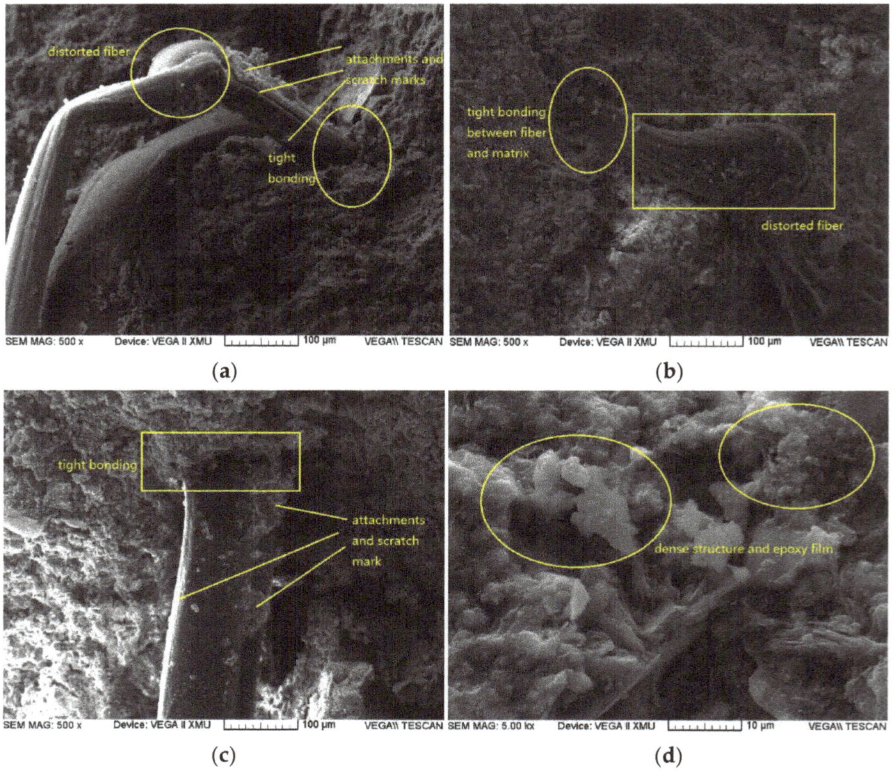

Figure 14. Microstructural images of PFRC-E15 by SEM. (**a**–**c**) PP fiber surfaces and interfacial bonding between PP fibers and the matrix, (**d**) the matrix.

4. Cost Analysis

Compared to ordinary concrete, the material cost of PFRCs has increased due to the existence of admixtures such as PP fibers and WBE. Therefore, the material cost per cubic meter of PFRC can be determined by:

$$C = C_0 + C_{pp} \times I_{pp} + C_{WBE} \times I_{WBE}, \tag{4}$$

where C and C_0 stand for the cost of PFRCs and the ordinary concrete per cubic meter, respectively; C_{pp} and C_{WBE} stand for the cost of PP fibers and WBE per unit kg, respectively; and I_{pp} and I_{WBE} stand for the content of PP fibers and WBE in PFRCs (kg), respectively.

According to the market survey, the price of PP fibers and WBE is 8.2 CNY/kg and 48 CNY/kg, respectively. Thus, compared to ordinary concrete, the material-cost increases per cubic meter for PFRC, PFRC-E5, PFRC-E10, PFRC-E15 and PFRC-E20 are CNY 7, CNY 1927, CNY 3847, CNY 5767 and CNY 7687, respectively. Considering that the thickness of the concrete slab is often around 25 cm, the material-cost increases per square meter for those five concretes are approximately CNY 1.8, CNY 482, CNY 962, CNY 1442 and CNY 1922, respectively. Although the material cost increases, it cannot reflect the total cost throughout the pavement life because the proportion of the construction, the management and the maintenance costs account for the larger part of the total cost. In addition, test results have indicated that the mechanical and durability properties were significantly enhanced by incorporating WBE; therefore, it can be reasonably expected that the service life of PFRCs can be extended, and the total cost can be lowered, which still needs to be carefully evaluated in a future study.

5. Conclusions and Future Work

According to the results obtained in this study, the following conclusions can be drawn:

1. The compressive strength of PFRC exhibited a steady increase with the content of WBE, reaching up to 10%, and then decreased gradually. The addition of 20% WBE even resulted in an inferior compressive strength compared to the control PFRC.
2. With an increasing content of WBE, the flexural strength of PFRC showed significant improvement until the WBE was beyond 15%, at which point a slight reduction in flexural strength was observed.
3. The dosage of WBE had a pronounced effect on the load versus deflection responses of PFRC. The deformability of PFRC was remarkably enhanced as the content of WBE increased, and the largest deformation was achieved in PFRC-E20, whereas the load-carrying capacity of PFRC-E20 decreased. Moreover, as the content of WBE increased, the fracture energy and fracture toughness of PFRC were increasingly improved.
4. The freeze–thaw resistance of PFRCs was notably enhanced through the addition of WBE. With the increasing content of WBE, the mass loss rate decreased, and the residual flexural strength showed a significant increase compared to the control PFRC. When the WBE content exceeded 15%, the improvement in the freeze–thaw resistance slowed down.
5. In practical pavement engineering, it is necessary to maximize the flexural properties of the concrete and ensure the stiffness and the durability of the concrete. Meanwhile, the cost effectiveness should be considered. Therefore, in this research, the WBE content of 15% was recommended.
6. The distorted PP fibers after fracture and the scratch marks on the surfaces of the PP fibers indicated that the bonding between PP fibers and their surrounding concrete matrix was effectively strengthened, and PP fibers effectively consumed part of the energy during the pull-out process. The dense microstructure and the attachments on the PP fiber surfaces also represented a stronger matrix to resist the initiation and propagation of cracks.

The present study mainly investigated the fracture properties and the freeze–thaw resistance of PFRCs. However, the tests were limited to one type of micro-PP fibers, while the combined use of different sizes of PP fibers may have better impact on the characteristics of concrete. In addition, the curing method can affect the polymerization process of polymer-modified concrete as well. Therefore, more experiments should be performed to obtain an in-depth understanding of the characteristics of PFRCs. In this regard, the influence of the combined use of micro-PP fibers and macro-PP fibers on the characteristics of WBE-modified concrete, and the effect of curing method on the characteristics of PFRCs, should be further researched concerning mechanical and durability properties. Furthermore, the anti-impact and fatigue properties of PFRCs should also be studied to better evaluate the performance of PFRCs.

Author Contributions: Conceptualization, H.Z., B.Z. (Bing Zhang) and L.W.; methodology, L.W. and Q.Z.; validation, H.Z. and Q.Z.; investigation, L.W., B.Z. (Bing Zhang) and B.Z. (Bendong Zhao); data curation, B.Z. (Bing Zhang) and B.W.; writing—original draft preparation, L.W.; writing—review and editing, M.S.; supervision, H.Z.; project administration, B.Z. (Bing Zhang) and B.W. All authors have read and agreed to the published version of the manuscript.

Funding: This research was supported by the Shandong Jianzhu University Doctoral Fund (No. X20081Z01).

Institutional Review Board Statement: Not applicable.

Informed Consent Statement: Not applicable.

Data Availability Statement: The data used to support this study are available within this article.

Acknowledgments: We would like to acknowledge numerous associated coworkers for providing specialized assistance with instrument examination.

Conflicts of Interest: The authors declare no conflict of interest.

References

1. Pujadas, P.; Cavalaro, S.H.P.; Aguado, A. Mives multicriteria assessment of urban-pavement conditions: Application to a case study in Barcelona. *Road Mater. Pavement Des.* **2019**, *20*, 1827–1843. [CrossRef]
2. Zhang, Z.; Yang, J.; Fang, Y.; Luo, Y. Design and performance of waterborne epoxy-SBR asphalt emulsion (WESE) slurry seal as under-seal coat in rigid pavement. *Constr. Build. Mater.* **2021**, *270*, 121467. [CrossRef]
3. Park, J.J.; Yoo, D.Y.; Park, G.J.; Kim, S.W. Feasibility of reducing the fiber content in ultra-high-performance fiber-reinforced concrete under flexure. *Materials* **2017**, *10*, 118. [CrossRef] [PubMed]
4. Quan, W.; Huang, W.; An, Y.; Miao, X.; Chen, Z. The effect of natural bamboo fiber and basalt fiber on the properties of autoclaved aerated concrete. *Constr. Build. Mater.* **2023**, *377*, 131153. [CrossRef]
5. Shen, D.; Liu, X.; Zeng, X.; Zhao, X.; Jiang, G. Effect of polypropylene plastic fibers length on cracking resistance of high performance concrete at early age. *Constr. Build. Mater.* **2020**, *244*, 117874. [CrossRef]
6. Li, D.; Liu, S. Macro polypropylene fiber influences on crack geometry and water permeability of concrete. *Constr. Build. Mater.* **2020**, *231*, 117128. [CrossRef]
7. Wang, J.; Dai, Q.; Si, R.; Guo, S. Mechanical, durability, and microstructural properties of macro synthetic polypropylene (PP) fiber-reinforced rubber concrete. *J. Clean. Prod.* **2019**, *234*, 1351–1364. [CrossRef]
8. Qin, Y.; Zhang, X.; Chai, J.; Xu, Z.; Li, S. Experimental study of compressive behavior of polypropylene-fiber-reinforced and polypropylene-fiber-fabric-reinforced concrete. *Constr. Build. Mater.* **2019**, *194*, 216–225. [CrossRef]
9. Akça, K.R.; Çakır, Ö.; İpek, M. Properties of polypropylene fiber reinforced concrete using recycled aggregates. *Constr. Build. Mater.* **2015**, *98*, 620–630. [CrossRef]
10. Das, C.S.; Dey, T.; Dandapat, R.; Mukharjee, B.B.; Kumar, J. Performance evaluation of polypropylene fibre reinforced recycled aggregate concrete. *Constr. Build. Mater.* **2018**, *189*, 649–659. [CrossRef]
11. Li, H.; Zhang, M.H.; Ou, J.P. Flexural fatigue performance of concrete containing nano-particles for pavement. *Int. J. Fatigue* **2007**, *29*, 1292–1301. [CrossRef]
12. Wang, D.; Ju, Y.; Shen, H.; Xu, L. Mechanical properties of high performance concrete reinforced with basalt fiber and polypropylene fiber. *Constr. Build. Mater.* **2019**, *197*, 464–473. [CrossRef]
13. Islam, G.S.; Gupta, S.D. Evaluating plastic shrinkage and permeability of polypropylene fiber reinforced concrete. *Int. J. Sustain. Built Environ.* **2016**, *5*, 345–354. [CrossRef]
14. Del Savio, A.A.; Esquivel, D.L.T.; de Andrade Silva, F.; Agreda Pastor, J. Influence of Synthetic Fibers on the Flexural Properties of Concrete: Prediction of Toughness as a Function of Volume, Slenderness Ratio and Elastic Modulus of Fibers. *Polymers* **2023**, *15*, 909. [CrossRef]
15. Gupta, S.; Kua, H.W.; Pang, S.D. Combination of polypropylene fibre and superabsorbent polymer to improve physical properties of cement mortar. *Mag. Concr. Res.* **2018**, *70*, 350–364. [CrossRef]
16. Hao, Y.; Cheng, L.; Hao, H.; Shahin, M.A. Enhancing fiber/matrix bonding in polypropylene fiber reinforced cementitious composites by microbially induced calcite precipitation pre-treatment. *Cem. Concr. Compos.* **2018**, *88*, 1–7. [CrossRef]
17. Akand, L.; Yang, M.; Wang, X. Effectiveness of chemical treatment on polypropylene fibers as reinforcement in pervious concrete. *Constr. Build. Mater.* **2018**, *163*, 32–39. [CrossRef]
18. Han, J.W.; Jeon, J.H.; Park, C.G. Bond characteristics of macro polypropylene fiber in cementitious composites containing nanosilica and styrene butadiene latex polymer. *Int. J. Polym. Sci.* **2015**, *2015*, 207456. [CrossRef]
19. Wang, L.; Zhang, H.; Zhao, B.; Wang, B.; Zhao, Q.; Sun, M. Experimental Investigation on Physical Properties of Concrete Containing Polypropylene Fiber and Water-Borne Epoxy for Pavement. *Coatings* **2023**, *13*, 452. [CrossRef]
20. Smarzewski, P. Influence of basalt-polypropylene fibres on fracture properties of high performance concrete. *Compos. Struct.* **2019**, *209*, 23–33. [CrossRef]
21. Liang, N.; Dai, J.; Liu, X.; Zhong, Z. Experimental study on the fracture toughness of concrete reinforced with multi-size polypropylene fibres. *Mag. Concr. Res.* **2019**, *71*, 468–475. [CrossRef]
22. Chen, Y.; Cen, G.; Cui, Y. Comparative study on the effect of synthetic fiber on the preparation and durability of airport pavement concrete. *Constr. Build. Mater.* **2018**, *184*, 34–44. [CrossRef]
23. Nam, J.; Kim, G.; Lee, B.; Hasegawa, R.; Hama, Y. Frost resistance of polyvinyl alcohol fiber and polypropylene fiber reinforced cementitious composites under freeze thaw cycling. *Compos. Part B Eng.* **2016**, *90*, 241–250. [CrossRef]
24. Richardson, A.E.; Coventry, K.A.; Wilkinson, S. Freeze/thaw durability of concrete with synthetic fibre additions. *Cold Reg. Sci. Technol.* **2012**, *83*, 49–56. [CrossRef]
25. Zhang, P.; Li, Q.F. Effect of polypropylene fiber on durability of concrete composite containing fly ash and silica fume. *Compos. Part B Eng.* **2013**, *45*, 1587–1594. [CrossRef]

26. Karahan, O.; Atiş, C.D. The durability properties of polypropylene fiber reinforced fly ash concrete. *Mater. Des.* **2011**, *32*, 1044–1049. [CrossRef]
27. Abadel, A.A.; Alghamdi, H. Effect of high volume tile ceramic wastes on resistance of geopolymer mortars to abrasion and freezing-thawing cycles: Experimental and deep learning modelling. *Ceram. Int.* 2023, *in press*. [CrossRef]
28. *JTG/T F30-2014*; Technical Guidelines for Construction of Highway Cement Concrete Pavements. Ministry of Transport of the People's Republic of China: Beijing, China, 2014.
29. *JTG E30-2005*; Test Methods of Cement and Concrete for Highway Engineering. Ministry of Transport of the People's Republic of China: Beijing, China, 2005.
30. *JCI-S-002-2003*; Method of Test for Load-Displacement Curve of Fiber Reinforced Concrete by Use of Notched Beam. Japan Concrete Institute Standard: Tokyo, Japan, 2003.
31. Anagnostopoulos, C.A.; Sapidis, G.; Papastergiadis, E. Fundamental properties of epoxy resin-modified cement grouts. *Constr. Build. Mater.* **2016**, *125*, 184–195. [CrossRef]
32. Moodi, F.; Kashi, A.; Ramezanianpour, A.A.; Pourebrahimi, M. Investigation on mechanical and durability properties of polymer and latex-modified concretes. *Constr. Build. Mater.* **2018**, *191*, 145–154. [CrossRef]
33. Eren, F.; Gödek, E.; Keskinateş, M.; Tosun-Felekoğlu, K.; Felekoğlu, B. Effects of latex modification on fresh state consistency, short term strength and long term transport properties of cement mortars. *Constr. Build. Mater.* **2017**, *133*, 226–233. [CrossRef]
34. Pang, B.; Zhang, Y.; Liu, G. Study on the effect of waterborne epoxy resins on the performance and microstructure of cement paste. *Constr. Build. Mater.* **2018**, *167*, 831–845. [CrossRef]
35. Peng, Y.; Zeng, Q.; Xu, S.; Zhao, G.; Wang, P.; Liu, X. BSE-IA reveals retardation mechanisms of polymer powders on cement hydration. *J. Am. Ceram. Soc.* **2020**, *103*, 3373–3389. [CrossRef]
36. Ma, H.; Li, Z. Microstructures and mechanical properties of polymer modified mortars under distinct mechanisms. *Constr. Build. Mater.* **2013**, *47*, 579–587. [CrossRef]
37. Li, L.; Wang, R.; Lu, Q. Influence of polymer latex on the setting time, mechanical properties and durability of calcium sulfoaluminate cement mortar. *Constr. Build. Mater.* **2018**, *169*, 911–922. [CrossRef]
38. Wang, M.; Wang, R.; Yao, H.; Farhan, S.; Zheng, S.; Wang, Z.; Du, C.; Jiang, H. Research on the mechanism of polymer latex modified cement. *Constr. Build. Mater.* **2016**, *111*, 710–718. [CrossRef]
39. Liu, J.; Chen, H.; Guan, B.; Liu, K.; Wen, J.; Sun, Z. Influence of mineral nano-fibers on the physical properties of road cement concrete material. *Constr. Build. Mater.* **2018**, *190*, 287–293. [CrossRef]
40. Kizilkanat, A.B.; Kabay, N.; Akyüncü, V.; Chowdhury, S.; Akça, A.H. Mechanical properties and fracture behavior of basalt and glass fiber reinforced concrete: An experimental study. *Constr. Build. Mater.* **2015**, *100*, 218–224. [CrossRef]
41. Yoo, D.Y.; Zi, G.; Kang, S.; Yoon, S. Biaxial flexural behavior of ultra-high-performance fiber-reinforced concrete with different fiber lengths and placement methods. *Cem. Concur. Compos.* **2015**, *63*, 51–66. [CrossRef]
42. Li, Y.; Li, W.; Deng, D.; Wang, K.; Duan, W.H. Reinforcement effects of polyvinyl alcohol and polypropylene fibers on flexural behaviors of sulfoaluminate cement matrices. *Cem. Concr. Compos.* **2018**, *88*, 139–149. [CrossRef]
43. Rilem, D.R. Determination of the fracture energy of mortar and concrete by means of three-point bend tests on notched beames. *Mater. Struct.* **1985**, *18*, 285–290.
44. Kazemi, M.T.; Golsorkhtabar, H.; Beygi, M.H.A.; Gholamitabar, M. Fracture properties of steel fiber reinforced high strength concrete using work of fracture and size effect methods. *Constr. Build. Mater.* **2017**, *142*, 482–489. [CrossRef]
45. Xu, F.; Li, T.; Li, C.; Li, Z.; Wang, S.; Zhao, N. Compressive Behavior, Microstructural Properties, and Freeze–Thaw Behavior of Tailing Recycled Aggregate Concrete with Waste Polypropylene Fiber Addition. *Materials* **2021**, *14*, 6712. [CrossRef] [PubMed]
46. Giustozzi, F. Polymer-modified pervious concrete for durable and sustainable transportation infrastructures. *Constr. Build. Mater.* **2016**, *111*, 502–512. [CrossRef]

coatings

MDPI

Article

Structural Fault Diagnosis Based on Static and Dynamic Response Parameters

Qiuwei Yang [1,2], Fengjiang Qin [3,*] and Xi Peng [1,2]

1 School of Civil and Transportation Engineering, Ningbo University of Technology, Ningbo 315211, China; yangqiuwei@nbut.edu.cn (Q.Y.); pengxi@nbut.edu.cn (X.P.)
2 Engineering Research Center of Industrial Construction in Civil Engineering of Zhejiang, Ningbo University of Technology, Ningbo 315211, China
3 Key Laboratory of New Technology for Construction of Cities in Mountain Area, Ministry of Education, School of Civil Engineering, Chongqing University, Chongqing 400045, China
* Correspondence: qinfengjiang@cqu.edu.cn

Abstract: Structural fault diagnosis is an important subject for ensuring the normal use of structures. More test data will help to improve the accuracy and reliability of structural fault diagnosis. Therefore, a structural fault detection algorithm based on static–dynamic mixed sensitivity analysis is proposed. The vibration parameters used were the vibration modes of some of the nodes in the structure measured by the vibration test system. The static response parameter used was the vertical displacement of the structure under the gravity load measured by the static test system. In particular, the gravity load and the structure were connected rigidly to form a new added-mass system. The vibration mode of the additional-mass system was measured again to obtain more equations for fault evaluation. Based on the static and dynamic measurement data, the failure coefficients of all components in the structure were calculated through the mixed sensitivity of the static displacement and vibration-mode shape. According to the calculated value of the failure coefficient, the failure state of all components in the structure could be finally evaluated. The main innovation of the proposed method was the use of the static load as a part of the new added-mass system to obtain more vibration parameters for the defect diagnosis. The implementation process and effect of this method were verified using a numerical truss structure and an experimental steel beam structure. Moreover, the defect diagnosis results of the proposed hybrid method were compared with those of a pure static algorithm and a pure dynamic algorithm to illustrate the advantages of the hybrid method. The research results showed that this method has the advantages of simple implementation and high diagnosis accuracy. Especially for symmetric structures, the proposed method can successfully avoid the possible missed diagnoses of the pure static algorithm and pure dynamic method. The algorithm provides a simple and feasible method for structural defect identification.

Keywords: fault diagnosis; static–dynamic test; hybrid sensitivity; symmetric structures; steel roof truss structure

1. Introduction

A structure will inevitably have faults due to the influence of environmental corrosion, material fatigue, disaster load, and so on. A fault in a structure will lead to a change in the structure's response parameters under a static or dynamic load [1–3]. In practice, the static displacement of a structure can be measured through an optical system, and the vibration modal data of a structure can be tested by a vibration test system. Therefore, the fault condition of a structure can be evaluated according to a change in the static or dynamic parameters. In recent decades, a lot of research has been carried out on structural fault diagnosis. These fault diagnosis methods can generally be divided into static methods and dynamic methods. Static methods use the change in the static displacement or strain of a structure to evaluate the state of the structural faults. Using flexibility disassembly perturbation, Yang and Sun [4–6] proposed a fast analysis method for the static displacement for structural damage detection. Using the static strain change, Yang et al. [7] developed a model-free method for damage localization for a grid structure. Bakhtiari-Nejad et al. [8] developed non-linear equations that relate static response variations to fault location and extent. Abdo [9] established a relationship between the structural fault characteristics and displacement curvature variations. Tian et al. [10] carried out fault diagnosis on a wind turbine blade using the static strain responses of the fiber Bragg grating. Using static testing data, Santos et al. [11] detected faults in stay cables using statistical and machine-learning methods. Viola and Bocchini [12] developed a fault diagnosis technique for truss structures based on static load tests. Using structural optimization concepts, Terlaje et al. [13] proposed a parameter identification method to accurately assess the stiffness in linear elastic models of civil structures. Boumechra [14] used a bridge's deflection under a moving load to determine the possible faults in the bridge's structure. Based on static displacement, Peng et al. [15] developed a method for determining the damage location in a beam's structure by using the redistribution of static shear energy. Chrysanidis [16] studied the main factors that affect the cracking of reinforced concrete columns and walls, such as load, tensile strain, etc. Based on the experimental results, the ranges, sizes, and locations of cracks in these components were summarized. The above research has shown the potential of static methods in diagnosing faults in beam-type structures. Dynamic methods use changes in structural vibration parameters, especially modal data, to evaluate the state of structural faults. Lu and Law [17] studied the sensitivity of dynamic parameters and proposed a sensitivity-based algorithm to detect faults in a structure. Feng [18] reported the applications of computer visions in structural modal parameter identification, model updating, fault diagnosis, and cable force estimation. Gentile et al. [19] proposed a dynamic-based technique for the continuous health monitoring of the tallest historic tower in Mantua, Italy. Almeida et al. [20] developed a fault diagnosis approach by considering both the time and frequency responses of structural dynamic responses. Yang et al. [21,22] studied a dynamic model reduction and carried out fault diagnosis using modal sensitivity based on a reduced model. Peng et al. [23] developed a dynamic flexibility method for fault diagnosis in steel truss structures. Weng et al. [24] reviewed dynamic sub-structuring methods for model correction and fault diagnosis in large-scale structures. Meixedo et al. [25] developed an unsupervised automatic data-driven methodology for detecting damage in railway bridges based on traffic-induced dynamic responses. These studies have shown the advantages of dynamic methods in damage diagnosis, without affecting the normal use of a structure.

Although significant progress has been made in fault diagnosis, there are still some challenges that need to be further addressed. The disadvantage of pure static methods is that some members in a structure may not be sensitive to vertical loading. Thus, it is difficult to diagnose the defect condition of these members using a pure static method since conventional static loading is carried out by gravity. For pure dynamic methods, a similar limitation exists since the faults in some components of a structure may not cause significant changes in the structural vibration parameters. Especially for symmetric structures, pure dynamic methods may lead to the misjudgment of defects because the damages in the symmetrical components may lead to similar changes in the vibration parameters. In view of this, a fault diagnosis technique based on static and dynamic hybrid measurements is more reliable in engineering applications since the missed diagnosis problem for symmetrical structures can be overcome by asymmetric gravity loading. For this purpose, a combination of static displacement and mode-shape sensitivities are employed in this work to calculate the damage parameters of all the components of a structure. The fault state of each component in a structure can be subsequently evaluated according to the calculation results. In addition, more test data will help to improve the accuracy and reliability of structural fault diagnosis. As a by-product of static loading, the gravity load and the structure can be connected rigidly to form a new added-mass system. The vibration mode of the added-mass system can be measured again so that more equations can be obtained for the fault evaluation. Combined with all static and dynamic test data, the proposed hybrid sensitivity method can obtain more accurate and reliable defect diagnosis results. The greatest innovation of the proposed method its use of static loading as a new added-mass system to obtain more vibration parameters for defect diagnosis. A steel roof truss structure was used to illustrate the implementation process and feasibility of this hybrid sensitivity method. It was shown that the proposed method can successfully avoid the possible missed diagnoses of pure static algorithms and pure dynamic methods. The results showed the potential application value of the proposed method in engineering applications. The framework of this article is as follows: Section 2 provides a detailed explanation of the calculation formula and operation process of the proposed hybrid sensitivity method. In Section 3, the proposed method is validated using a numerical model of a truss structure. In Section 4, an experimental beam is used to further verify the proposed approach. Section 5 presents the main conclusions of this work.

2. The Hybrid Sensitivity Method for Structural Fault Diagnosis

As stated before, some component faults in a structure are only sensitive to gravity loading and some component faults are only sensitive to structural vibration. Therefore, using a hybrid algorithm based on static and dynamic measurements to diagnose faults in a structure will be a more reliable method. To this end, a hybrid sensitivity method based on static and dynamic test data was developed to evaluate the fault states in a structure. The main formulas of this method were derived as follows.

2.1. Static Displacement Sensitivity

Static displacement is the most commonly used response parameter for structural defect diagnosis. In practice, static displacement can be measured using devices such as a dial gauge, a wire displacement gauge, or an optical system. The static displacement sensitivity formulas were derived to establish the relationship between the structural defect parameters and the static displacement changes. For a structure with and without faults, the static displacement under the gravity load vector f can be obtained from the finite element model (FEM) by:

$$S_u d_u = f, \tag{1}$$

$$S_d d_d = f, \tag{2}$$

$$S_d = S_u - \Delta S, \text{ and} \tag{3}$$

$$d_d = d_u + \Delta d, \tag{4}$$

where S_u is the stiffness matrix of the intact structure, d_u is the displacement vector under the load f, S_d is the stiffness matrix of the structure with faults, d_d is the corresponding displacement vector, and ΔS and Δd are the variations in the stiffness matrix and the displacement due to the faults. From Equations (1)–(4), we can obtain

$$S_u d_u + S_u \Delta d - \Delta S d_u - \Delta S \Delta d = f. \tag{5}$$

Equation (5) can be approximately rewritten as

$$S_u \Delta d = \Delta S d_u. \tag{6}$$

Using Equations (1) and (6), we can obtain

$$S_u \Delta d = \Delta S S_u^{-1} f. \tag{7}$$

From Equation (7), the displacement variation Δd can be expressed as

$$\Delta d = S_u^{-1} \Delta S S_u^{-1} f. \quad (8)$$

According to FEM theory, the variation ΔS in the stiffness matrix can be written as the sum of the elementary stiffness matrix multiplied by a fault coefficient, that is:

$$\Delta S = \sum_{i=1}^{N} \varepsilon_i S_i, \ (\varepsilon_i \subset [0,1]), \quad (9)$$

where S_i and ε_i are the i-th elementary stiffness matrix and the fault coefficient in the structure and N is the number of elements in the structural FEM. Using Equation (9), Equation (8) can be further expressed as

$$\Delta d = \sum_{i=1}^{N} \varepsilon_i \xi_i \text{ and} \quad (10)$$

$$\xi_i = S_u^{-1} S_i S_u^{-1} f, \quad (11)$$

where ξ_i is called as the displacement sensitivity vector of the i-th element.

2.2. Vibration Mode Sensitivity

Vibration modes are the most commonly used dynamic parameters for structural defect diagnosis. In practice, vibration modes can be measured using accelerometers and spectrum analysis software. The vibration mode sensitivity formulas were derived to establish the relationship between the structural defect parameters and the vibration mode changes. It is known that the free vibration modes of an intact structure can be obtained by solving the following generalized eigenvalue equations:

$$(S_u - \lambda_r M)\varphi_r = 0 \text{ and} \quad (12)$$

$$\varphi_r^T M \varphi_r = 1, \quad (13)$$

where λ_r and φ_r are the r-th eigenvalue and mode shape of structural free vibration, respectively, and M is the mass matrix of the structural FEM. Generally, M remains unchanged for a structure before and after damage. By taking the partial derivative of Equation (12) with respect to ε_i, we can obtain

$$(S_u - \lambda_r M)\frac{\partial \varphi_r}{\partial \varepsilon_i} = (\frac{\partial \lambda_r}{\partial \varepsilon_i} M - S_i)\varphi_r, \quad (14)$$

where $\frac{\partial \lambda_r}{\partial \varepsilon_i}$ and $\frac{\partial \varphi_r}{\partial \varepsilon_i}$ are the first-order eigenvalue and mode shape sensitivities, respectively. According to the FEM theory, S_u and M are both symmetric matrices. From Equation (12), we can also obtain

$$\varphi_r^T (S_u - \lambda_r M) = 0. \quad (15)$$

Using Equations (14) and (15), the eigenvalue sensitivity can be obtained by multiplying Equation (14) by φ_r^T, as follows:

$$\frac{\partial \lambda_r}{\partial \varepsilon_i} = \varphi_r^T S_i \varphi_r. \quad (16)$$

The mode shape sensitivity $\frac{\partial \varphi_r}{\partial \varepsilon_i}$ can be calculated using the formula proposed by Yang et al. [26], as follows:

$$\frac{\partial \varphi_r}{\partial \varepsilon_i} = \Theta^{-1} \Omega \varphi_r, \quad (17)$$

$$\Omega = \frac{\partial \lambda_r}{\partial \varepsilon_i} M - S_i, \text{ and} \tag{18}$$

$$\Theta = S_u - \lambda_r M + \lambda_r \varphi_r \varphi_r^T M. \tag{19}$$

If we assume that $\varphi_{d,r}$ is the measured mode shape of a damaged structure, due to the faults in the structure, the mode shape variation $\Delta\varphi_r$ can be obtained using

$$\Delta\varphi_r = \varphi_{d,r} - \varphi_r. \tag{20}$$

On the other hand, $\Delta\varphi_r$ can be approximated using Taylor's series expansion, as follows:

$$\Delta\varphi_r = \sum_{i=1}^{N} \varepsilon_i \frac{\partial \varphi_r}{\partial \varepsilon_i} \tag{21}$$

2.3. The Hybrid Sensitivity

As stated before, more test data will help to improve the accuracy and reliability of structural fault diagnosis. As a by-product of static loading, the gravity load and the structure can be connected rigidly to form a new added-mass system. We can measure the vibration mode of the added-mass system again so that more modal equations can be obtained for the fault evaluation, if necessary. Similar to Equation (21), the mode shape sensitivity equation of the additional-mass system with the gravity load is

$$\Delta\varphi_r^a = \sum_{i=1}^{N} \varepsilon_i \frac{\partial \varphi_r^a}{\partial \varepsilon_i}, \tag{22}$$

where $\Delta\varphi_r^a$ denotes the change of the r-th mode shape of the additional-mass system with the gravity load. The mode shape sensitivity $\frac{\partial \varphi_r^a}{\partial \varepsilon_i}$ of the added-mass system can be calculated by

$$\frac{\partial \varphi_r^a}{\partial \varepsilon_i} = \Theta_a^{-1} \Omega_a \varphi_r^a, \tag{23}$$

$$\Omega_a = \frac{\partial \lambda_r^a}{\partial \varepsilon_i} M_a - S_i, \tag{24}$$

$$\Theta_a = S_u - \lambda_r^a M_a + \lambda_r^a \varphi_r^a (\varphi_r^a)^T M_a, \tag{25}$$

$$\frac{\partial \lambda_r^a}{\partial \varepsilon_i} = (\varphi_r^a)^T S_i \varphi_r^a, \text{ and} \tag{26}$$

$$M_a = M + D(f), \tag{27}$$

where $D(f)$ represents a diagonal matrix whose diagonal element is the mass corresponding to the load vector f. Finally, Equations (10), (21), and (22) can be combined to construct the hybrid sensitivity system, as follows:

$$\Delta p = \Pi \cdot \alpha, \tag{28}$$

$$\Delta p = \begin{Bmatrix} \Delta d \\ \Delta\varphi_r \\ \Delta\varphi_r^a \end{Bmatrix}, \tag{29}$$

$$\Pi = \begin{bmatrix} \xi_1 & \cdots & \xi_N \\ \frac{\partial \varphi_r}{\partial \varepsilon_1} & \cdots & \frac{\partial \varphi_r}{\partial \varepsilon_N} \\ \frac{\partial \varphi_r^a}{\partial \varepsilon_1} & \cdots & \frac{\partial \varphi_r^a}{\partial \varepsilon_N} \end{bmatrix}, \text{ and} \tag{30}$$

$$\alpha = (\varepsilon_1, \cdots, \varepsilon_N)^T. \tag{31}$$

From Equation (28), the fault coefficient $\varepsilon_i (i = 1 \sim N)$ can be calculated by

$$\alpha = \Pi^+ \Delta p. \tag{32}$$

We note that the linear approximation was adopted in the above sensitivity analysis. Reference [27] discusses the influence of the linear approximation on fault diagnosis results. It was found that the fault coefficients calculated from the linear approximation equations are always larger than the true values. When the degree of damage is severe, this error caused by the linear approximation cannot be ignored. The research results of Reference [27] found that 0.176 is the critical value for considering this error. For damage parameters with a calculated value greater than 0.176, another correction operation must be performed. The formula of the correction operation for those calculated fault coefficient values greater than 0.176 is

$$\varepsilon_i^a = \frac{\varepsilon_i}{1 + \varepsilon_i}. \tag{33}$$

Thus far, the fault condition of all components in a structure can be evaluated based on the calculation results ε_i and ε_i^a. Figure 1 shows the flow chart of the entire technique to explain the process more clearly.

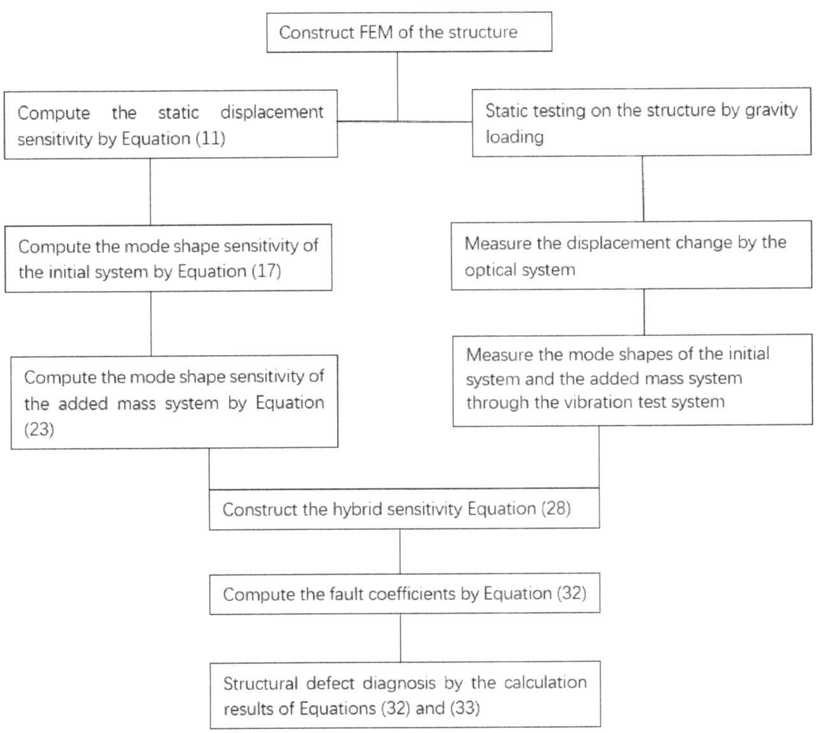

Figure 1. Flowchart of the entire technique.

3. Numerical Example

As presented in Figure 2, the hybrid sensitivity method is illustrated by a steel roof truss structure with 25 bar elements. Note that the number in the circle in Figure 2 denotes the bar number in the FEM. We note that the connections of the bar elements were pin-jointed connections at all joints. The diagonal elements making an X-shape were not connected or truncated at the intersection points. Obviously, this structure could basically be regarded as a geometrically symmetric structure. The main physical parameters of the structure were as follows: the elastic modulus was 200 GPa, the density was 7800 kg/m^3, and the cross-sectional area of each bar was 2.109×10^{-4} m^2.

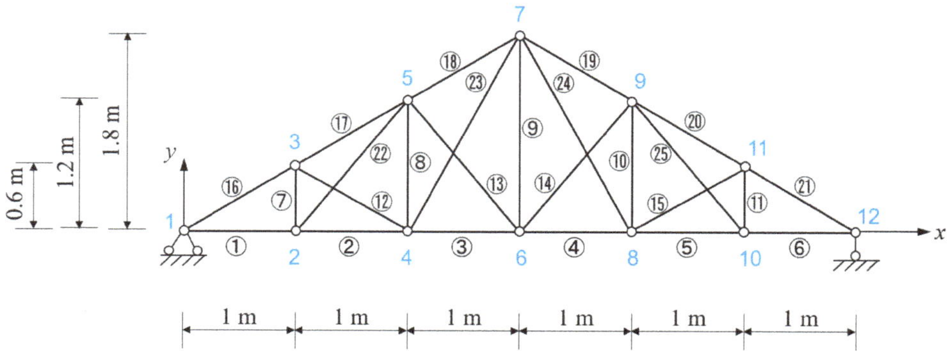

Figure 2. A steel roof structure.

As stated before, the asymmetric gravity loading can overcome the possible missed diagnosis problem in the diagnosis of symmetrical structural defects. For this purpose, the gravity loading, as shown in Figure 3, applied a concentrated force of 20 kN at node 4 in the structure. The number in the circle in Figure 3 denotes the bar number in the FEM. As a by-product of static loading, the gravity load and the structure were assumed to be connected rigidly to form a new added-mass system. In the static test, the vertical displacements of nodes 2–11 were observed through the optical measurement system. In the dynamic test, the horizontal accelerations of nodes 2–11 were observed through the vibration test system. We assumed that the accelerometers were arranged in the horizontal direction of nodes 2–11 to extract the vibration mode data in the horizontal direction. Figure 3 also provides the layout points marked in yellow for the acceleration sensors in the dynamic test. In the next numerical experiment, the fault in the structure was simulated by reducing the elastic modulus of some bar elements. The measured static and dynamic data were simulated from the FEMs of the undamaged and damaged structure.

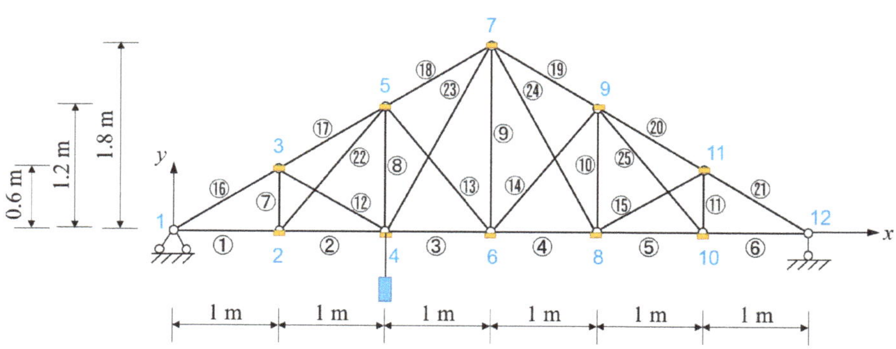

Figure 3. Layout of the loading and the measuring points.

The first defect case assumed that the elastic modulus of bar element 3 was reduced by 20%. Figures 4–6 present the calculation results of the fault coefficients using a pure static algorithm, a pure dynamic algorithm, and the proposed hybrid method, respectively. We note that the pure static algorithm computed the fault coefficients using Equation (10) with only static data. The pure dynamic algorithm computed the fault coefficients using Equation (21) with only dynamic data. From Figures 4–6, one can find that the fault coefficients obtained by the hybrid method more clearly indicated that the fault was located in bar element 3. Specifically, the fault extents obtained by the pure static method, the pure dynamic method, and the hybrid method were 7.5%, 18.4%, and 19.6%, respectively. Obviously, the fault extent (19.6%) calculated by the hybrid method was the closest to the assumed value (20%). It showed that the diagnosis result of the proposed method was the most reliable and accurate.

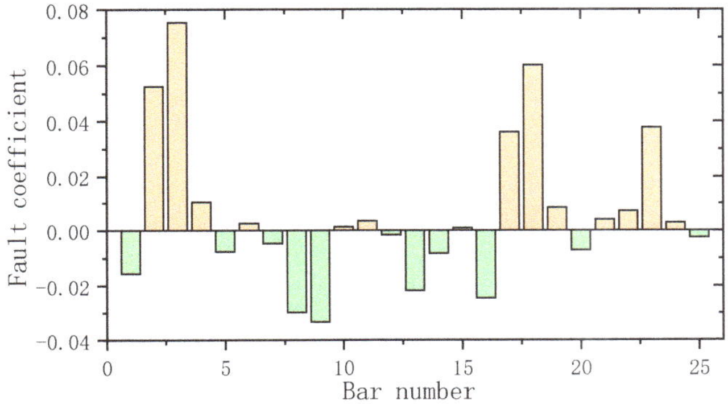

Figure 4. Fault coefficients obtained using a pure static algorithm (bar 3 is damaged).

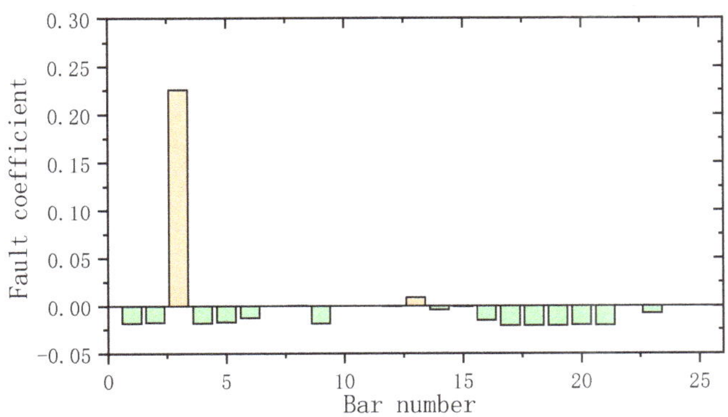

Figure 5. Fault coefficients obtained using a pure dynamic algorithm (bar 3 is damaged).

The second defect case assumed that the elastic modulus of bar element 8 was reduced by 15%. Figures 7–9 present the calculation results of the fault coefficients using a pure static algorithm, a pure dynamic algorithm, and the proposed hybrid method, respectively. As seen in Figure 8, the pure dynamic method missed the diagnosis because it did not indicate that the defect was located in bar element 8. By comparing Figure 7 to Figure 9, one can find that the fault coefficients obtained by the hybrid method more clearly indicated that the fault was located in bar element 8. Specifically, the fault extents obtained by the pure static method, the pure dynamic method, and the hybrid method were 13.2%, 0%, and 15.2%, respectively. Clearly, the fault extent (15.2%) calculated by the hybrid method was the closest to the assumed value (15%). This showed again that the diagnosis result of the proposed method was the most reliable and accurate.

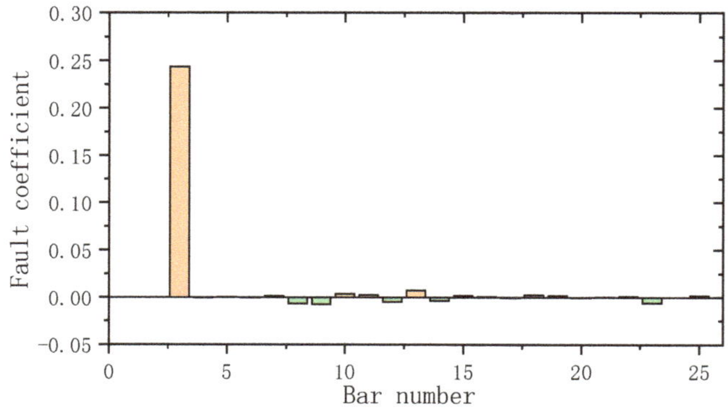

Figure 6. Fault coefficients obtained using the proposed hybrid algorithm (bar 3 is damaged).

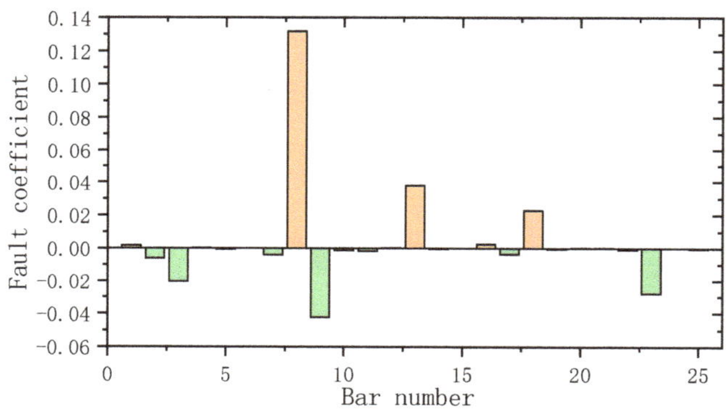

Figure 7. Fault coefficients obtained using a pure static algorithm (bar 8 is damaged).

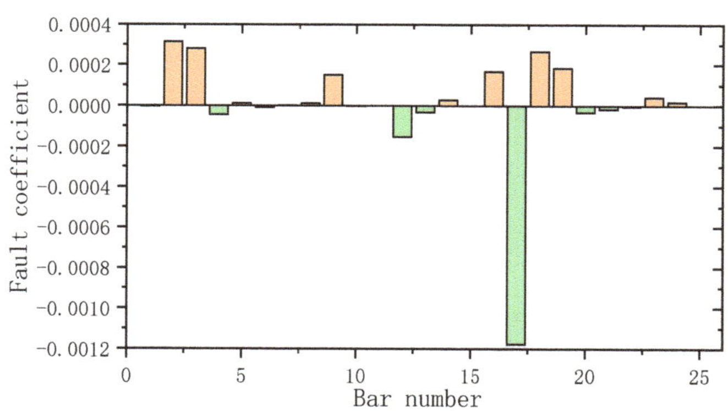

Figure 8. Fault coefficients obtained using a pure dynamic algorithm (bar 8 is damaged).

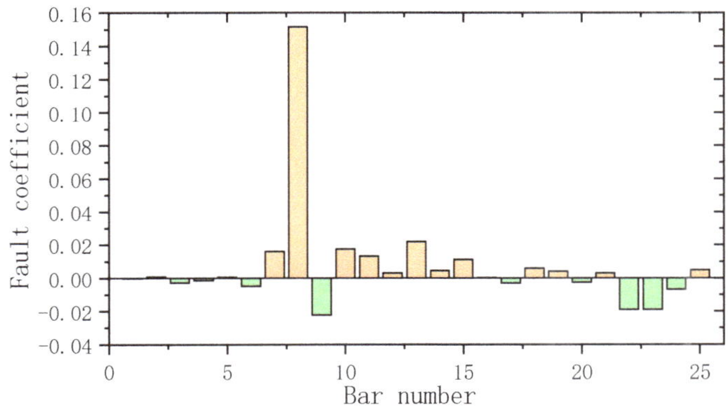

Figure 9. Fault coefficients obtained using the proposed hybrid algorithm (bar 8 is damaged).

The third defect case assumed that the elastic moduli of bar elements 12 and 20 were reduced by 15% and 20%, respectively. Figures 10–12 present the calculation results for this fault case using a pure static algorithm, a pure dynamic algorithm, and the proposed hybrid method, respectively. From Figures 10–12, one can find that only the fault coefficients obtained using the hybrid method could clearly indicate that the faults were located in bar elements 12 and 20. Both the pure static method and the pure dynamic method could only indicate that there was defect in bar 20 and the fault in bar 12 was missed. Specifically, the fault extents of element 12 obtained by the pure static method, the pure dynamic method, and the hybrid method were 0%, 0.2%, and 14.8%, respectively. The fault extents of element 20 obtained by the pure static method, the pure dynamic method, and the hybrid method were 16.4%, 17.3%, and 19%, respectively. Clearly, the fault extents (14.8% and 19%) obtained by the proposed hybrid method were the closest to the assumed values (15% and 20%). These results showed that the proposed method did not miss diagnosis and had high calculation accuracy.

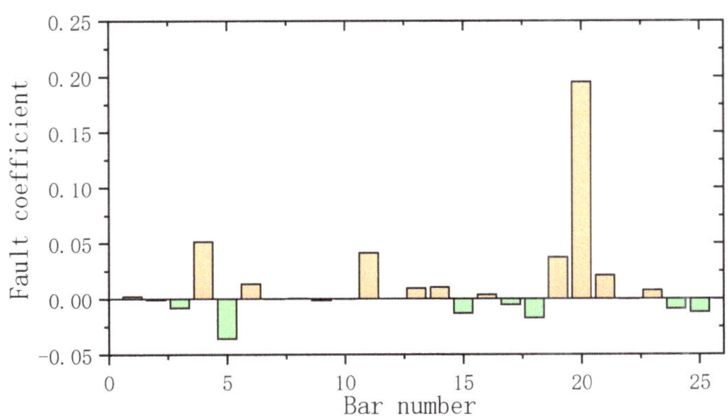

Figure 10. Fault coefficients obtained using a pure static algorithm (bars 12 and 20 are damaged).

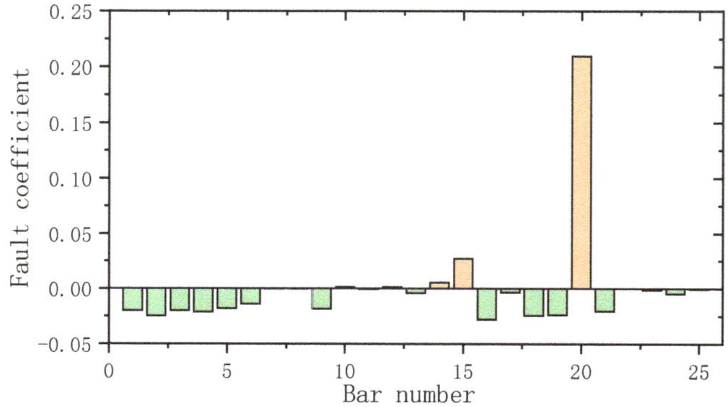

Figure 11. Fault coefficients obtained using a pure dynamic algorithm (bars 12 and 20 are damaged).

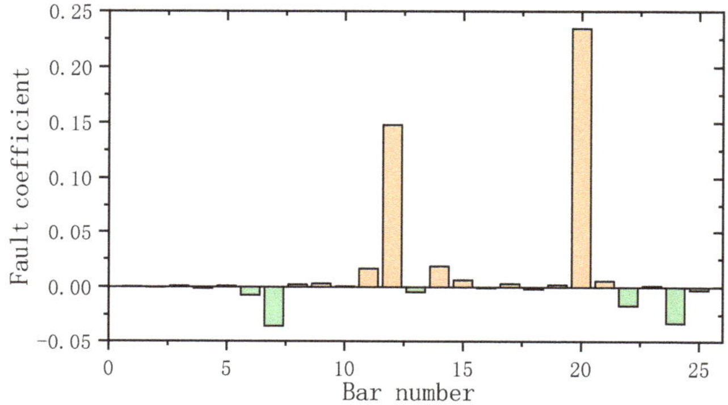

Figure 12. Fault coefficients obtained using the proposed hybrid algorithm (bars 12 and 20 are damaged).

4. Experimental Verification

As presented in Figure 13a, the proposed hybrid sensitivity method was further verified using an experimental steel beam with fixed ends. Obviously, this beam structure could basically be regarded as a geometrically symmetric structure. The main physical parameters of the structure were as follows: the elastic modulus was 193 GPa, the density was 7850 kg/m^3, the length of the beam was 0.8 m, and the width and height of the cross-section were 15 mm and 3 mm, respectively. The geometric parameters of the beam and the arrangement of measurement points are shown in Figure 13b. In the static test, the vertical displacements of nodes 3–5 were observed through the dial indicators. The reason for selecting these static measurement points was that they were all located in the mid-span area, and the corresponding vertical displacements under the given gravity load were relatively large and easy to measure. In the dynamic test, the accelerometers were arranged in the vertical direction of nodes 1, 2, 6, and 7 to extract the first-order vibration mode using the vibration test system. The purpose of selecting these dynamic measurement points was to monitor the entire structure as they were located in the areas not measured during the static testing.

As stated before, the asymmetric gravity loading could overcome the possible missed diagnosis problem in the defect diagnosis of the symmetrical structure. For this purpose, the gravity loading, as shown in Figure 14, applied a concentrated force of 10 N at node 3 of the structure, i.e., the location that was a distance of 300 mm from the left end. As a by-product of static loading, the gravity load and the structure were connected using strong adhesive to form a new added-mass system. Both the original structure and the added-mass structure underwent dynamic testing to obtain the first-order vibration modal data, as shown in Figure 15. Figure 16 shows the damaged beam with a notch. As shown in Figure 16, the cut width (approximately 3 + 3 = 6 mm) accounted for 40% of the entire cross-sectional width of 15 mm. For the fault diagnosis, the undamaged FEM with eight elements, as shown in Figure 17 for the beam, was established using MATLAB software of 7.0 version to represent the undamaged beam. Note that the number in the circle in Figure 17 denotes the element number in the FEM. Obviously, the damage location was related to the fifth element in this FEM with eight elements, as shown in Figure 17. Figures 18–20 present the calculation results of the fault coefficients using a pure static algorithm, a pure dynamic algorithm, and the proposed hybrid method, respectively. From Figures 18–20, one can find that the fault coefficients obtained using the hybrid method more clearly indicated that the fault was located in element 5. Specifically, the fault coefficients of element 5 obtained using the pure static method, the pure dynamic method, and the hybrid method were 6.55%, 13.6%, and 24.1%, respectively.

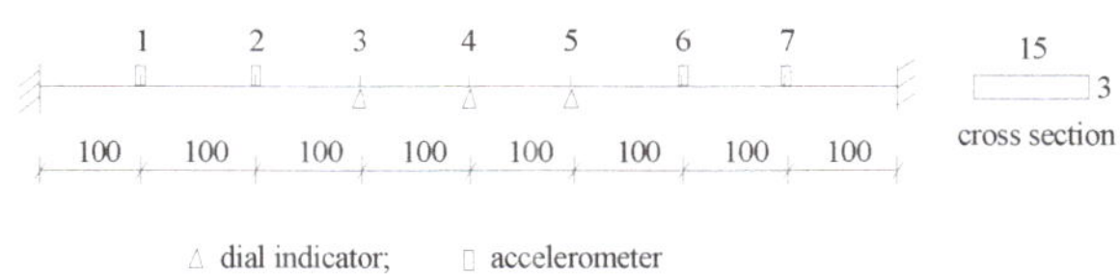

Figure 13. (**a**) An experimental steel beam with fixed ends. (**b**) Geometric parameters and measurement locations (unit: mm).

Figure 14. Static testing.

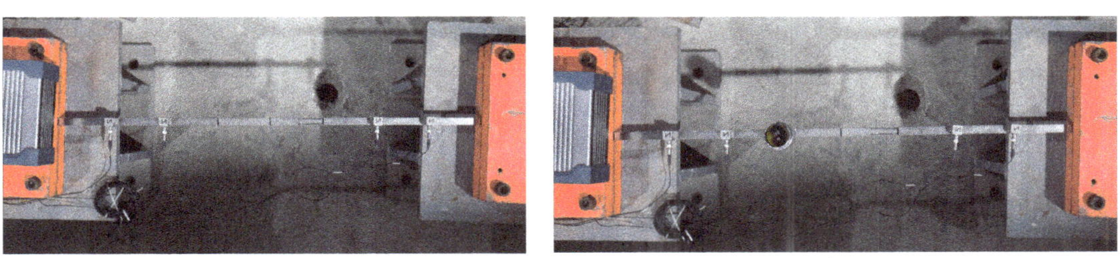

Figure 15. (**a**) Dynamic testing for the original system. (**b**) Dynamic testing for the added-mass system.

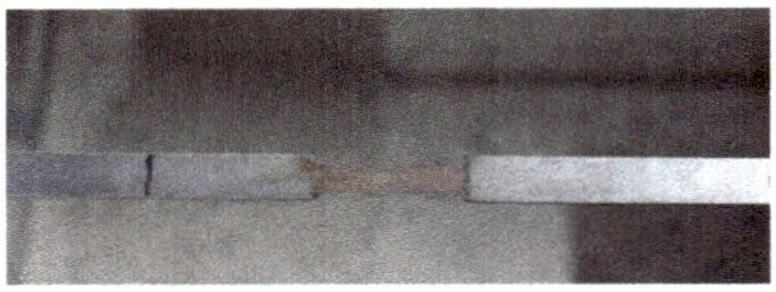

(a)

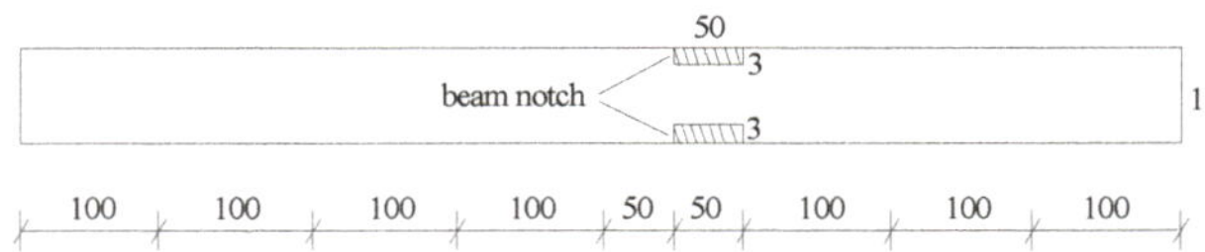

(b)

Figure 16. (**a**) The damaged beam with a notch. (**b**) Size of the notch (unit: mm).

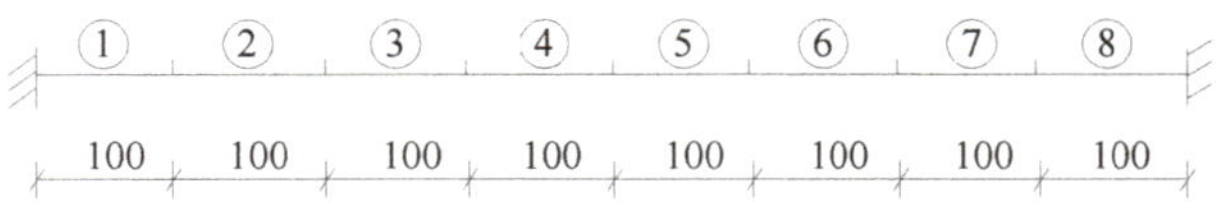

Figure 17. FEM with eight elements (unit: mm).

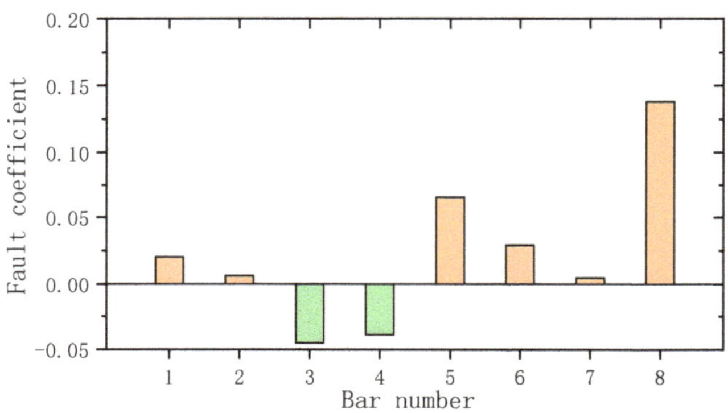

Figure 18. Fault coefficients of the FEM with eight elements obtained using the pure static algorithm (element 5 is damaged).

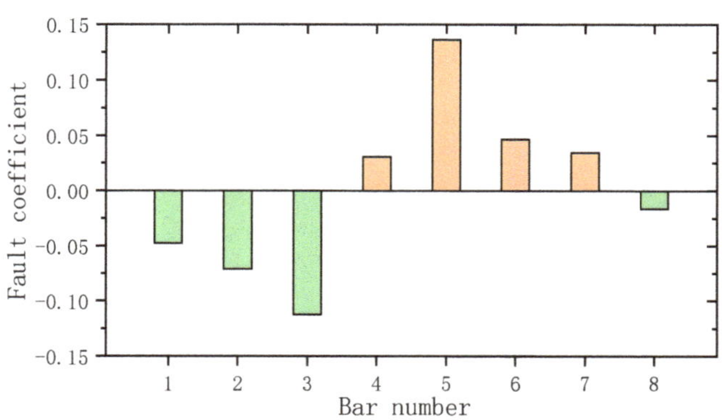

Figure 19. Fault coefficients of the FEM with eight elements obtained using the pure dynamic algorithm (element 5 is damaged).

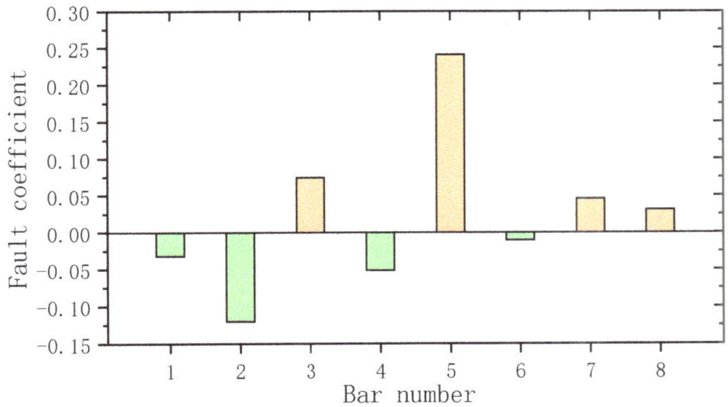

Figure 20. Fault coefficients of the FEM with eight elements obtained using the proposed hybrid algorithm (element 5 is damaged).

Next, a more refined FEM with sixteen elements, as shown in Figure 21, was established to discuss the impact of FEM mesh density on the damage diagnosis results. Note that the number in the circle in Figure 21 denotes the element number in the refined FEM. Obviously, the damage location was related to the tenth element in this refined FEM with sixteen elements, as shown in Figure 22. Compared with the FEM of eight elements, the number of unknown fault coefficients in the FEM with sixteen elements that needed to be solved had doubled from 8 to 16. Based on this refined FEM, Figures 23–25 present the calculation results of the fault coefficients using a pure static algorithm, a pure dynamic algorithm, and the proposed hybrid method, respectively. One can find that the fault coefficients obtained using the hybrid method more clearly indicated that the fault was located in element 10. Specifically, the fault coefficients of element 10 obtained using the pure static method, the pure dynamic method, and the hybrid method were 7.7%, 9.3%, and 33.2%, respectively. Obviously, the calculated fault coefficient (33.2%) obtained using the hybrid method was the closest to the true value (approximately 40%). It showed that the diagnosis result of the proposed method was the most reliable and accurate. The above results indicated that a more refined FEM would contribute to a more accurate damage localization result. However, the meshing of the FEM was not necessarily as dense as possible since an FEM with more elements clearly increased the number of unknown fault coefficients. This meant that an FEM with more elements may have required more test data to successfully compute the fault coefficients for the damage identification. It should also be noted that the damage extents calculated by these two FEMs were different because the lengths of the elements changed. As shown in Figure 22, the fault coefficient corresponding to the same notch changed for the different element grids. The fault coefficients calculated by an FEM with sparse grid partitioning are usually smaller than those obtained by an FEM with dense grid partitioning.

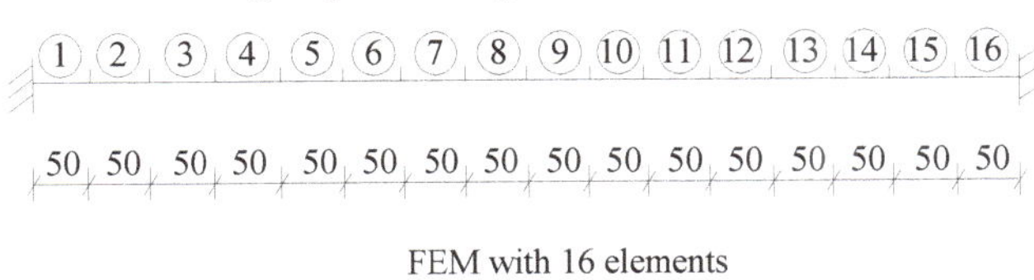

Figure 21. A refined FEM with sixteen elements (unit: mm).

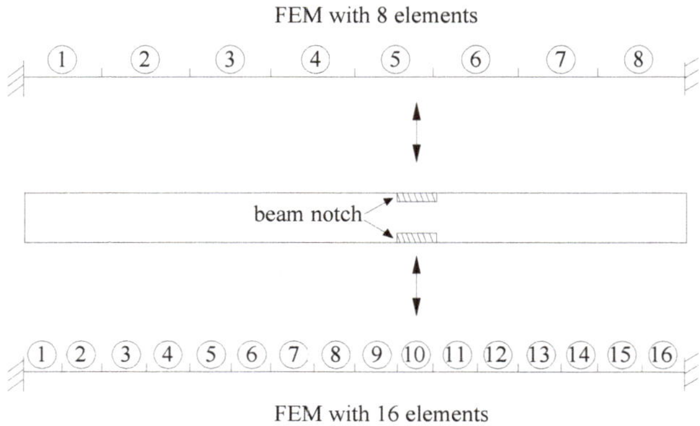

Figure 22. Defect location and the corresponding element number in different FEMs (the number in the circle denotes the element number in the FEM).

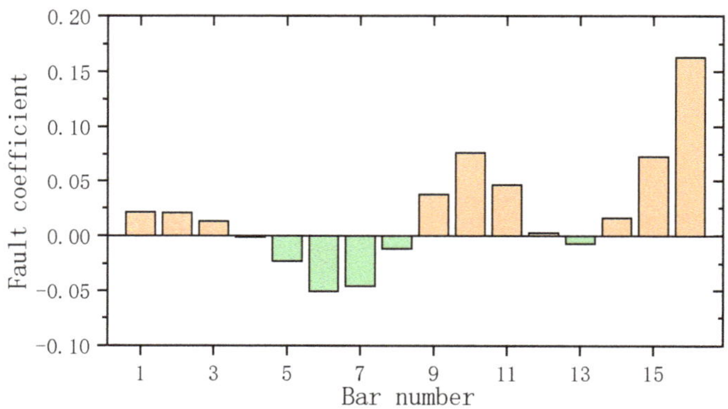

Figure 23. Fault coefficients of the FEM with sixteen elements obtained using the pure static algorithm (element 10 is damaged).

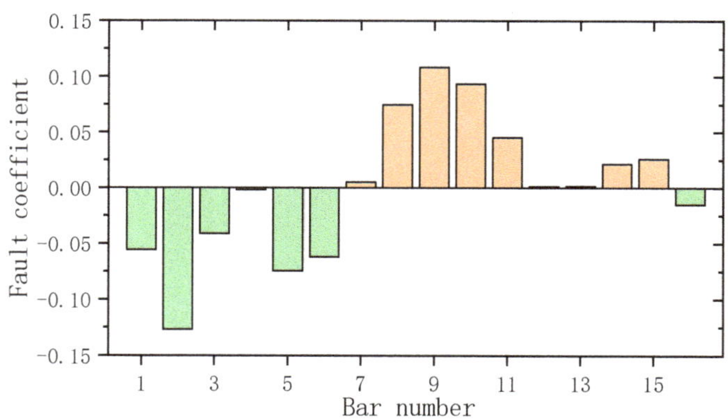

Figure 24. Fault coefficients of the FEM with sixteen elements obtained using the pure dynamic algorithm (element 10 is damaged).

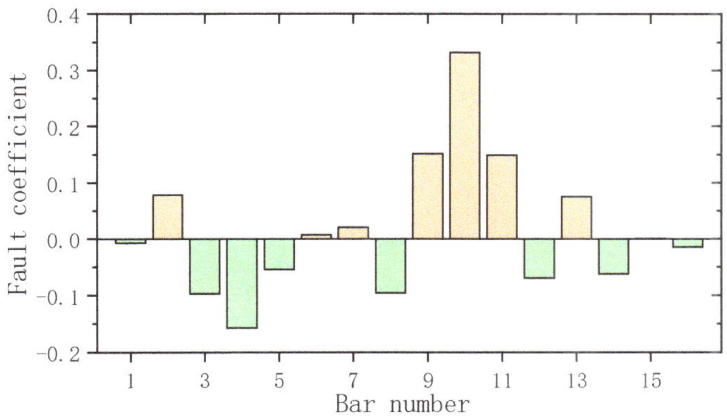

Figure 25. Fault coefficients of the FEM with sixteen elements obtained using the proposed hybrid algorithm (element 10 is damaged).

5. Conclusions

In this article, a hybrid sensitivity method was developed by combining the static displacement and the dynamic mode shape to find a fault's location and extent in a structure. The greatest innovation of the proposed method was its use of static loading as a new added-mass system to obtain more vibration parameters for defect diagnosis. The proposed method can avoid the shortcoming of the possible missed diagnosis, which is seen when using pure static and pure dynamic methods, and it obtained more reliable and accurate fault diagnosis results. A steel roof truss structure with 25 bar elements was used as a numerical example to illustrate the feasibility and advantages of the proposed method. An experimental steel beam was further used to verify the proposed method. By comparing the pure static algorithm, the pure dynamic algorithm, and the hybrid algorithm, the following main conclusions could be obtained: (1) The symmetry of a structure may lead to a missed diagnosis when using pure static or pure dynamic methods. The adverse effect of structural symmetry on defect diagnosis can be overcome by using asymmetric gravity loading. Among the three methods, the diagnosis result of the proposed hybrid method was the most reliable and accurate. (2) For a single defect condition, the three methods may be able to indicate the location of the defect, but the accuracy of the damage extent obtained by the proposed method was the highest. (3) For multiple defect conditions, only the proposed hybrid method could accurately indicate the locations of all the defects while the pure static method and the pure dynamic method missed diagnoses. The proposed method can provide a reference for structural defect diagnosis in engineering practice, especially for symmetric structures. We note that this work mainly elaborated on the application of this hybrid method in diagnosing structural on-site faults. The proposed method may also be used for desk research during design, and further research in this area can be carried out in the future.

Author Contributions: Conceptualization, software and writing—original draft preparation, Q.Y.; methodology, validation and formal analysis, F.Q.; experimental testing and investigation, X.P. All authors have read and agreed to the published version of the manuscript.

Funding: This research was funded by the Natural Science Foundation of China (52192663 and 52008215), the Zhejiang Public Welfare Technology Application Research Project (LGF22E080021), the Major Special Science and Technology Project (2019B10076) of "Ningbo Science and Technology Innovation 2025", and the Natural Science Foundation of Zhejiang Province, China (LQ20E080013).

Institutional Review Board Statement: Not applicable.

Informed Consent Statement: Not applicable.

Data Availability Statement: The data used to support the findings of this study are included in this study and also available from the corresponding author upon request.

Conflicts of Interest: The authors declare no conflict of interest.

References

1. Sun, Y.; Yang, Q.; Peng, X. Damage Identification for Shear-Type Structures Using the Change of Generalized Shear Energy. *Coatings* **2022**, *12*, 192. [CrossRef]
2. Sun, Y.; Yang, Q.; Peng, X. Structural Damage Assessment Using Multiple-Stage Dynamic Flexibility Analysis. *Aerospace* **2022**, *9*, 295. [CrossRef]
3. Fang, R.; Wu, Y.; Wei, W.; Na, L.; Biao, Q.; Jiang, P.; Yang, Q. An Improved Static Residual Force Algorithm and Its Application in Cable Damage Identification for Cable-Stayed Bridges. *Appl. Sci.* **2022**, *12*, 2945. [CrossRef]
4. Yang, Q.; Sun, B. Structural damage localization and quantification using static test data. *Struct. Health Monit.* **2010**, *10*, 381–389. [CrossRef]
5. Yang, Q.W. Fast and Exact Algorithm for Structural Static Reanalysis Based on Flexibility Disassembly Perturbation. *AIAA J.* **2019**, *57*, 3599–3607. [CrossRef]
6. Yang, Q.; Peng, X. A Fast Calculation Method for Sensitivity Analysis Using Matrix Decomposition Technique. *Axioms* **2023**, *12*, 179. [CrossRef]
7. Yang, Q.; Wang, C.; Li, N.; Luo, S.; Wang, W. Model-Free Method for Damage Localization of Grid Structure. *Appl. Sci.* **2019**, *9*, 3252. [CrossRef]
8. Bakhtiari-Nejad, F.; Rahai, A.; Esfandiari, A. A structural damage detection method using static noisy data. *Eng. Struct.* **2005**, *27*, 1784–1793. [CrossRef]
9. Abdo, M.A.B. Parametric study of using only static response in structural damage detection. *Eng. Struct.* **2012**, *34*, 124–131. [CrossRef]
10. Tian, S.; Yang, Z.; Chen, X.; Xie, Y. Damage Detection Based on Static Strain Responses Using FBG in a Wind Turbine Blade. *Sensors* **2015**, *15*, 19992–20005. [CrossRef]
11. Santos, J.; Cremona, C.; Orcesi, A.; Silveira, P.; Calado, L. Static-based early-damage detection using symbolic data analysis and unsupervised learning methods. *Front. Struct. Civ. Eng.* **2015**, *9*, 1–16. [CrossRef]
12. Viola, E.; Bocchini, P. Non-destructive parametric system identification and damage detection in truss structures by static tests. *Struct. Infrastruct. Eng.* **2013**, *9*, 384–402. [CrossRef]
13. Terlaje, A.S., III; Truman, K.Z. Parameter identification and damage detection using structural optimization and static response data. *Adv. Struct. Eng.* **2007**, *10*, 607–621. [CrossRef]
14. Boumechra, N. Damage detection in beam and truss structures by the inverse analysis of the static response due to moving loads. *Struct. Control. Health Monit.* **2016**, *24*, e1972. [CrossRef]
15. Peng, X.; Yang, Q. Damage detection in beam-like structures using static shear energy redistribution. *Front. Struct. Civ. Eng.* **2022**, *16*, 1552–1564. [CrossRef]
16. Chrysanidis, T. Experimental and numerical research on cracking characteristics of medium-reinforced prisms under variable uniaxial degrees of elongation. *Eng. Fail. Anal.* **2023**, *145*, 107014. [CrossRef]
17. Lu, Z.; Law, S. Features of dynamic response sensitivity and its application in damage detection. *J. Sound Vib.* **2007**, *303*, 305–329. [CrossRef]
18. Feng, D.; Feng, M.Q. Computer vision for SHM of civil infrastructure: From dynamic response measurement to damage detection—A review. *Eng. Struct.* **2018**, *156*, 105–117. [CrossRef]
19. Gentile, C.; Guidobaldi, M.; Saisi, A. One-year dynamic monitoring of a historic tower: Damage detection under changing environment. *Meccanica* **2016**, *51*, 2873–2889. [CrossRef]
20. Cardoso, R.D.A.; Cury, A.; Barbosa, F. Automated real-time damage detection strategy using raw dynamic measurements. *Eng. Struct.* **2019**, *196*, 109364. [CrossRef]
21. Yang, Q.W.; Peng, X. A highly efficient method for structural model reduction. *Int. J. Numer. Methods Eng.* **2022**, *124*, 513–533. [CrossRef]
22. Yang, Q.; Peng, X. Sensitivity Analysis Using a Reduced Finite Element Model for Structural Damage Identification. *Materials* **2021**, *14*, 5514. [CrossRef] [PubMed]
23. Peng, X.; Yang, Q.-W. Sensor Placement and Structural Damage Evaluation by Improved Generalized Flexibility. *IEEE Sensors J.* **2021**, *21*, 11654–11664. [CrossRef]
24. Weng, S.; Zhu, H.; Xia, Y.; Li, J.; Tian, W. A review on dynamic substructuring methods for model updating and damage detection of large-scale structures. *Adv. Struct. Eng.* **2019**, *23*, 584–600. [CrossRef]
25. Meixedo, A.; Santos, J.; Ribeiro, D.; Calçada, R.; Todd, M. Damage detection in railway bridges using traffic-induced dynamic responses. *Eng. Struct.* **2021**, *238*, 112189. [CrossRef]

26. Yang, Q.; Peng, X. An Exact Method for Calculating the Eigenvector Sensitivities. *Appl. Sci.* **2020**, *10*, 2577. [CrossRef]
27. Yang, Q.; Sun, B.; Lu, C. An improved spectral decomposition flexibility perturbation method for finite element model updating. *Adv. Mech. Eng.* **2018**, *10*, 1687814018814920. [CrossRef]

coatings

Article

Mechanical Strengths of Alkali-Activated Blast Furnace Slag Powder with Different Alkali Activators and Plant Fibers

Jing Zhu [1,*], Lizhuo Song [1], Zijian Qu [1], Xiaopeng Wang [1], Zijie Wen [1], Xiaodong Liu [2,*] and Hui Wang [3,*]

[1] College of Civil Engineering and Architecture, Harbin University of Science and Technology, Harbin 150080, China; songlizhuo12@163.com (L.S.); 1715020417@stu.hrbust.edu.cn (Z.Q.); m13048993423@163.com (Z.W.)
[2] School of Energy and Built Environment, Guilin University of Aerospace Technology, Guilin 541004, China
[3] School of Civil and Environmental Engineering, Ningbo University, Ningbo 315000, China
* Correspondence: zhujing@hrbust.edu.cn (J.Z.); liuxd@hrbust.edu.cn (X.L.); wanghui4@nbu.edu.cn (H.W.)

Abstract: In this paper, the influence of water glass types, the modulus of water glass, the alkali content, the water consumption, and plant fibers on the mechanical strengths of alkali-activated blast furnace slag powder (BFS) is investigated. Moreover, the fiber types and pretreatment on the plant fibers and the measuring temperature on the performance of alkali-activated BFS are further considered. Results indicate that BFS activated by potassium silicate shows higher mechanical strengths than that activated by sodium silicate. The alkali-activated BFS with alkali treatment on fibers is the most advantageous. The modulus of alkali leads to decreasing the compressive strength. A total of 35% water consumption is the most beneficial to the specimens' flexural and compressive strengths. Samples with 14% potassium silicate show the maximum mechanical strength. Alkali-activated BFS with 1% wheat straw fibers in addition by total volume represents the maximum mechanical strength. The alkali-activated BFS with alkali treatment on fibers is the most advantageous. The addition of potassium silicate can improve the flexural and compressive strengths by the maximum values of 30.4% and 16.8% compared to specimens with sodium silicate. A total of 35% water consumption can increase the flexural and compressive strengths by 33.8% and 32.7%.

Keywords: water glass; alkali-activated; blast furnace slag powder; mechanical strengths; potassium silicate

1. Introduction

Ordinary Portland Cement (OPC) is one of the most widely used cement-based building materials, which is consumed a lot in engineering construction [1]. The annual CO_2, SO_2, and NO_x emissions from OPC production around the world account for about 8% of the global CO_2 emissions [2]. During the entire production process of OPC, approximately 3.2 GJ (gigajoule) per ton is consumed [3–7]. The residual energy and hazardous gas can accelerate the greenhouse effect. In order to postpone the acceleration, some new energy-saving and environmentally protected cementitious materials need to be developed.

In recent years, fly ash, slag powder, and volcanic ash are used as cementitious materials in concrete [8–16]. These mineral admixtures can improve the mechanical strengths when the curing temperature is higher than 40 °C and the curing age is higher than 28 days [17]. However, when the curing temperature and the curing age are lower than 40 °C and 28 days, respectively, the mechanical strengths are decreased by the addition of mineral admixtures [18]. The alkali slag cement-based material has become a popular green and environmentally friendly material studied at home and abroad [19,20]. This kind of material shows excellent mechanical strengths when cured in several curing environments and curing ages [21,22]. Prior researchers have developed alkali-activated mineral admixtures with ultra-high mechanical strengths. As pointed out by prior researchers, the mechanical performance of alkali-activated blast furnace slag (BFS) shows higher mechanical strength and durability than the alkali-activated fly ash [23–27]. Therefore, in this paper, the BFS is applied in manufacturing the alkali-activated binder materials and the corresponding properties should be intensively studied. In previous studies, sodium silicate is a common activator in the alkali-activated cementitious materials. However, little attention has been paid to the influence of types of activators on the properties of alkali-activated cementitious materials [28,29].

Micro-fine steel fibers have been used for reinforcing the mechanical strengths of alkali-activated BFS by prior researchers. However, most of the steel fibers sink to the bottom of the samples during manufacture, leading to the uneven distribution of steel fibers [30]. Mastali et al. [31] have reported that the polyvinyl alcohol fibers, basalt fibers, and polypropylene fibers can improve the mechanical strengths of alkali-activated BFS, but the corresponding heat resistance is poor.

Plant fibers can not only significantly improve the toughness of the matrix material, inhibit the shrinkage of the material, and limit the crack propagation, but also make a large use of industrial and agricultural wastes such as slag and wheat straw to recycle resources [32]. As obtained in Zhu's research, alkali-activated BFS with plant fibers shows excellent heat resistance [33]. Although multiple performances of alkali-activated mineral admixtures have been studied by several scholars [34–36], little attention has been paid to the types of silicate, the type, dosages, and treatment methods of reinforced fibers.

In this study, the influence of silicate types and contents, the fibers' volume, and the effects of the fibers' type on the flexural and compressive strength of alkali-activated BFS are investigated. Moreover, the treatment method of the fibers and the modulus of the silicate are considered. The scanning electron microscope is used for revealing the inner mechanism of the mechanical properties. This research will provide a reference for developing alkali-activated mineral admixtures with excellent mechanical performance and durability in the future. As an inorganic adhesive, the alkali-activated BFS can be used in the repair of road, beam, and column in the future.

2. Experimental

2.1. Raw Materials

The alkali-activated BFS powder in our study is composed of the BFS powder, sodium silicate ($Na_2O{\cdot}nSiO_2$), potassium silicate ($K_2O{\cdot}nSiO_2$), sodium hydroxide (NaOH), and water. Potassium silicate ($K_2O{\cdot}nSiO_2$) and sodium silicate ($Na_2O{\cdot}nSiO_2$) are provided by Huida Chemical Plant (Tianjin, China). The baume degree, initial modulus, and density of potassium silicate are 46.3, 2.76, and 1.465 g/cm^3, respectively. The content of SiO_2 and K_2O of potassium water glass are 28.15% and 15.98%, respectively. Sodium silicate shows the baume degree, initial modulus, and density of 48.0, 2.3, and 1.495 g/cm^3, respectively. At the same time, the corresponding SiO_2 and K_2O of sodium water glass are 31.10% and 13.51%, respectively. NaOH is provided by Harbin Chemical Reagent Factory (Harbin, China), with the purity of 99.3%.

NaOH and KOH are provided by Harbin Chemical Reagent Factory (Harbin, China), with the purity of 99.4% and 99.1%, respectively.

The wheat straw fibers, rice straw fibers, and corn straw fibers are produced by Anshan Changhong Fiber Factory (Anshan, China). The diameters of the fibers are 5, 3, and 7 mm, while the lengths of the fibers are 3, 2.5, and 2 cm showing the aspect ratios of 18, 16.7, and 11.4, respectively. The densities of the fibers are 1.55, 1.0, and 0.25 g/cm^3, with the tensile strength of the fibers being 20, 8, and 12 MPa, respectively.

Polypropylene fibers (PP)s produced by Shandong Senhong Co., Ltd. (Taian, China) have a diameter of 18~45 um, an average length of 12 mm, and show an aspect ratio of 267. The density, the elastic modulus, and the tensile strength of PPs are 0.91 g/cm^3, 42 GPa, and 1600 MPa, respectively. Micro-steel fibers (MS) produced by Anshan Changhong Steel Fiber Factory (Anshan, China) show a diameter of 0.22 mm, a length of 13 mm, and perform an aspect ratio of 59.1. Additionally, the density and tensile strength of MS are 7.8 g/cm^3 and 2870 MPa, respectively. Coarse steel fibers (CS) produced by Huaixing Steel Fiber

Processing Plant in Taocheng District of Hengshui City (Hengshui, China) have a diameter of 2.5 mm and a length of 2~3 cm. At the same time, the aspect ratio, the density, and the tensile strength of CS are 10, 7.8 g/cm^3, and higher than 2150 MPa, respectively. The morphology of fibers is shown in Figure 1. The technical indicators of applied fibers are measured and provided by the manufacturers.

(**a**) wheat straw fiber (**b**) rice straw fiber (**c**) corn straw fiber

(**d**) polypropylene fiber (**e**) micro-steel fiber (**f**) coarser steel fiber

Figure 1. The fibers used in alkali-activated BFS.

The S95 BFS powder with specific surface area of 475 m^2/kg is provided by Liaoyuan Jingang Company (Liaoyuan, China). The main activity indicators, the mass coefficient, alkaline coefficient, and activity coefficient of BFS powder are 1.91, 1.03, and 0.44, respectively. The main chemical components of raw materials are shown in Table 1.

Table 1. Main chemical composition of BFS (%).

Types	SiO_2	AL_2O_3	CaO	MgO	Fe_2O_3	TiO_2	MnO	K_2O
BFS	36.9	15.66	37.57	9.3	0.36	0.18	0.16	0.25

2.2. Specimens' Preparation

The mass ratios of potassium silicate used in this study are 11.7%, 12.7%, 14.1%, 15.8%, 16.9%, and 17.8% by the total mass of alkali-activated BFS (see Table 2). At the same time, the corresponding mass ratios of NaOH are 3.9%, 3.1%, 2.5%, 1.6%, 0.9%, and 0.4%, respectively. First, NaOH is slowly poured into the water glass and stirred with a glass rod until there are no obvious floccules. The mixture is allowed to stand for 1.5 h until the heat release is finished. After that, the BFS powder and fibers are added to the cement slurry mixer, and the water glass with adjusted modulus is added to the mixer and stirred at the speed of 140 r/min for 1 min. Then, the weighted water is added to the mixer and stirred at the same stirring speed for 6 min. When the mixing is finished, the mixture is poured into steel molds of sizes 40 mm × 40 mm × 160 mm and 100 mm × 100 mm × 100 mm, and then all specimens are formed by high-frequency vibration on the concrete vibration table. The specimens are cured in a room with 20 ± 1 °C and relative humidity of 52.1% for 24 h. Finally, all samples are moved to a standard curing room with a relative humidity of 98.1% and a temperature of 20 ± 2 °C for the curing ages of 3, 7, and 28 days. The experimental details can be found in Feng's research [37].

Table 2. The mixing proportions of alkali-activated BFS powder.

Modulus	Alkali Content (%)	BFS (kg)	Water Glass Types	Water Glass (kg)	NaOH (kg)	Water–Binder Ratio	Fiber Types	Fiber Volume (%)	Fiber Treatment Method
1.0	14	1562.5	Potassium	339.6	84.2	32	-	-	-
1.0	14	1562.5	Sodium	393	97.3	32	-	-	-
0.8	12	1562.5	Potassium	267.7	89.2	35	-	-	-
1.0	12	1562.5	Potassium	291	72.1	35	-	-	-
1.2	12	1562.5	Potassium	323.7	57.2	35	-	-	-
1.6	12	1562.5	Potassium	361.7	35.7	35	-	-	-
2.0	12	1562.5	Potassium	388.9	21.1	35	-	-	-
2.4	12	1562.5	Potassium	409.5	8.3	35	-	-	-
1.0	8	1562.5	Potassium	199.1	47.7	35	-	-	-
1.0	10	1562.5	Potassium	248.8	59.6	35	-	-	-
1.0	12	1562.5	Potassium	291	72.1	35	-	-	-
1.0	14	1562.5	Potassium	348.4	83.4	35	-	-	-
1.0	16	1562.5	Potassium	398.2	95.3	35	-	-	-
1.0	18	1562.5	Potassium	447.9	107.2	35	-	-	-
1.0	20	1562.5	Potassium	497.7	241.2	35	-	-	-
1.0	22	1562.5	Potassium	547.5	131.1	35	-	-	-
1.0	12	1562.5	Potassium	291	72.1	28	-	-	-
1.0	12	1562.5	Potassium	291	72.1	32	-	-	-
1.0	12	1562.5	Potassium	291	72.1	35	-	-	-
1.0	12	1562.5	Potassium	291	72.1	38	-	-	-
1.0	12	1562.5	Potassium	291	72.1	42	-	-	-
1.0	12	1562.5	Potassium	291	72.1	35	Wheat straw	1	-
1.0	12	1562.5	Potassium	291	72.1	35	Rice	1	-
1.0	12	1562.5	Potassium	291	72.1	35	Corn	1	-
1.0	12	1562.5	Potassium	291	72.1	35	PP	1	-
1.0	12	1562.5	Potassium	291	72.1	35	XS	1	-
1.0	12	1562.5	Potassium	291	72.1	35	CS	1	-
1.0	12	1562.5	Potassium	291	72.1	35	Wheat	1	Acid
1.0	12	1562.5	Potassium	291	72.1	35	Wheat	1	Acid
1.0	12	1562.5	Potassium	291	72.1	35	Wheat	1	Alkali
1.0	12	1562.5	Potassium	291	72.1	35	Wheat	1	Alkali
1.0	12	1562.5	Potassium	291	72.1	35	Wheat	1	-
1.0	12	1562.5	Potassium	291	72.1	35	Wheat	0.2	Alkali
1.0	12	1562.5	Potassium	291	72.1	35	Wheat	1	Alkali
1.0	12	1562.5	Potassium	291	72.1	35	Wheat	3	Alkali
1.0	12	1562.5	Potassium	291	72.1	35	Wheat	5	Alkali

The treatment method on the plant fibers is described as follows. The sulfuric acid treatment and hydrochloric acid treatment are provided. The fibers are immersed in 10% sulfuric acid solution or hydrochloric acid solution for 60 min, and then the fibers are treated in deionized water for 2 min. After all these are finished, the treated plant fiber is dried naturally in a cool place. The fibers with alkali treatment are shown as follows. The plant fibers are immersed in 10% NaOH solution for 60 min, and then are moved in water for 2 min.

2.3. Measurement Methods

Measurement of Mechanical Strength

The YAW-300 cement flexural and compressive constant stress testing machine produced by Beijing Longchen Weiye (Beijing, China), is used for the measurement of mechanical strengths. The maximum compressive strength and flexural strength were 300 and 10 KN, respectively. The measurement accuracy ranges from 1% to 100%. The loading speeds for the compressive and flexural strengths are 2.4 and 0.05 kN/s, respectively. First, the mortar specimen is placed on the YAW-300 automatic compression-folding testing machine and folded into two halves. Then, half of the fractured prism specimen is placed in the middle of the compressive clamp, and the compression surface is the two sides of the specimen during forming. The mechanical strengths of alkali-activated BFS are measured in accordance with Chinese standard GB/T 17671-2021 [38]. The measuring process is shown in Figure 2.

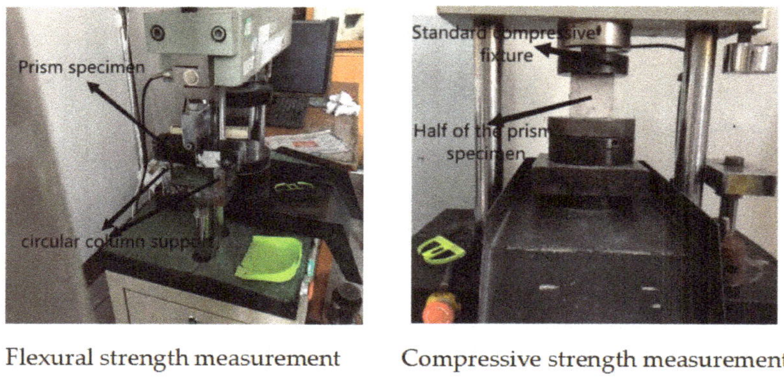

Figure 2. Measurement of mechanical strengths.

2.4. Steps of Micro-Performance

The central crushing part of the sample is selected for scanning electron microscope (SEM) measurement. The fragments are cut into small particles of about 5 mm, regular shape and smooth surface by pliers. On a fixed metal plate, the sample is fixed with conductive adhesive and vacuumed in a vacuum machine. After this, the samples are used to obtain photographs of the scanning electron microscope by the FEI Sirion model scanning electron microscope which is provided by Philips of Amsterdam, The Netherlands. Instrument resolution is demonstrated by the following: when the voltage > 10 kV, 1.5 nm; when the voltage is 1 kV, 2.5 nm; and when the voltage is 500 kV, 3.5 nm. The acceleration voltage is 200 V~30 kV, the sample chamber diameter is 284 mm, and the analysis working distance is 5 mm. This measurement is carried out following GB/T 17362-2008 [39]. Figure 3 shows the measuring process of the SEM. Table 3 shows the test device specifications and standards corresponding to different experiments.

Table 3. The test device specifications and standards corresponding to different experiments.

Parameter	Mechanical Strength	Micro-Performance
Equipment	YAW-300	FEI Sirion
Specification	Flexural strength: 0~10 KN Compressive strength: 0~300 KN	Magnification: 74×~300,000×
Manufacturer	Beijing Longchen Weiye, Beijing, China	Philips of The Netherlands
Standard	GB/T 17671-2021	GB/T 17362-2008

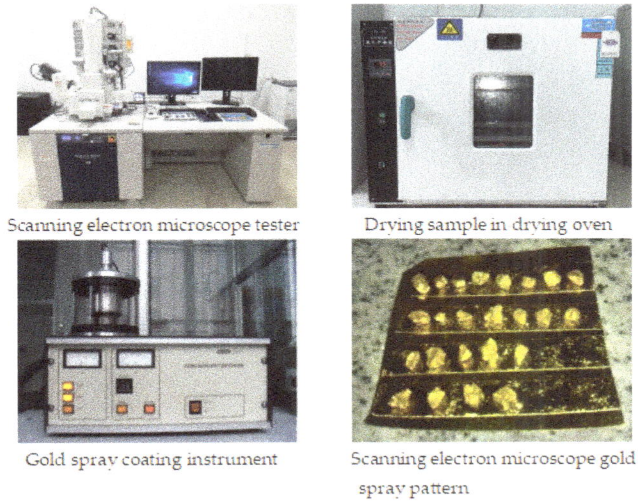

Figure 3. Measurement of SEM.

3. Results and Discussion

3.1. Influence of Silicate Type

Potassium silicate and sodium silicate are used as alkali activators to excite BFS. The flexural strength and compressive strength of the specimens are measured after standard cured for 3, 7, and 28 days. The measuring results are illustrated in Figure 4. As shown in Figure 4, the flexural strengths of the specimen mixed with potassium silicate are 3%, 24%, and 30.4% higher than that of the specimen mixed with sodium silicate. The compressive strength of the specimens mixed with potassium silicate cured for 3, 7, and 28 days is 4.4%, 3.3%, and 16.8% higher than that of the specimens mixed with sodium silicate. It is obvious that potassium silicate is superior to sodium silicate in strengthening the mechanical properties of the alkali-activated BFS [40,41]. When the amount of sodium silicate is small, the reaction and the potential activity of slag cannot be thoroughly and fully excited. Excessive dosage of sodium silicate will lead to a high concentration of OH^- ions, and the hydration products generated by the rapid reaction on the surface of slag particles form a protective film, which prevents the further progress of the reaction and leads to slow development of the strength in the later stage [42]. In the follow-up experimental results, potassium water glass is selected as alkali activator.

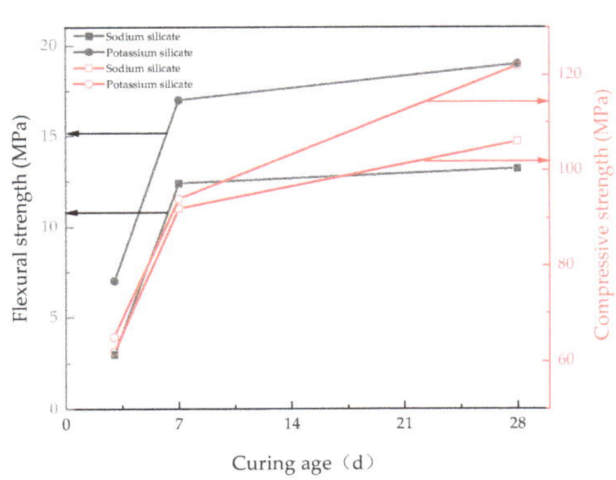

Figure 4. The mechanical strengths of alkali-activated BFS with different alkali activators.

3.2. Influence of Modulus

As reported in prior research, unreasonable water consumption will bring inconvenience to the construction of alkali-activated BFS. According to prior research [43], the modulus 0.8~2.4 is reasonable to the mechanical properties of alkali-activated BFS. Therefore, in this study, this modulus range is taken as the potassium silicate, and the compressive strength of specimens is determined. The compressive strength of the specimens cured in the standard environment for 28 days is shown in Figure 5. The compressive strength decreases with the increasing modulus. The compressive strength of specimens with the modulus of 0.8 is 53% higher than the specimens with the modulus of 2.4. This is ascribed to the fact that when the modulus increases, the alkali solution can not be able to fully activate the slag activity due to insufficient alkalinity, resulting in the reduction of compressive strength [44].

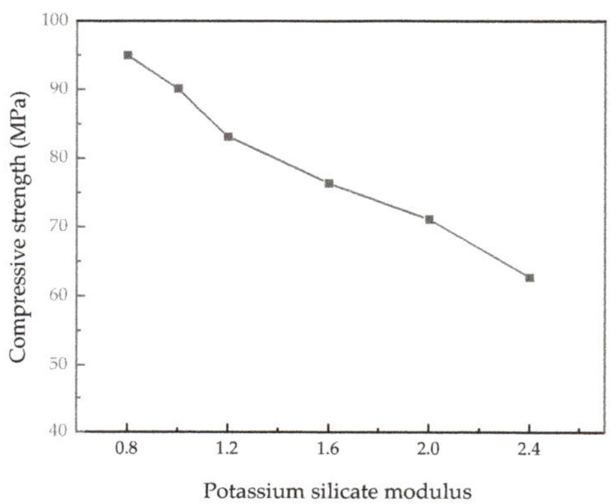

Figure 5. The compressive strength of alkali-activated BFS with different modulus.

Figure 6 shows the flexural and compressive strengths of potassium silicate with modulus of 1.0 and 2.0. The curing age ranges from 3 to 28 days. As depicted in Figure 6, the flexural strengths of the specimens with modulus of 1.0 cured for 3, 7, and 28 days are 24.2%, 14.6%, and 15.9% higher than that of the specimen whose modulus is 2.0. Moreover, as illustrated in Figure 6, specimens with modulus of 1.0 cured for 3, 7, and 28 days show the compressive strength 10.1%, 27.1%, and 12.9% higher than that of the specimen with modulus of 2.0. This is due to the improved hydration degree by the increased curing age [45].

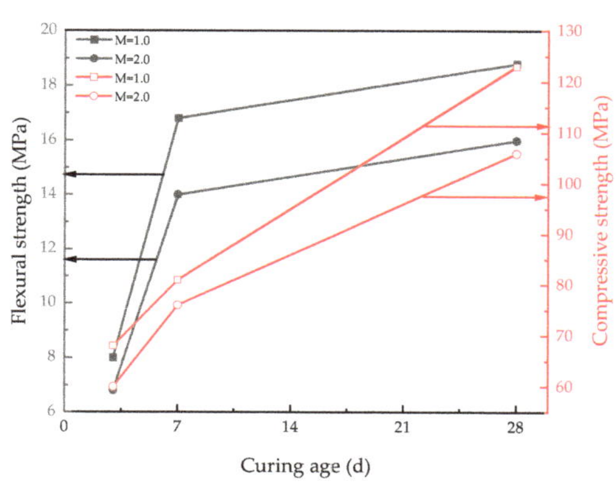

Figure 6. Mechanical strengths of alkali-activated BFS with different curing age and modulus.

3.3. Effect of Alkali Content

Figure 7 shows the flexural and compressive strengths of potassium silicate with different dosages of alkali. As found in Figure 7, the flexural strength and compressive strengths of the specimens with 14% alkali are 33.8% and 32.7% higher than those with 8% alkali. Meanwhile, the flexural strength and compressive strength of the specimens with 14% alkali are 21.2% and 17.1% higher than those with 18% alkali content. Additionally, the flexural strength and compressive strength of the specimens with 14% alkali content are 20.3% and 11.9% higher than those with 12% alkali. This is because the alkali is beneficial in stimulating the activity of slag, enhancing the dissolution and recombination ability of slag, forming more calcium silicate hydrate gel, and filling some pores' inner specimens. Therefore, the mechanical strengths are improved. When the alkali content is 14%, the alkali solution is sufficient to stimulate the activity of slag. However, with the continuous increase of alkali content, the production of calcium silicate hydrate gel will be inhibited, leading to the increase of the number and sizes of pores, thus decreasing the mechanical strength.

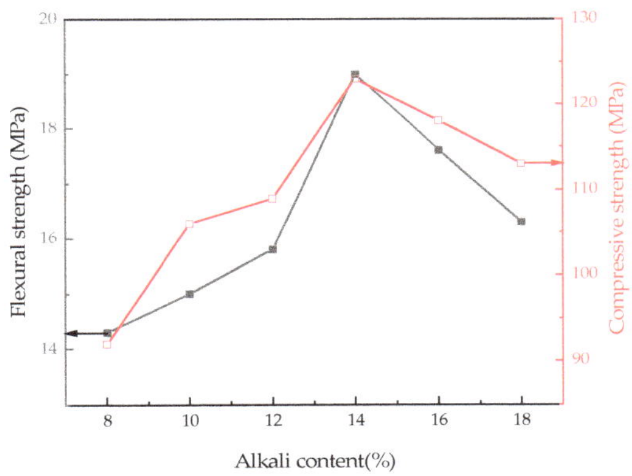

Figure 7. Mechanical strengths of alkali-activated BFS with different alkali contents.

The compressive strength of alkali-activated BFS with different dosages of potassium silicate is shown in Figure 8. As illustrated in Figure 8, the compressive strength of alkali-activated BFS first increases and then decreases with the increasing dosages of potassium silicate. The compressive strength reaches the maximum value when the dosage of potassium silicate is 12%. This can be ascribed to the reason that a small amount of potassium silicate will lead to an incomplete reaction and the potential activity of BFS can not be fully activated in the process of hydration [46,47]. Therefore, the compressive strength is increased by the addition of potassium silicate, when the amount ranges from 8% to 12%. However, when the amount of potassium silicate increases from 12% to 22%, the compressive strength decreases. The maximum value of compressive strength of the specimens with 12% potassium silicate is 1.7 multiples of the specimens with 22% potassium silicate. This is ascribed to the fact that with the increasing dosage of potassium silicate, the activity of BFS is improved by adding more potassium silicate. However, when the dosage is higher than 12%, because the excessive alkali reacts with CO_2, forming carbonates on the surface of hydration, this prevents a further reaction. Consequently, the mechanical strengths decrease with the dosage of potassium silicate ranging from 12% to 22%.

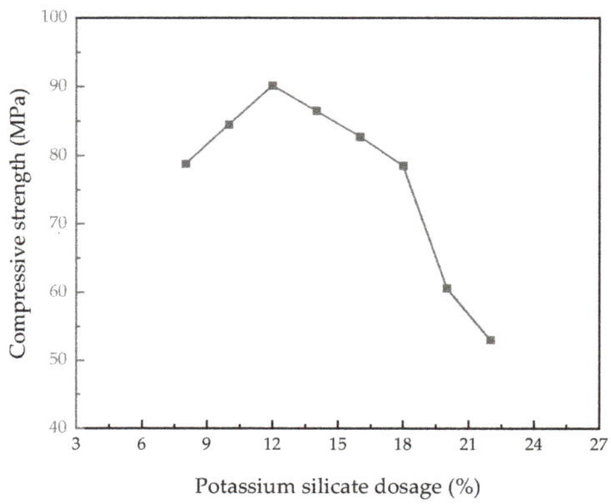

Figure 8. The compressive strength of alkali-activated BFS with different potassium silicate.

3.4. *Effect of Water Consumption*

Figure 9 shows the mechanical strengths of alkali-activated BFS with water consumption increasing from 28% to 42%. As demonstrated in Figure 9, the flexural and compressive strengths of alkali-activated BFS with 35% water consumption reach the highest, while specimens with 42% water consumption show the lowest mechanical strengths. The flexural and compressive strengths of specimens with water consumption of 35% are 313% and 35.5% of the specimens with 42% water consumption. This is ascribed to the fact that the addition of deionized water can increase fluidity of fresh alkali-activated BFS, leading to increasing the transport of free alkali ions. More silicon oxide tetrahedron and aluminum oxide tetrahedron formed, resulting in increasing the mechanical strengths. However, the volume and quantity of pores' inner alkali-activated BFS are increased by excessive water, which decreases the mechanical strengths [48].

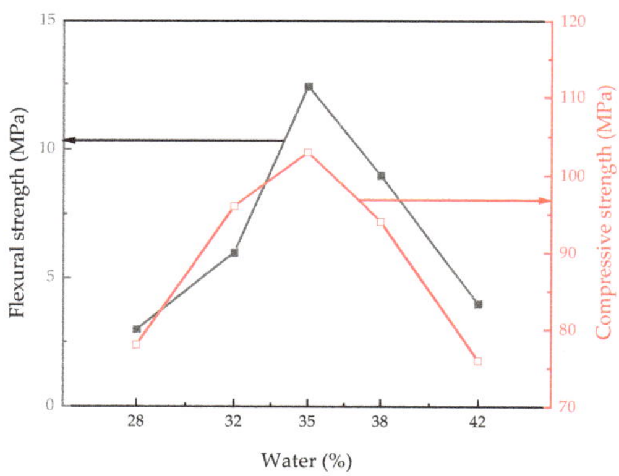

Figure 9. The mechanical strengths of alkali-activated BFS with different potassium silicate.

3.5. *Effect of Plant Fiber Types*

Figure 10a shows the mechanical strengths of alkali-activated BFS with wheat straw fibers, rice straw fibers, or corn straw fibers. As illustrated in Figure 10a, the mechanical strengths increase with the rising curing ages due to the increased hydration degree by curing age. The mechanical strengths of the specimens decrease in this order: alkali-activated BFS with wheat straw > alkali-activated BFS with corn straw > alkali-activated BFS without fiber > alkali-activated BFS with rice straw fibers. This is because the wheat straw is well occluded with the substrate, which enhances the contact interface between fibers and the alkali-activated BFS matrix [49]. However, the density of corn stalk is lower than the alkali-activated BFS matrix; therefore, the corn stalk floats on the surface of alkali-activated BFS. Consequently, alkali-activated BFS with corn stalk shows the worst mechanical strengths.

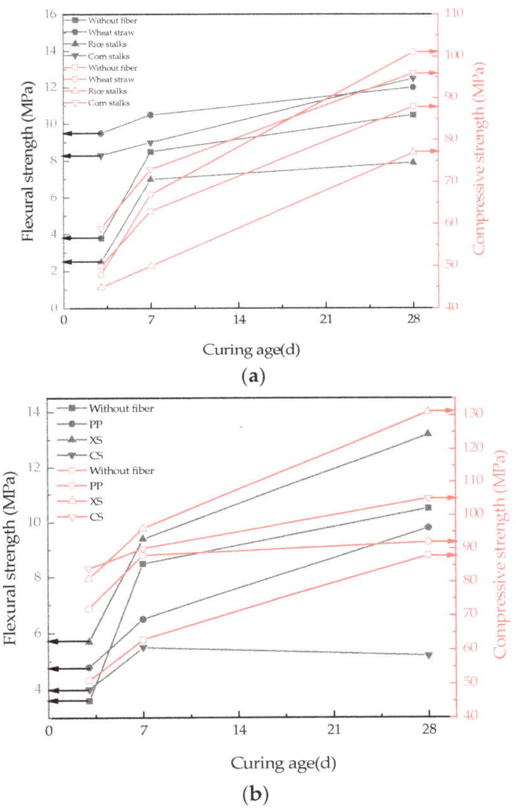

Figure 10. Mechanical strengths of alkali-activated BFS with different fibers. (**a**) Plant fibers; (**b**) Polypropylene fiber (PP), fine steel fiber (XS), and coarser steel fiber (CS).

Figure 10b shows the mechanical strengths of alkali-activated BFS with PP, XS, or CS fibers. As illustrated in Figure 10b, the mechanical strengths increase with the rising curing ages due to the increased hydration degree by curing age. The mechanical strengths of the specimens decrease in this order: alkali-activated BFS with XS > alkali-activated BFS with CS > alkali-activated BFS with PP > alkali-activated BFS without fiber. XS and CS have higher mechanical strength than wheat straw. However, due to their high density and heavy mass, most of the steel fibers sink to the bottom of the specimen during the coagulation process of alkali-activated BFS. The distribution is uneven, resulting in great differences in the specimens.

3.6. Effects of Different Treatment Methods on Plant Fiber

The mechanical strengths of alkali-activated BFS with different treatment methods on the plant fibers are shown in Figure 11. As observed in Figure 11, the mechanical strengths show an ascending trend with the increasing curing age due to the increased hydration degree. The mechanical strengths of alkali-activated BFS with different treatment methods on the plant fibers show the order of alkali-activated BFS with alkali treatment > alkali-activated BFS with acid treatment > alkali-activated BFS without any treatment. This is attributed to the fact that alkali treatment and acid treatment can improve the bonding performance of fibers and alkali-activated BFS, leading to improve the mechanical strengths [50]. Moreover, the alkali treatment and acid treatment lead to improving the dispersing degree of fibers and enhancing the mechanical strengths of alkali-activated BFS. In addition, the dispersing effect of alkali treatment on fibers is higher than the acid treatment, resulting in increasing the mechanical strength [51].

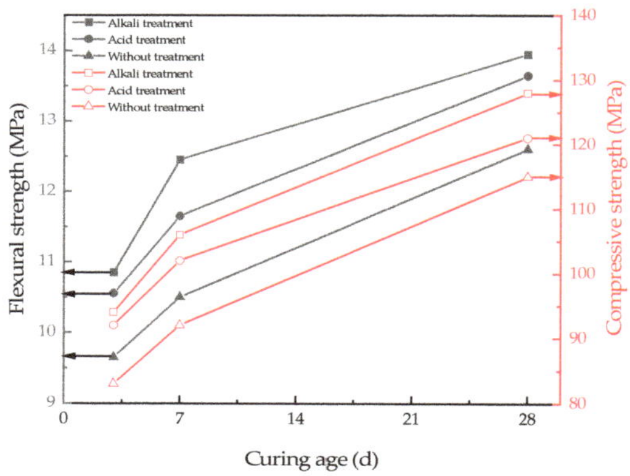

Figure 11. Mechanical strengths of alkali-activated BFS with different treatment methods.

3.7. *Effects of Plant Fiber Content*

Figure 12 shows the mechanical strengths of alkali-activated BFS with wheat straw. The wheat straw is treated by alkali and the specimens are cured in the standard environment for 28 days. As depicted in Figure 12, the specimens with 1% fibers show the maximum flexural and compressive strengths. When the volume ratio of fibers increases from 0.2% to 1%, the flexural and compressive strengths increase by 26.9% and 16.3%, respectively. Meanwhile, when the volume ratio of fibers ranges from 1% to 5%, the flexural and compressive strengths increase by 58.3% and 66.1% respectively. This is ascribed to the fact that when the volume of fibers is less than 1%, the increasing dosages of fibers can effectively improve mechanical strength by bridging the inner cracks. However, excessive fibers can agglomerate, leading to decreasing the mechanical strengths [52,53].

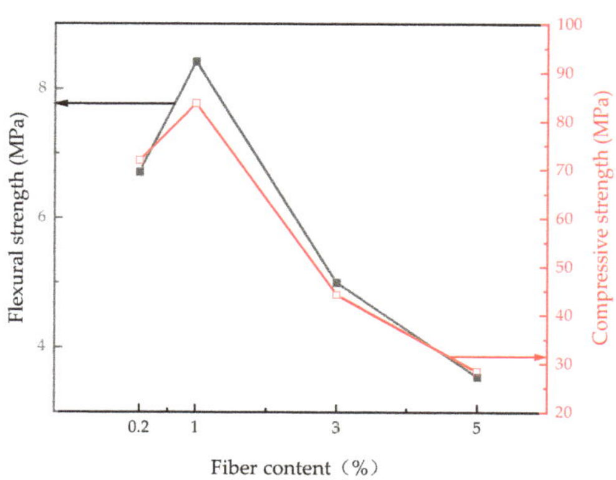

Figure 12. Mechanical strengths of alkali-activated BFS with different dosages of fibers.

3.8. Scanning Electron Microscope

Figure 13 shows the alkali-activated BFS with 1% wheat straw fibers, rice straw fibers, and corn stalk fibers, respectively. As illustrated in Figure 13a, strip substance can be found in the SEM photos of specimens with wheat straw fibers. As shown in Figure 13b,c, the serrated substances are observed in the micro-structures. When the wheat straw fibers are mixed, the strip substances are well occluded with the other hydration substrate, the contact surface is serrated, and there is no gap at the junction of the transition zone. However, many serrated substances are found in the SEM photos of specimens with rice straw and corn stalk, due to the fact that this type of fibers is prone to float on the surface of the matrix. Consequently, alkali-activated BFS with wheat straw fibers shows the optimum mechanical strengths.

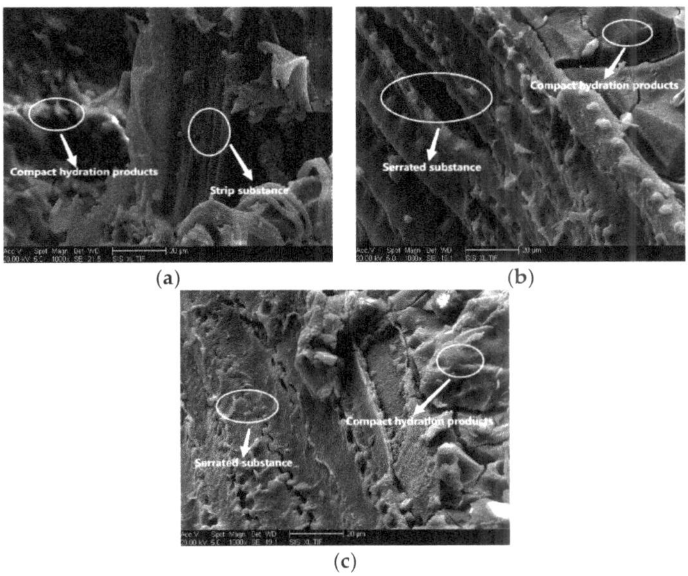

Figure 13. SEM of specimens. (**a**) With wheat straw; (**b**) With rice straw; (**c**) With corn stalk.

4. Conclusions

This study investigates the various influencing factors on the mechanical strengths of alkali-activated BFS powder. The research conclusions can be derived as follows.

Alkali-activated BFS powder with potassium silicate exhibits better mechanical strengths than the sodium silicate alkali-activated BFS powder. The flexural strengths of potassium silicate alkali-activated BFS cured for 3, 7, and 28 days are 3%, 24%, and 30.4% higher than those of the specimen mixed with sodium silicate. The compressive strengths of sodium silicate alkali-activated BFS cured for 3, 7, and 28 days are 4.4%, 3.3%, and 16.8%, respectively.

The modulus of alkali demonstrates a negative effect on the mechanical strengths of alkali-activated BFS powder. Alkali-activated BFS with 14% alkali shows the highest mechanical strengths. The flexural strength and compressive strength of the specimens with 14% alkali are 33.8% and 32.7% higher than those with 8% alkali. Meanwhile, the flexural strength and compressive strength of the specimens with 14% alkali are 21.2% and 17.1% higher than those with 18% alkali content.

The mechanical strengths of specimens with 35% water consumption are the highest, while specimens with 42% water consumption show the lowest mechanical strengths. The flexural and compressive strengths of specimens with water consumption of 35% are 313% and 35.5% of the specimens with 42% water consumption.

An amount of 1% wheat straw fibers by volume of the total alkali-activated BFS powder and alkali treatment on fibers are the most favorable to the mechanical strength.

Strip substance has been found in the SEM photos of samples with wheat straw fibers, while the serrated substances are observed in the micro-structures of specimens with rice straw and corn stalk. The micro-structures of specimens with wheat straw fibers are the densest.

The alkali-activated BFS with the optimum mixing proportion and the manufacturing method can be applied in the repair of damaged buildings in the future. Nowadays, this kind of material is prone to cracking when applied in actual engineering. The rubber powder can be used to hinder the development of cracks. Such research will be conducted in the future.

Author Contributions: Conceptualization, J.Z., X.L. and H.W.; methodology, J.Z. and X.W.; validation, H.W. and L.S.; investigation, J.Z., H.W. and L.S.; data curation, X.L. and Z.Q.; writing—original draft preparation, J.Z.; writing—review and editing, Z.W.; supervision, H.W.; project administration, X.L. and Z.Q. All authors have read and agreed to the published version of the manuscript.

Funding: This research was sponsored by the Natural Science Foundation of Heilongjiang (LH2019E066), the Natural Science Foundation of China (51508140) and the Innovation and the College Students Innovation and Entrepreneurship Project (202110214006x, 202110214252). This support is hereby acknowledged.

Institutional Review Board Statement: Not applicable.

Informed Consent Statement: Not applicable.

Data Availability Statement: Not applicable.

Conflicts of Interest: The authors declare no conflict of interest.

References

1. Poudyal, L.; Adhikari, K. Environmental sustainability in cement industry: An integrated approach for green and economical cement production. *Curr. Opin. Environ. Sustain.* **2021**, *4*, 100024. [CrossRef]
2. Provis, J.L.; Bernal, S.A. Geopolymers and related alkali-activated materials. *Adv. Mater. Res.* **2014**, *44*, 299–327. [CrossRef]
3. Jegan, M.; Annadurai, R.; Kannan Rajkumar, P.R. A state of the art on effect of alkali activator, precursor, and fibers on properties of geopolymer composites. *Constr. Mater.* **2023**, *18*, e01891. [CrossRef]
4. Van den Heede, P.; De Belie, N. Environmental impact and life cycle assessment (LCA) of traditional and 'green'concretes: Literature review and theoretical calculations. *Cem. Concr. Compos.* **2012**, *34*, 431–442. [CrossRef]
5. Bhutta, A.; Farooq, M.; Zanotti, C. Pull-out behavior of different fibers in geopolymer mortars: Effects of alkaline solution concentration and curing. *Mater. Struct.* **2017**, *50*, 80. [CrossRef]
6. Hasanbeigi, A.; Price, L.; Lin, E. Emerging energy-efficiency and CO_2 emission-reduction technologies for cement and concrete production: A technical review. *Renew. Sustain. Energy Rev.* **2012**, *16*, 6220–6238. [CrossRef]
7. Guo, Y.; Luo, L.; Liu, T.; Hao, L.; Li, Y.; Liu, P.; Zhu, T. A review of low-carbon technologies and projects for the global cement industry. *J. Environ. Sci.* **2024**, *136*, 682–697. [CrossRef]
8. Zhang, Z.; Zhang, J.; Yang, X.; Ning, Y.; Ji, T. Effect mechanism of slag activity on the basic tensile creep of alkali-activated slag mortar. *J. Build. Eng.* **2023**, *68*, 105903. [CrossRef]
9. Koppoju, M.; Mudimby, A.; Abhinay, A. Fracture parameters of flyash and GGBS based Alkali activated concrete. *Mater. Today* **2022**, *65*, 2053–2059. [CrossRef]
10. Mohanty, S.; Roy, N.; Singh, S.P.; Sihag, P. Strength and durability of flyash, GGBS and cement clinker stabilized dispersive soil. *Cold Reg. Sci. Technol.* **2021**, *191*, 103358. [CrossRef]
11. Varadharajan, S. Determination of mechanical properties and environmental impact due to inclusion of flyash and marble waste powder in concrete. *Structures* **2020**, *25*, 613–630. [CrossRef]
12. Nishanth, L.; Patil, N.N. Experimental evaluation on workability and strength characteristics of self-consolidating geopolymer concrete based on GGBFS, flyash and alccofine. *Mater. Today* **2022**, *59*, 51–57. [CrossRef]
13. Liu, H.; Wang, C.; Liu, X.; Wu, D.; Yang, H.; Zhang, Z.; Jan Khan, K.U. Deformation characteristics and prediction of unbound volcanic ash pavement based on the mechanistic-empirical design guide. Constr. *Build. Mater.* **2022**, *327*, 126975. [CrossRef]
14. Guler, S.; Akbulut, Z.F. Workability & mechanical properties of the single and hybrid basalt fiber reinforced volcanic ash-based cement mortars after freeze–thaw cycles. *Structures* **2023**, *48*, 1537–1547.

15. Liu, H.; Wang, C.; Wu, D.; Liu, X.; Zhang, Z. Deformation and critical dynamic stress for compacted volcanic ash subjected to monotonic and dynamic loads. *Constr. Build. Mater.* **2022**, *358*, 129454. [CrossRef]
16. Gobin, M.; Yasufuku, N.; Liu, G.; Watanabe, M.; Ishikura, R. Small strain stiffness, microstructure and other characteristics of an allophanic volcanic ash. *Eng. Geol.* **2023**, *313*, 106967. [CrossRef]
17. Zhang, H.; Ai, J.; Ren, Q.; Zhu, X.; He, B.; Jiang, Z. Understanding the strength evolution of alkali-activated slag pastes cured at subzero temperature. *Cem. Concr. Compos.* **2023**, *138*, 104993. [CrossRef]
18. Khaled, Z.; Mohsen, A.; Soltan, A.M.; Kohail, M. Optimization of kaolin into Metakaolin:Calcination Conditions, mix design and curing temperature to develop alkali activated binder. *Ain Shams Eng. J.* **2023**, *14*, 102142. [CrossRef]
19. Xu, Z.; Guo, Z.; Zhao, Y.; Li, S.; Luo, X.; Chen, G.; Liu, C.; Gao, J. Hydration of blended cement with high-volume slag and nano-silica. *J. Build. Eng.* **2023**, *64*, 105657. [CrossRef]
20. Jawad Ahmed, M.; Santos, W.F.; Brouwers, H.J.H. Air granulated basic Oxygen furnace (BOF) slag application as a binder: Effect on strength, volumetric stability, hydration study, and environmental risk. *Constr. Build. Mater.* **2023**, *367*, 130342. [CrossRef]
21. Yao, W.; Xia, K.; Liu, Y.; Shi, Y.; Peterson, K. Dependences of dynamic compressive and tensile strengths of four alkali-activated mortars on the loading rate and curing time. *Constr. Build. Mater.* **2019**, *202*, 891–903. [CrossRef]
22. Fang, S.; Shu Lam, E.S.; Li, B.; Wu, B. Effect of alkali contents, moduli and curing timeon engineering properties of alkali activated slag. *Constr. Build. Mater.* **2020**, *249*, 118799. [CrossRef]
23. Zhang, S.; Niu, D. Hydration and mechanical properties of cement-steel slag system incorporating different activators. *Constr. Build. Mater.* **2023**, *363*, 129981. [CrossRef]
24. Shi, J.; Liu, B.; Chu, S.H.; Zhang, Y.; Zhang, Z.; Han, K. Recycling air-cooled blast furnace slag in fiber reinforced alkali-activated mortar. *Powder Technol.* **2022**, *407*, 117686. [CrossRef]
25. Chen, X.; Chen, J.; Li, M.; Wang, J.; Zhou, Z.; Du, P.; Zhang, X. Synthesis of kaliophilite from high calcium fly ash: Effect of alkali concentration. *Constr. Mater.* **2022**, *17*, e01542. [CrossRef]
26. Raj, R.S.; Arulraj, P.; Anand, N.; Kanagaraj, B.; Lubloy, E.; Naser, M.Z. Nanomaterials in geopolymer composites: A review. *Dev. Built Environ.* **2023**, *13*, 100114.
27. Tian, B.; Ma, W.; Li, X.; Jiang, D.; Zhang, C.; Xu, J.; He, C.; Niu, Y.; Dan, J. Effectof ceramic polishing waste on the properties of alkali-activatedslag pastes: Shrinkage, hydration and mechanical property. *J. Build. Eng.* **2023**, *63*, 105448. [CrossRef]
28. Ionescu, B.A.; Barbu, A.-M.; Lăzărescu, A.-V.; Rada, S.; Gabor, T.; Florean, C. The Influence of Substitution of Fly Ash with Marble Dust or Blast Furnace Slag on the Properties of the Alkali-Activated Geopolymer Paste. *Coatings* **2023**, *13*, 403. [CrossRef]
29. Jamaludin, L.; Razak, R.A.; Al Bakri Abdullah, M.M.; Vizureanu, P.; Sandu, A.V.; Abd Rahim, S.Z.; Ahmad, R. Solid-to-Liquid Ratio Influenced on Adhesion Strength of Metakaolin Geopolymer Coating Paste Added Photocatalyst Materials. *Coatings* **2023**, *13*, 236. [CrossRef]
30. Cheng, H.; Chen, P.; Rong, X.; Cao, S.; Zhao, W. Effect of steel fibre and manufactured sand on mechanical properties of alkali-activated slag green cementitious material after high temperature. *Constr. Mater.* **2023**, *18*, e01919. [CrossRef]
31. Mastali, M.; Mohammad, S.K.; Abdollahnejad, Z. Towards sustainable bricks made withfiber-reinforced alkali-activated desulfurization slag mortars incorporating carbonated basic oxygenfurnace aggregates. *Constr. Build. Mater.* **2020**, *232*, 117258. [CrossRef]
32. Santana, H.A.; Amorim Júnior, N.S.; Ribeiro, D.V.; Cilla, M.S.; Dias, M.R. Vegetable fibers behavior in geopolymers and alkali-activated cement based matrices: A review. *J. Build. Eng.* **2021**, *44*, 103291. [CrossRef]
33. Zhu, J.; Feng, S.; Liang, S. Experimental research on compressive performances of masonry after beiing subjected to high temperature. *Ind. Constr.* **2021**, *51*, 170–176.
34. Gao, K.; Cui, Y.; Zhang, P.; Zhao, T. Research Progress on Reinforcement Corrosion in AlkaliActivated Concrete under Chloride Attack. *Bull. Chin. Ceram. Soc.* **2020**, *39*, 3070–3077.
35. Shi, J.; Guan, X.; Ming, J.; Zhou, X. Improved corrosion resistance of reinforcing steel in mortars containing red mud after long-term exposure to aggressive environments. *Cem. Concr. Compos.* **2022**, *130*, 104522. [CrossRef]
36. Vu, T.H.; Dang, L.C.; Kang, G.; Sirivivatnanon, V. Chloride induced corrosion of steel reinforcement in alkali activated slag concretes: A critical review. *Constr. Mater.* **2022**, *16*, e01112.
37. Feng, L.; Jin, K.; Wang, H. Research on the thermal conductivity and water resistance of foamed phosphogypsum. *Coatings* **2021**, *11*, 802. [CrossRef]
38. *GB/T 17671-2021*; Test Method of Cement Mortar Strength (ISO Method). The State Administration for Market Regulation: Beijing, China, 2021.
39. *GB/T 17362-2008*; Analysis Method for Gold Products with X-ray EDS in SEM. The Standardization Administration: Beijing, China, 2008.
40. Zhang, B.; Yu, T.; Deng, L.; Li, Y.; Guo, H.; Zhou, J.; Li, L.; Peng, Y. Ion-adsorption type rare earth tailings for preparation of alkali-based geopolymer with capacity for heavy metals immobilization. *Cem. Concr. Compos.* **2022**, *134*, 104768. [CrossRef]
41. Serdar, A.; Bulent, B. Effect of activator type and content on properties of alkali-activated slagmortars. *Ann. Compos.* **2014**, *57*, 166–172.
42. Shi, Z.; Shi, C.; Zhang, J.; Wan, S.; Zhang, Z.; Ou, Z. Alkali-silica reaction in waterglass-activated slag mortars incorporating fly ash and metakaolin. *Cem. Concr. Res.* **2018**, *108*, 10–19. [CrossRef]

43. Huang, W.; Wang, Y.; Zhang, Y.; Zheng, W. Experimental study of high-temperature resistance of alkali-activated slag crushed aggregate mortar. *Mater. Technol.* **2023**, *23*, 3961–3973. [CrossRef]
44. Ma, Q.; Yang, J.; Zhu, X.; Sreejith, V.; Huang, L. Influence of alkali concentration and modulus on compressive strength of alkali activated slag concrete. *New. Build. Mater.* **2019**, *46*, 1–4.
45. Shen, J.; Li, Y.; Lin, H.; Lv, J.; Feng, S.; Ci, J. Early properties and chemical structure analysis of alkali-activated brick geopolymer with varied alkali dosage. *J. Build. Eng.* **2022**, *60*, 105186. [CrossRef]
46. Zhang, S.; Ghouleh, Z.; Mucci, A.; Bahn, O.; Provençal, R.; Shao, Y. Production of cleaner high-strength cementing material using steel slag under elevated-temperature carbonation. *J. Cleaner Prod.* **2022**, *342*, 130948. [CrossRef]
47. Garanayak, L. Strength effect of alkali activated red mud slag cement in ambient condition. *Mater. Today* **2021**, *44*, 1437–1443. [CrossRef]
48. Wang, Q.; Sun, S.; Yao, G.; Wang, Z.; Lyu, X. Preparation and characterization of an alkali-activated cementitious material with blast-furnace slag, soda sludge, and industrial gypsum. *Constr. Build. Mater.* **2022**, *340*, 127735. [CrossRef]
49. Zhu, J.; Lesley, H.; Huang, Y.; Sun, Y. Experimental research on mechanical properties of different fibers reinforced alkali-activated slag cementitious material after high temperature. *J. Wuhan Univ. Technol.* **2020**, *44*, 97–102.
50. Tong, W.; Chu, Q.; Li, J.; Xie, X.; Wang, J.; Jin, Y.; Wu, S.; Hu, J.; Song, K. Insight into understanding sequential two-stage pretreatment on modifying lignin physiochemical properties and improving holistic utilization of renewable lignocellulose biomass. *Renew. Energy* **2022**, *187*, 123–134. [CrossRef]
51. Garlock, R.J.; Balan, V.; Dale, B.E. Comparative material balances around pretreatment technologies for the conversion of switchgrass to soluble sugars. *Bioresour. Technol.* **2011**, *102*, 11063–11071. [CrossRef]
52. Zhu, J.; Zheng, W.; Lessley, H. Mechanical properties of plant fibers reinforced alkali-activated slag cementitious material at hightemperature. *Ann. Chim. Sci. Mat.* **2019**, *43*, 240–255. [CrossRef]
53. Sellami, A.; Merzoud, M.; Amziane, S. Improvement of mechanical properties of green concrete by treatment of the vegetals fibers. *Constr. Build. Mater.* **2013**, *47*, 1117–1124. [CrossRef]

coatings

MDPI

Article

Mathematical Model for Early-Aged UHPFRC Compressive Strength Changes

Xi Peng [1,2], Qiuwei Yang [1,2,*], Hongfei Cao [1,3] and Haozhen Wang [1,3]

[1] School of Civil and Transportation Engineering, Ningbo University of Technology, Ningbo 315211, China
[2] Engineering Research Center of Industrial Construction in Civil Engineering of Zhejiang, Ningbo University of Technology, Ningbo 315211, China
[3] Faculty of Mechanical Engineering and Mechanics, Ningbo University, Ningbo 315211, China
* Correspondence: yangqiuwei@nbut.edu.cn

Abstract: Compressive strength is the most important mechanical index of ultra-high performance fiber-reinforced concrete (UHPFRC). The rule of changes in compressive strength in early-aged UHPFRC is of great significance to guide concrete curing, formwork removal, and prestress stretching. Therefore, it is very necessary to study an accurate mathematical model to describe the change in compressive strength of UHPFRC at an early age. For this purpose, a new mathematical model of compressive strength age is proposed in this work for predicting the long-term strength of UHPFRC according to a few test data from early-aged UHPFRC. This new model can overcome the shortcomings of the existing models, such as the exponential model, logarithmic model, and polynomial model. The proposed model is first demonstrated by using four groups of compressive strength test data compiled from previous research studies. Subsequently, an experiment of early-aged UHPFRC compressive strength was carried out to further verify the proposed mathematical model. The mixed proportion used in the UHPFRC compressive strength test was 10.87:0.82:1 (powder:steel fiber:water), and the design strength grade was 120 MPa. Based on the UHPFRC experimental data, it was shown that the average fitting error and standard deviation of the new model were about 10%~20% of that of the logarithmic model and the polynomial model. The proposed model can precisely predict the compressive strength of UHPFRC, with a determination coefficient (R^2) of 0.9974. The research results show that the average fitting error and standard deviation of this new model were significantly reduced when compared to the existing models, and the predicted compressive strength by the new model on the 60th day is the closest to the actual design strength grade of concrete. The greatest advantage of the proposed method lies in its simple formula, fast implementation, and no need for complex mathematical operations. It has been shown that the proposed model is superior to the existing models due to its higher fitting accuracy and prediction accuracy, and it can be better used to predict the later strength of UHPFRC by using only a few compressive strength test data taken at the early age stage.

Keywords: ultra-high performance concrete; age; compressive strength; exponential model; logarithmic model; polynomial model

1. Introduction

Concrete is widely used in bridge structure, beam, slab, column, and pile foundation engineering in building construction [1–4]. In recent years, ultra-high performance fiber reinforced concrete (UHPFRC) has become an important development direction of cement-based composites due to its ultra-high strength and excellent durability. Compressive strength is the most important mechanical parameter of UHPFRC. Yang et al. [5] found that the compressive and bending strength of the UHPFRC samples cured at 20 °C decreased by about 20% and 10% compared with the samples cured at 90 °C. Kwon et al. [6] found that the proper volume fraction of the longer fiber is essential in improving the various mechanical properties of UHPFRC, including compressive strength, strain capacity, and energy absorption capacity. Yu et al. [7] discussed the influence of UHPFRC mix design on compressive strength. They used the improved Andreasen & Andersen particle filling model to carry out the mix design to achieve the goal of a dense cement matrix. These studies revealed that curing temperature, mix ratio, and steel fiber content have an important influence on the development of the compressive strength of early-aged UHPFRC. In other words, these factors complicate the development law of UHPFRC compressive strength at an early age. The rule of changes in UHPFRC early compressive strength is of great significance to guide concrete curing, formwork removal, prestress tensioning, and later curing. In view of this, many studies have been carried out on the development law and estimation method of concrete compressive strength in the last few decades. These studies can be broadly divided into two types. The first type is the estimation methods based on multiple factors, such as the curing temperature, water-cement ratio, steel fiber content, curing time, etc. Yeh [8] studied the possibility of artificial neural network (ANN) predictions of compressive strength in high-performance concrete. It was found that the strength model based on the ANN was more accurate than the model based on regression analysis. Bui et al. [9] developed an expert system based on an ANN model and an improved firefly algorithm, which could better predict the property of high-performance concrete. Han et al. [10] studied the prediction of the compressive strength of high-performance concrete based on machine learning. By selecting the appropriate variables based on the importance of random forest measurements, the accuracy and efficiency of the proposed model prediction can be improved. Anyaoha et al. [11] adopted a new integrated computing technology-boosting smooth transition regression tree (BooST) to predict the compressive strength of high-performance concrete. The research results show that BooST is superior to other methods in prediction accuracy, with the smallest error, and is more in line with the laboratory experimental results. Hameed et al. [12] studied the ability of ANNs combined with principal component analysis (PCA) and cross-validation (CV) technology to predict the compressive strength of high-performance concrete. When compared with other models published in previous studies, the performance of the proposed model achieved the expected advantages and more stable prediction. Emad et al. [13] developed four soft computing technologies to predict the compressive strength of UHPFRC, including linear relationships, pure quadratics, M5P trees, and ANNs. The research results show that the ANN model had high prediction accuracy for the compressive strength of UHPFRC. Based on the optimized ANN model, a closed equation is proposed to quickly estimate the compressive strength of concrete. Ali et al. [14] used six models: linear regression (LR), multiple linear regression (MLR), nonlinear regression (NLR), pure quadratic (PQ), interaction (IA), and full quadratic (FQ) to predict the compressive strength of modified concrete with different nano-silica contents in different mix proportions. The research results showed that the most effective and accurate model was the FQ model. Ibrahim et al. [15] studied four models, including LR, NLR, PQ, and FQ, to evaluate the impact of recycled coarse aggregate on the long-term compressive strength of concrete under different mix proportions and different compressive strength ranges. Mohammed et al. [16] tested the effect of three water-reducing agent additives on the workability and compressive strength of concrete during the curing period of 1 to 28 days. A nonlinear M5P mathematical model is established to predict the compressive strength of concrete and to study the influence of the mix ratio on the compressive strength of concrete. Moradi et al. [17] developed a robust and time-saving method based on machine learning (ML) to predict the compressive strength of concrete with binary supplementary cementitious materials (SCMs) at different ages. It was found that the ML-based model could accurately predict the compressive strength of concrete containing any SCM at any age. Rezvan et al. [18] studied the benefits of using polyethylene terephthalate (PET) fibers as reinforcement elements in concrete by the ML method. The results showed that the content of PET fiber in concrete has a significant impact on the concrete compressive strength. Imran et al. [19] developed a new hybrid model for predicting the compressive strength of UHPFRC by using a cascaded forward neural network (CFNN) and artificial bee colony (ABC) optimization. It was found that the optimal neural network structure selected by the ABC method consists of two layers, and the developed model (CFNN-ABC) can accurately predict the compressive strength of UHPFRC. Other similar studies [20–26] reported the successful application of a BP neural network, probabilistic neural network, and a depth neural network in the early age strength prediction of high-strength concrete. In general, most of the first types of methods are ANN-based prediction methods. Its advantage is that it can consider the impact of various factors on compressive strength, and its limitation is that it requires a large number of test data to train the neural network.

The second type of method is to predict the later strength of concrete according to a function curve between concrete strength and age. Based on the test data from an early age, the concrete compressive strength age function equation can be obtained by data fitting, and then the concrete strength at any time can be calculated. The advantage of this kind of method is that only a few test data are needed to predict the concrete strength at any time, but the disadvantage is that only the change rule of strength with time is considered. At present, the commonly used concrete compressive strength–age curve models are the exponential model [27–30], logarithmic model [31,32], and polynomial model [33–36]. Yang et al. [27] tested the compressive strength of high-strength concrete cylinder blocks on days 1, 3, 5, 7, 10, 14, and 28. They fitted the data using the exponential model and obtained the empirical formula to predict the strength of high-strength concrete. Zhao et al. [28] examined the compressive strength of C50 concrete at different ages and generated the curve of compressive strength with age using an exponential model. Yang et al. [29] established a diachronic variation model of axial compressive strength of C60 concrete in the early aging period using the regression fitting method and combining the exponential model. Wang et al. [30] analyzed the development law of compressive strength of high-strength concrete with age. Zhao [31] discussed the development law of high-strength concrete aging strength and obtained the corresponding logarithmic fitting formula. Li et al. [32] examined the compressive strength of the C60 concrete cube on days 3, 7, 14, 28, and 60 and fitted the data using a logarithmic model to predict the later strength of concrete. Li [33] and Jin [34] developed the diachronic variation models of early-aged concrete axial-compressive strength by using a polynomial and the regression fitting method. Ling [35] analyzed the influence of concrete composition on compressive strength and established a prediction model for the compressive strength of mechanical sand concrete using a third-order nonlinear function. Jia et al. [36] examined the compressive strength of ordinary Portland concrete from the first to the 28th day and fitted the data using the cubic polynomial model to obtain the concrete compressive strength–time curve model. In general, these existing curve models may be inconsistent with the strength growth law of concrete in some aspects. For example, the strength of concrete slowly increases and tends to stabilize after 28 days of pouring. However, the polynomial model cannot conform to this rule, and the exponential model and logarithmic model cannot indicate this feature well. In addition, the strength of high-strength concrete increases rapidly in the ultra-early stage (i.e., before 7 days after pouring), which will lead to the poor fitting accuracy of the exponential model and logarithmic model at the ultra-early stage. Therefore, it is still necessary to study a new mathematical curve model to overcome the defects of the existing curve models so as to more accurately describe the change rule of UHPFRC compressive strength.

The work of this paper also belongs to the second type of method. The research significance of this work is to develop a new mathematical curve model that can accurately predict the compressive strength of UHPFRC at any time based on only a few early-age test data. For this purpose, the advantages and disadvantages of the three existing curve models are discussed first, and then a new model that can overcome the shortcomings of the existing model is proposed in Section 2. The implementation effects of the new model and the old models are compared to illustrate the feasibility and superiority of the new model by using four groups of compressive strength test data compiled from previous research studies in Section 3. Moreover, a UHPFRC early compressive strength test was carried out to further verify the proposed mathematical model in Section 4. The research results show that the average fitting error and standard deviation of the proposed model are significantly smaller than those of the existing models, and the concrete compressive strength predicted by the proposed model on the 60th day is also the closest to the design strength grade of concrete. The proposed model provides a simple and feasible way to predict the UHPFRC compressive strength with only a few early-age test data in engineering practice.

2. Compressive Strength-Time Mathematical Models

As stated before, the method of concrete strength evaluation based on a certain assumed mathematical curve is a simple and easy way in engineering practice because it only requires a few early-age test data of compressive strength. The main steps of this type of method are shown in Figure 1.

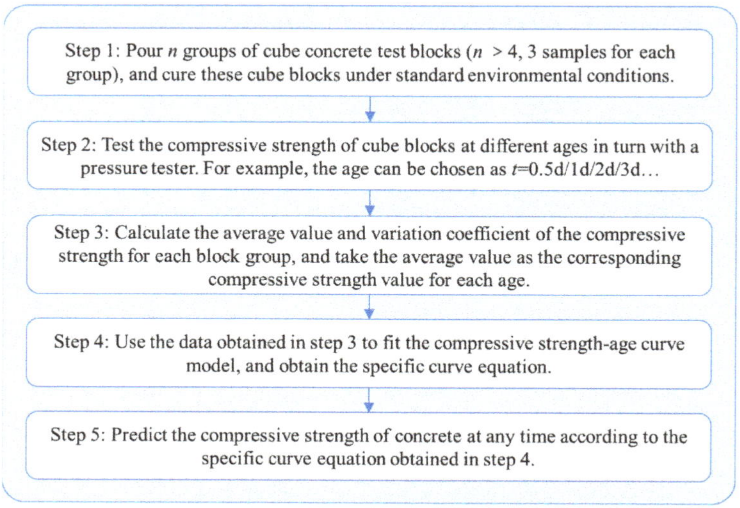

Figure 1. The flow chart of concrete compressive strength evaluation based on strength-age curve model.

It is clear that the proper mathematical curve model is the key to determining whether such methods can be implemented successfully. This section first reviews several of the existing mathematical models of changes in concrete compressive strength with time, discusses the advantages and disadvantages of these models, and then proposes a new compressive strength–time mathematical model. Assuming that t represents the age (time) after concrete pouring and $s(t)$ represents the compressive strength of concrete corresponding to the moment, t, the exponential model of the compressive strength-time is expressed as:

$$s(t) = x_0 \cdot e^{-x_1/t} \tag{1}$$

In Equation (1), x_0 and x_1 denote two non-negative fitting parameters that should be obtained by curve fitting based on the strength test data of concrete at an early age. Equation (1) indicates that the advantage of the exponential model is that the function is an increasing function. $s(t)$ increases with an increase in t, which conforms to the law that the strength of concrete increases with its age in engineering practice. Moreover, when $t \to \infty$, $s(t) \to x_0$, which is also consistent with the law that the strength of concrete remains almost stable after 28 days. The disadvantage of the exponential model is that its definition domain lacks the point $t = 0$. In practice, the concrete strength is 0 when $t = 0$. It is noteworthy that the exponential model cannot contain this situation. Furthermore, the equation contains only two fitting parameters, so the accuracy of the fitting is low for data with greater dispersion.

The logarithmic model of diachronic change in compressive strength is written as:

$$s(t) = x_0 + x_1 \cdot \ln(t) \tag{2}$$

In Equation (2), x_0 and x_1 represent two fitted parameters that are obtained based on partial test data combined with curve fitting. The comparison between Equations (1) and (2) suggests that the logarithmic model can be regarded as a form of change obtained by taking logarithms on both sides of the exponential model equation. Accordingly, the advantages and disadvantages of the logarithmic model are similar to the exponential model. To be specific, a detailed advantage is that the function is an increasing function, which conforms to the law that the strength of concrete increases with its age in engineering practice. The disadvantage is that its definition domain also lacks the point of $t = 0$, and when $t \to \infty$, the growth rate slows down but does not converge, which is not consistent with the actual growth law of concrete strength. The equation comprises only two fitted parameters, and the fitting accuracy is also low for data with greater dispersion.

Compressive strength–age polynomial model generally refers to the quadratic power function model and cubic power function model, which are expressed as follows:

$$s(t) = x_0 + x_1 t + x_2 t^2 \tag{3}$$

$$s(t) = x_0 + x_1 t + x_2 t^2 + x_3 t^3 \tag{4}$$

Equation (3) is a quadratic polynomial model, Equation (4) is a cubic polynomial model, and x_0, x_1, x_2, and x_3 are the fitting parameters. The advantage of the polynomial model is that the definition domain contains the point $t = 0$, and the obtained function after fitting is an increasing function, which is more consistent with the actual strength growth law of high-strength concrete within 28 days. The equation contains three or four fitted parameters, so the fitting accuracy is higher than an exponential or logarithmic model with only two parameters. The disadvantage of the polynomial model is that when $t \to \infty$, $s(t)$ will increase rapidly, which is not consistent with the law that the strength of concrete at the later stage remains nearly unchanged in practice, so it cannot be used to predict the later strength of concrete.

In summary, it is very necessary to propose a novel mathematical model for the change of compressive strength with time to better describe the growth law of concrete compressive strength in practice. This study refers to the mathematical form of the polynomial model, maintains its advantages and overcomes its disadvantages, and proposes the following new compressive strength-time curve model:

$$s(t) = x_0 + x_1 \cdot \left(\frac{1}{t+1}\right) + x_2 \cdot \left(\frac{1}{t+1}\right)^2 \tag{5}$$

where x_0, x_1, and x_2 represent the three fitting parameters. Obviously, the novel model Equation (5) can be considered an improved form of Equation (3), and its independent variable t is modified as $\frac{1}{t+1}$. The novel model shows the following advantages. (1) The definition domain contains the point $t = 0$, and $s(t) = 0$ can be determined when $t = 0$ through parameter fitting, thus overcoming the disadvantage of the lack of definition domains for the exponential model and logarithmic model. (2) When $t \to \infty$, $s(t) \to x_0$, consistent with the law that the concrete strength at the later stage remains almost unchanged in practice, thus overcoming the limitation that the polynomial model does not converge when $t \to \infty$. As a result, the novel model can be adopted to predict the later strength of concrete. (3) The equation comprises three fitting parameters, and the fitting accuracy is high. The cubic term can be added to further improve the fitting accuracy if necessary, which will not be repeated here.

Next, the calculation method of the fitted parameters in the above mathematical model is expressed. The corresponding linear equations should first be established based on the compressive strength data of different ages obtained through the test to calculate the fitting parameters. With the proposed model Equation (5) as an example, the corresponding system of linear equations is written as follows:

$$y = C \cdot x \tag{6}$$

$$y = \begin{Bmatrix} s(t_1) \\ \vdots \\ s(t_n) \end{Bmatrix} \tag{7}$$

$$x = \begin{Bmatrix} x_0 \\ x_1 \\ x_2 \end{Bmatrix} \tag{8}$$

$$C = \begin{bmatrix} 1 & \frac{1}{t_1+1} & \left(\frac{1}{t_1+1}\right)^2 \\ \vdots & \vdots & \vdots \\ 1 & \frac{1}{t_n+1} & \left(\frac{1}{t_n+1}\right)^2 \end{bmatrix} \tag{9}$$

where $s(t_n)$ represents the compressive strength value of concrete obtained by t_n time test. Obviously, the vector y and the coefficient matrix C are known quantities, and x is the unknown fitted parameter vector. In general, the coefficient matrix of Equation (6) is a rectangular matrix with a greater number of rows than columns, i.e., the number of examined data (known quantities) is greater than the number of fitted parameters (unknowns). Equation (6) can generally be solved using the least squares method, for which both sides of Equation (6) are simultaneously multiplied by the transposed matrix C^T of matrix C yields:

$$C^T y = (C^T C) \cdot x \tag{10}$$

From Equation (10), the solution of x is expressed as:

$$\hat{x} = (C^T C)^{-1} C^T y \tag{11}$$

The calculated $\hat{x}$ refers to the least squares estimate of x.

The specific curve equation of the fitted prediction model can be obtained based on the calculated fitting parameters, and the superiority of the model can be examined in accordance with the difference between the model prediction value and the experimental test value. Therefore, the difference between the model prediction and the experimental test value is calculated as follows:

$$\{\delta\} = |C \cdot \hat{x} - y| \tag{12}$$

where $\{\delta\} = (\delta_1, \cdots, \delta_n)^T$ is called the residual vector. To be specific, δ_i is the residuals between the ith experimental data and the predicted value. Further calculation of the mean $\overline{\delta}$ and standard deviation σ of the residual vector is as follows:

$$\overline{\delta} = \frac{\delta_1 + \cdots + \delta_n}{n} \tag{13}$$

$$\sigma = \sqrt{\frac{\sum_{i=1}^{n} (\delta_i - \overline{\delta})^2}{n}} \tag{14}$$

The mean $\overline{\delta}$ and standard deviation σ of the residual vector are called the fitting mean error (absolute error) and standard deviation of error, respectively. The smaller $\overline{\delta}$ and σ, the higher the fitting accuracy of the model and the more reliable the model will be.

To check the model's predictability, the coefficient of determination (R^2) is calculated along with root mean square error ($RMSE$). These performance indexes are calculated to measure how close are the predicted values from the experimental ones. This helps in better evaluation of the model capabilities for compressive strength prediction. The equations for R^2 and RMSE are given by:

$$R^2 = 1 - \frac{\sum_{i=1}^{n} (y_e - y_p)^2}{\sum_{i=1}^{n} (y_e - \overline{y}_e)^2} \tag{15}$$

$$RMSE = \sqrt{\frac{1}{n}\sum_{i=1}^{n} (y_e - y_p)^2} \tag{16}$$

where, y_e = experimental value, $\overline{y}_e$ = mean of experimental values, y_p = predicted value, and n = total number of observations.

3. Model Verification and Comparison by the Test Data Compiled from Previous Literature

Four sets of experimental data (called examples 1 to 4) from literature [28,29,36,37] are used to validate the new model, and the results are compared with the existing logarithmic model and cubic polynomial model fitting results to illustrate the advantages of the proposed model.

In example 1 (the literature [28]), the concrete cube test blocks with a side length of 150 mm were designed and fabricated according to the C50 strength grade, and the compressive strengths on days 1, 2, 3, 5, 7, 11, 14, 28, and 40 were examined under standard curing conditions, respectively, as shown in Figure 2 (expressed by symbol "o"). These data are fitted using the logarithmic model, the cubic polynomial model, and the model proposed, and the obtained fitting curves are also presented in Figure 2. The specific fitting equations for these models are expressed as follows:

$$s(t) = 27.2818 + 8.9263 \cdot \ln(t) \tag{17}$$

$$s(t) = 26.0318 + 3.6070t - 0.1449t^2 + 0.0019t^3 \tag{18}$$

$$s(t) = 58.4528 - 95.1937 \cdot (\frac{1}{t+1}) + 53.3113 \cdot (\frac{1}{t+1})^2 \tag{19}$$

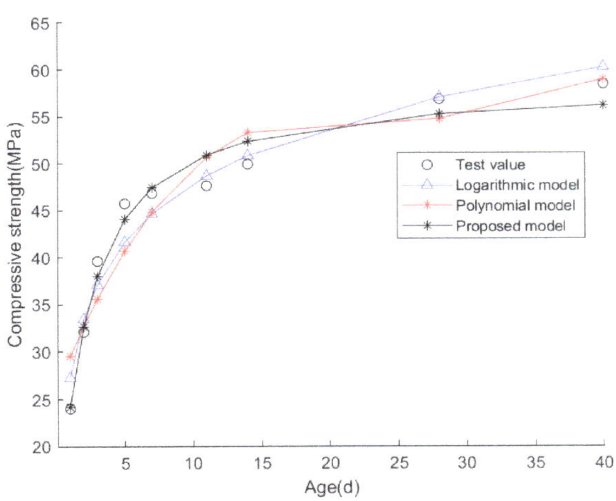

Figure 2. The fitting results for the compressive strength test data in example 1.

Table 1 lists the mean fitting error, standard deviation of error, R^2, RMSE, and the compressive strength value on day 60 predicted by these models.

Table 1. The mean fitting errors, standard deviations, and the predicted 60th-day compressive strengths by the three models based on experimental data in example 1.

Index	Logarithmic Model	Polynomial Model	Proposed Model
Mean fitting error $\overline{\delta}$	0.1907	0.2950	0.1133
Fitted standard deviation of error σ	0.0410	0.0670	0.0194
R^2	0.9541	0.8976	0.9694
$RMSE$	2.2453	3.3528	1.8317
Predicted day 60 compressive strength	63.83 MPa	127.39 MPa	56.91 MPa

It can be concluded from Table 1 that the average fitting error of the novel model is about 59% of the logarithmic model and about 38% of the polynomial model. The fitting standard deviation of the novel model is nearly 47% of the logarithmic model and about 29% of the polynomial model. The mean fitting error and mean standard deviation of the new model are significantly smaller than the logarithmic model and the cubic polynomial model. For the strength values predicted by the three models on the 60th day, only the 56.91 MPa predicted by the novel model is closest to the C50 intensity grade. It has been shown that the novel model is more accurate and reliable than the logarithmic model and the polynomial model.

In example 2 (the literature [29]), 150 mm × 150 mm × 300 mm concrete prismatic test blocks were designed and fabricated according to the C60 strength grade, and the compressive strengths on days 3, 5, 6, 7, 14, and 28 were examined, respectively, as illustrated in Figure 3 (represented by the symbol "o"). These data were fitted using the logarithmic model, the cubic polynomial model, and the model proposed, and the obtained fitting curves are presented in Figure 3. The specific fitting equations are as follows:

$$s(t) = 29.9666 + 11.2588 \cdot \ln(t) \tag{20}$$

$$s(t) = 23.3841 + 6.6224t - 0.3747t^2 + 0.0068t^3 \tag{21}$$

$$s(t) = 70.6181 - 151.4078 \cdot (\frac{1}{t+1}) + 112.5353 \cdot (\frac{1}{t+1})^2 \tag{22}$$

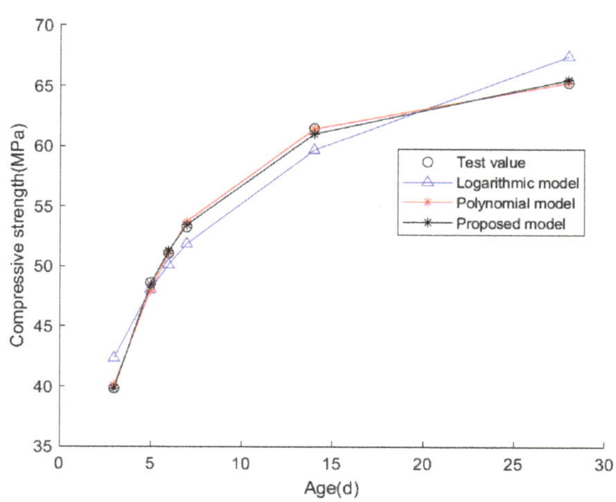

Figure 3. The fitting results for compressive strength test data in example 2.

The mean fitting error, standard deviations of error, R^2, and RMSE are shown in Table 2. The compressive strength values on day 60 predicted by these models are also listed in Table 2.

Table 2. The mean fitting errors, standard deviations, and the predicted 60th-day compressive strengths by the three models based on experimental data in example 2.

Index	Logarithmic Model	Polynomial Model	Proposed Model
Mean fitting error $\bar{\delta}$	0.0847	0.0170	0.0101
Fitted standard deviation of error σ	0.0179	0.0054	0.0023
R^2	0.9579	0.9984	0.9991
RMSE	1.7148	0.339	0.2442
Predicted day 60 compressive strength	76.06 MPa	550.14 MPa	68.17 MPa

It can be concluded from Table 2 that the average fitting error of the novel model is about 12% of the logarithmic model and about 59% of the polynomial model. The fitting standard deviation of the novel model is about 13% of the logarithmic model and about 43% of the polynomial model. The mean fitting error and mean standard deviation of the proposed model are significantly smaller than the logarithmic model and the cubic polynomial model. For the strength values predicted by the three models on the 60th day, only the 68.17 MPa predicted by the novel model is the closest to the C60 intensity grade. This shows that the novel model is more accurate and reliable than the logarithmic model and the polynomial model.

In example 3 (the literature [36]), 50 mm × 50 mm × 150 mm Portland concrete cuboid test blocks were designed, produced, and cured under standard curing conditions, and the compressive strength was examined for 1–28 days, as shown in Figure 4 (expressed by the symbol "o"). Based on the above experimental data, the fitting curves are also presented in Figure 4. The specific fitting equations are written as follows:

$$s(t) = 10.2190 + 21.3746 \cdot \ln(t) \quad (23)$$

$$s(t) = 2.6400 + 10.8638t - 0.5936t^2 + 0.0111t^3 \quad (24)$$

$$s(t) = 85.9810 - 273.5529 \cdot \left(\frac{1}{t+1}\right) + 229.0393 \cdot \left(\frac{1}{t+1}\right)^2 \quad (25)$$

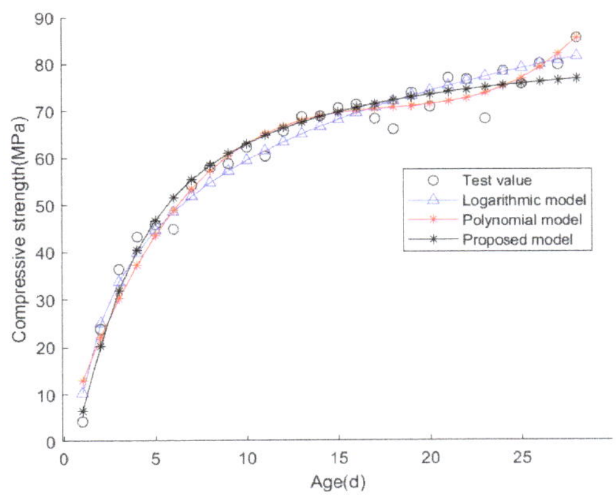

Figure 4. The fitting results for compressive strength test data in example 3.

The mean fitting error, standard deviation of error, R^2, RMSE, and the predicted compressive strength on day 60 are shown in Table 3.

Table 3. The mean fitting errors, standard deviations, and the predicted 60th-day compressive strengths by the three models based on experimental data in example 3.

Index	Logarithmic Model	Polynomial Model	Proposed Model
Mean fitting error $\bar{\delta}$	1.4240	2.0567	0.6167
Fitted standard deviation of error δ	0.2581	0.3783	0.0988
R^2	0.9672	0.9639	0.9623
$RMSE$	3.2636	3.4255	3.4973
Predicted day 60 compressive strength	97.73 MPa	917.90 MPa	81.56 MPa

From Table 3, it can be concluded that the average fitting error of the novel model is about 43% of the logarithmic model and about 30% of the polynomial model. The fitting standard deviation of the novel model is about 38% of the logarithmic model and about 26% of the polynomial model. For the strength values predicted by the three models on the 60th day, only the 81.56 MPa predicted by the novel model is closest to the C80 strength grade. These results show that the novel model is more accurate and reliable than the logarithmic model and the polynomial model.

In example 4 (the literature [37]), the 150 mm × 150 mm × 150 mm high-strength concrete cube test blocks were designed and fabricated according to the C80 strength grade, and the compressive strengths of days 0.5, 1, 3, 7, 14, 28, and 90 are examined respectively, as shown in Figure 5 (represented by the symbol "o"). The fitting curves obtained by the three models are also shown in Figure 5. The specific fitting equations are given as follows:

$$s(t) = 47.5587 + 9.8023 \cdot \ln(t) \tag{26}$$

$$s(t) = 41.6676 + 4.8059t - 0.1582t^2 + 0.0012t^3 \tag{27}$$

$$s(t) = 83.3955 - 49.5567 \cdot \left(\frac{1}{t+1}\right) - 37.5315 \cdot \left(\frac{1}{t+1}\right)^2 \tag{28}$$

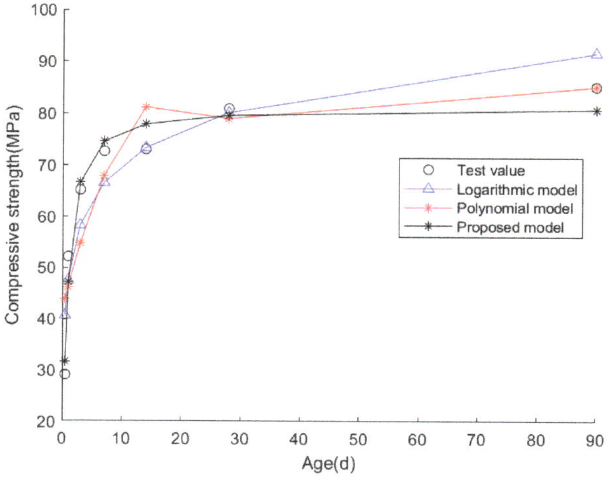

Figure 5. The fitting results of compressive strength test data in example 4.

Note that the coefficients in Equations (26)–(28) are obtained by fitting the test data, so different fitting coefficients will be obtained for the concrete test blocks with different components and water-cement ratios. Table 4 lists the mean fitting error, standard deviation of error, R^2, RMSE, and the predicted compressive strength on day 60.

Table 4. The mean fitting errors, standard deviations, and the predicted 60th-day compressive strengths by the three models based on experimental data in example 4.

Index	Logarithmic Model	Polynomial Model	Proposed Model
Mean fitting error $\overline{\delta}$	0.4344	0.5604	0.1571
Fitted standard deviation of error σ	0.1325	0.1712	0.0314
R^2	0.8738	0.7948	0.9639
RMSE	6.3454	8.0907	3.3948
Predicted day 60 compressive strength	87.69 MPa	24.89 MPa	80.57 MPa

Similar conclusions can be drawn from Table 4 that the mean fitting error and standard deviation of the proposed model are significantly smaller than the logarithmic model and the cubic polynomial model. For the strength values predicted by the three models on the 60th day, only the 80.57 MPa predicted by the novel model is the closest to the C80 strength grade. This means that the proposed model is more reasonable and reliable than the logarithmic model and the polynomial model.

4. UHPFRC Experimental Validation

In this section, the compressive strength test of UHPFRC at a very early age is analyzed to further investigate the applicability of the proposed model. In the experiment, 36 UHPFRC cube blocks with a side length of 100 mm were designed and manufactured according to the strength level of 120 MPa. As shown in Figure 6, the main materials used in the UHPFRC include powder, steel fiber, and water. The mix proportion and basic physical parameters of the UHPFRC are listed in Table 5. The powder is composed of cement, silica fume, silica sand, and superplasticizer. The diameter of the steel fiber used was 0.2 mm, the length of the fibers includes 16 mm and 12 mm, and the ratio of the long fiber-to-short fiber was 1:2. Note that the main purpose of this study is to verify the applicability of the proposed curve model to the prediction of concrete compressive strength, so only one mix design was used in this experiment. More relevant experiments are needed in the future to further verify the applicability of the proposed model for predicting the early compressive strength of UHPFRC under different mix proportions and curing temperatures.

Under the standard curing conditions, the compressive strength of the test block from half a day to the 14th day was tested. The curing temperature and relative humidity were about (20 ± 2) °C and 95%, respectively. The compression strength experiment was carried out by the STYE-3000E automatic pressure tester. The loading speed was 0.6 N/mm^2 per second. The test process was in accordance with GB/T50081-2002 Test Methods for Mechanical Properties of Ordinary Concrete [38]. The measured compressive strengths of these cubes at different times are listed in Table 6.

Table 5. Mix proportion and basic physical parameters of UHPFRC.

Mix Proportion			Density	Initial Setting Time	Final Setting Time
Powder 2130 kg/m^3	Steel fiber 160 kg/m^3	Water 196 kg/m^3	2488 kg/m^3	≥240 min	≤900 min

Figure 6. Experimental powder, steel fiber, test blocks, and equipment. (a) Powder; (b) Steel fiber; (c) High-strength concrete block; (d) Compressive strength test; (e) Some damaged test blocks.

Table 6. The measured compressive strengths of the UHPFRC cubes (MPa).

Age (Unit: Day)	Cube 1	Cube 2	Cube 3	Mean Value
0.5	44.596	41.413	43.899	43.303
1	65.842	71.954	75.241	71.012
2	101.524	100.801	97.419	99.915
3	106.935	110.032	109.759	108.909
4	120.471	109.423	112.927	114.274
5	122.200	116.100	104.400	114.300
6	125.919	107.592	122.186	118.566
7	120.030	123.060	116.28	119.790
8	121.192	122.166	121.963	121.774
9	124.069	116.635	125.644	122.116
10	117.506	122.822	128.429	122.919
14	121.747	136.977	119.324	126.016

As mentioned in the previous section, based on the experimental data, the logarithmic model, the polynomial model, and the model proposed were used to fit the data, and the resulting fitting curves are shown in Figure 7. The specific fitting equations for each model are as follows:

$$s(t) = 72.7723 + 24.0246 \cdot \ln(t) \tag{29}$$

$$s(t) = 38.3404 + 25.5434t - 3.9534t^2 + 0.1489t^3 \quad (30)$$

$$s(t) = 129.3477 - 58.5280 \cdot \left(\frac{1}{t+1}\right) - 107.3955 \cdot \left(\frac{1}{t+1}\right)^2 \quad (31)$$

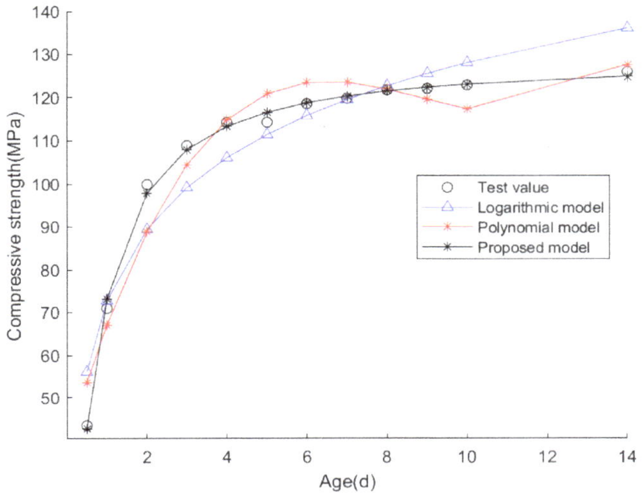

Figure 7. Curve-fitting results using the three models for the UHPFRC blocks.

Note that the coefficients in Equations (29)–(31) were obtained by fitting the test data, so different fitting coefficients will be obtained for those concrete test blocks with different components and water-cement ratios. The mean fitting error, standard deviation of error, R^2, RMSE, and predicted compressive strength of UHPFRC on the 60th day by the three models are shown in Table 7.

Table 7. The mean fitting errors, standard deviations, and the predicted 60th-day compressive strengths by the three models based on UHPFRC experimental data.

Index	Logarithmic Model	Polynomial Model	Proposed Model
Mean fitting error $\bar{\delta}$	0.3503	0.2877	0.0480
Fitted standard deviation of error σ	0.0797	0.0649	0.0095
R^2	0.9130	0.9437	0.9974
$RMSE$	7.0595	5.6791	1.2304
Predicted day 60 compressive strength	171.13 MPa	1192 MPa	128.36 MPa

It can be concluded from Table 7 that the average fitting error of the new model is about 13.7% of the logarithmic model and about 16.7% of the polynomial model. The fitted standard deviation of the new model is about 11.9% of the logarithmic model and about 14.6% of the polynomial model. Obviously, the mean fitting error and mean standard deviation of the proposed new model are much smaller than the logarithmic model and the cubic polynomial model. The proposed model has the largest R^2 and the smallest RMSE. For the compressive strength values predicted by the three models on the 60th day, only the 128.36 MPa predicted by the new model is closest to the design intensity grade of 120 MPa. It has been shown that the new model is more accurate and reliable than the logarithmic model and the polynomial model.

5. Conclusions

In this paper, a new compressive strength–age curve model is proposed, which can accurately predict the later strength of concrete with only a few early-age test data. The proposed model was validated first by using four sets of compressive strength test data collected from previous studies. Subsequently, UHPFRC early compressive strength tests were carried out to further verify the proposed mathematical model. According to the comparative analysis of the calculation results, the following main conclusions can be drawn:

(1) The existing compression strength–time curve models are not ideal in terms of fitting accuracy or prediction accuracy. The proposed new model can well describe the characteristics of compressive strength increases with time at the early age stage. From the comparison results, it was found that the new model is more accurate and reliable than the logarithmic model and the polynomial model;
(2) From the UHPFRC experiment, it can be seen that the compressive strength of UHPFRC increases very rapidly at the ultra-early age stage, and the compressive strength was very close to that of the strength design grade on about the 7th day. After the 7th day, the compressive strength of UHPFRC increases slowly and gradually maintains a stable value;
(3) Based on the UHPFRC experimental data, it has been shown that the average fitting error and standard deviation of the new model are about 10%~20% of the logarithmic model and the polynomial model. The proposed model has the largest R^2 of 0.9974 and the smallest RMSE of 1.2304. The 60-day compressive strength predicted by the proposed model (i.e., 128.36 MPa) is closest to the strength design grade of UHPFRC (i.e., 120 MPa).

It should be pointed out that the curing temperature, mix ratio, and steel fiber content have an important impact on the compressive strength of UHPFRC at an early age. More relevant experiments need to be carried out in the future to further verify the growth law of UHPFRC compressive strength at an early age with different components and water-cement ratios. The proposed method is a quick method to predict the compressive strength of concrete, which can be implemented using only a few early-age test data. The greatest advantage of the proposed method lies in its simple formula, fast implementation, and no need for complex mathematical operations. However, the proposed method can not consider the influence of other factors (except for age) on compressive strength, which is possible using the surrogate model or ML-based method. In the following research, a comparative study between the proposed model and the surrogate model or ML-based algorithm can be carried out to better apply it to the strength prediction of UHPFRC in engineering practice.

Author Contributions: Conceptualization and software, X.P.; methodology, validation and formal analysis, Q.Y.; experimental testing and investigation, H.C.; resources, X.P.; writing—original draft preparation, H.W.; writing—review and editing, X.P and Q.Y.; visualization, H.W.; project administration, X.P.; funding acquisition, X.P and Q.Y. All authors have read and agreed to the published version of the manuscript.

Funding: This research was funded by the Natural Science Foundation of China (52008215), the National Key R&D Program of China (2021YFF0501004), Zhejiang Public Welfare Technology Application Research Project (LGF22E080021), the Major Special Science and Technology Project (2019B10076) of "Ningbo Science and Technology Innovation 2025", Ningbo Natural Science Foundation Project (202003N4169), and Natural Science Foundation of Zhejiang Province, China (LQ20E080013).

Institutional Review Board Statement: Not applicable.

Informed Consent Statement: Not applicable.

Data Availability Statement: The data used to support the findings of this study are included in this study and also available from the corresponding author upon request.

Conflicts of Interest: The authors declare that there are no conflict of interest. The funders had no role in the collection, analysis, or interpretation of the data or in the decision to publish the results.

References

1. Yang, Q.W.; Sun, Y.; Peng, X. Experimental study on mechanical properties of concrete at super-early age. *Materials* **2022**, *15*, 7582. [CrossRef]
2. Yang, Q.W.; Peng, X.; Sun, Y. Shear capacity evaluation of the recycled concrete beam. *Materials* **2022**, *15*, 3693. [CrossRef] [PubMed]
3. Yang, Q.W.; Ba, Z.; Zhao, Z.; Peng, X.; Sun, Y. Stability evaluation method of hole wall for bored pile under blasting impact. *Symmetry* **2022**, *14*, 79. [CrossRef]
4. Peng, X.; Yang, Q. Damage detection in beam-like structures using static shear energy redistribution. *Front. Struct. Civ. Eng.* **2023**, *16*, 1552–1564. [CrossRef]
5. Yang, S.L.; Millard, S.G.; Soutsos, M.N.; Barnett, S.; Le, T. Influence of aggregate and curing regime on the mechanical properties of ultra-high performance fibre reinforced concrete (UHPFRC). *Constr. Build. Mater.* **2009**, *23*, 2291–2298. [CrossRef]
6. Kwon, S.; Nishiwaki, T.; Kikuta, T.; Mihashi, H. Development of ultra-high-performance hybrid fiber-reinforced cement-based composites. *ACI Mater. J.* **2014**, *111*, 309. [CrossRef]
7. Yu, R.; Spiesz, P.; Brouwers, H.J.H. Mix design and properties assessment of ultra-high performance fibre reinforced concrete (UHPFRC). *Cem. Concr. Res.* **2014**, *56*, 29–39. [CrossRef]
8. Yeh, I.C. Modeling of strength of high-performance concrete using artificial neural networks. *Cem. Concr. Res.* **1998**, *28*, 1797–1808. [CrossRef]
9. Bui, D.-K.; Nguyen, T.; Chou, J.S.; Nguyen-Xuan, H.; Ngo, T. A modified firefly algorithm-artificial neural network expert system for predicting compressive and tensile strength of high-performance concrete. *Constr. Build. Mater.* **2018**, *180*, 320–333. [CrossRef]
10. Han, Q.; Gui, C.; Xu, J.; Lacidogna, G. A generalized method to predict the compressive strength of high-performance concrete by improved random forest algorithm. *Constr. Build. Mater.* **2019**, *226*, 734–742. [CrossRef]
11. Anyaoha, U.; Zaji, A.; Liu, Z. Soft computing in estimating the compressive strength for high-performance concrete via concrete composition appraisal. *Constr. Build. Mater.* **2020**, *257*, 119472. [CrossRef]
12. Hameed, M.M.; AlOmar, M.K.; Baniya, W.J.; AlSaadi, M.A. Incorporation of artificial neural network with principal component analysis and cross-validation technique to predict high-performance concrete compressive strength. *Asian J. Civ. Eng.* **2021**, *22*, 1019–1031. [CrossRef]
13. Emad, W.; Mohammed, A.S.; Kurda, R.; Ghafor, K.; Cavaleri, L.; Qaidi, S.M.A.; Hassan, A.; Asteris, P.G. Prediction of concrete materials compressive strength using surrogate models. *Structures* **2022**, *46*, 1243–1267. [CrossRef]
14. Ali, R.; Muayad, M.; Mohammed, A.S.; Asteris, P.G. Analysis and prediction of the effect of Nanosilica on the compressive strength of concrete with different mix proportions and specimen sizes using various numerical approaches. *Struct. Concr.* **2022**, 12. [CrossRef]
15. Ibrahim, A.K.; Dhahir, H.Y.; Mohammed, A.S.; Omar, H.A.; Sedo, A.H. The effectiveness of surrogate models in predicting the long-term behavior of varying compressive strength ranges of recycled concrete aggregate for a variety of shapes and sizes of specimens. *Arch. Civ. Mech. Eng.* **2023**, *23*, 61. [CrossRef]
16. Mohammed, A.S.; Emad, W.; Sarwar Qadir, W.; Kurda, R.; Ghafor, K.; Kadhim Faris, R. Modeling the impact of liquid polymers on concrete stability in terms of a slump and compressive strength. *Appl. Sci.* **2023**, *13*, 1208. [CrossRef]
17. Moradi, N.; Tavana, M.H.; Habibi, M.R.; Amiri, M.; Moradi, M.J.; Farhangi, V. Predicting the compressive strength of concrete containing binary supplementary cementitious material using machine learning approach. *Materials* **2022**, *15*, 5336. [CrossRef] [PubMed]
18. Rezvan, S.; Moradi, M.J.; Dabiri, H.; Daneshvar, K.; Karakouzian, M.; Farhangi, V. Application of machine learning to predict the mechanical characteristics of concrete containing recycled plastic-based materials. *Appl. Sci.* **2023**, *13*, 2033. [CrossRef]
19. Imran, M.; Khushnood, R.A.; Fawad, M. A Hybrid data-driven and metaheuristic optimization approach for the compressive strength prediction of high-performance concrete. *Case Stud. Constr. Mater.* **2023**, *18*, e01890. [CrossRef]
20. Li, Z.J.; Song, Y.H.; Li, S.H. Prediction of metakaolin high-performance concrete compressive strength. *Bull. Chin. Ceram. Soc.* **2017**, *36*, 2963–2968.
21. Zhang, J.L.; Liu, X.; Zhu, H.B.; Li, C.; Yuan, Y. Model for predicting the mechanical behavior of UHPC based on BP neural network. *Concrete* **2013**, *281*, 18–25.
22. Chen, Q.; Ma, R.; Jiang, Z.W.; Wang, H. Compressive Strength Prediction and Mix Proportion Design of UHPC Based on GA-BP Neural Network. *J. Build. Mater.* **2020**, *23*, 176–183.
23. Zhang, J.; Zhao, Y. Prediction of compressive strength of ultra-high performance concrete (UHPC) containing supplementary cementitious materials. In Proceedings of the 2017 International Conference on Dmart Grid and Electrical Automation (ICSGEA), Changsha, China, 27–28 May 2017; pp. 522–525.
24. Abuodeh, O.R.; Abdalla, J.A.; Hawileh, R.A. Assessment of compressive strength of Ultra-high Performance Concrete using deep machine learning techniques. *Appl. Soft Comput.* **2020**, *95*, 106552. [CrossRef]
25. Wu, X.G.; Liu, P.C.; Chen, H.Y.; Zeng, T.M.; Xu, W. Characteristic screening and prediction of high-performance concrete compressive strength based on random forest method. *Concrete* **2022**, *387*, 17–24.

26. Roberson, M.M.; Inman, K.M.; Carey, A.S.; Howard, I.L.; Shannon, J. Probabilistic neural networks that predict compressive strength of high strength concrete in mass placements using thermal history. *Comput. Struct.* **2022**, *259*, 106707. [CrossRef]
27. Yang, Y.; Chen, F.Y. Strength development and evaluation of high-strength concrete at early age. *China Concr. Cem. Prod.* **2004**, *137*, 1–4.
28. Zhao, S.B.; Zhao, Y.J.; Chen, J.H.; Liu, S.; Sun, Y.P. Application and study on time-dependent compressive strength and elastic modulus of C50 pumping concrete. *Concrete* **2015**, *303*, 98–101.
29. Yang, W.J.; Wang, Y. Diachronic variation model of compressive strength and elastic modulus of concrete at early age. *J. China Foreign Highw.* **2007**, *27*, 149–151.
30. Wang, M.L.; Wang, X.A. Study on strength development law of high strength concrete at early age in high altitude environment. *Build. Struct.* **2021**, *51*, 1311–1315.
31. Zhao, Z. Proportion design and stregth regularity of high strength concrete. *Concrete* **2011**, *264*, 101–103.
32. Li, X.F.; Ma, W.L.; Pei, S.W.; Sun, S.N. Experimental study on compressive strength of commercial concrete at early age. *J. Zhongyuan Inst. Technol.* **2006**, *17*, 59–61.
33. Li, X.Z.; Han, L.J.; Liu, M.Q. Analysis of early aging prediction method of strength state of high-performance concrete. *Shanxi Archit.* **2007**, *33*, 167–168.
34. Jin, X.Y.; Shen, Y.; Li, Z.J. Experimental study on early aging characteristics of high-strength concrete. *China Concr. Cem. Prod.* **2003**, *133*, 5–7.
35. Ling, J.J. Compressive strength and nonlinear prediction model of high-grade basalt manufactured sand concrete. *J. Water Resour. Archit. Eng.* **2021**, *19*, 124–130.
36. Jia, M.; Han, Z.B. Analysis of compressive test of silicate concrete based on bridge estimation. *J. North China Univ. Sci. Technol. (Nat. Sci. Ed.)* **2019**, *41*, 84–89.
37. Wang, D.Y.; Wang, M.M.; Yang, Z.K.; Wu, Y.G.; Yuan, Q.; Liao, J. Study on mechanical properties of C80 concrete at different ages. *Ind. Constr.* **2014**, *44*, 106–111.
38. Ministry of Construction of China. *Test Method Standard for Mechanical Properties of Common Concrete: GB/T 50081-2002 [M]*; China Construction Industry Press: Beiging, China, 2003. (In Chinese)

coatings

MDPI

Article

The Influence of Assembly Unit of Fibers on the Mechanical and Long-Term Properties of Reactive Powder Concrete

Zhangjie Cai [1], Jie Ren [2], Guangming Shen [3], Changhong Jin [2], Xingqing Gu [4], Wenjie Cheng [1] and Hui Wang [5,*]

[1] Civil Engineering Department, Yancheng Institute of Technology, Yancheng 224051, China
[2] China Construction Infrastructure Co., Ltd., Beijing 100029, China
[3] Jiuzhou Engineering Design Co., Ltd., Zhengzhou 451162, China
[4] College of Civil Aviation, Nanjing University of Aeronautics and Astronautics, Nanjing 210016, China
[5] School of Civil Engineering and Geographic Environment, Ningbo University, Ningbo 315000, China
* Correspondence: huiwang123@aliyun.com

Abstract: The corrosion of concrete structures by chloride salt is very significant in coastal environments. In order to improve the durability of marine concrete structures, cement-based materials with high durability need to be developed. In this investigation, the influence of NaCl freeze–thaw cycles (FT-C) and NaCl dry-wet alternations (DW-A) on the flexural and compressive strengths of reactive powder concrete (RPC) with an assembly unit of basalt fibers and steel fibers is studied. Additionally, the mass loss rate, the relative dynamic modulus of elasticity (RDEM), the chloride ion migration coefficient (CMC) and the impact toughness are measured after the NaCl FT-C and DW-A action. Our findings show that the RDEM, mass loss, and mechanical strength loss of RPC are increased by the ascending NaCl FT-C and DW-A. Meanwhile, the RDEM and the impact toughness are decreased by the NaCl FT-C and DW-A. The RPC with 0.5% basalt fibers and 1.5% steel fibers by volume of RPC shows the optimum mechanical performance and resistance to NaCl FT-C and DW-A. However, RPC with 3% steel fibers shows the worst resistance to NaCl erosion. The maximum mass loss rates, RDEM, flexural strength loss rate, compressive strength loss rate, CMC and impact toughness of all specimens after 300 NaCl FT-C and 30 NaCl DW-A are 4.5%, 91.7%, 28.1%, 29.3%, 3.2×10^{-12} (m^2/s) and 2471 J. Meanwhile, the corresponding minimum values are 1.62%, 83.2%, 20.4%, 15.7%, 1.1×10^{-12} (m^2/s) and 625 J. The researching findings will provide an optimum mix ratio of RPC with an assembly unit of basalt fibers and steel fibers, which can be applied in the marine engineering environment.

Keywords: mechanical strength; assembly unit; basalt fibers; steel fibers; reactive powder concrete; NaCl erosion action; impact toughness

1. Introduction

Cement concrete is one of the most widely used materials [1,2]. Reactive powder concrete (RPC) is a kind of cement-based material which is manufactured with cement, mineral admixtures, quartz sand, water and high-performance water-reducing agent (WRA) [3,4]. This kind of material shows high compactness and mechanical performance. When the cement concrete is used in the coastal urban environment, serious deterioration in the concrete occurs. For coastal civil engineering construction materials, the chloride erosion effect is the most severe. Ordinary cement concrete is prone to corrosion during service in the marine environment. Meanwhile, the application of RPC can improve the service life of construction buildings [5–9].

Steel fibers, plant fibers, basalt fibers, carbon fibers, and carbon nanotubes have been applied in reinforcing the performance of RPC. However, steel fibers are easy to corrode when applied in the marine environment [10,11]. The freeze–thaw cycles and dry-wet alternations of seawater can accelerate the deterioration of cement concrete [12–14]. When the steel fibers are mixed in RPC, the corroded steel fibers can induce more serious cracks in the cement concrete, thus decreasing the mechanical performance of cement concrete. The carbon fibers and carbon nanotubes show positive excellent corrosion resistance [15–17]. These fibers are extremely expensive, and not suitable for application in cement-based materials. The plant fibers shows low cost and good corrosion resistance. However, their reinforcing effect on mechanical performance is limited.

Basalt fiber is a kind of continuous fiber manufactured by natural basalt after melting in a high temperature environment of 1450~1500 °C and drawing at high speed through a platinum rhodium alloy wire drawing bushing. The tensile ratio of basalt fiber is as high as 15%~18%, which can be uniformly distributed in cement concrete. The addition of basalt fibers can prevent the sink of aggregate and improve the uniformity of RPC. Moreover, the segregation can be decreased and the workability can be improved through the addition of basalt fibers [18–20]. Additionally, the tensile strain of cement concrete can be limited by basalt fibers; meanwhile, concrete shrinkage can be decreased by basalt fibers. Li et al. [2], reported that 0.5% of the total volume of RPC is the optimum fiber volume of basalt fibers. Furthermore, basalt fibers can effectively improve the impact toughness and provide excellent resistance to freeze–thaw and carbonation of RPC. The copper-plated fine steel fibers exhibit high corrosion resistance and perfect mechanical performance. The RPC with an assembly unit of basalt fibers and copper-plated fine steel fibers may exhibit better mechanical and durable properties. However, little attention has been paid to the effect of NaCl erosion on the performances of RPC with assembly units of basalt fibers and copper-plated fine steel fibers [21,22].

In this study, the mass loss rate, the relative dynamic modulus of elasticity (RDME), the flexural loss rate, the compressive strength loss rate and the impact toughness of RPC with an assembly unit of basalt fibers and copper-plated fine steel fibers are measured. Freeze–thaw cycles (FT-C) and the dry-wet alternations (DW-A) of NaCl are exerted on the RPC specimens. This research will provide a new high-performance concrete material which may be applicable to coastal environment.

2. Materials and Methods

2.1. Raw Materials

Ordinary Portland cement (OPC) provided by Aosheng Construction Co., Ltd., Yanchen, China is selected. The cement has a density of 3.05 g/cm^3, a compressive strength of 42.5 MPa, and an initial setting time and final setting time of 107 min and 361 min. The blast furnace slag (BFS) shows a density of 2.89 g/cm^3, a specific surface area of 437.5 m^2/g and a thermogravimetric loss rate of 2.3%. Silica fume (SF) with a specific surface area and density of 15 m^2/g and 2.2 g/cm^3 is used for this research. SF is produced by the Shanggao Mingzheng Plastic Factory, Yichun, China. Meanwhile, the ultra-fine fly ash (FA) exhibits a density of 2.11 g/cm^3 and a specific surface area of 387 m^2/g. The chemical composition and the particle size distributions of the binder materials are shown in Tables 1 and 2. Quartz sand with particle sizes of 1.1~0.70 mm, 0.61~0.34 mm and 0.148~0.298 mm is used as the aggregate in this experiment. Liquid poly-carboxylic acid water-reducing agent (WRA) with a 40% water-reducing rate is used for adjusting the working performance of fresh RPC paste. Calcium formate, polyether surfactants and dry powdered DF-04 defoamer are used as early strength agent and defoamer. Basalt fibers (BFs) with excellent acid and alkali resistance are provided by Shandong Taicheng Fiber Factory, Taian, China, showing a density of 2.71 g/cm^3, a length of 15–18 mm, and a diameter of 18–48 μm, a the fracture strength of 1050 MPa, and an average modulus of elasticity of 3781 MPa. Copper-plated steel fibers (SFs), offered by Anshan Cobit, Anshan, China, with an average length of 1.5 cm, a diameter of 0.2 mm, a density of 7.85 g/cm^3 and a tensile strength of 2812 MPa are used in the measurement.

Table 1. Particle passing percentage of raw materials (%).

Types' Particle Size (%)	0.3 μm	0.6 μm	1 μm	4 μm	8 μm	64 μm	360 μm
OPC	0	0.3	2.7	15.0	28.8	93.6	100
BFS	0.04	0.11	3.6	19.7	35.1	98.0	100
Sand	0.01	0.02	0.02	0.02	0.04	20	100
FA	12.3	41.7	66.2	100	100	100	100
SF	31.3	58.4	82.2	94.6	99.4	99.4	100

Table 2. The compositions (%).

Types' Compositions	SiO_2	Al_2O_3	Fe_xO_y	MgO	CaO	SO_3	K_2O	Na_2O	Ti_2O	Loss on Ignition
OPC	20.9	5.5	3.9	1.7	62.2	2.7	-	-	-	3.1
BFS	34.1	14.7	0.2	9.7	35.9	0.2	3.5	-	-	-
Sand	99.6	-	0.02	-	-	-	-	-	-	-
FA	55.00	20.00	6.00	10.20	4.50	0.11	1.26	2.13	0.06	0.74
SF	90.0	0.2	0.6	0.2	0.4	0	7.4	-	-	-

2.2. Sample Preparation and Measurement Methods

2.2.1. Sample Manufacturing

The samples are prepared with the mixing proportions shown in Table 3, including water, OPC, FA, SF, BFS, quartz sand, WR, BFs and SFs. All dry powder materials are mixed in a UJZ-15 mortar mixer for 30 s. Then, the steel fibers and basalt fibers are mixed evenly in the dry materials for another 30 s. Finally, the water-reducing agent and water are stirred uniformly for the last 5 min. After the stirring, the fresh RPC paste is poured, forming specimens with sizes of 40 × 40 × 160 mm^3, 100 × 100 × 400 mm^3 and Φ 100 × 50 mm^3. All the specimens are cured in the standard curing environment for 28 days before measurement. The standard curing environment can be divided into two steps. Firstly, specimens are cured in the environment with a temperature of 20 ± 2 °C and a relative humidity of 60% for 1 day before demolding. Then, the specimens are moved to the standard curing room with a temperature of 20 ± 2 °C and a relative humidity of 98% ± 1%. Figure 1 shows the manufacturing and measuring processes.

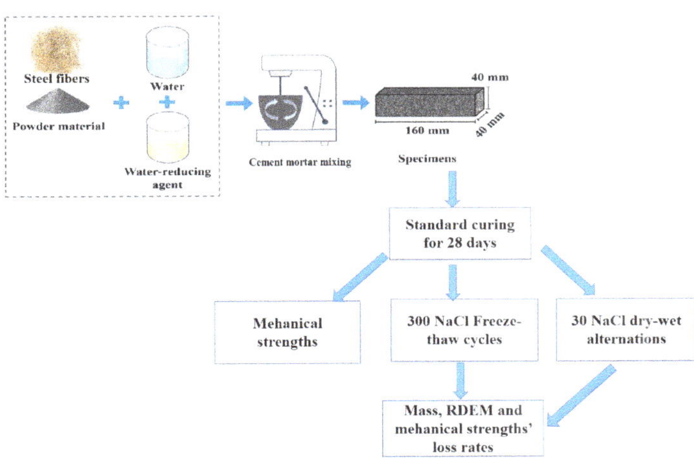

Figure 1. The manufacturing and measuring processes.

Table 3. Mix proportion of RPC per unit volume (kg/m^3).

Water	OPC	FA	SF	BFS	Sand	WR	BFs	SFs
183.3	740.7	148.12	222.18	111.1	977.9	16.3	0	157
183.3	740.7	148.12	222.18	111.1	977.9	16.3	0	235.5
183.3	740.7	148.12	222.18	111.1	977.9	16.3	54.2	0
183.3	740.7	148.12	222.18	111.1	977.9	16.3	81.3	0
183.3	740.7	148.12	222.18	111.1	977.9	16.3	13.55	117.75

2.2.2. Measurement of Mechanical Strengths

Specimens with a size of 40 × 40 × 160 mm^3 are made for measuring the flexural and compressive strengths. The mechanical strengths are measured with a YDW-100C microcomputer full-automatic universal tester provided by Jinan Xulian Instrument and Equipment Co., Ltd., Jinan, China. The loading speeds for the flexural and compressive strengths are 0.05 and 2.4 kN/s. Figure 2 shows the measuring process of flexural and compressive strengths.

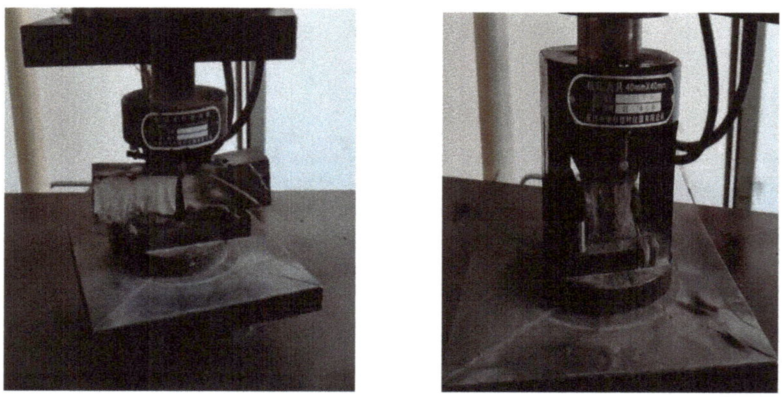

Broken specimens **Crushed specimens**

Figure 2. The measuring process of flexural and compressive strengths.

2.2.3. Measurement of Impact Toughness

The JLW-100 drop weight impact test measured impact toughness. The test method proposed by the American Concrete Institute (ACI) 544 committee is used to measure the impact toughness of concrete beams. The specimen is 100 × 100 × 400 mm^3 in size, the drop hammer weighs 1.5 kg, the clear span of the beam is 360 mm, and the span of the upper loading section is 120 mm.

A four-point loading mode is adopted to apply impact load on the specimen. The measuring equipment is exhibited in Figure 3. In the study, three replicates are used for each group. The average impact toughness is considered as the index of concrete impact resistance. The strain gauge is fixed at the axial position of the bottom center of the beam, and the dynamic strain gauge is used to monitor the concrete. If a sudden increase in strain happens during impact, cracks may start to generate in the beam. With an increase in impact times, the number of impact times virtually stays the same. The impact toughness can be calculated by Equation (1).

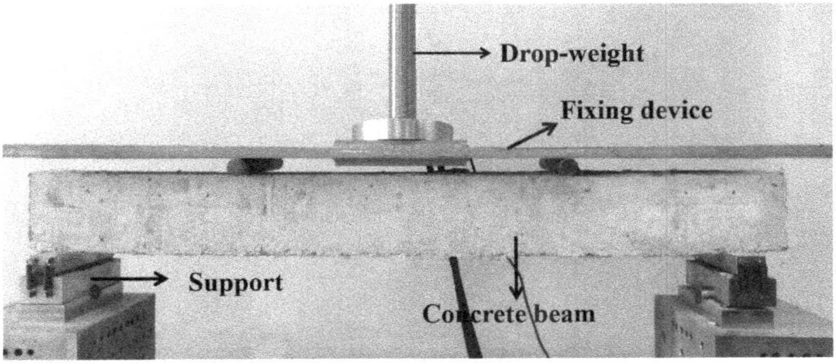

Figure 3. The test process and loading mode of the impact test.

The specimen is simply supported, and its calculated span is 340 mm. In this section, the impact toughness index and impact energy dissipation performance of different beams are obtained using a drop weight bending impact test for nine groups of different mixing proportions.

In this impact test, a 2.5 kg drop hammer (5 kg rod weight) is used to impact the side of the forces specimen from a height of 400 mm. The crack propagation width of the specimen surface is observed by the specimens' crack width detector (as shown in Figure 2), which is used to determine the impact times at the initial crack. The impact test is repeated until the specimen fails.

The impact toughness index (β) for drop hammer impact testing is defined as the ratio of the difference between the failure impact number (N_2) and the initial crack impact number (N_1) to the initial crack impact number (N_1). The impact toughness index represents the ratio of the impact energy consumption after the crack appears to the impact energy consumption before the initial crack, and reflects the toughness of the material after the crack. Among them, the impact times of the initial crack are the initial stage of impact. For each test, the experimental steps are carried out as follows. Firstly, the bottom of the specimen is wiped with alcohol, and then the impact times are recorded and detected. The impact times of failure are recorded as soon as the crack width detector identifies the first crack that has a width of 0.05 mm. In the whole impact test, the impact energy consumption (W) of the specimen is the work accumulation of each impact. It can be calculated as follows:

$$W = mghN \tag{1}$$

where N represents the total number of impact failures, m notes the mass of the whole drop-weight, which in this study is 7.5 kg, h is the dropping height, in this case 400 mm, g stands for gravitational acceleration, in this case 9.8 m/s^2. The impact toughness index (β) can be calculated by Equation (2) as follows:

$$\beta = (N_2 - N_1)/N_1 \tag{2}$$

2.2.4. NaCl Erosion Environment

Previously, NaCl (FT-C) and NaCl (DW-A) specimens with sizes of 40 × 40 × 100 mm^3 were immersed in 3% NaCl solution for 4 days after standard curing for 24 days. Then, the specimens are subjected to the experiments of NaCl FT-C by the concrete rapid FT-C test box. A DW-A can be described following these steps. The specimens immersed in NaCl solution for 4 days are air dried at room temperature for 2 h. Then, the specimens are dried in the vacuum drying oven for 36 h and moved to room temperature and cooled for 2 h. Finally, all specimens are immersed in NaCl solution for the residual 10 h.

3. Results and Discussion

3.1. Action of NaCl FT-C

3.1.1. Mass Loss Rate and RDEM after Different NaCl FT-C

Figure 4 shows the mass loss rate of specimens during FT-C. It can be observed from figure that the mass loss rate of RPC increases with the increasing number of FT-C. The mass loss rate of RPC with 2% steel fibers is lower than that of RPC with 3% steel fibers. Meanwhile, the RPC with 3% basalt fibers shows a lower mass loss rate than the RPC with 2% basalt fibers and 3% steel fibers, but higher than that with 2% steel fibers. The RPC with an assembly unit of basalt fibers and steel fibers shows the lowest mass loss rate of all mix proportions. The main reason for this result is that the frost-heaving stress inside the RPC and the crystallization stress of NaCl during freezing and thawing cycles cause the RPC to bulge and crack, resulting in peeling. Therefore, the mass of RPC continues to decline during freezing and thawing. However, excessive steel fibers (3%) can accelerate the corrosion of steel fibers, and the corroded steel fiber is prone to cause cracking and peeling of RPC, thus accelerating its mass loss. Therefore, when the steel fiber content is 3%, the mass loss rate of RPC is the highest. The reinforcement effect of RPC mixed with 0.5% basalt fiber and 1.5% steel fiber is the best, leading to an optimum crack resistance effect of RPC. Consequently, the mass loss rate of RPC during FT-C is the smallest. The mass loss rates after 300 FT-C vary from 1.62%~4.5%. It can be observed from Figure 4 that the mass loss of RPC is less than 5%, which indicates that this type of RPC has excellent anti-NaCl freeze–thaw effect [23].

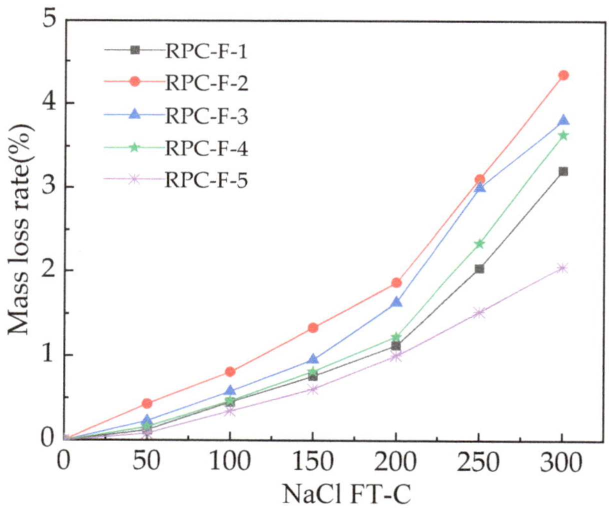

Figure 4. Mass loss rate during NaCl FT-C.

The relative dynamic modulus of elasticity (RDME) after exposure to a different number of FT-C is shown in Figure 5. As illustrated in Figure 5, the RDME decreases with the increasing FT-C. This is mainly because the cracks in the inner RPC increase with the increase of FT-C, and therefore the RDME of RPC decreases gradually during FT-C [24]. After suffering FT-C, the RDME is in accordance with the following order: 0.5% basalt fiber + 1.5% steel fiber RPC > 2% steel fiber RPC > 2% steel fiber RPC > 3% steel fiber RPC > 3% basalt fiber RPC > 2% basalt fiber RPC. This is due to the fact that the reinforcing effect of steel fibers on RPC is higher than that of basalt fibers, and the RDME of steel fiber-reinforced RPC with the same content is higher than that of the RPC with basalt fibers [25]. However, steel fibers will rust under the action of NaCl FT-C, which will cause acceleration of cracking, thus reducing the RDME of RPC. The fiber-reinforced effect of the

RPC mixed with 0.5% basalt fibers and 1.5% steel fibers is the best, hence the RDME of RPC is the highest at this time. The RDME after 300 NaCl FT-C ranges from 84.1%~91.7%, which is higher than 80%, thus showing excellent resistance to NaCl FT-C.

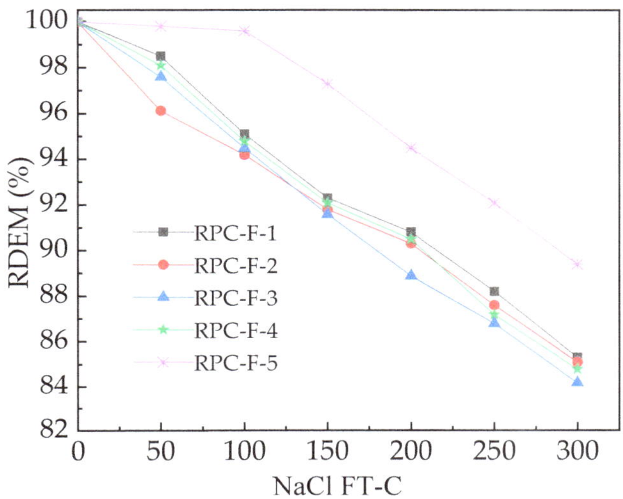

Figure 5. The RDEM of RPC during NaCl FT-C.

3.1.2. Mechanical Strength Loss Rates after NaCl FT-C

Figure 6 shows the flexural strength and compressive strength loss rates after different NaCl FT-C. It can be observed from Figure 6 that the mechanical strength loss rates of RPC gradually increase with the ascending number of FT-C. The mechanical strength loss rates of all specimens increase in the following order: RPC with 0.5% basalt fibers + 1.5% steel fibers < RPC with 2% steel fibers < RPC with 3% basalt fibers < RPC with 2% basalt fibers < RPC with 3% steel fibers. This is because the increased cracks during the FT-C lead to decreased mechanical strength [26]. Therefore, the increased number of FT-C will cause a continuous decrease in RPC strength. We can see from our results that a certain number of steel fibers and basalt fibers can enhance the mechanical and resistance of FT-C. However, excessive steel fibers will aggravate their corrosion, thus reducing the tensile property of steel fiber and leading to the reduction in mechanical strength. Additionally, steel fibers' corrosion leads to RPC expansion cracking, further reducing the mechanical strength. The assembly unit of basalt fibers and steel fibers possesses the advantages of two kinds of fibers, thus weakening their shortcomings. Therefore, the mechanical strength loss of RPC with 0.5% basalt fiber + 1.5% steel fiber is at a minimum. It can be seen in Figure 6 that the loss of compressive strength is lower than the flexural strength loss. This is because the cracks of RPC increase and widen due to FT-C. The crack growth rate of RPC in tension is higher than that in compression. Therefore, the reduction rate of flexural strength is higher than that of compressive strength. The flexural and compressive strengths loss rates are 22.3%~36.7% and 15.7%~28.1%.

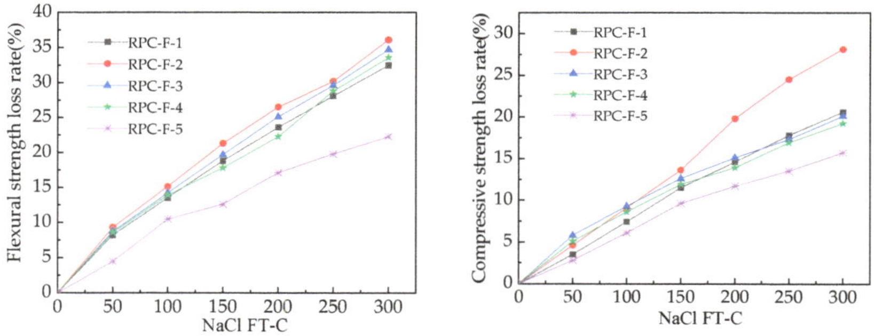

(**a**) (**b**)

Figure 6. Mechanical strength loss rates after NaCl FT-C. (**a**) Flexural strength loss rate (**b**) Compressive strength loss rate.

3.1.3. CMC during NaCl FT-C

The CMC of RPC during different FT-C is exhibited in Figure 7. It can be seen in Figure 7 that the CMC decreases in the following order: RPC with 3% steel fibers > RPC with 2% steel fibers > RPC with 2% basalt fibers > RPC with 0.5% basalt fibers + 1.5% steel fibers, before the NaCl FT-C. As shown in Figure 7, the CMC descends obviously with the increasing number of NaCl FT-C. This is mainly ascribed to the fact that the cracking expansion of RPC is accelerated by the FT-C, leading to an increase in the CMC. Consequently, the CMC of RPC increases with the increasing FT-C [27]. After the FT-C, the CMC descends in the same order as that before NaCl FT-C. The CMC after 300 FT-C ranges from 1.1×10^{-12} (m^2/s)~3.2×10^{-12} (m^2/s).

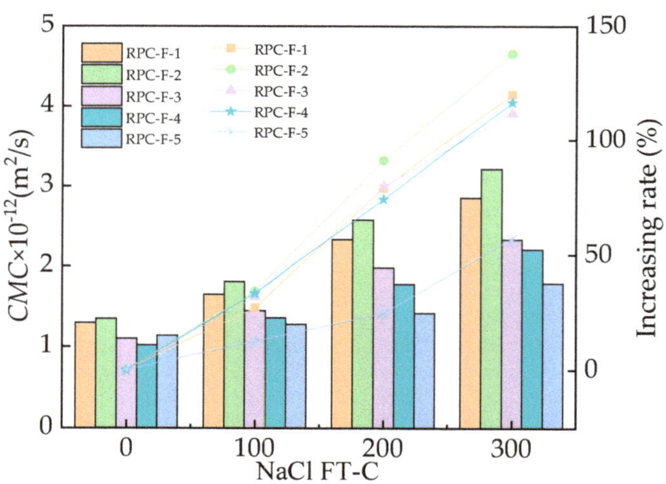

Figure 7. The CMC of RPC during NaCl FT-C.

3.1.4. Impact Toughness during NaCl FT-C

Figure 8 shows the impact toughness of RPC after different NaCl FT-C. As depicted in Figure 8, the initial impact toughness (impact toughness before NaCl corrosion) of RPC increases with the increasing fiber contents. The RPC with steel fibers shows higher impact toughness than the RPC with basalt fibers. This is mainly attributed to the fact that the fibers can consume the impact energy during the impact process. Besides, the fibers' pullout effect limits the extension of RPC cracks during impact; therefore, the impact toughness of RPC is increased by adding fibers [28]. Meanwhile, the tensile toughness of steel fibers is higher than that of basalt fibers, leading to the result that the RPC with steel fibers shows higher impact toughness than the RPC with basalt fibers. RPC with 0.5% basalt fibers + 1.5% steel fibers performs the highest impact toughness. This is because this combination of fibers performs the highest the mechanical strength and the fiber dispersion, thus improving the impact toughness [29]. Moreover, as depicted in Figure 8, increasing NaCl FT-C results in decreasing impact toughness. This is attributed to increased cracks in the inner RPC during NaCl FT-C action [30]. Hence, the impact toughness is reduced by the NaCl FT-C. The impact toughness of RPC after being subjected to NaCl FT-C decreases in the following order: RPC with 3% steel fibers > 2% steel fibers RPC > 2% basalt fibers RPC > 3% basalt fibers > 0.5% basalt fibers mixed with 1.5% steel fiber RPC. When 300 NaCl FT-C are finished, the impact toughness is 625~2456 J.

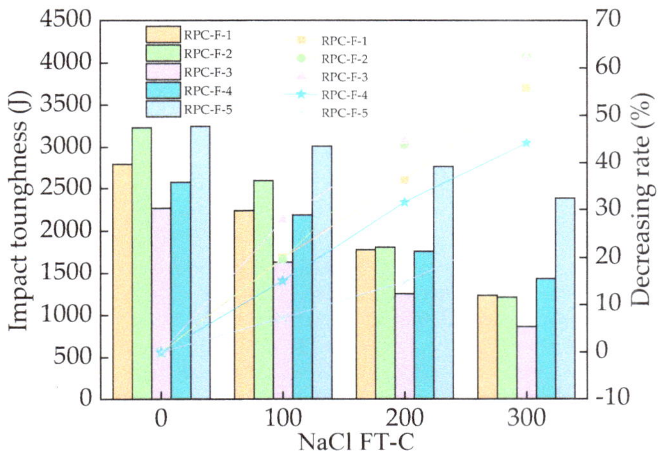

Figure 8. The impact toughness of RPC during NaCl FT-C.

Comparing our results with those of other researchers [25,27], the mass loss rates, RDEM, mechanical strengths loss rates and CMC of RPC with an assembly unit of steel fibers and basalt fibers after 300 NaCl FT-C are 77.1%~85.3%, 118.4%~126.7%, 82.5%~86.2%, and are 91.1%~92.3% in the RPC with mono-doped basalt fibers or steel fibers.

3.2. Action of NaCl DW-A

3.2.1. Mass Loss Rate and RDEM after Different NaCl DW-A

The mass loss rates of RPC specimens after exposure to 0, 10, 20 and 30 NaCl DW-A are shown in Figure 9. As depicted in Figure 9, the mass loss rates of RPC increase with the rising number of DW-A. This is mainly due to the fact that NaCl will continuously crystallize and precipitate when the NaCl DW-A are exerted on the specimens [31]. As a result, the internal cracking and surface peeling of RPC will increase the mass loss of RPC. As found in Figure 9, the mass loss rates increase in the following order: RPC with 0.5% basalt fibers + 1.5% steel fibers < RPC with 2% steel fibers < RPC with 3% basalt fibers < RPC with 2% basalt fibers < RPC with 3% steel fibers. This is mainly because the addition of basalt fibers and steel fibers can limit crack propagation in the inner RPC, while steel fiber has a stronger limiting ability. However, during the process of DW-A, more Cl^- and O_2 will enter the RPC to accelerate the corrosion of steel fibers, thus accelerating the crack propagation and peeling of the RPC [32]. Therefore, RPC with 3% steel fibers has the highest mass loss rate. Compared with steel fibers, basalt fibers have higher corrosion resistance. Therefore, the assembly unit of steel fibers and basalt fibers combine the advantages of the excellent mechanical properties of steel fibers and the high corrosion resistance of basalt fibers. Consequently, the RPC with 0.5% basalt fibers + 1.5% steel fibers shows the lowest mass loss rate. The mass loss rate after 30 NaCl DW-A is 2.1%~3.2%, indicating high resistance to NaCl DW-A.

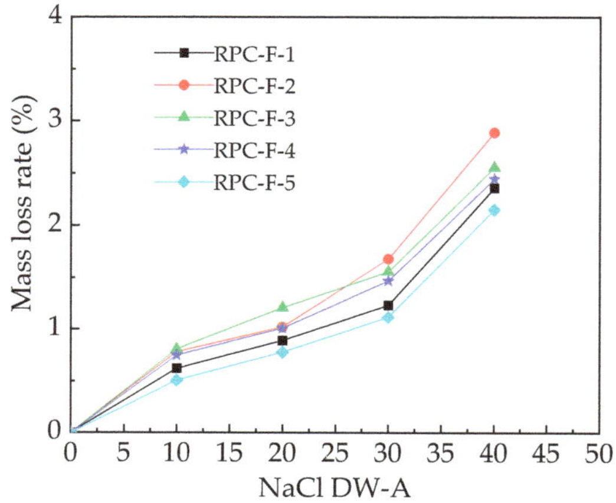

Figure 9. Mass loss rate during NaCl DW-A.

The RDEM of RPC after suffering from 0, 10, 20, and 30 NaCl DW-A is illustrated in Figure 10. It can be observed in Figure 10, the RDEM descends with the increasing number of NaCl DW-A. The main reason is attributed to the increasing and widening of cracks in the inner RPC by NaCl DW-A resulting in the decreased RDEM [33]. After the NaCl DW-A, the RDEM decreases in the following order: RPC with 0.5% basalt fibers + 1.5% steel fibers > RPC with 2% steel fibers > RPC with 3% basalt fibers > RPC with 2% basalt fibers > RPC with 3% steel fibers. This is mainly because the mechanical properties of steel fibers are higher than basalt fibers [34]. However, excessive steel fibers will accelerate the steel fibers' corrosion and increase cracks, leading to decreasing the RDEM. Therefore, when the steel fiber content is 3%, the RDEM of RPC is the lowest, while the RPC with 0.5% basalt fibers + 1.5% steel fibers shows the highest RDEM. The RDEM of RPC after suffering 30 NaCl DW-A is 82.1%~90.2%, showing perfect performance after NaCl DW-A action.

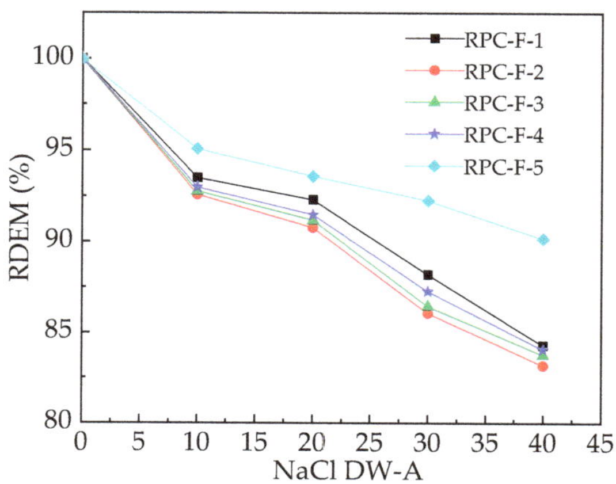

Figure 10. The RDEM of RPC during NaCl DW-A.

3.2.2. Mechanical Strength Loss Rates after NaCl DW-A

The mechanical strength loss rates after different NaCl DW-A are shown in Figure 11. As observed in Figure 11, mechanical strength loss rates increase with the number of NaCl DW-A. The mechanical strengths' loss rates are in accordance with the following order: RPC with 0.5% basalt fibers + 1.5% steel fibers < RPC with 2% steel fibers < RPC with 3% basalt fibers < RPC with 2% basalt fibers < RPC with 3% steel fibers. The reason for this phenomenon is that the assembly unit of basalt fibers and steel fibers is the best for reinforcing the mechanical strengths of RPC. The tensile strength of steel fibers is higher than that of basalt fibers. However, massive dosages of steel fibers accelerate the corrosion of steel fibers, leading to a reduction in the RPC's mechanical strength. As can be observed from Figure 11, the flexural strength loss rate of RPC is higher than the compressive strength loss rate of RPC. Comparing the mass loss rates and mechanical strength loss rates of the RPC exposed to NaCl DW-A with that of the RPC exposed to NaCl FT-C, the RPC shows higher resistance to NaCl FT-C than resistance to NaCl DW-A. The flexural and compressive strength loss rates are 20.4 %~28.5% and 18.2 %~22.1%.

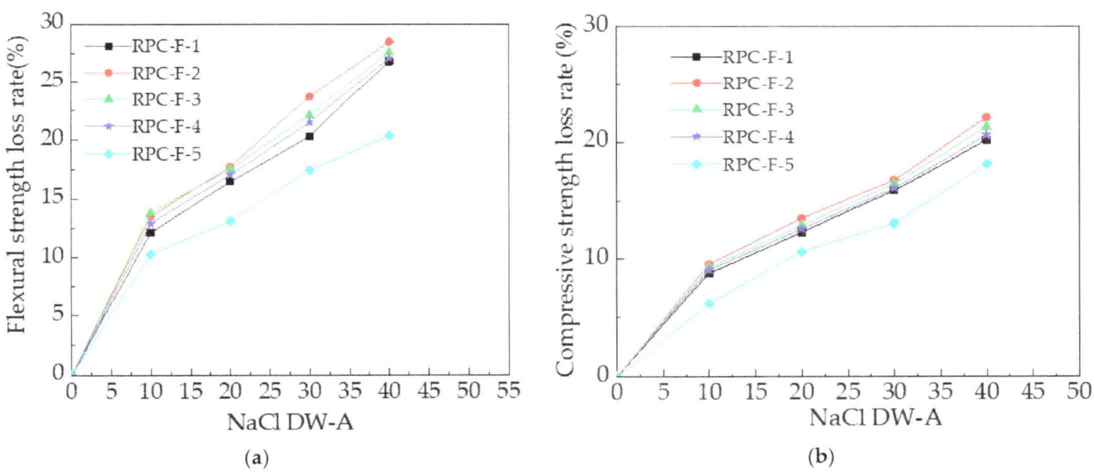

Figure 11. Mechanical strength loss rates after NaCl DW-A. (**a**) Flexural strength loss rate (**b**) Compressive strength loss rate.

3.2.3. CMC during NaCl DW-A

The CMC of RPC after 0, 10, 20, and 30 NaCl DW-A is shown in Figure 12. As obtained from Figure 12, the CMC increases with the increasing NaCl DW-A. It is mainly because the cracks caused by the NaCl DW-A increase the probability of chloride ion diffusion, leading to the increase in CMC. The CMC after NaCl DW-A varies in the following order: RPC with 3% steel fibers > RPC with 2% steel fibers > RPC with 2% basalt fibers > RPC with 3% basalt fibers > RPC with 0.5% basalt fibers + 1.5% steel fibers. When 30 NaCl DW-A are finished, the CMC is 1.6×10^{-12}~3.1×10^{-12} (m^2/s).

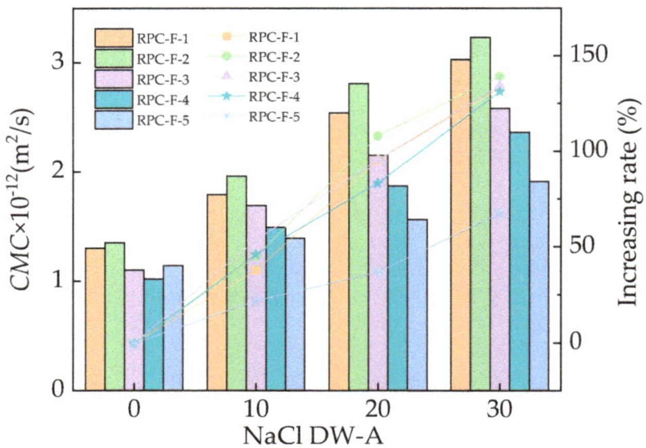

Figure 12. The CMC of RPC during NaCl DW-A.

3.2.4. Impact Toughness during NaCl DW-A

The impact toughness of RPC after different NaCl DW-A is illustrated in Figure 13. As depicted in Figure 13, the impact toughness decreases obviously with the increasing NaCl DW-A. This is mainly because that the cyclic crystallization stress of RPC is generated in the process of NaCl DW-A [35]. Hence, the cracks in RPC will expand, resulting in reducing the impact toughness. The reduction rate of impact toughness due to the NaCl DW-A coincides with the following order: RPC with 3% steel fibers >RPC with 2% steel fibers > RPC with 2% basalt fibers > RPC with 3% basalt fibers > RPC with 0.5% basalt fibers + 1.5% steel fibers. This is mainly because the passive film of steel fibers is corroded by chloride ions under the alternating action of NaCl DW-A, which leads to a rapid reduction in impact toughness [36]. Comparing Figures 8 and 13, the impact toughness of RPC after NaCl FT-C is higher than that after NaCl DW-A. After 30 NaCl DW-A, the impact toughness is 678~2471 J.

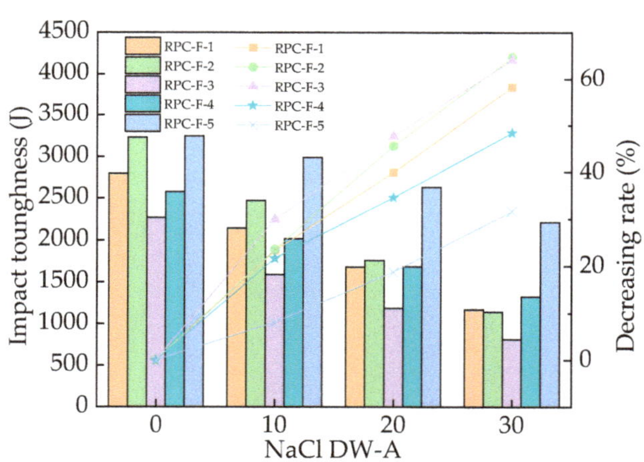

Figure 13. The impact toughness of RPC during NaCl DW-A.

Comparing the results of our research with those of other researchers [25,27], the mass loss rates, RDEM, mechanical strength loss rates and CMC of the RPC with an assembly unit of steel fibers and basalt fibers after 30 NaCl DW-A are 79.8%~86.1%, 123.1%~135.2%, 83.6%~88.1% and are 92.3%~94.5% in the RPC with mono-doped basalt fibers or steel fibers.

4. Conclusions

The conclusion of the research can be summarized as follows.

The mass loss rates, the mechanical strength loss rates and CMC are increased by the actions of NaCl FT-C and DW-A. Meanwhile, the NaCl FT-C and DW-A lead to decreased RDEM and impact toughness.

The addition of BFs and SFs can improve mechanical performances and resistance to NaCl actions. The NaCl erosion resistance of the RPC accords with the following order: RPC with 0.5% basalt fibers + 1.5% steel fibers > RPC with 2% steel fibers > RPC with 3% basalt fibers > RPC with 2% basalt fibers > RPC with 3% steel fibers. Specimens with 0.5% basalt fibers and 1.5% steel fibers show the optimum mechanical strength and the resistance to NaCl erosion.

The mass loss rates, RDEM, flexural strength loss rate, compressive strength loss rate, CMC and impact toughness of all specimens after 300 NaCl FT-C are 1.62%~4.5%, 83.2%~91.7%, 22.3%~36.7% and 15.7%~28.1%, 1.1×10^{-12}~3.2×10^{-12} (m^2/s) and 625~2456 J, while the corresponding parameters of specimens after 40 NaCl DW-A are 2.1%~3.2%, 82.1%~90.2%, 20.4%~28.5%, 18.2%~22.1%, 1.6×10^{-12}~3.1×10^{-12} (m^2/s) and 678~2471 J. According to comprehensive comparisons, these properties decay more seriously after 300 NaCl FT-C.

The goal of this study is to provide a kind of concrete material with ultra-high mechanical strength and durability for marine structures.

Author Contributions: Conceptualization, Z.C. and H.W.; methodology, J.R.; software, C.J.; validation, Z.C., H.W. and G.S.; formal analysis, X.G.; investigation, W.C.; resources, H.W.; data curation, Z.C.; writing—original draft preparation, Z.C.; writing—review and editing, H.W.; visualization, J.R.; supervision, Z.C.; project administration, H.W.; funding acquisition, H.W. All authors have read and agreed to the published version of the manuscript.

Funding: This research was funded by Zhejiang Provincial Natural Science Foundation [No. LY22E080005].

Institutional Review Board Statement: Not applicable.

Informed Consent Statement: Not applicable.

Data Availability Statement: Not applicable.

Conflicts of Interest: The authors declare no conflict of interest.

References

1. Dong, S.F.; Li, L.W.; Ashour, A.; Dong, X.F.; Han, B.G. Self-assembled 0D/2D nano carbon materials engineered smart and multifunctional cement-based composites. *Constr. Build. Mater.* **2021**, *272*, 121632. [CrossRef]
2. Li, Y.; Li, X.; Tan, Y. Effect of aging on fatigue performance of cement emulsified asphalt repair material. *Constr. Build. Mater.* **2021**, *292*, 123417. [CrossRef]
3. Ju, Y.; Zhao, J.; Wang, D.; Song, Y. Experimental study on flexural behaviour of reinforced reactive powder concrete pole. *Constr. Build. Mater.* **2021**, *312*, 125399. [CrossRef]
4. Rajasekar, A.; Arunachalam, K.; Kottaisamy, M. Durability of ultra-high strength concrete with waste granite sand as partial substitute for aggregate. *J. Comput. Theor. Nanosci.* **2018**, *15*, 446–452. [CrossRef]
5. Shen, W.G.; Liu, Y.; Cao, L.G.; Huo, X.J.; Yang, Z.G.; Zhou, C.C.; He, P.T.; Lu, Z.L. Mixing design and microstructure of ultra-high strength concrete with manufactured sand. *Constr. Build. Mater.* **2017**, *143*, 312–321. [CrossRef]
6. Shan, B.; Lai, D.D.; Xiao, Y.; Luo, X.B. Experimental research on concrete-filled RPC tubes under axial compression load. *Eng. Struct.* **2018**, *155*, 358–370. [CrossRef]
7. Haselbach, L.; Valavala, S.; Montes, F. Permeability predictions for sand clogged Portland cement pervious concrete pavement systems. *J. Environ. Manag.* **2005**, *81*, 42–49. [CrossRef] [PubMed]
8. Li, S.Q.; Feng, Y.C.; Yang, J.B.; Cerny, R. Expansion mechanism and properties of magnesium oxide expansive hydraulic cement for engineering applications. *Adv. Mater. Sci. Eng.* **2021**, *2021*, 5542072. [CrossRef]
9. Yoo, D.Y.; Banthia, N.; Lee, J.Y.; Yoon, Y.S. Effect of fiber geometric property on rate dependent flexural behavior of ultra-high-performance cementitious composite. *Cem. Concr. Compos.* **2018**, *86*, 57–71. [CrossRef]

10. Dong, S.; Wang, Y.; Ashour, A.; Han, B.G.; Ou, J.P. Uniaxial compressive fatigue behavior of ultra-high performance concrete reinforced with super-fine stainless wires. *Int. J. Fatigue* **2021**, *142*, 105959. [CrossRef]
11. Heng, K.; Jia, P.C.; Xu, J.P.; Li, R.W.; Wu, H. Vehicular impact resistance of highway bridge with seismically-designed UHPC pier. *Eng. Struct.* **2022**, *252*, 113635. [CrossRef]
12. Zhu, J.; Liu, S.; Song, L.; Qu, Z.; Wang, H. Influence of Carbon Dioxide Curing on the Corrosion Resistance of Reinforced Cement Mortar under the External Erosion of NaCl Freeze–Thaw Cycle. *Appl. Sci.* **2022**, *12*, 5061. [CrossRef]
13. Tan, C.J.; Zhang, Y.; Zhao, H.; Zhang, B.; Du, T. Study on Shear-Lag Effect of Steel–UHPC Ribbed Slab Composite Structures Using Bar Simulation Method. *Buildings* **2022**, *12*, 1884. [CrossRef]
14. Wang, H.; Liu, X.; Yue, Q.; Wang, N.; Zheng, M.Z. Longitudinal shear behavior and design method of UHPC connection for prefabricated slabs in composite beams. *Eng. Struct.* **2023**, *277*, 115386. [CrossRef]
15. Wang, H.; Zhang, A.; Zhang, L.; Wang, Q.; Yang, X.; Gao, X.; Shi, F. Electrical and piezoresistive properties of carbon nanofiber cement mortar under different temperatures and water contents. *Constr. Build. Mater.* **2020**, *265*, 120740. [CrossRef]
16. Zhao, X.; Cai, L.; Ji, X.H.; Zeng, W.; Liu, J.T. Mechanical Properties of Polyethylene Fiber Reinforced Ultra High Performance Concrete (UHPC). *Materials.* **2022**, *15*, 8374. [CrossRef]
17. Zhang, Y.; Li, X.L.; Zhu, Y.P.; Shao, X.D. Experimental study on flexural behavior of damaged reinforced concrete (RC) beam strengthened by toughness-improved ultra-high performance concrete (UHPC) layer. *Composites Part B* **2020**, *186*, 107834. [CrossRef]
18. Sohn, Y.S.; Lee, J.H.; Lee, S.H. Fire resistance of hybrid fibre-reinforced, ultra-high-strength concrete columns with compressive strength from 120 to 200 MPa. *Mag. Concr. Res.* **2012**, *64*, 539–550.
19. Wang, H.; Gao, X.; Liu, J.; Ren, M.; Lu, A. Multi-functional properties of carbon nanofiber reinforced reactive powder concrete. *Constr. Build. Mater.* **2018**, *187*, 699–707. [CrossRef]
20. Alhoubi, Y.; Mahaini, Z.; Abed, F. The Flexural Performance of BFRP-Reinforced UHPC Beams Compared to Steel and GFRP-Reinforced Beams. *Sustainability* **2022**, *14*, 15139. [CrossRef]
21. Sun, M.; Visintin, P.; Bennett, T. The effect of specimen size on autogenous and total shrinkage of ultra-high performance concrete (UHPC). *Constr. Build. Mater.* **2022**, *327*, 126952. [CrossRef]
22. Huang, G.; Wang, H.; Shi, F. Coupling Effect of Salt Freeze-Thaw Cycles and Carbonation on the Mechanical Performance of Quick Hardening Sulphoaluminate Cement-Based Reactive Powder Concrete with Basalt Fibers. *Coatings* **2021**, *11*, 1142. [CrossRef]
23. Wang, H.; Jin, K.; Zhang, A.; Zhang, L.; Han, Y.; Liu, J.; Shi, F.; Feng, L. External erosion of sodium chloride on the degradation of self-sensing and mechanical properties of aligned stainless steel fiber reinforced reactive powder concrete. *Constr. Build. Mater.* **2021**, *287*, 123028. [CrossRef]
24. Shi, X.; Fay, L.; Peterson, M.; Yang, Z. Freeze-thaw damage and chemical change of a portland cement concrete in the presence of diluted deicers. *Mater. Struct.* **2010**, *43*, 933–946. [CrossRef]
25. Cai, Z.; Wang, H. Influence of NaCl Freeze–thaw Cycles on the Mechanical Strength of Reactive Powder Concrete with the Assembly Unit of Sulphoaluminate Cement and Ordinary Portland Cement. *Coatings* **2021**, *11*, 1238. [CrossRef]
26. Cao, H.; Liang, Z.; Peng, X.; Cai, X.; Wang, K.; Wang, H.; Lyu, Z. Research of Carbon Dioxide Curing on the Properties of Reactive Powder Concrete with Assembly Unit of Sulphoaluminate Cement and Ordinary Portland Cement. *Coatings* **2022**, *12*, 209. [CrossRef]
27. Hong, X.; Wang, H.; Shi, F. Influence of NaCl freeze thaw cycles and cyclic loading on the mechanical performance and permeability of sulphoaluminate cement reactive powder concrete. *Coatings* **2020**, *10*, 1227. [CrossRef]
28. Qiang, L.; Ibrahim, L.; Zhou, W.; Zhang, M.; Yuan, Z. Treatment methods for plant fifibers for use as reinforcement in cement-based materials. *Cellulose* **2021**, *28*, 5257–5268.
29. Zhong, S.; Fu, J. Surface modifification of copper-plated steel fiber and its corrosion resistance and bonding properties. *J. Build. Mater.* **2019**, *22*, 606–610.
30. Liang, X.; Kang, J.; Chen, J.; Geng, L.; You, Z. Mechanical properties test of copper-coated steel fifiber lightweight aggregate concrete. *J. North China Univ. Sci. Technol.* **2021**, *43*, 32–37.
31. Zhu, J.; Qu, Z.; Liang, S.; Li, B.; Du, T.; Wang, H. The macro-scopic and microscopic properties of cement paste with carbon dioxide curing. *Materials* **2022**, *15*, 1578. [CrossRef] [PubMed]
32. Wang, H.; Zhang, A.; Zhang, L.; Liu, J.; Han, Y.; Wang, J. Research on the Influence of Carbonation on the Content and State of Chloride Ions and the Following Corrosion Resistance of Steel Bars in Cement Paste. *Coatings* **2020**, *10*, 1071. [CrossRef]
33. Liang, J.; Zhu, H.; Chen, L.; Han, X.; Guo, Q.; Gao, Y.; Liu, C. Rebar corrosion investigation in rubber aggregate concrete via the chloride electro-accelerated test. *Materials* **2019**, *12*, 862. [CrossRef]
34. Xu, Q.; Ji, T.; Yang, Z.; Ye, Y. Steel rebar corrosion in artifificial reef concrete with sulphoaluminate cement, sea water and marine sand. *Constr. Build. Mater.* **2019**, *227*, 116685. [CrossRef]

35. Wang, J.; Chen, Z.; Wu, K. Properties of calcium sulfoaluminate cement made ultra-high performance concrete: Tensile performance, acoustic emission monitoring of damage evolution and microstructure. *Constr. Build. Mater.* **2019**, *208*, 767–779. [CrossRef]
36. Keshavarzian, F.; Saberian, M.; Li, J. Investigation on mechanical properties of steel fifiber reinforced reactive powder concrete containing nano-SiO_2: An experimental and analytical study. *J. Build. Eng.* **2021**, *44*, 102601. [CrossRef]

coatings

Article

Reusing Fine Silty Sand Excavated from Slurry Shield Tunnels as a Sustainable Raw Material for Synchronous Grouting

Rui Wang [1,*], Haoqing Xu [2,*], Yi Liu [3], Pengming Jiang [2] and Aizhao Zhou [2]

[1] School of Transportation Engineering, Shandong Jianzhu University, Jinan 250101, China
[2] School of Architecture and Civil Engineering, Jiangsu University of Science and Technology, Zhenjiang 212003, China
[3] Jinan Rail Transit Group Co., Ltd., Jinan 250014, China
* Correspondence: rui_wang1985@163.com (R.W.); hank1nxu@just.edu.cn (H.X.)

Abstract: Using the Nanjing Dinghuaimen Yangtze River Tunnel project as a case study, we proposed a method to reuse the excavated silty-fine sand by adjusting the proportion of the waste sand to replace the commercial sand. This would address the issue of recycling the significant amount of waste sand generated when the slurry shield passes through the silty-fine sand stratum. Moreover, we have evaluated grout indicators such as density, fluidity, consistency, bleeding rate, volumetric shrinkage, setting time, and unconfined compressive strength and examined how the particle size and distribution of the sand affected the grout's performance. The findings show that as the replacement ratio increases, the grout's density, fluidity, consistency, and bleeding rate gradually increase; meanwhile, the volumetric shrinkage increases initially before decreasing; the setting time decreases gradually; the unconfined compressive strength initially decreases before increasing. The key factor altering the grout's performance when the replacement ratio is less than 50% is the weakening of the adsorption effect of fine sand particles on water due to the increase in the sand's fineness modulus. When it is greater than 50%, the particle size of the sand tends to be distributed nonuniformly and fine particles fill the voids between larger particles, thus contributing to the changes in grout properties.

Keywords: excavated fine silty sand; recycling; slurry shield tunnel; shield synchronous grouting materials; engineering properties

1. Introduction

In recent years, the slurry shield tunnelling method has been widely used in the construction of river-crossing tunnels and submarine tunnels under regions with complex hydrogeology [1–3]. Examples include the Tokyo Bay Tunnel [4], Nanjing Yangtze River Tunnel [3,5], Tuen Mun-Chek Lap Kok Tunnel [4,6] and other projects [7–9]. With the slurry shield method, slurry pressure is applied to effectively balance the stratum soil pressure and water pressure and maintain the stability of the excavation surface [10–12]. A large volume of soil is cut by the shield cutterhead, mixed with the slurry, and then discharged to ground slurry treatment equipment via the slurry system. After treatment, the discharged soil and slurry are separated; the slurry is recycled and the excess soil is discarded and stacked. In addition to consuming a lot of land resources, this also readily leads to secondary pollution, which is bad for the environment. The waste residue itself is a form of resource, and the methods of transportation and landfill have resulted in significant resource waste. Therefore, one of the project's main concerns is how to implement the environmentally friendly treatment and resource usage of a substantial amount of waste residue.

Generally, the abovementioned discharged soil is disposed of by stacking it in a yard at relatively high cost, which depletes land resources [13–15]. In addition, the properties of the slurry have a great impact on shield tunnelling. To obtain a high-quality slurry, polymer materials are added. Despite the slurry treatment system, residues remain in the discharged soil, which adversely affect the surrounding environment. The fine particles (mean diameter < 0.075 mm) in the discharged soil after slurry separation and treatment can be used as slurry materials, while larger particles (such as fine silty sand) can be used as a sustainable raw material for synchronous grouting materials [13,16,17]. Such recycling has economic and environmental significance. Therefore, it is important to study the engineering properties of recycled discharged soil and its application in synchronous grouting for tunnel construction.

Most research on shield synchronous grouting materials has focused on its engineering properties and performance during the processes of (1) injection of synchronous grouting material into a shield tail void and (2) volumetric deformation of the grout itself after injection. Traditional shield synchronous grouting materials [18,19] have been improved by adding polymers or other materials. To understand the deformation induced by consolidation after the grout is injected into a shield tail void, the variations in grout pressure and volume in the shield tail void have been studied by field testing/monitoring [20,21] and laboratory tests [22,23]. The consolidation equations of grouts have been derived thusly. These studies have promoted our understanding of synchronous grouting techniques and their functions and provide a theoretical basis for optimizing grout for shield tunnel construction. With the extensive engineering application of shield tunnelling, the disposal of excavated soil is becoming an even more urgent issue. Some studies [24–26] have focused on the recycling of such discharged soil, and corresponding recycling schemes have been proposed.

Research on the recycling of excavated soil from shield tunnelling and its applications in shield synchronous grouting material has mainly focused on the factors affecting the engineering properties and composition of grout (such as its water–binder and soil-sand ratios) when excavated soils are used as a raw material. Sand is used as a fine aggregate in shield synchronous grouting material and comprises >50% of the material's volume. Hence, the characteristics of sand have a significant impact on the engineering performance of the grout. The main sand characteristics are the particle size and its distribution; however, there is little research on the influence of such characteristics on shield synchronous grouting material properties. Given the similarities between shield synchronous grouting material and cement mortar, research on the effects of sand properties—particularly the fineness modulus and particle size distribution—on the performance of cement mortar was used as a reference. Previous studies on cement mortar have shown that the fineness modulus of sand has a significant impact on the water retention capability and strength of the mortar [27]. Sand with different fineness moduli and relative specific surface areas are correlated with the water consumption of the mortar [28,29]; meanwhile, the fineness modulus of sand also affects the rheological properties of concrete [30]. Despite the similarities between the shield synchronous grouting material and the cement mortar [31] and the grout used in jet grouting methods [32], the construction technology of shield synchronous grouting material requires higher workability and pumpability; and the main function is to limit the ground settlement to an acceptable level, which requires it to have good filling property.Therefore, the influence of sand characteristics on the engineering properties of shield synchronous grouting material and the volumetric change after its injection into a shield tail void require further study.

This study investigates the feasibility and technical aspects of recycling excavated silty sand from the silty sand stratum of the Nanjing Dinghuaimen Yangtze River Tunnel. This study proposed a specific method to reuse the waste sand as the raw material for synchronous grouting and analyzed the influence of sand properties on the grout's performance. The findings may serve as a foundation for the design of synchronous grouting and serve as a guide for similar projects.

2. Project Overview

2.1. Project Site

Nanjing is located in the Yangtze River Delta, China. The Yangtze River divides the city into two parts. Figure 1 shows the location of the Nanjing Dinghuaimen Yangtze River Tunnel project. Located 5.0 km downstream of the Nanjing Yangtze River Tunnel and 4.5 km upstream of the Nanjing Yangtze River Bridge, the tunnel connects the main urban area of Nanjing and the planned new urban centre of Pukou. The tunnel adopts an eight-lane X-shaped scheme and was designed as a double-layer bidirectional eight-lane tunnel. The tunnel is divided into two lines: a northern line with a total shield section length of 3537 m and a southern line with a total shield section length of 4135 m. Two slurry shields with diameter of 14.96 m were used to conduct tunnel excavation in the same direction from an originating well to the north of the Yangtze River. The total construction cost is RMB 5.2 billion. It is one of most challenging large-diameter slurry shield tunnelling projects crossing the Yangtze River.

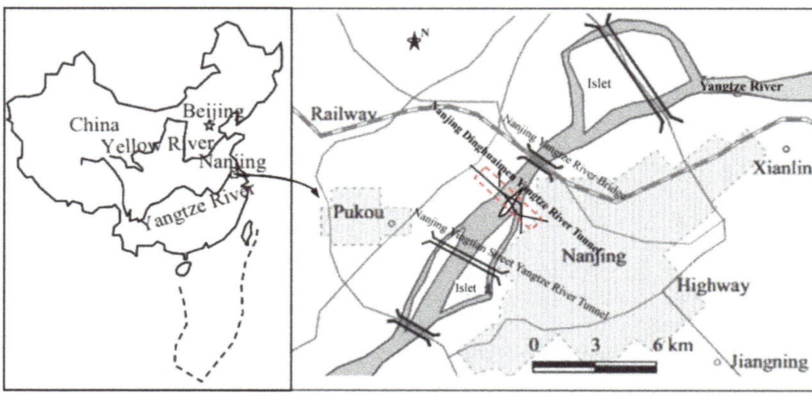

Figure 1. Location of the Nanjing Dinghuaimen Yangtze River Tunnel project.

2.2. Geological Conditions

The tunnel runs through the Yangtze River's alluvial plain, which mostly consists of the river's water area and central bar as well as its waterside beach, high landside floodplain, and low landside floodplain. Except for the levees on both sides of the Yangtze River, the proposed tunnel does not pass through any major roadways; all of the crossing areas are farmland and levees. The Nanjing Yangtze River levee, a high-level levee, is the main water conservancy facility that the tunnel passes through. The stability of the bank slope and levee is ensured by the employment of dry masonry slope protection and mortar masonry toe protection close to the waterside of the levee. In the passage area of the tunnel site, there are no fractures or fracture zones.

The upper part of the stratum through which the shield tunnel passes is filled with soil and quaternary Holocene alluvial soil consisting of muddy-silty clay, silty clay, silt, and silty sand. The middle stratum is composed of medium-to-dense silty-fine sand of the Quaternary Holocene. The lower stratum is dense gravel sand and round gravel of the Upper Pleistocene. The rock stratum is mainly Cretaceous mudstones and silty mudstone, as shown in Figure 2. The geological profile of the northern line is shown in Figure 2; the specific length of the southern line shield passing through each stratum is shown in Table 1. It can be seen that the shield section passing through the silty-fine sand stratum comprises about 45% of the total length. The grain size distribution of the sand in the silty-fine sand stratum is shown in Figure 3; 85% of particles are <1.25 mm, while the content of clay particles is only 2%. The particle size distribution is suitable for use as shield synchronous grouting material.

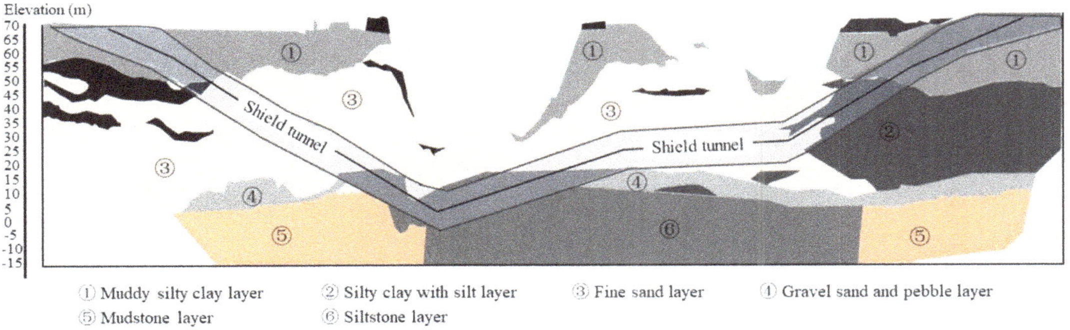

Figure 2. Geological profile of the northern line.

Table 1. Lengths of the southern line shield passing through each stratum (m).

Soil Stratum	Muddy-Silty Clay	Muddy-Silty Clay and Silty-Fine Sand	Silty-Fine Sand	Silty-Fine Sand, Gravel Sand, and Pebbles	Gravel Sand and Pebbles	Gravelly Sand, Pebbles, and Moderately Weathered Sandstone
Length	870	345	1670	330	380	540

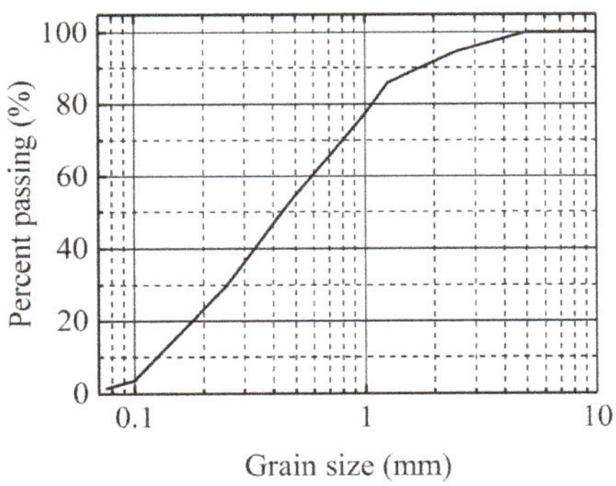

Figure 3. Grain size distribution of the sand in the silty-fine sand stratum.

3. Materials and Experimental Methods

3.1. Materials

The test materials included pulverized limestone, fly ash, sand, water, bentonite, and a water reducer. The content of calcium hydroxide in the hydrated lime was 89% by dry weight. The fly ash was grade-III fly ash from Nanjing Xiaguan Power Plant. The bentonite was grade-I sodium bentonite from Nanjing Tangshan Bentonite Co., Ltd. (Nanjing, China). The water reducer was HLC-NAF high-efficiency water reducer from Nanjing R&D Hi-tech Co., Ltd. (Nanjing, China). The main chemical compositions of the cement, fly ash and bentonite, are shown in Table 2, and the characteristics of the cement and fly ash are shown in Tables 3 and 4. The grain size distribution of the original sand used in the project, the excavated silty-fine sand, and the excavated silty-fine sand passing through a 1.25 mm sieve are shown in Figure 4.

The shield synchronous grouting material used in the northern line was selected as the test material. Its composition is shown in Table 5. This grout is inert because it does not contain cement.

Table 2. Main chemical compositions of the cement, fly ash, and bentonite (%).

Chemical	SiO_2	Al_2O_3	Fe_2O_3	CaO	MgO	SO_3
Cement	21.50	5.80	2.70	63.00	2.09	3.55
Fly ash	52.70	30.50	4.32	6.38	1.46	0.72
Bentonite	70.76	14.93	1.89	0.96	1.77	-

Table 3. Characteristics of the cement.

Requirement of Normal Consistency (%)	Setting Time (h:min)		Flexural Strength (28 Days; MPa)	Compressive Strength (28 Days; MPa)
	Initial Setting Time	Final Setting Time		
25.0	2:35	3:50	8.1	38.4

Table 4. Characteristics of the fly ash.

Fineness	45 μm Sieve Residue	Loss on Ignition (%)	Water Demand Ratio (%)
25.6	14.6	1.53	101

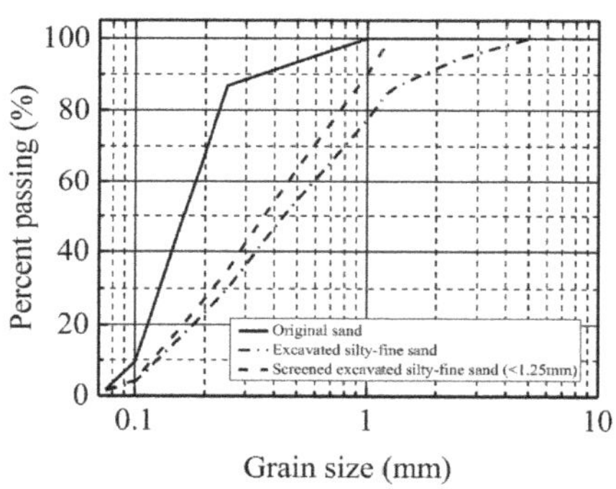

Figure 4. Grain size distribution of different sands.

Table 5. Composition of shield tunnel grout (kg/m^3).

Pulverized Limestone	Fly Ash	Sand	Bentonite	Water Reducer	Water
60.0	390.0	950.0	120.0	4.0	417.0

3.2. Test Program

The excavated sand was first screened through a 1.25 mm sieve and then dried in an oven for 24 h for standby. The excavated sand was mixed with the original sand used in the northern line at different ratios by weight, and the water–binder ratio was adjusted by adding different amounts of tap water. Then, the mixture was stirred uniformly with a stirrer to form the grouts. All the steps in this process were performed at room temperature (25 ± 2 °C). Fifteen groups of shield synchronous grouting materials were prepared according to the test plan presented in Table 6. The water–binder ratio is defined as the mass ratio of water to cementitious material (the sum of pulverized limestone and fly ash). The density, consistency (0 h, 3 h), setting time, fluidity (0 h, 3 h), bleeding rate, volume shrinkage (under 0.3 MPa), and unconfined compressive strength (days 7, 28) of the grouts were tested.

Table 6. Test plan.

No.	Replacement Ratio	Water–Binder Ratio	Fineness Modulus	Test Item
A1 A2 A3	0	0.92 0.85 0.80	0.781	Density; Consistency (0 h, 3 h); Setting time; Fluidity (0 h, 3 h); Bleeding rate; Volume shrinkage (under 0.3 MPa); Unconfined compressive strength (days 7, 28).
B1 B2 B3	0.25	0.92 0.85 0.80	1.046	
C1 C2 C3	0.50	0.92 0.85 0.80	1.210	
D1 D2 D3	0.75	0.92 0.85 0.80	1.372	
E1 E2 E3	1.00	0.92 0.85 0.80	1.652	

3.3. Test Methods

The fluidity of the shield synchronous grouting material was measured following Chinese Standard GB/T2419—2005 [33]. The consistency, unconfined compressive strength, and setting time of the grouts were measured according to Chinese Standard JGJ/T70—2009 [34]. To investigate the variation in the engineering properties of different grouts, their consistency and fluidity were tested at ages of 0 h and 3 h. The bleeding rate was determined following Chinese Standard GB/T 50080—2002 [35]. The volume shrinkage of the grout was measured by one-dimensional consolidation tests, as shown in Figure 5. The consolidation tests were carried out on specimens with a diameter of 75 mm under a pressure of 300 kPa. The maximum consolidation volume compression was measured, and its ratio in relation to the total volume was taken as the volumetric shrinkage of the shield synchronous grouting material.

Figure 5. Volume shrinkage test device for shield synchronous grouting material.

4. Results

4.1. Changes in Grout Properties

The density of grout is an important influence on the stability of a shield tunnel at the early stage. Figure 6 shows that the densities of the 15 groups of grouts in the test were basically within the range of 1.90–2.00 g/cm^3. At a given water–binder ratio (w/b), the density increased with the replacement ratio. If the proportion of excavated silty-fine sand was <50%, the density remained almost constant, while a slight increase was observed at replacement ratios >50%. As the water–binder ratios used in this study only varied slightly, the ratio had little influence on grout density.

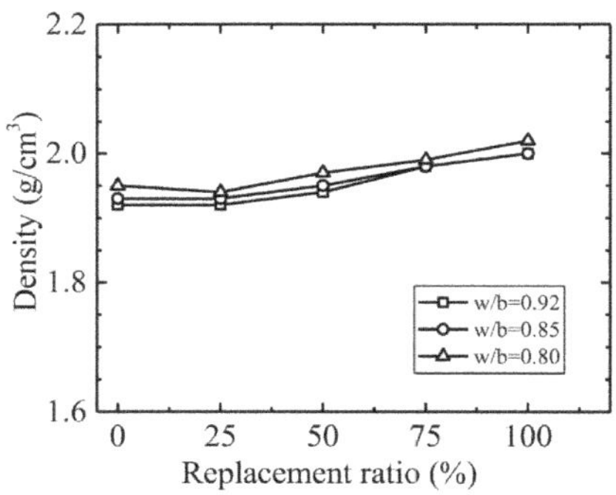

Figure 6. Relationship between grout density and replacement ratio.

Fluidity and consistency can be used to evaluate the potential to pump shield synchronous grouting material. In this study, the initial fluidity, fluidity after 3 h, initial consistency, and consistency after 3 h of the shield synchronous grouting were measured, as shown in Figures 7 and 8. The initial fluidity of the grout was relatively high, being >23.0 cm for all 15 groups of test materials. As the replacement ratio increased, the fluidity tended to increase. The growth rate in fluidity is faster at replacement ratios >50% than at ratios <50%. The grout fluidity was measured 3 h after preparation. As shown in Figure 7, the grout fluidity at this time was lower than the initial fluidity. Moreover, the variation in fluidity with replacement ratio exhibited a similar pattern to that of the initial fluidity. As shown in Figure 8, the variation in grout consistency with increases in the replacement ratio was similar to that of fluidity.

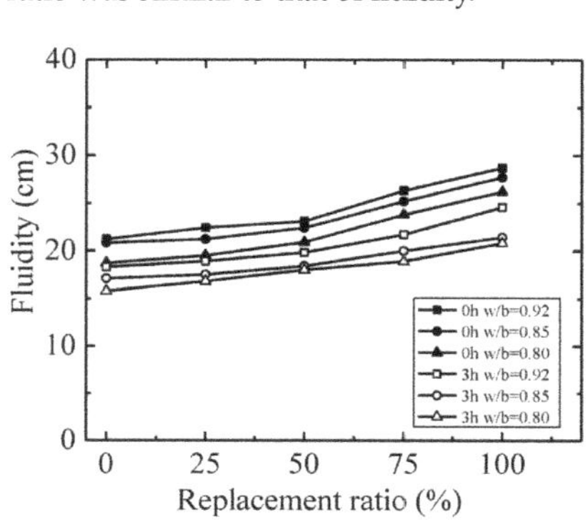

Figure 7. Relationship between grout fluidity and replacement ratio.

The bleeding rate is an index reflecting the solid–liquid separating property of a shield synchronous grouting material. The smaller the bleeding rate, the better the stability of the grout. As shown in Figure 9, the bleeding rates of the prepared grout were <5%. At a given water–binder ratio, the bleeding rate of the grout increased linearly with the replacement ratio of excavated fine-silty sand. At a given replacement ratio, the grout bleeding rate increased with the water–binder ratio.

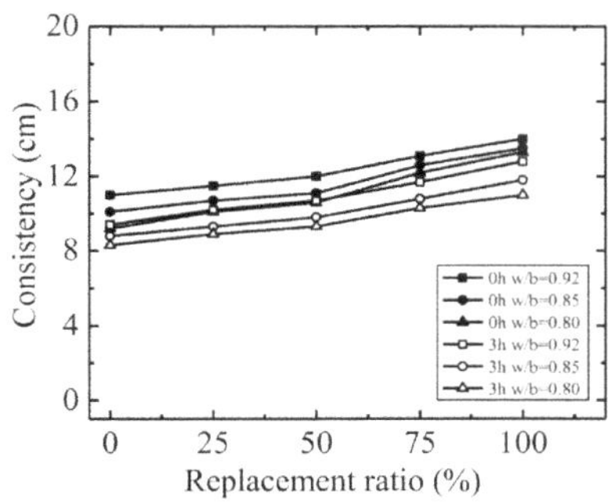

Figure 8. Relationship between grout consistency and replacement ratio.

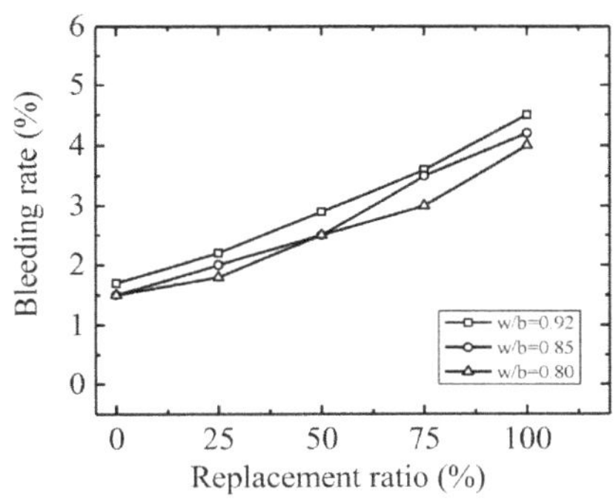

Figure 9. Relationship between grout bleeding rate and replacement ratio.

Volumetric shrinkage is an index for evaluating grout shrinkage under pressure. A lower grout volume shrinkage also reflects lower ground subsidence. Figure 10 shows the shrinkage deformation rate of the shield synchronous grouting material. The volumetric shrinkage of the prepared grouts was <20%, which is low. At a given water–binder ratio, the grout volumetric shrinkage increased with excavated silty-fine sand replacement ratios of <50% but decreased with ratios >50%.

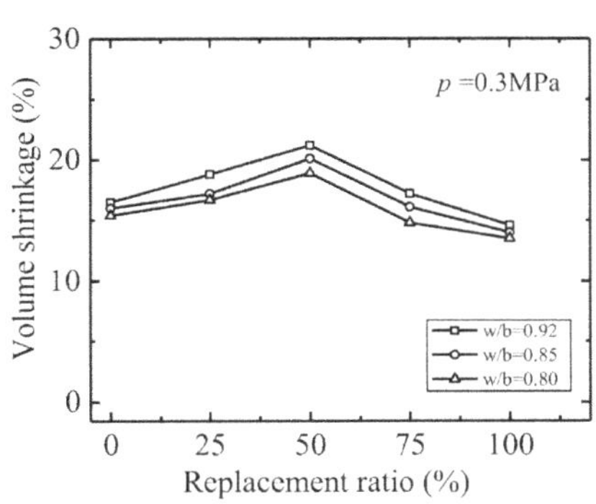

Figure 10. Relationships between grout volumetric shrinkage and replacement ratio at different water–binder ratios.

The setting time is an index of the time it takes for the grout to become solid. It should be determined according to the stratum, for example, for a high-permeability stratum, it should be short. If the setting time is too short, pipe blocking or incomplete filling might occur, while a long setting time is not conducive to the stability of a shield tunnel at the early stage. The variations in the setting times of the prepared grouts are shown in Figure 11. Due to the grout being inert (no cement component), the setting times are 45–65 h. The replacement ratio of the excavated silty-fine sand influences the setting time of the shield synchronous grouting material. At a given water–binder ratio, the setting time decreases with the replacement ratio.

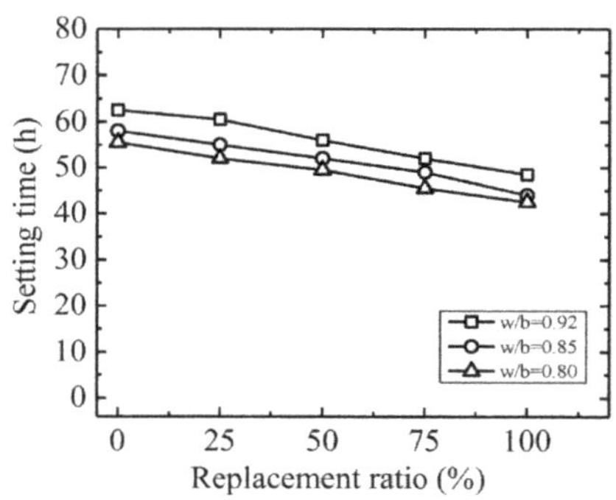

Figure 11. Relationships between grout setting time and replacement ratio at different water–binder ratios.

The strength soon after injection into a shield tail void is an important property of shield synchronous grouting material and is of great significance to the stability and antiseepage properties of a shield tunnel. Therefore, shield synchronous grouting material of a certain strength is required at the early stage. Generally, it is advisable to use shield synchronous grouting material with an unconfined compressive strength (UCS) close to that of the surrounding rock–soil mass. The UCS of the grout was tested on days 7 and 28, as shown in Figure 12. The UCSs of the prepared grouts were low, being lower on days 7 than on days 28, while they did not differ much for grouts with different water–binder ratios of the same age. The UCSs of grouts with the same water–binder ratio decreased first and then increased with increases in the replacement ratio of excavated silty-fine sand.

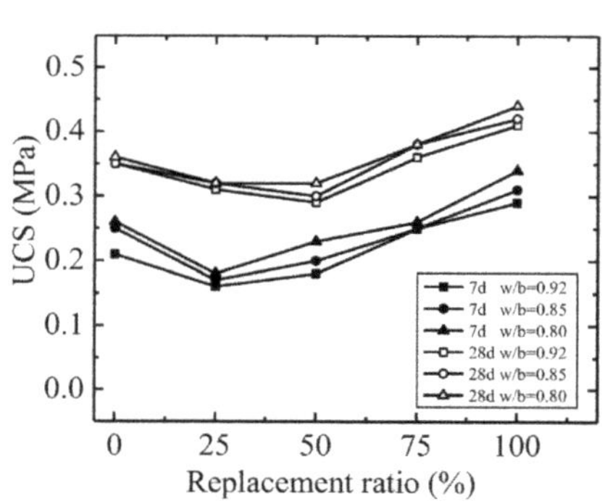

Figure 12. Relationship between the unconfined compressive strength (UCS) of grout and the replacement ratio.

4.2. Selection of Grout Formula

According to the test results, when the proportion of excavated silty-fine sand accounted for 25% of the total sand (B1, B2, B3), the UCS (days 7, 28) was less than that of the original-formula grout (A1, A2, A3). The volume shrinkage was greater than that of the original, while the other engineering properties were similar. With 50% excavated silty-fine sand (C1, C2, C3), properties such as fluidity and consistency were higher than those of the original formula, while the volume shrinkage, bleeding rate, and strength were lower. At 75% (D1, D2, D3), the engineering properties other than bleeding rate were close to or slightly higher than those of the original formula. When all the original sand was replaced by excavated silty-fine sand (100%; E1, E2, E3), the bleeding rate of the grout was high, the setting time was short, and the other engineering properties were close to or slightly higher than those of the original formula. Therefore, grout D was selected as the basic grout and one of the three grouts was selected as the new grout formula.

The engineering performance of grouts A1, D1, D2, and D3 are shown in Table 7. Although the consistency and fluidity of grout D3 were the lowest among the four grouts, the differences were small, and the engineering requirements are still satisfied. The bleeding rate and volume shrinkage of grout D3 were also the lowest among the three grouts, which is very important for shield synchronous grouting materials. Therefore, grout D3 (75% excavated silty-fine sand) was selected as the optimum grout formula.

Table 7. Engineering performance of grouts A1, D1, D2, and D3.

No.	Density (g/cm^3)	Consistency (cm)		Fluidity (cm)		Bleeding Rate (%)	Setting Time (h)	Volume Shrinkage (%)	UCS (MPa)	
		Initial	3 h	Initial	3 h				7 Days	28 Days
A1	1.92	11.0	9.4	21.2	18.7	1.7	62.5	16.5	0.21	0.35
D1	1.96	13.1	11.7	26.3	21.7	3.6	52.0	17.2	0.25	0.36
D2	1.96	12.6	10.7	25.2	20.0	3.6	49.0	16.1	0.25	0.38
D3	1.97	12.2	10.1	23.8	18.9	3.0	45.5	14.8	0.26	0.38

5. Discussion

This study found the main influences on the engineering properties of shield synchronous grouting material to include the sand's fineness and particle size distribution and the water–binder ratio. The former two are varied by adjusting the proportion of excavated silty-fine sand, while the latter is varied by adjusting the water consumption in the original formula.

5.1. Influence of Fineness Modulus

The mixed sand with a different proportion of excavated silty-fine sand, which screened through a 1.25 mm sieve, was screened to calculate the fineness modulus. Figure 13 shows the relationship between the fineness modulus of the mixed sand and the replacement ratios. It is indicated that the fineness modulus of mixed sand increases with the replacement ratio.

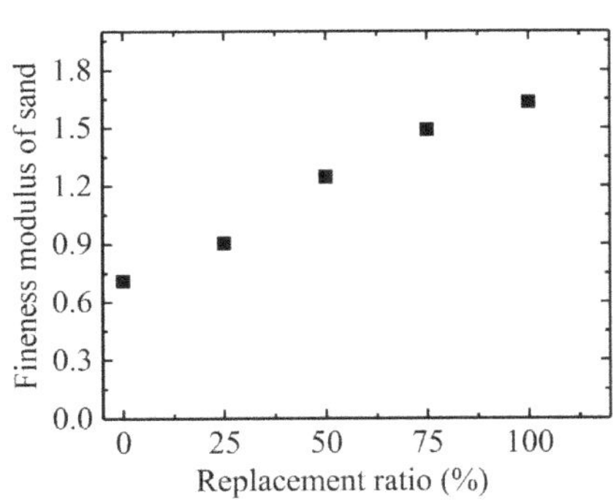

Figure 13. Relationship between the fineness modulus of mixed sand and the replacement ratio.

The replacement ratio (x-axis in Figure 13) could be directly replaced by the fineness modulus for discussion in Figures 6–12. As the fineness modulus of the sand increases, the density of the grout tended to increase, while the range of variation in the density of the grout is not high. This indicates that the fineness modulus of the sand has little influence on the density of the shield synchronous grouting material.

The fluidity, consistency, and bleeding rate of the grout increase with the proportion of excavated silty-fine sand in the total sand. However, the growth rates in fluidity and consistency are different in different regimes (replacement ratios < or >50%). This is because, with increases in the proportion of excavated silty-fine sand, the sand particle size increases, the fineness modulus of the sand increases, the total surface area of particles reduces, the adsorption of water by particles decreases, and the free water around particles increases so that the fluidity and consistency of the grout tend to increase. With <50% excavated silty-fine sand, the sand grain size is relatively uniform so the increases in the grout's fluidity and consistency are mainly affected by the sand's fineness. At >50%, the sand grain size tends to be nonuniform and fine particles fill the voids between larger particles. This reduces friction between them, thus further increasing the fluidity and consistency of the shield synchronous grouting material. Due to the range of variation in the grout bleeding rate being small, only the fineness modulus has a significant influence on this variation.

With increases in the proportion of sand that is excavated silty-fine sand, the volumetric shrinkage of grout (under a pressure of 0.3 MPa) first increases and then decreases, while the UCS first decreases and then increases. At an excavated silty-fine sand ratio of <50%, with increases in fineness modulus, the grain size of sand particles increases, which leads to increases in void spaces between larger particles. The fine particles cannot effectively fill these pores, so the volumetric shrinkage of grout increases and the UCS decreases. When the ratio reaches 50%, more fine particles fill the voids between larger particles. The increased contact points between larger particles form a skeleton so that the volumetric shrinkage of grout gradually decreases and the UCS gradually increases. The grouts prepared in this study did not contain cement or other materials that could induce a rapid hydration reaction. The fly ash contents were the same in different shield synchronous grouting materials, so the influence of the grout's chemical reactions on setting time can be ignored. With increasing sand particle size, the liquid resistance of particles per unit volume decreases, so the contacts between particles are increased, resulting in a shorter setting time for the shield synchronous grouting material.

5.2. *Influence of Grain Size Distribution*

Based on the above analyses, the fineness of the sand in the shield synchronous grouting material (fineness modulus) has an obvious influence on the grout's engineering properties. The sand used in the original formula of shield synchronous grouting material in the project was fine sand with a relatively uniform particle size. The silty sand excavated from the stratum has a certain size distribution.

Mixed sand samples with different amounts of excavated silty-fine sand from the silty-fine sand stratum were screened to obtain their grain size distributions (Figure 14). The figure shows that the grain size distribution of the sand in the grout changed gradually as the amount of added excavated silty-fine sand increased. The nonuniformity coefficient C_u was calculated for five mixed sand samples (Table 8). The C_u value of the sand used in original grout formula was 1.817, and that of the excavated silty-fine sand with particles >1.25 mm removed was 3.853. As the amount of added excavated silty-fine sand increased, the C_u value of the mixed sand gradually increased within this range. Based on the calculation of the curvature coefficient C_c, the C_c values of all sand samples were <1.0, which indicates that the grain size distributions of the sand became increasingly uneven. However, since the C_u values are <5.0, the grain sizes of the mixed sands still have poor grading and lack particles of intermediate sizes.

In summary, the sand used in the original grout formula was fine with a relatively low fineness modulus and uniform particle size. When a small amount of excavated silty-fine sand was added, the fineness modulus of the mixed sand increased, and the engineering properties of the grout changed accordingly. The grain size distribution of the sand samples did not differ much from that of the original sand. Therefore, changes in the grout's engineering performance were mainly influenced by increases in the sand particle size. As the amount of excavated silty-fine sand increased, the grain size distribution of the mixed sand gradually became uneven. The number of large particles in the mixed sand gradually increased and formed voids, while small particles in the sand filled the voids between large particles and reduced rolling friction, thus intensifying the changes in the engineering performance of the shield synchronous grouting material.

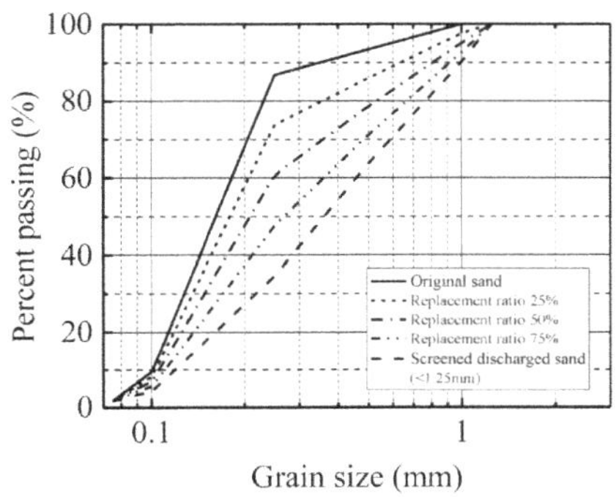

Figure 14. Grain size distribution of mixed sand samples.

Table 8. Indexes of fine sand samples.

Sample	Fineness Modulus	C_u	C_c
A	0.781	1.817	0.90
B	1.046	1.961	0.91
C	1.210	2.352	0.85
D	1.372	3.263	0.73
E	1.652	3.853	0.86

5.3. Economic Analysis and Environmental Benefits

Generally, the cost of the synchronous grouting material has become a concern for the construction unit. The calculation parameters for Equation (1): outer diameter of shield D_1 = 14.96 m; outer diameter of segment D_2 = 14.5 m; ring width B = 2 m; injection rate α = 150%; and injection volume of a single ring = 31.9 m^3. In other words, the grouted weights per ring of pulverized limestone, fly flash, bentonite, river sand, water reducer, and water were approximately 1.914 t, 12.444 t, 3.829 t.302 t, 30.314 t, 0.128 t, and 13.306 t, respectively, according to the local market price of the raw material (river sand about RMB 250 per ton). Thus, this method decreases the cost of raw materials by RMB 5683.9. Moreover, taking into account the collection and treatment costs of the excavated silty-fine sand (about RMB 40 per ton), this method decreases the collection and treatment cost by RMB 909.42. As a result, the total costs are decreased by 6593.32 RMB/ring by taking the recycling option for the excavated silty-fine sand.

$$V = \frac{\pi}{4}\left(D_1^2 - D_2^2\right)B\alpha \tag{1}$$

At the time of the study, there were about 1200 rings left, and the total injection volume was about 38,280 m^3. Considering the collection and treatment costs of the excavated silty-fine sand, the replacement according to the $D3$ scheme was expected to save about RMB 7.911 million.

Furthermore, the recycling of the excavated silty-fine sand from the construction site enhances a portion of the environmental benefits, and it includes the following three aspects: (1) the accumulated excavated silty-fine sand will occupy the construction site, which will influence the smell of the surrounding environment and blockages of the drainage system; (2) the transport of excavated silty-fine sand to spoil area will causes adverse an environmental impact to the surroundings, such as CO_2 emissions, soil splash, and raised dust; and (3) the accumulated excavated silty-fine sand in the spoil area will result in a potential source of landslides and debris flows [36].

Therefore, it is significant to reuse the excavated silty-fine sand from the silty-fine sand stratum as the raw material for shield synchronous grouting material.

6. Conclusions

The following conclusions can be drawn:

(1) The engineering performance of the shield synchronous grouting material in which 75% of the sand in the original formula was replaced by excavated fine-silty sand screened through a 1.25 mm sieve, and in which the water/binder ratio was adjusted to 0.8, is close to or slightly higher than that of the original formula. Hence, it can be used in the recycling scheme.

(2) As the replacement ratio increases, the grout's density, fluidity, consistency, and bleeding rate gradually increase; meanwhile, the volumetric shrinkage increases initially before decreasing; the setting time decreases gradually

(3) The sand used in the original formula was fine sand with a small fineness modulus and uniform particle size. At a mixing amount of <50%, a small change in the grain size distribution does not affect the engineering performance of the shield synchronous grouting material. However, the fineness modulus is the main influencing factor. At mixing amounts >50%, the grain size distribution changes greatly and has a leading influence on the shield synchronous grouting material's performance.

(4) Recycling the excavated silty-fine sand as shield synchronous grouting material can not only greatly reduce engineering costs but also significantly reduce environmental pollution, thereby promoting sustainable development.

Author Contributions: Conceptualization, R.W.; methodology, R.W. and H.X.; validation, H.X. and Y.L.; formal analysis, R.W. and H.X.; investigation, H.X.; resources, P.J. and A.Z.; data curation, R.W. and H.X.; writing—original draft preparation, R.W. and H.X.; writing—review and editing, P.J.; supervision, Y.L. and H.X.; project administration, P.J. and Y.L. All authors have read and agreed to the published version of the manuscript.

Funding: This work was financially supported by Shandong Provincial Natural Science Foundation, China (Grant No. ZR2020QE267).

Institutional Review Board Statement: Not applicable.

Informed Consent Statement: Not applicable.

Data Availability Statement: All data that support the findings of this study are included within the article.

Acknowledgments: We would like to recognize numerous coworkers, students, and research facility associates for giving specialized assistance on instrument examination.

Conflicts of Interest: The authors declare no conflict of interest.

References

1. Wang, R.; Xu, H.; Liu, Y.; Jiang, P.; Zhou, A.; Lv, Y. Study on the shear strength characteristics and source mechanism of early-age shield synchronous grouting materials. *Acta Geotech.* **2023**. [CrossRef]
2. Geng, Z.; Jin, D.; Yuan, D. Face stability analysis of cohesion-frictional soils considering the soil arch effect and the instability failure process. *Comput. Geotech.* **2023**, *153*, 105050. [CrossRef]

3. Feng, K.; He, C.; Qiu, Y.; Zhang, L.; Wang, W.; Xie, H.; Zhang, Y.; Cao, S. Full-scale tests on bending behavior of segmental joints for large underwater shield tunnels. *Tunn. Undergr. Space Technol.* **2018**, *75*, 100–116. [CrossRef]
4. Funasaki, T.; Izumi, Y.; Masuda, T.; Kanai, M. The Tokyo Bay Tunnel of the Tram-Tokyo Bay Highway. In Proceedings of the 7th International Offshore and Polar Engineering Conference Proceedings (ISOPE-97), Honolulu, HI, USA, 25–30 May 1997; pp. 6–15.
5. Min, F.; Zhu, W.; Lin, C.; Guo, X. Opening the excavation chamber of the large-diameter size slurry shield: A case study in Nanjing Yangtze River Tunnel in China. *Tunn. Undergr. Space Technol.* **2015**, *46*, 18–27. [CrossRef]
6. Morris, M.; Yang, M.W.W.; Tsang, C.K.; Hu, A.Y.M.; Shut, D.S.C. An overview of subsea tunnel engineering in Hong Kong. *Proc. Inst. Civ. Eng.-Civ. Eng.* **2016**, *169*, 9–15. [CrossRef]
7. Min, F.; Du, J.; Zhang, N.; Chen, X.; Lv, H.; Liu, L.; Yu, C. Experimental study on property change of slurry and filter cake of slurry shield under seawater intrusion. *Tunn. Undergr. Space Technol.* **2019**, *88*, 290–299. [CrossRef]
8. Jin, D.; Shen, Z.; Yuan, D. Effect of spatial variability on disc cutters failure during TBM tunneling in hard rock. *Rock Mech. Rock Eng.* **2020**, *53*, 4609–4621.
9. Geng, Z.; Yuan, D.; Wang, D.; Zhao, Y.; Xie, T. Study on Longitudinal Stress Relaxation Effect and Reinforcement Technology of Segment Lining during Shield Docking. *Appl. Sci.* **2022**, *12*, 2831. [CrossRef]
10. Min, F.; Song, H.; Zhang, N. Experimental study on fluid properties of slurry and its influence on slurry infiltration in sand stratum. *Appl. Clay Sci.* **2018**, *161*, 64–69. [CrossRef]
11. Wang, D.; Min, F.; Lyu, H.; Chen, J.; Wang, B. Recycling waste sand from slurry shield tunneling: A sustainable filter aid for waste slurry dehydration. *J. Clean. Prod.* **2023**, *383*, 135387. [CrossRef]
12. Liu, M.-B.; Liao, S.-M.; Shi, Z.-H.; Liu, H.-K.; Chen, L.-S. Analytical study and field investigation on the effects of clogging in slurry shield tunneling. *Tunn. Undergr. Space Technol.* **2023**, *133*, 104957. [CrossRef]
13. Zhang, C.; Yang, J.; Fu, J.; Wang, S.; Yin, J.; Xie, Y. Recycling of discharged soil from EPB shield tunnels as a sustainable raw material for synchronous grouting. *J. Clean. Prod.* **2020**, *268*, 121947. [CrossRef]
14. Wang, J.; Cai, Y.; Ni, J.; Geng, X.; Xu, F. Effect of sand on the vacuum consolidation of dredged slurry. *Mar. Georesour. Geotechnol.* **2018**, *36*, 238–244. [CrossRef]
15. Ooishi, Y.; Murakawa, S.; Hatakoshi, A.; Mori, T. Development of discharged soil measuring device for shield tunneling machines. *J. Terramech.* **1992**, *29*, 465–475. [CrossRef]
16. Song, W.; Zhu, Z.; Pu, S.; Wan, Y.; Huo, W.; Song, S.; Hu, L. Synthesis and characterization of eco-friendly alkali-activated industrial solid waste-based two-component backfilling grouts for shield tunnelling. *J. Clean. Prod.* **2020**, *266*, 121974. [CrossRef]
17. Zhang, C.; Chen, K.; Yang, J.; Fu, J.; Wang, S.; Xie, Y. Reuse of Discharged Soil from Slurry Shield Tunnel Construction as Synchronous Grouting Material. *J. Constr. Eng. Manag.* **2022**, *148*, 04021193. [CrossRef]
18. Koyama, Y.; Okano, N.; Sato, Y.; Shimizu, M. Back-fill grouting model test for shield tunnel. *Railw. Tech. Res. Inst. Q. Rep.* **1998**, *39*, 35–39.
19. Peila, D.; Borio, L.; Pelizza, S. The behaviour of a two-component back-filling grout used in a tunnel-boring machine. *Acta Geotech. Slov.* **2011**, *8*, 5–15.
20. Bezuijen, A.; Talmon, A.; Kaalberg, F.; Plugge, R. Field Measurements of Grout Pressures During Tunnelling of the Sophia Rail Tunnel. *Soils Found.* **2004**, *44*, 39–48. [CrossRef]
21. Talmon, A.; Bezuijen, A. Simulating the consolidation of TBM grout at Noordplaspolder. *Tunn. Undergr. Space Technol.* **2009**, *24*, 493–499. [CrossRef]
22. Zhang, S.S.; Dai, Z.; Bai, Y. Research on the dissipation law of grout pressure during the simultaneous grouting of shield tunnel. *China Railw. Sci.* **2012**, *33*, 40–48.
23. Liang, Y.; Yang, J.S.; Wang, S.Y.; Zeng, X.Y. A study on grout consolidation and dissipation mechanism during shield backfilled grouting with considering time effect. *Rock Soil Mech.* **2015**, *36*, 3373–3380.
24. Zhou, S.; Li, X.; Ji, C.; Xiao, J. Back-fill grout experimental test for discharged soils reuse of the large-diameter size slurry shield tunnel. *KSCE J. Civ. Eng.* **2017**, *21*, 725–733. [CrossRef]
25. Zhang, J.; Lu, S.; Feng, T.; Yi, B.; Liu, J. Research on reuse of silty fine sand in backfill grouting material and optimization of backfill grouting material proportions. *Tunn. Undergr. Space Technol.* **2022**, *130*, 104751. [CrossRef]
26. Cui, Y.; Tan, Z. Experimental Study of High Performance Synchronous Grouting Materials Prepared with Clay. *Materials* **2021**, *14*, 1362. [CrossRef] [PubMed]
27. Lim, S.K.; Tan, C.S.; Chen, K.P.; Lee, M.L.; Lee, W.P. Effect of different sand grading on strength properties of cement grout. *Constr. Build. Mater.* **2013**, *38*, 348–355. [CrossRef]
28. De Schutter, G.; Poppe, A.-M. Quantification of the water demand of sand in mortar. *Constr. Build. Mater.* **2004**, *18*, 517–521. [CrossRef]
29. Haach, V.G.; Vasconcelos, G.; Lourenço, P.B. Influence of aggregates grading and water/cement ratio in workability and hardened properties of mortars. *Constr. Build. Mater.* **2011**, *25*, 2980–2987. [CrossRef]
30. Westerholm, M.; Lagerblad, B.; Silfwerbrand, J.; Forssberg, E. Influence of fine aggregate characteristics on the rheological properties of mortars. *Cem. Concr. Compos.* **2008**, *30*, 274–282. [CrossRef]
31. Farhangi, V.; Karakouzian, M.; Geertsema, M. Effect of Micropiles on Clean Sand Liquefaction Risk Based on CPT and SPT. *Appl. Sci.* **2020**, *10*, 3111. [CrossRef]

32. Roshani, M.; Kargar, S.; Farhangi, V.; Karakouzian, M. Predicting the Effect of Fly Ash on Concrete's Mechanical Properties by ANN. *Sustainability* **2021**, *13*, 1469. [CrossRef]
33. *GB/T 2419-2005*; Test Method for Fluidity of Cement Mortar. General Administration of Quality Supervision. Inspection and Quarantine of the People's Republic of China: Beijing, China, 2005.
34. *JGJ/T 70-2009*; Standard for Test Method of Performance on Building Mortar. Ministry of Housing and Urban-Rural Development of the People's Republic of China: Beijing, China, 2009.
35. *GB/T 50080-2016*; Standard for Test Method of Performance on Ordinary Fresh Concrete. General Administration of Quality Supervision. Inspection and Quarantine of the People's Republic of China: Beijing, China, 2016.
36. Liu, W.; Yan, S.; He, S. Landslide damage incurred to buildings: A case study of Shenzhen landslide. *Eng. Geol.* **2018**, *247*, 69–83. [CrossRef]

coatings

Article

The Influence of Aspect Ratio of Steel Fibers on the Conductive and Mechanical Properties of Compound Cement Reactive Powder Concrete

Zhao Liang [1,2], Xi Peng [2,3,*] and Hui Wang [1,*]

1 School of Civil Engineering and Geographical Environment, Ningbo University, Ningbo 315000, China
2 School of Civil Transportation Engineering, Ningbo University of Technology, Ningbo 315000, China
3 Engineering Research Center of Industrial Construction in Civil Engineering of Zhejiang, Ningbo University of Technology, Ningbo 315000, China
* Correspondence: pengxi@nbut.edu.cn (X.P.); wanghui4@nbu.edu.cn (H.W.)

Abstract: The performance of steel fibers reinforced RPC has a great relationship with the aspect ratio of the fibers. It is necessary to systematically study the relationship between the aspect ratio and the mechanical properties of RPC and the force-electricity response. In this paper, the flexural strength, the compressive strength and the flexural toughness and AC electrical resistance and AC impedance spectroscopy of reactive powder concrete are investigated. The flexural strength, compressive strength, and flexural toughness with the curing age of 1 day and 28 days are tested. The cement is the compound cement in this study, and the mass ratio of ordinary portland cement and sulphoaluminate cement is 1:1, and two kinds of steel fibers with aspect ratios of 30(AR-30) and 75(AR-75) are used. The fiber content is 2% of the concrete volume. Scanning electron microscope (SEM) and scanning electron microscope energy spectrum analysis (EDS) are investigated to analyze the composition of the hydration products. It can be obtained from the research results that the flexural strength and the compressive strength of reactive powder concrete decrease with the increase of steel fibers content with an aspect ratio of 30. The decreasing rates of the flexural strength and the compressive strength with the curing age of 1 d and 28 d are 14.93%~83.26% and 0.40%~46.36% with the incorporation of steel fibers with an aspect ratio of 30. The flexural toughness decreases in the form of a quadratic function with the mass ratio of steel fibers with an aspect ratio of 30. The electrical resistance of reactive powder concrete increases with the increase of steel fibers with an aspect ratio of 30. The maximum decreasing rate of electrical conductivity is 91.16%. The AC impedance spectrum is obtained, and the electric circuit of reactive powder concrete accords with the series conduction model, which parallel electrical resistance and capacitance of pore solution, rapid strength RPC and steel fibers respectively connect with a contact resistance of the electrode and specimens. It is found that the steel fibers network with an aspect ratio of 75 is closer by studying the equivalent circuit diagram.

Keywords: steel fibers; concretes; aspect ratio; electrical property; mechanical property; micro analysis

1. Introduction

Ordinary portland cement (OPC) has a history of nearly 200 years since it is manufactured. The invention of OPC has brought significant changes to the building materials industry. The performance of the concrete has also been greatly improved by the changes in the type, amount of cement, and admixtures. With the rapid repairing of damaged structures in engineering, special requirements for early strength, and rapid setting of cement are required, and OPC can not meet the requirements in terms of technology and performance [1].

Coastal cities are economically developed. At the same time, trade requires a lot of transportation. The bridge is an important part of transportation. However, due to the erosion of the sea-crossing bridge in the marine environment and traffic loading [2], the pavement of the bridge is more prone to cracking, and the scene of repairing the damaged bridge will not have too much time to close. For example, Beijing Xizhimen Interchange has a large number of cracks due to chloride erosion and the freeze-thaw cycle. Many port projects in some coastal countries will appear cracking and peeling in a short period of time. The decks of 9700 bridges in Wisconsin, USA, were damaged due to chloride corrosion [3]. The damaged roads should be repaired in time [4–6]. Furthermore, the repairing materials of the bridge deck should have good corrosion resistance.

The electric conductivity of concrete has been used for intelligent detection and evaluation of hydration and mechanical properties [7,8]. Wang's article points out that using electrical properties to characterize mechanical properties is an effective method for the nondestructive evaluation of concrete mechanical properties [9]. This nondestructive evaluation method has been applied to bridges such as Qinglong Bridge in Ningbo City, Zhejiang Province [10].

Sulfoaluminate cement (SAC) has been used for repairing roads and bridges due to its fast-hardening characteristics in recent years [11,12]. Feng's article points out that SAC mortar reinforced with ultra-fine steel fibers displays a maximum compressive strength (f_{cu}) of 19 MPa and flexural strength (f_t) of 5.3 MPa after curing for 3 h. SAC has excellent impermeability and corrosion resistance [13]. However, SAC has a high cost and the phenomenon of late strength recession.

Reactive powder concrete (RPC) is a special cement concrete material with ultra-high strength, high toughness, and durability [14,15]. Moreover, as described in Hui's research that RPC with the silica fume and fly ash demonstrates f_{cu} of 124.2 MPa and f_t of 19.2 MPa. Additionally, after standard curing, the f_{cu} and f_t of steel fibers reinforced RPC can reach 130 MPa and 52.9 MPa. The autogenous shrinkage rate decreases by 29.9% [16,17].

The compound cement has more advantages than single cement [18,19]. Scholars have optimized the deficiency of single cement by compositing SAC and OPC. Hong's article points out that the initial setting time and final setting time after the combination of SAC and OPC are 30 min and 55 min respectively. The f_{cu} and f_t of 5 h are 34.8 MPa and 3.7 MPa. The strength decline in the later stage will also be improved [20].

The addition of fibers has a significant influence on mechanical performance and durability [21–23]. Several types of research about the fibers reinforced cement concrete have been reported. Scholars have reported that steel fibers, plant fibers, and polypropylene fibers are effective to improve the mechanical strength of RPC [24,25]. Plant fibers and polypropylene fibers have excellent corrosion resistance and cracking resistance. However, they are difficult to disperse in cement-based materials [26–29]. Meanwhile, the effect of plant fibers and polypropylene fibers on the mechanical strength and toughness of RPC is not obvious. The steel fibers have been proven to improve the mechanical strength of RPC [30]. As reported in some research, the increasing rates of the splitting tensile strength, f_t, and tension compression ratio by steel fibers are 74%, 119%, and 75% respectively [31,32]. The dispersion and aspect ratio of steel fibers will have a considerable impact on the performance of RPC. At present, few people have carried out systematic research on the area of the dispersion and aspect ratio of steel fibers.

In this experiment, the mechanical properties (f_t, f_{cu}, and the flexural toughness (α)) and the electrical properties (AC electrical resistance (R_e) and AC impedance spectrum) of reactive powder concrete with 2% fibers content of steel fibers with different ratio of the aspect ratios of 30 and 75 are determined. In addition, the mechanical properties of RPC with different aspect ratio fibers are characterized by electrical properties. Scanning electron microscope (SEM) and scanning electron microscope energy spectrum analysis (EDS) are investigated to reveal the mechanism of the macro performances. This paper will provide a reference for rapid repairing materials for bridge decks by adjusting the aspect ratio of steel fibers in the future.

2. Experimental

2.1. Raw Materials

Rapid hardening SAC and OPC with strength grades of 42.5 MPa are employed in the research. The initial setting time and final setting time of SAC are 20 min and 205 min

(GBT1346-2011). OPC and SAC are produced by Anhui Conch Group Co., Ltd. (Wuhu, China) The specific surface area of SAC is 365 m^2/g. Silica fume (SF) with a specific surface area and density of 15 m^2/g and 2.2 g/cm^3 is used for this research. SF is produced by Shanggao Mingzheng Plastic Co., Ltd. The density and specific surface area of granulated blast furnace slag powder (GGBS) are 2.9 g/cm^3 and 436 m^2/g respectively. GGBS is provided by Lingshou Qiangdong Mineral Products Processing Plant. Tables 1 and 2 show the particle size distribution and chemical composition of the cementitious materials. Aggregate is the quartz sand produced by Ling Shou County Yongshun Fruit Products Processing Co., Ltd., LingShou, China. The particle sizes are 0.99~0.72 mm, 0.60~0.34 mm, and 0.14~0.299 mm. And the mass ratio of sand is 1:1.5:0.8. The copper-plated steel fibers produced by Chongqing Yingzhi Building Materials Co., Ltd., Jiulongpo District, China are used in the experiment. The average length and diameter of steel fibers with an aspect ratio of 75(AR-75) are 1.5 cm and 0.2 mm. The average length and diameter of steel fibers with an aspect ratio of 30(AR-30) are 1.5 cm and 0.5 mm. The tensile strength and the nonbreaking rate at one bending of 90° of steel fibers with AR-30 and AR-75 are 3000 MPa and 98%, respectively. The density and the resistivity of steel fibers are 7.85 g/cm^3 and 9.78×10^{-8} Ω·m. The polycarboxylate-based (SP) that has a water-reducing rate of 40% provided by Shenteng Co., Ltd., Lingshou, China is used as a water-reducing agent. Li_2SO_4, tartaric acid, and polyether defoamer (The purity of these drugs is 99%) are regarded as early-strength agent, retarder and defoamer in this research. The densities of Li_2SO_4, tartaric acid, and polyether defoamer are 2.06 g/cm^3, 1.90 g/cm^3, and 1.00 g/cm^3, respectively. The mixing proportion of specimens is shown in Table 3. "C-RPC" represents the composite reactive powder concrete group.

Table 1. The passing rate of the particle size distribution (%).

Types \ Particle Size (%)	0.3	0.6	1	4	8	64	360
OPC	0	0.34	2.67	15.00	28.76	93.60	100
SAC	0	0.34	1.90	16.37	30.13	95.16	100
GGBS	0.026	0.2	3.54	19.65	35.05	97.7	100
SF	31.1	58.5	82.1	100	100	100	100

Table 2. Chemical composition (%).

Types	Chemical Composition (%)						
	SiO_2	Al_2O_3	Fe_2O_3	MgO	CaO	SO_3	Ti_2O
OPC	20.90	5.46	3.93	1.71	62.22	5.78	/
SAC	13.96	22.45	2.66	2.93	42.01	14.33	1.64
GGBS	35.62	14.75	0.25	9.72	35.91	0.24	3.50

Table 3. Mixing proportion of specimens (g).

Types	H_2O	OPC	SAC	SF	GGBS	Sand	Water-Reducing Agent	Li_2SO_4	Calcium Formate	Tartaric Acid	Defoamer	Aspect Ratio of 75 Steel Fibers	Aspect Ratio of 30 Steel Fibers
C-RPC-1	244.42	740.7	740.7	370.3	111.1	977.9	20.3	0.6	2.6	1.9	0.6	15.8	0
C-RPC-2	244.42	740.7	740.7	370.3	111.1	977.9	20.3	0.6	2.6	1.9	0.6	11.85	3.95
C-RPC-3	244.42	740.7	740.7	370.3	111.1	977.9	20.3	0.6	2.6	1.9	0.6	7.9	7.9
C-RPC-4	244.42	740.7	740.7	370.3	111.1	977.9	20.3	0.6	2.6	1.9	0.6	3.95	11.85
C-RPC-5	244.42	740.7	740.7	370.3	111.1	977.9	20.3	0.6	2.6	1.9	0.6	0	15.8

2.2. Specimen Preparation

All weighed raw materials are poured into a stirring UJZ-15 mortar stirring pot, stirring in the mixture for 480 s. The specimens are prepared according to GB/T 50081-2019. The preparation of the specimens can be divided into the following steps.

First of all, the dry materials are stirred in the UJZ-15 for 30 s. The materials are mixed in a mixture of water-reducing agents and water. Mixing materials are continuously stirred in the UJZ-15 for 480 s.

After the materials are stirred, the mixed concrete is poured into the oiled plastic molds obtaining the specimens with sizes 40 mm × 40 mm × 160 mm and 50 mm × 50 mm × 50 mm. These two sizes of specimens are used for measurements of mechanical and electrical properties. The numbers of specimens for measurement of mechanical and electrical parameters are 3 and 6, respectively. The difference between the data and the average value exceeds ±10% of the average value, which is regarded as unreasonable data (GB/T17671-1999). The average values after removing the unreasonable data are treated as the testing values. All specimens are cured at a temperature of 20 ± 2 °C and a relative humidity above 95% (GB50204-2015) after demolding.

2.3. Measurement Method

2.3.1. The Experiments of Mechanical Strengths

f_{cu} and f_t of specimens are tested by the YAW-300 automatic pressure testing machine with the hydraulic pump rated pressure of 25 kN provided by Jinan Rutong Testing Technology Factory. The control pattern is force control, and the 2400 N/s and 60 N/s are used for compressive and flexural tests. The specific operation process refers to GB/T 17671-1999 Chinese standard [33]. Figure 1 shows the photos of the compressive and flexural tests.

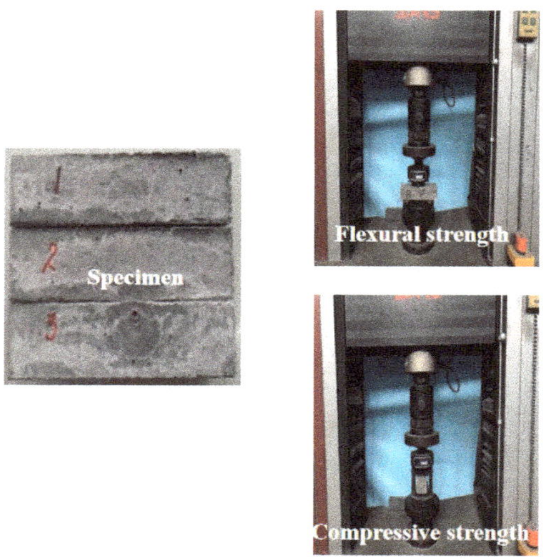

Figure 1. The diagram of f_t and f_{cu}.

2.3.2. The Experiments of Flexural Toughness

CMT5205 microcomputer-controlled electronic universal testing machine is employed to measure α. The specimens with a size of 40 mm × 40 mm × 160 mm are used for the measurement of α. Linearly varying displacement transducer (LVDT) produced by Zhengzhou Muchen Automation Technology Co., Ltd., Zhengzhou, China is applied to measuring the mid-span displacement. The loading speed is 0.5 mm/min in the experiment as shown in Figure 2 (The force exerted by the press is expressed in "F". The bottom edge width and thickness of the sample are expressed in "b" and "h" respectively). The detailed measurement steps can be followed in Wang's paper [34]. In the force-displacement curve, the integral value from 0 to the highest of the force is used as α. Figure 2 is the schematic diagram for measuring α.

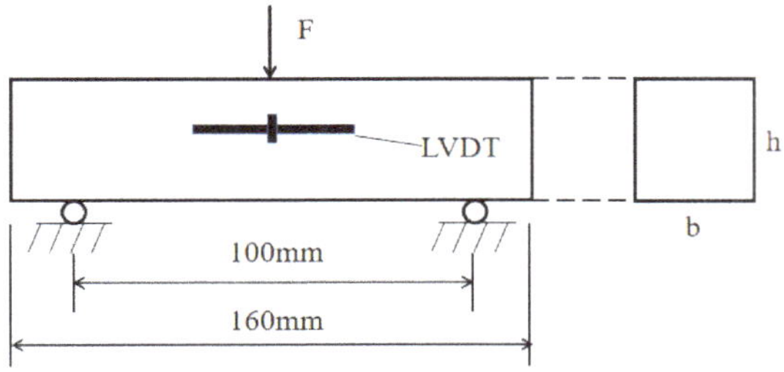

Figure 2. The schematic diagram of the loading regime and electrode layout.

2.3.3. The Experiments of Electrical Properties

TH2810D LCR digital electric bridge with a testing standard frequency of 100 Hz, 120 Hz, 1 kHz, and 10 kHz provided by Changzhou Tonghui Factory is used for measuring AC R_e. Additionally testing voltage and frequency of TH2810D are 1 V and 10 kHz, respectively. The specific connection and method refer to Wang's article [34]. For the reason that the AC R_e measurement of concrete can minimize R_e instability caused by polarization, in this study, so the two electrodes are applied in the measurement of R_e.

Parstat3000 electrochemical workstation with a testing frequency of 1~10^5 Hz provided by Shanghai Princeton Instrument Co., Ltd., Shanghai, China is used for recording the AC impedance spectrum. Figure 3 is the schematic diagram of the connection between the specimen and TH2810D LCR digital electric bridge and Parstat3000 electrochemical workstation respectively.

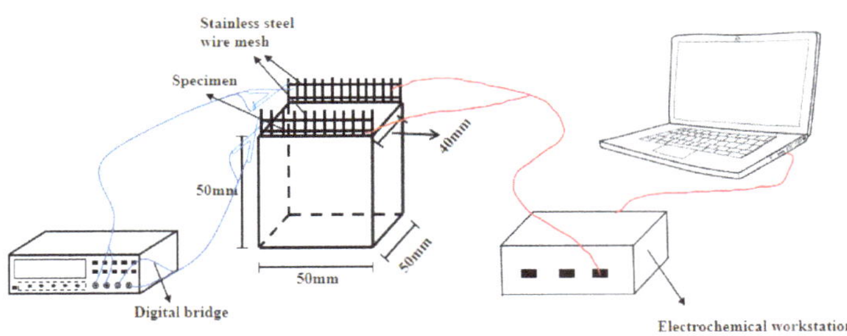

Figure 3. Electrical parameters measurement.

2.3.4. The Experiments of Micro Analysis

The following is the procedure of SEM and EDS.

After the samples are cured, the samples are soaked in absolute ethanol for 5 days to prevent cement hydration. YZF-6050 vacuum drying oven with an output power of 1200 W (Shanghai yaoshi Co., Ltd., Shanghai, China) is used to dry the samples at 105 °C for 2 days. Then, the bean-sized and pebble-free particles are removed from the oven-dried samples. The collected particles are pasted on the aluminum plate with conductive tape. The particle surfaces are sputtered with a layer of gold film in a vacuum environment. The prepared samples are used for measuring SEM and EDS. The analysis method of the hydration products analyzed by results of SEM and EDS is based on Wang's article [9].

3. Results and Discussions

3.1. Mechanical Performance Analysis

Figure 4 shows f_{cu} and f_t of specimens with different steel fiber content with AR-30. It can be found that the mechanical performance decreases with the increase of steel fibers with AR-30 (The maximum rate of decline is 80%). Meanwhile, the decline in f_{cu} with the curing age of 28 d is more obvious than f_{cu} with the curing age of 1 d. However, the decline in f_t with the curing age of 28 d is lower than f_t with the curing age of 1 d. The dramatic declines in f_{cu} with the curing age of 28 d and f_t with the curing age of 1 d are observed with the ratio of the steel fibers with AR-30 from 50% to 75%. When the total amount of fibers remains unchanged, the increase of steel fibers content with AR-30 will reduce the number of fibers. Thus affecting the bridging effect of fibers in RPC. In addition, steel fibers with AR-30 tend to sink to the bottom, resulting in uneven dispersion of fibers [35]. The degree of hydration of the specimen is low after curing for 1 day, and f_t is mainly provided by steel fibers. Therefore, f_t decreases with the increase of steel fibers content with AR-30.

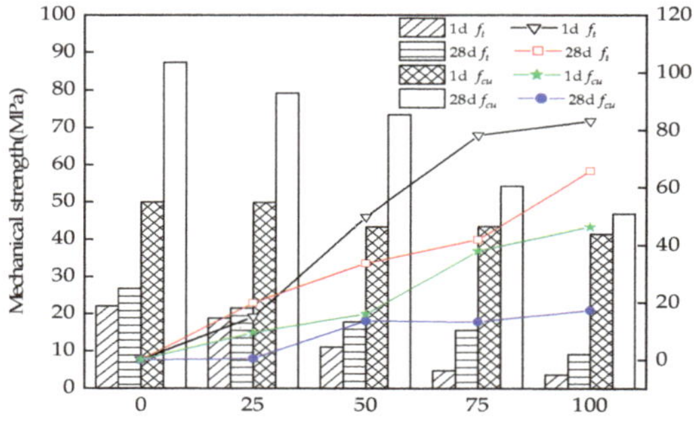

Figure 4. Mechanical strength is a relation of the ratio of fibers with an aspect ratio of 30.

3.2. Flexural Toughness Analysis

Figures 5 and 6 show the flexural stress (P_t) and mid-span deflection (w_m) of specimens. It can be found from Figures 5 and 6, w_m decreases with the ratio of steel fibers with AR-30 from 0% to 100%. At the same time, the maximum P_t decreases. Figures 7 and 8 show the fitting curves of P_t, w_m, and the fitting results of P_t and w_m curves. The curves of P_t and w_m with the curing age of 1 d extremely fit a quadratic parabola. α in Figure 9 is calculated. It can be found from Figure 9, α decreases with the increase of steel fibers with AR-30. Meanwhile, a dramatic decrease in α is observed with the ratio of steel fibers with AR-30 from 0% to 25%. However, the decreasing rate decreases with the increase of steel fibers with AR-30, which can be attributed to the fact that the decrease of α will slow down when the ratio of steel fibers with AR-30 reaches a value. Furthermore, the number of steel fibers with AR-75 will be more at the same ratio, which is more uniformly distributed in the RPC. The fibers are more tightly embedded in the concrete, resulting in improving α of RPC [36]. Additionally, f_t with a curing age of 28 d is higher than f_t with a curing age of 1 d, indicating the accuracy of experimental data.

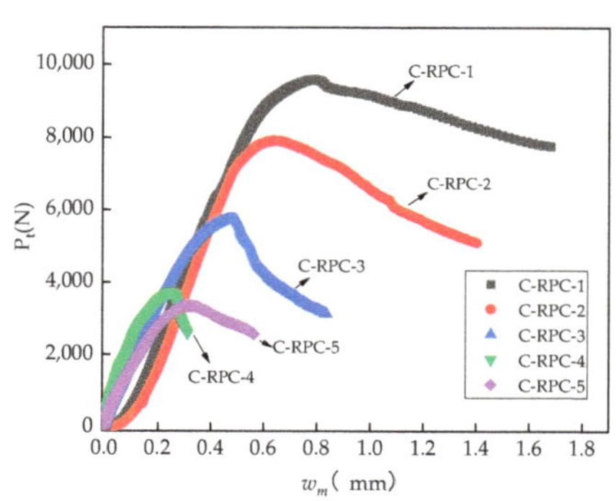

Figure 5. P_t and w_m curve for 1 d.

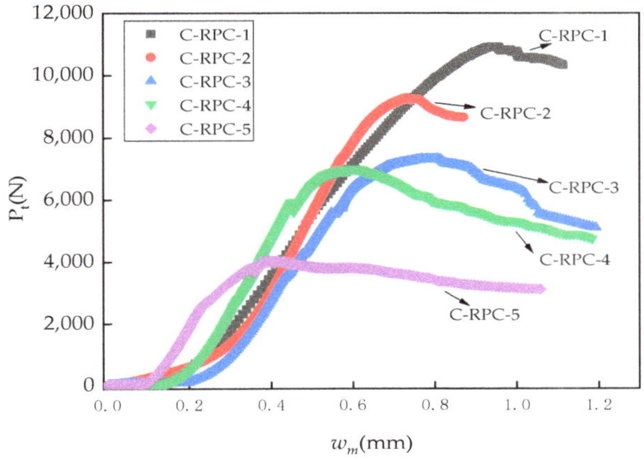

Figure 6. P_t and w_m curve for 28 d.

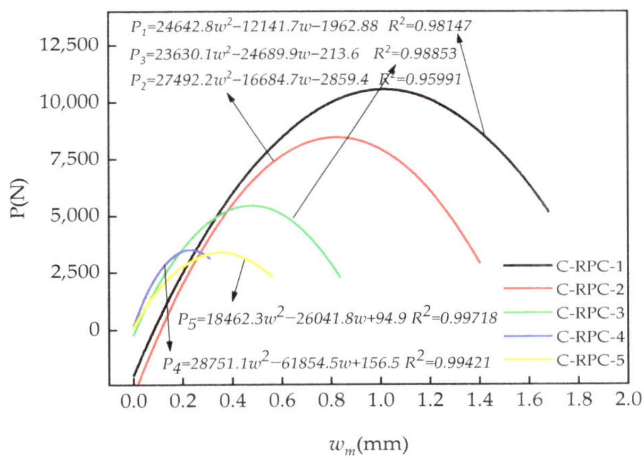

Figure 7. The fitting curve of 1 d P_t and w_m.

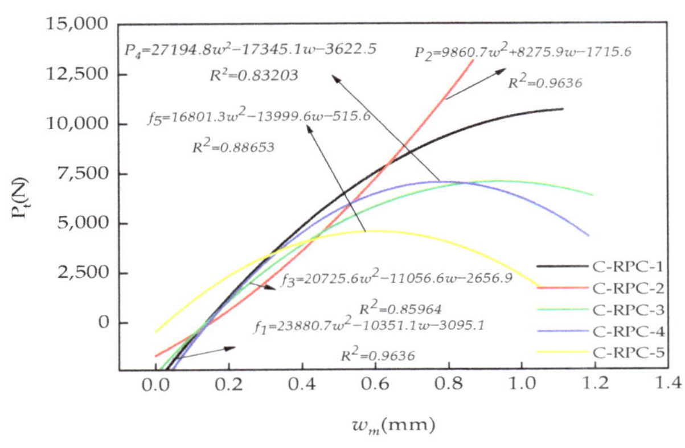

Figure 8. The fitting curve of 28 d P_t and w_m.

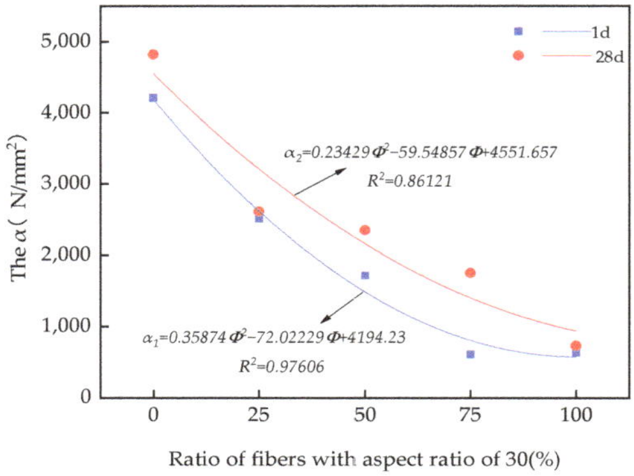

Figure 9. α of specimens.

3.3. Electrical Performance Analysis

Figure 10 shows R_e, the increasing rate, and the ratio of steel fibers with AR-30. The steel fibers with AR-30 induce the ascending in R_e. However, the increase of R seems to be irregular. R_e of specimens with the curing age of 1 d and 28 d are measured reaching the maximum value at 25% and 75%, respectively. The increase of steel fibers with AR-30 will reduce the tightness of the electric grid composed of fibers in RPC. However, the gaps increase with the increase of steel fibers with AR-30. The free ions in the specimens produce electrophoresis during electrification, and the electrophoresis phenomenon leads to the enhancement of the electric conductivity of the specimens, leading to the instability of R_e.

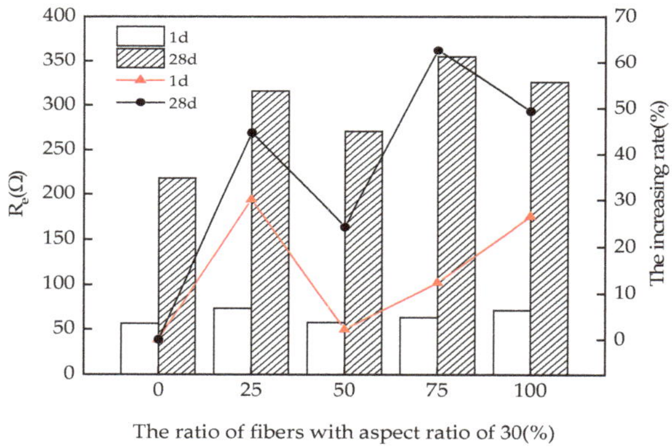

Figure 10. R_e of specimens.

Figure 11 shows the electrical resistance of specimens with different content of steel fibers with AR-30. As shown in Figure 11, the electrical resistance of pore solution (R_p) decreases with the increase of steel fibers with AR-30, and its decrease is gradual. The steel fibers with AR-30 lead to increasing inner pores' volume, so the pore solution increases, resulting in the decline of R_p. Moreover, when the fiber content is 0%~25%, the electrical resistance of steel fibers (R_s) increases slowly. However, when the fiber content exceeds 50%, R_s increases rapidly. The increasing ratio of steel fibers with AR-30 will lead to the decreasing dispersion of steel fibers in RPC. Furthermore, the number of steel fibers decreases with the increase of steel fibers content with AR-30 leading to a sparse electrical conductive network. R_s decreases [37]. The fitting curves fit the quadratic function, and the fitting degrees are 0.99823 and 0.97846, respectively, indicating that the fitting function is very reasonable.

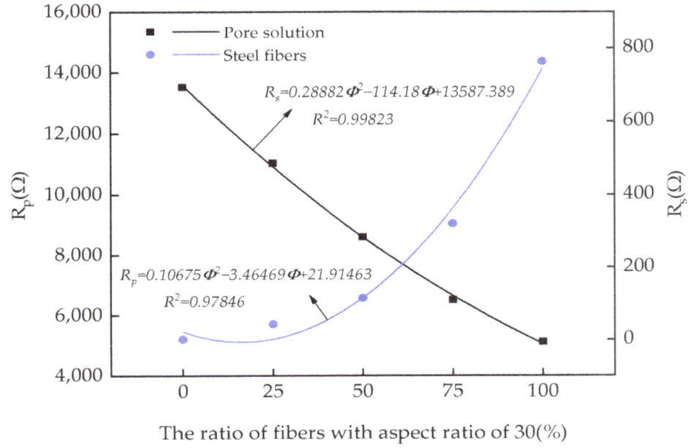

Figure 11. The relationship between R_p, R_s, and steel content of RPC.

Figure 12 displays the electrochemical impedance spectroscopy of specimens. The imaginary component (-Zi) and the real component (Zr) represent electrical reactance and R_e respectively. Table 4 shows the abscissa of the extreme point after removing individual points. It can be found that R_e increases with the increasing ratio of steel fibers with AR-30. Moreover, the distribution of points becomes dispersed. The reason for this phenomenon may be that the increase of steel fibers with AR-30 will reduce the number of steel fibers, leading to the decrease of dispersion of steel fibers. The electrical conductivity decreases in RPC [38].

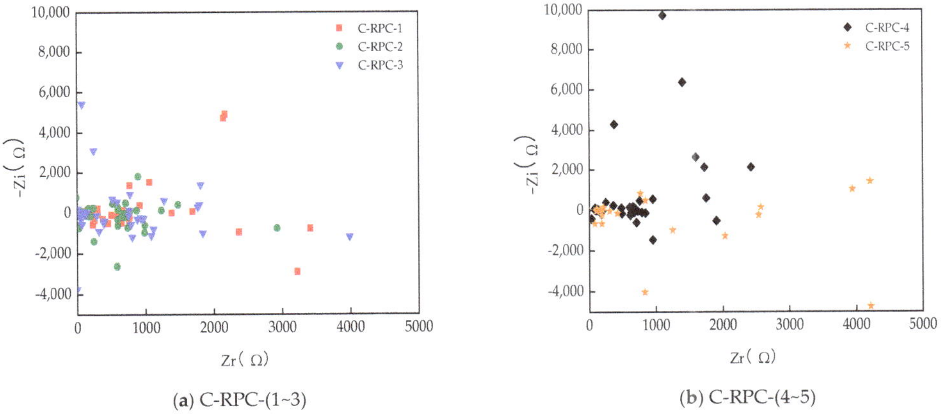

(**a**) C-RPC-(1~3) (**b**) C-RPC-(4~5)

Figure 12. Electrochemical impedance spectroscopy.

Table 4. The abscissa of the extreme point after removing individual points.

Tpyes	C-RPC-1	C-RPC-2	C-RPC-3	C-RPC-4	C-RPC-5
Zr	230.11	582.2	803.2	953.5	2033.9

Figure 13 shows the circuit schematic and equivalent circuit diagram of electrochemical impedance spectroscopy. It can be seen that the electric circuit diagram of RPC with compound cement accord with the series conduction model. R_e and capacitance include steel fibers, plain RPC, and pore solution, and these three parts respectively connect with contact resistances of the electrode and specimens [39]. The Chi-squared is less than 2.324×10^{-3}, indicating that the equivalent circuit is reasonable.

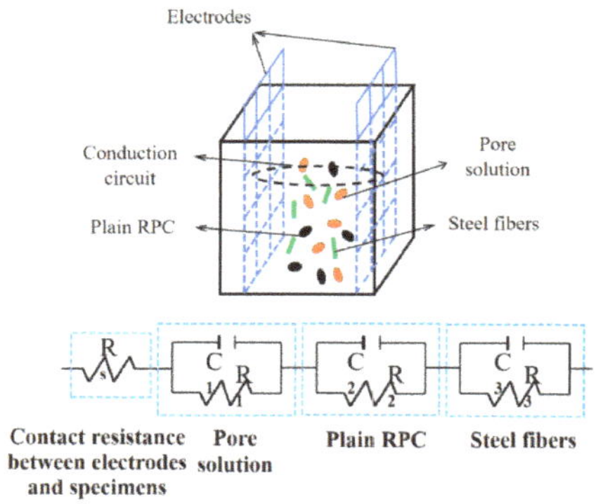

Figure 13. The equivalent circuits of RPC.

3.4. Micro Analysis

Figure 14 shows SEM-EDS photos of samples cured for 28 days. The samples are plain samples without any steel fibers. The dense hydration products, the flocculent hydration products, and the agglomerate hydrate can be observed in the photos. Table 5 shows the element distribution obtained by EDS. As depicted in Figure 14 that the element of carbon (C), oxygen (O), silicon (Si), sulfur (S), aluminium (Al), calcium (Ca), tellurium (Te), kalium (K), and magnesium (Mg) in Table 5 can be found. As observed in Table 5, the content of the elements decreases in this order of O > Si > S. This is attributed to the fact that the contents of S and Si in the part A and part B are low indicating that the parts are mainly other substances. Furthermore, the content of Si is significantly higher than the content of S in part C and part D indicating that the parts are mainly composed of hydrated calcium silicate and other substances [9].

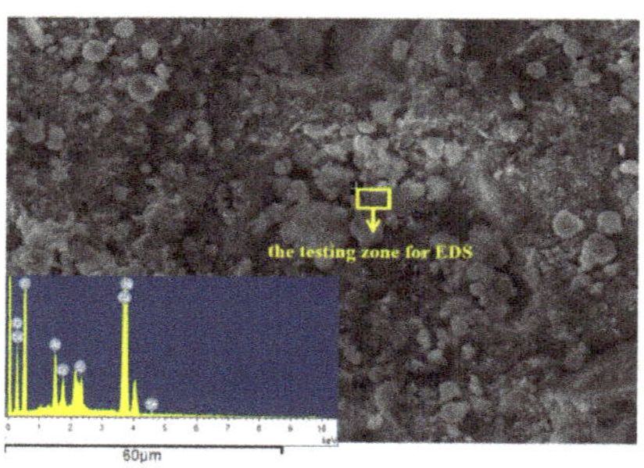

(**a**) Part A

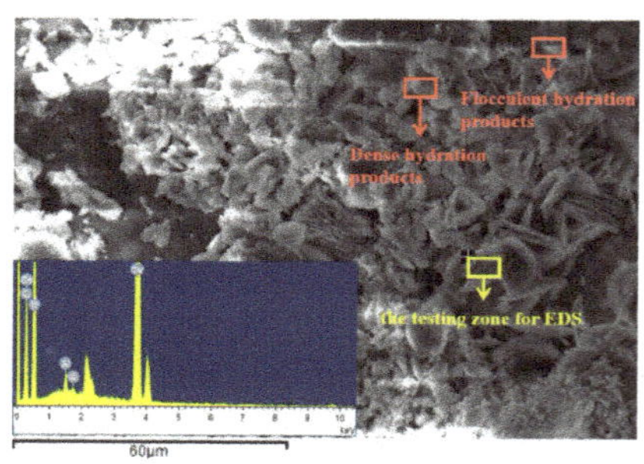

(**b**) Part B

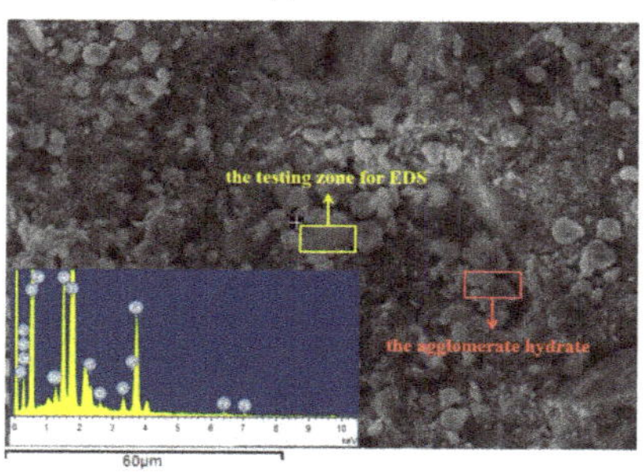

(**c**) Part C

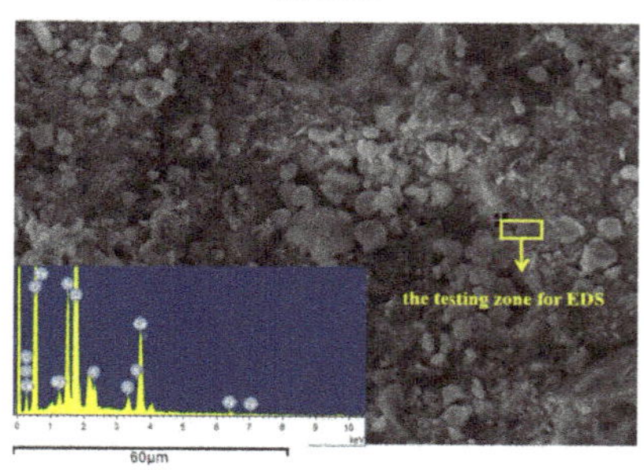

(**d**) Part D

Figure 14. SEM-EDS of samples for 28 d.

Table 5. The element distribution obtained by EDS (%).

Part	C	O	Si	S	Al	Ca	Te	K	Mg
A	11.57	48.44	1.29	1.40	2.73	30.08	4.49	-	-
B	12.42	53.71	0.22	-	0.72	32.93	-	-	-
C	7.12	49.93	25.08	0.90	5.03	9.36	0.74	1.20	0.64
D	7.12	49.93	25.08	0.90	5.03	10.10	-	1.20	0.64

4. Conclusions

In this experiment, the mechanical properties and electrical parameters of RPC with sulfoaluminate compound cement are studied. The experimental conclusions are as follows.

The max f_t and f_{cu} of 1 d are 22.1 MPa and 50 MPa. The max f_t and f_{cu} of 28 d are 26.8 Mpa and 87.32 MPa. The increase of steel fibers with an aspect ratio of 30 fibers leads to the decline of the mechanical properties of RPC. The decreasing rates of f_t and f_{cu} with the curing age of 1 d and 28 d are 14.93%~83.26% and 0.40%~46.36%.

The addition of steel fibers with an aspect ratio of 30 can weaken α of RPC. The decreasing rates with the curing age of 1 d and 28 d are 85.48% and 84.77%, respectively.

The increase of steel fibers with an aspect ratio of 30 leads to the decrease of electrical conductivity in RPC. The max R_c of 28 d is 2033.9 Ω. The electrical conductivity of specimens of steel fibers with an aspect ratio of 30 is 1/10 of the electrical conductivity of specimens of steel fibers with an aspect ratio of 75.

The electric circuit of reactive powder concrete accords with a series conduction model, which parallel electrical resistance and capacitance of pore solution, rapid strength RPC and steel fibers respectively connect with the contact resistance of the electrode and specimens.

The main hydration product of SAC and OPC composite cement is hydrated calcium silicate.

In this study, the effect of steel fibers' aspect ratio on the electrical performance and mechanical strength has been reported. However, the influence of steel fibers' aspect ratio on the durability needs further investigation in the future.

Author Contributions: Conceptualization, Z.L. and H.W.; methodology, X.P.; software, Z.L.; validation, H.W., X.P. and Z.L.; formal analysis, X.P.; investigation, H.W.; resources, X.P.; data curation, H.W.; writing—original draft preparation, Z.L.; writing—review and editing, H.W.; visualization, Z.L.; supervision, X.P.; project administration, H.W.; funding acquisition, H.W. All authors have read and agreed to the published version of the manuscript.

Funding: This research was funded by [the Natural Science Foundation of China] grant number (52008215) and [the Natural Science Foundation of Zhejiang Province, China] grant number [LY22E080005].

Institutional Review Board Statement: Not applicable.

Informed Consent Statement: Not applicable.

Data Availability Statement: The data used to support the findings of this study are available.

Conflicts of Interest: The authors declare no conflict of interest.

References

1. Sun, J.; Wang, X.; Yin, X.; Wang, W.; Chen, B.; Xiong, J. Future trends and development characteristics of cement industry. *China Cement.* **2021**, *9*, 66–69.
2. Zhang, G. Emergency Repair treatment of cement concrete bridge deck open to traffic damage. *Commun. Sci. Technol. Heilongjiang* **2016**, *39*, 160.
3. Li, C.; Ye, Q. Corrosion status of concrete bridges at home and abroad. *Highw. Transp. Technol.* **2016**, *2*, 31–36.
4. Nemati, M.K.; Uhlmeyer, J.S. Accelerated construction of urban intersections with Portland Cement Concrete Pavement (PCCP). *Case Stud. Constr. Mater.* **2021**, *14*, e00499. [CrossRef]
5. Liang, R.; Zhu, C. Study on basic mechanical properties of hybrid fiber recycled concrete for rapid pavement repair. *Jilin Water Resour.* **2021**, *6*, 38–41.
6. Fu, T.; Guo, B.; Luo, Y.; Zhang, J. Study on ultra high early strength sulphoaluminate cement concrete in concrete pavement repair. *Concrete* **2018**, *02*, 140–144.

7. Aku, S.; Milad, H.; Mohammad, P.; Qiu, X. Electrical impedance tomography-based sensing skin. *Dam Saf.* **2016**, *03*, 77–82.
8. He, L.; Chen, Q.; Jiang, Z. Hydration Based Dynamic Calculation Model for Electric Conductivity of Hardened Cement Paste. *J. Build. Mater.* **2022**, *09*, 99–100.
9. Wang, H.; Cai, X.; Rao, C.; Wang, K.; Wang, J. Mechanical and Electrical Properties of Rapid-Strength Reactive Powder Concrete with Assembly Unit of Sulphoaluminate Cement and Ordinary Portland Cement. *Materials* **2022**, *15*, 3371. [CrossRef]
10. Ren, X. Application of nondestructive testing technology in road and bridge detection. *Heilongjiang Sci.* **2019**, *09*, 90–91.
11. Sirtoli, D.; Wyrzykowski, M.; Riva, P.; Tortelli, S.; Marchi, M.; Lura, P. Shrinkage and creep of high-performance concrete based on calcium sulfoaluminate cement. *Cem. Concr. Compos.* **2019**, *98*, 61–73. [CrossRef]
12. Gołaszewska, M.; Klemczak, B.; Gołaszewski, J. Thermal Properties of Calcium Sulphoaluminate Cement as an Alternative to Ordinary Portland Cement. *Materials* **2021**, *14*, 7011. [CrossRef] [PubMed]
13. Feng, H.; Chen, G.; Hadi, M.; Sheikh, M.; Zhou, B. Mechanical behaviour of micro-fifine steel fifibre reinforced sulphoaluminate cement composite. *Constr. Build. Mater.* **2018**, *170*, 91–100. [CrossRef]
14. Zemir, I.; Debieb, F.; Kenai, S.; Ouldkhaoua, Y.; Irki, I. Strengthening of ordinary vibrated concrete using steel fibers self-compacting concrete. *J. Adhes. Sci. Technol.* **2020**, *34*, 1556–1571. [CrossRef]
15. Ju, Y.; Zhao, J.; Wang, D.; Song, Y. Experimental study on flexural behaviour of reinforced reactive powder concrete pole. *Constr. Build. Mater.* **2021**, *312*, 125399. [CrossRef]
16. Kou, J.; Liu, Y.; Zhang, B. Experimental study on mechanical property and durability of reactive powder concrete. *Build. Struct.* **2018**, *48*, 48–54+47.
17. Wang, Y.; Zhong, C.; Zhang, Y.; Wang, J. Early cracking resistance of reactive powder concrete mixed with steel fiber. *J. Jilin Jianzhu Univ.* **2022**, *39*, 12–16.
18. Li, K.; Li, D.; Liu, Y. Meso-scale investigations on the effective thermal conductivity of multi-phase materials using the finite element method. *Int. J. Heat Mass Transf.* **2020**, *151*, 119383. [CrossRef]
19. Wang, Y.; He, Y.; He, S.; Hong, W.; He, W. Effects of the mixed optimization of PCE on pore structure evolution of cement mortar. *New Build. Mater.* **2021**, *48*, 100–105.
20. Hong, J.; Qiu, J. Performance Research on the Sulfoaluminate Cement Based Rapid Repairing Materials. *Sci. Technol. Ind.* **2021**, *21*, 270–273.
21. Wang, H.; Gao, X.; Liu, J.; Ren, M.; Lu, A. Multi-functional properties of carbon nanofiber reinforced reactive powder concrete. *Constr. Build. Mater.* **2018**, *187*, 699–707. [CrossRef]
22. Han, B.; Zhang, L.; Zhang, C.; Wang, Y.; Yu, X.; Ou, J. Reinforcement effect and mechanism of carbon fibers tomechanical and electrically conductive properties of cement-based materials. *Constr. Build. Mater.* **2016**, *125*, 479–489. [CrossRef]
23. Yavaş, A.; Hasgul, U.; Turker, K.; Birol, T. Effective fiber type investigation on the shear behavior of ultrahigh-performance fiber-reinforced concrete beams. *Adv. Struct. Eng.* **2019**, *22*, 1591–1605. [CrossRef]
24. Qian, X. Study on improving frost thaw resistance and corrosion resistance of bridge concrete with carbon nanofibers. *Synth. Fiber China* **2022**, *51*, 54–57+62.
25. Ma, F.; Chang, Q.; Wang, X. Application of polyacrylonitrile fiber in improving concrete performance. *Chem. Ind.* **2011**, *29*, 26–28.
26. Wang, J. Study on the influence of different amount of steel fiber on mechanical properties of concrete. *Jiangxi Build. Mater.* **2021**, *7*, 35–36.
27. Zhu, H.; Zhou, H.; Gou, H. Evaluation of carbon fiber dispersion in cement-based materials using mechanical properties, conductivity, mass variation coefficient, and microstructure. *Constr. Build. Mater.* **2021**, *266*, 120891. [CrossRef]
28. Jia, E.; Mou, H.; Liu, Z.; Wang, J.; Zeng, L.; Yang, X.; Liu, P. Surface hydrophilic modification of polypropylene fibers and their application in fiber-reinforced cement-based materials. *J. Macromol. Sci.* **2020**, *60*, 286–298. [CrossRef]
29. Hu, L. Research on factors affecting the strength of high performance concrete. *Jiangxi Build. Mater.* **2015**, *19*, 177.
30. Tan, L. Application and research progress of steel fiber reinforced concrete. *China Concr. Cem. Prod.* **2018**, *24*, 160.
31. Niu, L.; Zhang, S.; Wei, Y. The influence of steel fiber content on mechanical properties of concrete. *China Concr. Cem. Prod.* **2019**, *3*, 51–54.
32. Ke, G.; Zhang, J.; Xie, S.; Pei, T. Rheological behavior of calcium sulfoaluminate cement paste with supplementary cementitious materials. *Constr. Build. Mater.* **2020**, *243*, 118234. [CrossRef]
33. *GB/T17671-1999*; Method of Testing Cements-Determination of Strength. The State Bureau of Quality and Technical Supervision: Beijing, China, 1919.
34. Wang, H.; Shi, F.; Shen, J.; Zhang, A.; Zhang, L.; Huang, H.; Liu, J.; Jin, K.; Feng, L.; Tang, Z. Research on the self-sensing and mechanical properties of aligned stainless steel fiber reinforced reactive powder concrete. *Cem. Concr. Compos.* **2021**, *119*, 104001. [CrossRef]
35. Li, F.; Li, X.; Tang, H.; Mu, B.; Jiang, Y.; Feng, L. Experimental Study on Mechanical Properties of Hybrid Fiber UHPC. *China Concr. Cem. Prod.* **2022**, *316*, 56–59.
36. Wang, Y. Experimental Research on Flexural Toughness of Steel Fiber Reinforced Concrete. *North. Commun.* **2022**, *356*, 21–25.
37. Wei, X.; Xiao, L.; Li, Z.; Sui, T. Conductive mechanism and hydration property of cement-based materials with steel fibers. *Concrete* **2006**, *4*, 11–13+51.

38. Wang, R.; He, F.; An, X.; Chen, C. Effect of Dispersion of AC Impedance Spectroscopy of Cement-based Materials with Fly Ash. *Bull. Chin. Ceram. Soc.* **2016**, *35*, 1877–1882.
39. Cui, L.; Wang, H. Research on the Mechanical Strengths and the Following Corrosion Resistance of Inner Steel Bars of RPC with Rice Husk Ash and Waste Fly Ash. *Coatings* **2021**, *11*, 1480. [CrossRef]

coatings

MDPI

Article

Influence of CO_2 Curing on the Alkali-Activated Compound Mineral Admixtures' Corrosion Resistance to NaCl Dry–Wet Alternations

Houchao Sun [1], Wenjie Cheng [2], Haoxin Xu [3], Zhangjie Cai [1], Minggan Yin [1] and Feiting Shi [1,*]

1 School of Civil Engineering, Yancheng Institute of Technology, Yancheng 224051, China
2 School of Mechanical Engineering, Yancheng Institute of Technology, Yancheng 224051, China
3 School of Chemistry & Chemical Engineering, Yancheng Institute of Technology, Yancheng 224051, China
* Correspondence: shifeiting@ycit.cn

Abstract: In this study, the influence of CO_2 curing on the corrosion resistance of reinforced alkali-activated compounds is investigated. Fly ash (FA) and blast furnace slag powder (BFS) are used as mineral admixtures. The specimens were subjected to dry–wet alternations with 3% NaCl, used to simulate a concrete structure under a corrosion environment. The ultrasonic velocity, mass loss rate, and electrical characteristics (such as electrical resistance, AC impedance spectra, and corrosion area rates determined by Tafel curves) are utilized to determine the degree of corrosion. Scanning electron microscopy (SEM) and X-ray diffraction (XRD) are used to analyze the corrosion mechanism. Results show that the corrosion resistance is decreased by the addition of FA but improved by CO_2 curing. When CO_2 curing is provided, the addition of BFS shows a higher enhancing effect on the corrosion resistance than that of FA. The equivalent circuit diagram of reinforced alkali-activated compound mineral admixtures obtained by AC impedance spectra is composed of three electrical elements (electrical resistance and capacitance in parallel) in series. The X-ray diffraction results show that adding BFS and CO_2 curing can decrease the rust's iron oxides on the steel bars' surface. Finally, as found in the SEM photos, BFS and CO_2 curing can effectively improve the compactness of specimens. Meanwhile, the roughness of hydration is increased by CO_2 curing.

Keywords: corrosion resistance; CO_2 curing; corrosion area rates; NaCl dry–wet alternation; electrical characteristics

1. Introduction

Cement-based materials have been broadly used in buildings such as houses, bridges, and highways for many years [1–3]. However, cement production also produces a great number of byproducts [4,5]. To alleviate this issue, mineral admixtures are usually applied to replace cement. Silica fume, fly ash, and slag powder are commonly used additives in cement materials. These solid industrial waste-derived mineral admixtures can increase cement concrete's mechanical strength and durability [6–8]. Meanwhile, cement-based materials with mineral admixtures can also benefit from high-temperature curing to increase their strength and durability [9–11]. Although mineral admixtures are easy to obtain, their incorporation often deteriorates the strength and durability of concrete under standard curing conditions [12,13]. Consequently, alkali-activated mineral admixtures can be an option to replace some cement materials.

Alkali-activated mineral admixtures are composed of mineral admixtures, sodium silicate, alkali, water, etc. [14–17]. It has been pointed out in prior research that this kind of material shows excellent mechanical properties. Cement-based materials with NaOH as an activator show ultra-high mechanical strengths (flexural strength of higher than 20 MPa and compressive strength of higher than 120 MPa) [18–20]. Several studies have been carried out on its mechanical performance, the use of different alkali-activated agents, and their high-temperature resistance and micro-properties [21]. Little attention has been paid to the influence of NaCl erosion on its performance.

Cementitious materials are always applied with reinforcing steel bars to sustain the load in service [22,23]. Alkali-activated mineral admixtures without steel bars cannot be applied in a real project. Meanwhile, concrete usually encounters chlorine salt erosion when it is used in a coastal environment. As reported in prior research, the alkali-activated cementitious material shows quite small porosity, thus improving the corrosion resistance of steel bars. Moreover, the surface of steel bars' passivation film can be ameliorated by the alkali-activated mineral admixtures, leading eventually to increasing the subsequent corrosion resistance [24,25]. However, NaCl dry–wet alternations usually cause serious erosion to the reinforcement inside cement-based materials [26]. Little attention has been paid to the corrosion resistance of reinforced alkali-activated mineral admixtures under the environment of NaCl dry–wet alternations.

CO_2 is the main cause of the greenhouse effect. The production of cement can increase the emissions of CO_2, which has been used for curing cement-based materials [27,28]. The reinforced cement matrix's mechanical characteristics and corrosion resistance can be enhanced by CO_2 curing, according to earlier studies [29]. However, the effect of CO_2 curing on mineral admixtures is still unknown [30–32]. Whether CO_2 curing can help improve the corrosion resistance of reinforced alkali-activated mineral admixtures should be investigated. The innovation of this study is to investigate the influence of CO_2 curing on the corrosion resistance of reinforced alkali-activated material under the action of NaCl dry–wet alternations. Multiple sets of electrical parameters and ultrasonic velocity, along with X-ray diffraction (XRD) and scanning electron microscopy (SEM), are used for the analysis of corrosion resistance. However, related studies are rarely reported.

In this study, the effect of CO_2 curing on mineral admixtures with alkali-activated compounds is investigated. The mass, the ultrasonic velocity, and the electrical resistance of reinforced CO_2-cured alkali-activated compound mineral admixtures are determined. Meanwhile, the mass loss rate and the corrosion area rate of steel bars are measured. Finally, the micro-parameters (the scanning electron microscopy (SEM) of the alkali-activated mineral admixtures and X-ray diffraction (XRD) of the rust) are obtained to reveal the corrosion mechanism of the reinforced CO_2-cured alkali-activated compound mineral admixtures. The outcomes of this study can provide a guideline for the application of CO_2-cured reinforced alkali-activated mineral admixtures, which may be applied in a NaCl erosion environment in the future.

2. Experimental Arrangement

2.1. Raw Materials

The sodium silicate applied in this research is made by Tongxiang Hengli Chemical Co., Ltd., Tongxiang, China. The sodium silicate shows a melting point of 40 °C~48 °C. S95 blast furnace slag powder (BFS) produced by HSBC New Materials Co., Ltd., Zhengzhou, China, is applied in this research. The BFS shows a specific surface area and an ignition loss of the blast furnace slag powder of 2.88 g/cm^3, 435.8 m^2/g, and 1.98%, respectively. Secondary fly ash made by the Shijiazhuang Shunli mineral products Co., Ltd., Shijiazhuang City, China, is used in this study. The density of fly ash is 2.4 g/cm^3. The sodium hydroxide is manufactured by Shanghai Dongmiao Chemical Technology Co., Ltd., Shanghai, China. Sodium hydroxide with a density of 2.13 g/cm^3, a boiling point of 1390 °C, and a purity of 99.9% is used in this research. The density, boiling point, and purity of sodium hydroxide are 2.13 g/cm^3, 1390 °C, and 99.9%, respectively.

Tables 1 and 2 display the raw materials' particle passing percentage and chemical composition, respectively. The particle passing percentage of raw materials is measured by negative pressure sieve analysis, according to the Chinese standard GB/T1345-2005 [33]. Meanwhile, the chemical compositions of BFS and FA in Table 2 are determined by X-ray fluorescence spectrometry. All data in Tables 1 and 2 are provided by the manufacturer of raw materials, which are within acceptable limits.

Table 1. Particle passing percentage of raw materials (%).

Types	0.3 μm	0.6 μm	1 μm	4 μm	8 μm	64 μm	360 μm
BFS	0.03	0.1	3.5	19.6	35.0	97.9	100
FA	12.3	41.7	66.2	100	100	100	100

Table 2. The chemical composition of cement (%).

Types	SiO_2	Al_2O_3	Fe_xO_y	MgO	CaO	SO_3	K_2O	Na_2O	Ti_2O	LI
BFS	34.1	14.7	0.2	9.7	35.9	0.2	3.5	—	—	—
FA	55.00	29.58	6.00	0.62	4.50	0.11	1.26	2.13	0.06	0.74

As shown in Table 1, the particle sizes of BFS average 0.3~64 μm, while the particle sizes of FA average 0.3~1 μm. It can be obtained from Table 1 that FA shows higher fineness than BFS. Moreover, as illustrated in Table 2, FA contains a higher content of Al_2O_3 and Fe_xO_y, indicating that FA shows higher activity. Additionally, the content of SiO_2 in FA is higher than that of BFS; hence, FA is more prone to secondary hydration.

2.2. Specimen Preparation

Table 3 lists the mixing proportions of all samples. To prepare the specimen, the sodium hydroxide and potash water glass are mixed using the NJ-160A cement paste mixer at the beginning. The cement paste is mixed at 140 r/min for 2 min and then stirred at 180 r/min for another 2 min. After that, the fresh paste is poured into molds with a size of 50 mm × 50 mm × 50 mm. Each specimen is embedded with a plain round bar showing a diameter of 8 mm and a piece of stainless-steel wire mesh with a 4.75 mm nominal aperture (mesh size). Specimens are stored in a standard curing chamber with a temperature of 20 °C and a relative humidity of 98.6%, and a TH-B concrete carbonization test box, provided by Tianjin Gangyuan test instrument factory, Tianjin, China, is used to provide a CO_2 environment with a concentration of 8%, respectively. The modulus of potassium silicate is 1.0. The modulus of potassium silicate means the amount of substance ratio of SiO_2:(K_2O + NaOH).

Table 3. The mixing proportions (kg/m^3).

Specimens	FA	BFS	Alkali Equivalent	W/B	Modulus	$Na_2O{\cdot}nSiO_2$	NaOH
A1	300	0	3%	0.3	1	29.83	6.69
A2	240	60	3%	0.3	1	29.83	6.69
A3	180	120	3%	0.3	1	29.83	6.69
A4	120	180	3%	0.3	1	29.83	6.69
A5	60	240	3%	0.3	1	29.83	6.69
A6	0	300	3%	0.3	1	29.83	6.69

2.3. Measurements

2.3.1. Macro-Performance

The JITAI990 ultrasonic detector is used to measure ultrasonic velocity. It can generate sound with a speed range of 1000 m/s to 9999 m/s. The sides of specimens are smooth and flat, and the probes should be stuck on the axis positions with Vaseline. After all these are finished, the signal of ultrasonic velocity is collected. Figure 1 shows the measuring process of the ultrasonic velocity. The detailed measuring process of the ultrasonic velocity can be found in Wang's research [34].

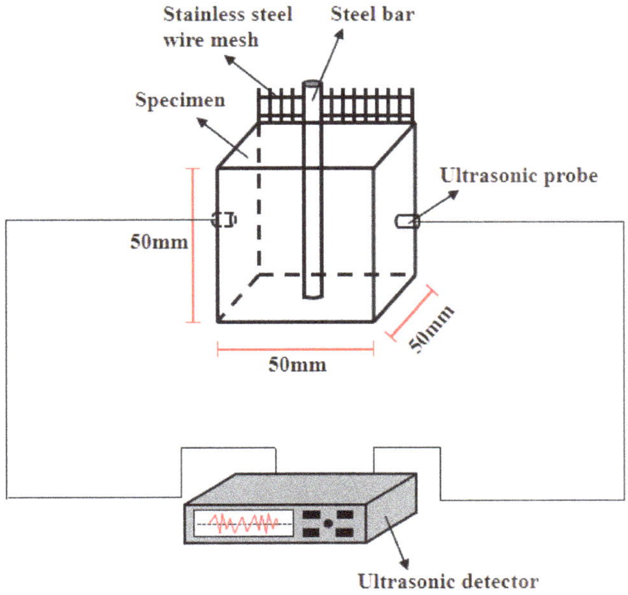

Figure 1. The ultrasonic velocity measurement.

The mass loss of the reinforcement can be measured by the following steps.

Firstly, the sanding process is used to remove the original rust on the surface of the steel bars, thus eliminating its effect on the final results. In this study, the original steel bars are polished by a sandblasting machine, and then the impurities on the surface of the steel bars are cleaned with citric acid and moved to be dried and weighed. After that, the standard cured samples with polished steel bars are applied in the corrosion environment. Then, the specimens are destroyed by a press machine, and then the steel bars are taken out and immersed in the citric acid solution for 4 h. Finally, the steel bars are sanded with sandpaper. Additionally, the steel bars are cleaned with water and dried in a vacuum oven at 40 °C before the test. The reinforcement and stainless-steel mesh serve as two electrodes. TH2810D LCR digital bridge with a testing voltage of 1 V and testing frequencies of 10^4 Hz is used for testing the AC electrical resistance. The Princeton electrochemical workstation with a testing voltage of 10 mV and testing frequencies of $0.01 \sim 10^5$ Hz is used to measure the AC impedance spectrum. The average value of six specimens of each group is the measuring value of each parameter. The Tafel method is applied to measure the corrosion area rate of the reinforced alkali-activated mineral admixtures. The voltage of the Tafel method increases from −0.15 V to 0.15 V. The increasing rate of the voltage is 50 mV/s. The corrosion area rate of the inner reinforcement can be calculated by the following Equation (1).

$$\mu = \frac{Mit}{Fr\rho} \quad (1)$$

where μ is the corrosion area rate, M is the molar mass of iron, i is the corrosion current density by the Tafel curves. Additionally, t, F, r, and ρ are the corrosion time, Faraday constant, radius of reinforcement, and density of reinforcement. All measuring detail processes can be found in prior research [35,36]. The measurement of AC electrical parameters is shown in Figure 2.

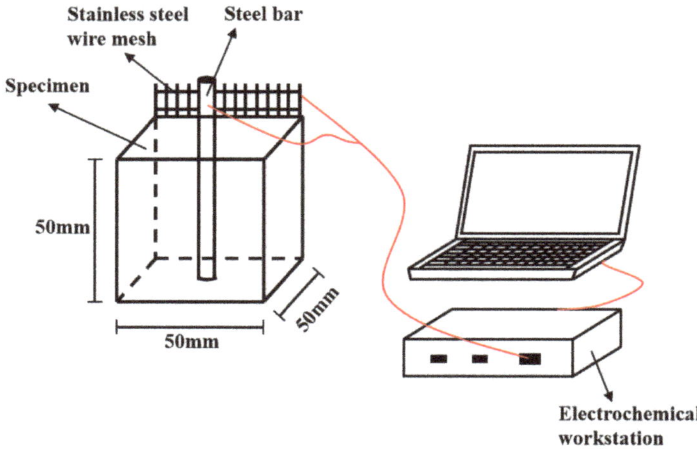

Figure 2. Measurement of AC electrical parameters.

2.3.2. Microscopic Performance

Samples the size of a soybean used for SEM measurement are firstly sprayed with gold in a vacuum environment. After that, all samples are moved to the JSM-IT800 high-resolution thermal field emission scanning electron microscope for measurement.

After the experiment of the macro-performance is finished, the reinforced specimens are destroyed. The alkali-activated cementitious matrix on the reinforcement surface is removed, and the rust on the surface of the steel bars is collected. The powdered rust is applied in the determination of X-ray diffraction by the DX-2700BH crystal structure XRD analyzer, provided by Suzhou Shipu Instrument Co., Ltd., Suzhou, China.

3. Results and Discussions

3.1. The Mass Loss Rate of the Reinforced Alkali-Activated Mineral Admixtures

When the reinforcement is corroded, the rusted reinforcement will expand, thus accelerating the development of cracks in specimens. Therefore, in this study, the mass loss rate of the reinforced alkali-activated mineral admixtures is measured to reflect the corrosion degree. Figure 3 shows the mass loss rate of the reinforced alkali-activated mineral admixtures. As shown in the figure, the mass loss rate is increased by the NaCl D–W alternations and the addition of FA. This can be attributed to the accelerated penetration of chloride ions in materials by NaCl D–W alternations, leading to increasing the corrosion rate of reinforcement [37,38]. More cracks in the specimens are formed by the increased corrosion rate, resulting in higher mass loss rate. Additionally, the mass loss rate is decreased by CO_2 curing, which is attributed to the higher content of calcium carbonate formed by the reaction of CO_2 and $Ca(OH)_2$. As shown in Figure 3, when FA is used, the decreasing rate of mass loss rate further increases. This is because FA shows a high degree of $[SiO_4]^{4-}$ polymerization in the vitreous structure network, which results in low reaction activity. FA is more difficult to be excited by alkali [39,40]. Therefore, the mass of alkali-activated mineral admixtures is decreased by the addition of FA.

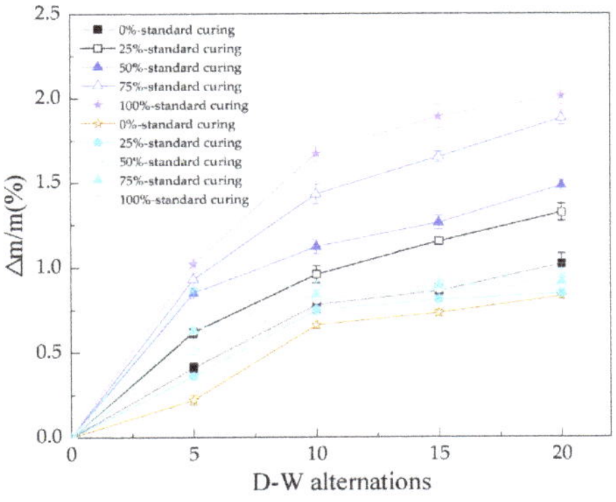

Figure 3. Δ m/m of the reinforced alkali-activated mineral admixtures during D–W alternations.

The mass loss rate of the reinforcement is exhibited in Figure 4. It can be observed that the mass loss rate of reinforcement increases with the increasing NaCl D–W alternations and the addition of BFS, due to the increased corrosion degree by D–W alternations and the corrosion resistance by BFS. Meanwhile, CO_2 curing demonstrates a positive effect on the corrosion resistance of the reinforcement [41,42]. The result of the reinforcement's mass loss rate is consistent with the mass loss rate of the reinforced alkali-activated mineral admixtures.

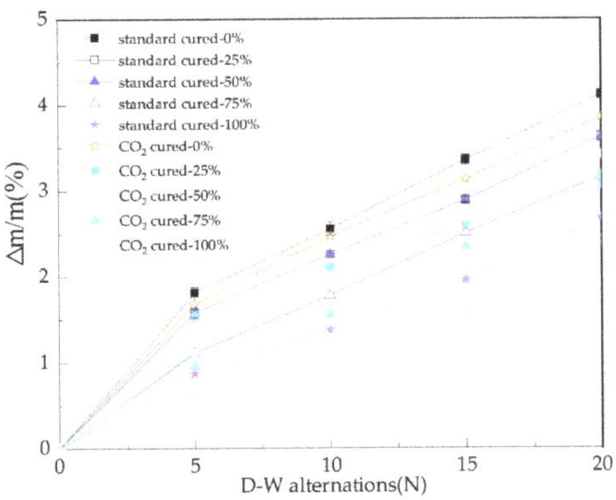

Figure 4. The mass loss rate of steel bars during D–W alternations.

3.2. *The Ultrasonic Velocity of the Reinforced Alkali-Activated Mineral Admixtures*

The ultrasonic velocity of the reinforced alkali-activated mineral admixtures is illustrated in Figure 5. It can be seen that the ultrasonic velocity decreases with the increasing NaCl D–W alternations and the decreasing dosage of BFS. This reduction can be attributed to the fact that the NaCl D–W alternations induce the inner cracks of the specimens [43,44]. Additionally, the NaCl D–W alternations accelerate the corrosion of reinforcement, increasing the number and width of cracks, thus decreasing the corresponding ultrasonic velocity. Furthermore, it can be noted that CO_2 curing increases the ultrasonic velocity. The more dosages of BFS, the higher the increase. CO_2 curing can accelerate the carbonization rate; therefore, the compactness of hydration products is improved, leading eventually to increasing the corresponding ultrasonic velocity. Moreover, the alkali-activated activity of BFS is higher than that of FA. As a consequence, BFS demonstrates a positive effect on the ultrasonic speed.

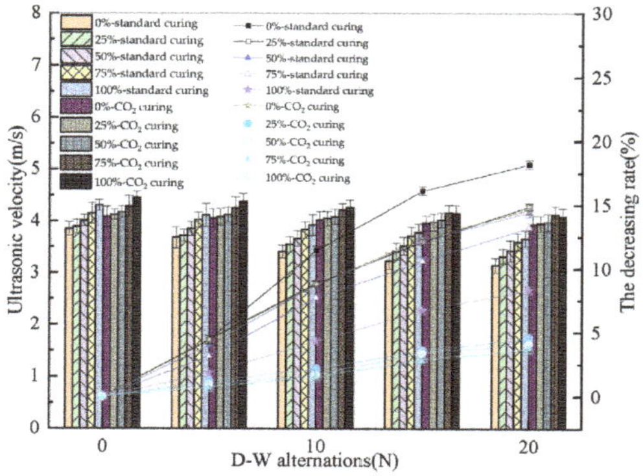

Figure 5. The ultrasonic velocity of reinforced alkali-activated mineral admixtures during D–W alternations.

3.3. The Electrical Parameters of Reinforced Alkali-Activated Mineral Admixtures

The electrical resistivity of reinforced alkali-activated mineral admixtures is depicted in Figure 6. It can be observed that the electrical resistivity continues to increase with more NaCl D–W alternations. This is because the NaCl D–W alternations can increase the inner cracks of the reinforced alkali-activated mineral admixtures [43,44]. The inner cracks block the electron migration, thus decreasing the electrical conduction and increasing the electrical resistivity. CO_2 curing reduces the increasing rate of electrical resistivity due to the improved compactness of the specimens. Additionally, the addition of BFS also increases the electrical resistivity of the reinforced alkali-activated mineral admixtures before D–W alternations. However, as the D–W alternations continue to increase, the increasing rate is decreased by adding BFS and CO_2 curing, thus improving the fact that the increasing dosage of BFS is effective in decreasing the corrosion rate of reinforced alkali-activated mineral admixtures and improving the corrosion resistance.

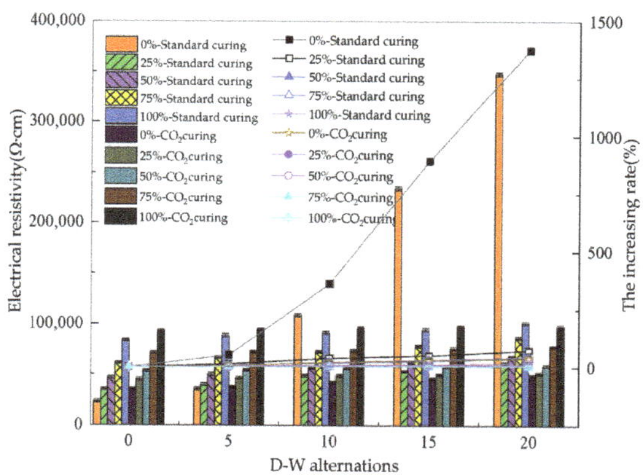

Figure 6. The electrical resistivity of reinforced alkali-activated mineral admixtures during D–W alternations.

The Tafel curves of the reinforced alkali-activated mineral admixtures after 20 D–W alternations are shown in Figure 7. It can be seen that the Tafel curves consist of two bifurcation curves. Previous studies reported that the bifurcation point could reflect the corrosion degree of inner reinforcement [35,36]. Additionally, it can be found that the addition of BFS can effectively decrease the potential of the bifurcation point, thus improving the corrosion resistance of concrete under the NaCl dry–wet alternations. This potential can be further reduced with CO_2 curing, which validates that CO_2 curing is effective in enhancing corrosion resistance. Table 4 shows the fitted corrosion current obtained using C-view software while measuring Tafel curves. Figure 8 shows the corrosion area rate of the reinforcement calculated by Table 4 and Equation (1).

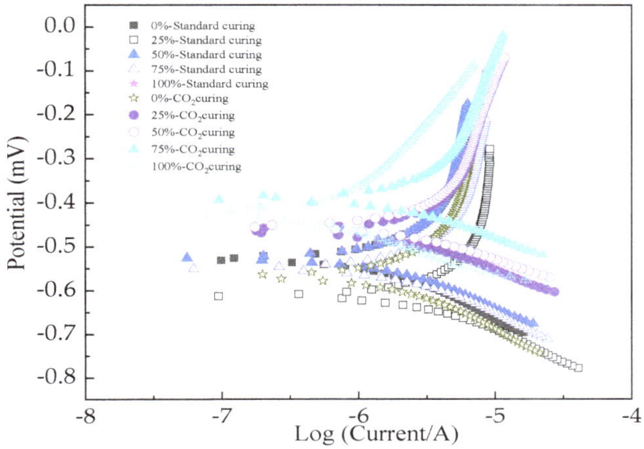

Figure 7. The Tafel curves of reinforced alkali-activated mineral admixtures after 20 D–W alternations.

Table 4. The corrosion current of reinforced alkali-activated mineral admixtures(10^{-6}A/cm^2).

Specimens	A_1	A_2	A_3	A_4	A_5
Standard curing—5 alternations	3.15	2.89	2.46	2.33	0.31
CO_2 curing—5 alternations	3.41	3.23	2.87	2.68	0.46
Standard curing—10 alternations	2.71	2.53	2.14	1.98	0.24
CO_2 curing—10 alternations	3.79	3.56	3.24	3.11	0.59
Standard curing—20 alternations	4.12	3.98	3.66	3.27	0.89
CO_2 curing—20 alternations	4.08	3.79	3.60	2.96	0.67

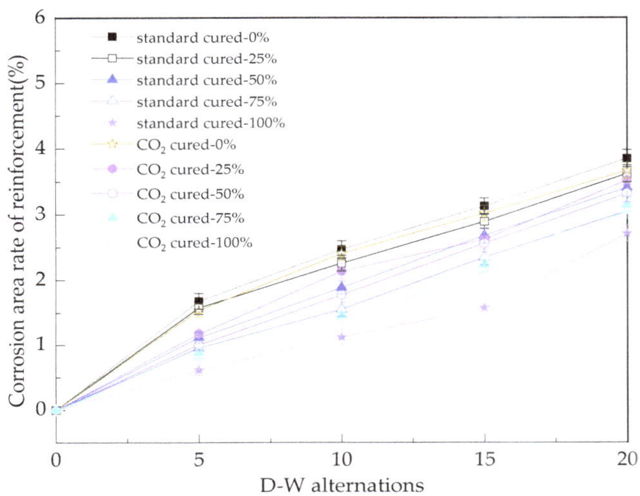

Figure 8. The corrosion rate of inner steel bars during D–W alternations.

The corrosion area rate of the reinforcement is shown in Figure 8. This parameter is obtained by testing the Tafel curves. It can be seen that the corrosion area rate of the reinforcement increases with the increasing NaCl D–W alternations and the decreasing dosage of BFS. Additionally, the corrosion area rate decreases with the CO_2 curing condition. This is because the NaCl D–W alternations accelerate the corrosion rate of reinforcement, thus increasing the corrosion area rate of the reinforcement. Meanwhile, CO_2 curing increases the compactness of the specimens. Hence, the corrosion resistance is improved, and, thus, eventually decreases the corrosion area rate.

In this study, the AC impedance spectrum is obtained to reflect the corrosion degree of the reinforced alkali-activated mineral admixtures. The $Z_i(\Omega)$ and $Z_r(\Omega)$ are the imaginary part and real part, which represent the electrical reactance and resistance of the AC impedance spectrum curves, respectively. Figure 9 demonstrates the AC impedance spectrum of the reinforced alkali-activated mineral admixtures. The minimum point of the AC impedance spectrum increases with the addition of BFS and CO_2 curing. This is owing to the fact that FA effectively decreases the electrical resistance of reinforced alkali-activated mineral admixtures, thus decreasing the minimum point of the AC impedance spectrum. In addition, CO_2 curing can diminish the number of free ions, leading to increased electrical resistance. Additionally, the minimum point of the AC impedance spectrum increases with the NaCl D–W alternations since the NaCl D–W alternations accelerate the corrosion. The increased rust can block the conductive channel; therefore, the electrical resistance is increased. Moreover, the increasing extent of the minimum point of AC impedance spectrum by NaCl D–W alternations is decreased with the increasing dosage of BFS and the CO_2 curing.

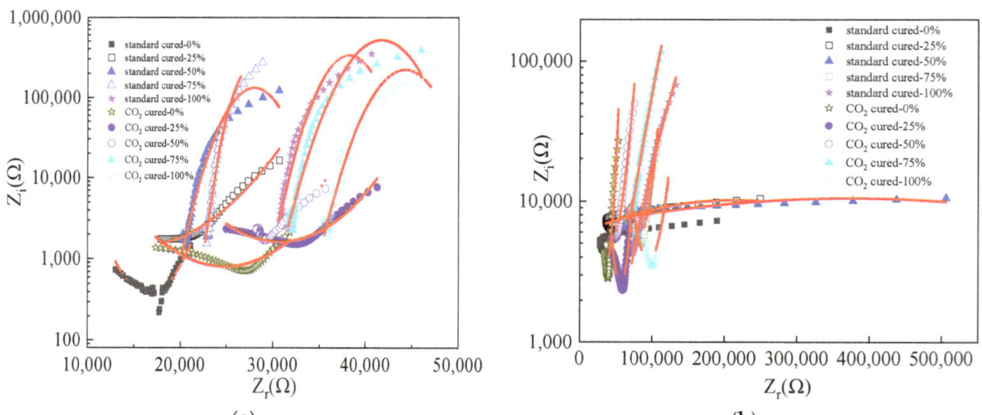

Figure 9. The AC impedance spectrum curves of reinforced alkali-activated mineral admixtures during D–W alternations. (**a**) Before NaCl corrosion. (**b**) After 20 NaCl D–W alternations.

All values of the AC impedance spectrum curves were input to ZSimpWin software and the fitting curves of the equivalent circuit diagrams were obtained (the red lines in Figure 9). Additionally, the equivalent circuit diagram of reinforced alkali-activated mineral admixtures is demonstrated in Figure 10. According to the fitting results and the prior studies [44–48], the electric circuit comprises the three electrical resistance and reactance of the rust, the matrix, and the pore solution. As shown in Figure 10, all these electrical elements are in series after a parallel connection (the electrical components of the same substance are in parallel, while the electrical components of different substances are in series). Table 5 shows the chi-square of the fitting curves. As observed in Table 5, the highest chi-squared value of all curves is 0.014, indicating the accuracy of the selected equivalent circuit diagram.

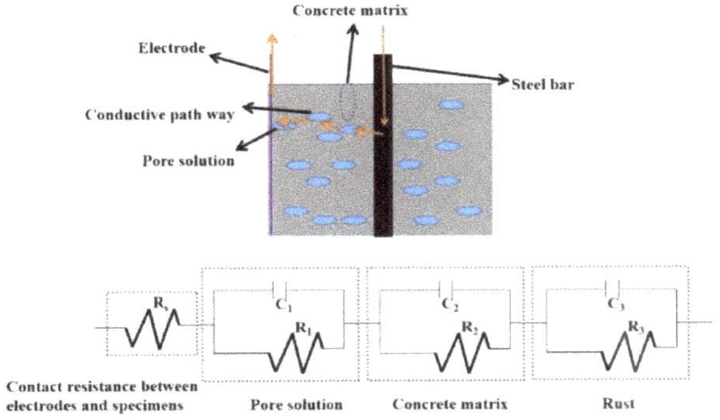

Figure 10. Corresponding equivalent circuits of reinforced alkali-activated mineral admixtures.

Table 5. The fitting results of the equivalent circuit diagram.

Specimens	Chi-Square
Standard cured-0%-Before NaCl D–W alternation	0.012
Standard cured-25%-Before NaCl D–W alternation	0.014
Standard cured-50%-Before NaCl D–W alternation	0.008
Standard cured-75%-Before NaCl D–W alternation	0.0056
Standard cured-100%-Before NaCl D–W alternation	0.013
CO_2-cured-0%-Before NaCl D–W alternation	0.0097
CO_2-cured-25%-Before NaCl D–W alternation	0.012
CO_2-cured-50%-Before NaCl D–W alternation	0.011
CO_2-cured-75%-Before NaCl D–W alternation	0.0089
CO_2-cured-100%-Before NaCl D–W alternation	0.0091
Standard cured-0%-After NaCl D–W alternation	0.0099
Standard cured-25%-After NaCl D–W alternation	0.012
Standard cured-50%-After NaCl D–W alternation	0.01
Standard cured-75%-After NaCl D–W alternation	0.0087
Standard cured-100%-After NaCl D–W alternation	0.0065
CO_2-cured-0%-After NaCl D–W alternation	0.0079
CO_2-cured-25%-After NaCl D–W alternation	0.0088
CO_2-cured-50%-After NaCl D–W alternation	0.0092
CO_2-cured-75%-After NaCl D–W alternation	0.0097
CO_2-cured-100%-After NaCl D–W alternation	0.0045

The electrical resistivity of the rust of reinforced alkali-activated mineral admixtures is illustrated in Figure 11. As demonstrated in Figure 11, the electrical resistivity of the rust of reinforced alkali-activated mineral admixtures decreases with the increasing dosage of BFS. It can be obtained from Figure 11 that the addition of BFS can help improve the corrosion resistance of reinforced alkali-activated mineral admixtures. Moreover, CO_2 curing can decrease the electrical resistance of rust. Therefore, as proved in Figure 11, the corrosion resistance of reinforced alkali-activated mineral admixtures is improved by CO_2 curing.

3.4. The Micro-Structure Research Results

This research was undertaken to investigate the corrosion resistance of reinforced alkali-activated mineral admixtures. The specimens are alkali-activated FA and alkali-activated BFS, which are cured in the standard curing environment and CO_2 curing environment, respectively. The X-ray diffraction (XRD) of the rust on the surface of the reinforcement is shown in Figure 12. Compared with the XRD results in other journals [45–47], the diffraction diagrams of (α-FeO(OH) and γ-Fe(OH)) are observed in the rust, and the height of the diffraction diagrams of γ-Fe(OH) increases in the order of CO_2-cured BFS < CO_2-cured FA < standard cured alkali-activated BFS < standard cured alkali-activated cured FA. This result confirms that adding BFS and CO_2 curing effectively improve corrosion resistance against NaCl D–W alternations.

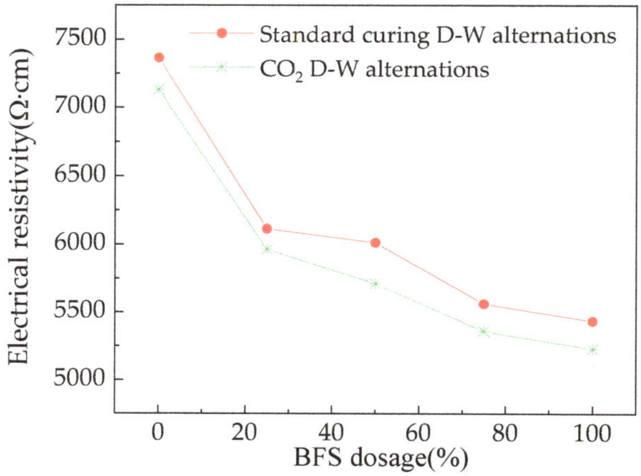

Figure 11. The electrical resistivity of the rust through an equivalent circuit diagram.

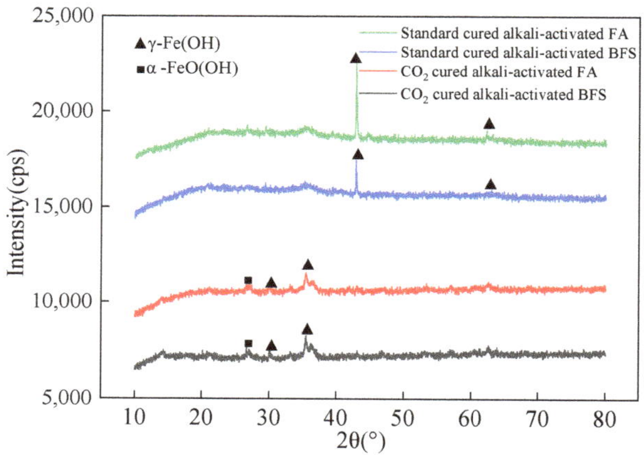

Figure 12. The X-ray diffraction (XRD) diagrams of the rust on the surface of steel bars.

The SEM micro-structure photos of standard cured alkali-activated FA, standard cured alkali-activated BFS, and CO_2-cured alkali-activated FA specimens are presented in Figure 13. As shown in Figure 13, more needle-like and loose hydration products are found. Additionally, as shown in Figure 13b, the hydration products of standard cured alkali-activated BFS are more compact than those of the standard cured alkali-activated FA, while as shown in Figure 13c, the hydration products of CO_2-cured alkali-activated BFS are coarser and more compact. Therefore, the SEM micro-structure photos of specimens further prove that adding BFS and CO_2 curing can effectively improve the corrosion resistance of reinforced alkali-activated mineral admixtures exposed to the NaCl D–W alternations.

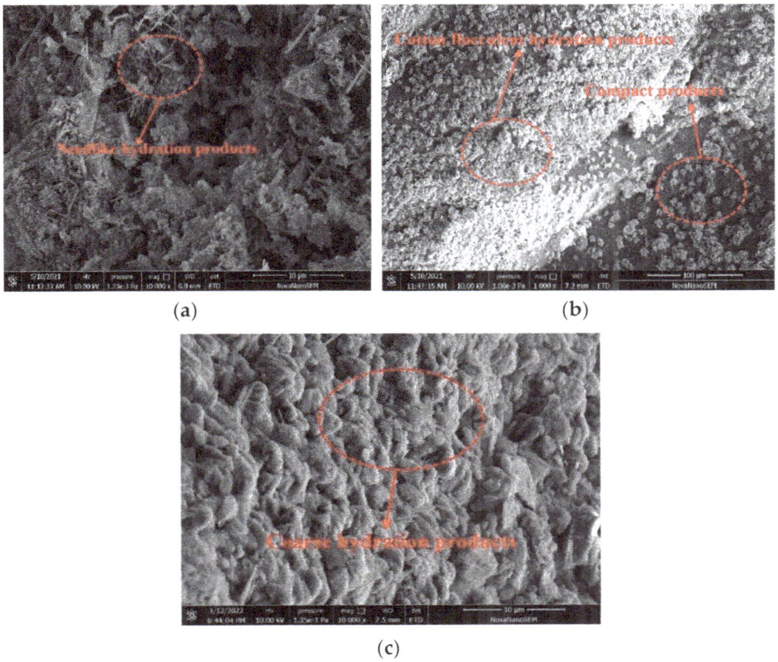

Figure 13. SEM micro-structure photos of specimens. (**a**) Standard cured alkali-activated FA. (**b**) Standard cured alkali-activated BFS. (**c**) CO_2-cured alkali-activated BFS.

4. Conclusions

The mass loss rates of the reinforced alkali-activated mineral admixtures and the inner steel bars increase with the increasing number of NaCl D–W alternations but decrease with the addition of BFS and CO_2 curing.

The electrical resistivity of the reinforced alkali-activated mineral admixtures increases with increasing D–W alternations, the addition of BFS, and CO_2 curing. Meanwhile, the increasing rate of electrical resistivity by D–W alternations decreases with more dosage of BFS and CO_2 curing.

The corrosion area rate of reinforced alkali-activated mineral admixtures and the electrical resistivity of the rust increase with NaCl D–W alternations but decrease with the increasing dosage of blast furnace slag powder and CO_2 curing. Adding BFS and CO_2 curing can improve the corrosion resistance of reinforced alkali-activated mineral admixtures.

This study will provide an alkali-activated mineral admixture with good corrosion resistance under a chloride corrosion environment.

Author Contributions: Conceptualization, F.S. and H.S.; methodology, H.S.; software, W.C.; validation, H.S., H.X. and Z.C.; formal analysis, H.S.; investigation, W.C.; resources, M.Y.; data curation, F.S.; writing—original draft preparation, H.S.; writing—review and editing, F.S.; visualization, H.S.; supervision, F.S.; project administration, F.S.; funding acquisition, H.S. All authors have read and agreed to the published version of the manuscript.

Funding: This research was funded by Science and Technology Projects of Ministry of Housing and Urban–Rural Development of China, grant number 2014-K4-024.

Institutional Review Board Statement: Not applicable.

Informed Consent Statement: Not applicable.

Data Availability Statement: The data used to support the findings of this study are available from the corresponding author upon request.

Acknowledgments: The authors gratefully acknowledge the financial support of YCIT, and the testing services from Analysis and Test center of YCIT.

Conflicts of Interest: The authors declare no conflict of interest.

References

1. Shah, J. Laboratory characterization of controlled low-strength material and its application to construction of flexible pipe drainage system. *Health Manpow. Manag.* **2000**, *19*, 30–59.
2. Haselbach, L.; Valavala, S.; Montes, F. Permeability predictions for sand clogged Portland cement pervious concrete pavement systems. *J. Environ. Manag.* **2005**, *81*, 42–49. [CrossRef] [PubMed]
3. Wei, Y.; Gao, X.; Zhang, Q. Evaluating performance of concrete pavement joint repair using different materials to reduce reflective cracking in asphalt concrete overlay. *Road Mater. Pavement Des.* **2014**, *15*, 966–976. [CrossRef]
4. Witoon, T.; Lapkeatseree, V.; Numpilai, T.; Cheng, C.; Limtrakul, J. CO_2 hydrogenation to light olefins over mixed Fe-Co-K-Al oxides catalysts prepared via precipitation and reduction methods. *Chem. Eng. J.* **2022**, *428*, 131389. [CrossRef]
5. Witoon, T.; Numpilai, T.; Nijpanich, S.; Chanlek, N.; Kidkhunthod, P.; Cheng, C.; Ng, K.; Vo, D.N.; Ittisanronnachai, S.; Wattanakit, C.; et al. Enhanced CO_2 hydrogenation to higher alcohols over K-Co promoted In_2O_3 catalysts. *Chem. Eng. J.* **2022**, *431*, 133211. [CrossRef]
6. Liska, M.; Jin, F.; Yi, Y.; Al-Tabbaa, A. Mechanism of reactive magnesia-ground granulated blastfurnace slag (GGBS) soil stabilization. *Can. Geotech. J.* **2016**, *53*, 773–782.
7. Li, Y.; Li, X.; Tan, Y. Effect of aging on fatigue performance of cement emulsified asphalt repair material. *Constr. Build. Mater.* **2021**, *292*, 123417. [CrossRef]
8. Chen, X.; Wang, S.; Zhou, Y.; Cheeseman, C.; Bi, W.; Zhang, T. Improved low-carbon magnesium oxysulfate cement pastes containing boric acid and citric acid. *Cem. Concr. Compos.* **2022**, *134*, 104813. [CrossRef]
9. Liu, M.; Tan, H.; He, X. Effects of nano-SiO_2 on early strength and microstructure of steam-cured high volume flfly ash cement system. *Constr. Build. Mater.* **2019**, *194*, 350–359. [CrossRef]
10. Celikten, S.; Sarldemir, M.; Deneme, I.O. Mechanical and microstructural properties of alkali-activated slag and slag+ fly ash mortars exposed to high temperature. *Constr. Build. Mater.* **2019**, *217*, 50–61. [CrossRef]
11. Huang, G.; Wang, H.; Shi, F. Coupling Effect of Salt Freeze-Thaw Cycles and Carbonation on the Mechanical Performance of Quick Hardening Sulphoaluminate Cement-Based Reactive Powder Concrete with Basalt Fibers. *Coatings* **2021**, *11*, 1142. [CrossRef]
12. Li, S.; Feng, Y.; Yang, J.; Cerny, R. Expansion mechanism and properties of magnesium oxide expansive hydraulic cement for engineering applications. *Adv. Mater. Sci. Eng.* **2021**, *2021*, 5542072. [CrossRef]
13. Li, S.; Yang, J.; Zhang, P. Hydration and hardening properties of reactive magnesia and Portland cement composite. *Constr. Build. Mater.* **2022**, *327*, 126779. [CrossRef]
14. Ozata, S.; Akturk, B.; Yuzer, N. Utilization of waste Cappadocia earth as a natural pozzolan in alkali activation: A parametric study. *Constr. Build. Mater.* **2022**, *329*, 127192. [CrossRef]
15. Yang, J.; Huang, J.; Su, Y.; He, X.; Tan, H.; Yang, W.; Strnadel, B. Eco-friendly treatment of low-calcium coal fly ash for high pozzolanic reactivity: A step towards waste utilization in sustainable building material. *J. Clean. Prod.* **2019**, *238*, 117962. [CrossRef]
16. Matějková, P.; Matějka, V.; Sabovčík, T.; Gryžbon, L.; Vlček, J. Alkali Activation of Ground Granulated Blast Furnace Slag and Low Calcium Fly Ash Using "One-Part" Approach. *J. Sustain. Metall.* **2022**, *8*, 511–521. [CrossRef]
17. Supriya, J.; Raut, A. Performance parameter analysis of magnesia based cement products-a review. *IOP Conf. Ser. Mater. Sci. Eng.* **2021**, *1197*, 012078. [CrossRef]
18. Glukhovsky, V.D.; Rostovskaja, G.S.; Rumyna, G.V. High strength slag—Alkaline cements. In Proceedings of the 7th International Congress on the Chemistry of Cement, Paris, France, 11–13 July 1980.
19. Alomayri, T.; Adesina, A.; Das, S. Influence of amorphous raw rice husk ash as precursor and curing condition on the performance of alkali activated concrete. *Case Stud. Constr. Mater.* **2021**, *15*, e00777. [CrossRef]
20. Alrefaei, Y.; Dai, J. Effects of delayed addition of polycarboxylate ether on one-part alkali-activated fly ash/slag pastes: Adsorption, reaction kinetics, and rheology. *Constr. Build. Mater.* **2022**, *323*, 126611. [CrossRef]
21. Feng, S.; Zhu, J.; Wang, R.; Qu, Z.; Song, L.; Wang, H. The Influence of CaO and MgO on the Mechanical Properties of Alkali-Activated Blast Furnace Slag Powder. *Materials* **2022**, *15*, 6128. [CrossRef]
22. Xian, X.; Zhang, D.; Lin, H.; Shao, Y. Ambient pressure carbonation curing of reinforced concrete for CO_2 utilization and corrosion resistance. *J. CO2 Util.* **2022**, *56*, 101861. [CrossRef]
23. Ju, Y.; Zhao, J.; Wang, D.; Song, Y. Experimental study on flexural behaviour of reinforced reactive powder concrete pole. *Constr. Build. Mater.* **2021**, *312*, 125399. [CrossRef]
24. Chi, M. Effects of dosage of alkali-activated solution and curing conditions on the properties and durability of alkali-activated concrete. *Constr. Build. Mater.* **2012**, *35*, 240–245. [CrossRef]

25. Zhu, H.; Zhang, Z.; Zhu, Y.; Tian, L. Durability of alkali-activated fly ash concrete: Chloride penetration in pastes and mortars. *Constr. Build. Mater.* **2014**, *65*, 51–59. [CrossRef]
26. Wang, H.; Jin, K.; Zhang, A.; Zhang, L.; Han, Y.; Liu, J.; Shi, F.; Feng, L. External erosion of sodium chloride on the degradation of self-sensing and mechanical properties of aligned stainless steel fiber reinforced reactive powder concrete. *Constr. Build. Mater.* **2021**, *287*, 123028. [CrossRef]
27. AL-Ameeri, A.S.; ImranRafiq, M.; Tsioulou, O.; Rybdylova, O. Impact of climate change on the carbonation in concrete due to carbon dioxide ingress: Experimental investigation and modelling. *J. Build. Eng.* **2021**, *44*, 102594. [CrossRef]
28. Qin, L.; Gao, X. Properties of coal gangue-Portland cement mixture with carbonation. *Fuel* **2019**, *245*, 1–12. [CrossRef]
29. Kamal, N.L.M.; Itam, Z.; Sivaganese, Y.; Beddu, S. Carbon dioxide sequestration in concrete and its effects on concrete compressive strength. *Mater. Today Proc.* **2020**, *31*, A18–A21. [CrossRef]
30. Adesina, A. Recent advances in the concrete industry to reduce its carbon dioxide emissions. *Environ. Chall.* **2020**, *1*, 100004. [CrossRef]
31. Zhu, J.; Qu, Z.; Liang, S.; Li, B.; Du, T.; Wang, H. The macro-scopic and microscopic properties of cement paste with carbon dioxide curing. *Materials* **2022**, *15*, 1578. [CrossRef]
32. Ashraf, W.; Olek, J. Carbonation behavior of hydraulic and non-hydraulic calcium silicates: Potential of utilizing low-lime calcium silicates in cement-based materials. *J. Mater. Sci.* **2016**, *51*, 6173–6191. [CrossRef]
33. *GB/T1345-2005;* The Test Sieving Method for Fineness of Cement. General Administration of Quality Supervision, Inspection and Quarantine of the People's Republic of China: Beijing, China, 2005.
34. Wang, H.; Shi, F.; Shen, J.; Zhang, A.; Zhang, L.; Huang, H.; Liu, J.; Jin, K.; Feng, L.; Tang, Z. Research on the self-sensing and mechanical properties of aligned stainless steel fiber reinforced reactive powder concrete. *Cem. Concr. Compos.* **2021**, *119*, 104001. [CrossRef]
35. Mei, K.; He, Z.; Yi, B.; Lin, X.; Wang, J.; Wang, H.; Liu, J. Study on electrochemical characteristics of reinforced concrete corrosion under the action of carbonation and chloride. *Case Stud. Constr. Mater.* **2022**, *17*, e01351. [CrossRef]
36. Zhang, L.; Zhang, A.; Wang, Q.; Han, Y.; Tang, Z. Corrosion resistance of wollastonite modified magnesium phosphate cement paste exposed to freeze-thaw cycles and acid-base corrosion. *Case Stud. Constr. Mater.* **2020**, *13*, e00421. [CrossRef]
37. Zhang, B.; Tan, H.; Shen, W.; Xu, G.; Ma, B.; Ji, X. Nano-silica and silica fume modifified cement mortar used as Surface Protection Material to enhance the impermeability. *Cem. Concr. Compos.* **2018**, *92*, 7–17. [CrossRef]
38. Jin, F.; Gu, K.; Al-Tabbaa, A. Strength and hydration properties of reactive MgO-activated ground granulated blastfurnace slag paste. *Cem. Concr. Compos.* **2015**, *57*, 8–16. [CrossRef]
39. Lei, L.; Wang, Q.; Xu, S.; Wang, N.; Zheng, X. Fabrication of superhydrophobic concrete used in marine environment with anti-corrosion and stable mechanical properties. *Constr. Build. Mater.* **2020**, *251*, 118946. [CrossRef]
40. Qin, L.; Gao, X.; Chen, T. Influence of mineral admixtures on carbonation curing of cement paste. *Constr. Build. Mater.* **2019**, *212*, 653–662. [CrossRef]
41. Cao, H.; Liang, Z.; Peng, X.; Cai, X.; Wang, K.; Wang, H.; Lyu, Z. Research of Carbon Dioxide Curing on the Properties of Reactive Powder Concrete with Assembly Unit of Sulphoaluminate Cement and Ordinary Portland Cement. *Coatings* **2022**, *12*, 209. [CrossRef]
42. Liang, J.; Zhu, H.; Chen, L.; Qinglin, X.; Ying, G. Rebar corrosion investigation in rubberaggregate concrete via the chloride electro-accelerated test. *Materials* **2019**, *12*, 862. [CrossRef]
43. Cui, L.; Wang, H. Research on the Mechanical Strengths and the Following Corrosion Resistance of Inner Steel Bars of RPC with Rice Husk Ash and Waste Fly Ash. *Coatings* **2021**, *11*, 1480. [CrossRef]
44. Wang, H.; Zhang, A.; Zhang, L.; Liu, J.; Han, Y.; Wang, J. Research on the Influence of Carbonation on the Content and State of Chloride Ions and the Following Corrosion Resistance of Steel Bars in Cement Paste. *Coatings* **2020**, *10*, 1071. [CrossRef]
45. Zhang, X.W.; Liu, Y.Y.; Li, C.Q.; Tian, L.; Yuan, F.; Zheng, S.L.; Sun, Z.M. Fast and lasting electron transfer between γ-FeOOH and g-C3N4/kaolinite containing N vacancies for enhanced visible-light-assisted peroxymonosulfate activation. *Chem. Eng. J.* **2022**, *429*, 132374. [CrossRef]
46. Jadhav, A.S.; Mokhtar, A.A.; Sunit, K.S.; Ahmed, S.A.; Amit, B.; Vadali, V.S.S.S.; Nitin, K.L. γ-FeOOH and γ-FeOOH decorated multi-layer graphene: Potential materials for selenium(VI) removal from water. *J. Water Process Eng.* **2020**, *37*, 101396. [CrossRef]
47. Chen, G.; Ma, C.G.; Fu, Z.H.; Wang, J.; Dai, P.B. Significantly enhancing fracture toughness of epoxy composite with promising γ-FeOOH@Fe2O3 hybrid nanoparticles by magnetic field assistanc. *Nano Mater. Sci.* **2022**, *4*, 139–150. [CrossRef]
48. Wang, H.; Zhang, A.; Zhang, L.; Liu, J.; Han, Y.; Shu, H.; Wang, J. Study on the influence of compound rust inhibitor on corrosion of steel bars in chloride concrete by electrical parameters. *Constr. Build. Mater.* **2020**, *262*, 120763. [CrossRef]

coatings

Article

The Influence of CO_2 Curing on the Mechanical Performance and the Corresponding Chloride Ion Resistance of Alkali-Activated Compound Mineral Admixtures

Ling Peng [1], Junchao Yang [2,*], Hui Wang [3,*] and Xiaoqing Jin [1,*]

1 Department of Mathematics and Physics, Hebei University of Architecture, Zhangjiakou 075000, China
2 School of Civil Engineering, Hebei University of Architecture, Zhangjiakou 075000, China
3 School of Civil and Environmental Engineering, Ningbo University, Ningbo 315000, China
* Correspondence: yangjunchao@aliyun.com (J.Y.); huiwang123@aliyun.com (H.W.); jinxiaoqing@aliyun.com (X.J.)

Abstract: In this paper, the mechanical properties (the flexural strength, compressive strength and the drying shrinkage rate) of CO_2-cured alkali-activated compound mineral admixtures (blast furnace slag powder (BFS) and fly ash (FA)) are investigated. In addition, the corresponding chloride ion mobility coefficient is measured. Additionally, the freeze–thaw cycles with an NaCl concentration of 3% is studied. Thermogravimetric analysis and scanning electron microscopy are applied in analyzing the mechanical properties. The curing ages of the alkali-activated compound mineral admixtures are 1 day, 3 days and 28 days. Results show that the mechanical strengths are decreased by the addition of FA and increased by the increasing curing age and CO_2 curing. The maximum reducing rates of flexural and compressive strengths by FA are 47.6% and 42.3%. Meanwhile, the corresponding increasing rates by CO_2 curing are 26.5% and 23.1%, respectively. The improving effect of alkali-activated BFS by CO_2 curing is higher than that of FA. Furthermore, the drying shrinkage rate is increased by the increasing dosages of BFS, the increasing curing ages and CO_2 curing. Additionally, CO_2 curing and the increasing dosage of BFS leads to decreasing the chloride ion mobility coefficient. Finally, CO_2 curing and the addition of BFS can effectively improve the resistance of NaCl freeze–thaw cycles. The compactness of the hydration products is improved by the addition of BFS and the roughness of hydration products is increased by CO_2 curing.

Keywords: CO_2 curing; alkali-activated; drying shrinkage rate; NaCl freeze–thaw cycles

1. Introduction

Cement-based materials are the most widely used building materials, having been invented hundreds years ago [1,2]. Housing construction, bridge construction, highway engineering, etc. have adopted cement-based materials for construction [3,4]. Cement-based materials show good mechanical properties, high durability, low cost and a wide application range [5,6]. However, a large amount of harmful gases are produced and a great quantity of energy is consumed in the production of cement. Therefore, more cementitious materials need to be developed to replace cement.

Blast furnace slag powder (BFS), fly ash (FA), silica fume, volcanic ash, etc. are solid wastes generated from industrial production and nature [7,8]. As reported in prior research, they possess a composition similar to cement. Therefore, these mineral admixtures have become excellent choices to replace cement. However, the mineral admixtures do not possess enough mechanical strength when cured in conventional curing processes [9,10]. Research points out that the addition of mineral admixtures can decrease the mechanical strength of cement-based materials when they are cured in a standard curing environment. Meanwhile, when the materials are cured in a high temperature curing environment, mechanical strengths are improved by the mineral admixtures.

Alkali-activated cementitious materials have the characteristics of high early strength, fast setting and hardening, excellent corrosion resistance, etc. [9]. BFS and FA are common industrial wastes in our daily life, showing high production and wide distribution [11]. A large amount of BFS and FA have polluted the atmosphere, causing smog. Moreover, long-term stacking will pollute rivers, lakes and soil and infiltrate into groundwater. Furthermore, massive accumulation will also seriously occupy land [12]. Alkali-activated mineral admixtures are manufactured by mixing mineral admixtures, sodium silicate, alkali, water, etc. [12–16]. The alkali-activated mineral admixtures present excellent mechanical strengths. In the 1950s Gluk hovsky et al. [17] used NaOH or sodium silicate as an activator to excite the mixture of crushed stone, boiler slag or blast furnace slag powder, quicklime, blast furnace slag and Portland cement, and prepared a cementitious material with strength up to 120 MPa [18,19]. Additionally, the compressive strengths of alkali-activated BFS and FA can reach values higher than 80 MPa and 100 MPa, respectively. Feng et al. found that the assembly unit of CaO and MgO can improve the mechanical strengths, high temperature resistance and compactness of the microstructure [20]. The compressive strength of ordinary concrete is less than 60 MPa while high performance concrete shows a compressive strength of 60 MPa–100 MPa; therefore, the compressive strength of alkali-activated BFS is close to the compressive strength of ultra-high performance concrete.

CO_2 is one of the main gases which causes the greenhouse effect. A large amount of CO_2 gas is produced during the production of cement. The cementitious materials cured with CO_2 can not only consume it, but also enhance the strength of materials, which is of great significance to alleviate greenhouse gases in the air [21,22]. CO_2 has been used for curing cement-based materials for several years [23,24]. As reported in prior research, CO_2 curing can improve the mechanical properties and the corrosion resistance of the reinforced cement matrix [25]. Additionally, CO_2 curing can effectively increase the mechanical strengths of reactive powder concrete by 30.1%–72.6%. The chloride ion mobility coefficient is decreased by 11.2%–41.6% with CO_2 curing. However, the effect of CO_2 curing on the mechanical strengths and chloride ion mobility coefficient of alkali-activated mineral admixtures is unknown. The increased use of mineral admixtures may reduce CO_2's emissions [26].

This paper aims to study the influence of CO_2 curing on the mechanical performance and the resistance to chloride ion permeability and freeze–thaw resistance of alkali-activated compound mineral admixtures (blast furnace slag powder and fly ash). The corresponding mechanism of the mechanical performance is revealed by thermogravimetric (TG) analysis and scanning electron microscopy (SEM). This research will provide a new idea for developing diverse cementitious materials and consuming CO_2 in the future.

2. Experimental

2.1. Raw Materials

The sodium silicate is produced by Tongxiang Hengli Chemical Co., Ltd., Tongxiang, China. The melting point of the sodium silicate ranges from 40 °C to 48 °C. S95 grade mineral powder is manufactured by Lingshou Anda mineral powder factory, Shijiazhuang City, China. The corresponding density, the specific surface area and the loss on ignition of the BFS are 2.88 g/cm^3, 435.8 m^2/g and 1.98%, respectively. The FA is made by the Shijiazhuang Shunli mineral products Co., Ltd., Shijiazhuang City, China. The density of FA is 2.4 g/cm^3. The sodium hydroxide is manufactured by Shanghai Dongmiao Chemical Technology Co., Ltd., Shanghai, China. The sodium hydroxide shows a density of 2.13 g/cm^3, a boiling point of 1390 °C and a purity of 99.9%. The particle passing percentage and the chemical composition of raw materials are shown in Tables 1 and 2, respectively.

Table 1. Particle passing percentage of raw materials (%).

Types	0.3 μm	1 μm	4 μm	8 μm	64 μm	360 μm
BFS	0.03	3.5	19.6	35.0	97.9	100
FA	12.3	66.2	100	100	100	100

Table 2. The chemical composition of cement (%).

Types	SiO_2	Al_2O_3	Fe_xO_y	MgO	CaO	SO_3	K_2O	Na_2O	Ti_2O	LI
BFS	34.1	14.7	0.2	9.7	35.9	0.2	3.5	-	-	-
FA	55.00	20.00	6.00	10.20	4.50	0.11	1.26	2.13	0.06	0.74

2.2. Specimen Preparation

The specimens are prepared following these steps: Table 3 indicates the mixing proportions of alkali-activated cementitious material. The NJ-160A cement paste mixer is applied in the mixing of the potash water glass and sodium hydroxide. The stirring speeds of the cement paste mixer are 140 r/min and 285 r/min. A 2 min stirring speed of 140 r/min and 2 min of 285 r/min are provided, respectively. Specimens are kept in the standard curing room (temperature of 20 ± 2 °C and relative humility of 98.6%) and TH-B concrete carbonization test box (CO_2 concentration of 8%) provided by Tianjin Gangyuan test instrument factory, Tianjin, China.

Table 3. The mixing proportions of alkali-activated cementitious material (kg/m^3).

Number	Water	P·O	SAC	SF	GGBS	Quartz Sand	Water-Reducer	Li_2SO_4	Calcium Formate	Tartaric Acid	Defoamer	Steel Fibers
1	244.4	370.5	370.5	370.5	111.1	978	20.3	0.6	2.6	1.9	0.6	78.5
2	244.4	370.5	370.5	370.5	111.1	978	24.4	0.6	2.6	1.9	0.6	117.75
3	244.4	370.5	370.5	370.5	111.1	978	12.2	0.6	2.6	1.9	0.6	157
4	244.4	370.5	370.5	370.5	111.1	978	16.3	0.6	2.6	1.9	0.6	196.25
5	244.4	370.5	370.5	370.5	111.1	978	20.3	0.6	2.6	1.9	0.6	235.5

2.3. Measurement

Specimens with size of 40 × 40 × 160 mm^3 are used for the measurement of the flexural and compressive strengths with the loading speeds of 2.4 kN/s and 0.05 kN/s, respectively. HYE-300B cement flexural and compressive constant stress testing machine is used for providing the loading. In this study, three specimens are used for the measurement of flexural strength, while six specimens are applied in the determination of compressive strength.

The Mitutoyo Japan Mitutoyo high precision handheld digital display produced by Shanghai Shoufeng Precision Instrument Co., Ltd., Shanghai, China is applied in the measurement of the dry shrinkage rate. Before the measurement, specimens are fixed in the BC-160 type length comparator. The measuring details of the dry shrinkage rate can be found in JGJT70-2009.

The dry shrinkage rate is calculated by Equation (1).

$$\varepsilon = \frac{L_1 - L_t}{L_t} \tag{1}$$

where ε is the dry shrinkage rate and L_1 and L_t are the lengths of specimen cured for 1 day and 28 days, respectively.

The steps of the measurement of the chloride ion permeability coefficient (CMC) are described as follows.

The specimens with a size of Φ100 × 50 mm^3 are saturated in a vacuum saturator for 2 days. After that, specimens are moved for the CMC test with the NELD-CCM550 concrete chloride ion diffusion coefficient tester provided by Beijing Neerde Intelligent Technology Co., Ltd., Beijing, China.

Specimens with size of 100 × 100 × 400 mm^3 are applied in the measurement of NaCl freeze–thaw cycles. Before the experiment, all samples are immersed in an NaCl solution with a concentration of 3%. After that, all specimens are moved to the DR-10A cement automatic quick freezing and thawing machine for the freeze–thaw measurement. The measuring details of determinations of CMC and NaCl freeze–thaw cycles are followed by the Chinese Standard GB/T 50082-2009. Three specimens of each group are used for the measurements of dry shrinkage rate, the CMC and the NaCl freeze–thaw cycles.

The central parts of the specimens are taken out and used for the measurements of the TG and SEM. The specimens with mung bean size are sprayed with gold under a vacuum environment. After the gold spraying is finished, the measurement of the scanning electron microscope photos is carried out. The powdered sample after crushing and screening is used for the thermal analysis. The temperatures on the TG curves are determined by the temperature sensor for automatic temperature recording. The accuracy of determining these values is ±0.8 °C, these descriptions have been added in the revised manuscript. The measuring details can be found in Ref. [27].

3. Results and Discussions

3.1. Influence of Steel Fibers on Working Performance

Figure 1 illustrates the slump flow of the fresh alkali-activated cementitious material. As shown in Figure 1, the slump flow decreases with increasing dosages of FA. The fact that the specific surface area of FA is larger than that of BFS leads to a decrease in the slump flow of fresh alkali-activated cementitious material. The values of all error bars are lower than 0.1, indicating the accuracy of the experimental results.

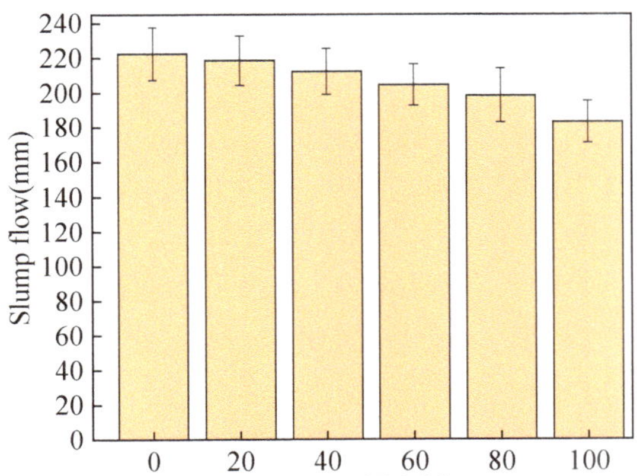

Figure 1. The slump flow of fresh alkali-activated compound mineral admixtures.

3.2. The Mechanical Strengths

Figure 2 shows the mechanical strengths of alkali-activated compound mineral admixtures. It is depicted in Figure 2 that the flexural and compressive strengths increase with increasing curing age and decrease with the increasing dosage of FA. Meanwhile, the reduction rate of flexural strength has a positive correlation with the dosages of FA and has a negative correlation with the curing age. Moreover, the addition of FA leads to increasing the reduction rate of compressive strength; furthermore, an increasing curing age results in a decrease of the reduction rate. This is ascribed to the fact that the FA shows a high degree of $[SiO_4]^{4-}$ polymerization in a vitreous structure network, resulting in low activity; therefore, FA is difficult to excite by alkali [10]. Consequently, the mechanical strengths of alkali-activated mineral admixtures are decreased by the addition of FA. The decreasing rates of flexural and compressive strengths of alkali-activated compound mineral admixtures induced by the addition of FA are 32.6%–47.6% and 35.4%–42.3%.

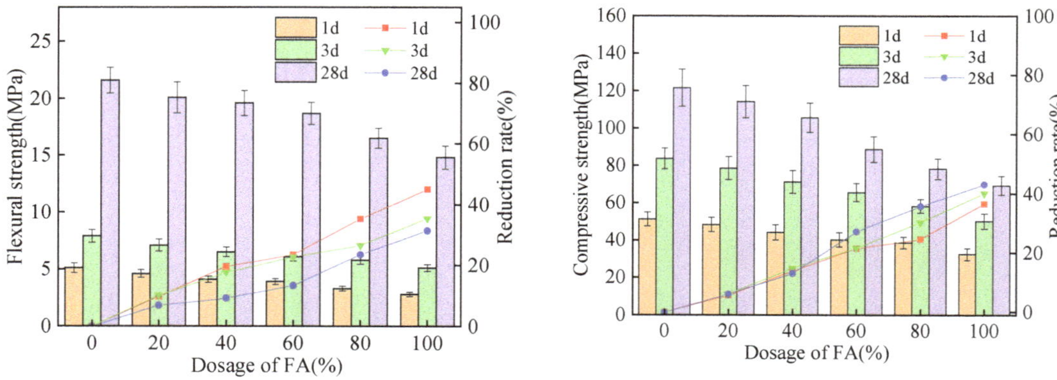

(a) (b)

Figure 2. The mechanical strengths of alkali-activated compound mineral admixtures. (**a**) flexural strength, (**b**) compressive strength.

The flexural and compressive strengths of CO_2-cured alkali-activated compound mineral admixtures are shown in Figure 3. As illustrated in Figure 3, the change rule of the flexural and compressive strengths of CO_2-cured alkali-activated compound mineral admixtures are similar to that of the standard cured alkali-activated compound mineral admixtures. Comparing Figures 2 and 3, CO_2 curing can improve the mechanical strengths of the alkali-activated compound mineral admixtures. Meanwhile, when more FA is added, the improvement effect is worse. This is ascribed to the fact that BFS possesses a higher content of CaO, which can react with more CO_2 forming a higher content of calcium carbonate leading to more compact products [28]. Comparing Figures 2 and 3, the increasing rates of flexural and compressive strengths by CO_2 curing are 16.7%–26.5% and 15.4%–23.1%, respectively.

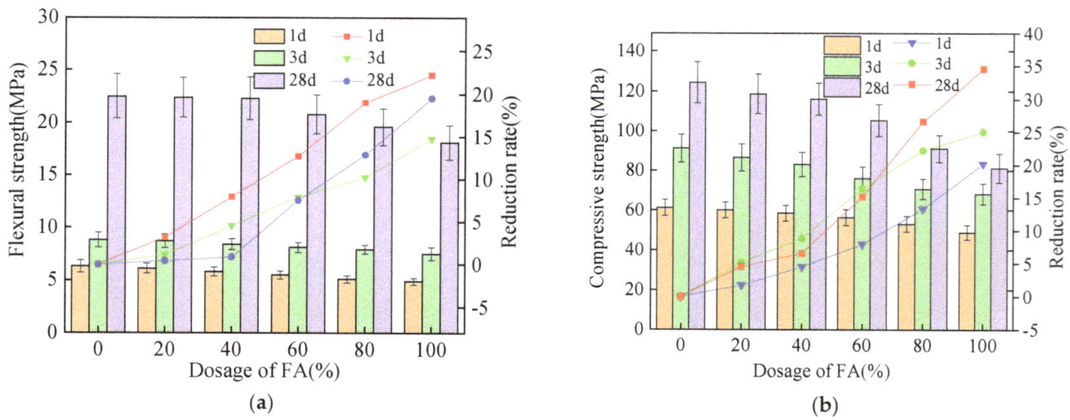

(a) (b)

Figure 3. The mechanical strengths of CO_2 -cured alkali-activated compound mineral admixtures. (**a**) flexural strength, (**b**) compressive strength.

Figure 4 shows the dry shrinkage rate of alkali-activated compound mineral admixtures. Table 4 shows the fitting results between the dry shrinkage rate and the mass ratio of FA. As depicted in Figure 4 and Table 4, the dry shrinkage rate increases in the form of a quadratic function with the mass ratio of BFS. Moreover, the dry shrinkage rate of alkali-activated compound mineral admixtures increases with the increasing curing age and CO_2 curing. This is ascribed to the fact that the hydration degree is increased with curing age, leading to a decrease in the free water inner specimens and increasing the dry shrinkage rate [29]. Moreover, CO_2 can react with $Ca(OH)_2$ forming $CaCO_3$; the increased amount of $CaCO_3$ and the porosity in the specimen is higher, resulting in an increase in the drying shrinkage rate [30]. Additionally, as depicted in Figure 4, the drying shrinkage rate increases with the increasing BFS dosage, due to higher alkalinity of BFS, thus forming more $CaCO_3$ and increasing the drying shrinkage rate. When the dosages of BFS increase from 0% to 100%, the drying shrinkage rate is decreased by 23.5%–32.1%. Whereas, with CO_2 curing, the drying shrinkage rate is increased by 36.5%–63.1%.

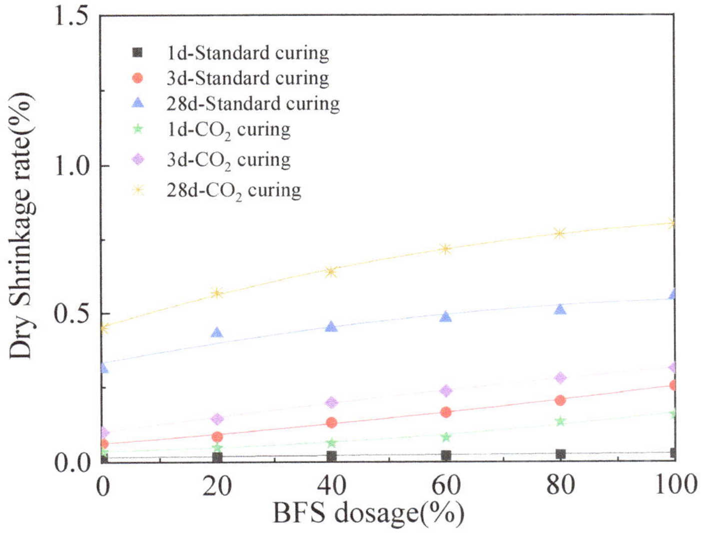

Figure 4. The dry shrinkage rate of alkali-activated compound mineral admixtures.

Table 4. The fitting results between the dry shrinkage rate and the mass ratio of FA.

Equation	FA Dosage (%)	a	b	c	R^2
$\frac{\Delta L}{L} = av^2 + bv + c$	0	-2.14×10^{-5}	0.0056	0.45	0.99
	20	-1.50×10^{-5}	0.0036	0.33	0.93
	40	-4.20×10^{-5}	0.0025	0.10	0.99
	60	4.28×10^{-5}	0.0014	0.06	0.99
	80	8.57×10^{-6}	0.0004	0.04	0.98
	100	-7.14×10^{-8}	1.20×10^{-4}	0.02	0.98

The CMC of the alkali-activated compound mineral admixtures. When the specimens are cured in the standard environment, the CMC increases with the increasing dosage of FA. As depicted in Figure 5, CO_2 curing leads to decreasing the CMC. When CO_2 curing is provided, the CMC first increases and then decreases with the increasing dosage of FA. This is due to the fact that the alkali excitation of BFS is higher than the FA, therefore, the compactness of alkali-activated compound mineral admixtures is improved by BFS [31,32]. Consequently, the CMC is increased by the increasing dosages of FA.

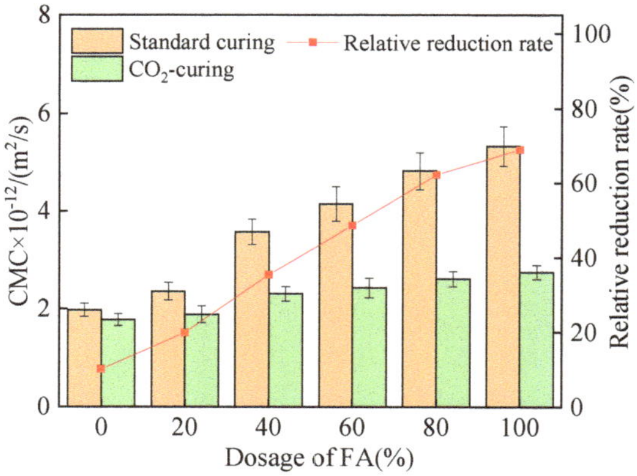

Figure 5. The CMC of the alkali-activated compound mineral admixtures.

The mass loss rate (△m/m) of alkali-activated compound mineral admixtures is illustrated in Figure 6. The △m/m increases with the increasing dosages of FA and the increasing NaCl freeze–thaw cycles. This is attributed to the fact the alkali-activated activity of FA is lower than that of BFS [33,34]. Therefore, alkali-activated FA shows lower compactness, leading to more serious NaCl freeze–thaw damage. Meanwhile, the NaCl freeze–thaw cycles can accelerate peeling of the outer surface of alkali-activated compound mineral admixtures, thus increasing the △m/m. Additionally, CO_2 curing can improve the compactness of alkali-activated compound mineral admixtures, resulting in reducing the surface peeling.

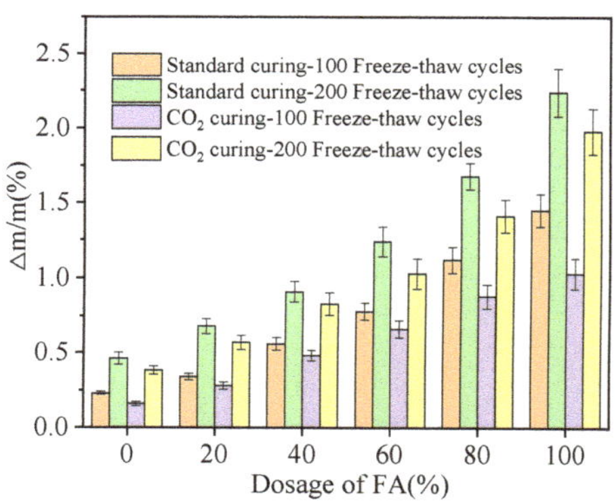

Figure 6. The △m/m of the alkali-activated compound mineral admixtures during NaCl freeze–thaw cycles.

The relative dynamic modulus of elasticity (RDME) of the alkali-activated mineral admixtures is illustrated in Figure 7. As observed in Figure 7, the RDME decreases with the increasing NaCl freeze–thaw cycles and the decreasing dosage of BFS. This is attributed to the fact that the NaCl freeze–thaw cycles induce inner cracks in the specimens. Moreover, as depicted in Figure 7, CO_2 curing leads to an increase in the RDME; when the addition of BFS dosage is increased, the improving effect is higher. CO_2 curing can accelerate the reaction between alkaline substances and CO_2, thus improving the compactness of alkali-activated mineral admixtures and increasing the corresponding RDME [35]. Moreover, the alkali-activated activity of BFS is higher than that of FA. As a consequence, the RDME is increased by an increasing dosage of BFS.

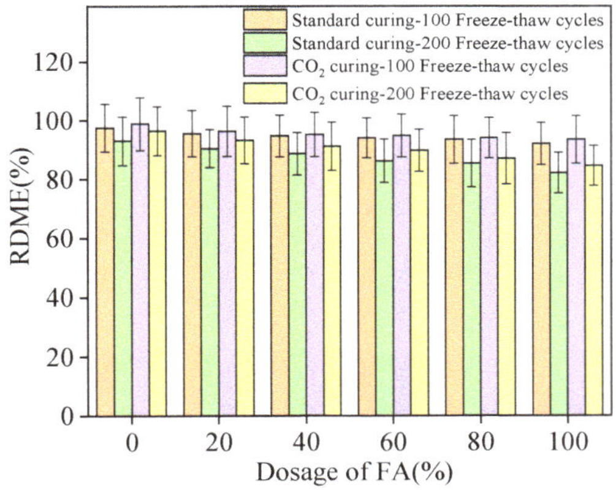

Figure 7. The RDME of the alkali-activated compound mineral admixtures during NaCl freeze–thaw cycles.

The TG and DTG curves of alkali-activated compound mineral admixtures are shown in Figure 8. As demonstrated in Figure 8, the TG values of alkali-activated compound mineral admixtures decreases in three steps. In the first step, the temperature varies from 30 °C to 156.7 °C, the TG values decrease due to the evaporation of free water. In addition, as illustrated in Figure 8, the second step can be observed, which is the temperature of 368.2 °C. In this step, the TG values decrease, which is ascribed to the decomposition of calcium silicate hydrate, the hydrated single sulfur calcium sulphoaluminate and hydrated polysulfur calcium sulphoaluminate [20,36]. As illustrated in Figure 8, the TG values are decreased by CO_2 curing, due to the fact that CO_2-cured alkali-activated compound mineral admixtures show less bound water than the alkali-activated compound mineral with standard curing, thus decreasing the reduction of TG values. Finally, when the temperature varies from 368.2 °C to 900 °C, the TG value decreases, which can be ascribed to the decomposition of hydrated calcium silicate and calcium carbonate. As observed in Figure 8, the addition of FA and CO_2 curing decreases the reduction of TG values. This is attributed to the lower alkali activity of FA, the increased calcium carbonate and the decreased of bound water by CO_2 curing [37].

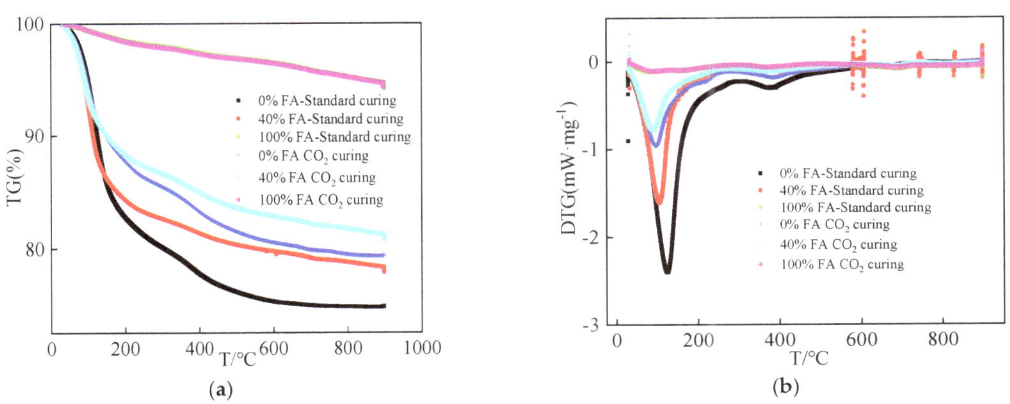

Figure 8. Thermogravimetric analysis curves. (**a**) TG curves, (**b**) DTG curves.

The scanning electron microscope (SEM) photos of the of the alkali-activated mineral admixtures are shown in Figure 9. The samples include the alkali-activated FA, alkali-activated BFS and CO_2-cured alkali-activated BFS. As observed in Figure 9, the alkali-activated BFS shows more compact hydration products than the alkali-activated FA. Moreover, as illustrated in Figure 9, the hydration products of CO_2-cured alkali-activated BFS are more compact than the alkali-activated FA. This is attributed to the fact that the alkali excitation degree of BFS is higher than that of FA. Meanwhile, as depicted in Figure 9, more flocculent and granular hydration products exist on the surface of the CO_2-cured alkali-activated BFS. This may be ascribed to the increased $CaCO_3$ content formed by the reaction of CO_2 and CaO [27]. Therefore, it is further confirmed that alkali-activated BFS shows higher mechanical strength than that of alkali-activated FA, and CO_2 curing can improve the mechanical strength.

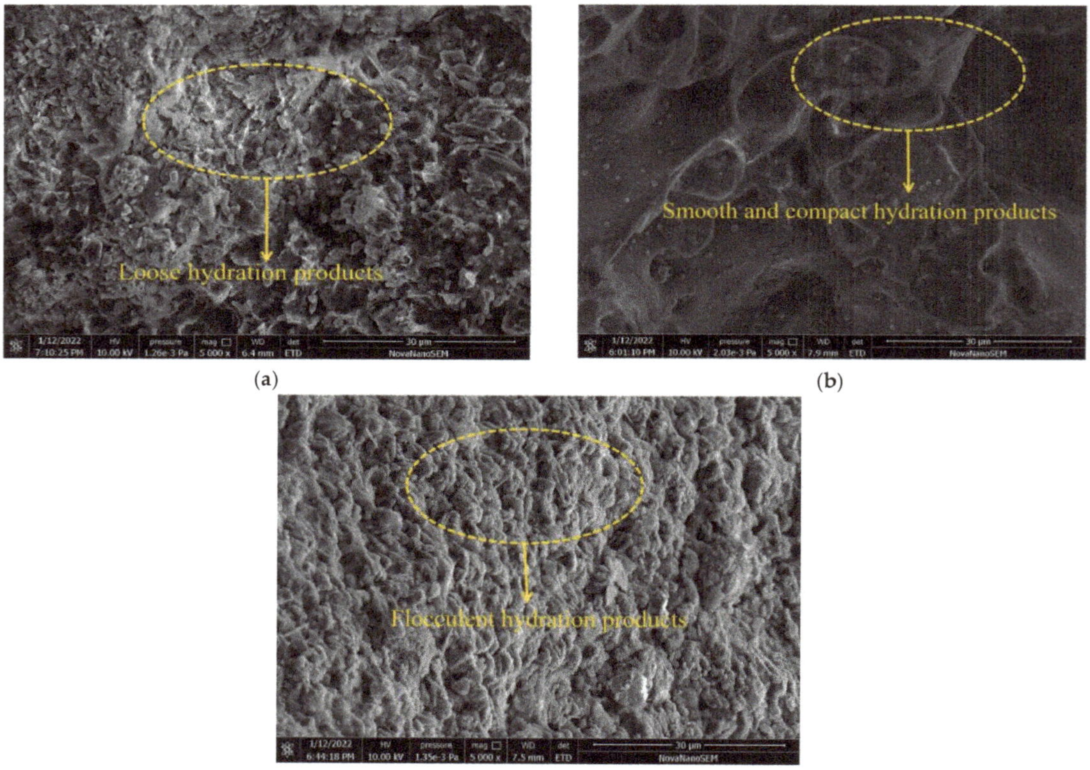

Figure 9. The SEM photos of alkali-activated mineral admixtures. (**a**) alkali-activated FA; (**b**) alkali-activated BFS; (**c**) CO_2-cured alkali-activated BFS.

4. Conclusions

In this paper, the effect of CO_2 curing on the mechanical properties of alkali-activated compound mineral admixtures is investigated. The conclusions are obtained as follows.

(1) The mechanical strengths of alkali-activated compound mineral admixtures are decreased by the addition of FA. Moreover, CO_2 curing is effective in improving the mechanical strength. The addition of FA can decrease the flexural and compressive strengths of alkali-activated compound mineral admixtures by 32.6%–47.6% and 35.4%–42.3%, respectively. Meanwhile, CO_2 curing can improve the flexural and compressive strengths of alkali-activated compound mineral admixtures by 16.7%–26.5%

and 15.4%–23.1%, respectively. When, the BFS content is higher, the improving effect of CO_2 curing is higher.

(2) The addition of BFS and CO_2 curing demonstrate an increasing effect on the drying shrinkage rate of alkali-activated compound mineral admixtures. Due to the addition of BFS, the drying shrinkage rate is decreased by 23.5%–32.1%, while, with CO_2 curing, the drying shrinkage rate is increased by 36.5%–63.1%.

(3) The chloride ion impermeability and resistance of NaCl freeze–thaw cycles of alkali-activated compound mineral admixtures are improved by the increasing dosages of BFS and CO_2 curing.

(4) CO_2 curing is able to increase the production of $CaCO_3$, while when the dosage of BFS is higher, the $CaCO_3$ product is higher. The addition of BFS can improve the compactness of the hydration products and CO_2 curing can increase the amount of rough hydration products.

Author Contributions: Conceptualization, J.Y.; Methodology, L.P.; Formal analysis, J.Y.; Investigation, L.P. and X.J.; Resources, J.Y. and H.W.; Data curation, L.P.; Writing—original draft, L.P.; Writing—review & editing, H.W.; Project administration, X.J.; Funding acquisition, H.W. All authors have read and agreed to the published version of the manuscript.

Funding: Central government guided local science and technology development fund project (Development and demonstration of key technologies for coordinated regulation of heat storage and heating) and self funded project of Hebei Provincial Department of Education (Grant No. ZC2022023).

Institutional Review Board Statement: Not applicable.

Informed Consent Statement: Not applicable.

Data Availability Statement: Data sharing is not applicable to this article.

Conflicts of Interest: The authors declare no conflict of interest.

References

1. Li, Y.; Li, X.; Tan, Y. Effect of aging on fatigue performance of cement emulsified asphalt repair material. *Constr. Build. Mater.* **2021**, *292*, 123417. [CrossRef]
2. Shah, J. Laboratory characterization of controlled low-strength material and its application to construction of flexible pipe drainage system. *Health Manpow. Manag.* **2000**, *19*, 30–59.
3. Haselbach, L.; Valavala, S.; Montes, F. Permeability predictions for sand clogged Portland cement pervious concrete pavement systems. *J. Environ. Manag.* **2005**, *81*, 42–49. [CrossRef] [PubMed]
4. Wei, Y.; Gao, X.; Zhang, Q. Evaluating performance of concrete pavement joint repair using different materials to reduce reflective cracking in asphalt concrete overlay. *Road Mater. Pavement Des.* **2014**, *15*, 966–976. [CrossRef]
5. Huang, G.; Wang, H.; Shi, F. Coupling Effect of Salt Freeze-Thaw Cycles and Carbonation on the Mechanical Performance of Quick Hardening Sulphoaluminate Cement-Based Reactive Powder Concrete with Basalt Fibers. *Coatings* **2021**, *11*, 1142. [CrossRef]
6. Ju, Y.; Zhao, J.; Wang, D.; Song, Y. Experimental study on flexural behaviour of reinforced reactive powder concrete pole. *Constr. Build. Mater.* **2021**, *312*, 125399. [CrossRef]
7. Liu, M.; Tan, H.; He, X. Effects of nano-SiO_2 on early strength and microstructure of steam-cured high volume flfly ash cement system. *Constr. Build. Mater.* **2019**, *194*, 350–359. [CrossRef]
8. Zhang, B.; Tan, H.; Shen, W.; Xu, G.; Ma, B.; Ji, X. Nano-silica and silica fume modified cement mortar used as Surface Protection Material to enhance the impermeability. *Cem. Concr. Compos.* **2018**, *92*, 7–17. [CrossRef]
9. Liska, M.; Jin, F.; Yi, Y.; Al-Tabbaa, A. Mechanism of reactive magnesia-ground granulated blastfurnace slag (GGBS) soil stabilization. *Can. Geotech. J.* **2016**, *53*, 773–782.
10. Celikten, S.; Sarldemir, M.; Deneme, I.O. Mechanical and microstructural properties of alkali-activated slag and slag+ fly ash mortars exposed to high temperature. *Constr. Build. Mater.* **2019**, *217*, 50–61. [CrossRef]
11. Li, S.; Feng, Y.; Yang, J.; Cerny, R. Expansion mechanism and properties of magnesium oxide expansive hydraulic cement for engineering applications. *Adv. Mater. Sci. Eng.* **2021**, *2021*, 5542072. [CrossRef]
12. Li, S.; Yang, J.; Zhang, P. Hydration and hardening properties of reactive magnesia and Portland cement composite. *Constr. Build. Mater.* **2022**, *327*, 126779. [CrossRef]
13. Ozata, S.; Akturk, B.; Yuzer, N. Utilization of waste Cappadocia earth as a natural pozzolan in alkali activation: A parametric study. *Constr. Build. Mater.* **2022**, *329*, 127192. [CrossRef]

14. Marple, M.; Koroglu, B.; Morrison, K.; Crowhurst, J.; Balachandra, A.; Soroushian, P.; Mason, H.E. Accelerated carbonation and structural transformation of blast furnace slag by mechanochemical alkali-activation. *Cem. Concr. Res.* **2022**, *156*, 106760. [CrossRef]
15. Matějková, P.; Matějka, V.; Sabovčík, T.; Gryžbon, L.; Vlček, J. Alkali Activation of Ground Granulated Blast Furnace Slag and Low Calcium Fly Ash Using "One-Part" Approach. *J. Sustain. Metall.* **2022**, *8*, 511–521. [CrossRef]
16. Supriya, J.; Raut, A. Performance parameter analysis of magnesia based cement products-a review. *IOP Conf. Ser. Mater. Sci. Eng.* **2021**, *1197*, 012078. [CrossRef]
17. Glukhovsky, V.D.; Rostovskaja, G.S.; Rumyna, G.V. High strength slag—Alkaline cements. In Proceedings of the 7th International Congress on the Chemistry of Cement, Paris, France, September 1980.
18. Alomayri, T.; Adesina, A.; Das, S. Influence of amorphous raw rice husk ash as precursor and curing condition on the performance of alkali activated concrete. *Case Stud. Constr. Materials* **2021**, *15*, e00777. [CrossRef]
19. Alrefaei, Y.; Dai, J. Effects of delayed addition of polycarboxylate ether on one-part alkali-activated fly ash/slag pastes: Adsorption, reaction kinetics, and rheology. *Constr. Build. Mater.* **2022**, *323*, 126611. [CrossRef]
20. Feng, S.; Zhu, J.; Wang, R.; Qu, Z.; Song, L.; Wang, H. The Influence of CaO and MgO on the Mechanical Properties of Alkali-Activated Blast Furnace Slag Powder. *Materials* **2022**, *15*, 6128. [CrossRef]
21. Xian, X.; Zhang, D.; Lin, H.; Shao, Y. Ambient pressure carbonation curing of reinforced concrete for CO_2 utilization and corrosion resistance. *J. CO_2 Util.* **2022**, *56*, 101861. [CrossRef]
22. Adesina, A. Recent advances in the concrete industry to reduce its carbon dioxide emissions. *Environ. Chall.* **2020**, *1*, 100004. [CrossRef]
23. Witoon, T.; Lapkeatseree, V.; Numpilai, T.; Cheng, C.; Limtrakul, J. CO_2 hydrogenation to light olefins over mixed Fe-Co-K-Al oxides catalysts prepared via precipitation and reduction methods. *Chem. Eng. J.* **2022**, *428*, 131389. [CrossRef]
24. Witoon, T.; Numpilai, T.; Nijpanich, S.; Chanlek, N.; Kidkhunthod, P.; Cheng, C.; Ng, K.; Vo, D.N.; Ittisanronnachai, S.; Wattanakit, C.; et al. Enhanced CO_2 hydrogenation to higher alcohols over K-Co promoted In_2O_3 catalysts. *Chem. Eng. J.* **2022**, *431*, 133211. [CrossRef]
25. AL-Ameeri, A.S.; ImranRafiq, M.; Tsioulou, O.; Rybdylova, O. Impact of climate change on the carbonation in concrete due to carbon dioxide ingress: Experimental investigation and modelling. *J. Build. Eng.* **2021**, *44*, 102594. [CrossRef]
26. Kamal, N.L.M.; Itam, Z.; Sivaganese, Y.; Beddu, S. Carbon dioxide sequestration in concrete and its effects on concrete compressive strength. *Mater. Today Proc.* **2020**, *31*, A18–A21. [CrossRef]
27. Zhu, J.; Qu, Z.; Liang, S.; Li, B.; Du, T.; Wang, H. The macro-scopic and microscopic properties of cement paste with carbon dioxide curing. *Materials* **2022**, *15*, 1578. [CrossRef]
28. Smirnov, V.G.; Manakov, A.Y.; Dyrdin, V.V.; Ismagilov, Z.R.; Mikhailova, E.S.; Rodionova, T.V. The formation of carbon dioxide hydrate from water sorbed by coals. *Fuel* **2018**, *228*, 123–131. [CrossRef]
29. Branston, J.; Das, S.; Kenno, S.; Taylor, C. Mechanical behaviour of basalt fifibre reinforced concrete. *Constr. Build. Mater.* **2016**, *124*, 878–886. [CrossRef]
30. Wei, B.; Cao, H.; Song, S. Tensile behavior contrast of basalt and glass fifibers after chemical treatment. *Mater. Des.* **2010**, *31*, 4244–4250. [CrossRef]
31. Shi, X.; Fay, L.; Peterson, M.; Yang, Z. Freeze-thaw damage and chemical change of a portland cement concrete in the presence of diluted deicers. *Mater. Struct.* **2010**, *43*, 933–946. [CrossRef]
32. Wang, H.; Gao, X.; Liu, J. Effects of salt freeze-thaw cycles and cyclic loading on the piezoresistive properties of carbon nanofibers mortar. *Constr. Build. Mater.* **2018**, *177*, 192–201. [CrossRef]
33. Yang, J.; Huang, J.; Su, Y.; He, X.; Tan, H.; Yang, W.; Strnadel, B. Eco-friendly treatment of low-calcium coal fly ash for high pozzolanic reactivity: A step towards waste utilization in sustainable building material. *J. Clean. Prod.* **2019**, *238*, 117962. [CrossRef]
34. Chao, L.; Duy, H.; Mitiku, D.; Vu, A. Improving the strength and engineering properties of alkali-activated slag–rice husk ash paste at the early ages with addition of various magnesium oxide content. *Int. J. Struct. Civ. Eng. Res.* **2019**, *8*, 210–214.
35. Qin, L.; Gao, X. Properties of coal gangue-Portland cement mixture with carbonation. *Fuel* **2019**, *245*, 1–12. [CrossRef]
36. Faridmehr, I.; Bedon, C.; Fahim, G.; Nikoo, M.; Baghban, M. Assessment of Mechanical Properties and Structural Morphology of Alkali-Activated Mortars with Industrial Waste Materials. *Sustainability* **2021**, *13*, 2062. [CrossRef]
37. Ashraf, W.; Olek, J. Carbonation behavior of hydraulic and non-hydraulic calcium silicates: Potential of utilizing low-lime calcium silicates in cement-based materials. *J. Mater. Sci.* **2016**, *51*, 6173–6191. [CrossRef]

coatings

Review

Review on Durability Deterioration and Mitigation of Concrete Structures

Jiwei Ma [1,2], Qiuwei Yang [1,2,*], Xi Peng [1,2,*] and Kangshuo Xia [3]

[1] School of Civil and Transportation Engineering, Ningbo University of Technology, Ningbo 315211, China; jiwei_ma@nbut.edu.cn
[2] Zhejiang Key Laboratory of Intelligent Construction and Operation & Maintenance for Deep-Sea Foundations, Ningbo University of Technology, Ningbo 315211, China
[3] Key Laboratory of New Technology for Construction of Cities in Mountain Area, School of Civil Engineering, Chongqing University, Chongqing 400045, China; xiakangshuo@cqu.edu.cn
* Correspondence: yangqiuwei@nbut.edu.cn (Q.Y.); pengxi@nbut.edu.cn (X.P.)

Abstract

Concrete bridges, as a vital component of modern transportation infrastructure, have their structural durability directly tied to safety and service life. In recent years, with the aging of bridge structures and increasingly complex environmental conditions, various durability-related deteriorations have become more prominent, significantly affecting structural performance and maintenance costs. This paper presents a systematic analysis of concrete carbonation as a key chemical process and its impact on durability-related pathologies. Particular attention is given to the formation mechanisms and influencing factors of critical deterioration modes such as cracking, reinforcement corrosion, and freeze–thaw damage. A multi-level prevention and mitigation strategy is proposed, encompassing optimized structural material design, strict construction quality control, and effective maintenance and repair techniques. The study concludes that the durability issues of concrete bridge structures exhibit a strong multi-factor coupling effect and proposes a core durability assurance framework. Finally, the paper briefly outlines emerging trends in intelligent monitoring and digital operation and maintenance, offering insights for future durability management of bridges.

Keywords: concrete bridges; durability deterioration; carbonation; cracking; reinforcement corrosion; freeze–thaw damage; mitigation strategies

1. Introduction

In the field of transportation engineering, the construction of concrete bridge structures holds a position of paramount importance, serving as an indispensable component of transportation infrastructure [1]. The structural durability of concrete bridges refers to their ability to maintain safety, functionality, and aesthetic integrity throughout their designed service life. This endurance persists despite exposure to environmental erosion, operational stresses, and material-related degradation—without necessitating extensive reinforcement or reconstruction. Neglecting the issue of durability can have severe consequences. Over time, many bridges may experience progressive structural deterioration due to insufficient durability, ultimately requiring premature demolition and replacement. More alarmingly, numerous bridges demand frequent maintenance and reinforcement well before reaching their intended service life, resulting in considerable financial costs and placing a heavy burden on public finances. These interventions also disrupt traffic flow, leading to widespread socio-economic consequences [2,3]. In-depth research into the durability of concrete bridge structures is therefore of strategic importance. Such studies are essential for the early identification and effective mitigation of structural damage, ensuring safe and stable bridge operation, prolonging service life, and reducing long-term maintenance costs. Moreover, they offer a robust scientific foundation for the planning, construction, and post-construction management of new bridges. This not only enhances durability design standards and drives technological innovation in bridge construction but also reinforces the reliability of infrastructure that supports sustainable economic and social development [4–8].

In the 20th century, both academia and the engineering community conducted comprehensive and in-depth research on the durability-related deterioration of concrete bridges and corresponding mitigation strategies, achieving a series of landmark breakthroughs. Researchers meticulously investigated the underlying causes of key durability issues such as reinforcement corrosion, concrete cracking, spalling damage, and alkali–aggregate reactions [9–13]. Building on this solid foundation, systematic strategies were proposed across multiple critical dimensions, including design concept optimization, the development of advanced materials, quality control during construction, and long-term maintenance management. These strategies specifically encompassed the enhancement of durability-oriented structural design principles, the development and application of high-performance concrete and corrosion-resistant reinforcement, rigorous quality assurance throughout construction processes, the establishment of scientific inspection and evaluation systems, and the adoption of efficient repair technologies for already deteriorated structures. The fundamental goal of these innovative initiatives was to significantly reduce the incidence of durability-related deterioration and ensure the long-term safety, stability, and reliability of concrete bridges. In doing so, they laid a solid foundation for the sustainable development of bridge engineering and provided valuable theoretical and practical guidance for future projects of a similar nature. However, researchers also faced numerous challenges in their efforts to explore durability deterioration and mitigation strategies. The underlying mechanisms of deterioration are highly complex, often interdisciplinary in nature, and influenced by multiple environmental factors, making them difficult to analyze comprehensively. Long-term data collection posed another major difficulty, as limitations in monitoring technologies restricted the ability to track internal structural conditions and environmental interactions over extended periods. In addition, discrepancies between laboratory simulations and real bridge environments—due to the inability to replicate complex environmental conditions and size effects—often led to deviations in experimental results. Furthermore, evaluating the effectiveness of mitigation strategies remains challenging, requiring multi-factor analyses that consider long-term performance as well as economic feasibility. These obstacles represented significant barriers to research at the time and continue to pose challenges to the advancement of durability studies in concrete bridge engineering [14,15].

In contemporary research on the durability of concrete bridge structures, there has been a marked advancement in both depth and precision. The application of cutting-edge microscopic analysis techniques and the development of sophisticated multi-field coupled models have significantly improved our understanding of the fundamental mechanisms behind deterioration processes. The integration of non-destructive testing (NDT) methods and long-term structural health monitoring systems has greatly enhanced the efficiency and accuracy of condition assessments, enabling reliable real-time evaluations of bridge performance. High-performance materials and a wide array of innovative repair technologies have been extensively applied in practice, leading to substantial improvements in structural durability. Meanwhile, the integration of durability-based design concepts with full life-cycle design approaches has become a core component of modern bridge engineering codes and standards. However, a range of formidable challenges remains. These include the complex interactions of multiple deterioration factors under harsh environmental conditions, the impacts of extreme weather events, the limited accuracy of long-term performance prediction models, insufficient accumulation of long-term monitoring data, uncertainties regarding the long-term aging behavior of new materials, and the urgent need to optimize maintenance decision-making and ensure the applicability of maintenance technologies. Together, these factors continue to impede progress in the field of bridge durability research [16,17]. To address these challenges, a comprehensive summary and analysis of deterioration mechanisms and corresponding mitigation strategies is essential. Such efforts can facilitate the development of more scientific evaluation standards [18], support the formulation of comprehensive maintenance strategies and technical solutions, and enable more adaptive responses to the specific demands of different bridge types and regional conditions. Ultimately, these advances will provide a solid foundation for ensuring the long-term safe and reliable service performance of concrete bridge structures.

2. Durability Degradation Mechanisms of Concrete Structures

2.1. Concrete Carbonation

The chemical process of concrete carbonation involves complex mass transfer and chemical equilibrium among multiphase media at the microscopic level. During cement hydration, a large amount of calcium hydroxide [$Ca(OH)_2$] is produced, forming a saturated solution within the pore structure that maintains a highly alkaline environment inside the concrete (pH 12.5–13.5). This alkaline system protects the embedded steel reinforcement through the formation of a passive film, a dense oxide layer composed primarily of Fe_2O_3 and Fe_3O_4 with a thickness of approximately 5–10 nm. This passive film enhances corrosion resistance by inhibiting the transport of oxygen and electrons. Studies have shown that the passive film remains stable as long as the pore solution pH stays above 11.5; however, once the pH drops below this critical threshold due to carbonation, the passive film dissolves, leading to an increased risk of steel reinforcement corrosion. This deterioration mechanism is especially pronounced in atmospheric environments, where CO_2 diffuses through the pore structure into the concrete and reacts with $Ca(OH)_2$ to form calcium carbonate ($CaCO_3$). This reaction reduces the alkalinity of the pore solution and alters the microstructure of the concrete [19,20], as described by the carbonation reaction Equations (1) and (2). Therefore, an in-depth understanding of the chemical kinetics and mass transfer processes during carbonation is essential for predicting the long-term service performance of concrete structures.

$$Ca(OH)_2 + CO_2 = CaCO_3 + H_2O \tag{1}$$

$$3CaO \cdot 2SiO_2 \cdot 3H_2O + 3CO_2 = 3CaCO_3 + 2SiO_2 + 3H_2O \tag{2}$$

The carbonation behavior of concrete is influenced by a combination of environmental, material, and mechanical factors, among which environmental conditions play a predominant role. Climatic variables such as temperature and relative humidity have significant effects on the carbonation process. Elevated temperatures accelerate both the diffusion of carbon dioxide (CO_2) and the rate of the chemical reactions involved. Relative humidity levels between 50% and 75% are considered optimal for carbonation, as CO_2 dissolves and diffuses most efficiently in the pore solution within this range, resulting in the highest carbonation rates. Conversely, excessively high or low humidity levels suppress the carbonation reaction. Air quality also exerts a notable influence. Acidic gases such as sulfur dioxide (SO_2) and nitrogen oxides (NO_x) in the atmosphere readily dissolve in the moisture retained on the surface and within the pore structure of concrete. Through a series of complex chemical reactions, strong acids such as nitric acid and sulfuric acid are ultimately formed. The accumulation of these acidic substances inside the concrete not only alters its chemical properties but also may induce material corrosion and structural deterioration, severely compromising the durability and service life of concrete structures such as bridges. Groundwater level fluctuations further impact carbonation dynamics. When the water table is high, concrete pores remain saturated for extended periods, hindering CO_2 diffusion and slowing carbonation. As the water level recedes, pore saturation decreases, enhancing gas exchange and allowing CO_2 to diffuse more rapidly, significantly increasing the carbonation rate. In terms of material properties, the concrete mix design and compressive strength are critical factors. The water-to-binder (w/b) ratio directly affects porosity and density: a lower w/b ratio leads to a denser microstructure, reducing CO_2 penetration. The type and dosage of mineral admixtures also influence the alkalinity and microstructure of concrete. For example, incorporating fly ash or ground granulated blast-furnace slag can enhance workability but may reduce the early-age alkalinity reserve, thereby increasing carbonation susceptibility. Concrete strength is closely related to carbonation resistance. High-strength concrete typically exhibits a denser microstructure and lower porosity, resulting in significantly slower carbonation under identical environmental conditions compared to low-strength concrete. It is important to note that these influencing factors do not operate independently; rather, they interact in complex ways. A comprehensive understanding of the synergistic effects among these factors is essential for accurately predicting and effectively controlling the carbonation process, thereby ensuring the long-term durability of concrete structures [21–25]. The specific impacts of carbonation on the durability of bridge structures are summarized in Table 1.

Table 1. Factors influencing concrete carbonation.

Category	Specific Factors	Influence Pattern
Material Properties	Water–cement ratio, cement content, aggregate size, admixtures	Carbonation rate increases with higher water–cement ratio and larger aggregate size; it decreases with higher cement content. Water-reducing agents and air-entraining agents tend to reduce carbonation rate.
Environmental Conditions	Relative humidity, temperature, CO_2 concentration	Carbonation proceeds most rapidly at a relative humidity of 70%–80%. Higher temperatures accelerate carbonation. The carbonation rate increases proportionally with CO_2 concentration.

The impact of concrete carbonation on the durability of bridge structures can be quantitatively assessed using the carbonation depth as a key indicator. For the calculation of carbonation depth, the Alexeyev carbonation model is commonly employed. Based on references [26,27], the mathematical expression of this model is given as follows:

$$X_c = \sqrt{\frac{2DC}{m}}\sqrt{t} \tag{3}$$

Based on Equation (3), Reference [28] replaces the carbon dioxide uptake per unit volume of concrete with the concentrations of $Ca(OH)_2$, CSH, C_3S, and C_2S. Building on this substitution, the study further improved and expanded the mathematical model, such as Equation (4), to better represent the carbonation process in the concrete matrix.

$$x = \sqrt{\frac{2DC}{C_{Ca(OH)_2} + 3C_{CSH} + 3C_{C3S} + 2C_{C2S}}\sqrt{t}} \quad (4)$$

Reference [29] improves upon Fick's law by incorporating several influencing factors, including the 28-day compressive strength of concrete, clinker content, carbon dioxide concentration, and the equivalent water absorption capacity of the aggregate mixture. On this basis, the following mathematical model is proposed:

$$x = K_C\sqrt{t} = (72.470 - 0.772f_c - 0.117C_0 + 4.617c + 1.594EWA) \cdot \sqrt{t} \quad (5)$$

Reference [30] primarily focuses on the effect of the water–cement ratio on the carbonation behavior of concrete. The corresponding mathematical expressions are presented as follows:

$$x = -0.56213 - \frac{8.792}{\sqrt{t}} + 17.8372(w/c) \quad (6)$$

The risk of reinforcement corrosion induced by concrete carbonation, along with the resulting deterioration in structural performance, poses a serious threat to the long-term durability of infrastructure. In recent years, numerous researchers have conducted in-depth investigations into carbonation-related issues. Studies have shown that carbonation-induced steel corrosion and degradation of concrete properties can cause the actual load-bearing capacity of bridge structures to fall below their design capacity over time. Under service loads such as vehicular traffic and wind, these weakened structures are more susceptible to deformation, crack propagation, and other failure mechanisms. For example, Reference [31] presents a case study of a Yangtze River bridge where the concrete has been exposed to atmospheric conditions for over 50 years without effective protective coatings. The carbonation process was significantly accelerated, and prolonged rainwater erosion further exacerbated surface degradation, posing serious safety risks. Moreover, carbonation-induced deterioration is often latent and progressive, making early detection difficult. Once damage becomes severe, repair becomes both technically challenging and costly and may even require traffic closures, resulting in substantial economic losses and social disruption. In coastal regions, carbonation compromises the protective properties of concrete, facilitating the ingress of chloride ions and accelerating steel corrosion. In cold climates, carbonation-induced cracking provides pathways for moisture ingress; during freeze–thaw cycles, this leads to spalling and fragmentation of concrete, severely impairing the bridge's functionality and structural safety [32–36]. Table 2 summarizes the concrete carbonation deterioration mechanism.

Table 2. Summary of concrete carbonation deterioration mechanism.

Reference	Research Content/Model	Main Factors	Mechanism/Pattern	Impact on Structural Durability
[19,20]	The chemical reaction process of carbonation (Equations (1) and (2)), emphasizing the loss of alkalinity leading to the dissolution of the steel passive film	Micro-scale chemical reactions, CO_2 diffusion	pH > 11.5: passive film remains stable; pH < 11.5: passive film deteriorates, increasing steel corrosion susceptibility	Higher risk of reinforcement corrosion and reduced structural durability
[21–25]	Influence of environmental factors on carbonation	Temperature, relative humidity, air pollutants, groundwater level fluctuation	Optimal RH: 50%–75%; higher temperature accelerates reactions; SO_2 and NO_x produce strong acids, corroding concrete; groundwater fluctuations alter pore saturation, affecting CO_2 diffusion	Accelerated carbonation, changes in chemical properties, and potential structural damage

Table 2. *Cont.*

Reference	Research Content/Model	Main Factors	Mechanism/Pattern	Impact on Structural Durability
[21–25]	Influence of material factors on carbonation	Water-to-binder ratio, mineral admixtures, strength grade	Higher w/b ratio → higher porosity → faster carbonation; mineral admixtures may reduce early-age alkalinity reserve; high-strength concrete → lower porosity → slower carbonation	Higher carbonation risk in concretes with high porosity and low alkalinity reserve
[26,27]	Alexeyev carbonation model (Equation (3))	Carbonation depth calculation	Predicts carbonation depth based on CO_2 uptake and time	Used for service life and durability prediction
[28]	Improved Alexeyev model (Equation (4))	Concentrations of $Ca(OH)_2$, CSH, C_3S, and C_2S	Replaces CO_2 uptake with concentrations of hydration products to better represent the carbonation process	Improves model accuracy and facilitates material optimization
[29]	Carbonation model based on improved Fick's law (Equation (5))	28-day compressive strength, clinker content, CO_2 concentration, aggregate water absorption capacity	Incorporates multiple influencing factors for more realistic carbonation depth estimation	More accurate reflection of service environment effects
[30]	Influence of water–cement ratio on carbonation (Equation (6))	Water–cement ratio	Higher w/c ratio → higher porosity → faster carbonation	Provides guidance for mix design optimization
[31]	Long-term carbonation case study (Yangtze River Bridge)	Atmospheric exposure, rainwater erosion	Over 50 years of exposure without protection → accelerated carbonation → surface degradation	Safety risks and high maintenance costs
[32–36]	Coupled effects of carbonation and other environmental factors	Chloride ingress, freeze–thaw cycles	Carbonation reduces protective capacity → facilitates chloride penetration; cracks + freeze–thaw cause spalling	Reduced load-bearing capacity and impaired functionality

2.2. Concrete Crack

Cracking is one of the most common durability-related defects in concrete bridges during long-term service, and it can be classified into various types based on their causes. Load-induced cracks result from mechanical actions, primarily static and dynamic loads. Cracks may develop during any phase of the bridge lifecycle, including investigation, design, construction, and operation. In the design stage, inaccuracies in load calculations, inappropriate modeling assumptions, or inadequate reinforcement design can cause discrepancies between actual and theoretical stress distributions. When the applied stress exceeds the tensile strength of concrete, cracking occurs. During construction, many factors can lead to cracking, with non-compliant construction practices being the most prominent. Improper placement of construction equipment, incorrect sequencing of construction steps, and alterations in the designed load-bearing mechanisms can all induce structural cracking. In the service stage, impacts from vehicles or ships, extreme weather events, and seismic activity are also significant contributors to cracking. Due to the varying magnitude and nature of applied loads, the distribution patterns of load-induced cracks also differ. Typically, such cracks appear in tensile zones, shear zones, and areas subject to strong vibrations [37–39]. Figures 1–4 present a summary of macroscopic images of concrete cracks as compiled by the author in reference [40].

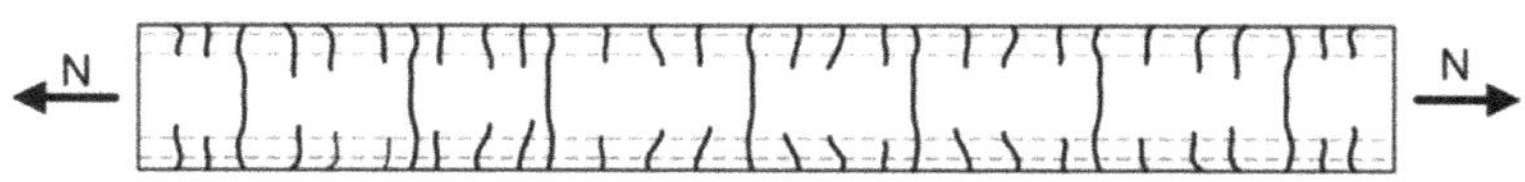

Figure 1. Cracks caused by tension; N—axial force [40].

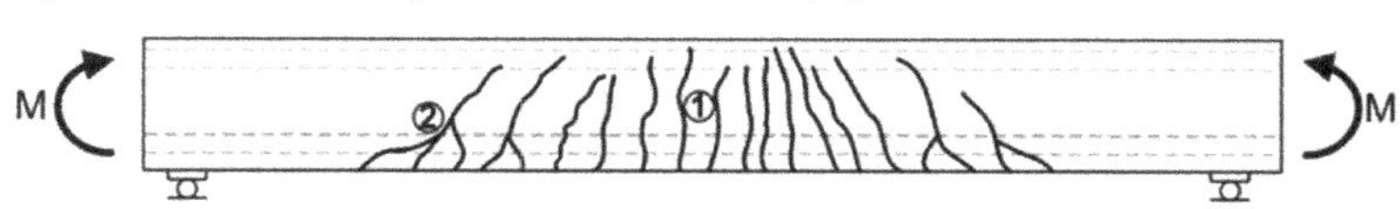

Figure 2. Flexural cracks; 1—single crack, 2—multiple cracks, M—bending moment [40].

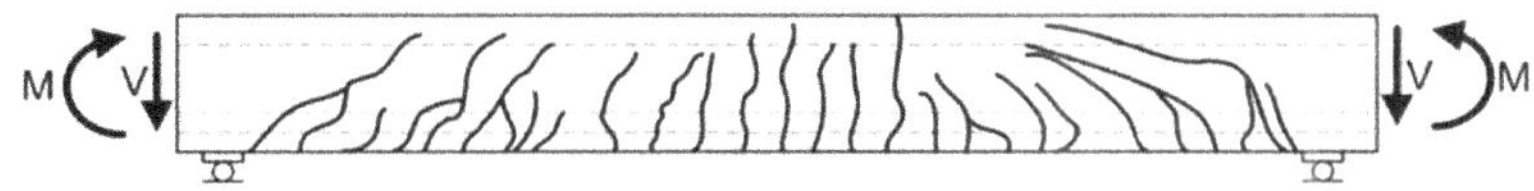

Figure 3. Shear-induced cracks; M—bending moment, V—shear force [40].

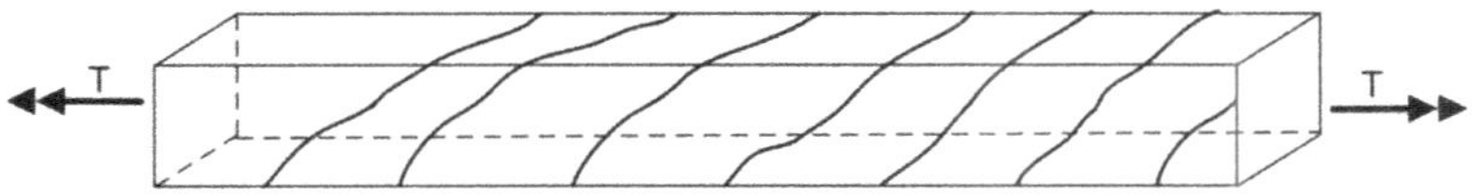

Figure 4. Torsion-induced cracks; T—torque [40].

Temperature variations are another major cause of cracking in concrete. Due to thermal expansion and contraction, concrete deforms with temperature changes. When exposed to significant temperature fluctuations in the external environment, differential temperatures can develop within the bridge structure. For instance, during hot summer periods, the surface of the concrete heats up faster than the interior. As the surface expands, the cooler inner concrete resists this movement, generating tensile stresses. Conversely, in cold winter conditions, the surface contracts more quickly than the inner layers, again inducing tensile stresses. When these thermally induced tensile stresses exceed the tensile strength of concrete, temperature-induced cracks can form [40–42]. In addition, the hydration of Portland cement releases a significant amount of heat. When there is a substantial temperature difference between the concrete and its surrounding environment, internal temperature gradients can develop, leading to non-uniform thermal expansion and contraction. This, in turn, may induce early-age thermal cracking. Such cracks are commonly observed in mass concrete structures or under construction conditions with sharp temperature fluctuations. If not properly controlled, they can adversely affect the overall durability of the structure.

Concrete shrinkage is also a key factor contributing to cracking [43–45]. During the hardening process, cement hydration consumes water, resulting in autogenous shrinkage. Simultaneously, water evaporation from the surface leads to drying shrinkage. If the shrinkage deformation is restrained—such as by abutments or piers—tensile stresses develop within the concrete, leading to shrinkage cracks.

Reinforcement corrosion is another significant cause of cracking. When steel bars embedded in concrete corrode, the resulting rust products expand. The volume of these corrosion products typically ranges from two to six times the original volume of the steel [46]. This expansion exerts radial pressure on the surrounding concrete. Once this pressure exceeds the tensile strength of the concrete, cracks propagate along the reinforcement, forming corrosion-induced cracks. The presence of such cracks facilitates the ingress of moisture, oxygen, and other aggressive agents, which in turn accelerates the corrosion process, creating a vicious cycle that severely undermines the durability of the bridge structure [47,48]. Table 3 summarizes the causes for concrete cracking.

Table 3. Summary of causes for concrete cracking.

Reference	Stage	Frost Damage	Plastic Stage Characteristics	Construction Activities	Physicochemical Factors	Structural Design Factors	Volume Change Factors
[40–42]	Before Concrete Hardening	Early-age frost damage	Plastic shrinkage cracking	Formwork displacement and deformation	—	—	—
		Surface scaling, map cracking	Plastic settlement cracking	Uneven foundation settlement	—	—	Autogenous volume shrinkage
[40,43]	After Concrete Hardening	Freeze–thaw cycles	—	—	Alkali–aggregate reaction	Design load and overloading	Drying shrinkage
		—	—	—	Reinforcement corrosion	Thermal stress	Thermal expansion and contraction
		—	—	—	—	Material fatigue	Creep deformation

2.3. Reinforcement Corrosion

Steel reinforcement corrosion primarily follows the principles of electrochemical corrosion. Within the concrete environment, due to the inherent heterogeneity of the steel material and the presence of electrolyte solutions in the concrete pores, potential differences tend to form on the steel surface. This results in the formation of anodic and cathodic regions. At the anodic site, iron undergoes oxidation, losing electrons to form ferrous ions (Fe^{2+}). These electrons are conducted through the steel to the cathodic site, where they react with dissolved oxygen in the pore water, resulting in the formation of hydroxide ions (OH^-). The ferrous ions then combine with hydroxide ions to form ferrous hydroxide [$Fe(OH)_2$], which is further oxidized in the presence of oxygen to form ferric hydroxide [$Fe(OH)_3$], commonly known as rust [49,50].

$$Fe \rightarrow Fe^{2+} + 2e^- \tag{7}$$

$$H_2O + 1/2O_2 + 2e^- \rightarrow 2OH^- \tag{8}$$

$$Fe^{2+} + OH^- \rightarrow Fe(OH)_2 \tag{9}$$

$$4Fe(OH)_2 + O_2 + 2H_2O \rightarrow 4Fe(OH)_3 \tag{10}$$

Equations (7)–(10) represent the primary chemical reactions involved in the corrosion of steel reinforcement:

Moisture, oxygen, and chloride ions are key environmental factors influencing the rate of steel reinforcement corrosion. Moisture acts as the medium for ion transport and is a fundamental prerequisite for electrochemical corrosion reactions. Oxygen, as the main reactant in the cathodic process, contributes to an increased corrosion rate with rising concentrations. Chloride ions are considered one of the most critical and aggressive agents in initiating depassivation and accelerating steel corrosion. They can disrupt the passive film on the steel surface and form localized corrosion cells within the concrete matrix, thereby facilitating electron migration and significantly increasing the corrosion rate. Once corrosion occurs, the steel cross-section is reduced and its load-carrying capacity is diminished. Additionally, corrosion products lead to the loss of bond strength between steel and concrete, ultimately undermining the structural integrity and durability of reinforced concrete bridges [51,52].

The impact of chloride ions on steel reinforcement corrosion can be assessed from multiple perspectives, including the time to corrosion initiation, defined as the moment when the chloride concentration at the steel surface reaches a critical threshold; the progressive reduction in steel bar diameter over time; and the degradation of mechanical properties of the reinforcement under chloride-induced attack. Based on Fick's diffusion theory and other relevant models, researchers have developed a range of mathematical formulations to quantitatively describe these processes. Drawing from references [53,54], the following mathematical models have been established:

Reference [54] proposed that if the critical chloride ion concentration on the surface of reinforcement at the initiation of corrosion is defined as C_{cr}, then the time of corrosion initiation can be derived based on the diffusion behavior of chloride ions in concrete. This is typically modeled using Fick's Second Law of Diffusion, as expressed in Equation (11):

$$T_{cor} = \left\{ \frac{d_c^2}{4k_e k_t k_c D_0 (t_0)^n} \left[erf^{-1}\left(\frac{C_s - C_{cr}}{C_s} \right) \right]^{-2} \right\}^{\frac{1}{1-n}} \tag{11}$$

In Equation (11), T_{cor} represents the time to corrosion initiation of the reinforcing steel (d), while d_c denotes the thickness of the concrete cover (mm). The coefficients k_e, k_t, and

k_c are correction factors accounting for environmental conditions, testing methods, and curing conditions, respectively. D_0 is the apparent diffusion coefficient of chloride ions in concrete (m^2/s), n is the time-dependent decay exponent reflecting the reduction in diffusion over time, and C_s is the chloride ion concentration at the surface of the reinforced concrete structure (kg/m^3).

Reference [55] pointed out that once reinforcement begins to be attacked by chloride ions and corrosion occurs, pitting will gradually appear on the steel surface, leading to a reduction in diameter and cross-sectional area, thereby causing a progressive decline in the structural resistance provided by the reinforcement. Assuming uniform corrosion of the steel, this process can be described by Equation (12):

$$d(t) = \begin{cases} d_0, & T \leq T_{cor} \\ d_0 - 2\lambda(T - T_{cor}), & T_{cor} < T \leq [T_{cor} + d_0/(2\lambda)] \\ 0, & T > [T_{cor} + d_0/(2\lambda)] \end{cases} \tag{12}$$

In Equation (12), $d(t)$ is the time-dependent reduction in the diameter of reinforcing steel (mm) and λ is the steel corrosion rate.

Reference [56] proposed that, based on the corrosion mechanism of reinforcement and the associated corrosion rate theory, and by considering the degradation of reinforcement material properties under chloride-induced corrosion conditions, a corresponding degradation model can be established, as shown in Equation (13):

$$\begin{cases} f_y = (1 - \beta_y \rho_s)\, f_{y0} \\ f_u = (1 - \beta_u \rho_s)\, f_{u0} \end{cases} \tag{13}$$

In Equation (13), f_y is the yield strength of reinforcing steel after corrosion (MPa), f_{y0} is the initial (uncorroded) theoretical yield strength of steel (MPa), β_y is the reduction coefficient of yield strength due to chloride-induced corrosion, ρ_s is the corrosion rate of reinforcing steel, f_u is the ultimate strength of reinforcing steel after corrosion (MPa), f_{u0} is the initial (uncorroded) theoretical ultimate strength of steel (MPa), and β_u is the reduction coefficient of ultimate strength due to chloride-induced corrosion.

As shown in the following table, Table 4 summarizes the steel reinforcement corrosion deterioration mechanism.

Table 4. Summary of steel reinforcement corrosion deterioration mechanism.

Reference	Research Content/Model Formula	Main Influencing Factors	Mechanism/Pattern	Impact on Structural Durability
[49,50]	The electrochemical corrosion principles of steel in concrete and the main reaction (Equations (7)–(10))	Electrolyte solution in concrete pores, steel material heterogeneity, moisture, oxygen	Formation of anodic and cathodic regions on the steel surface, with electron transfer triggering corrosion reactions	Reduction of steel cross-section, loss of bond strength between steel and concrete, decreased structural load-bearing capacity, and durability
[51,52]	Destructive effects of chloride ions on steel reinforcement	Chloride ions, moisture, oxygen	Chloride ions disrupt the passive film, form localized corrosion cells, and accelerate electron migration and corrosion rate	Accelerated loss of steel cross-section and reduced bond performance, leading to premature structural deterioration
[54]	Prediction model for time to corrosion initiation based on Fick's Second Law (Equation (11))	Chloride ion diffusion coefficient, concrete cover thickness, environmental correction factors, surface chloride concentration	Calculates time to corrosion initiation when chloride concentration at the steel surface reaches the critical threshold	Used for service life assessment and durability design
[55]	Corrosion model describing uniform reduction in steel diameter over time (Equation (12))	Steel corrosion rate, time	Assumes uniform corrosion; steel diameter decreases linearly with time	Reduction in cross-sectional area, progressive loss of load-bearing capacity
[56]	Material property degradation model of steel reinforcement under chloride-induced corrosion (Equation (13))	Yield strength, ultimate strength, corrosion rate, strength reduction coefficients	Corrosion leads to reduction in yield and ultimate strength; the model predicts deterioration based on corrosion rate	Decrease in steel load-bearing capacity and reduction in structural safety margin

2.4. Freeze–Thaw Damage

In cold climate conditions, concrete bridge structures often suffer from durability degradation due to freeze–thaw cycles. When the ambient temperature drops below 0 °C, the free water within the concrete pores begins to freeze. Because water expands by approximately 9% upon freezing, significant tensile stresses are exerted on the pore walls. Due to the heterogeneous distribution of pores in concrete, water in larger capillary pores freezes first, forming so-called "ice lenses". As the temperature continues to fall, water in smaller pores gradually freezes, further expanding the ice volume. This expansion drives unfrozen water toward the freezing front, resulting in the accumulation of hydrostatic pressure within the pore structure. When the internal freezing pressure exceeds the tensile strength of the concrete, microcracks may form within the material. As the temperature rises and the ice melts, water is drawn back into the cracks and pores by capillary action. Under repeated freeze–thaw cycles, these microcracks progressively propagate and interconnect, causing ongoing internal damage to the concrete structure [32,57,58]. Studies have shown that the moisture content of concrete is one of the critical factors influencing its freeze–thaw resistance. The higher the moisture content, the more ice forms during freezing, leading to increased internal stresses and a significantly higher risk of freeze–thaw damage. Moreover, the number of freeze–thaw cycles has a cumulative effect on the extent of deterioration. As the number of cycles increases, concrete strength declines markedly, often accompanied by surface scaling, cracking, and spalling, all of which can severely compromise the structural safety and service life of bridge structures [59–61].

Freeze–thaw damage in concrete occurs only when two fundamental conditions are met: the material is in a near-saturated state, and it undergoes repeated cycles of freezing and thawing. Salt freeze damage, a particular form of freeze–thaw deterioration, shares certain mechanistic similarities with ordinary freeze–thaw processes but also exhibits notable differences. Mechanistically, salt freeze damage still relies on the freezing of pore water. However, in saline environments, the increased concentration of salt ions in the pore solution raises the osmotic pressure, which partially suppresses the freezing process. Despite this inhibition, salts continuously precipitate and crystallize within the pores during moisture evaporation, generating substantial crystallization pressure. This crystallization pressure exacerbates internal structural damage in the concrete. Thus, salt freeze damage essentially results from the combined effects of freeze–thaw action and salt erosion, and its destructive impact is often significantly greater than that of freeze–thaw alone.

Currently, many bridge structures are located not only in cold climates but also in high-salinity environments such as mining areas, salt lakes, or coastal regions, which further aggravates concrete durability degradation [62,63]. Xu et al. [64] demonstrated that the damage to concrete after multiple freeze–thaw cycles exhibits a pronounced cumulative effect. With increasing freeze–thaw cycles, the relative dynamic elastic modulus of concrete decreases continuously. Notably, C30 concrete shows a more rapid deterioration in freeze–thaw resistance compared to C40 and C50 concretes, highlighting the significant influence of strength grade on freeze–thaw performance. Related studies also indicate that freeze–thaw cycling substantially weakens the mechanical properties of concrete. This deterioration process involves not only physical aspects, such as pore expansion and crack formation, but also chemical degradation of cement hydration products and ionic migration, making the mechanism complex and slowly progressive.

In practice, various factors affecting the durability of concrete bridges often interact and exacerbate each other. The most common manifestation of freeze–thaw damage is crack propagation. These cracks not only compromise the overall structural integrity of the concrete but also create pathways for external corrosive agents (e.g., chlorides) to penetrate, accelerating reinforcement corrosion and significantly reducing the load-bearing capacity and service life of structural elements. Under prolonged freeze–thaw exposure, the rate of concrete deterioration accelerates markedly. In real service conditions, the lower parts of concrete bridges are particularly susceptible to freeze–thaw damage. Components such as piers and pile foundations are often located within the freezing zone influenced by the interaction of surface water and groundwater. Subjected to cyclical temperature variations, these areas frequently experience freezing and thawing cycles, making them high-risk zones for freeze–thaw damage. Inspection data reveal that, after approximately 10 years in service, bridges in severe cold regions commonly exhibit varying degrees of freeze–thaw deterioration in their lower structures. Typical damage includes crack propagation, surface scaling, and exposed corroded reinforcement [65–67]. These deterioration phenomena reduce the effective load-bearing cross-section of the elements, leading to decreased structural capacity and safety risks in mild cases and potentially causing component instability or failure in severe cases, thereby posing a serious threat to the overall structural safety and service life of the bridge. Table 5 summarizes the deterioration mechanism of freeze-thaw damage.

Table 5. Summary of freeze–thaw damage deterioration mechanism.

Reference	Research Content/Model Formula	Main Influencing Factors	Mechanism/Pattern	Impact on Structural Durability
[32,57,58]	Under freeze–thaw cycles, pore water in concrete freezes and expands, generating tensile stresses. Ice lenses first form in larger pores, followed by gradual freezing in smaller pores, driving unfrozen water toward the freezing front, causing capillary pressure buildup and microcrack formation.	Ambient temperature, pore structure, moisture content	Freezing expansion pressure exceeds tensile strength of concrete → microcrack formation → crack propagation under repeated freeze–thaw cycles	Accumulation of internal microcracks, reduction in structural strength, surface scaling, and cracking
[59–61]	High moisture content and numerous freeze–thaw cycles lead to cumulative deterioration, significant strength loss, and occurrences of surface scaling, cracking, and spalling.	Moisture content, number of freeze–thaw cycles	Higher moisture content → more ice → greater internal stresses; clear cumulative effect of cycles	Reduction in load-bearing capacity, diminished durability, increased structural safety risks
[62,63]	Mechanism and hazards of salt freeze damage	Saline environments (salt lakes, mining areas, coastal regions), temperature variation	Salt freeze damage combines freeze–thaw action and salt attack; salt ions increase osmotic pressure, partially suppressing freezing, but salt crystallization during evaporation generates crystallization pressure → exacerbates structural damage	More harmful than freeze–thaw alone, severe damage to pore structure
[64]	Cumulative freeze–thaw damage effects in concretes of different strength grades	Concrete strength grade, number of freeze–thaw cycles	Increasing cycles → continuous decrease in relative dynamic elastic modulus; C30 deteriorates faster than C40/C50	Strength grade significantly affects frost resistance
[65–67]	Substructures of bridges (piers, pile foundations) in freezing zones are prone to freeze–thaw damage, often showing cracks, scaling, and exposed reinforcement after 10 years	Location (influenced by surface water/groundwater), service time, temperature variation	Freeze–thaw induces crack propagation, providing pathways for chlorides and other corrosive agents → accelerates steel reinforcement corrosion	Reduction of effective load-bearing cross-section; in severe cases, component instability or failure, threatening overall structural safety and service life

3. Durability Prevention Strategies for Concrete Structures

3.1. Carbonation

To address the durability deterioration of concrete structures caused by carbonation, Li et al. analyzed actual engineering conditions and proposed corresponding anti-corrosion strategies. They emphasized the need to improve the effectiveness of protection measures during construction. Two solutions were proposed: firstly, applying protective coatings to the concrete surface, a common physical barrier method that forms a layer to reduce the carbonation-induced corrosion of reinforcement; secondly, adding inhibitors directly into the concrete mixture, which proved to be more practical and efficient in real-world applications. Comparative analysis showed that the first method has a lower cost-effectiveness and is technically challenging for large-scale implementation. In contrast, the second method offers easier application and better anti-carbonation performance, making it more widely adopted in construction [68].

Li further theorized that concrete carbonation occurs primarily because constituents within the concrete readily react with carbon dioxide in the air to form carbonation products. Thus, applying a surface coating can act as a physical shield to block direct contact between concrete and air, effectively slowing down the carbonation process. However, this approach has clear drawbacks—it is costly and requires skilled labor for execution. Based on theoretical and practical studies, introducing corrosion inhibitors during the concrete mixing stage has emerged as a more mature and efficient strategy. Through extensive experimentation and optimization, it has been demonstrated that, when the type and dosage of inhibitors are properly controlled, concrete can remain largely unaffected by carbonation over extended periods [69].

Yang [70] emphasized that delaying carbonation and improving structural durability require comprehensive consideration during the design and application of concrete materials. Firstly, the type of cement should match the environmental conditions in which the structure operates. Empirical studies show that, under the same cement content, Portland cement demonstrates the slowest carbonation rate, making it an ideal choice. Secondly, aggregate properties—including type and particle size—also affect the carbonation rate. For instance, if volcanic ash is present in the aggregate, it may react with siliceous aggregates under heat curing conditions in an alkali–aggregate reaction, which consumes alkaline reserves and accelerates carbonation. Thirdly, controlling the water–cement ratio is critical, as it directly influences the internal pore structure of concrete. A higher water–cement ratio increases porosity and the effective diffusion coefficient of CO_2, thereby accelerating carbonation.

Research by L and PAM et al. found that increasing sand content can reduce the water–cement ratio and complicate permeability pathways, thus decreasing both carbonation depth and air permeability [71].

Zhou and Gencturk et al. observed that using well-graded fine aggregates in concrete is more effective in delaying CO_2 ingress than using sand alone. Well-graded aggregates help block diffusion pathways, thereby enhancing carbonation resistance. In addition, applying thicker concrete cover or plaster coatings has been shown to delay CO_2 penetration into reinforced concrete structures, extending their service life [72].

Huang and Jiang et al. proposed that CO_2 diffusion and carbonation reactions mainly occur within the mortar phase, which comprises water, cement, and sand. This implies that reducing the amount of mortar or using coarser aggregates can help lower carbonation depth. Moreover, using smaller-sized coarse aggregates further reduces carbonation depth due to their larger surface area, which creates a longer and more tortuous path for CO_2 diffusion [73]. Table 6 presents a comparative analysis of protective measures against concrete carbonation.

Although significant progress has been made in the study of carbonation mechanisms, current models are mostly based on simplified assumptions and have not fully accounted for the nonlinear effects of environmental disturbances such as temperature and humidity variations and wet–dry cycles on the carbonation process. In addition, the coupling effects between carbonation and other deterioration factors (e.g., steel corrosion, freeze–thaw damage) remain insufficiently and systematically modeled, which limits the accuracy of predictions. Future research should focus on developing coupled carbonation–corrosion models under multi-factor interactions, incorporating both field data and accelerated testing to improve the reliability and applicability of durability assessments.

Table 6. Comparative analysis of protective measures against concrete carbonation.

Reference	Main Findings/Insights	Protective Measures	Advantages	Disadvantages
[68]	Based on field analysis, two corrosion protection schemes were proposed	1. Surface coating of concrete (physical barrier) 2. Addition of buffering agents (e.g., corrosion inhibitors)	Adding buffering agents is simple and highly effective in protection	Surface coatings have low cost-effectiveness and high construction difficulty
[69]	Carbonation is caused by reactions between concrete components and atmospheric substances	Coating to isolate air; addition of corrosion inhibitors during mixing	Corrosion inhibitor strategies are mature, economical, and efficient	Coating method is costly and requires high technical skill
[70]	Concrete material design should comprehensively consider multiple factors to delay carbonation	1. Selection of suitable cement types (e.g., silicate cement) 2. Control of aggregate properties 3. Proper water–cement ratio control	Silicate cement slows carbonation; low water–cement ratio reduces porosity	High water–cement ratio accelerates carbonation; poor aggregates may cause alkali–aggregate reaction
[71]	Increasing sand content can reduce water–cement ratio and improve protection	Increasing sand reduces carbonation depth and permeability	Reduces carbonation depth and results in a denser structure	Sand content must be controlled properly to avoid affecting other properties
[72]	Well-graded fine aggregates perform better than single sand	Use of well-graded fine aggregates; application of thick protective or plaster layers	Delays CO_2 penetration and extends service life	Requires strict control of grading during construction
[73]	Diffusion and carbonation mainly occur in the mortar phase	Reduce mortar content; use small particle coarse aggregates	Decreases carbonation depth; provides longer diffusion paths	Precise selection of small particle coarse aggregates needed; increases mixing difficulty

3.2. Crack

Cracking is one of the primary durability-related defects in bridge structures during their service life. It not only compromises the overall load-bearing capacity of the structure but also accelerates reinforcement corrosion and material degradation, thereby posing a serious threat to the long-term safety and service life of bridges. In recent years, researchers have conducted multi-dimensional and multi-level studies on various types of bridge cracking and proposed corresponding prevention and repair technologies. Chen et al. conducted on-site surveys and analytical assessments of bridge cracks and systematically developed a treatment strategy for existing cracks in box girders. Their study indicated that cracks with widths less than 0.15 mm should be sealed using epoxy mortar, whereas those with widths equal to or greater than 0.15 mm should be repaired using the "Bekaphen Method" (i.e., pressure injection of epoxy grout). This method has been shown to effectively mitigate the impact of cracks on structural performance [74]. To prevent longitudinal cracking in box girder segments, Ruan et al. proposed a set of integrated measures [75]: (1) strictly control the age difference of concrete between adjacent segments during casting [76,77]; (2) enhance structural crack resistance by adding anti-crack reinforcement meshes in the bottom slab and web to reduce temperature and shrinkage-induced stresses; (3) optimize the concrete curing process, with special attention to bottom slab maintenance—ensuring water curing for no less than 7 days—and controlling the internal–external temperature gradient during curing, while appropriately extending the formwork removal period. These measures significantly improve the crack resistance of precast box girder segments. Han addressed the issue of thermal shrinkage cracking in mass concrete abutments caused by heat of hydration and identified excessive internal–external temperature differentials as the primary cause. Through the use of thermal insulation layers, optimized casting times, and controlled cement content, thermal control techniques were successfully implemented to suppress cracking and improve the thermal adaptability of the structure [78]. Focusing on abutment sidewall cracking, Liu analyzed the causes from the perspectives of asynchronous concrete creep, accumulation of hydration heat, and inadequate curing practices. He proposed a grout-based repair method using mineral admixtures and achieved effective restoration of structural performance by optimizing the repair mix design and grouting process [79]. Wu concentrated on deep structural cracks in road and bridge concrete. Given the limitations of traditional repair methods in penetrating deep into narrow cracks, he adopted a combined approach using polymer-modified cement grout, high-pressure injection, and rotary jet grouting. This method effectively sealed deep cracks and significantly enhanced the structure's shear and flexural capacities [80]. Tan conducted a systematic analysis of cracks in concrete bridge structures caused by common factors such as loading, temperature variation, foundation deformation, and reinforcement corrosion. The study examined crack patterns, depths, and propagation trends and proposed a graded repair standard based on crack width. Specifically, cracks with a width less than 0.1 mm may be left untreated temporarily. However, once the width exceeds 0.1 mm, appropriate remedial measures must be taken. Repair and reinforcement techniques such as surface bonding, shotcreting, and carbon fiber strengthening can be employed. The crack treatment standard is illustrated in Table 7 [81]. Together, these studies offer valuable theoretical foundations and engineering insights into crack identification, cause analysis, and repair and reinforcement technologies. They provide important references for the prevention and management of structural deterioration throughout the entire life cycle of bridges. Table 8 presents a comparative analysis of crack prevention and repair technologies for concrete structures.

Table 7. Standards for bridge crack treatment.

Crack Width (x/mm)	Treatment Method
$x < 0.1$	Leave untreated
$0.1 \leq x \leq 0.2$	Seal with adhesive
$x > 0.2$	Repair (structural)

Table 8. Comparative analysis of crack prevention and repair technologies for concrete structures.

Reference	Research Object	Cause Analysis	Repair/Preventive Measures	Features
[74]	Cracks in precast box girders	–	1. Cracks < 0.15 mm: sealed with epoxy mortar 2. Cracks > 0.15 mm: epoxy grout injection (Wikona method)	Crack width grading allows for precise repair and improved efficiency
[75]	Segmental cracks in box girders	1. Differences in concrete age; high thermal and shrinkage stress 2. Inadequate curing	1. Control concrete age differences 2. Install anti-crack reinforcement mesh at the bottom and web 3. Enhance curing (≥7 days), control internal–external temperature differences	Improved crack resistance from construction, structural, and curing perspectives
[78]	Cracks in mass concrete abutments	Large internal–external temperature difference due to heat of hydration	1. Install thermal insulation layers 2. Optimize pouring schedule 3. Control cement content	Effective mitigation of thermal shrinkage cracks through temperature control
[79]	Cracks in abutment sidewalls	1. Unsynchronized creep and heat buildup 2. Inadequate curing measures	Grouting with mineral admixtures; optimize material proportions and grouting process	Restores structural performance; suitable for structural crack repair
[80]	Deep structural cracks in road bridges	Cracks are deep; traditional methods lack penetration	1. Polymer cement slurry 2. High-pressure grouting with rotary jetting technique	Reinforces and seals deep cracks; enhances shear and flexural strength
[81]	Common bridge cracks (various causes)	Load, temperature variation, foundation deformation, corrosion, etc.	1. Graded crack treatment 2. Surface bonding, shotcreting, and CFRP reinforcement	Proposes a systematic crack identification, strengthening, and repair technology system

The formation and propagation of concrete cracks are influenced by various factors, including environmental conditions, applied loads, and construction defects. These cracks often exhibit coupled interactions with other forms of deterioration, such as carbonation and chloride-induced corrosion, mutually accelerating structural degradation. Most existing studies focus on the identification of static cracks while lacking dynamic descriptions of crack evolution over time. Additionally, certain crack prevention measures or repair materials show limited adaptability under complex service conditions. Future research should focus on elucidating the mechanisms linking crack development to structural durability, exploring AI-based methods for crack identification and evolution prediction, and advancing the application of intelligent self-healing materials in bridge crack repair.

3.3. Reinforcement Corrosion

Corrosion of reinforcement is one of the primary durability issues affecting concrete bridge structures, with chloride-induced corrosion being particularly critical. As a result, extensive research has been conducted to investigate the mechanisms of chloride ion ingress and effective mitigation strategies. Suo [82] recommended incorporating 5%–10% fly ash and 3%–5% silica fume into concrete mixtures to enhance impermeability. Additionally, he emphasized the importance of strictly controlling the water-to-binder ratio and improving concrete density to effectively slow down chloride ion diffusion and thereby inhibit steel corrosion. Dhir and Jones [83] proposed the use of low-lime fly ash to refine pore structure and improve the chloride-binding capacity of concrete, leading to the development of mixtures with superior chloride resistance. They further demonstrated that blending cement with fly ash, silica fume, or metakaolin resulted in optimal resistance to chloride ingress. Hossain et al. [84] found that the inclusion of ultrafine fly ash (UFFA) significantly enhanced both the compressive strength and chloride penetration resistance of concrete, with silica fume playing a particularly prominent role. Hariharan et al. [85] also reported that ternary blends—where 30%–50% of cement is replaced with Class C fly ash and 6%–10% silica fume—effectively improved compressive strength and reduced chloride permeability. Thomas and Bamforth [86] developed models to predict chloride ion diffusion. Their data indicated that although the early-age resistance to chloride ingress may not be substantially improved by the addition of fly ash or ground granulated blast furnace slag (GGBS), long-term exposure over several years reveals significant performance benefits.

In terms of material protection, in the marine splash zone, KOBAYASHI and TAKEWAKA compared the test results of epoxy-coated reinforcement with untreated galvanized steel. The results indicated that epoxy-coated reinforcement exhibited significantly superior corrosion resistance compared to galvanized steel [87]. Similarly, SWAMY and KOYAMA discussed the engineering performance of epoxy-coated bars, showing that when exposed to natural and harsh marine environments, the epoxy coating demonstrated excellent corrosion resistance. Even when the coating was damaged and left unrepaired, the extent of corrosion was still much lower than that observed in uncoated or galvanized steel bars [88]. Many engineers consider GFRP (Glass Fiber Reinforced Polymer) to be one of the most innovative materials capable of overcoming the inherent vulnerability of steel-reinforced concrete structures to corrosion in aggressive environments. ALSAYED et al. investigated GFRP bars and concluded that they possess higher corrosion resistance, greater tensile strength, and lighter weight. However, their low elastic modulus and lack of ductility are recognized as major engineering drawbacks. Therefore, extensive laboratory and field testing is still required to validate their long-term performance [89].

In terms of improving concrete performance, the use of mineral admixtures can effectively mitigate steel reinforcement corrosion caused by alkali–silica reaction (ASR) [90–92]. Papadakis [28] found that partially replacing aggregates with silica fume, low-calcium fly ash, or high-calcium fly ash can enhance the carbonation resistance of concrete. Gonen and Yazicioglu [93] reported that replacing 10% of cement with silica fume and 20% with fly ash significantly improved the concrete's resistance to wet–dry cycles, reduced capillary water absorption, and enhanced carbonation resistance. Furthermore, Shashiprakash and Thomas [94] demonstrated that mortar incorporating low-calcium fly ash or appropriate amounts of ultrafine fly ash exhibited good resistance to sulfate attack. These findings, from various perspectives, underscore the critical role of mineral admixtures in the prevention and control of steel corrosion. Although mineral admixtures may slightly slow the early-age strength development of concrete, studies have shown that they do not adversely affect the long-term mechanical performance. On the contrary, they may even enhance the long-term strength development. Concrete mixtures incorporating high volumes of fly ash or blended with ground granulated blast-furnace slag (GGBS) exhibit excellent workability, high compressive strength, and superior durability, thereby significantly inhibiting reinforcement corrosion [95,96]. Table 9 presents a comparative analysis of reinforcement corrosion prevention measures.

Table 9. Comparative analysis of reinforcement corrosion prevention measures.

Scholar/Research Team	Research Focus	Type and Dosage of SCMs	Technical Highlights	Conclusions and Effects
[82]	Chloride diffusion control	Fly ash 5%–10%, silica fume 3%–5%	Controlled water–binder ratio, enhanced compactness	Significantly improved impermeability, delayed chloride diffusion, inhibited corrosion
[83]	Pore structure improvement and chloride binding	Low-lime fly ash + silica fume/metakaolin	Optimized mix design	Cement–mineral blends exhibited the best chloride resistance
[84]	Role of ultra-fine fly ash	Ultra-fine fly ash (UFFA), silica fume	Enhanced chloride resistance and strength	UFFA notably improved early-age strength and chloride resistance
[85]	Performance of ternary SCM systems	Class C fly ash 30%–50%, silica fume 6%–10%	Partial cement replacement	Enhanced compressive strength and reduced chloride permeability
[86]	Chloride diffusion modeling	Fly ash, slag	Long-term performance modeling	Limited early improvement, but significant long-term chloride resistance
[87–89]	Reinforcement protection	Epoxy coating	Surface coating for corrosion protection	Demonstrated excellent anti-corrosion performance
[28]	Carbonation resistance	Silica fume and low/high-calcium fly ash replacing aggregate	Altered alkalinity and carbonation path	Improved carbonation resistance, indirectly reducing corrosion risk
[93]	Dry–wet cycling and capillary absorption	Silica fume 10% + fly ash 20% (cement replacement)	Enhanced surface density	Significantly enhanced resistance to carbonation and water ingress
[94]	Sulfate attack resistance	Low-calcium fly ash, UFFA	Reduced deterioration under aggressive environment	Improved chemical resistance and overall durability
[95,96]	Long-term performance and strength development	High-volume fly ash + slag	Combined SCM optimization	Slightly lower early strength, but excellent long-term performance and corrosion control

Although the initiation mechanisms of steel reinforcement corrosion are relatively well understood, significant deviations still exist in predicting corrosion rates under the combined influence of multiple factors, such as the synergistic effects of chloride ingress and carbonation [97]. Moreover, current protective measures (e.g., coatings and admixtures) exhibit limited durability and stability in extreme environments, and there is a lack of standardized evaluation systems. The accuracy and adaptability of existing structural health monitoring techniques for early-stage corrosion detection also require improvement. Future research should focus on the integrated study of corrosion–crack–structural performance degradation mechanisms, promote long-term validation of novel protective materials [98] in real bridge applications, and develop online corrosion monitoring technologies that can be embedded within structural systems.

3.4. Freeze–Thaw Damage

The complexity of freeze–thaw controlling factors presents significant challenges in the design and construction of bridge structures within civil engineering projects. In cold and permafrost regions, structural frost heave and shrinkage induced by fluctuating environmental temperatures have been widely documented, causing various structural damages and even safety incidents in practical engineering. To minimize the risk of structural failures triggered by freeze–thaw (FT) cycles, current research primarily focuses on the mechanisms by which FT cycles affect the strength of bridge materials, component durability, and overall service performance. This includes investigating the formation and progression of freeze–thaw damage as well as its long-term impacts on bridge safety. To effectively mitigate concrete deterioration caused by freeze–thaw cycles, researchers have proposed various material modification strategies, among which controlling the internal pore structure is a key approach. Current techniques for regulating pore structure mainly include introducing air-entraining agents to form a stable microbubble system; using mineral admixtures (such as fly ash, silica fume, and slag) to optimize the distribution of capillary pores; incorporating nanomaterials (e.g., nano-SiO_2 and nano-Al_2O_3) to fill microvoids; and adjusting the water-to-binder ratio and curing methods to reduce overall porosity. These methods improve the compactness and impermeability of concrete from different aspects, providing diverse technological pathways to develop a pore structure favorable for frost resistance.

Among various methods, the use of air-entraining agents has been extensively studied and widely applied due to their convenience and remarkable effectiveness. Bao et al. [99] demonstrated that under the combined action of sustained compressive loading and freeze–thaw cycling, the incorporation of AEA significantly reduced concrete mass loss and the decline in relative dynamic elastic modulus (RDEM). Zhang [100] found that adding AEA during concrete mixing produces numerous uniformly distributed, disconnected microscopic air bubbles within the concrete matrix, which effectively reduce the hydrostatic pressure caused by freezing pore water in the early stages of freezing, thereby significantly improving concrete's freeze–thaw durability. Kosior-Kazberuk and Berkowski [101] further reported that air-entrained concrete subjected to combined salt freeze–thaw cycles and flexural loading exhibited a 62.8% reduction in mass loss compared to ordinary concrete, demonstrating superior resistance under complex environmental conditions. Additionally, the introduction of uniformly distributed, closed microbubbles by AEA provides internal cushioning space for ice expansion, which inhibits CO_2 diffusion and the development of freeze–thaw-induced cracking. Consequently, this significantly enhances concrete durability under the coupled effects of carbonation and freeze–thaw cycling [102].

Under complex environmental conditions such as freeze–thaw cycles, concrete inevitably experiences a certain degree of structural degradation. In such scenarios, fiber materials do not directly prevent the initiation of damage but rather play a crucial role after cracking by bridging and stabilizing the damaged matrix through their excellent crack resistance and toughening capabilities. As a result, fibers significantly enhance the integrity and durability of the overall structure, demonstrating their importance as an effective material for improving concrete durability under multiple environmental stressors. Yuan et al. [103] demonstrated that the incorporation of polypropylene fibers (PPF) and polyvinyl alcohol fibers (PVAF) significantly reduced the loss of relative dynamic elastic modulus (RDEM) of concrete under the combined effects of freeze–thaw cycling and compressive loading, with the loss reduced by 77.5%. Ren and Lai [104] further found that concrete containing 0.2% PPF exhibited a 32.4% reduction in compressive strength loss after undergoing 120 cycles of sulfate freeze–thaw exposure. Experimental results from Kosior-Kazberuk and Berkowski [101] also showed that under 90 cycles of salt freeze–thaw cycling combined with flexural loading, steel fiber (SF) reinforced concrete experienced an 80% reduction in mass loss compared to ordinary concrete. Notably, when air-entraining agents are combined with steel fibers in concrete, the composite material exhibits even greater improvements in freeze–thaw resistance under external loading conditions [97].

In terms of binder material substitution, pozzolanic mineral admixtures have also demonstrated significant potential for enhancing freeze–thaw resistance. Besheli et al. [105] reported that after 55 freeze–thaw (FT) cycles in a 4% NaCl solution, concrete partially replaced with 15% metakaolin or zeolite exhibited an 86.7% reduction in mass loss compared to plain concrete, indicating a substantial improvement in freeze–thaw durability. Tian et al. [106] pointed out that fly ash (FA) and blast furnace slag (BFS) generate secondary hydration products through pozzolanic reactions, thereby refining the pore structure and enhancing overall density and impermeability, which effectively slows down deterioration caused by the coupling of freeze–thaw and salt exposure. Furthermore, Li et al. [107] conducted 300 freeze–thaw cycles in a 5% NaCl solution and found that concrete with 25% FA replacement showed reductions of 2.3% and 3.2% in the loss of relative dynamic elastic modulus (RDEM) and compressive strength, respectively, compared to ordinary concrete. This improvement is primarily attributed to the higher glass content and purer reactive components in FA, which promote densification of the interfacial transition zone (ITZ), thereby enhancing resistance to freeze–thaw damage. In summary, pozzolanic mineral admixtures, when used at appropriate dosages and under favorable reactivity conditions, can enhance the mechanical properties and durability of concrete subjected to freeze–thaw cycles and chloride-induced deterioration. This improvement is mainly achieved through mechanisms such as refining the concrete's microstructure, increasing the density of the interfacial transition zone, and delaying crack propagation. However, the actual freeze–thaw resistance imparted by these mineral admixtures strongly depends on their specific types, reactive indices, dosage, and particle morphology. Improper use may even adversely affect concrete performance. Therefore, in cold climate regions—especially those at high risk of severe freeze–thaw damage—it is not recommended to use cements with a high content of non-clinker components. Scientific selection and validation based on the local environment and material characteristics are essential.

In addition to commonly used modification methods such as air-entraining agents, fiber materials, and mineral admixtures, various novel approaches have recently shown promising effects in enhancing freeze–thaw durability. Research indicates that the incorporation of viscosity-modifying agents (VMA) can optimize the rheological properties of fresh concrete mixtures, leading to the formation of a more uniform microporous structure during curing, thereby effectively improving freeze–thaw resistance [108]. Moreover, glazed hollow beads (GHB), a type of solid air-entraining material, have demonstrated significant advantages in resisting salt freeze–thaw environments. Experimental results showed that after 100 cycles of salt freeze–thaw, concrete incorporating GHB exhibited reductions of 74.6% in mass loss and 8.5% in relative dynamic elastic modulus (RDEM) loss [109]. Pang et al. [110] confirmed that partially replacing conventional coarse aggregates with carbonated steel slag aggregates (CSA) effectively enhances concrete's resistance to salt freeze–thaw damage. After 85 freeze–thaw cycles in a 3% chloride solution, CSA concrete achieved a mass loss reduction of up to 75.8%. Additionally, early-stage carbonation curing techniques have attracted extensive attention. Introducing a CO_2-rich environment during the initial setting phase promotes uniform precipitation of calcium carbonate ($CaCO_3$), which improves the pore structure and densifies the interfacial transition zone (ITZ), thereby significantly mitigating deterioration caused by carbonation [111]. These findings collectively suggest that the integrated application of novel materials and curing methods provides expanded possibilities for improving concrete freeze–thaw durability under complex environments, offering both theoretical support and practical pathways for the design and maintenance of future high-performance bridge structures [112–116]. Table 10 presents methods for mitigating concrete deterioration under freeze-thaw damage.

Most existing freeze–thaw cycle tests are conducted under standardized laboratory conditions, which fail to accurately capture the coupled effects of temperature gradients, moisture migration, and chloride action in real service environments. In addition, freeze–thaw deterioration often interacts with other damage processes such as cracking and carbonation, potentially forming feedback mechanisms that accelerate structural degradation and complicate durability modeling. It is recommended that future research establish in-situ monitoring systems on bridges located in representative climate zones and integrate multi-scale studies of freeze–thaw damage mechanisms. This will support the development of evaluation methods for freeze–thaw deterioration in extreme environments and facilitate the design and validation of new freeze-resistant concrete materials.

Table 10. Methods to mitigate concrete deterioration under freeze–thaw damage.

Reference	Mitigation Method	Additive Type	Experimental Condition	Experimental Result
[99–101]	Air-Entraining Agent	—	Compressive load and freeze–thaw cycle	Reduced concrete mass loss and RDEM loss [99]
		—	Concrete mixing	Reduced hydrostatic pressure inside concrete pores [100]
		Saponified liquid resin	Salt freeze–thaw cycle and flexural load	Reduced concrete mass loss by 62.8% [101]
[101–103]	Fibers	Polypropylene fibers (PPF) and polyvinyl alcohol fibers (PVAF)	Freeze–thaw cycle and compressive load	Reduced RDEM loss by 77.5% [102]
		Polypropylene fibers (PPF)	Salt freeze–thaw cycle	Reduced compressive strength loss by 32.4% [103]
		Steel fibers (SF)	Salt freeze–thaw cycle and flexural load	Reduced mass loss of SF concrete by 80% [101]
[105–107]	Pozzolanic Materials	Metakaolin or zeolite	Salt freeze–thaw cycle	Reduced concrete mass loss by 86.7% [105]
		Fly ash (FA) and blast furnace slag (BFS)	Salt freeze–thaw cycle	Improved concrete density and impermeability [106]
		Fly ash (FA)	Salt freeze–thaw cycle	Reduced RDEM loss by 2.3% and compressive strength loss by 3.2% [107]
[108–110]	Other Methods	Viscosity-modifying agent (VMA)	Flexural load	Reduced RDEM loss [108]
		Glazed hollow beads (GHB)	Salt freeze–thaw cycle	Reduced mass loss by 74.6% and RDEM loss by 8.5% [109]
		Carbonated steel slag aggregate (CSA)	Salt freeze–thaw cycle	Reduced concrete mass loss by 75.8% [110]

3.5. *Evaluation Criteria for Durability Prevention Strategies*

Unlike the singular performance indicators commonly used in laboratory studies, practical engineering demands a more comprehensive assessment of how various prevention strategies perform under complex service conditions. Therefore, when assessing the applicability of durability prevention strategies in practical engineering projects, as indicated in Reference [18], it is crucial to establish more detailed evaluation criteria that reflect the multifaceted requirements for durability in real-world engineering scenarios. Based on the recommendations of some existing literature [18,81,97,98,110], the following six aspects can be adopted as evaluation criteria for durability prevention strategies.

3.5.1. Specificity to Deterioration Mechanisms

Each type of deterioration—carbonation, corrosion, cracking, and freeze–thaw damage—has its own unique cause and progression pathway. Therefore, a core standard is whether a technique effectively targets the critical stages of a given degradation process. For instance, surface treatments can efficiently slow CO_2 penetration and prevent carbonation but offer limited benefit once corrosion has occurred [81]. On the other hand, cathodic protection is highly effective for halting steel corrosion but irrelevant for crack sealing. This criterion helps avoid the misapplication of solutions to problems for which they are not designed.

3.5.2. Long-Term Performance in Field Conditions

Durability by nature emphasizes performance over time. Thus, the second criterion focuses on how well a method maintains its effectiveness under long-term service conditions, including fluctuating humidity, chloride ingress, repetitive loads, and seasonal temperature variations [97]. Some methods may perform well in short-term tests but fail to withstand the cumulative environmental stresses found in real-world applications, limiting their practical value.

3.5.3. Structural Compatibility

In bridge maintenance and rehabilitation, it is crucial that new materials and technologies are compatible with existing structural components in terms of mechanical behavior, volumetric changes, and chemical properties. Incompatibilities can lead to issues like interface cracking or secondary damage, compromising the repair outcome [98]. Hence, material compatibility and the integration capacity with existing structures are vital evaluation factors.

3.5.4. Monitoring and Quantifiability

With the growing adoption of structural health monitoring technologies, it is increasingly desirable that repair or protection measures allow for quantifiable, traceable performance data. For parameters such as corrosion rate, carbonation depth, or crack propagation, a lack of effective monitoring makes it difficult for engineers to plan maintenance cycles or interventions. Therefore, the adaptability of a technique to real-time or periodic monitoring is an important consideration.

3.5.5. Economic Feasibility and Constructability

Given limited maintenance budgets, cost-effectiveness and ease of application are major factors influencing the viability of a solution. Techniques that require specialized labor, expensive equipment, or complicated procedures may be less attractive despite strong technical performance. Thus, aspects such as material cost, labor demand, installation difficulty, and disruption to traffic should all be taken into account.

3.5.6. Environmental Impact and Regulatory Compliance

As sustainability becomes a central goal in infrastructure management, the environmental implications of repair strategies must also be addressed [110]. Whether a material contains hazardous substances, generates pollutants during use, or complies with local environmental and construction regulations are key factors in modern durability evaluation.

Together, these six dimensions provide a comprehensive framework for evaluating and comparing different durability interventions. This structured approach also supports more informed decision-making in bridge maintenance planning and reflects the broader trend toward data-driven and systematic infrastructure management.

4. Durability Enhancement Strategies at the Whole Life Cycle Level for Concrete Structures

Based on the analyses of concrete carbonation, cracking, steel bar corrosion, and freeze–thaw damage presented in Sections 2 and 3, Table 11 summarizes a multi-level durability enhancement framework from the perspective of the entire life cycle of concrete structures. The table categorizes a comprehensive set of strategies proposed by researchers in recent years for preventing and mitigating concrete deterioration at each stage, according to the four stages of the concrete structure's life cycle. It reflects the latest research advancements on durability-related issues since the release of important codes and guidelines, such as the Eurocode EN 1992-1-1 Durability Guide [117] and relevant approaches from the American Concrete Institute (ACI). Compared with existing review literature [118–124], this review systematically integrates various common durability distresses and their prevention and mitigation strategies within a unified framework, facilitating the development of more targeted and synergistic durability enhancement strategies from a whole life cycle perspective.

Table 11. Comparison of strategies for enhancing durability of concrete structures at different life cycle stages.

Reference	Durability Distress	Design Stage	Construction Stage	Service Stage	Repair Stage
[68–72,118,119]	Carbonation [118,119]	1. Preferably select cement types (e.g., Portland cement has a slower carbonation rate) [70]. 2. Use well-graded fine aggregates and coarse aggregates [71,72]. 3. Reasonably control the water–cement ratio to reduce porosity and CO_2 diffusion channels [70,71]. 4. Increase the thickness of the concrete cover [72].	1. Strengthen wet curing to avoid early drying cracks. 2. Minimize the use of materials with high carbonation rates. 3. Add an appropriate amount of corrosion inhibitors during material mixing to enhance anti-carbonation capability [68,69].	1. Regularly inspect the carbonation depth of concrete and CO_2 concentration. 2. Apply protective coatings on the concrete surface or construct physical barriers to isolate air contact [68]. 3. Perform spray-type realkalization treatment when necessary.	1. Remove severely carbonated areas and re-pour or repair. 2. Adopt electrochemical realkalization or deep coating penetration technology. 3. Implement local anti-corrosion treatment for damaged rebars in the area [68,69].
[74–81,119,120]	Cracking [119,120]	1. Reasonably arrange the structure, optimize reinforcement design, and set anti-crack rebar nets at the bottom and web plates. 2. Select materials with good temperature control performance to reduce heat of hydration concentration [75].	1. Control the age difference of segments to avoid early shrinkage-induced cracks. 2. Optimize pouring sequence and time, and use insulation layers to control internal and external temperature differences [75,78].	1. Monitor cracks and assess their grades to identify crack development trends and impacts. 2. Set repair grades and cycles according to crack width classification [81].	1. For cracks < 0.15 mm: seal with epoxy putty. 2. For cracks ≥ 0.15 mm: inject with "Bike method" epoxy mortar. 3. For deep structural cracks: use high-pressure grouting, polymer cement, or rotary jet grouting for reinforcement. 4. For cracks caused by loads or corrosion: use carbon fiber reinforcement, the shotcrete method, or surface bonding [74,79–81].
[82–96,122]	Steel Bar Corrosion [122]	1. Optimize concrete mix ratio, control water–cement ratio, and enhance compactness. 2. Add mineral admixtures (such as 10% coal fly ash, 3%–5% silica fume, or metakaolin) to disperse and retard chloride ion diffusion. 3. Adopt anti-chloride ion permeable mixture systems (such as UFFA, slag, and ternary systems). 4. Reasonably set the thickness of the steel bar protective layer [82–86,95,96].	1. Strictly control construction quality to prevent honeycombs, pits, and voids. 2. Precisely control raw material ratios and mixing time to improve uniformity and compactness. 3. Ensure sufficient vibration and compaction, and reduce permeable channels [82,83].	1. Regularly monitor the steel bar potential, chloride ion concentration, and carbonation depth. 2. Surface treatment: spray anti-chloride ion agents or chloride ion retarders. 3. If the structure is in a heavy chloride salt pollution area, adopt electrochemical chlorine removal or migration technology [87–89,93,94].	1. Locally apply high-performance repair mortar in damaged areas. 2. For severely corroded components, implement electrochemical protection or cathodic protection technology. 3. If necessary, replace damaged steel bars and re-apply protective layers [87–89].

Table 11. *Cont.*

Reference	Durability Distress	Design Stage	Construction Stage	Service Stage	Repair Stage
[99–116,123,124]	Freeze–thaw Damage [123,124]	1. Add air-entraining agent (AEA) to form closed air bubbles and relieve freezing expansion pressure [99–102]. 2. Incorporate polypropylene fiber (PPF), steel fiber (SF), etc., to enhance crack resistance and freeze–thaw resistance [105–107]. 3. Add volcanic ash mineral admixtures (such as coal fly ash, metakaolin, zeolite, and BFS) to improve pore structure and enhance freeze–thaw and anti-permeability performance. 4. Optimize water–cement ratio, improve concrete compactness, and transition zone (ITZ) performance. 5. Air-entrained materials such as glazed hollow beads (GHB) can be used to enhance salt-freeze resistance [101–104].	1. Incorporate viscosity-modifying admixtures (VMA) to improve the fluidity and deformability of the mixture and form a uniform micro-pore structure [108]. 2. Use carbonized steel slag materials (CSA) to partially replace ordinary coarse aggregates and improve freeze–thaw resistance [110]. 3. Ensure a suitable construction environment temperature to reduce early freeze damage.	1. Regularly detect the concrete relative dynamic elastic modulus (RDEM) and mass loss rate, and monitor the freeze–thaw damage evolution trend. 2. Set drainage slopes and systems to prevent water retention and penetration.	1. For freeze–thaw damaged areas, apply surface waterproof coatings and seal with air-retarding agents. 2. Locally repair and reuse high-freeze-resistance performance repair materials or fiber-reinforced mortar. 3. For severe freeze–thaw cracks, combine surface sealing and internal grouting technology for repair [109,111–116].

5. Conclusions

Concrete bridge structures are an essential component of modern transportation infrastructure. Their durability during service life is directly related to structural safety, service lifespan, and subsequent maintenance and repair costs. This paper systematically reviews the common durability-related deteriorations in concrete bridges, analyzing the underlying mechanisms, influencing factors, and manifestations in engineering practice across four major categories of damage. Based on current research findings and practical experience, corresponding prevention and control strategies are proposed. At the end of each damage section, the paper summarizes the existing mitigation measures and their limitations and offers suggestions for future research. The main conclusions regarding the types of deterioration discussed in this study are as follows:

1. Based on existing studies, concrete carbonation, as a long-standing and widely studied durability deterioration mechanism, fundamentally involves the reaction between atmospheric carbon dioxide and alkaline substances within the concrete. This process reduces the alkalinity of the concrete, thereby weakening the stability of the passive film on the reinforcement surface and indirectly triggering reinforcement corrosion. The carbonation rate and its impact on structural performance are influenced by multiple factors, including the water-to-binder ratio, cover thickness, environmental humidity, and temperature. In current engineering practice, carbonation is regarded as one of the critical factors affecting the long-term service performance of reinforced concrete structures and serves as an important basis for durability design, material selection, and subsequent maintenance strategies.
2. Cracking not only indicates the degradation of the structure's mechanical integrity but also serves as a pathway for the ingress of aggressive agents such as moisture, chloride ions, and carbon dioxide. This accelerates carbonation, freeze–thaw damage, and

reinforcement corrosion. Crack formation is complex and multi-factorial, commonly associated with material shrinkage, construction practices, and load-induced stresses.

3. Reinforcement corrosion is one of the most prevalent and severe durability issues in concrete bridges. The expansive corrosion products can cause concrete cover spalling, crack propagation, and a significant reduction in load-bearing capacity, thus posing a direct threat to structural safety. Corrosion is primarily driven by chloride ingress and carbonation, with a concealed and time-delayed progression that makes it difficult to detect in early stages.
4. Freeze–thaw damage predominantly occurs in cold or seasonally freezing regions. The expansion of pore water during freezing disrupts the concrete's density and integrity, promoting crack initiation and propagation and accelerating overall structural degradation. Enhancing the freeze–thaw resistance of concrete requires a comprehensive approach involving material design, optimization of air-void structure, and waterproofing protection.
5. In response to these durability challenges, this study proposes a holistic prevention and control framework encompassing the design, construction, operation, and maintenance phases. Studies have shown that novel materials such as ultra-high-performance concrete (UHPC) have been widely applied in bridges and related structures, significantly enhancing their durability and service life. The appropriate selection of these advanced materials, in combination with optimized structural detailing, strict construction quality control, and the integration of intelligent monitoring technologies, remains a key approach to improving the long-term performance of bridges. Although existing durability models have partially incorporated the unique properties of new materials such as UHPC—particularly in terms of modified diffusion coefficients, permeability, and damage-healing mechanisms—most remain at the stage of theoretical exploration or experimental fitting and lack mature, widely accepted generalized models.

Author Contributions: Conceptualization, J.M., Q.Y., X.P. and K.X.; investigation, J.M., Q.Y., X.P. and K.X.; writing—original draft preparation, J.M., Q.Y., X.P. and K.X. All authors have read and agreed to the published version of the manuscript.

Funding: This research received no external funding.

Data Availability Statement: No new data were created or analyzed in this study.

Conflicts of Interest: The authors declare that there are no conflicts of interest. The funders had no role in the collection, analyses, or interpretation of data or in the decision to publish the results.

References

1. Collura, D.; Nascimbene, R. Comparative Assessment of Variable Loads and Seismic Actions on Bridges: A Case Study in Italy Using a Multimodal Approach. *Appl. Sci.* **2023**, *13*, 2771. [CrossRef]
2. Zhong, X.; Jin, W.; Zhang, B. Durability design method for concrete structures in chloride salt environments. *J. Build. Mater.* **2016**, *19*, 544–549.
3. Wang, J.; Wang, M.; Men, P.; Qin, F.; Peng, X.; Di, J. Experimental investigation on mechanical behavior of Push-out Perfobond connectors. *Eng. Struct.* **2025**, *341*, 120828. [CrossRef]
4. Zhao, D. Durability design of highway concrete structures based on carbonation numerical model. *Shanghai Highw.* **2006**, 25–28. [CrossRef]
5. Jiang, T.; Ai, L.; Yang, Z. Analysis of typical durability diseases and causes in concrete bridges. *World Build. Mater.* **2013**, *34*, 23–26.
6. Zhou, X.; Zhang, X. Thoughts on the Development of Bridge Technology in China. *Engineering* **2019**, *5*, 1120–1130. [CrossRef]
7. Qin, F.; Han, Y.; Wei, X.; Wang, X.; Zhang, Z.; Zhang, X. Flexural behavior of engineered cementitious composites (ECC) slabs with different strength grades. *Materials* **2025**, *18*, 2047. [CrossRef]

8. Wei, X.; Li, X.; Di, J.; Qin, F.; Zhang, Z.; Liang, F. Flexural performance of U-shaped high-strength ECC perma-nent formwork concrete composite beams. *Structures* **2025**, *74*, 108641. [CrossRef]
9. Aktan, A.E.; Farhey, D.N.; Brown, D.L.; Dalal, V.; Helmicki, A.J.; Hunt, V.J.; Shelley, S.J. Condition Assessment for Bridge Management. *J. Infrastruct. Syst.* **1996**, *2*, 108–117. [CrossRef]
10. Vu, K.A.T.; Stewart, M.G. Structural reliability of concrete bridges including improved chloride-induced corrosion models. *Struct. Saf.* **2000**, *22*, 313–333. [CrossRef]
11. Stewart, M.G.; Rosowsky, D.V. Structural Safety and Serviceability of Concrete Bridges Subject to Corrosion. *J. Infrastruct. Syst.* **1998**, *4*, 146–155. [CrossRef]
12. Wang, H.; Wang, A.; Wang, J.; Zhang, J.; Du, T.; Shen, J. Optimizing quaternary chemical additives and steel fiber dispersion in CSA cement based HPC for rapid repair engineering. *Constr. Build. Mater.* **2025**, *489*, 142265. [CrossRef]
13. Men, P.; Chen, F.; Qin, F.; Peng, X.; Di, J.; Jiao, H. Behavior of composite beams with UHPC-concrete composite slabs under negative bending moment. *J. Constr. Steel Res.* **2025**, *227*, 109415. [CrossRef]
14. Vassie, P.; TRRL. Reinforcement Corrosion and the Durability of Concrete Bridges. *Proc. Inst. Civ. Eng.* **1984**, *76*, 713–723. [CrossRef]
15. Sarja, A. Durability design of concrete structures—Committee report 130-CSL. *Mater. Struct.* **2000**, *33*, 14–20. [CrossRef]
16. Lei, J. Detection and Evaluation of Disease in Jinghuai Bridge. Master's Thesis, Hefei University of Technology, Hefei, China, 2008.
17. Tang, D.; Zhang, X. Diagnosis and treatment measures of durability defects in in-service concrete bridges. *Heilongjiang Sci. Technol. Inf.* **2010**, *225*, 236.
18. Abdel-Jaber, H.; Glisic, B. Monitoring of prestressing forces in prestressed concrete structures—An overview. *Struct. Control Health Monit.* **2019**, *26*, e2374. [CrossRef]
19. Rozière, E.; Loukili, A.; Cussigh, F. A performance based approach for durability of concrete exposed to carbonation. *Constr. Build. Mater.* **2009**, *23*, 190–199. [CrossRef]
20. Glasser, F.P.; Marchand, J.; Samson, E. Durability of concrete—Degradation phenomena involving detrimental chemical reactions. *Cem. Concr. Res.* **2008**, *38*, 226–246. [CrossRef]
21. Chen, X. Factors Affecting the Durability of Bridge Structures and Their Optimization Design. *Eng. Constr. Des.* **2023**, 85–87. [CrossRef]
22. Chen, Z. Study on influencing factors and optimization design of bridge structure durability. *Transp. Sci. Manag.* **2023**, *4*, 143–145.
23. Ding, W.; Huang, W. Discussion on field detection of durability of bridge concrete structures. *Transp. World* **2022**, 99–101. [CrossRef]
24. Cusson, D.; Lounis, Z.; Daigle, L. Durability Monitoring for Improved Service Life Predictions of Concrete Bridge Decks in Corrosive Environments. *Comput.-Aided Civ. Infrastruct. Eng.* **2011**, *26*, 524–541. [CrossRef]
25. Hooton, R.D.; Bickley, J.A. Design for durability: The key to improving concrete sustainability. *Constr. Build. Mater.* **2014**, *67*, 422–430. [CrossRef]
26. Li, C. Study on Factors Affecting Bridge Structure Durability and Its Optimization Design. *Transp. Sci. Manag.* **2023**, *4*, 147–149.
27. Liu, B. Analysis of common diseases and reinforcement of expressway bridges. *Transp. Sci. Manag.* **2023**, *4*, 75–77.
28. Papadakis, V.G. Effect of supplementary cementing materials on concrete resistance against carbonation and chloride ingress. *Cem. Concr. Res.* **2000**, *30*, 291–299. [CrossRef]
29. Silva, R.V.; Silva, A.; Neves, R.; Brito, J. Statistical Modeling of Carbonation in Concrete Incorporating Recycled Aggregates. *J. Mater. Civ. Eng.* **2016**, *28*, 04015082. [CrossRef]
30. Czarnecki, L.; Woyciechowski, P. Concrete Carbonation as a Limited Process and Its Relevance to Concrete Cover Thickness. *ACI Mater. J.* **2012**, *109*, 275–282. [CrossRef]
31. Li, X.; Wu, Z.; Li, D. Design and key construction technologies of protective coating systems for concrete bridge durability. *Highway* **2021**, *66*, 178–183.
32. Wang, W.; Yang, L.; Wu, F.; He, X.; Tan, G.; Liu, B.; Wang, H.; Lv, D.; Liu, H. Compressive behavior deterioration mechanism of concrete under carbonization and freeze-thaw cycles and time-varying structural reliability evaluation. *Constr. Build. Mater.* **2025**, *466*, 140359. [CrossRef]
33. Zheng, B. Analysis of disease treatment techniques in road and bridge engineering construction. *Hous. Real Estate* **2021**, 235–236. [CrossRef]
34. Parameswaran, L.; Kumar, R.; Sahu, G.K. Effect of Carbonation on Concrete Bridge Service Life. *J. Bridge Eng.* **2008**, *13*, 75–82. [CrossRef]
35. Song, X. Discussion on bridge structure defects and maintenance strengthening measures. *Urban Constr. Theory Res. (Electron. Ed.)* **2020**, 62. [CrossRef]
36. Ann, K.Y.; Pack, S.W.; Hwang, J.P.; Song, H.W.; Kim, S.H. Service life prediction of a concrete bridge structure subjected to carbonation. *Constr. Build. Mater.* **2010**, *24*, 1494–1501. [CrossRef]

37. Du, M. Comprehensive Analysis and Remedial Strategies for Crack Occurrences in Concrete Bridge Structural Systems. *Heilongjiang Transp. Sci. Technol.* **2021**, *44*, 108–109. [CrossRef]
38. Wang, J. Causes and treatment technologies of cracks in concrete beam bridges on expressways. *Transp. World* **2019**, 131–132. [CrossRef]
39. Chen, S.; Duffield, C.; Miramini, S.; Nasim Khan Raja, B.; Zhang, L. Life-cycle modelling of concrete cracking and reinforcement corrosion in concrete bridges: A case study. *Eng. Struct.* **2021**, *237*, 112143. [CrossRef]
40. Golewski, G.L. The Phenomenon of Cracking in Cement Concretes and Reinforced Concrete Structures: The Mechanism of Cracks Formation, Causes of Their Initiation, Types and Places of Occurrence, and Methods of Detection—A Review. *Buildings* **2023**, *13*, 765. [CrossRef]
41. Cai, C.; Chen, S.; Liu, L. Detection of Fatigue Cracks for Concrete Structures by Using Carbon Ink-Based Conductive Skin and Electrical Resistance Tomography. *Sensors* **2023**, *23*, 8382. [CrossRef] [PubMed]
42. Pal, M.; Palevičius, P.; Landauskas, M.; Orinaitė, U.; Timofejeva, L.; Ragulskis, M. An Overview of Challenges Associated with Automatic Detection of Concrete Cracks in the Presence of Shadows. *Appl. Sci.* **2021**, *11*, 11396. [CrossRef]
43. Abdel-Qader, I.; Abudayyeh, O.; Kelly, M.E. Analysis of Edge-Detection Techniques for Crack Identification in Bridges. *J. Comput. Civ. Eng.* **2003**, *17*, 255–263. [CrossRef]
44. Gribniak, V.; Kaklauskas, G.; Bacinskas, D. State-Of-Art Review of Shrinkage Effect on Cracking and Deformations of Concrete Bridge Elements. *Balt. J. Road Bridge Eng.* **2007**, *2*, 183–193.
45. Flah, M.; Suleiman, A.R.; Nehdi, M.L. Classification and quantification of cracks in concrete structures using deep learning image-based techniques. *Cem. Concr. Compos.* **2020**, *114*, 103781. [CrossRef]
46. Li, Z. Study on formation and repair methods of concrete structural cracks. *Jiangxi Build. Mater.* **2022**, 95–97. [CrossRef]
47. Xue, C. Control of cast-in-place box girder concrete cracks in bridge structures. *Urban Constr. Theory Res. (Electron. Ed.)* **2024**, 151–153. [CrossRef]
48. Issa, M.A. Investigation of Cracking in Concrete Bridge Decks at Early Ages. *J. Bridge Eng.* **1999**, *4*, 116–124. [CrossRef]
49. Raupach, M. Models for the propagation phase of reinforcement corrosion—An overview. *Mater. Corros.* **2006**, *57*, 605–613. [CrossRef]
50. Chen, D.; Mahadevan, S. Chloride-induced reinforcement corrosion and concrete cracking simulation. *Cem. Concr. Compos.* **2008**, *30*, 227–238. [CrossRef]
51. Capozucca, R. Damage to reinforced concrete due to reinforcement corrosion. *Constr. Build. Mater.* **1995**, *9*, 295–303. [CrossRef]
52. Ahmad, S. Reinforcement corrosion in concrete structures, its monitoring and service life prediction—A review. *Cem. Concr. Compos.* **2003**, *25*, 459–471. [CrossRef]
53. Khan, M.U.; Ahmad, S.; Al-Gahtani, H.J. Chloride-Induced Corrosion of Steel in Concrete: An Overview on Chloride Diffusion and Prediction of Corrosion Initiation Time. *Int. J. Corros.* **2017**, *2017*, 5819202. [CrossRef]
54. Cheng, Y.; Zhang, Y.; Wu, C.; Jiao, Y. Experimental and Simulation Study on Diffusion Behavior of Chloride Ion in Cracking Concrete and Reinforcement Corrosion. *Adv. Mater. Sci. Eng.* **2018**, *2018*, 8475384. [CrossRef]
55. Li, Z. Seismic vulnerability analysis of large-span high-pier rigid frame bridges under chloride ion erosion. *Fujian Transp. Sci. Technol.* **2024**, 31–35.
56. Bažant, Z.P. Physical Model for Steel Corrosion in Concrete Sea Structures—Application. *J. Struct. Div.* **1979**, *105*, 1155–1166. [CrossRef]
57. Zheng, Z. Study on deterioration mechanisms and protection measures of reinforced concrete bridge durability. *Fujian Transp. Sci. Technol.* **2021**, *90–94*, 125.
58. Ren, J. Investigation into the Influencing Factors and Optimization Design of Bridge Structure Durability. *Transp. World* **2024**, 205–207. [CrossRef]
59. Yu, D. Experimental Study on Durability of Precast Segmental Beam Joints of Bridges Under Freeze-Thaw and Carbonation in Northern Hebei. Master's Thesis, Hebei University of Architecture, Zhangjiakou, China, 2020. [CrossRef]
60. Shi, C.; Zhao, R.; Wang, W.; Zhang, S. Effect of load on bonding properties and salt freeze-thaw resistance of bridge expansion joint concrete. *Constr. Build. Mater.* **2024**, *411*, 134680. [CrossRef]
61. Ma, Z.; Zhu, F.; Ba, G. Effects of freeze-thaw damage on the bond behavior of concrete and enhancing measures. *Constr. Build. Mater.* **2019**, *196*, 375–385. [CrossRef]
62. Tian, J.; Hong, J. Influence of freeze-thaw cycles on the service life of concrete pavements. *Shanxi Archit.* **2008**, 287–289. [CrossRef]
63. Peng, R.; Qiu, W.; Teng, F. Investigation on seawater freeze-thaw damage deterioration of marine concrete structures in cold regions from multi-scale. *Ocean. Eng.* **2022**, *248*, 110867. [CrossRef]
64. Xu, Y.; Li, X.; Zhang, P. Influence of freeze-thaw cycles on the service life of concrete bridges. *J. Hebei Univ. Technol.* **2009**, *38*, 107–110.
65. Xue, F.; Zhao, S.; Liu, J. Effects of freeze-thaw erosion on concrete bridges in plateau mountainous areas and protective measures. *Urban Roads Bridges Flood Control* **2018**, 199-201+231+21-22. [CrossRef]

66. Sun, W.; Zhang, Y.M.; Yan, H.D.; Mu, R. Damage and damage resistance of high strength concrete under the action of load and freeze-thaw cycles. *Cem. Concr. Res.* **1999**, *29*, 1519–1523. [CrossRef]
67. Liu, D.; Tu, Y.; Sas, G.; Elfgren, L. Freeze-thaw damage evaluation and model creation for concrete exposed to freeze–thaw cycles at early-age. *Constr. Build. Mater.* **2021**, *312*, 125352. [CrossRef]
68. Li, X.; Guo, M.; Man, G. Analysis of pavement diseases and maintenance technologies in road and bridge construction. *Urban Hous.* **2019**, *26*, 173–174.
69. Li, Y. Common diseases and protective measures of urban roads and bridges. *Smart City* **2017**, *3*, 154. [CrossRef]
70. Yang, F. Common diseases and durability analysis of concrete bridges. *Eng. Constr.* **2016**, *48*, 95-97+101. [CrossRef]
71. Basheer, L.; Basheer, P.A.M.; Long, A.E. Influence of coarse aggregate on the permeation, durability and the microstructure characteristics of ordinary Portland cement concrete. *Constr. Build. Mater.* **2005**, *19*, 682–690. [CrossRef]
72. Zhou, Y.; Gencturk, B.; Willam, K.; Attar, A. Carbonation-Induced and Chloride-Induced Corrosion in Reinforced Concrete Structures. *J. Mater. Civ. Eng.* **2015**, *27*, 04014245. [CrossRef]
73. Huang, Q.; Jiang, Z.; Zhang, W.; Gu, X.; Dou, X. Numerical analysis of the effect of coarse aggregate distribution on concrete carbonation. *Constr. Build. Mater.* **2012**, *37*, 27–35. [CrossRef]
74. Chen, Z.; Wu, D.; Dong, X. Analysis of causes for cracking of bottom slabs during construction of prestressed concrete continuous box girder bridges. *World Bridges* **2021**, *49*, 103–107.
75. Ruan, X.; Wang, B.; Wu, J.; Zhao, X.; Chen, Y. Identification of spalling and rebar exposure defects in reinforced concrete bridges based on deep learning. *World Bridges* **2020**, *48*, 88–92.
76. Yang, H.; Pan, J. Crack control technology during construction of concrete continuous beams of Hong Kong-Zhuhai-Macao Bridge. *Bridge Constr.* **2017**, *47*, 106–111.
77. Hu, Z.; Yu, S.; Li, G. Cracking analysis in negative moment regions of prestressed steel-concrete box composite continuous beams. *Bridge Constr.* **2020**, *50*, 66–71.
78. Han, W. Study on temperature and shrinkage crack control of large volume concrete in bridge abutment foundations. *China Highw.* **2021**, 96–98. [CrossRef]
79. Liu, A. Research on repair technology of cracks in highway bridge abutment sidewalls based on mineral admixtures. *Transp. World* **2023**, 130–132. [CrossRef]
80. Wu, Z. Application and performance testing of polymer cement mortar in crack repair of road and bridge concrete structures. *Aging Appl. Compos. Mater.* **2022**, *51*, 88-90+10. [CrossRef]
81. Tan, Q. Analysis and Treatment Methods for Cracks in Concrete Bridge Structures. *Eng. Constr. Des.* **2024**, 233–235. [CrossRef]
82. Suo, D. Durability design analysis of highway bridge structures based on multi-factor analysis. *Transp. Sci. Manag.* **2025**, *6*, 118–120.
83. Dhir, R.K.; Jones, M.R. Development of chloride-resisting concrete using fly ash. *Fuel* **1999**, *78*, 137–142. [CrossRef]
84. Hossain, A.B.; Shrestha, S.; Summers, J. Properties of Concrete Incorporating Ultrafine Fly Ash and Silica Fume. *Transp. Res. Rec.* **2009**, *2113*, 41–46. [CrossRef]
85. Hariharan, A.R.; Santhi, A.S.; Ganesh, G.M. Effect of ternary cementitious system on compressive strength and resistance to chloride ion penetration. *Int. J. Civ. Struct. Eng.* **2011**, *1*, 695–706.
86. Thomas, M.D.A.; Bamforth, P.B. Modelling chloride diffusion in concrete: Effect of fly ash and slag. *Cem. Concr. Res.* **1999**, *29*, 487–495. [CrossRef]
87. Kobayashi, K.; Takewaka, K. Experimental studies on epoxy coated reinforcing steel for corrosion protection. *Int. J. Cem. Compos. Lightweight Concr.* **1984**, *6*, 99–116. [CrossRef]
88. Swamy, R.; Koyama, S. Epoxy coated rebars the panacea for steel corrosion in concrete. *Constr. Build. Mater.* **1989**, *3*, 86–91. [CrossRef]
89. Alsayed, S.H.; Al-Salloum, Y.A.; Almusallam, T.H. Performance of glass fiber reinforced plastic bars as a reinforcing material for concrete structures. *Compos. Part B Eng.* **2000**, *31*, 555–567. [CrossRef]
90. Shehata, M.H.; Thomas, M.D.A. The role of alkali content of Portland cement on the expansion of concrete prisms containing reactive aggregates and supplementary cementing materials. *Cem. Concr. Res.* **2010**, *40*, 569–574. [CrossRef]
91. Taha, B.; Nounu, G. Using lithium nitrate and pozzolanic glass powder in concrete as ASR suppressors. *Cem. Concr. Compos.* **2008**, *30*, 497–505. [CrossRef]
92. Xu, G.J.Z.; Watt, D.F.; Hudec, P.P. Effectiveness of mineral admixtures in reducing ASR expansion. *Cem. Concr. Res.* **1995**, *25*, 1225–1236. [CrossRef]
93. Gonen, T.; Yazicioglu, S. The influence of mineral admixtures on the short and long-term performance of concrete. *Build. Environ.* **2007**, *42*, 3080–3085. [CrossRef]
94. Shashiprakash, S.G.; Thomas, M.D.A. Sulfate resistance of mortars containing high-calcium fly ashes and combinations of highly reactive pozzolans and fly ash. *Spec. Publ.* **2001**, *199*, 221–238.

95. Collepardi, S.; Corinaldesi, V.; Moriconi, G.; Bonora, G.; Collepardi, M. Durability of high-performance concretes with pozzolanic and composite cements. *Spec. Publ.* **2000**, *192*, 159–172.
96. Shi, X.; Xie, N.; Fortune, K.; Gong, J. Durability of steel reinforced concrete in chloride environments: An overview. *Constr. Build. Mater.* **2012**, *30*, 125–138. [CrossRef]
97. Qu, F.; Li, W.; Dong, W.; Tam, V.; Tao, Y. Durability deterioration of concrete under marine environment from material to structure: A critical review. *J. Build. Eng.* **2021**, *35*, 102074. [CrossRef]
98. Niu, F.; Liu, Y.; Xue, F.; Sun, H.; Liu, T. Ultra-high performance concrete: A review of its material properties and usage in shield tunnel segment. *Case Stud. Constr. Mater.* **2025**, *22*, e04194. [CrossRef]
99. Bao, J.; Xue, S.; Zhang, P.; Dai, Z.; Cui, Y. Coupled effects of sustained compressive loading and freeze–thaw cycles on water penetration into concrete. *Struct. Concr.* **2021**, *22*, E944–E954. [CrossRef]
100. Zhang, Q. Analysis of structural damage of bridge concrete under environmental factors. *Shanxi Archit.* **2011**, *37*, 177–178. [CrossRef]
101. Kosior-Kazberuk, M.; Berkowski, P. Surface Scaling Resistance of Concrete Subjected to Freeze-thaw Cycles and Sustained Load. *Procedia Eng.* **2017**, *172*, 513–520. [CrossRef]
102. He, Z.; Tang, S.W.; Zhao, G.S.; Chen, E. Comparison of three and one dimensional attacks of freeze-thaw and carbonation for concrete samples. *Constr. Build. Mater.* **2016**, *127*, 596–606. [CrossRef]
103. Yuan, Y.; Zhao, R.; Li, R.; Wang, Y.; Cheng, Z.; Li, F.; Ma, Z. Frost resistance of fiber-reinforced blended slag and Class F fly ash-based geopolymer concrete under the coupling effect of freeze-thaw cycling and axial compressive loading. *Constr. Build. Mater.* **2020**, *250*, 118831. [CrossRef]
104. Ren, J.; Lai, Y. Study on the durability and failure mechanism of concrete modified with nanoparticles and polypropylene fiber under freeze-thaw cycles and sulfate attack. *Cold Reg. Sci. Technol.* **2021**, *188*, 103301. [CrossRef]
105. Ebrahimi Besheli, A.; Samimi, K.; Moghadas Nejad, F.; Darvishan, E. Improving concrete pavement performance in relation to combined effects of freeze–thaw cycles and de-icing salt. *Constr. Build. Mater.* **2021**, *277*, 122273. [CrossRef]
106. Tian, J.; Wang, W.; Du, Y. Damage behaviors of self-compacting concrete and prediction model under coupling effect of salt freeze-thaw and flexural load. *Constr. Build. Mater.* **2016**, *119*, 241–250. [CrossRef]
107. Li, Y.; Wang, R.; Li, S.; Zhao, Y.; Qin, Y. Resistance of recycled aggregate concrete containing low- and high-volume fly ash against the combined action of freeze–thaw cycles and sulfate attack. *Constr. Build. Mater.* **2018**, *166*, 23–34. [CrossRef]
108. Li, N.; Long, G.; Fu, Q.; Wang, X.; Ma, K.; Xie, Y. Effects of freeze and cyclic flexural load on mechanical evolution of filling layer self-compacting concrete. *Constr. Build. Mater.* **2019**, *200*, 198–208. [CrossRef]
109. Hao, L.; Liu, Y.; Wang, W.; Zhang, J.; Zhang, Y. Effect of salty freeze-thaw cycles on durability of thermal insulation concrete with recycled aggregates. *Constr. Build. Mater.* **2018**, *189*, 478–486. [CrossRef]
110. Pang, B.; Zhou, Z.; Cheng, X.; Du, P.; Xu, H. ITZ properties of concrete with carbonated steel slag aggregate in salty freeze-thaw environment. *Constr. Build. Mater.* **2016**, *114*, 162–171. [CrossRef]
111. Monkman, S.; Shao, Y. Carbonation Curing of Slag-Cement Concrete for Binding CO_2 and Improving Performance. *J. Mater. Civ. Eng.* **2010**, *22*, 296–304. [CrossRef]
112. Wang, R.; Zhang, Q.; Li, Y. Deterioration of concrete under the coupling effects of freeze–thaw cycles and other actions: A review. *Constr. Build. Mater.* **2022**, *319*, 126045. [CrossRef]
113. Luo, S.; Bai, T.; Guo, M.; Wei, Y.; Ma, W. Impact of Freeze–Thaw Cycles on the Long-Term Performance of Concrete Pavement and Related Improvement Measures: A Review. *Materials* **2022**, *15*, 4568. [CrossRef]
114. Tian, J.; Wu, X.; Zheng, Y.; Hu, S.; Ren, W. Investigation of damage behaviors of ECC-to-concrete interface and damage prediction model under salt freeze-thaw cycles. *Constr. Build. Mater.* **2019**, *226*, 238–249. [CrossRef]
115. Cheng, Y.; Zhang, Y.; Jiao, Y.; Yang, J. Quantitative analysis of concrete property under effects of crack, freeze-thaw and carbonation. *Constr. Build. Mater.* **2016**, *129*, 106–115. [CrossRef]
116. Wang, R.; Hu, Z.; Li, Y.; Wang, K.; Zhao, H. Review on the deterioration and approaches to enhance the durability of concrete in the freeze–thaw environment. *Constr. Build. Mater.* **2022**, *321*, 126371. [CrossRef]
117. Beeby, A.W.; Narayanan, R.S. *Designers' Guide to Eurocode 2: Design of Concrete Structures: Designers' Guide to EN 1992-1-1 and EN 1992-1-2 Eurocode 2: Design of Concrete Structures Design of Concrete Structures General Rules and Rules for Buildings and Structural Fire Design*; Thomas Telford Publishing: London, UK, 2005.
118. Wang, X.; Yang, Q.; Peng, X.; Qin, F. A Review of Concrete Carbonation Depth Evaluation Models. *Coatings* **2024**, *14*, 386. [CrossRef]
119. Ibrahim, A.; Abdelkhalek, S.; Zayed, T.; Qureshi, A.H.; Mohammed Abdelkader, E. A comprehensive review of the key deterioration factors of concrete bridge decks. *Buildings* **2024**, *14*, 3425. [CrossRef]
120. Fan, C.; Ding, Y.; Liu, X.; Yang, K. A review of crack research in concrete structures based on data-driven and intelligent algorithms. *Structures* **2025**, *75*, 108800. [CrossRef]

121. Liu, Y.; Feng, C.; Chen, W.; Fan, J. A Review of Research on Bridge Apparent Defect Detection Based on Machine Vision Methods. *China J. Highw. Transp.* **2024**, *37*, 1–15. [CrossRef]
122. Huang, J.; Lu, W.; Yin, C.; Fu, C. A Review of Research on the Impact of Corrosion on the Bonding Performance Between Steel Bars and Concrete. *Mater. Rep.* **2024**, *38*, 228–239.
123. Dong, H.; Li, H.; Yang, Z.; Wen, J.; Huang, F.; Wang, Z.; Yi, Z. Mechanism of Freeze-Thaw Damage in Concrete and Methods for Life Prediction. *Mater. Rep.* **2024**, *38*, 143–153.
124. Guo, J.; Sun, W.; Xu, Y.; Lin, W.; Jing, W. Damage Mechanism and Modeling of Concrete in Freeze–Thaw Cycles: A Review. *Buildings* **2022**, *12*, 1317. [CrossRef]

coatings

MDPI

Review

Review of Research on Prediction Models for Residual Life of Concrete Structures

Linyuan Qi [1], Xi Peng [1,2,3,*], Qiuwei Yang [1,2,*], Kangshuo Xia [4] and Bin Xu [3]

1 School of Civil and Transportation Engineering, Ningbo University of Technology, Ningbo 315211, China; qilinyuan@nbut.edu.cn
2 Engineering Research Center of Industrial Construction in Civil Engineering of Zhejiang, Ningbo University of Technology, Ningbo 315211, China
3 Ningbo Roaby Technology Industrial Group Co., Ltd., Ningbo 315800, China; 5632374376@aliyun.com
4 Key Laboratory of New Technology for Construction of Cities in Mountain Area, School of Civil Engineering, Chongqing University, Chongqing 400045, China; xiakangshuo@cqu.edu.cn
* Correspondence: pengxi@nbut.edu.cn (X.P.); yangqiuwei@nbut.edu.cn (Q.Y.)

Abstract: The performance degradation of concrete structures directly impacts their safety. As such, accurately predicting their remaining service life is critical for effective operation and maintenance management. This paper reviews the key factors influencing the degradation of concrete structures, providing a comprehensive summary of current research on deterioration mechanisms, steel corrosion, crack propagation, and chloride ion penetration. It also compares the advantages and limitations of physical, empirical, statistical, and machine learning models used for life prediction. A critical aspect highlighted in this paper is the importance of model validation based on real-world field data, which can more effectively determine the applicability of prediction models in actual engineering practice. Model validation incorporates evaluation metrics like sensitivity analysis to gauge how fluctuations in input parameters, such as temperature, influence life prediction models and thereby reveal the uncertainties inherent in complex engineering environments. Currently, life prediction models are widely applied to infrastructure projects like bridges and tunnels. By incorporating environmental factors such as chloride ion concentration, temperature, and humidity, as well as real-time monitoring data, these models effectively predict the remaining service life, aiding engineers in developing optimized maintenance strategies. However, current models still face challenges in terms of data requirements and accuracy. Future research should focus on the integration of hybrid models and intelligent technologies. By leveraging the combined strengths of physical and data-driven approaches, hybrid models can enhance prediction accuracy. Additionally, intelligent technologies and real-time monitoring will help dynamically update and optimize models, resulting in more precise and efficient life predictions.

Keywords: concrete structures; performance degradation; life prediction; evaluation metrics; hybrid models

1. Introduction

Concrete structures find widespread application in the field of civil engineering, especially in the construction of buildings, bridges, tunnels, and other forms of infrastructure development, where they serve a fundamental function. The lifetime of such structures needs to be predicted to maintain public safety, minimize costs of maintenance, and improve the efficiency of resource allocation. Concrete is the reigning material in infrastructure construction as the primary building material. Research shows that nearly half of the global production of ordinary Portland cement goes towards the production of about 11 billion tons of concrete [1]. With such a vast demand, the assessment of the durability of concrete structures has become a more pertinent subject of research with far-reaching implications.

The durability of concrete structures over the long term is especially critical under severe environments, such as those characterized by marine and cold climates. Research has established that, in these specific environments, the rate of corrosion affecting concrete structures significantly exceeds the rate under normal conditions, leading to increased deterioration. For instance, the high level of chloride ions characteristic of marine environments promotes the corrosion of reinforcing bars, thereby compromising the structural integrity of the concrete [2,3]. In cold environments, low temperatures cause the internal moisture in concrete to freeze. As the temperature fluctuates, the expansion and contraction of the ice generate stress within the concrete, particularly under prolonged exposure to low temperatures. While cold environments do not directly cause freeze–thaw cycles, extended periods of low temperatures can significantly reduce the physical properties of concrete, thereby affecting its structural stability [4]. In contrast, freeze–thaw cycles refer to the repeated freezing and melting of moisture within concrete due to temperature fluctuations. During freezing, the moisture expands, and during melting, it contracts. These cycles generate internal expansion stresses within the concrete, which gradually enlarge microcracks, ultimately leading to surface spalling and crack propagation. The effect of freeze–thaw cycles on concrete's durability is especially pronounced [5,6].

The degradation of concrete structures is a complicated multivariable phenomenon [7,8] involving numerous factors such as corrosion of crack propagation, chloride ion infiltration, and material fatigue. In particular, chloride-induced corrosion causes rust on the surface of the rebar, initiating crack propagation that accelerates the reduction in the structure's strength and service life. Thus, the precise prediction of the residual service life of concrete structures is crucial to avert structural failure and effectively extend their service life.

The development of machine learning and artificial intelligence in recent years has enabled the creation of lifespan prediction models that involve real-time monitoring data and multi-factor integration, thereby rendering it a forefront research area. As opposed to the conventional physical models, the novel approaches utilize big datasets and machine learning algorithms to optimally address complicated nonlinear relationships, thereby greatly improving the accuracy of predictions [9,10].

This paper offers an extensive overview of current lifespan prediction models for concrete structures and their merits and drawbacks, and proposes ways for enhancement. First, we review the fundamental principles and application domains of physical and statistical models. Next, we examine how machine learning-based models, by processing real-time monitoring data, enable dynamic updates that enhance prediction accuracy. Finally, we discuss future research directions, particularly the integration of multi-factor coupling with intelligent models, and how real-time data feedback can further optimize prediction outcomes. As technology advances, future lifespan prediction models for concrete structures are expected to become increasingly accurate and efficient, offering enhanced support for structural health management in civil engineering.

2. Aging and Failure Mechanisms of Concrete Structures

The aging and failure mechanisms of concrete structures are crucial aspects of concrete durability research. Over time, chloride ions penetrate into concrete, eroding the passive film on the steel reinforcement, initiating electrochemical corrosion and leading to the expansion of corrosion products. This results in concrete cracking, accelerates the ongoing corrosion of the steel reinforcement, significantly reduces the mechanical properties of the concrete, and severely compromises its service life. This process involves the coupling of multiple factors, including rebar corrosion, crack growth, and environmental effects. As corrosion and cracking advance, the concrete's strength and stability diminish, potentially resulting in catastrophic structural failure.

2.1. Rebar Corrosion and the Decline of Concrete's Mechanical Properties

Rebar corrosion is a primary factor contributing to the degradation of concrete structures, particularly in environments with chlorides, moisture, and temperature variations, where the corrosion rate accelerates significantly. The expansion of corroded rebar generates internal stress, causing surface cracking in the concrete, which, in turn, accelerates the decline of its mechanical properties. A thorough understanding of rebar corrosion and its effects on the mechanical behavior of concrete is essential for providing scientific guidance in the design, repair, and maintenance of concrete structures.

For instance, Tian et al. [11] reviewed the key characteristics of rebar corrosion in concrete under natural environmental conditions. The study found that, while the corrosion rate decreases over time, crack propagation accelerates, highlighting the importance of establishing a correlation between laboratory accelerated tests and natural corrosion, as well as developing a durability assessment system for structures after repair. Furthermore, to evaluate the corrosion risk of rebar in marine environments, Ju et al. [12] proposed a time-dependent model for the surface chloride ion concentration and diffusion coefficient. This model can predict the variation in chloride ion concentration in concrete under various heights and exposure durations. Zhang et al. [13] developed a probabilistic model for rebar corrosion based on the Markov chain model and Faraday's law. Their experimental data showed that, as current density increased from 50 $\mu A/cm^2$ to 100 $\mu A/cm^2$, the corrosion rate rose from 6.645×10^{-5} mm/h to 13.290×10^{-5} mm/h. The increase in current density accelerated the corrosion process, providing valuable insights for the evaluation and durability prediction of rebar corrosion in concrete structures. As corrosion advances, the concrete cover deteriorates, exposing the rebar to external environmental conditions. Sun et al. [14] proposed an enhanced model for concrete carbonation and rebar corrosion to evaluate crack vulnerability induced by corrosion. This model accurately predicts crack evolution in concrete bridges under corrosion and offers guidance for bridge maintenance. Zhang and Su [15] examined non-uniform crack propagation due to rebar corrosion and developed a corner crack model to describe the degradation of concrete cover. This model provides theoretical support for the durability design, service life prediction, and crack width estimation of concrete structures. To assess the durability of concrete structures exposed to marine environments for extended periods, Wang et al. [16] introduced a chloride ion diffusion model that accounts for the coupling effects of coarse aggregate and rebar. The study demonstrated that, as the volume fraction of coarse aggregate increased from 0 to 0.3, the chloride ion diffusion coefficient decreased from 9.37×10^{-12} m^2/s to 5.98×10^{-12} m^2/s, a reduction of 20%. This suggests that coarse aggregate improves the concrete's resistance to chloride ion penetration by modifying its pore structure, thus enhancing durability. Hajkova et al. [17] proposed a chemical–mechanical model for chloride-induced rebar corrosion and examined the impact of corrosion on the serviceability of concrete structures. Their study provides valuable theoretical support for assessing the serviceability of structures affected by rebar corrosion, underscoring the importance of early diagnosis and repair. Wang and Lu [18] investigated the corrosion process of rebar in ordinary concrete and concrete with varying fly ash contents under constant climatic conditions. The results showed that increasing the fly ash substitution ratio from 0% to 45% significantly reduced the corrosion rate of rebar, thereby improving concrete durability. Finally, Ann et al. [19] analyzed chloride ion penetration in concrete using four different surface chloride ion accumulation models. The study found that the constant surface chloride ion model overestimated corrosion risk, whereas the optimized accumulation model more accurately predicted chloride ion distribution and corrosion life. This approach provides a reliable foundation for structural maintenance and durability assessment.

2.2. Crack Propagation and Structural Damage

Rebar corrosion not only threatens the strength and durability of concrete, but also significantly compromises the overall safety of the structure. Common corrosion products include carbide, chloride, and sulfate compounds. Table 1 presents a comparison of the effects of these three typical corrosion products on the corrosion process. As observed from Table 1, among the three corrosion products, chloride ions have the most significant impact on both the concrete protection layer and the corrosion process.

Table 1. Impact of three common corrosion products on concrete protection and the corrosion process.

Corrosion Product	Conditions for Corrosion Occurrence	Effect on Concrete Cover	Effect on Corrosion Process	Relative Effect During Continuous Corrosion Process
Carbonate Products [20]	Reaction of carbon dioxide with calcium hydroxide, forming calcium carbonate	Reduces the alkalinity of the concrete, compromising the concrete cover of the reinforcement	Lowers the pH of concrete; the steel protection layer gradually fails, promoting steel corrosion	Moderate
Chloride Products [21]	Chloride ions penetrate, particularly in marine or high-salinity environments	Increases the permeability of concrete, diminishing its protective capacity	Directly breaks down the steel passivation layer, initiates corrosion, causes pitting (localized corrosion)	High
Sulfate Products [22]	Sulfates react with cement hydration products to form expansive compounds	Causes expansion and cracking, damaging the concrete structure	Expanding reactions create cracks, increasing permeability and accelerating corrosion	Low

As corrosion progresses, the expansion of the rebar generates internal cracks in the concrete. The study in reference [23] primarily investigates the influence of different types of cement and supplementary cementitious materials (SCMs) on the corrosion rate of reinforcement steel, particularly under conditions with cover depths of 20 mm and 40 mm, and crack widths of 0.2 mm and 0.7 mm. The experimental data in Table 2 are derived from the research presented in reference [23], with the corrosion rates obtained under accelerated corrosion conditions in a laboratory setting. These conditions involve a wet–dry cycling process and exposure to a 5% sodium chloride solution. The corrosion rate is calculated by applying a small electrical current and monitoring the changes in the reinforcement steel potential, which allows for determining the polarization resistance of the steel. The corrosion rate is then calculated using the Stern–Geary equation. According to Table 2, significant differences in corrosion rates are observed for different types of cement; ordinary Portland cement (PC) shows higher corrosion rates under both 20 mm and 40 mm cover depths at 2.65 $\mu A/cm^2$ and 1.20 $\mu A/cm^2$, respectively. This is primarily due to its lower resistivity, which allows chloride ions to more easily penetrate the steel surface. In contrast, cements containing fly ash (FA) and ground granulated blast furnace slag (GGBS) significantly reduce the corrosion rate. In the case of fly ash concrete, the corrosion rates are 0.64 $\mu A/cm^2$ for 20 mm cover depth and 0.50 $\mu A/cm^2$ for 40 mm cover depth, indicating that fly ash increases the concrete resistivity and inhibits chloride ion ingress. GGBS exhibits the lowest corrosion rates at 0.35 μA/cm^2 and 0.24 μA/cm^2, respectively, due to improved concrete density and resistivity, which reduces chloride ion penetration. Silica fume shows considerable fluctuation in corrosion rates, especially at 40 mm cover depth and larger crack widths, where the corrosion rate reaches 1.03 μA/cm^2. This may be attributed to the rapid hydration characteristics of silica fume, leading to an unstable microstructure, which exposes the steel to corrosive environments at the crack sites and accelerates the corrosion process.

Table 2. Influence of crack width on corrosion rate (μA /cm^2).

Corrosion rate at 40 mm cover (μA/cm^2)			
Cement Type	Crack width		Percentage change
	0.2 mm	0.7 mm	
Ordinary Portland Cement	1.20	1.48	23%
7% Silica Fume Blended Cement	0.59	1.03	75%
30% Fly Ash Blended Cement	0.39	0.50	28%
50% Ground Granulated Blast Furnace Slag Blended Cement	0.35	0.53	51%
Corrosion rate at 20 mm cover (μA/cm^2)			
Cement Type	Crack width		Percentage change
	0.2 mm	0.7 mm	
Ordinary Portland Cement	2.65	3.23	22%
7% Silica Fume Blended Cement	0.67	1.12	67%
30% Fly Ash Blended Cement	0.64	0.71	11%
50% Ground Granulated Blast Furnace Slag Blended Cement	0.39	0. 51	31%

Therefore, supplementary materials such as fly ash and ground granulated blast furnace slag significantly improve the corrosion resistance of concrete by enhancing its resistivity and density. In contrast, ordinary Portland cement and silica fume exhibit different corrosion rate variations due to their distinct microstructural characteristics.

Several studies have investigated crack propagation and structural damage. For instance, Guo et al. [24] proposed a new fatigue life prediction model for ordinary Portland cement (PC) beams that takes into account the spatial variability of pitting corrosion. The model uses a probabilistic approach to predict fatigue damage resulting from rebar corrosion. Wu et al. [25], based on Fick's second law and Paris' law, developed a model to predict the lifespan of reinforced concrete (RC) beams under chloride ion corrosion and fatigue. Their study revealed that crack initiation significantly affects fatigue life, which decreases with lower stress ratios, longer service times, and higher environmental temperatures, offering valuable insights for structural safety assessments. As corrosion progresses, cracks in the concrete propagate, which usually leads to the deterioration of the structure and, in extreme cases, the premature failure of the reinforced concrete elements. Ge and Kim [26] introduced a probabilistic update method based on Bayesian inference, integrating multiple inspection results using Markov chain Monte Carlo sampling. This approach enhances the prediction accuracy of remaining life under corrosion–fatigue interactions in reinforced concrete structures, providing critical support for preventive maintenance and life prediction. Bhargava et al. [27] developed a model to evaluate concrete cover cracking caused by rebar corrosion. This model considers the effect of corrosion on the concrete cover and is capable of predicting the probability of crack propagation and structural deterioration. Choe et al. [28] also introduced a probabilistic model integrating chloride ion corrosion, time-dependent corrosion rates, and shear capacity analysis for examining the vulnerability of reinforced concrete columns. Sensitivity analysis was employed to identify the most dominant parameters, giving a theoretical basis for reinforced concrete structure life prediction and lifecycle cost estimation. These studies highlight the fact that corrosion is not just a parameter influencing concrete strength, but, by inducing crack propagation and structural damage accumulation, eventually compromises the safety of the whole structure. As the corrosion process continues, the structural integrity is lost, highlighting the need for early detection and evaluation to ensure the long-term safety of concrete structures.

2.3. Chloride Ion Penetration and Concrete Corrosion

Concrete structures, particularly those exposed to marine and cold climates, are prone to corrosion from chloride ion intrusion. Chloride ions are one of the primary causes of rebar corrosion initiation, and their diffusion and penetration significantly affect concrete durability. It is important to distinguish between corrosion efficiency and degree of reinforcement corrosion. Corrosion efficiency refers to the effectiveness of the electrochemical reaction in causing the dissolution of steel, while the degree of reinforcement corrosion describes the overall extent of material loss and damage to the rebar due to corrosion. Figure 1, derived from reference [29], illustrates that the corrosion efficiency of the 5% NaCl solution is highest at a current density of 3 mA/cm^2, suggesting that the corrosion rate of the rebar increases at lower current densities. This also highlights the variation in the corrosion efficiency of rebar due to chloride ions under different environmental conditions.

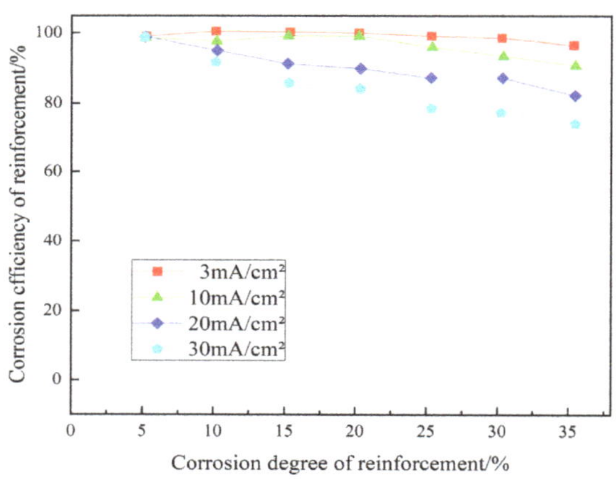

Figure 1. The corrosion efficiency of reinforcement under 5% NaCl solution.

In recent years, extensive research has been directed towards chloride ion diffusion models in concrete, corrosion initiation mechanisms, and the corrosion prediction of reinforcing bars. For example, Yu et al. [30] performed long-term experimental investigations to examine the correlation between the chloride ion diffusion coefficient, exposure time, and curing age. The results obtained showed that the diffusion coefficient is a power function of curing age, but its relation to exposure time was not found to be significant. This observation has provided a scientific basis for estimating the durability of concrete structures. Chen et al. [31] have proposed a probabilistic multi-factor predictive model in which various aggregate sizes (i.e., dmax = 10 mm, 20 mm, 25 mm) are taken into consideration when estimating the chloride ion diffusion coefficient. The results proved a significant increase in the coefficient of diffusion for chloride ions relating to the increased sizes of aggregates. Furthermore, an analysis of a case study of Shenzhen-Zhongshan Bridge with a concrete cover thickness of 78 mm provided evidence that raising the cover thickness reduced the penetration depth of the chloride ions and, hence, enhanced the durability of the structure. Their study highlights the relevance of both the cover thickness of the concrete and the diffusion coefficient of the chloride ions as key factors deciding durability. In practical applications, the concentration of chloride ions plays a role in assessing the durability of concrete structures. To solve this issue, Guo et al. [32] proposed a random forest–fuzzy logic hybrid model, in which the random forest algorithm was utilized to enhance training samples and forecast surface chloride concentrations in concrete. The model effectively deals with complicated environmental conditions. In the same way, Taffese and Espinosa-Leal [33] used five various machine learning models with the goal of predicting chloride ions' permeability in concrete, thereby developing novel approaches for assessing the lifespan and durability of construction materials. Liu et al. [34] analyzed how precipitation affects chloride ions' transport dynamics in concrete and established that higher rainfall intensity and duration decreased chloride ion concentration at the surface considerably. For chloride-induced corrosion in concrete, Shao et al. [35] proposed a combined evaluation approach with both deterministic and probabilistic methodologies for evaluating the corrosion initiation life of reinforced concrete hollow piles in chloride environments. Safehian and Ramezanianpour [36] also analyzed chloride ion penetration in a dock exposed to harsh marine environments, taking into account the influence of construction methods and exposure conditions. Their findings underscored the significant influence that microenvironments have on the penetration of chloride ions. The research shows that, with the progress in investigating corrosion mechanisms triggered by chloride, increasingly predictive models and assessment methods are being created, thus giving valuable support for concrete maintenance.

In conclusion, concrete structure ageing and deterioration are complex and multi-factorial processes involving rebar corrosion, cracking, chloride ion ingress, and fatigue. As artificial intelligence and machine learning have advanced in recent times, real-time monitoring-data-driven lifespan prediction models and multi-factor coupling models have gained popularity in research on concrete structure ageing. Future research must further investigate the relationship between these variables, with a particular focus on how real-time monitoring and feedback systems can enhance predictive modeling. This would, therefore, allow for more precise and effective assistance in the maintenance and management of concrete structures.

3. Classification of Lifespan Prediction Models

This section outlines different types of life prediction models and categorizes them into four types: physical models, empirical models, statistical models, and machine learning-based models. It also provides a detailed discussion on each model type, analyzing their principles, application areas, advantages, and limitations.

3.1. Physical Lifespan Prediction Models

Physical lifespan prediction models of structures assess their durability by simulating the interaction between concrete and the environment and physical–chemical processes. Founded on sound theoretical principles, these models yield credible lifespan predictions for varying environmental conditions. Carbonation models and chloride ion penetration models are representative models in this group.

3.1.1. Carbonation Depth Model

Carbonation is a major cause of concrete degradation. Carbon dioxide migrates through the surface of the concrete, reducing its pH and harming the protective coating of the rebar, leading to the corrosion of the rebar. The Fick model, used routinely for the estimation of carbonation depth, assumes that carbonation depth varies with the square root of time [37]. The equation is given as follows.

$$h = x_0\sqrt{t} \tag{1}$$

In this Equation (1), h denotes the carbonation depth, or the extent to which carbon dioxide penetrates into the concrete, t represents time, and x_0, known as the carbonation coefficient, indicates the carbonation rate of the concrete.

The Fick model's linear nature and strong physical basis make it amenable to particular uses, namely the initial estimation of carbonation depth. Its drawbacks are, however, exposed under complicated environmental circumstances, for instance, where the carbonation rate changes with high humidity and, hence, departs from the Fick model's assumptions [38]. As a solution to this issue, Cao et al. [39] suggested an improved variant of the Fick model by adding a correction term to the basic equation.

$$h = x_0\sqrt{t} + x_1\frac{e^t - e^{-t}}{e^t + e^{-t}} \tag{2}$$

In this Equation (2), x_1 denotes an unknown fitting parameter in the correction term introduced.

The data in Table 3 come from reference [39]. As demonstrated in Table 3, the improved model not only reduces prediction errors, but also minimizes the variability in the results, thereby enhancing the applicability and reliability of the Fick model for carbonation depth prediction.

Table 3. Average relative error and standard deviation of the fitted carbonation test results at 20 °C and 30 °C.

Carbonation Conditions	Statistical Indicators	Fick Model	New Model
20 °C Carbonation	Mean error value $\bar{e}$ (%)	1.80%	0.78%
	Standard deviation σ (%)	1.91%	0.35%
30 °C Carbonation	Mean error value $\bar{e}$ (%)	6.67%	2.67%
	Standard deviation σ (%)	7.70%	3.39%

3.1.2. Chloride Ion Penetration Model

Chloride ion penetration is a significant cause of corrosion in reinforced concrete structures, particularly in marine and industrial environments. Chloride ions diffuse through the concrete's surface and penetrate the structure, compromising the rebar's passive film, which leads to corrosion. The prediction of chloride ion penetration is generally based on Fick's second law. Yu et al. [40] applied Fick's second law model to describe the diffusion process of chloride ions in concrete. The basic Equation (3) of the model is as follows:

$$C(x,t) = C_0 + (C_s - C_0)\left[1 - erf\left(\frac{x}{2\sqrt{tD_{cl}}}\right)\right] \tag{3}$$

$C(x,t)$ represents the chloride ion concentration at time t and depth x,
C_0 is the initial concentration, typically the chloride ion concentration in the concrete,

C_s is the surface concentration, corresponding to the concentration at the concrete surface exposed to water or air,
D_{cl} is the chloride diffusion coefficient,
erf denotes the error function.

In practical applications, Fick's law assumes uniform diffusion. However, in concrete structures, the diffusion coefficient is typically not constant due to variations in environmental factors such as temperature and humidity. As a result, numerous studies have modified Fick's law to account for the time dependence of the diffusion coefficient and other influencing factors. The improved Equation (4) [41] is as follows:

$$C(x,t) = C_s \left[1 - \operatorname{erf}\left(\frac{x}{2\sqrt{\frac{D_0}{1-n}\left(\frac{t_0}{t}\right)^n t}} \right) \right] \quad (4)$$

D_0 represents the initial chloride ion diffusion coefficient,
t_0 is the reference time,
n is the coefficient for time-dependent decay.

Equation (4) has been evaluated in both laboratory and field environments. Under laboratory conditions, the model tends to show a larger and more stable diffusion coefficient, near the ideal time exponent, and comparably low computational complexity. For field applications, the diffusion coefficient is significantly affected by external conditions, so there are larger fluctuations. The time exponent can be different from the ideal, and other factors must be taken into account. Although this adds computational complexity, the model provides a more realistic description of actual nonlinear diffusion processes and exhibits great benefits in practical applications. Chen et al. [42] also presented a variable-order fractional diffusion model that has the ability to describe the chloride ion diffusion–absorption process more accurately by introducing a concentration influence factor k. The study shows that the new model describes experimental data more precisely than the classic Fick's second law and can forecast the distribution of chloride ion concentration for 200 days.

Physical models predict the lifespan of concrete structures by simulating the interaction between environmental factors and physical–chemical processes. Examples are the Fick model and the chloride ion diffusion model. The Fick model estimates carbonation depth as a function of the square root of time and is, thus, suitable for application during the early age, but accuracy declines under complicated environments. The improved model incorporates a time-dependent diffusion coefficient for increased accuracy. The diffusion model of chloride ions using Fick's second law simulates the rebar corrosion caused by chloride ion intrusion. With the influence of environmental variations on the diffusion coefficient, researchers have proposed more advanced models, such as those with fractional derivatives, to better fit actual conditions. In conclusion, although physical models offer a sound theoretical basis for lifespan predictions, optimization is needed under complex environments.

3.2. Empirical Models

The estimation of the lifespan of concrete structures has the highest priority in civil engineering, especially their durability for long periods under changing environmental conditions and loading. By combining historical data with information about environmental effects and changes in loading, researchers have developed a series of empirical formulas and data fitting models to precisely estimate the lifespan of concrete structures. This section presents an exhaustive overview of empirical models, such as their applicability, limitations, and future directions.

3.2.1. Empirical Formulas

Empirical formulas are predictive models that have been created by analyzing large amounts of historical data, as well as environmental factors such as temperature, humidity, and corrosive substances, along with loading conditions. Empirical formulas are a scientifically based platform for the construction, building, and maintenance of concrete structures.

For example, Lu et al. [43] compared seven empirical models on the basis of 156 experimental datasets and introduced a new model that comprises variables such as concrete resistivity, temperature, humidity, corrosion time, and chloride ion content. This model effectively predicts corrosion rates and, through error probability distribution analysis, provides theoretical support for the reliability-based lifespan prediction of reinforced concrete structures. Chen et al. [44] developed a simplified empirical model using regression analysis, enabling the rapid evaluation of reinforced concrete structure service life and offering a practical computational tool for engineering applications. While empirical models are convenient and user-friendly, they generally rely on historical data and lack the detailed consideration of structural specifics and complex environmental changes.

3.2.2. Applicability and Limitations of Empirical Models

While empirical models are simple to use and computationally efficient, their application presents several challenges. First, these models often fail to account for the impact of specific environmental factors, such as climate change and various corrosive media, which can lead to biases in predictions. Second, empirical models rely heavily on historical data, which limits their predictive accuracy in situations with scarce data or unique environmental conditions. Table 4 summarizes several chloride diffusion models, which are examples of empirical models used to predict chloride penetration in concrete. These models, although computationally efficient, also face challenges such as simplifying assumptions and reliance on specific conditions. The Table 4 provides a comparative overview of these models, highlighting their advantages and disadvantages.

It is noteworthy that, while the empirical models in Table 4 provide practical methods for predicting chloride ion ingress into concrete, these models must be used with great caution, particularly in situations where the data are limited or the environmental conditions are complex.

Table 4. Comparison of some empirical models.

Model	Disadvantages	Advantages
The model in reference [45]: $C_{x,t}^{f} = C_0^{f}\left[1 - \text{erf}\left(\frac{x}{2\sqrt{D_c{}^{f} t}}\right)\right]$ $C_{x,t}^{f}$: chloride concentration in the pore fluid at depth x from the concrete surface after time t of exposure to a chloride environment; C_0^{f}: free chloride concentration at the concrete surface; **erf**: error function; $D_c{}^{f}$: the coefficient of Cl^- diffusion.	1. Assumes chloride diffusion is uniform, ignoring the pores and cracks in concrete, which affects accuracy. 2. Assumes the diffusion coefficient remains constant, whereas it actually changes with time and environmental conditions. 3. In extreme environments such as high or low temperatures, the model's applicability and accuracy decrease.	1. The formula is based on Fick's second law, with a simple structure that is easy to apply. 2. It is consistent with experimental data for short-term and medium-term exposures (90 to 520 days). 3. Suitable for different types of concrete and exposure environments.
The model in reference [46]: $C = C_0\left\{1 - \text{erf}\left(\frac{x}{2\sqrt{\frac{D_1}{1-m}t^{(1-m)}}}\right)\right\}$ C: chloride concentration at distance x and time t; C_0: chloride concentration at the concrete surface; erf: error function; D_1: the initial chloride diffusion coefficient; m: the empirical coefficient that represents the time-dependent variation of the diffusion coefficient.	1. The m coefficient depends on the mix ratio and environment, requiring calibration. 2. Extensive experimental data are needed for validation and adjustment. 3. Long-term predictions (e.g., decades) face uncertainties regarding diffusion coefficient behavior.	1. Incorporating time-dependent diffusion coefficients enhances long-term chloride concentration predictions. 2. Predictions align well with actual data, especially in marine environments. 3. Suitable for various concrete mixes and environmental conditions.

Table 4. *Cont.*

Model	Disadvantages	Advantages
The model in reference [47]: $\frac{C}{C_s} = 1 - \mathrm{erf}\left(\frac{x}{2\sqrt{D_0 \frac{1}{1-n}\left[\left(1+\frac{t_{ex}}{t}\right)^{1-n} - \left(\frac{t_{ex}}{t}\right)^{1-n}\right]\left(\frac{t_0}{t}\right)^n t}}\right)$ C: chloride concentration inside the concrete; C_s: chloride concentration at the surface of the concrete; x: depth of chloride penetration; t: exposure time; D_0: initial chloride diffusion coefficient; n: age factor representing how diffusion changes with concrete age; t_{ex}: initial exposure time (when the concrete starts being exposed); t_0: reference time for the initial measurements.	1. Assumes chloride binding is time-independent and linear, which may not reflect actual behavior. 2. When the age factor is high, it may underestimate chloride penetration depth, leading to overly conservative predictions. 3. For short-term exposure, the model may overestimate the diffusion coefficient, leading to inaccurate predictions.	1. Introduces a time-dependent diffusion coefficient, more accurately reflecting the change in chloride diffusion over time. 2. Provides a more accurate prediction of chloride penetration, especially under long-term exposure conditions like marine environments. 3. Aligns better with measured chloride penetration profiles in real-world applications.
The model in reference [48]: $C(x,t) =$ $A_1 D_{max} k_1 \left(1 - \mathrm{erf}\left(\frac{x}{4KD_0 t}\right) - e^{-\frac{x^2}{4KD_0 t}} Re\left[e^{-g^2}\mathrm{erfc}(-ig)\right]\right)$ $C(x,t)$: the chloride ion concentration at position x and time t; D_{max}: the stabilized surface chloride ion concentration; k_1: the coefficient representing the influence of water–cement ratio on the stabilized surface chloride ion concentration; A_1: the impact of sulfate ions on the chloride ion concentration on the concrete surface; erf: error function; D_0: the chloride diffusion coefficient of the concrete measured at the hydration age.	1. The model features complex mathematical expressions, requires large computational resources, and is highly sensitive to input parameters, requiring precise measurements. 2. In real marine environments, the diffusion coefficient undergoes dynamic changes due to the inhomogeneity and variability of the medium, and the boundary conditions are uncertain, which adds challenges to its field application.	1. The model is particularly suitable for describing diffusion processes involving dynamic changes in both time and space variables. 2. It can be applied to diffusion processes under complex boundary conditions, taking into account the changes in substances over time.

3.2.3. Integrating Empirical Models with Real-Time Data

Advancements in sensor technology and real-time monitoring systems offer significant potential to enhance the predictive accuracy of empirical models. By integrating real-time data, such as temperature, humidity, and chloride ion concentration, into these models, parameters can be dynamically updated, improving prediction precision. For instance, Ashrafian et al. [49] used 642 field data points and identified the water-to-cement ratio, annual average temperature, and exposure time as key factors in predicting chloride ion concentration on the surface of concrete in marine environments. Wang et al. [50] applied extensive experimental data from marine tidal zone field studies, reported in the literature, to predict long-term chloride ion diffusion in concrete and developed a structural lifespan assessment method suitable for marine environments.

Empirical models, known for their simplicity and efficiency, are widely applicable in predicting the lifespan of concrete structures. However, their accuracy and applicability heavily depend on the quality of historical data and variations in environmental conditions. In practical application, it is recommended that environmental conditions, real-time monitoring, and the strengths and weaknesses of other modeling techniques be integrated to enable a more in-depth analysis that leads to the enhanced accuracy and reliability of lifespan estimation.

3.3. Statistical Lifespan Prediction Models

The prediction of the lifespan of concrete structures has emerged as a key component of civil engineering, specifically in the context of sustainability and maintaining infrastructure over extended durations. Specifically, regression analysis and Bayesian networks, in particular, are widely used to quantify the relationship between various factors and the expected lifespan of concrete structures.

3.3.1. Regression Analysis Models

Linear and nonlinear regression analyses are commonly used to predict the life of concrete structures by defining the interrelations among contributing parameters and durability responses. For instance, Prieto et al. [51] used multiple linear regression analysis with over 650 data points of rebar bond tests, demonstrating that their suggested model is effective and reliable compared to other models for bond strength assessment, and it is suitable for assessing the structural integrity of reinforced concrete elements subjected to corrosion.

Additionally, Jin et al. [52] used nonlinear and hybrid regression analyses, showing that these models offer higher accuracy and better R^2 values compared to linear methods. Similarly, Hanif et al. [53] reviewed the latest developments in nonlinear-model-based approaches for structural system damage assessment, identified limitations in field applications, and proposed new methods for damage assessment. While regression analysis is a powerful and widely used tool, it has limitations, particularly in handling nonlinear relationships, poor data quality, and multicollinearity. In practice, additional validation and optimization methods are required to ensure the model's accuracy and stability.

3.3.2. Bayesian Networks for Lifespan Prediction

Bayesian networks offer a powerful statistical framework to integrate uncertainty and complex dependencies between influencing factors. These models are particularly useful in predicting the lifespan of concrete structures, as various uncertainty factors interact over time. HSPBN is a type of Bayesian network that models both discrete and continuous variables. Figure 2 [54] illustrates an example of an HSPBN. Discrete nodes can only have other discrete nodes as parents, while continuous nodes can have any combination of parent types. The conditional distribution of discrete variables is usually represented as a categorical distribution (CPT), while continuous variables are represented by parameterized or non-parameterized conditional probability distributions (CPDs). When the CPDs for all continuous variables are parameterized, the HSPBN becomes a special case of CLGBN. However, the model is more flexible when non-parameterized CPDs are used.

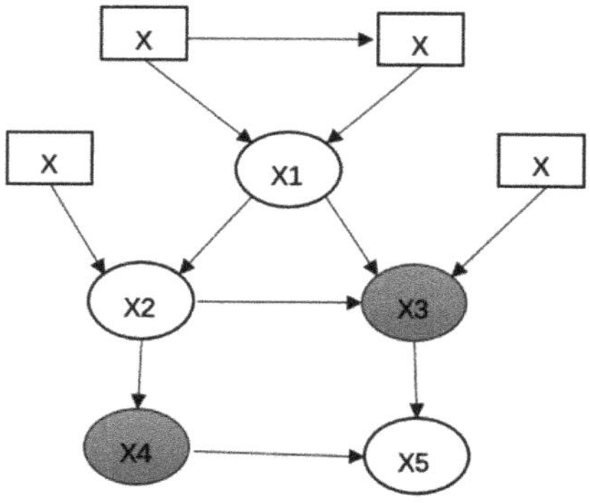

Figure 2. Example representing the structure of an HSPBN. Discrete variable nodes are represented with rectangles and continuous variable nodes with ellipses. Parametric CPD nodes are represented with white nodes and non-parametric CPDs nodes are represented with gray-shaded nodes.

Bayesian networks use probability distributions to model relationships between factors and predict the probability distribution of structural lifespan. This approach integrates expert knowledge with empirical data to address uncertainties in material properties, environmental exposure, and operational conditions. Guo and Dong [55] developed a dynamic Bayesian network-based framework for assessing reinforced concrete structures, utilizing the finite difference method to focus on the impact of chloride ion transport on damage evaluation. In addition, Guo et al. [56] presented an MBN-based reliability analysis framework for reinforced concrete structures with modules to assess tension, to compute the load-carrying capacity, and to conduct time-dependent reliability calculations. In accordance with their study, it has been concluded that early inspection may overestimate failure probabilities because of the high occurrence of cracks, while neglecting temporal distinctions, environmental effects, and chloride ion transport could result in the underestimation of impending failures.

Statistical models, specifically regression analysis and Bayesian networks, are the most important methods for the prediction of concrete structure lifespan. Regression analysis provides an easy model for the relationship between different factors that affect the life of concrete. Bayesian networks can handle uncertainty in addition to the interaction among numerous variables and, thus, provide more accurate and reliable predictions of lifespan.

3.4. Machine Learning-Based Lifespan Prediction Models

Due to the limitations that are inherent in classic models used for predicting lifespan, especially with respect to complicated environmental factors and large datasets, machine learning algorithms have become effective tools in the area of the longevity prediction of concrete structures. With the computational capabilities of artificial intelligence, these algorithms process nonlinear relationships and large datasets efficiently, thereby enabling more accurate lifespan predictions. In this section, we will go through various popular machine learning models and their uses in predicting concrete lifespan.

3.4.1. Artificial Neural Networks (ANNs)

ANNs are a powerful technique for lifespan prediction, and the reason for this lies in their ability to readily identify nonlinear and complex patterns. The schematic diagram in Figure 3 [57] of a computational neuron in an ANN system is provided with three inputs and a single output. The neuron computes the weighted sum of the inputs as a product of input features I and weights w and compares it to a threshold term called the "bias" to produce the output o. When the sum is greater than or equal to the bias, it fires the output signal; otherwise, it does not.

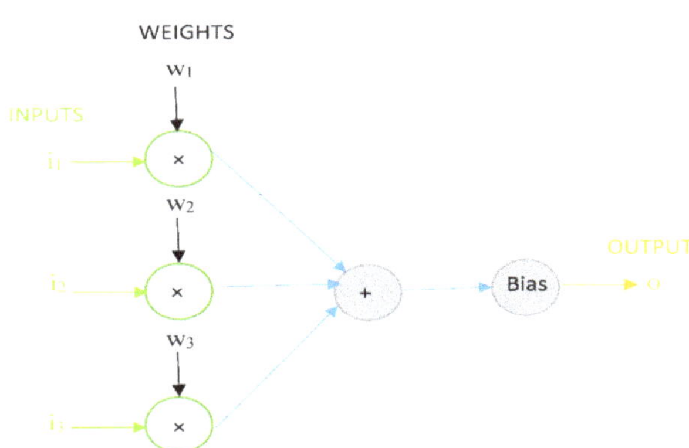

Figure 3. A computational neuron in the artificial neural network (ANN) model [52].

With the growth of research on neural networks, scientists have increasingly utilized them to estimate the lifespan of concrete structures. Kellouche et al. [58] developed an ANN-based model for the prediction of carbonation depth in fly ash concrete, which accurately predicted it. Chen [59] began an ANN-based new approach to estimate the compressive strength of high-performance concrete and obtained high accuracy. To improve the neural networks' predictive accuracy, BKA et al. [60] compared various architectures. While the first model with one hidden layer comprising 50 neurons yielded a mean squared error (MSE) of 34.81, the second model with two hidden layers with 50 and 10 neurons each lowered the MSE to 28.33, indicating that the prediction accuracy is improved with an increase in the hidden layers and neurons. Other studies, such as Zhang et al. [61], have expanded ANN applications, proposing an interpretable machine learning framework that combines chloride ion erosion and mechanical aging characteristics to predict the lifespan of medium- and short-span bridges.

3.4.2. Support Vector Machines (SVMs)

SVMs are a key tool for concrete lifespan prediction, particularly due to their ability to handle small sample sizes and high-dimensional data. The basic structure of an SVM is illustrated in Figure 4 [62].

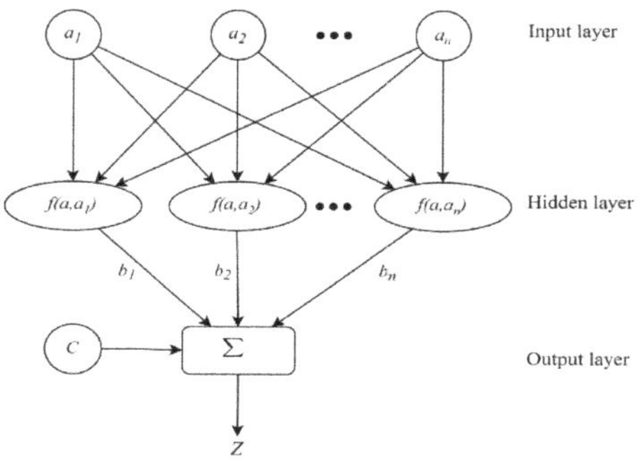

Figure 4. The basic structure of an SVM.

In concrete lifespan prediction, an SVM effectively handles complex nonlinear relationships. Hariri-Ardebili and Pourkamali-Anaraki [63] applied an SVM to assess the structural response of concrete dams. When comparing an SVM with traditional reliability analysis methods, the three commonly used methods are first-order reliability method (FORM), second-order reliability method (SORM), and Monte Carlo simulation (MCS). The failure probability estimated by the SVM closely matched other methods but with higher computational efficiency. Fan [64] proposed a Hybrid Machine Learning (HML) approach, combining critical analysis and an SVM, which is valuable for rapidly evaluating concrete structure damage. Additionally, for predicting concrete strength degradation in complex marine environments, Ling et al. [65] developed an SVM model optimized using K-Fold cross-validation. The results showed that, by optimizing input parameters, the prediction accuracy of the SVM model was significantly improved. Compared to the original model, the average relative error decreased from 34.8% to 27.6%, and the median relative error reduced from 24.7% to 20.8%, further proving the superiority of the optimized SVM model.

3.4.3. Random Forests and Decision Trees

Random forest and decision tree models are pivotal in lifespan prediction due to their ability to perform multivariable analysis and their interpretability. Golafshani et al. [66] applied random forest models to analyze the effects of chloride ion diffusion and environmental exposure on lifespan, thereby enhancing the durability assessment of concrete structures. The model demonstrated strong predictive performance. Random forests are also applied in concrete material design. For example, Liu et al. [67] showcased the potential of random forests in predicting concrete wear resistance and optimizing mix proportions.

Concha and Oreta [68] combined random forests with decision trees to assess the impact of rebar corrosion on bond strength, improving the understanding of concrete structure deterioration. Additionally, random forests have been integrated into hybrid models. Ahmad et al. [69] demonstrated how to combine random forests with other machine learning techniques to predict chloride ion concentration accurately, providing a scientific basis for concrete lifespan management. Table 5 summarizes the advantages, disadvantages, and applicability of ANNs, SVMs, random forests, and decision trees.

Table 5. Comparison of ANN, SVM, random forest, and decision tree models for lifespan prediction.

Reference	Type	Advantages	Disadvantages	Applicable Conditions
Reference [70]	Neural Networks	Strong adaptability, no gradient information required, optimization potential	Sensitive to network architecture, difficult to optimize, risk of overfitting	Complex and nonlinear problems with large datasets
Reference [65]	SVMs	High prediction accuracy, adaptable to complex problems, strong optimization capability	Requires many parameters, computationally complex	Small high-dimensional datasets
Reference [71]	Random Forests	Handles high-dimensional data, easy to apply	Lack of interpretability, long computation time	Exploratory analysis when relationships are unclear
Reference [72]	Decision Trees	Can handle multiple data types, requires little data preprocessing	Instability, prone to bias	Data need clear classification, applicable to structured data

The machine learning models discussed above (ANN, SVM, random forest, etc.) are capable of handling complex multi-factor data in concrete lifespan prediction. In practice, an ANN is well suited for large-scale, complex, and nonlinear data, while an SVM excels in small sample sizes and high-dimensional datasets. Random forest offers strong multivariable analysis capabilities and high interpretability. However, these models still face challenges, including the need for large datasets, high computational resource requirements, and the risk of overfitting.

4. Model Validation and Prediction

In civil engineering, model validation is required to ensure both scientific accuracy and practical applicability. This section discusses four significant aspects: validation methods, typical evaluation metrics, sensitivity analysis, and uncertainty analysis.

4.1. Validation Methods

Concrete structure durability models are typically examined using two main approaches: experimental validation and field-data-based verification. Experimental validation provides early observations using simulation procedures, including accelerated corrosion testing and freeze–thaw cycling tests. In contrast, field-data-based verification provides a more effective determination of the suitability of the model in real practice in the engineering field.

As an example, Ann et al. [73] reported that concrete that was naturally exposed for a period of 18 years had a model-predicted carbonation depth of 8.3 mm, while the experimental measurement was 11.62 mm with a standard deviation of 2.45 mm, demonstrating a strong correlation between model predictions and empirical observation, which shows near conformity between calculated model values and experimental values. Papadakis and Demis [74] performed an in-depth examination of carbonation depths using various water-to-cement ratios (w/c) and various cement types with very small experimental measurement errors, which provided a mean calculated value of 7.6%. This observation indicates the high accuracy of the model in experimental use. Xu et al. [75] studied Aramid Fiber Reinforced Polymer-reinforced concrete exposed to multiple freeze–thaw cycles in an alkaline environment, where the measured compressive strength was 38.100 Mpa and the back propagation (BP) model predicted 42.263 Mpa. The relative error of 10.927% suggests room for optimization in the model's prediction stability. While experimental validation can assess model performance under controlled conditions, it cannot fully capture the complexity of actual service environments; thus, further validation with field monitoring data is necessary. For example, Wang et al. [76] proposed a validation framework using both experimental and long-term field monitoring data to assess the durability of concrete structures in marine environments. Similarly, Fan et al. [77] validated the applicability of concrete durability models in complex corrosive environments by integrating experimental and numerical simulation results. To further illustrate the validation based on field data, Table 6 summarizes several cases of concrete durability models validated by field data. These cases demonstrate the comparison between model predictions and actual field data, highlighting the importance of field data in model validation.

Table 6. Studies on the applicability of concrete durability models validated by field data.

	Context	Model Prediction Accuracy	Importance of Field Data
Reference [78]	Field validation of chloride ion diffusion in UK harbors	At 1.17 and 6.17 years, model predictions closely matched the monitoring data. At year 18, the model slightly underestimated chloride ion diffusion at deeper layers, with about 10% error.	Highlights the role of field data in validating the model's performance over time, showing discrepancies at deeper layers at later years.
Reference [79]	Study of chloride ion diffusion coefficients in aging structures along the South China Sea	For the HPW6 structure, the measured chloride ion surface concentration was 1.78%, while the model predicted 4.13%.	Demonstrates how field data from actual aging structures provides critical validation for model predictions, emphasizing real-world applications.
Reference [80]	Geothermal power plant concrete exposed to chloride- and sulfate-rich geothermal water	Field data show that, after 18 years, chloride ions had penetrated into deeper layers in normal concrete and began corroding the reinforcement; ultra-high-durability concrete performed better at the same time, although the diffusion occurred slightly earlier, but the overall trend was consistent with the model.	Field data confirmed that the actual chloride diffusion rate was faster than predicted in normal concrete, validating the importance of monitoring for accurate durability modeling.
Reference [81]	Lifecycle modeling of concrete cracking and reinforcement corrosion in Changfeng Creek Bridge, China	The study predicts, through lifecycle modeling, that reducing the maintenance cycle (e.g., shortening it from 12 years to 4 years) can significantly extend the service life of the bridge. Field data from regular monitoring revealed an extension of approximately 15 years.	This case highlights the importance of field monitoring in ensuring the long-term durability of infrastructure and optimizing models.

The cases in Table 5 clearly demonstrate the differences between model predictions and actual data, proving the importance of field data in validating the accuracy of a model. While laboratory simulations provide initial validation, field data can more accurately assess the applicability of the model in real-world environments. Future research should continue to strengthen the integration of experimental data and field monitoring to improve the models' predictive capabilities.

4.2. Expansion and Application of Evaluation Metrics

In concrete structure lifespan prediction modeling, the traditional performance measures such as MSE, Coefficient of Determination (R^2), Standard Error (SE), and Prediction Error (PE) have been widely utilized. However, existing studies indicate that the sole utilization of a single measure of assessment is not normally adequate to generate an extensive judgment of model performance, especially in the context of complex engineering problems [82–84]. To resolve this, researchers have augmented conventional measures by combining various assessment techniques, incorporating precision, stability, and generalizability into models. Although MSE and R^2 remain cornerstone measures, Yu et al. [85] found that, in addressing complicated nonlinear relationships, MSE alone may not account for extreme error impacts. MSE is, thus, sometimes combined with other error metrics, such as Absolute Error or Maximum Error. This approach skillfully recognizes outlier values and unusual cases in prediction analyses, particularly in the estimation of concrete structure service life and explaining abrupt changes due to factors like corrosion and cracking. SE is widely used to assess model reliability, mostly in longitudinal research or repeated experiments. For instance, Kostić et al. [86] combined SE with experimental design methods in the analysis of slope stability. The SE ranged from 0.06 to 0.12 through the comparison of the model predictions and experimental values, showing a very small error of prediction and enhancing the stability and reliability of the model.

In addition, Kefei and Torrent [87] presented a Hong Kong-Zhuhai-Macao Bridge concrete structure lifespan prediction model using electric flux and accelerated chloride ion penetration tests. A confidence interval for the lifespan estimation with a concrete cover thickness of SE = 3.9 mm was employed, which indicated the variability range of model inaccuracy under standard exposure conditions. While SE alone can measure deviations of model error, researchers now combine SE with sensitivity analysis to examine the effects of environmental variables (e.g., chloride ion concentration, temperature, and humidity changes) on predictions concerning concrete durability. The hybrid analytical strategy enhances the understanding of the model's robustness and stability under variable conditions. Sarmiento et al. [88] carried out a reliability analysis of the lifecycle of prestressed concrete bridges that had reached a high age and observed that the Safety Evaluation Index increased from 1.8×10^{-2} to 3.7×10^{-2} in the course of a 50-year service life, illustrating the significant effect of failures in safety-critical elements, such as external steel strands, on the safety of the infrastructure as a whole. Moreover, the structural robustness index fell from 0.74 to 0.60, indicating a decrease of approximately 18.92% in the structure's ability to withstand cascading failures when damaged. In contrast, relying solely on the reliability index β (from 1.95 to 1.78) reveals changes in overall risk levels but does not capture the effects on local components or system redundancy. Therefore, combined evaluation methods offer greater diagnostic power for identifying structural vulnerabilities and assessing lifecycle safety. Moreover, Carevic and Ignjatovic [89] incorporated various evaluation metrics into their green concrete carbonation prediction model, predicting the service life by establishing a relationship between compressive strength and carbonation resistance.

4.3. Sensitivity Analysis and Uncertainty Analysis

Both sensitivity analysis and uncertainty analysis are crucial in life prediction models of concrete structures for determining model performance. Sensitivity analysis entails the identification of critical input variables by examining how changes in input parameters affect the output and, hence, allowing for the identification of parameters that have a notable effect on accuracy in predictions of lifespan. Conversely, uncertainty analysis displays the risk and stability range of model predictions, thereby offering insights into model reliability and associated risks. For example, Yu et al. [90] discovered that cover depth is a critical parameter influencing durability. An increase in cover depth from 40 mm to 70 mm increased the 100-year reliability index β from 0.3696 to 2.1042, which implies that a higher cover depth considerably enhances structural durability and safety. This, again, underscores the significance of cover thickness in lifespan-related predictions. In addition, Guo et al. [56] utilized a Hybrid Bayesian Network model to investigate the sensitivity of key parameters in relation to the reliability of reinforced concrete structures in different environments. Strauss et al. [91] proved that sensitivity analysis based on monitoring data together with probabilistic prediction models more precisely identifies the critical variables affecting the durability of concrete.

Prediction uncertainty is a critical issue in engineering practice. Monte Carlo simulation is a commonly used method to address this uncertainty. For instance, Franceschini et al. [92] proposed a method to evaluate the remaining service life of prestressed concrete beams under corrosion conditions. The study focused on the failure mechanisms caused by corrosion and quantitatively assessed the structural degradation process using statistical methods. Additionally, Vishwanath and Banerjee [93] enhanced the reliability of bridge pier lifespan prediction by introducing the uncertainty analysis of chloride corrosion when evaluating the seismic vulnerability of aging bridge piers. Cho et al. [94] combined random variable simulations in the study of a fly ash concrete carbonation, systematically quantifying the confidence intervals of model predictions. Furthermore, Taffese and Espinosa-Leal [33] used machine learning techniques to assess the confidence intervals of concrete durability prediction models, demonstrating that, in large-scale data-driven contexts, the uncertainty in model predictions can be significantly reduced.

5. Application Cases of Models

With the increasing complexity and variability of engineering environments, traditional single models are becoming insufficient. In modern engineering practice, many models combine various factors such as environmental influences, load variations, and material degradation. This is particularly crucial for lifespan prediction in key infrastructure like bridges, tunnels, and high-rise buildings, where model accuracy and adaptability are essential. Below, we review the practical applications of representative lifespan prediction models, categorized by typical civil engineering structures such as bridges, underground facilities, and high-rise buildings. As a vital part of transportation infrastructure, bridges bear substantial loads and long-term usage pressure, making the accuracy of their lifespan prediction models critical. Ultra-high-performance concrete (UHPC) is widely applied in bridge construction. For example, Shafikhani and Chidiac [95] measured the chloride ion diffusion coefficient through concrete materials. The diffusion coefficient for UHPC ranged from 6×10^{-15} m^2/s to 5.9×10^{-13} m^2/s, while for ordinary concrete, it ranged from 2×10^{-13} m^2/s to 7.97×10^{-12} m^2/s, demonstrating UHPC's superior corrosion resistance. Additionally, Liang et al. [96] analyzed the chloride-induced lifespan of the Linbian and Dajin bridges in Keelung, comparing the lifespan predictions of three models: the Langmuir isotherm model, the linear binding model, and the Fick diffusion model [97,98]. In low-chloride conditions, the Langmuir isotherm model predicted a lifespan of 33.12 years for the bridge deck, which was higher than the linear binding model (31.55 years) and the Fick diffusion model (31.54 years).

The findings indicate that the Langmuir isotherm model predicts lifespan and longevity more precisely than other models, particularly in low-chloride-concentration environments. Moreover, Xue and Mao [99] modified the chloride ion diffusion model by introducing variables such as the time-dependent decay of the diffusion coefficient, chloride ion binding capacity, type of concrete, and service environment. The researchers confirmed the model's accuracy with an application study on the Huangpu Bridge, proving that environmental parameters have a great impact on chloride ion diffusion rates and the corrosion process. The revised model provides a more precise method for predicting bridge durability and service life. Chen et al. [100] performed a lifecycle cost analysis on reinforced concrete bridge edge beams, effectively controlling crack width and improving resistance to chloride ion corrosion, resulting in a 58% extension of edge beam lifespan. Yamamoto et al. [101] found that the relationship between the apparent chloride ion diffusion coefficient and the air permeability coefficient showed different diffusion trends in road bridges exposed to deicing salts and in marine environments or laboratory tests. Their study highlighted the critical role of environmental factors in concrete evaluation. Furthermore, Li and Feng [102] studied the fatigue life of bridge cables, validating the model's reliability using real-world monitoring data. Fathalla et al. [103] proposed a fatigue life prediction model for concrete bridge decks, further advancing bridge lifespan evaluation methods. Underground facilities are subjected to complex environmental conditions, which increase the demands on lifespan prediction models. Balestra et al. [104] introduced a nonlinear lifespan model that incorporates environmental factors such as chloride concentration, temperature, and humidity to predict the corrosion rate of underground structures. For example, the corrosion of buried cast iron pipes is a complex process [105]. Li and Mahmoodian [106], using corrosion data from pipes in the UK that have been in service for up to 150 years, combined a corrosion pit growth model to calculate the failure probability of pipes with varying diameters and wall thicknesses under both internal and external corrosion effects. Dasar et al. [107] experimentally investigated the natural corrosion of beams with different cover layers and rebar diameters, validating the long-term corrosion performance prediction accuracy of their model. Chen et al. [108] applied a nonlinear chloride ion transport model to simulate chloride ion intrusion in underwater concrete tunnels, and the simulation results closely matched experimental data, providing a useful tool for assessing chloride ion corrosion in cylindrical tunnels. Additionally, Keßler [109] proposed a probabilistic corrosion state assessment method for tunnel structures based on half-cell potential measurements, significantly improving lifespan prediction accuracy. High-rise buildings, airport runways, and similar infrastructure are subjected to heavy loads and complex environmental conditions. Meng et al. [110] evaluated the performance of nano-silica-modified recycled concrete in marine environments and proposed a lifespan prediction model based on material modification for environmental adaptation. Canestro et al. [111] developed a multiscale modeling method to accurately assess the long-term performance of prestressed concrete structures. For airport runway lifespan prediction, Zheng and Pornsing [112] introduced a novel empirical optimization light-gradient-boosting machine model, enhancing prediction accuracy through detailed parameter optimization, feature selection, and model integration. Mortagi and Ghosh [113] presented a vulnerability assessment method for highway bridges under climate change scenarios, offering a new perspective for building lifespan prediction.

6. Analysis of Advantages and Disadvantages of Existing Models

In predicting the lifespan of concrete structures, different models offer distinct advantages and limitations. The selection of an appropriate model depends on the specific engineering environment, data availability, and prediction needs.

Table 7 provides an analysis of the advantages and disadvantages of several commonly used models.

Table 7. Analysis of advantages and disadvantages of common concrete structure lifespan.

Model Type	Advantages	Disadvantages	Applicable Scenarios
Physical Models	Strong theoretical foundation, capable of simulating long-term degradation processes, incorporates environmental factors	Computationally complex, requires accurate data, limited applicability	Suitable for simpler controlled environments, such as laboratory settings or less complex engineering projects
Empirical Models	Simple, fast, cost-effective, highly adaptable	Limited precision, reliant on past data, not applicable to all situations	Ideal for scenarios with available data, especially for quick preliminary assessments
Statistical Models	Based on data-driven predictions, adaptable to complex environments, capable of handling uncertainty	Requires high-quality data, difficult to interpret	Best for data-rich complex environments with numerous variables and uncertainties
Machine Learning Models	High accuracy, capable of handling complex data and problems, real-time prediction updates	Requires large amounts of data and computational resources, difficult to interpret model mechanisms	Best suited for environments that require high-precision predictions, real-time monitoring, and substantial data support
Hybrid Models	Strong theoretical foundation, capable of simulating long-term degradation processes, incorporates environmental factors	Complex models that require multidisciplinary knowledge for integration	Ideal for engineering projects that need to consider multiple factors, complex structures, or dynamic environments

In summary, each model has its own strengths and limitations in practical applications. To achieve more accurate predictions, it is essential to choose the most appropriate model based on the specific needs of the project. With technological advancements, hybrid models and machine learning models have proven especially advantageous in handling complex problems, particularly for real-time monitoring and dynamic updates. The continuous improvement and integration of these models will help better address the diverse needs of engineering projects.

7. Research Progress and Development Trends

As civil engineering technology continues to advance, concrete structure lifespan prediction models are progressing toward greater intelligence and precision. Recent developments, particularly in the application of high-performance materials and the rapid development of intelligent technologies, have significantly enhanced the accuracy and reliability of predictions of the life of concrete. Jeon et al. [114] presented a maintenance strategy based on a health index for concrete bridges, which combines real-time monitoring data with changes in environmental conditions, thus greatly increasing prediction accuracy.

Multiscale modeling has become a necessity for lifespan prediction. Ying et al. [115], for example, employed multiscale modeling to examine the behavior of concrete under high-temperature conditions and revealed the influences of environmental conditions on the aging process. Through integrated multiscale analysis approaches, researchers are able to analyze concrete material aging and its influence on the overall structure performance more accurately. Kaveh and Alhajj [116] proposed a multiscale model based on the development of microcracks, emphasizing the central role of crack evolution in the life of concrete. Wang and Maekawa [117] developed a multiscale modeling framework that predicts the life of concrete based on moisture transport, temperature gradients, and electrochemical coupling under different environmental conditions.

The fast evolution of data-driven technologies has led to the increased application of intelligent approaches in predicting the life of concrete. The combination of machine learning and big data analytics not only improves the accuracy of predictions, but also facilitates real-time monitoring. Azimi et al. [118] conducted a review of the latest advances of deep learning for structural health monitoring, pointing out that the coupling of real-time data with machine learning improves diagnostic capability, thereby rendering life predictions more accurate.

In the future, concrete structure life prediction models will continue to evolve with the help of even more advanced technologies. Incorporating smart technologies, real-time feedback, and multivariable modeling approaches is particularly of interest, as this will significantly enhance model precision and ease of use. With the deeper integration of interdisciplinary technologies, future research will focus on enhancing prediction accuracy through real-time data feedback and dynamic updates, providing more accurate and efficient support for the maintenance of civil engineering projects.

8. Conclusions

Concrete structures find extensive applications in civil engineering, and it is essential to guarantee their long-term durability and safety for public welfare and the effective utilization of resources. With changing environmental conditions and the degradation of materials under the influence of a multitude of factors, the durability assessment of concrete infrastructures is a critical problem. This review presents a summary of lifespan prediction models such as physical, empirical, statistical, and machine learning models and evaluates their applicability and limitations in accordance with different environmental conditions.

With the development of technology, models based on real-time monitoring data and integrating multiple factors are given more prominence in research forums. While physical models offer a good theoretical model, they are computationally intensive and restrictive when used in complex environments. Empirical models are extremely simple and interactive but are highly reliant on historical data and could overlook certain elements of environmental complexity. Statistical modeling based on regression analysis and Bayesian networks is aptly suited for addressing complex environments and diverse uncertainties but requires large amounts of data. Machine learning models such as artificial neural networks (ANNs) and support vector machines (SVMs) exhibit an enhanced capability of addressing nonlinear complexities and can offer enhanced predictive accuracy when supplemented with large datasets.

Despite these advances, current models still face many challenges, particularly with respect to updating data in real time, dynamic model optimization, and the integration of various influencing factors. Future studies can focus on the following areas:

1. Combining real-time monitoring data with advanced machine learning algorithms to improve prediction accuracy and robustness;
2. Exploring hybrid models that integrate the strengths of both physical and data-driven models to enhance adaptability in complex environments;
3. Developing more precise nonlinear and multiscale modeling techniques to accommodate environmental and material variations;
4. Optimizing model validation and evaluation methods to improve their reliability and applicability in practical engineering projects;

5. By combining experimental data with field monitoring data, future research should continue to explore how to effectively integrate these two sources in order to further optimize the predictive capability of models and improve the accuracy of concrete structure durability assessments.

By continuously advancing technological innovations and improving model capabilities, future concrete structure lifespan prediction models will be more accurate and efficient, providing stronger support for structural health management in civil engineering.

Funding: This work was supported by the Key Projects of Yongjiang Science and Technology Innovation 2035 in Ningbo (2024Z090) and the Major Science and Technology Project of Ningbo High-tech Zone in 2023 (2023CX050001).

Conflicts of Interest: Authors Xi Peng and Bin Xu were employed by the company Ningbo Roaby Technology Industrial Group Co., Ltd. The remaining authors declare that the research was conducted in the absence of any commercial or financial relationships that could be construed as a potential conflict of interest.

References

1. Imbabi, M.S.; Carrigan, C.; McKenna, S. Trends and developments in green cement and concrete technology. *Int. J. Sustain. Built Environ.* **2012**, *1*, 194–216. [CrossRef]
2. Shi, X.; Xie, N.; Fortune, K.; Gong, J. Durability of steel reinforced concrete in chloride environments: An overview. *Constr. Build. Mater.* **2012**, *30*, 125–138. [CrossRef]
3. Athibaranan, S.; Karthikeyan, J.; Rawat, S. Investigation on service life prediction models of reinforced concrete structures exposed to chloride laden environment. *J. Build. Pathol. Rehabil.* **2022**, *7*, 1–15. [CrossRef]
4. Peng, R.; Qiu, W.; Jiang, M. Frost resistance performance assessment of concrete structures under multi-factor coupling in cold offshore environment. *Build. Environ.* **2022**, *226*, 109733. [CrossRef]
5. Zhang, K.; Guo, A.; Yu, Y.; Yang, B.; Yu, B.; Xie, C. Freeze–thaw damage degradation model and life prediction of air-entrained concrete in multi-year permafrost zone. *Materials* **2023**, *16*, 7703. [CrossRef]
6. Chen, J.; Li, Y.; Li, Y.; Wen, L.; Guo, H. Effects of curing conditions with different temperature and humidity on damage evolution of concrete during freeze–thaw cycling. *Mater. Struct.* **2022**, *55*, 80. [CrossRef]
7. Zhang, X.G.; Wang, J. Probabilistic Model of Corrosion in Reinforced Concrete Structures Considering Coupling Effects of Influence Factors. *Adv. Mater. Res.* **2013**, *671*, 740–745. [CrossRef]
8. Zhang, P.; Cong, Y.; Vogel, M.; Liu, Z.; Müller, H.S.; Zhu, Y.; Zhao, T. Steel reinforcement corrosion in concrete under combined actions: The role of freeze-thaw cycles, chloride ingress, and surface impregnation. *Constr. Build. Mater.* **2017**, *148*, 113–121. [CrossRef]
9. Zhou, L.; Pan, S.; Wang, J.; Vasilakos, A.V. Machine learning on big data: Opportunities and challenges. *Neurocomputing* **2017**, *237*, 350–361. [CrossRef]
10. Dinesh, A.; Prasad, B.R. Predictive models in machine learning for strength and life cycle assessment of concrete structures. *Autom. Constr.* **2024**, *162*, 105412. [CrossRef]
11. Tian, Y.; Zhang, G.; Ye, H.; Zeng, Q.; Zhang, Z.; Tian, Z.; Jin, X.; Jin, N.; Chen, Z.; Wang, J. Corrosion of steel rebar in concrete induced by chloride ions under natural environments. *Constr. Build. Mater.* **2023**, *369*, 130504. [CrossRef]
12. Ju, X.; Wu, L.; Lin, C.; Yang, X.; Yang, C. Prediction of chloride concentration with elevation in concrete exposed to cyclic drying-wetting conditions in marine environments. *Constr. Build. Mater.* **2021**, *278*, 122370. [CrossRef]
13. Zhang, W.; Chen, J.; Yu, Q.; Gu, X. Corrosion evolution of steel bars in RC structures based on Markov chain modeling. *Struct. Saf.* **2021**, *88*, 102037. [CrossRef]
14. Sun, B.; Xiao, R.; Ruan, W.; Wang, P. Corrosion-induced cracking fragility of RC bridge with improved concrete carbonation and steel reinforcement corrosion models. *Eng. Struct.* **2020**, *208*, 110313. [CrossRef]
15. Zhang, Y.; Su, R.K.L. Corner cracking model for non-uniform corrosion-caused deterioration of concrete covers. *Constr. Build. Mater.* **2020**, *234*, 117410. [CrossRef]
16. Wang, Y.; Gong, X.; Wu, L. Prediction model of chloride diffusion in concrete considering the coupling effects of coarse aggregate and steel reinforcement exposed to marine tidal environment. *Constr. Build. Mater.* **2019**, *216*, 40–57. [CrossRef]
17. Hájková, K.; Šmilauer, V.; Jendele, L.; Červenka, J. Prediction of reinforcement corrosion due to chloride ingress and its effects on serviceability. *Eng. Struct.* **2018**, *174*, 768–777. [CrossRef]
18. Wang, W.; Lu, C. Time-varying law of rebar corrosion rate in fly ash concrete. *J. Hazard. Mater.* **2018**, *360*, 520–528. [CrossRef]

19. Ann, K.Y.; Ahn, J.H.; Ryou, J.S. The importance of chloride content at the concrete surface in assessing the time to corrosion of steel in concrete structures. *Constr. Build. Mater.* **2009**, *23*, 239–245. [CrossRef]
20. Zhou, Y.; Gencturk, B.; Willam, K.; Attar, A. Carbonation-induced and chloride-induced corrosion in reinforced concrete structures. *J. Mater. Civ. Eng.* **2015**, *27*, 04014245. [CrossRef]
21. Saeki, T. Effect of carbonation on chloride penetration in concrete. In Proceedings of the Third International RILEM Workshop on Testing and Modelling Chloride Ingress into Concrete, Madrid, Spain, 9–10 September 2002; pp. 1–14.
22. Zhou, Y.; Tian, H.; Sui, L.; Xing, F.; Han, N. Strength deterioration of concrete in sulfate environment: An experimental study and theoretical modeling. *Adv. Mater. Sci. Eng.* **2015**, *2015*, 951209. [CrossRef]
23. Scott, A.; Alexander, M.G. The influence of binder type, cracking and cover on corrosion rates of steel in chloride-contaminated concrete. *Mag. Concr. Res.* **2007**, *59*, 495–505. [CrossRef]
24. Guo, Z.; Wang, L.; Ma, Y.; Wu, B.; Ren, L.; Gui, S.; Su, X. Probabilistic fatigue life prediction of corroded PC beams considering spatial variability of pitting corrosion. *Eng. Struct.* **2024**, *306*, 117721. [CrossRef]
25. Wu, J.; Xu, J.; Diao, B.; Jin, L.; Du, X. Fatigue life prediction for the reinforced concrete (RC) beams under the actions of chloride attack and fatigue. *Eng. Struct.* **2021**, *242*, 112543. [CrossRef]
26. Ge, B.; Kim, S. Probabilistic service life prediction updating with inspection information for RC structures subjected to coupled corrosion and fatigue. *Eng. Struct.* **2021**, *238*, 112260. [CrossRef]
27. Bhargava, K.; Ghosh, A.K.; Mori, Y.; Ramanujam, S. Model for cover cracking due to rebar corrosion in RC structures. *Eng. Struct.* **2006**, *28*, 1093–1109. [CrossRef]
28. Choe, D.E.; Gardoni, P.; Rosowsky, D.; Haukaas, T. Probabilistic capacity models and seismic fragility estimates for RC columns subject to corrosion. *Reliab. Eng. Syst. Saf.* **2008**, *93*, 383–393. [CrossRef]
29. Jianyu, Y.; Xin, Y.; Cong, J.; Weijun, Y. Effect of NaCl solution and simulated concrete pore solution environment on the efficiency of steel bar energized corrosion. *Materials* **2022**, *15*, 7040. [CrossRef] [PubMed]
30. Yu, H.; Li, L.; Ma, H.; Lu, Y.; Hu, D. Study on time range of definite integral in derivation process of chloride ion diffusion theoretical model of concrete. *Constr. Build. Mater.* **2024**, *428*, 135876. [CrossRef]
31. Chen, D.; Guo, W.; Quan, X.; Duan, B.; Guo, L. Mesoscopic statistics-based probability characteristics of chloride transport and reliability-based corrosion initiation life of bridge tower. *Thin-Walled Struct.* **2024**, *198*, 111680. [CrossRef]
32. Guo, Z.; Guo, R.; Lin, S. Multi-factor fuzzy prediction model of concrete surface chloride concentration with trained samples expanded by random forest algorithm. *Mar. Struct.* **2022**, *86*, 103311. [CrossRef]
33. Taffese, W.Z.; Espinosa-Leal, L. Prediction of chloride resistance level of concrete using machine learning for durability and service life assessment of building structures. *J. Build. Eng.* **2022**, *60*, 105146. [CrossRef]
34. Liu, J.; Liao, C.; Jin, H.; Jiang, Z.; Zhou, H. Numerical and experimental research on the effect of rainfall on the transporting behavior of chloride ions in concrete. *Constr. Build. Mater.* **2021**, *302*, 124160. [CrossRef]
35. Shao, W.; Nie, Y.; Liang, F.; Shi, D. A novel comprehensive evaluation method for the corrosion initiation life of RC hollow piles in chloride environments. *Constr. Build. Mater.* **2020**, *249*, 118801. [CrossRef]
36. Safehian, M.; Ramezanianpour, A.A. Assessment of service life models for determination of chloride penetration into silica fume concrete in the severe marine environmental condition. *Constr. Build. Mater.* **2013**, *48*, 287–294. [CrossRef]
37. Papadakis, V.G.; Vayenas, C.G.; Fardis, M.N. Fundamental modeling and experimental investigation of concrete carbonation. *Mater. J.* **1991**, *88*, 363–373.
38. Wang, D.; Tan, Q.; Wang, Y.; Liu, G.; Lu, Z.; Zhu, C.; Sun, B. Carbonation depth prediction and parameter influential analysis of recycled concrete buildings. *J. CO2 Util.* **2024**, *85*, 102877. [CrossRef]
39. Cao, H.; Xu, Z.; Peng, X. An Improved Fick Model for Predicting Carbonation Depth of Concrete. *Coatings* **2024**, *14*, 1345. [CrossRef]
40. Yu, H.; Sun, W.; Ma, H. Research on Theoretical Models of Chloride Ion Diffusion in Concrete I—Non-steady State Homogeneous and Non-homogeneous Diffusion Problems Based on Infinite Bodies. *Nanjing Univ. Aeronaut. Astronaut.* **2009**, *41*, 276–280.
41. Yu, H.; Sun, W.; Ma, H. Research on Theoretical Models of Chloride Ion Diffusion in Concrete II—Non-steady State Homogeneous and Non-homogeneous Diffusion Problems Based on Finite Bodies. *Nanjing Univ. Aeronaut. Astronaut.* **2009**, *41*, 408–413.
42. Chen, W.; Zhang, J.; Zhang, J. A variable-order time-fractional derivative model for chloride ions sub-diffusion in concrete structures. *Fract. Calc. Appl. Anal.* **2013**, *16*, 76–92. [CrossRef]
43. Lu, Z.H.; Lun, P.Y.; Li, W.; Luo, Z.; Li, Y.; Liu, P. Empirical model of corrosion rate for steel reinforced concrete structures in chloride-laden environments. *Adv. Struct. Eng.* **2019**, *22*, 223–239. [CrossRef]
44. Chen, X.; Ming, Y.; Fu, F.; Chen, P. Numerical and empirical modeling for the assessment of service life of RC structures in marine environment. *Int. J. Concr. Struct. Mater.* **2022**, *16*, 11. [CrossRef]
45. Mangat, P.S.; Molloy, B.T. Prediction of free chloride concentration in concrete using routine inspection data. *Mag. Concr. Res.* **1994**, *46*, 279–287. [CrossRef]
46. Mangat, P.S.; Molloy, B.T. Prediction of long term chloride concentration in concrete. *Mater. Struct.* **1994**, *27*, 338–346. [CrossRef]

47. Luping, T.; Gulikers, J. On the mathematics of time-dependent apparent chloride diffusion coefficient in concrete. *Cem. Concr. Res.* **2007**, *37*, 589–595. [CrossRef]
48. Ma, J.; Yang, Q.; Wang, X.; Peng, X.; Qin, F. Review of Prediction Models for Chloride Ion Concentration in Concrete Structures. *Buildings* **2025**, *15*, 149. [CrossRef]
49. Ashrafian, A.; Panahi, E.; Salehi, S.; Amiri, M.J.T. On the implementation of the interpretable data-intelligence model for designing service life of structural concrete in a marine environment. *Ocean. Eng.* **2022**, *256*, 111523. [CrossRef]
50. Wang, Y.; Wu, L.; Wang, Y.; Li, Q.; Xiao, Z. Prediction model of long-term chloride diffusion into plain concrete considering the effect of the heterogeneity of materials exposed to marine tidal zone. *Constr. Build. Mater.* **2018**, *159*, 297–315. [CrossRef]
51. Prieto, M.; Tanner, P.; Andrade, C. Multiple linear regression model for the assessment of bond strength in corroded and non-corroded steel bars in structural concrete. *Mater. Struct.* **2016**, *49*, 4749–4763. [CrossRef]
52. Jin, R.; Chen, Q.; Soboyejo, A.B.O. Non-linear and mixed regression models in predicting sustainable concrete strength. *Constr. Build. Mater.* **2018**, *170*, 142–152. [CrossRef]
53. Hanif, M.U.; Ibrahim, Z.; Ghaedi, K.; Hashim, H.; Javanmardi, A. Damage assessment of reinforced concrete structures using a model-based nonlinear approach–A comprehensive review. *Constr. Build. Mater.* **2018**, *192*, 846–865. [CrossRef]
54. Atienza, D.; Larrañaga, P.; Bielza, C. Hybrid semiparametric Bayesian networks. *TEST* **2022**, *31*, 299–327. [CrossRef]
55. Guo, H.; Dong, Y. Dynamic Bayesian network for durability of reinforced concrete structures in long-term environmental exposures. *Eng. Fail. Anal.* **2022**, *142*, 106821. [CrossRef]
56. Guo, H.; Dong, Y.; Bastidas-Arteaga, E. Mixed Bayesian Network for reliability assessment of RC structures subjected to environmental actions. *Struct. Saf.* **2024**, *106*, 102392. [CrossRef]
57. Aneja, S.; Sharma, A.; Gupta, R.; Yoo, D.Y. Bayesian regularized artificial neural network model to predict strength characteristics of fly-ash and bottom-ash based geopolymer concrete. *Materials* **2021**, *14*, 1729. [CrossRef]
58. Kellouche, Y.; Ghrici, M.; Boukhatem, B. Service life prediction of fly ash concrete using an artificial neural network. *Front. Struct. Civ. Eng.* **2021**, *15*, 793–805. [CrossRef]
59. Chen, L. Hybrid structured artificial network for compressive strength prediction of HPC concrete. *J. Appl. Sci. Eng.* **2022**, *26*, 989–999.
60. BKA, M.A.R.; Ngamkhanong, C.; Wu, Y.; Kaewunruen, S. Recycled aggregates concrete compressive strength prediction using artificial neural networks (ANNs). *Infrastructures* **2021**, *6*, 17. [CrossRef]
61. Zhang, B.; Wang, K.; Lu, G.; Guo, W.; Liu, J.; Zhang, N.; Yang, C. Life-cycle seismic performance analysis of an offshore small-to-medium span bridge based on interpretable machine learning. In *Structures*; Elsevier: Amsterdam, The Netherlands, 2024; Volume 70, p. 107511.
62. Wang, X.; Yang, Q.; Peng, X.; Qin, F. A review of concrete carbonation depth evaluation models. *Coatings* **2024**, *14*, 386. [CrossRef]
63. Hariri-Ardebili, M.A.; Pourkamali-Anaraki, F. Support vector machine based reliability analysis of concrete dams. *Soil Dyn. Earthq. Eng.* **2018**, *104*, 276–295. [CrossRef]
64. Fan, C.L. Detection of multidamage to reinforced concrete using support vector machine-based clustering from digital images. *Struct. Control. Health Monit.* **2021**, *28*, e2841. [CrossRef]
65. Ling, H.; Qian, C.; Kang, W.; Liang, C.; Chen, H. Combination of Support Vector Machine and K-Fold cross validation to predict compressive strength of concrete in marine environment. *Constr. Build. Mater.* **2019**, *206*, 355–363. [CrossRef]
66. Golafshani, E.M.; Kashani, A.; Arashpour, M. Chloride diffusion modeling of concrete using tree-based forest models. *Struct. Concr.* **2023**, *24*, 5614–5634. [CrossRef]
67. Liu, Q.; Andersen, L.V.; Wu, M. Prediction of concrete abrasion depth and computational design optimization of concrete mixtures. *Cem. Concr. Compos.* **2024**, *148*, 105431. [CrossRef]
68. Concha, N.C.; Oreta, A.W.C. Investigation of the effects of corrosion on bond strength of steel in concrete using neural network. *Comput. Concr.* **2021**, *28*, 77–91.
69. Ahmad, A.; Farooq, F.; Ostrowski, K.A.; Śliwa-Wieczorek, K.; Czarnecki, S. Application of novel machine learning techniques for predicting the surface chloride concentration in concrete containing waste material. *Materials* **2021**, *14*, 2297. [CrossRef]
70. Ding, S.; Li, H.; Su, C.; Yu, J.; Jin, F. Evolutionary artificial neural networks: A review. *Artif. Intell. Rev.* **2013**, *39*, 251–260. [CrossRef]
71. Rothacher, Y.; Strobl, C. Identifying informative predictor variables with random forests. *J. Educ. Behav. Stat.* **2024**, *49*, 595–629. [CrossRef]
72. Karbassi, A.; Mohebi, B.; Rezaee, S.; Lestuzzi, P. Damage prediction for regular reinforced concrete buildings using the decision tree algorithm. *Comput. Struct.* **2014**, *130*, 46–56. [CrossRef]
73. Ann, K.Y.; Pack, S.W.; Hwang, J.P.; Song, H.W.; Kim, S. Service life prediction of a concrete bridge structure subjected to carbonation. *Constr. Build. Mater.* **2010**, *24*, 1494–1501. [CrossRef]

74. Papadakis, V.G.; Demis, S. Estimation and validation of concrete strength and service life using software packages based on predictive models. In Proceedings of the 12th International Conference on Durability of Building Materials and Components, Porto, Portugal, 12–15 April 2011.
75. Xu, W.; Li, W.; Wang, D.; Ji, Y. Mechanical performance degradation investigation on FRP reinforced concrete based on neural network design method. *Sci. Rep.* **2024**, *14*, 11807. [CrossRef]
76. Wang, S.; Wu, R.; Gong, F.; Xia, J.; Zhao, Y.; Zeng, B. Service life evaluation of marine concrete structures considering spatial and temporal characteristics: A framework based on multi training-MCS-NLS. *Eng. Struct.* **2025**, *322*, 119193. [CrossRef]
77. Fan, J.; Zhu, H.; Yi, C.; Li, Z. Research on durability evaluation of concrete in complex attack environment. *AIP Adv.* **2019**, *9*, 115019. [CrossRef]
78. Nanukuttan, S.V.; Basheer, P.A.M.; McCarter, W.J.; Tang, L.; Holmes, N.; Chrisp, T.M.; Starrs, G.; Magee, B. The performance of concrete exposed to marine environments: Predictive modelling and use of laboratory/on site test methods. *Constr. Build. Mater.* **2015**, *93*, 831–840. [CrossRef]
79. Pang, L.; Li, Q. Service life prediction of RC structures in marine environment using long term chloride ingress data: Comparison between exposure trials and real structure surveys. *Constr. Build. Mater.* **2016**, *113*, 979–987. [CrossRef]
80. Al-Obaidi, S.; Bamonte, P.; Luchini, M.; Mazzantini, I.; Ferrara, L. Durability-based design of structures made with ultra-high-performance/ultra-high-durability concrete in extremely aggressive scenarios: Application to a geothermal water basin case study. *Infrastructures* **2020**, *5*, 102. [CrossRef]
81. Chen, S.; Duffield, C.; Miramini, S.; Raja, B.N.K.; Zhang, L. Life-cycle modelling of concrete cracking and reinforcement corrosion in concrete bridges: A case study. *Eng. Struct.* **2021**, *237*, 112143. [CrossRef]
82. Sharp, M.E. Simple metrics for evaluating and conveying prognostic model performance to users with varied backgrounds. In Proceedings of the Annual Conference of the PHM Society, San Diego, CA, USA, 27 September–1 October 2013; Volume 5.
83. Bennett, N.D.; Croke, B.F.W.; Guariso, G.; Guillaume, J.H.A.; Hamilton, S.H.; Jakeman, A.J.; Marsili-Libelli, S.; Newham, L.T.H.; Norton, J.P.; Perrin, C.; et al. Characterising performance of environmental models. *Environ. Model. Softw.* **2013**, *40*, 1–20. [CrossRef]
84. Gupta, H.V.; Clark, M.P.; Vrugt, J.A.; Abramowitz, G.; Ye, M. Towards a comprehensive assessment of model structural adequacy. *Water Resour. Res.* **2012**, *48*, W08301. [CrossRef]
85. Yu, J.; Wang, N.; Wang, M.; Zhang, J. Hou, D. Recyclable rubber-cement composites produced by interfacial strengthened strategy from polyvinyl alcohol. *Constr. Build. Mater.* **2020**, *264*, 120541. [CrossRef]
86. Kostić, S.; Vasović, N.; Sunarić, D. Slope stability analysis based on experimental design. *Int. J. Geomech.* **2016**, *16*, 04016009. [CrossRef]
87. Kefei, L.; Torrent, R. *Analytical and Experimental Service Life Assessment of Hong Kong-Zhuhai-Macau Link*; Civil Engng. Dept., Tsinghua University: Beijing, China, 2016.
88. Sarmiento, S.; Thöns, S.; Gonzalez, J.; Björnsson, I.; Sas, G. Comparison and Joint Evaluation of Importance, Redundancy and Robustness Indicators Applied to Aging Prestressed Concrete Bridges. *Nord. Concr. Res.* **2024**, *70*, 147–171. [CrossRef]
89. Carević, V.; Ignjatović, I. Evaluation of concrete cover depth for green concretes exposed to carbonation. *Struct. Concr.* **2021**, *22*, E1009–E1021. [CrossRef]
90. Yu, B.; Ning, C.; Li, B. Probabilistic durability assessment of concrete structures in marine environments: Reliability and sensitivity analysis. *China Ocean. Eng.* **2017**, *31*, 63–73. [CrossRef]
91. Strauss, A.; Vidovic, A.; Zambon, I.; Grossberger, H.; Bergmeister, K. Monitoring information and probabilistic-based prediction models for the performance assessment of concrete structures. *J. Perform. Constr. Facil.* **2016**, *30*, 04015081. [CrossRef]
92. Franceschini, L.; Vecchi, F.; Belletti, B.; Andrade, C.; Peiretti, H.C. Analytical method for the evaluation of the residual service life of prestressed concrete beams subjected to corrosion deterioration. *Struct. Concr.* **2022**, *23*, 121–137. [CrossRef]
93. Vishwanath, B.S.; Banerjee, S. Considering uncertainty in corrosion process to estimate life-cycle seismic vulnerability and risk of aging bridge piers. *Reliab. Eng. Syst. Saf.* **2023**, *232*, 109050. [CrossRef]
94. Cho, H.K.; Lee, H.S.; Wang, X.Y.; Ismail, M.; Park, W.J. Evaluation of CO_2 emission–absorption of fly-ash-blended concrete structures using cement-hydration-based carbonation model. *Mater. Struct.* **2015**, *48*, 3949–3963. [CrossRef]
95. Shafikhani, M.; Chidiac, S.E. Quantification of concrete chloride diffusion coefficient—A critical review. *Cem. Concr. Compos.* **2019**, *99*, 225–250. [CrossRef]
96. Liang, M.T.; Chang, J.J.; Yu, T.W. Studies of the effect of diffusion-induced chloride binding on chlorination life predictions for existing reinforced concrete bridges. *J. Mar. Sci. Technol.* **2012**, *20*, 10.
97. Martın-Pérez, B.; Zibara, H.; Hooton, R.D.; Thomas, M.D.A. A study of the effect of chloride binding on service life predictions. *Cem. Concr. Res.* **2000**, *30*, 1215–1223. [CrossRef]
98. Sergi, G.; Yu, S.W.; Page, C.L. Diffusion of chloride and hydroxyl ions in cementitious materials exposed to a saline environment. *Mag. Concr. Res.* **1992**, *44*, 63–69. [CrossRef]

99. Xue, P.F.; Mao, D.L. Corrected diffusion model of chloride in concrete and its engineering application. *Adv. Mater. Res.* **2011**, *189*, 1886–1891. [CrossRef]
100. Chen, E.; Berrocal, C.G.; Löfgren, I.; Lundgren, K. Comparison of the service life, life-cycle costs and assessment of hybrid and traditional reinforced concrete through a case study of bridge edge beams in Sweden. *Struct. Infrastruct. Eng.* **2022**, *19*, 39–57. [CrossRef]
101. Yamamoto, M.; Nakarai, K.; Yoshizumi, Y.; Torrent, R. Relationship between chloride ingress and concrete cover quality of inland structures exposed to deicing salts. *Case Stud. Constr. Mater.* **2024**, *20*, e03075. [CrossRef]
102. Li, J.A.; Feng, D. Fatigue life evaluation of bridge stay cables subject to monitoring traffic and considering road roughness. *Eng. Struct.* **2023**, *293*, 116572. [CrossRef]
103. Fathalla, E.; Tanaka, Y.; Maekawa, K. Fatigue lifetime prediction of newly constructed RC road bridge decks. *J. Adv. Concr. Technol.* **2019**, *17*, 715–727. [CrossRef]
104. Balestra, C.E.T.; Reichert, T.A.; Vizzotto, A.L.P.; Pansera, W.A.; Savaris, G. General model for service life analysis of reinforced concrete structures subject to chloride penetration. *Constr. Build. Mater.* **2021**, *305*, 124727. [CrossRef]
105. Shreir, L.L. (Ed.) *Corrosion: Metal/Environment Reactions*; Newnes-Butterworths: London, UK, 1976.
106. Li, C.Q.; Mahmoodian, M. Risk based service life prediction of underground cast iron pipes subjected to corrosion. *Reliab. Eng. Syst. Saf.* **2013**, *119*, 102–108. [CrossRef]
107. Dasar, A.; Patah, D.; Hamada, H.; Yamamoto, D.; Sagawa, Y. Life performance of 40-year-old RC beams with different concrete covers and bar diameters in natural corrosion environments. In *Structures*; Elsevier: Amsterdam, The Netherlands, 2022; Volume 46, pp. 2031–2046.
108. Chen, D.; Mei, G.; Xiao, L. Analysis of chloride invasion process in undersea RC circular lined tunnel based on nonlinear ADE theory: Analytical solution, numerical verification and experimental prediction. *Constr. Build. Mater.* **2022**, *343*, 128140. [CrossRef]
109. Keßler, S. Probabilistic corrosion condition assessment of a tunnel structure. *Struct. Concr.* **2020**, *21*, 1345–1355. [CrossRef]
110. Meng, T.; Yu, H.; Zhao, Y.; Ruan, S. Strength/Service life-oriented environmental assessment of normal and recycled concrete with nano-SiO2 in a natural and a marine environment: A pioneer case study in Zhejiang Province. *J. Build. Eng.* **2023**, *64*, 105623. [CrossRef]
111. Canestro, E.; Strauss, A.; Sousa, H. Multiscale modelling of the long-term performance of prestressed concrete structures–Case studies on T-Girder beams. *Eng. Struct.* **2021**, *231*, 111761. [CrossRef]
112. Zheng, L.; Pornsing, C. Prediction of airport road service life based on concrete structure characteristics. *Case Stud. Constr. Mater.* **2024**, *20*, e02814. [CrossRef]
113. Mortagi, M.; Ghosh, J. Climate change considerations for seismic vulnerability assessment of aging highway bridges. *ASCE-ASME J. Risk Uncertain. Eng. Syst. Part A Civ. Eng.* **2020**, *6*, 04020005. [CrossRef]
114. Jeon, C.H.; Kwon, T.H.; Kim, J.; Jung, K.S.; Park, K.T. Health Index-Based Maintenance of Prestressed Concrete Bridges Considering Building Information Modeling Application. *Buildings* **2024**, *14*, 4032. [CrossRef]
115. Ying, J.; Chen, W.; Wang, W.; Tian, Z.; Liang, L.; Chen, B. Multiscale study on high-temperature mechanical behaviors and durability of recycled concrete reinforced by polycarboxylate-modified 3D porous graphene. *Constr. Build. Mater.* **2024**, *450*, 138676. [CrossRef]
116. Kaveh, H.; Alhajj, R. Recent advances in crack detection technologies for structures: A survey of 2022-2023 literature. *Front. Built Environ.* **2024**, *10*, 1321634. [CrossRef]
117. Wang, Z.; Maekawa, K. Lifetime assessment of structural concrete–multi-scale integrated hygro-thermal-chemo-electrical-mechanistic approach and statistical evaluation. *Struct. Infrastruct. Eng.* **2022**, *18*, 933–949. [CrossRef]
118. Azimi, M.; Eslamlou, A.D.; Pekcan, G. Data-driven structural health monitoring and damage detection through deep learning: State-of-the-art review. *Sensors* **2020**, *20*, 2778. [CrossRef] [PubMed]

coatings

MDPI

Review

Review of Research on Tire–Pavement Contact Behavior

Zhenlong Gong [1,2], Yinghao Miao [1,*] and Claudio Lantieri [2,*]

1 National Center for Materials Service Safety, University of Science and Technology Beijing, Beijing 100083, China; gongzl@xs.ustb.edu.cn

2 Department of Civil, Chemical, Environmental and Material Engineering (DICAM), University of Bologna, 40136 Bologna, Italy

* Correspondence: miaoyinghao@ustb.edu.cn (Y.M.); claudio.lantieri2@unibo.it (C.L.)

Abstract: This article presents the latest progress in research on tire–pavement contact behavior. Firstly, the tire–pavement contact characteristics and their influencing factors are summarized. Then, the measurement methods and theoretical research on tire–pavement contact behavior are reviewed, and the advantages and shortcomings of different methods are compared and analyzed. Finally, analysis in the field of pavement engineering is summarized based on contact behavior. This article suggests a few key research directions: Tire–pavement contact behavior is influenced by multiple factors; therefore, multi-physical field-coupling analyses need to be carried out. Tire–pavement contact tests are mostly static and non-standardized, and it is a future trend to develop high-precision, low-cost, and standardized instruments that can measure dynamic contact. Theoretical research models rarely involve environmental factors; a contact model of the tire, pavement, and environment needs to be constructed that can truly describe the contact process. There is a relationship between contact characteristics and pavement performance; pavement performance evaluation indexes need to be established based on tire–pavement contact characteristics in the future.

Keywords: tire–pavement; contact behavior; pavement engineering; measurement method; theoretical research; pavement performance

1. Introduction

The tire–pavement contact problem is one of the important problems in the field of pavement engineering [1–3]. Tire–pavement contact behavior is essential to understanding issues such as pavement skid resistance [4,5], noise [6], rolling resistance [7], and driving safety and comfort [8,9]. Tire–pavement contact involves tribology [10], acoustics [11], contact mechanics [12,13], tire dynamics, and other multidisciplinary fields [14] and has attracted widespread interest in the field of pavement engineering. The tire–pavement contact process is also the process of pavement skid resistance deterioration [15,16], pavement disease formation [17], and tire performance decline [18]. Therefore, the study of tire–pavement contact behavior is of great significance for improving pavement performance, enhancing pavement life, and enhancing vehicle energy saving and environmental protection [19,20]. Furthermore, only with a clear understanding of the interaction between the tire and the pavement can we obtain real contact stress and pavement load distribution data. Only then can we design more economical and durable pavements as well as carry out friction analyses.

Currently, there are two main types of research methods for tire–pavement contact behavior: One is contact experimental research, which analyzes tire–pavement contact characteristics through tests, but these tests are difficult to perform, and only in recent years has there been related research. Some researchers use pressure sensors [21], carbon paper [22], and pressure-sensitive materials [23] to test the pressure distribution. The second type of research method is theoretical research, in which researchers analyze tire–pavement contact behavior through mathematical models. Later, with the development of computer technology, many researchers have studied the interaction between tires and pavements through numerical analysis techniques based on theoretical research [24–26]. The finite element analysis method can build tire and pavement models quickly and efficiently and analyze the tire–pavement contact accurately and cost-effectively, but there are differences between simulation and reality [27]. Experimental research can truly reflect the tire–pavement contact behavior, but these tests have problems such as high costs and external environmental influences [1,13]. Although tire–pavement contact behavior and its influence on pavements have been understood to some extent, contact characteristics are still controversial, and the influence law on the pavement is still unclear. Some researchers have carried out numerical simulations and experimental research on the tire–pavement contact behavior, but there is a lack of detailed and complete summary and generalization.

The objective of this article is to conduct a comprehensive review of the latest progress in the study of tire–pavement contact behavior. In order to achieve this objective, tire–pavement contact characteristics, factors influencing contact behavior, contact measurement methods, and theoretical research are reviewed. Firstly, the research progress on tire–pavement contact behavior is examined based on both geometric characteristics and mechanical characteristics. Secondly, the influencing factors of contact behavior are analyzed based on three aspects—the tire, pavement, and contact environment. Then, experimental research on tire–pavement contact behavior and measurement methods are summarized, and the advantages and disadvantages of different measurement methods are evaluated. Finally, tire–pavement models and computer simulations are reviewed, and the advantages and disadvantages of different research approaches are discussed. This article also outlines the analysis of tire–pavement contact behavior to study noise, skid resistance, rolling resistance, and pavement damage. Future research directions are also envisioned. The framework of this article is shown in Figure 1.

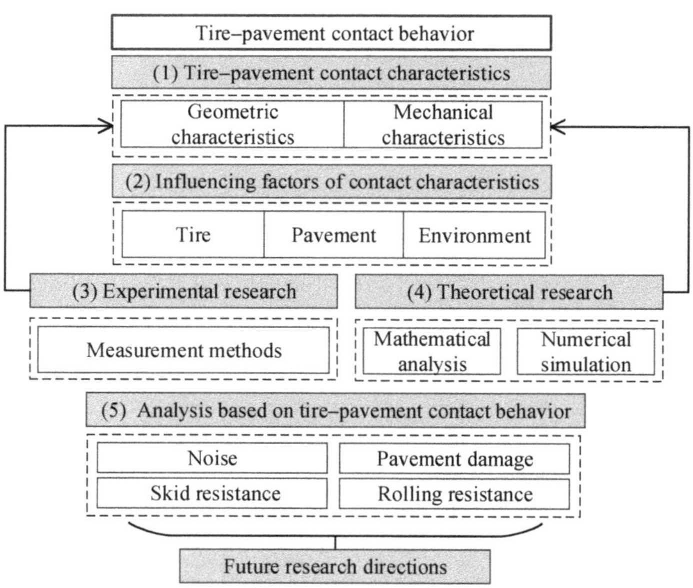

Figure 1. Article framework.

2. Tire–Pavement Contact Characteristics

Tire–pavement contact behavior is understood to help optimize tire design and pavement structure design. This can help improve pavement performance and the overall performance of tires. Researchers have characterized tire–pavement contact behavior through tire–pavement contact characteristics [2,28]. Tire–pavement contact characteristics are mainly geometric characteristics defined according to the contact area and mechanical characteristics defined according to the stress distribution [29,30]. The definition, understanding, and application of tire–pavement contact characteristics vary among researchers because of differences in the study subjects and methods.

2.1. Geometric Characteristics

The geometric characteristics of tire–pavement contact mainly refer to the contact area and contact shape. There is no clear definition of contact characteristics in the field of pavement engineering; however, the geometric characteristics of tire–pavement contact are summarized and defined in the field of tires, and "Tire Terms and Their Definitions" (GB/T 6326-2005) [31] define the tread contact length (*L*), tread contact width (*W*), coefficient of contact (*L/W*), contact area (*C-A*), and footprint area (*F-A*), as shown in Figure 2.

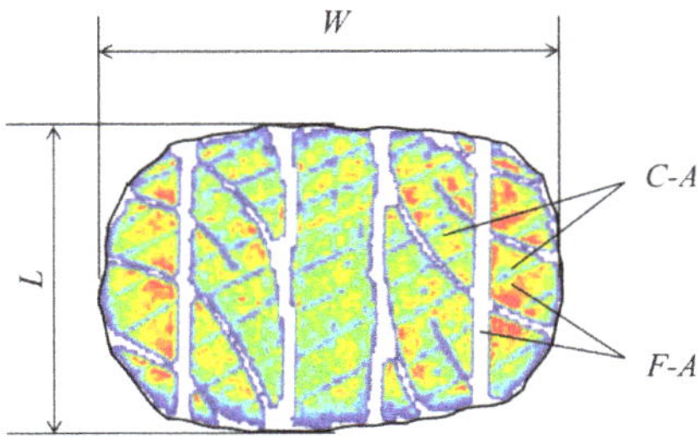

Figure 2. Geometric characteristics of tire–pavement contact.

The tire–pavement contact area is usually assumed to be circular or rectangular with uniform pressure distribution. This can simplify mechanical calculations and reflects tire–pavement contact behavior to a certain extent, and is widely used in pavement design [32]. However, measurements and modeling data show that the vertical stress distribution at the tire–pavement interface is not uniform, and the contact area is not regularly circular [29] or rectangular [33,34]. Tire pattern and pavement texture influences are ignored in this assumption, so researchers have conducted extensive research on contact geometric characteristics [35,36]. Pillai et al. [37] used the tire footprint to determine the tire–pavement contact area and obtained two simple predictive equations for tire–pavement contact area based on tire deflection and tire and wheel dimensional parameters found on the tire sidewall, as shown in Equations (1) and (2), with an error between the calculated and measured footprint areas within 15%. Ge and Wang [38] established a tire–pavement contact finite element model using ABAQUS to simulate the tire–pavement contact process and obtain a contact area equation based on tire pressure and wheel load, as shown in Equation (3). A quantitative expression of tire contact characteristics under vehicle load was obtained, as shown in Figure 3.

$$A = 2.17dr \tag{1}$$

$$A = 1.85d^{0.67}r^{0.33}(2r - s)/2a \tag{2}$$

where A is tire footprint area; r, s, and a are tire radius, wheel diameter, and aspect ratio, respectively; and d is tire deflection.

$$\delta = 0.1233p^{-0.9580}F^{0.9511} \tag{3}$$

where δ is tire footprint area, 10^3 mm; p is tire pressure, Mpa; and F is wheel load, kN.

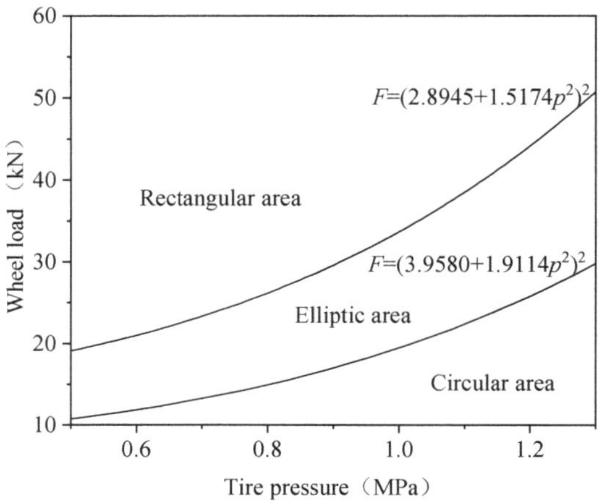

Figure 3. Distribution map of the tire contact area.

Tielking et al. [39] found that the contact area between the tire and the pavement increases as the wheel load increases, and the effect of a single wheel load on the contact area is lower. Based on experiments, Weissman et al. [40] found that the width of the tire footprint remained essentially constant due to the stiffness constraint of the tire sidewall, and that the wheel load mainly affected the length of the tire footprint. Yu et al. [2] conducted a static tire–pavement contact test using super-low-pressure (LLW) Fujifilm Holdings Corporation (FUJI) pressure film paper. As the mean profile depth (MPD) increases, the actual tire–pavement contact area decreases, and as the tire load increases, the contact area gradually increases. However, when the tire load exceeds 250 N, the tire–pavement contact area does not broadly change and even shows a slight decrease. This is mainly because the contact width becomes larger when the load is too large but the contact length is significantly reduced. In summary, it can be seen that the geometric characteristics of tire–pavement contact are related to the wheel load, pavement texture, tire type, etc. Research has mostly used the footprint method for static tire–pavement contact geometric characteristic analysis, lacking dynamic contact studies, while more direct simplification is used in pavement design.

2.2. Mechanical Characteristics

Tire–pavement contact creates stresses and forces. These tire–pavement contact stresses and forces provide the driving force when the vehicle accelerates, the lateral force when it rotates, and the braking force when it decelerates, and it can be seen that the contact behavior is directly related to vehicle performance [41] and pavement performance [42,43]. Therefore, it is significant to conduct contact mechanical characterization studies. When a tire is in contact with pavement, the tire is subjected to a complex load through the complex deformation of the tread and carcass. This complex load is generally described in the study of tire–pavement contact dynamics as the six tire forces, which are vertical force, longitudinal force, lateral force, aligning torque, lateral tilt moment, and rolling resistance moment. Some of these forces are shown in Figure 4. Due to the large deformation of a tire in mutual contact with pavement coupled with the environment and load, these forces are coupled with each other and affect each other [39,44].

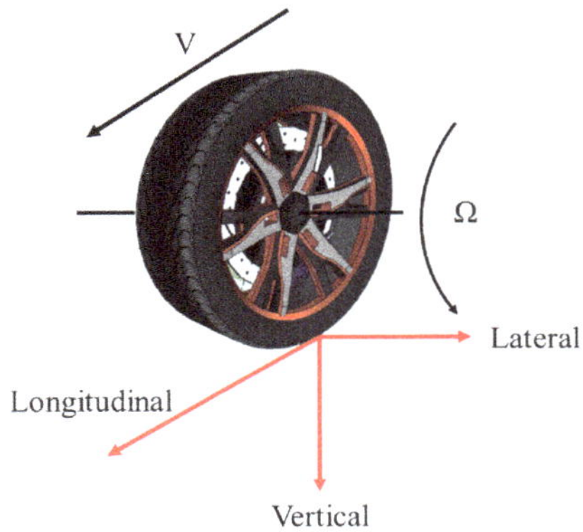

Figure 4. Schematic diagram of contact stress.

The vertical forces are of most concern in pavement studies. Marshek et al. [45] attempted to test the pressure distribution footprint of diagonal truck tires using pressure-sensitive films, and their results showed that the contact pressure was non-uniformly distributed and that the contact pressure at the tire shoulder was significantly greater than the tire air pressure. Wang et al. [46] analyzed the non-uniformity of vertical contact stresses and found that the non-uniformity of stress distribution decreases with increasing load but increases with increasing inflation pressure. Gong et al. [47] used a high-precision pressure sensor to test the tire–pavement contact pattern and stress under different forces, as shown in Figure 5. It can be seen that with an increase in force, the contact area and the region of stress concentration gradually increase. The distribution of vertical contact stresses is not uniform.

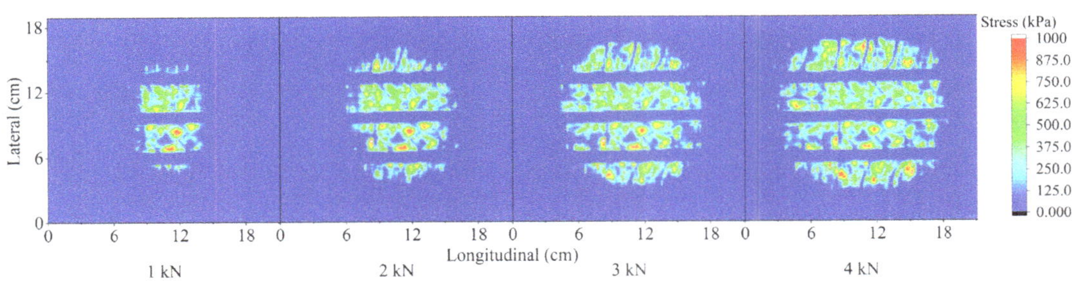

Figure 5. Contact stress distribution images.

For longitudinal forces, Tielking et al. [39] analyzed longitudinal contact stresses and vertical contact stresses as a function of vehicle speed. They found that vehicle speed had almost no effect on the vertical contact stress but had a significant influence on the longitudinal contact stress; the longitudinal stress on the pavement and the tire changed direction twice from the stationary state to the rolling state. Douglas et al. [48] tested the distribution of vertical and longitudinal pressures between the tire and the pavement under different tire pressures and loads. Their test was a full-scale laboratory test and confirmed Tielking's analysis of the contact stresses. For lateral forces, Hugo et al. [49] found that lateral forces can cause pavement cracking. Research on the aligning torque, lateral tilt moment, and rolling resistance moment has mostly been conducted in the field of vehicle engineering. In summary, it can be seen that the mechanical characteristics of tire–pavement contact are closely related to pavement performance and vehicle performance, and there are certain rules. Therefore, we can understand tire–pavement contact behavior better by characterizing these mechanical characteristics.

3. Factors Influencing Contact Characteristics

Many factors influence tire–pavement contact characteristics, but the main ones are tire factors, pavement factors, and environmental factors, as shown in Figure 6. Tire factors such as tire tread, type, and pressure all have an effect on contact area and contact stress. The pavement and the environment are equally important in influencing the contact characteristics.

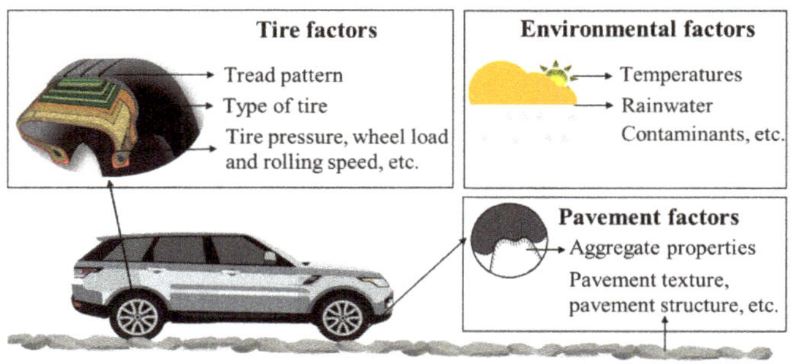

Figure 6. Factors influencing contact characteristics.

3.1. Tire

In the case of tires, the type of tire, tire pressure, tread pattern, wheel load, and the rolling speed of the tire all impact tire–pavement contact. Xie and Yang [30] analyzed the effect of tire type on contact stress by establishing a tire–pavement three-dimensional finite element model. The results showed that with the increase of vertical load, the position of the maximum contact stress between the diagonal tire and the pavement changed from the inner side of the tread to the outer side, while the maximum contact stress between the radial tire and the pavement always appeared in the middle of the tread. Under the same working conditions, the maximum contact stress between radial tires and pavement is greater than the maximum contact stress between diagonal tires and pavement. Furthermore, the change in lateral contact stress under different ribs of diagonal tires is smaller than that of radial tires. In this paper, the contact area of different tires was quantified, as shown in Figure 7. It can be seen that there are differences in the contact areas of different tire types, although this broadly conforms to the law that the larger the tire load, the larger the contact area. All the points in Figure 7 are linearly fitted, which achieves a good fitting effect. According to the 95% prediction band, the contact area of different tires also varies within a certain range.

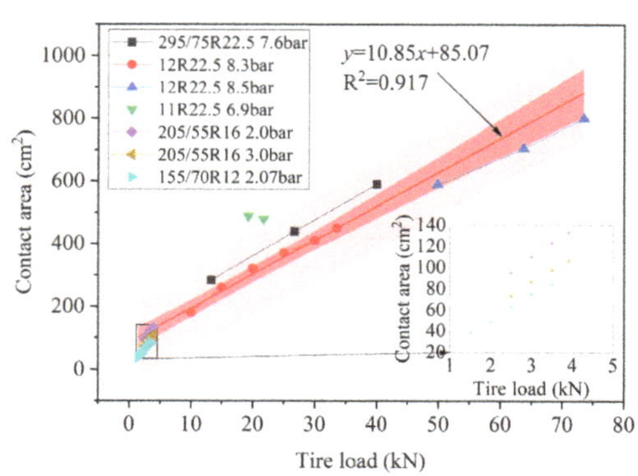

Figure 7. The contact area of different tires [3,21,29,50–52].

Muniandy et al. [36] selected seven different tread-pattern tires for footprint image analysis, and their results showed that the tread pattern affected the tire–pavement contact area second only to tire load. Machemehl et al. [53] used measured tire–pavement contact stress data to study the effect of tire pressure on pavements. Tire pressure was found to be significantly related to the tensile strain at the bottom of the asphalt concrete layer and the stress near the pavement surface for both thick and thin pavement structures. Wang et al. [54] analyzed the tire–pavement interaction by measuring contact stress for different loads and constant tire pressures. It was found that an increase in the wheel load mainly increased the contact stress at the edge of the tire and the corresponding shear strain and octahedral shear stress. Zheng et al. [3] established a three-dimensional tire–asphalt pavement contact model based on the Lagrangian algorithm and investigated the effects of internal inflation pressure and wheel load on the tire–asphalt pavement contact characteristics under static load and antilock braking system (ABS) states, respectively. The results show that a tire is more prone to wear at the shoulder position and tread-center area during emergency braking, and the center area has the highest contact pressure. With the same wheel load and internal inflation pressure, the mean contact pressure increases by 8.4%, and the contact area decreases by 7.7% when the tire is under ABS compared to static load. In summary, it can be seen that tire changes cause changes in tire–pavement contact behavior, and there are certain patterns of effect.

3.2. Pavement

Pavement factors include pavement structure, pavement texture, aggregate and asphalt characteristics, etc. Driving pavements are mainly asphalt and cement concrete pavements, which have significant differences. The Permanent International Association of Road Congresses (PIARC) classifies pavement textures into four categories—microtexture, macrotexture, mega texture, and unevenness—and tire–pavement contact studies are mainly conducted at the macrotexture and microtexture levels [55,56]. Pavement texture directly affects tire–pavement contact behavior. In addition, asphalt pavements are mainly composed of aggregates and asphalt, which inevitably affects tire–pavement contact behavior.

3.2.1. Pavement Type

Research on tire–pavement contact behavior has been more specific to one type of pavement, and few studies have compared the effects of asphalt and concrete pavements on contact behavior. Sheng et al. [57] used 3D printing technology to produce pavements with three different textures to analyze the effect of variations in pavement texture on stress distribution through numerical simulation. It was found that different textures lead to different locations of maximum stress distribution. He et al. [58] also used ABAQUS to simulate asphalt pavement–tire contact behavior, which differed from the results of Wei et al. This was mainly due to the difference in stiffness between asphalt and concrete pavements, which leads to different simulation results, especially for a pavement's internal stresses, which can show large differences. It can be seen that pavement type affects the stress distribution, and pavement topography directly affects the distribution of stress concentrations. Thus, most research on tire–pavement contact behavior focuses on pavement texture rather than pavement type.

3.2.2. Texture

Asphalt and cement pavements are shown in Figure 8; it can be seen that there is little difference in the appearance of the two pavements but notable differences in their textures [59]. Asphalt pavement texture is random and irregular, but the mean texture depth (MTD) of asphalt pavement for a specific gradation is stable. Cement pavements are very flat due to the fact that the grooves are kept in a specific pattern. Thus, the mean texture depth is broadly the same at different locations on a pavement. Texture differences can lead to different tire–pavement contact behavior. Yu et al. [2] used FUJI pressure film paper to test the differences in the tire–pavement contact area for different

MPDs, and the results showed that the pavement contact area decreased with increasing MPD. Gong et al. [47] analyzed the contact stress between asphalt pavement with different MTDs and tires and found that pavement texture had an effect on the contact behavior and showed a relationship with the MTD value. However, some researchers have ignored pavement texture during their studies, arguing that texture has a limited effect on contact stresses [32,60]. Zheng et al. [3] considered pavement texture during their modeling and analysis process and found that pavement texture had a significant effect on contact stresses.

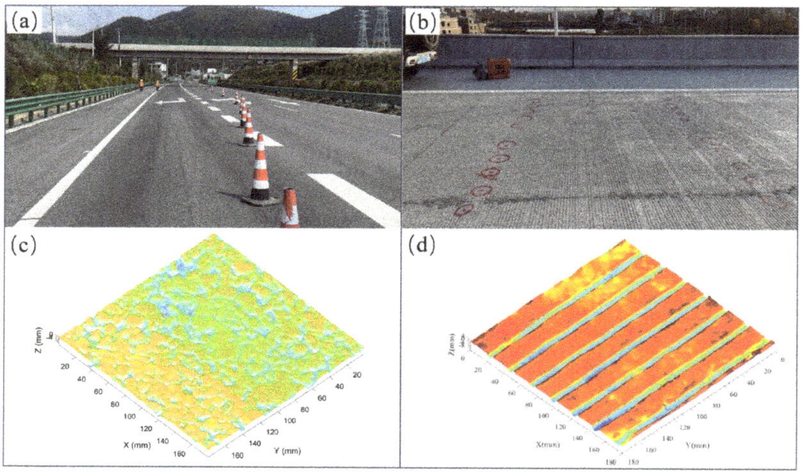

Figure 8. Asphalt and cement pavements: (**a**) Asphalt pavement; (**b**) Cement pavement; (**c**) Asphalt pavement texture; (**d**) Cement pavement texture.

Tire–pavement friction is determined by microtexture and macrotexture [61]. Most existing research is on macrotexture as it is more difficult to measure microtexture and, consequently, it is more difficult to analyze the effect of microtexture on tire–pavement contact behavior. At this stage, the impact on pavement performance is mostly determined by analyzing the morphological characteristics of the aggregates [62]. Sengoz et al. [63] determined the effect of different shapes of aggregates on the tire–pavement interface by experimentation. It was found that there is a relationship between the shape characteristics of the aggregates and the surface properties of asphalt concrete. However, their research did not examine tire–pavement contact behavior. Nevertheless, tire–pavement contact behavior has an impact on many pavement performance indicators, such as noise and skid resistance. Some relevant research has considered the effect of microtexture [64,65].

3.3. *Environment*

Tire–pavement contact behavior is not only an interaction between the tire and the pavement but is also influenced by environmental factors. Environmental factors such as temperature, rain, snow, and contaminations all affect tire–pavement contact behavior. Researchers have been most concerned with the effects of various weather and contaminations on contact behavior.

3.3.1. Climatic

Both tires and pavement are affected by the surrounding environment, and climate is one important environmental variable [61,66]. Pavements are affected by the climate, and there are various states such as dry, wet, ice, etc., that will make a significant difference in tire–pavement contact. Zheng et al. [67] used a combination of field measurements and finite element simulations to analyze tire–pavement adhesion properties in the dry, wet, lubricated, and ponding states of a pavement. It was found that under the dry road state, when the slip rate was around 15%, the adhesion coefficient reached the peak value, while the adhesion coefficient peaked at a slip rate of 11.5% for wet pavement. Gerthoffert et al. [68] analyzed friction in four states—dry, wet (due to water), compacted snow, and ice. By constructing a model consisting of tire, pavement, and contaminant components, it was found that friction varied greatly between states. Wies et al. [69] analyzed pavement skid resistance, noise, and rolling resistance based on tire imprints under winter and summer environmental conditions. Tires should be designed according to seasonal differences to improve driving. Contact behavior can be affected by temperature, which is directly related to climate. Tang et al. [70] developed a tire–pavement thermal–mechanical interaction model to research the effect of temperature on pavement skid resistance. They found that higher environmental temperatures led to lower skid resistance and higher braking distances. Hernandez et al. [71] used the finite element method to analyze the effect of temperature and speed on tire contact area, deflection, and three-dimensional contact stresses. It was found that the effect of temperature on contact stress is greater than that of vehicle speed. In summary, it can be seen that the climate has a significant effect on tire–pavement contact behavior: Higher temperatures increase the tire–pavement contact area and reduce the contact stress; and a wet pavement will reduce the contact area due to the lifting effect, thus affecting vehicle braking performance.

3.3.2. Contaminants

The range of contaminants is broad, including rain and snow as previously mentioned. Gerthoffert et al. [68] constructed a model to study the problem of the friction coefficient degradation of contaminated pavements to establish the relationship between contaminants and friction coefficients. The friction performance of four pavement states (dry, wet, snow, and ice) was compared. The results show that their constructed model fitted the experimental data from tire and pavement friction measurement devices. Contaminants also include rubber deposits, dust particles, jet fuel, oil spills, etc. [72]. Pérez-Fortes and Giudici [73] noted that contaminants are extremely important for pavement performance and categorized contaminants into four categories based on source—those produced by vehicles, pavement embankment deterioration, road equipment, and maintenance operations. Contaminant sources are not limited to these four categories, however; for example, the surrounding soil and plants can also produce contaminants that affect pavements. Lantieri et al. [74] analyzed the effect of contaminants spilled from vehicle accidents on the skid resistance of pavements. Tire–pavement contact is bound to change when a pavement is contaminated. Thus, contamination can also affect tire–pavement contact behavior. Thorpe et al. (2008) summarized non-exhaust emission contaminants, which are the main type of contaminants affecting pavements and are known to be increasing [75]. Pomoni et al. [76] found, experimentally, that loose contaminants reduce the tire–pavement contact area. In addition, debris from worn pavement markings, snowmelt residue, and other contaminants can affect the contact behavior of tires with the road surface, affecting driving performance [77,78].

In summary, it can be seen that the tire, pavement, and environment have an important influence on the contact area and contact stress, while tire–pavement contact will also react to the tire and pavement. Therefore, analyses should be carried out with the coupling of multi-physical fields such as stress, fluid, and temperature. This can better reflect the actual tire–pavement contact behavior. However, this is difficult, so researchers have mostly analyzed tire–pavement contact behavior through experimental and simulation research.

4. Experimental Research

4.1. Measurement Methods

Experimental research was initially only able to analyze the footprint between tires and pavements using carbon paper, and so there was less experimental research conducted on tire–pavement contact behavior. With the advancement of measurement methods and equipment, more measurement methods are now available, meaning that tire–pavement contact behavior research is gaining widespread attention [13,37]. Measurement methods mainly include the imprinting, pressure-sensitive, pressure sensor, and optical methods. The common tire–pavement contact measurement methods are summarized in Table 1.

Table 1. Measurement methods for tire–pavement contact behavior.

Measurement Methods		Principle Testing Methods	Features	Equipment View
Imprinting method	Carbon paper [22]	Break of microcapsules due to pressure; colorless hidden dyes spill out and become colored by acid	Easy to operate; low cost; only applicable for measuring contact area	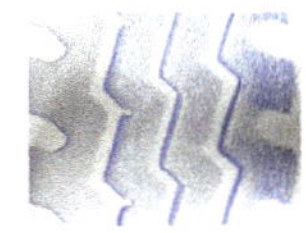
	Pressure plates [79]	Analysis of pressure magnitude and distribution based on the size of the imprint produced by the conical particles	Very low accuracy; can measure stress distribution	
Pressure-sensitive method	Pressure-sensitive film [45]	Color intensity is proportional to the applied pressure	Lower accuracy; non-reusable	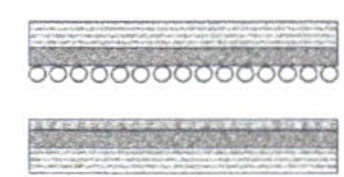
	FUJI pressure film [1]	Same as pressure-sensitive membranes	Measurement accuracy up to 0.125 × 0.125 mm; non-reusable	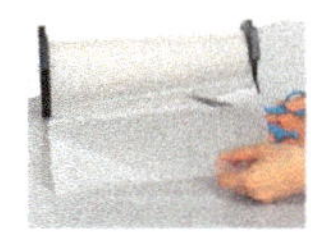
Pressure sensor method	Piezoresistive sensors [35]	Electrical resistance varies proportionally to the contact pressure	Easy to operate; intelligent; accuracy of 1.1 mm × 1.1 mm	
	Capacitive sensors [20]	Capacitive contact principle	Good flexibility; high sensitivity; can measure dynamic contact	
	Piezoelectric Sensors [80]	Piezoelectric effect	High reliability; high sensitivity; high dynamic measurement accuracy	
	Stress-In-Motion system [81,82]	Multiple principles	Measures lateral and longitudinal forces; low accuracy	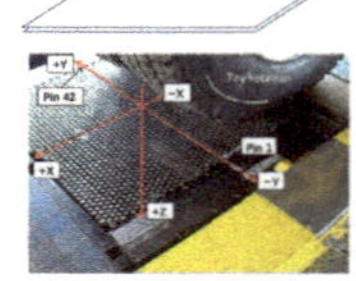
Optical method	Light absorption [83]	Based on the total reflection of light	High resolution; frequent calibrations required	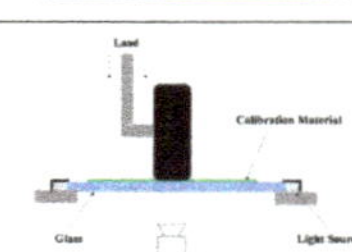

The imprint method is the earliest tire–pavement contact test method. Researchers have been conducting skid resistance and rolling resistance studies by analyzing tire footprints as early as the 20th century [84]; however, this method cannot measure pressure levels. Yu et al. [79] used a vulcanized rubber plate with conical particles and a calibrated pressure–imprint diameter relationship curve to obtain pressure data; However, the accuracy of this method is very low and does not reflect the contact between the tire and the pavement. The optical method also has the problem that the pavement is not realistically represented [85]. This method generates images based on the phenomenon of frustrated total internal reflection in the pressure area and processes the images to show the contact area as a color contour map of pressure, which creatively introduces optical technology into the study of tire–pavement contact behavior, but this can only be used to carry out static contact studies. Liang et al. [83] improved the method by addressing the influence of ambient light by optimizing the test bench. Consequently, they were able to measure not only static pressure distributions but also pressure under dynamic conditions.

The pressure-sensitive method and flexible pressure sensor method can truly reflect tire–pavement contact behavior. Marshek et al. [45] attempted to test the pressure distribution of tire–pavement contact using pressure-sensitive films, which consist of two layers of film—one containing capsules of developer and the other providing a developing layer on which the developer acts. Capsules are of various sizes corresponding to different breakage pressures. The pressure distribution is obtained by counting the distribution of different breakage capsules. Subsequently, FUJI Japan developed a commercial film and pressure image analysis system based on this principle, and the accuracy and intelligence of this technique were greatly improved when used in pavement research [1,86]. As several manufacturers have developed sophisticated flexible pressure sensor products, researchers have begun to use these products to obtain pressure distribution data during tire–pavement contact to build more accurate tire–pavement contact models, measure tire grounding performance, pavement skid resistance, etc. [13,20]. The method is easy to operate, can be controlled by a computer program, and has high accuracy, but pressure sensors are costly and easily damaged. A future development direction is to improve the measurement accuracy of pressure sensors and make them more applicable to various test environments.

4.2. Current Status and Trends in Measurement Methods

In summary, it can be seen that with the improvement in the accuracy of test equipment and system improvement, several measurement methods can be applied in pavement performance and tire performance studies. However, there are different problems with each technique at this stage. The first is the cost issue; except for the imprinting method, other methods are relatively expensive. Therefore, now and in the future, there are likely to be many studies that still use the imprinting method to calculate contact stress and contact area [22]. The second issue is that of dynamic acquisition. The pressure-sensitive method has been continuously updated (from an initial accuracy of a few millimeters to 0.125×0.125 mm) and is capable of converting pressure into a pressure cloud diagram. This has great advantages with respect to measurement accuracy and the visualization of results [1]; however, the inability for dynamic continuous acquisition remains a major limitation. At the same time, this method suffers from the problem that high accuracy and range cannot be realized. In general, a key development direction is obtaining high-precision data under static conditions. The third problem is the inability to reflect the true pavement surface. The optical method mostly ignores the real pavement and is limited to tire performance studies. The pressure sensor method is available for static and dynamic tests. The United States Tekscan company, Canada Xseneor, and others have developed mature, flexible sensor equipment. This technology is simple to operate, has computer program control, provides high accuracy, and, in recent years, has shown continuous improvement, although the cost of these pressure sensors is high and they are easily damaged. The development of high-accuracy, low-cost, standardized equipment that can measure dynamic contact is a trend of the future.

5. Theoretical Research

5.1. Mathematical Analysis of Tire–Pavement Contact Behavior

The study of tire–pavement contact behavior is not limited to experimental studies. Many researchers have constructed different contact models and carried out analyses [87–89]. One of the most classic contact models is the elastic contact model proposed by Hertz, which is also the basis for many subsequent classic models [90]. For the first time, Hertz treated the elastomer as a quadric and analyzed its deformation and contact problems, as shown in Figure 9. The contact radius and contact stress can be calculated directly from the model, as shown in Equations (4) and (5).

$$a = (\frac{3WR}{4E^*})^{1/3} \quad (4)$$

$$P(r) = \frac{3W}{2\pi a^2}[1 - (\frac{r}{a})^2]^{1/2} \quad (5)$$

where W is the applied load; R is the compound curvature, $\frac{1}{R} = \frac{1}{R_1} + \frac{1}{R_2}$; E^* is the contact modulus, $\frac{1}{E^*} = \frac{1-v^2}{E_1} + \frac{1-v^2}{E_2}$; E is the elastic modulus; v is the Poisson ratio; a is the radius of the contact circle; $P(r)$ is the contact pressure; and r is the distance from the point to the contact center.

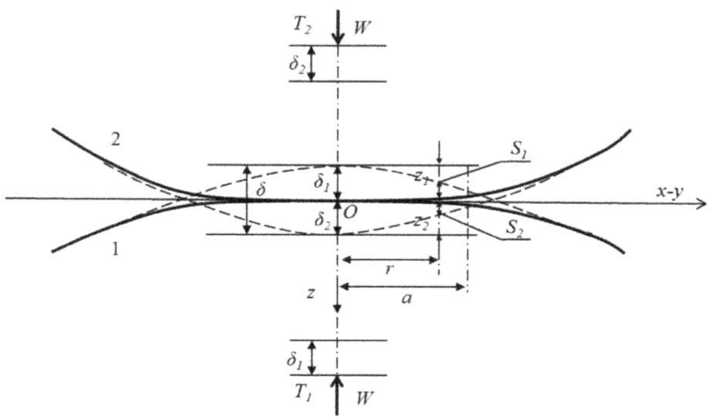

Figure 9. Frictionless static contact of two convex bodies.

Subsequent research based on this model adjusted the geometric parameters to obtain a model of a cylinder and an elastic half-space, as shown in Figure 10, and the contact radius and contact stress, as shown in Equations (6) and (7). Wu [91] used this model to analyze the effect of tire and pavement parameters on tire–pavement contact behavior.

$$a = (\frac{4FR}{E^*\pi})^{1/2} \quad (6)$$

$$P(x) = \frac{2F}{\pi a^2}[a - x^2]^{1/2} \quad (7)$$

where x is the distance from the point to the contact center; L is the length of the cylinder; F is the line load; and R is the radius of the cylinder.

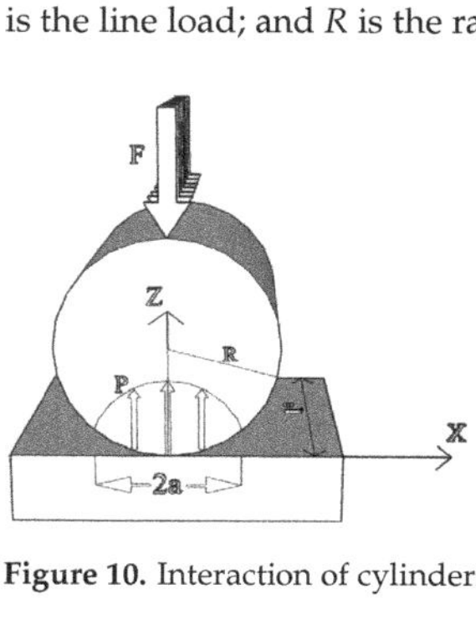

Figure 10. Interaction of cylinders and elastic half-space bodies.

In 1966, Greenwood and Williamson [92] proposed a mixed elastic and elasto-plastic contact model between rough and smooth surfaces based on statistical analysis, i.e., the G–W model, as shown in Figure 11. Their results showed that the actual area of objects in contact with each other, the number of micro-convex bodies, and the load generated by the contact are all related to the probability density function of the height of the surface form profile. The actual contact area is shown in Equation (8), and the actual average pressure is shown in Equation (9), where $p(z)$ is a function of the probability density.

$$A_{re} = n\pi R(z-d) = N\pi R\int_{d}^{\infty}(z-d)p(z)dz \tag{8}$$

$$p_r = \frac{4NE^*R_p^{1/2}}{3A_{re}}\int_{d}^{\infty}(z-d)^{3/2}p(z)dz \tag{9}$$

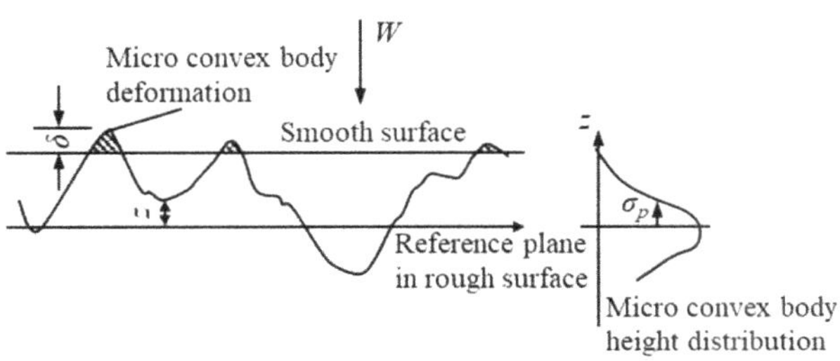

Figure 11. Contact between a rough surface and a smooth surface.

To further improve the accuracy of the tire–pavement model, Majumdar and Bhushan [93] introduced the fractal geometry theory and proposed the M–B fractal contact model using the W–M fractal function. This model, compared with the G–W model, includes fractal characteristics of the rough surface and can provide a quantitative representation of the relationship between surface roughness and the true contact area and the elastic and plastic contact area, respectively, as well as reflecting the relationship between load and contact area. Subsequently, Persson [94] proposed the surface fractal friction theory, which is based on the fractal characteristics of the pavement surface and the power spectrum density to model the tire–pavement contact. This model is widely used for tire–pavement friction analysis and reflects the skid resistance of the pavement by calculating the dynamic friction factor, which is calculated as shown in Equation (10). The contact state is shown in Figure 12, where $C(q)$ is the road surface spectral density; and $p(q)$ and $G(q)$ are the fractal characteristic functions, where q is the wavelength and is related to the amplification factor ζ. Persson analyzed experimental data on tire–pavement contact using Persson contact mechanics and rubber friction theory with better results [95,96].

$$\mu = \frac{1}{2}\int_{q_L}^{q_1} \mathrm{d}qq^3C(q)p(q) \times \int_{0}^{2\pi} d\phi\cos(\phi)\cdot\mathrm{Im}\frac{E(qv\cos\phi)}{(1-v^2)\sigma_0} \tag{10}$$

In summary, it can be seen that the Hertz contact model considers surface contact as elastic contact and single-point contact, and the model is used in tire–pavement contact studies to oversimplify the characteristics of the tire and the pavement. The G–W contact model takes into account the textural properties of the contact surface but assumes the micro-convex body as a uniform semicircle, which is different from the real situation. Meanwhile, both the Hertz contact model and the G–W contact model ignore the interactions of different contact areas, while Persson's model of tire–pavement contact takes into account the scale-dependent deformation response of rubber to surface roughness to make up for this deficiency. Furthermore, the friction theory system of this model fits better with the actual situation and is widely used in the study of the pavement skid-resistance mechanism.

However, this model assumes the conservation of energy dissipated by tires and work done by friction, ignoring energy transfer and dissipation. Therefore, further refinement is still needed.

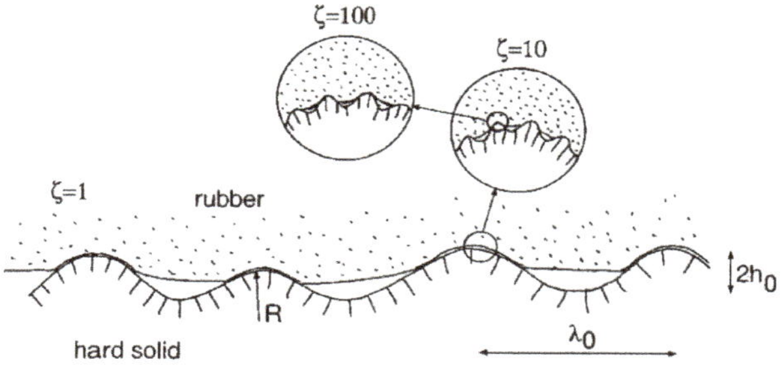

Figure 12. Rough pavement and tire contact.

5.2. Numerical Simulation Analysis of Tire–Pavement Contact Behavior

Although various tire–pavement contact models are available to analyze tire–pavement contact behavior and even the influence of pavement surface properties on the contact behavior, these models all assume the geometry and height distribution of a micro-convex body while ignoring the interaction between different contact zones; therefore, the usability of these models is very limited. With the development of computer technology, some researchers have begun to simulate tire–pavement contact behavior on computers and calculate tire–pavement contact stresses and contact areas via numerical simulation [14,60,97]. The modeling process is shown in Figure 13; contact behavior analysis is carried out by constructing models of tires and road surfaces, and some models even consider the water factor and can carry out water-wading calculations.

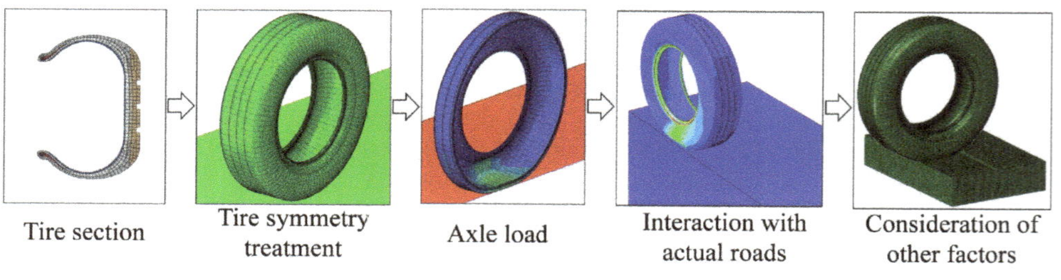

Figure 13. The modeling process.

Wang et al. [46] simulated tire–pavement interactions to predict contact stresses at static and various rolling conditions, using an arbitrary Lagrangian–Eulerian formulation for the steady-state rolling process of tires, and their model results were consistent with previous measurements. Their analysis results show that the non-uniformity of vertical contact stress decreases with the increase of load and increases with the increase of expansion pressure. However, vehicle-maneuvering behavior significantly affects the tire–pavement contact stress distribution. Oubahdou et al. [13] used a semi-analytical model to predict vertical contact stresses at the tire–pavement interface under different loads and inclination angles and compared the experimental measurements with Tekscan flexible sensors to confirm the reliability of the numerical analysis. Reynaud et al. [98] used the Hertz model for tire–pavement contact analysis and obtained a tire grounding pressure model using the boundary element method and semi-analytical method. Using their model, the contact differences were analyzed for different tire inflation pressures and different loads, and the contact stresses measured by using a Stress-In-Motion device were compared with the simulation calculation results. It was shown that the model could accurately analyze the tire–pavement contact behavior. Zheng et al. [3] established a three-dimensional tire–asphalt pavement contact model using the finite element method. The effects of internal inflation pressure and wheel load on the tire–pavement contact characteristics were explored. Their results show that when a tire is under emergency braking, the shoulder position and the center region of the tread are more prone to wear, and the contact pressure in the center region is the largest. The average contact pressure when the tire was in the ABS condition increased by 8.4% and the contact area decreased by 7.7% compared to a static load. In summary, it can be seen that the use of numerical simulation has improved the analysis of tire–pavement contact behavior.

5.3. Current Status and Trends of Theoretical Research

At this stage, numerical simulation techniques mostly construct tire–pavement contact models by analyzing pavements and tires, with little consideration of environmental factors; only a relatively few studies have analyzed the effects of temperature and wetness on tire–pavement contact. Bai et al. [99] established a three-dimensional finite element model based on non-uniformly distributed tire–pavement contact pressure, full interfacial layer bonding conditions, and viscoelastic properties of an asphalt layer. According to this model, it was found that higher temperatures could lead to a smoother stress distribution on the wheel track, and the top of the asphalt concrete pavement showed a tendency to become damaged under the shear stress in the vertical direction as the temperature increased. Zheng et al. [67] analyzed the adhesion characteristics of the tire–pavement contact interface using numerical simulation and revealed tire–pavement adhesion characteristics under different pavement conditions. The adhesion coefficient of tire–pavement increased with the increase of MPD value in both dry and wet conditions, but the rate of change was greater under wet pavement conditions. Tires and pavements are subject to complex environmental conditions, with not only temperature changes occurring in the pavement but also dust and weather extremes affecting the tire–pavement contact behavior. Therefore, the analysis of tire–pavement contact behavior under multi-physical field coupling should be carried out; only by constructing contact models for tires, pavements, and the environment can the contact process between tires and pavements be truly described. However, this puts higher requirements on model construction and computational power. Therefore, currently, existing simulation analyses still have many shortcomings.

6. Analysis Based on Tire–Pavement Contact Behavior

The purpose of tire–pavement contact behavior research is to analyze pavement performance and tire performance. The pavement engineering and tire fields have research needs, but older testing methods are difficult to employ and offer low accuracy. Researchers can only simulate and mathematically calculate tire–pavement contact behavior to analyze pavement and tire performance. With the development of equipment, researchers have gradually begun conducting contact-behavior tests to evaluate road performance and tire performance. Currently, there are many types of pavement-performance and tire-performance analyses based on tire–pavement contact behavior, such as skid resistance, noise, pavement damage, rolling resistance, etc.

6.1. Skid Resistance

Firstly, pavement skid resistance has been previously studied. Peng et al. [100] developed a skid-resistance simulation framework based on a finite element method using in situ 3D pavement texture and skid resistance data. The relationship between each contact feature and the interface friction parameters was quantified by principal component analysis and regression analysis. The results of the study provided better friction parameter inputs for finite element simulations. Najafi et al. [101] concluded that friction is reduced and vehicle crash rates are increased under wet weather conditions due to the water on the pavement reducing the contact area between the tire and the pavement. Yun et al. [5] also realized the problem of friction change and studied the effect of contact area on the friction coefficient. The relationship between pavement texture and contact area was investigated; the surface textures of 29 pavements were scanned; and the skid resistance of 23 of them and the tire–pavement contact area of some were also measured. The results showed that the contact area is influenced by the roughness of the pavement; the contact area decreases as a power function of the root-mean-square surface height; and the low-speed tire–pavement friction coefficient under dry conditions is not influenced by the real contact area, while the friction coefficient under wet conditions is weakly positively correlated with the real contact area. Yu et al. [102] used a pressure-distribution-mapping system, FPD-8010E, and a tire–pavement dynamic friction analyzer to test the contact area and friction coefficient of different pavements. They found a close linear correlation between the actual tire–pavement contact area and the dynamic friction coefficient (DFC), as shown in Figure 14. It can be seen that the DFC gradually increases as the actual area of contact increases, but this correlation varies with pavement type and aggregate type. Skid resistance is certainly related to tire–pavement contact behavior, but it is also influenced by many other factors, such as aggregate type [103], pavement gradation [104], etc. In summary, the analysis of pavement skid resistance requires a comprehensive understanding of tire–pavement contact behavior.

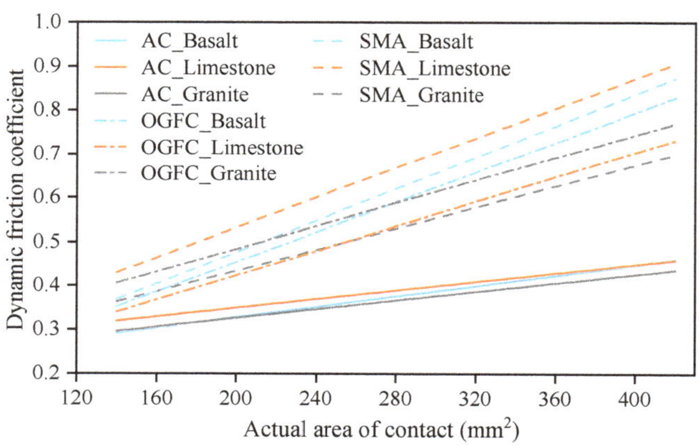

Figure 14. Fitting graph of the dynamic friction coefficient to the actual contact area.

6.2. Noise

In addition to pavement skid resistance studies based on contact behavior, researchers have also conducted noise studies. Ding [27] established a coupled finite element and boundary element analysis method (FEM/BEM) for tire–pavement interaction noise simulation to quantify the effect of pavement characteristics on the generation and propagation of tire vibration noise. It was found that tire–pavement noise increased with increasing texture level and that overall noise decreased with increasing porosity at the same texture level. Noise research has mostly used a combination of noise and stress equipment to analyze the effect of contact behavior on noise, as shown in Figure 15. Cesbron et al. [105] used a dynamic pressure-measurement system and a close-proximity (CPX) noise-measurement system to measure the contact force and noise levels for a slick tire rolling on six different pavement surfaces between 30 and 50 km/h. The results showed that contact forces and close-proximity noise measured at 30 km/h were correlated. Later, Cesbron et al. [106] summarized a relationship function between noise level and contact force level based on tire–pavement contact data and CPX rolling noise measurements, as shown in Equation (11). However, this relationship is valid only at low frequencies (315–1250 Hz for a slick tire and 315–800 Hz for a patterned tire). Huang et al. [107] used pressure film to measure the open void ratio at the tire–pavement contact interface and judged pavement noise according to the open void ratio. Their study showed that as the open void ratio increased, the noise between the tire and the pavement decreased.

$$L_N^e = a_i L_F^e(f_i) + b_i \tag{11}$$

where L_N^e is the contact force level; $L_F^e(f_i)$ is the noise level; and a_i and b_i are the coefficients of the linear regression using a statistical method.

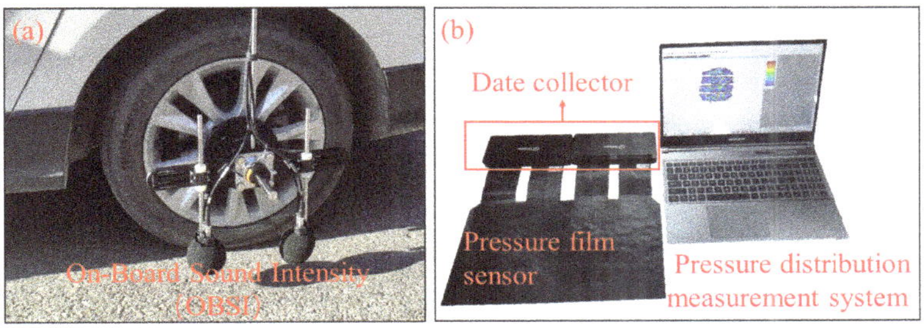

Figure 15. Noise test and contact stress test equipment: (**a**) On-Board Sound Intensity (OBSI) test method; (**b**) Pressure distribution measurement equipment.

6.3. Pavement Damage

Pavement damage is also closely related to tire–pavement contact behavior, and uneven tire–pavement contact stresses play an important role in the sprouting of top-down cracks. Studies of this phenomenon point out that high surface stresses may be the origin of the top-down crack creation. Manyo et al. [108] constructed a simulation model of tire–pavement contact pressure and used flexible sensor data to verify its reliability. They then analyzed the effect of tire–pavement contact behavior on the pavement structure. Their analysis showed that the maximum tensile strain was capable of causing top-down cracking. This strain is higher than the level considered in current pavement design models [109]. Hajj et al. [42] used a quartz sensor array to measure tire–pavement contact stress distributions and calculated pavement responses to predict bottom-up fatigue cracking using the Mechanistic–Empirical Pavement Design Guide (MEPDG). It was found that equivalent circular and square contact stress distributions overestimated bottom-up fatigue cracking by 4% to 23%, and the elliptical stress distributions resulted in similar or underestimated fatigue cracking by up to 17%. Contact behavior does not only affect pavement cracking but is also related to rutting damage. Ronald et al. [110] tested contact stresses with the Vehicle–Road Surface Pressure Transducer Array System and found that high contact stresses tend to cause rutting. Moazami and Muniandy [111] found that the actual contact area and stresses were not different from the design, which would cause the pavement to be highly susceptible to rutting damage. In summary, it can be seen that pavement damage is closely related to tire–pavement contact behavior. Understanding the true contact stresses will help design economical and durable pavements.

6.4. Rolling Resistance

Tire–pavement contact tests also have an important role in the field of automotive tires, where contact characteristics are closely related to rolling resistance. Guo et al. [15] established a three-dimensional tire–pavement model to analyze the changing characteristics of the contact stress and the relationship between the rolling resistance and tire–pavement operating conditions. Their results show that the strength of the relationship between transverse and longitudinal contact stresses is related to rolling conditions, and a method was proposed that can be used to predict the relationship between truck–bus tire conditions and rolling resistance. Kawakami et al. [7] reported tire–pavement pressure distributions using a pressure-measuring film and analyzed the relationship between rolling resistance and pavement contact characteristics. Their results for the correlation between the area percentage and the rolling resistance coefficient at 20 km/h are shown in Figure 16. It can be seen that the distribution of the contact pressure correlates with the rolling resistance coefficient at low vehicle speeds, and so the contact pressure distribution can be used to indirectly evaluate the rolling resistance. Liang Chen et al. [112] used the Tekscan pressure-measurement system to obtain the grounding pressure distribution, which was used to construct a parametric evaluation system expressing geometric and mechanical information in the ground area. Their results showed that the regression equation fitted better when tire–pavement characteristic parameters were used to evaluate the tire rolling resistance, and the maximum relative fitting error was only 2.367%; the corresponding equation is given as Equation (12).

$$f_1 = 1.324 - 0.028\overline{\Delta Z_{\mathrm{I}}'} + 0.021E_1 - 0.031x_9 - 0.038\overline{\Delta Z_{\mathrm{II}}'} + 0.100\sigma E_{y_{\mathrm{II}}} + 0.020E_{\mathrm{IV}} - 0.294x_{14} \quad (12)$$

where f_1 is the rolling resistance coefficient and the other parameters represent the outer tire shoulder area and transition area tire–pavement contact characteristics.

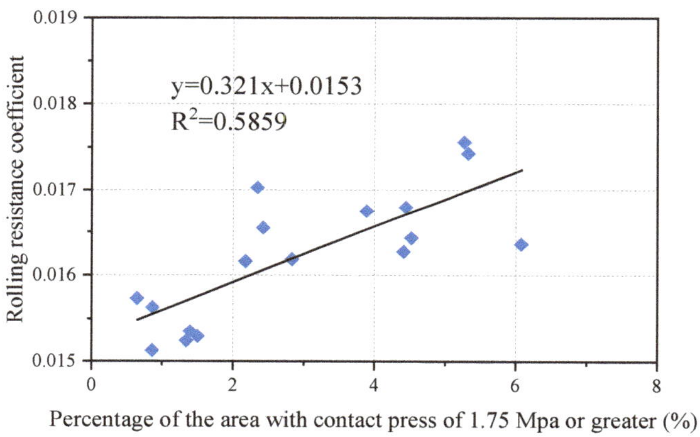

Figure 16. Relationship between the rolling resistance coefficient and the percentage of the area [7].

In addition to the above-described tire and pavement performance studies, contact behavior studies can also be used to analyze structural stresses in pavement [108], pavement durability [113], and other issues [42]. Therefore, it is crucial to define real tire–pavement contact behaviors. When contact behaviors are identified as such, road design and tire design can be optimized based on contact stresses, the extent of pavement damage by stress can be determined, changes in pavement performance can be analyzed based on changes in contact characteristics, and pavement damage processes can be understood. In summary, many pavement- and tire-performance analyses can be performed based on contact behavior.

7. Concluding Remarks

This paper has reviewed the research progress on tire–pavement contact behavior in recent years from the perspectives of tire–pavement contact, contact behavior experimental research, theoretical research, and analyses based on tire–pavement contact behavior. It was found that tire–pavement contact behavior research suffers from low experimental accuracy, high costs, and the limited applicability of contact models. Therefore, future in-depth research should be conducted mainly on the following aspects:

(1) Defining tire–pavement contact characteristics. Tire–pavement contact characteristics are mainly geometric and mechanical. There is a lack of definitions for specific contact

characteristics in the pavement field, and the influences of different characteristics on pavement performance are poorly understood and require further research.

(2) Many factors influence tire–pavement contact characteristics. Tire–pavement contact behavior in complex environments is inherently more complex, especially when rain, dust, and pollutants are present. This means that the comprehensive analysis of geometric characteristics, mechanical characteristics, and environmental factors remains limited and will be the focus of future research.

(3) The existing geometric characteristic test lacks the contact depth test and is limited to imprint measurement, and the mechanical feature test is only a vertical stress test, excluding transverse and longitudinal stresses. Tire–pavement contact tests are mostly static and non-standardized, and it is a future trend to develop high-precision, low-cost, and standardized instruments that can comprehensively measure dynamic contact stresses.

(4) In terms of theoretical research, finite element numerical simulation has more development potential than traditional tire–pavement contact calculation model research. Tire–pavement models need to be more realistic and coupled with multiple environmental factors. Model calculations will always differ from reality, and the more realistic the tire path model, the more complex the calculations.

(5) The performance of most pavement changes over time, and this change process is directly related to tire–pavement contact behavior while also being influenced by the pavement environment. Therefore, pavement performance evaluation indexes need to be established in the future based on tire–pavement contact characteristics.

Author Contributions: Conceptualization, Y.M. and Z.G.; investigation, Y.M., C.L. and Z.G.; writing, Z.G. and C.L. All authors have read and agreed to the published version of the manuscript.

Funding: This work was jointly supported by the National Natural Science Foundation of China (No. 51978048) and the China Scholarship Council (No. 202306460065).

Data Availability Statement: Data are contained within the article.

Conflicts of Interest: The authors declare no conflicts of interest.

References

1. Chen, B.; Zhang, X.; Yu, J.; Wang, Y. Impact of contact stress distribution on skid resistance of asphalt pavements. *Constr. Build. Mater.* **2017**, *133*, 330–339. [CrossRef]
2. Yu, M.; Xu, X.; Wu, C.; Li, S.; Li, M.; Chen, H. Research on the prediction model of the friction coefficient of asphalt pavement based on tire-pavement coupling. *Adv. Mater. Sci. Eng.* **2021**, *2021*, 6650525. [CrossRef]
3. Zheng, B.; Chen, J.; Zhao, R.; Tang, J.; Tian, R.; Zhu, S.; Huang, X. Analysis of contact behaviour on patterned tire-asphalt pavement with 3-D FEM contact model. *Int. J. Pavement Eng.* **2022**, *23*, 171–186. [CrossRef]
4. Yu, M.; You, Z.; Wu, G.; Kong, L.; Liu, C.; Gao, J. Measurement and modeling of skid resistance of asphalt pavement: A review. *Constr. Build. Mater.* **2020**, *260*, 119878. [CrossRef]
5. Yun, D.; Hu, L.; Tang, C. Tire–road contact area on asphalt concrete pavement and its relationship with the skid resistance. *Materials* **2020**, *13*, 615. [CrossRef] [PubMed]
6. Li, T.; Burdisso, R.; Sandu, C. Literature review of models on tire-pavement interaction noise. *J. Sound Vibr.* **2018**, *420*, 357–445. [CrossRef]
7. Kawakami, A.; Ishigaki, T.; Shirai, Y.; Terada, M.; Kubo, K. Evaluation method of pavement surface characteristics for rolling resistance. *Road Mater. Pavement Des.* **2017**, *18* (Suppl. 2), 2–11. [CrossRef]
8. Maia, M.M.A.S.; Dinis-Almeida, M.; Martinho, F.C.G. The influence of the affinity between aggregate and bitumen on the mechanical performance properties of asphalt mixtures. *Materials* **2021**, *14*, 6452. [CrossRef] [PubMed]
9. Yu, L.; Hu, J.; Li, R.; Yang, Q.; Guo, F.; Pei, J. Tire-pavement contact pressure distribution analysis based on ABAQUS simulation. *Arab. J. Sci. Eng.* **2022**, *47*, 4119–4132. [CrossRef]
10. Belghith, S.; Mezlini, S.; BelhadjSalah, H.; Ligier, J. Modeling of contact between rough surfaces using homogenisation technique. *Comptes Rendus Mécanique* **2010**, *338*, 48–61. [CrossRef]
11. Ren, W.; Han, S.; Fwa, T.F.; Zhang, J.; He, Z. A new laboratory test method for tire-pavement noise. *Measurement* **2019**, *145*, 137–143. [CrossRef]
12. Hernandez, J.A.; Al-Qadi, I.L. Tire–pavement interaction modelling: Hyperelastic tire and elastic pavement. *Road Mater. Pavement Des.* **2016**, *18*, 1067–1083. [CrossRef]

13. Oubahdou, Y.; Wallace, E.; Reynaud, P.; Picoux, B.; Dopeux, J.; Petit, C.; Nélias, D. Effect of the tire—Pavement contact at the surface layer when the tire is tilted in bend. *Constr. Build. Mater.* **2021**, *305*, 124765. [CrossRef]
14. Ma, B.; Xu, H. Vehicle unsteady dynamics characteristics based on tire and road features. *Adv. Mech. Eng.* **2013**, *5*, 153257. [CrossRef]
15. Guo, M.; Li, X.; Ran, M.; Zhou, X.; Yan, Y. Analysis of contact stresses and rolling resistance of truck-bus tyres under different working conditions. *Sustainability* **2020**, *12*, 10603. [CrossRef]
16. Rasol, M.; Schmidt, F.; Ientile, S.; Adelaide, L.; Nedjar, B.; Kane, M.; Chevalier, C. Progress and monitoring opportunities of skid resistance in road transport: A critical review and road sensors. *Remote Sens.* **2021**, *13*, 3729. [CrossRef]
17. Vázquez, V.; Terán, F.; Huertas, P.; Paje, S. Surface aging effect on tire/pavement noise medium-term evolution in a medium-size city. *Coatings* **2018**, *8*, 206. [CrossRef]
18. Lorenz, B.; Oh, Y.R.; Nam, S.K.; Jeon, S.H.; Persson, B. Rubber friction on road surfaces: Experiment and theory for low sliding speeds. *J. Chem. Phys.* **2015**, *142*, 194701. [CrossRef]
19. Ueckermann, A.; Wang, D.; Oeser, M.; Steinauer, B. A contribution to non-contact skid resistance measurement. *Int. J. Pavement Eng.* **2014**, *16*, 646–659. [CrossRef]
20. Woodward, D.; Millar, P.; Lantieri, C.; Sangiorgi, C.; Vignali, V. The wear of Stone Mastic Asphalt due to slow speed high stress simulated laboratory trafficking. *Constr. Build. Mater.* **2016**, *110*, 270–277. [CrossRef]
21. Hamlat, S.; Hammoum, F.; Kerzreho, J. Evaluation of the distribution of local pressures and the real contact area between the tyre and the road surface. *Int. J. Pavement Eng.* **2014**, *16*, 832–841. [CrossRef]
22. Wang, W.; Zhao, K.; Li, J.; Luo, R.; Wang, L. Characterization of dynamic response of asphalt pavement in dry and saturated conditions using the full-scale accelerated loading test. *Constr. Build. Mater.* **2021**, *312*, 125355. [CrossRef]
23. Riehm, P.; Unrau, H.; Gauterin, F.; Torbrügge, S.; Wies, B. 3D brush model to predict longitudinal tyre characteristics. *Veh. Syst. Dyn.* **2019**, *57*, 17–43. [CrossRef]
24. Jiang, X.; Zeng, C.; Gao, X.; Liu, Z.; Qiu, Y. 3D FEM analysis of flexible base asphalt pavement structure under non-uniform tyre contact pressure. *Int. J. Pavement Eng.* **2017**, *20*, 999–1011. [CrossRef]
25. Kuraishi, T.; Takizawa, K.; Tezduyar, T.E. Space–time computational analysis of tire aerodynamics with actual geometry, road contact, tire deformation, road roughness and fluid film. *Comput. Mech.* **2019**, *64*, 1699–1718. [CrossRef]
26. Wang, G.; Roque, R. Impact of wide-based tires on the near-surface pavement stress states based on three-dimensional tire-pavement interaction model. *Road Mater. Pavement Des.* **2011**, *12*, 639–662. [CrossRef]
27. Ding, Y. Numerical Simulation of Tire-Pavement Interaction For Noise, Safety, and Rolling Resistance. Doctoral Thesis, Rutgers The State University of New Jersey, School of Graduate Studies, New Brunswick, NJ, USA, 2018.
28. Liang, C.; Wang, G.; An, D.; Ma, Y. Tread wear and footprint geometrical characters of truck bus radial tires. *Chin. J. Mech. Eng.* **2013**, *26*, 506–511. [CrossRef]
29. Liu, Q.; Shalaby, A. Simulation of pavement response to tire pressure and shape of contact area. *Can. J. Civ. Eng.* **2013**, *40*, 236–242. [CrossRef]
30. Xie, Y.; Yang, Q. Tyre–pavement contact stress distribution considering tyre types. *Road Mater. Pavement Des.* **2018**, *20*, 1899–1911. [CrossRef]
31. *GB/T 6326-2014*; Tyre Terms and Definitions. General Administration of Quality Supervision, Inspection and Quarantine of the People's Republic of China, Standardization Administration of the P.R.C: Beijing China, 2014.
32. Guo, M.; Zhou, X. Tire-pavement contact stress characteristics and critical slip ratio at multiple working conditions. *Adv. Mater. Sci. Eng.* **2019**, *2019*, 5178516. [CrossRef]
33. Blab, R.; Harvey, J.T. Modeling measured 3D tire contact stresses in a viscoelastic FE pavement model. *Int. J. Geomech.* **2002**, *2*, 271–290. [CrossRef]
34. Prozzi, J.A.; Luo, R. Quantification of the joint effect of wheel load and tire inflation pressure on pavement response. *Transp. Res. Rec.* **2005**, *1919*, 134–141. [CrossRef]
35. Cesbron, J.; Anfosso-Lédée, F.; Yin, H.P.; Duhamel, D.; Le Houédec, D. Influence of Road Texture on Tyre/Road Contact in Static Conditions. *Road Mater. Pavement Des.* **2011**, *9*, 689–710. [CrossRef]
36. Muniandy, R.; Moazami, D.; Hamid, H.; Hassim, S. Characterization of effective tire contact area for various tread patterns. *Instrum. Sci. Technol.* **2013**, *42*, 15–26. [CrossRef]
37. Pillai, P.S.; Fielding-Russell, G.S. Empirical equations for tire footprint area. *Rubber Chem. Technol.* **1986**, *59*, 155–159. [CrossRef]
38. Ge, S.; Wang, B. Analysis and Test Verification on Contact Characteristics of Tire. *Mach. Des. Manuf.* **2014**, *7*, 233–235+239.
39. Tielking, J.T.; Roberts, F.L. Tire contact pressure and its effect on pavement strain. *J. Transp. Eng.* **1987**, *113*, 56–71. [CrossRef]
40. Weissman, S.L. Influence of tire-pavement contact stress distribution on development of distress mechanisms in pavements. *Transp. Res. Rec.* **1999**, *1655*, 161–167. [CrossRef]
41. Linhui, Z.; Zhiyuan, L.; Hong, C. Vehicle state and friction force estimation using nonlinear observer strategies. In Proceedings of the 27th Chinese Control Conference, Kunming, China, 1 January 2008.
42. Hajj, E.Y.; Thushanthan, P.; Sebaaly, P.E.; Siddharthan, R.V. Influence of tire–pavement stress distribution, shape, and braking on performance predictions for asphalt pavement. *Transp. Res. Rec. J. Transp. Res. Board* **2012**, *2306*, 73–85. [CrossRef]
43. Luo, R.; Prozzi, J.A. Effect of measured three-dimensional tire–pavement contact stress on pavement response at asphalt surface. *Transp. Res. Rec. J. Transp. Res. Board* **2007**, *2037*, 115–127. [CrossRef]

44. Yang, S.; Zhang, J.; Lu, Y.; Li, S. Research progress of vehicle-pavement interaction. *J. Mech. Eng.* **2021**, *57*, 1–17.
45. Marshek, K.M.; Chen, H.H.; Connell, R.B.; Hudson, R.W. Experimental determination of pressure distribution of truck tire-pavement contact. *Transp. Res. Rec.* **1986**, *1070*, 9–14.
46. Wang, H.; Al-Qadi, I.L.; Stanciulescu, I. Simulation of tyre–pavement interaction for predicting contact stresses at static and various rolling conditions. *Int. J. Pavement Eng.* **2012**, *13*, 310–321. [CrossRef]
47. Gong, Z.; Miao, Y.; Li, W.; Yu, W.; Wang, L. Analysis of tyre-pavement contact behaviour of heavy truck in full-scale test. *Int. J. Pavement Eng.* **2023**, *24*, 2235630. [CrossRef]
48. Douglas, R.A.; Woodward, W.D.H.; Woodside, A.R. Road contact stresses and forces under tires with low inflation pressure. *Can. J. Civ. Eng.* **2000**, *27*, 1248–1258. [CrossRef]
49. Hugo, F.; Kennedy, T.W. Surface cracking of asphalt mixtures in southern africa. In Proceedings of the Association of Asphalt Paving Technologists, St Paul, MN, USA, 7–11 January 1985.
50. Chae, S. Nonlinear Finite Element Modeling and Analysis of a Truck Tire. Ph.D. Thesis, Pennsylvania State University, State College, PA, USA, 2006.
51. Moazami, D.; Muniandy, R. Determination of relative damage of asphalt pavement from reduced tire contact area. *Int. J. Pavement Eng.* **2016**, *19*, 553–563. [CrossRef]
52. Wang, T.; Dong, Z.; Xu, K.; Ullah, S.; Wang, D.; Li, Y. Numerical simulation of mechanical response analysis of asphalt pavement under dynamic loads with non-uniform tire-pavement contact stresses. *Constr. Build. Mater.* **2022**, *361*, 129711. [CrossRef]
53. Machemehl, R.B.; Wang, F.; Prozzi, J.A. Analytical study of effects of truck tire pressure on pavements with measured tire–pavement contact stress data. *Transp. Res. Rec.* **2005**, *1919*, 111–120. [CrossRef]
54. Wang, H.; Al-Qadi, I.L. Combined effect of moving wheel loading and three-dimensional contact stresses on perpetual pavement responses. *Transp. Res. Rec. J. Transp. Res. Board* **2009**, *2095*, 53–61. [CrossRef]
55. Chen, B.; Xiong, C.; Li, W.; He, J.; Zhang, X. Assessing Surface Texture Features of Asphalt Pavement Based on Three-Dimensional Laser Scanning Technology. *Buildings* **2021**, *11*, 623. [CrossRef]
56. PIARC "Report of the committee on surface characteristics. In Proceedings of the XVII World Road Congress, Brussels, Belgium, 13–19 September 1987.
57. Wei, S.; Wang, Y.; Cheng, H.; Wang, D. Stress distributions in the textures of prefabricated pavement surface created with the assistance of 3D printing technology. *Int. J. Pavement Eng.* **2021**, *24*, 1–17. [CrossRef]
58. He, H.; Li, R.; Yang, Q.; Pei, J.; Guo, F. Analysis of the tire-pavement contact stress characteristics during vehicle maneuvering. *Ksce J. Civ. Eng.* **2021**, *25*, 2451–2463. [CrossRef]
59. Yu, J.; Zhang, B.; Long, P.; Chen, B.; Guo, F. Optimizing the Texturing Parameters of Concrete Pavement by Balancing Skid-Resistance Performance and Driving Stability. *Materials* **2021**, *14*, 6137. [CrossRef] [PubMed]
60. Al-Qadi, I.L.; Wang, H. Prediction of tire pavement contact stresses and analysis of asphalt pavement responses: A decoupled approach. In Proceedings of the Asphalt Paving Technology-Proceedings Association of Asphalt Technologists, Tampa, FL, USA, 27–30 March 2011.
61. Santos Maia, R.; Lacerda Costa, S.; Craveiro Cunto, F.J.; Castelo Branco, V.T.F. Relating weather conditions, drivers' behavior, and tire-pavement friction to the analysis of microscopic simulated vehicular conflicts. *J. Transp. Eng. Part B Pavements* **2021**, *147*, 04021037. [CrossRef]
62. Wang, F.; Xiao, Y.; Cui, P.; Mo, L.; Fang, M. Review on the characterization of aggregate morphology and its influence on asphalt mixture properties. *Mater. Rep.* **2022**, *17*, 21030063.
63. Sengoz, B.; Onsori, A.; Topal, A. Effect of aggregate shape on the surface properties of flexible pavement. *Ksce J. Civ. Eng.* **2014**, *18*, 1364–1371. [CrossRef]
64. Kokkalis, A.G.; Tsohos, G.H.; Panagouli, O.K. Consideration of Fractals Potential in Pavement Skid Resistance Evaluation. *J. Transp. Eng.* **2002**, *128*, 591–595. [CrossRef]
65. Wei, D.; Li, B.; Zhang, Z.; Han, F.; Zhang, X.; Zhang, M.; Li, L.; Wang, Q. Influence of surface texture characteristics on the noise in grooving concrete pavement. *Appl. Sci.* **2018**, *8*, 2141. [CrossRef]
66. Gong, Z.; Zhang, L.; Wu, J.; Xiu, Z.; Wang, L.; Miao, Y. Review of regulation techniques of asphalt pavement high temperature for climate change adaptation. *J. Infrastruct. Preserv. Resil.* **2022**, *3*, 9. [CrossRef]
67. Zheng, B.; Tang, J.; Chen, J.; Zhao, R.; Huang, X. Investigation of adhesion properties of tire—Asphalt pavement interface considering hydrodynamic lubrication action of water film on road surface. *Materials* **2022**, *15*, 4173. [CrossRef]
68. Gerthoffert, J.; Cerezo, V.; Bouteldja, M.; Do, M. Modeling aircraft braking performance on wet and snow/ice-contaminated runways. *Proc. Inst. Mech. Eng. Part J J. Eng. Tribol.* **2015**, *229*, 1065–1078. [CrossRef]
69. Wies, B.; Roeger, B.; Mundl, R. Influence of pattern void on hydroplaning and related target conflicts. *Tire Sci. Technol.* **2009**, *37*, 187–206. [CrossRef]
70. Tang, T.; Anupam, K.; Kasbergen, C.; Kogbara, R.; Scarpas, A.; Masad, E. Finite element studies of skid resistance under hot weather condition. *Transp. Res. Rec. J. Transp. Res. Board* **2018**, *2672*, 382–394. [CrossRef]
71. Hernandez, J.A.; Al-Qadi, I.L. Contact phenomenon of free-rolling wide-base tires: Effect of speed and temperature. *J. Transp. Eng.* **2016**, *142*, 295. [CrossRef]
72. Ashtiani, R.S.; Pullen, A.B.; Hammons, M.I. Comparative Study of Water-Blasting Equipment for Airfield Surface Decontamination. *J. Mater. Civ. Eng.* **2016**, *28*, 04016053. [CrossRef]

73. Pérez-Fortes, A.P.; Giudici, H. A recent overview of the effect of road surface properties on road safety, environment, and how to monitor them. *Environ. Sci. Pollut. Res.* **2022**, *29*, 65993–66009. [CrossRef] [PubMed]
74. Lantieri, C.; Ghasemi, N.; Cotignoli, L.; Vignali, V.; Simone, A. The evaluation of the effects of hazardous spills by road accidents on the surface performance of an asphalt-wearing course. *Int. J. Pavement Eng.* 2022, *ahead-of-print*. [CrossRef]
75. Thorpe, A.; Harrison, R.M. Sources and properties of non-exhaust particulate matter from road traffic: A review. *Sci. Total Environ.* **2008**, *400*, 270–282. [CrossRef]
76. Pomoni, M.; Plati, C.; Loizos, A. How can sustainable materials in road construction contribute to vehicles' braking? *Vehicles* **2020**, *2*, 55–74. [CrossRef]
77. Duong, T.T.T.; Lee, B. Determining contamination level of heavy metals in road dust from busy traffic areas with different characteristics. *J. Environ. Manag.* **2011**, *92*, 554–562. [CrossRef]
78. Ruiz-Llata, M.; Martín-Mateos, P.; López, J.R.; Acedo, P. Remote optical sensor for real-time residual salt monitoring on road surfaces. *Sens. Actuators B Chem.* **2014**, *191*, 371–376. [CrossRef]
79. Yu, Q.; Dai, Y.; Zhang, K. Study on determination of tire contact area pressure distribution under static load. *Tire Ind.* **1999**, *19*, 11–15.
80. Zhang, T.; Wang, M.; Wang, S.; Zhou, X.; Li, Q. A Dynamic Tire and Road Ground Stress Measurement Device and Method. China Patents CN201711320599.7, 2017.
81. De Beer, M.; Fisher, C. Stress-In-Motion (SIM) system for capturing tri-axial tyre–road interaction in the contact patch. *Measurement* **2013**, *46*, 2155–2173. [CrossRef]
82. Hernandez, J.A.; Gamez, A.; Al-Qadi, I.L.; De Beer, M. Analytical approach for predicting three-dimensional tire–pavement contact load. *Transp. Res. Rec. J. Transp. Res. Board* **2014**, *2456*, 75–84. [CrossRef]
83. Liang, C.; Zhu, D.; Wang, G.; Shan, M. Experimental study on tire-road dynamic contact pressure distribution using FTIR imaging. *Int. J. Automot. Technol.* **2021**, *22*, 1305–1317. [CrossRef]
84. Horne, W.B. *Influence of the Tire Tread Pattern and Runway Surface Condition on Breaking Friction and Rolling Resistance of a Modern Aircraft Tire*; National Aeronautics and Space Administration: Washington, DC, USA, 1962.
85. Gentle, C.R. Optical mapping of pressures in tyre contact areas. *Opt. Lasers Eng.* **1983**, *4*, 167–176. [CrossRef]
86. Zhang, X.; Liu, T.; Liu, C.; Chen, Z. Research on skid resistance of asphalt pavement based on three-dimensional laser-scanning technology and pressure-sensitive film. *Constr. Build. Mater.* **2014**, *69*, 49–59. [CrossRef]
87. Ghaednia, H.; Wang, X.; Saha, S.; Xu, Y.; Sharma, A.; Jackson, R.L. A review of elastic–plastic contact mechanics. *Appl. Mech. Rev.* **2017**, *69*, 060804. [CrossRef]
88. Király, T.; Primusz, P.; Tóth, C. Simulation of static tyre–pavement interaction using two FE models of different complexity. *Appl. Sci.* **2022**, *12*, 2388. [CrossRef]
89. Wollny, I.; Kaliske, M. Numerical simulation of pavement structures with inelastic material behaviour under rolling tyres based on an arbitrary Lagrangian Eulerian (ALE) formulation. *Road Mater. Pavement Des.* **2013**, *14*, 71–89. [CrossRef]
90. Johnson, K.L. *Contact Mechanics*; Cambridge University Press: Cambridge, UK, 1987.
91. Wu, X. *Study on the Contact Mechanics Model of Heavy Tire-Asphalt Pavement*; Chang'an University: Xi'An, China, 2018.
92. Greenwood, J.A.; Williamson, J.P. Contact of nominally flat surfaces. *Proc. R. Soc. Lond. Ser. A Math. Phys. Sci.* **1966**, *295*, 300–319.
93. Majumdar, A.; Bhushan, B. Role of fractal geometry in roughness characterization and contact mechanics of surfaces. *J. Tribol.* **1990**, *2*, 205–226. [CrossRef]
94. Persson, B.N.J. Theory of rubber friction and contact mechanics. *J. Chem. Phys.* **2001**, *115*, 3840–3861. [CrossRef]
95. Lorenz, B.; Persson, B.N.J.; Dieluweit, S.; Tada, T. Rubber friction: Comparison of theory with experiment. *Eur. Phys. J. E* **2011**, *34*, 1–11. [CrossRef] [PubMed]
96. Tolpekina, T.V.; Persson, B.N.J. Adhesion and friction for three tire tread compounds. *Lubricants* **2019**, *7*, 20. [CrossRef]
97. Wang, Y.; Lu, Y.J.; Si, C.D. Tire-pavement coupling dynamic simulation under tire high-speed-rolling condition. *Int. J. Simul. Model* **2016**, *15*, 236–248. [CrossRef]
98. Reynaud, P.; Nasr, S.B.; Allou, F.; Chaise, T.; Nelias, D.; Petit, C. 3D modelling of tyre-pavement contact pressure. *Eur. J. Environ. Civ. Eng.* **2016**, *21*, 712–729. [CrossRef]
99. Bai, T.; Cheng, Z.; Hu, X.; Fuentes, L.; Walubita, L.F. Viscoelastic modelling of an asphalt pavement based on actual tire-pavement contact pressure. *Road Mater. Pavement Des.* **2021**, *22*, 2458–2477. [CrossRef]
100. Peng, Y.; Li, J.Q.; Zhan, Y.; Wang, K.C.P.; Yang, G. Finite element method-based skid resistance simulation using in-situ 3D pavement surface texture and friction data. *Materials* **2019**, *12*, 3821. [CrossRef]
101. Najafi, S.; Flintsch, G.W.; Medina, A. Linking roadway crashes and tire–pavement friction: A case study. *Int. J. Pavement Eng.* **2015**, *18*, 119–127. [CrossRef]
102. Yu, M.; Xiao, B.; You, Z.; Wu, G.; Li, X.; Ding, Y. Dynamic friction coefficient between tire and compacted asphalt mixtures using tire-pavement dynamic friction analyzer. *Constr. Build. Mater.* **2020**, *258*, 119492. [CrossRef]
103. Pomoni, M.; Plati, C. Skid Resistance Performance of Asphalt Mixtures Containing Recycled Pavement Materials under Simulated Weather Conditions. *Recycling* **2022**, *7*, 47. [CrossRef]
104. Fwa, T.F.; Choo, Y.S.; Liu, Y. Effect of Aggregate Spacing on Skid Resistance of Asphalt Pavement. *J. Transp. Eng.* **2003**, *129*, 420–426. [CrossRef]

105. Cesbron, J.; Anfosso-Lédée, F.; Duhamel, D.; Ping Yin, H.; Le Houédec, D. Experimental study of tyre/road contact forces in rolling conditions for noise prediction. *J. Sound Vibr.* **2009**, *320*, 125–144. [CrossRef]
106. Dubois, G.; Cesbron, J.; Yin, H.P.; Anfosso-Lédée, F.; Duhamel, D. Statistical estimation of low frequency tyre/road noise from numerical contact forces. *Appl. Acoust.* **2013**, *74*, 1085–1093. [CrossRef]
107. Huang, Z.; Zhang, X.; Chen, B.; Yu, J.; Chen, F. Impact of tire contact texture on noise reduction characteristics of asphalt pavement. *Sci. Technol. Eng.* **2018**, *18*, 123–128.
108. Manyo, E.Y.; Reynaud, P.; Picoux, B.; Tautou, R.; Nelias, D.; Allou, F.; Petit, C. Towards fast modelling of the tire-pavement contact. *Eur. J. Environ. Civ. Eng.* **2021**, *25*, 2396–2412. [CrossRef]
109. Manyo, E.Y.; Reynaud, P.; Picoux, B.; Tautou, R.; Allou, F.; Petit, C.; Nélias, D. Tire–pavement tractive rolling contact under turning conditions: Towards pavement top-down cracking. *Int. J. Pavement Eng.* **2022**, *23*, 841–850. [CrossRef]
110. Ronald, B. Introducing improved loading assumptions into analytical pavement models based on measured contact stresses of tires. In Proceedings of the International Conference on Accelerated Pavement Testing, Reno, NV, USA, 18–20 October 1999.
111. Moazami, D.; Muniandy, R. Determination of rutting performance of asphalt pavements considering realistic tire-pavement contact area. *Int. J. Pavement Res. Technol.* **2021**, *14*, 764–770. [CrossRef]
112. Liang, C.; Wang, G.; Yu, K.; Mei, Y. Evaluation method of tire rolling resistance and ground-grip performance based on ground contact characteristics. *Automot. Eng.* **2020**, *42*, 1679–1687.
113. Wang, D.; Wang, G.; Li, Z.; Shao, S.; Sun, Y. Evaluation of anti-sliding durability of asphalt mixture based on pressure film technology. *China J. Highw. Transp.* **2017**, *30*, 1–9.